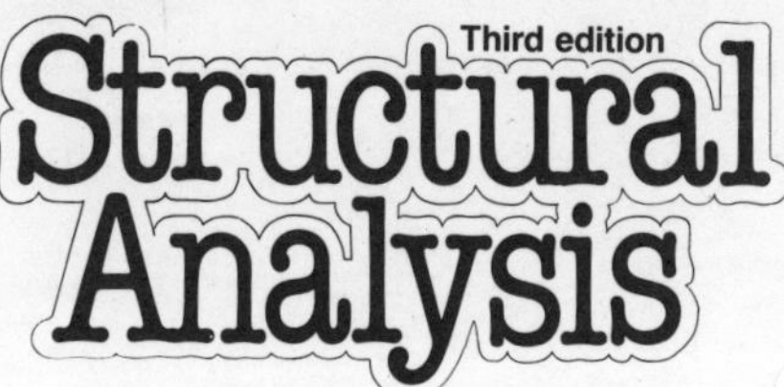
Third edition
Structural
Analysis

J.G.
D.W.

D1330590

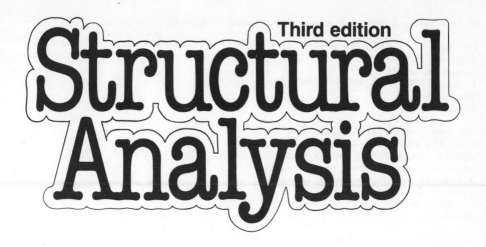

Third edition

Structural Analysis

R. C. Coates

Emeritus Professor of Civil Engineering,
University of Nottingham

M. G. Coutie

Senior Lecturer in Civil Engineering,
University of Nottingham

F. K. Kong

Professor of Structural Engineering,
University of Newcastle upon Tyne

International

First published in Great Britain 1972
Reprinted 1975, 1977
Second edition 1980
Reprinted 1980, 1981, 1982, 1984, 1985, 1986
Third edition 1988
Reprinted 1988

Published by Van Nostrand Reinhold (UK) Co. Ltd
Molly Millars Lane, Wokingham, Berkshire, England

Illustrations by Cecil Misstear Associates
and Hedgehog Design

Printed and bound in Hong Kong

Coates, R. C.
Structural analysis.—3rd ed.
1. Structures, Theory of
I. Title II. Coutie, M. G. III. Kong, F. K.
624.4'71 TA645

ISBN 0-278-00035-5

Contents

5 Stiffness and flexibility

6 Moment distribution

7 Matrix stiffness method

8 Matrix flexibility method

9 Instability of struts and frameworks

10 Structural dynamics

11 Elasticity problems and the finite difference method

Preface

The preface provides an author with the opportunity both to defend the choice of his material and to explain particular omissions. In our case we feel the reader might properly ask what reasons dictated yet another text on structural analysis. Simply, we believed it to be desirable. We have been jointly concerned for several years and in several places with the teaching of structural mechanics. We encountered it first in the heyday of moment distribution and of the application of energy methods, and saw it develop into a series of particular tricks each appropriate for the solution of some particular problem. The structural student learnt these tricks (as we did ourselves) and developed the appropriate response to particular stimuli, but frequently failed to appreciate the overall significance of what he was about—and great changes were taking place! The true place of Castigliano's theorems in the spectrum of energy equations was gradually understood; the special relationship of moment-distribution methods to the more generally applicable relaxation methods appeared relatively early; and the recognition in the 1950s of the importance of stability equations for compression members carrying moment was a major step. Overall, however, was the continuing development of computer-based analysis and it seems to us that no structural engineer of the future will resort to the various pretty little calculations which have been beloved of examiners till now, nor yet be far from a practicable machine. It will give both another dimension in his design thinking (and we readily admit that the primary purpose of analysis is to assist design) and confidence in his ability to produce solutions.

There have been, of course, other changes which may prove equally revolutionary. None can deny the far-reaching importance of work on the ultimate load-carrying capacity of structures, or of work on the probability assessments of loads and fitness for purpose, and we are conscious that little appears in this book about either. What we have tried to do is to produce a text suitable for the undergraduate, intended to give him a grounding in structural mechanics and the use of the computer in linear analysis, both of the skeletal structure and of the continuum. We could not contemplate the coverage of the entire subject in a single volume and, in any event, recent publications in the field of plastic theory have more than adequately filled the need.

Many may feel that the inclusion of Chapter 6 on moment-distribution methods is an anachronism, but it provides a relatively simple approach to the solution of problems for which the computer would prove uneconomical, and is less likely to give an incorrect answer than the orthodox methods available for the solution of small numbers of simultaneous equations. Moreover, it can provide some qualitative indication of structural behaviour, and we know that no quality is more valuable than an appreciation of deflected shapes and structural appropriateness—the computer alone could readily blunt both common sense and aesthetic discrimination.

We have encountered all the problems of earlier authors, and feel we have solved none, although the convention of signs used throughout is put forward as providing some guidance in a murky situation. The times in which we live have given us one additional difficulty not encountered by our predecessors, since the units of dimension used in the English-speaking world seem to be more diverse than ever before. In the United Kingdom it has been officially stated (but not agreed by all) that SI units will be used in future, whereas in the United States the kip still seems supreme and several authors appear ignorant of the essential differences between SI and other systems. We believe that, ultimately, SI will become the language of the international engineering community, although we do not pretend to know when that will be, and have written examples based on both current systems.

We have learnt a great deal whilst writing, not merely of structural analysis but also of each other, but hope that we shall not be the only beneficiaries. From those who read we shall be pleased to receive comments and suggestions and, inevitably, for it would be idle to hope there were none, to have our attention drawn to those errors which must exist. For those errors none is responsible save us, but for the great help in preparation which has been given us, we should like to acknowledge the debt we owe to Dr P. J. Robins (for some problems) and Mrs S.A. Howett and Mrs D.M. Wragg who typed our drafts.

Preface to the second edition

When we originally considered the content of this book, limitations of space led us to exclude any but passing reference to the ultimate load-carrying capacity—or plastic behaviour and design—of structures. It has become increasingly plain, however, that such an exclusion was undesirable if the text was to appeal to the generality of undergraduates, and we have inserted a chapter on this topic and brief sections on the analysis of two pinned arches, and the use of instantaneous centres, and ideas of symmetry and anti-symmetry in moment distribution.

We have taken advantage of the opportunity given us in the preparation of this new edition to recast all problems in SI units, since these are increasingly widely accepted universally, and to eliminate some of our earlier mistakes, to which our friends and readers have drawn our attention. Such is human frailty that we can scarcely hope that no errors will remain—we should be glad to learn of them and to rectify them on a subsequent occasion.

Preface to the third edition

Since the book was first published, increasingly we have been aware of its deficiencies, some of which arise from deliberate omissions on our part, and some from changing needs and practices in structural engineering itself. We were pleased, therefore—and somewhat flattered—when our (new) publisher invited us to contemplate a third edition.

We have included a new section on basic mechanics—a topic with which the present-day undergraduate appears increasingly to have difficulties—and have extended the sections devoted to certain structural concepts, and to instability. We are conscious that developments in computer technology and the ready availability of relatively small and cheap machines have increased the popularity of analytical methods dependent on simple routines, irrespective of the increase in computer time this might imply. Accordingly we have reduced the space devoted to matrix flexibility methods, which require more sophisticated judgement and knowledge of analysis than the average student can command. Finally, in our own thinking we find an increasing respect for the simplicity and enormous scope of the Principle of Virtual Work in its application to a great range of structural problems.

Since first publication we have separated ourselves, and, since our working areas are no longer adjacent, we cannot arrange an informal authors' meeting at an hour or two's notice. Our discussions have taken place by correspondence and, occasionally, by 'phone, but the physical separation has had little effect on the unanimity of our views. We must acknowledge the great help given to us by Dr G. Davies regarding sections of Chapter 9, and by Miss Rachel Ramsden, who typed all our drafts—both of the Civil Engineering Department of the University of Nottingham. We also give our grateful thanks to the Department's head, Professor P.S. Pell. They cannot be blamed, of course, for the errors which, we fear, are still likely to have occurred. We should be pleased to learn of these and to have an opportunity to correct them.

R.C.C.
M.G.C.
F.K.K.

Chapter 1
Definitions and introductory concepts

1.1 Sign convention for coordinate axes

In this book frequent reference will be made to the cartesian or rectangular system of mutually perpendicular axes Ox, Oy, and Oz. Only *right-handed systems* will be used. By this is meant that having chosen the positive directions for any two of the axes, say x and y, the positive direction for the z-axis will be that in which a right-handed screw would advance when turned from the x-axis to the y-axis through the right angle between the positive directions of these two axes.

The systems in Fig. 1.1–1 are all right-handed systems.

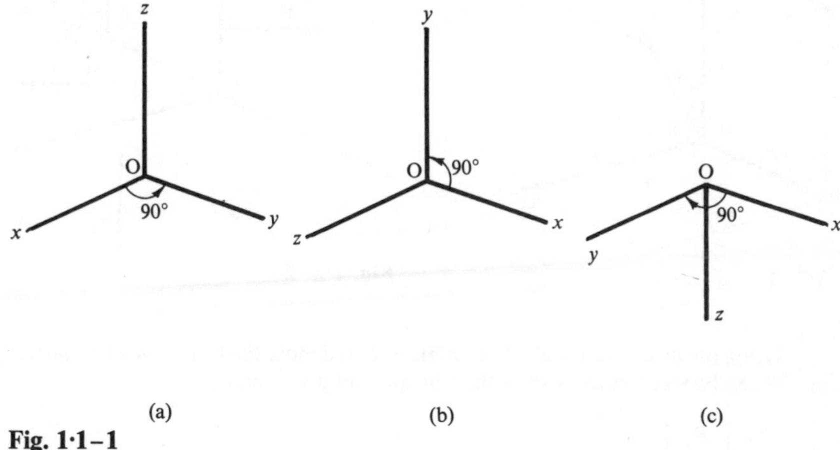

Fig. 1·1–1

1.2 Forces and moments

Consider a force **P**† of magnitude P and of direction as defined by the angles α, β, and γ which the force makes with the x-, y-, and z-axes of a cartesian coordinate system. In Fig. 1.2–1, the force **P** is represented in magnitude and direction by the vector OA, which

† Throughout this book, where a single letter is used for a vector quantity such as the force **P** here (or for a matrix, such as **N** in Eqn 1.7–8), it is printed in bold-face type.

is the diagonal of a rectangular prism of sides P_x, P_y, and P_z. P_x, P_y, and P_z are the projections of **P** on the *x*-, *y*-, and *z*-axis respectively, i.e. they are the components of **P** in the three coordinate directions. From Fig. 1.2–1,

$$P_x = P\cos\alpha; \qquad P_y = P\cos\beta; \qquad P_z = P\cos\gamma \tag{1.2–1}$$

The quantities cos α, cos β, cos γ are called the **direction cosines of P**, and are usually denoted by *l, m, n* respectively:

$$l = \cos\alpha; \qquad m = \cos\beta; \qquad n = \cos\gamma \tag{1.2–2}$$

The components P_x, P_y, and P_z completely specify the magnitude and direction of **P** and it is often convenient to write

$$\mathbf{P} = \begin{bmatrix} P_x \\ P_y \\ P_z \end{bmatrix} \tag{1.2–3}$$

In Eqn 1.2–3, **P** is expressed as a matrix of order 3 × 1 (i.e. a three-component column matrix), whose elements are the components of **P** in the three coordinate directions. This column matrix is often referred to as a **force vector**.. A force vector has magnitude and direction and can be represented by a straight line in a three-dimensional space, as in Figs 1.2–1 and 1.2–2.

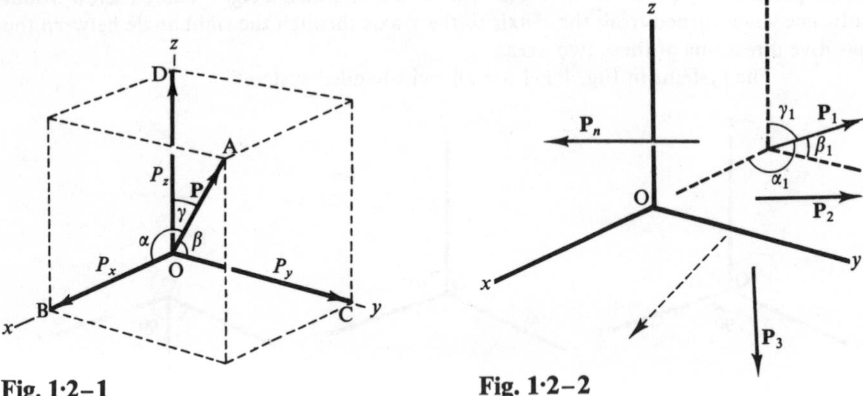

Fig. 1·2–1

Fig. 1·2–2

Using the usual matrix-algebra notation *T* to denote the transpose of a matrix, the force **P** can be represented also as the transpose of a row matrix:

$$\mathbf{P} = \begin{bmatrix} P_x \\ P_y \\ P_z \end{bmatrix} = [P_x \ P_y \ P_z]^T \tag{1.2–4}$$

Much printing space can be saved by using transposed row matrices to represent column matrices. Such transposed row matrices will be freely used throughout the book.

Note that the components P_x, P_y, and P_z depend only on the magnitude and direction of **P** but not on its point of application. Referring to Fig. 1.2–1, if **P** does not act at the origin O of the coordinate system, but at another arbitrary point, its components would still be as given by Eqn 1.2–1 as long as its magnitude and direction remain unchanged.

If there are a number of forces, $\mathbf{P}_1, \mathbf{P}_2, \mathbf{P}_3 \ldots \mathbf{P}_n$ as in Fig. 1.2–2, then,

$$\mathbf{P}_1 = \begin{bmatrix} P_{x1} \\ P_{y1} \\ P_{z1} \end{bmatrix}; \quad \mathbf{P}_2 = \begin{bmatrix} P_{x2} \\ P_{y2} \\ P_{z2} \end{bmatrix}; \quad \ldots; \quad \mathbf{P}_n = \begin{bmatrix} P_{xn} \\ P_{yn} \\ P_{zn} \end{bmatrix} \tag{1.2–5}$$

where $P_{x1} = P_1 \cos \alpha_1$, $P_{y1} = P_1 \cos \beta_1$, $P_{z1} = P_1 \cos \gamma_1$, and $P_{x2} = P_2 \cos \alpha_2$, and so on. The resultant of these n forces is the vector sum

$$\mathbf{P}_1 + \mathbf{P}_2 + \cdots + \mathbf{P}_n = \begin{bmatrix} P_{x1} \\ P_{y1} \\ P_{z1} \end{bmatrix} + \begin{bmatrix} P_{x2} \\ P_{y2} \\ P_{z2} \end{bmatrix} + \cdots + \begin{bmatrix} P_{xn} \\ P_{yn} \\ P_{zn} \end{bmatrix}$$

$$= \begin{bmatrix} P_{x1} + P_{x2} + \cdots + P_{xn} \\ P_{y1} \quad P_{y2} \quad\quad P_{yn} \\ P_{z1} \quad P_{z2} \quad\quad P_{zn} \end{bmatrix}$$

$$= \begin{bmatrix} \sum P_x \\ \sum P_y \\ \sum P_z \end{bmatrix} \tag{1.2–6}$$

If the resultant force vanishes, then, from Eqn 1.2–6,

$$\begin{bmatrix} \sum P_x \\ \sum P_y \\ \sum P_z \end{bmatrix} = 0 \tag{1.2–7}$$

That is, each of the components $\sum P_x$, $\sum P_y$, and $\sum P_z$ of the resultant must be zero.

Equation 1.2–7, however, does not ensure that there is no resultant moment. Consider, for example, the moments about the coordinate axes produced by force $\mathbf{P}_1$ (Fig. 1.2–3). The moment of $\mathbf{P}_1$ about any axis, say the x-axis, is equal to the sum of the

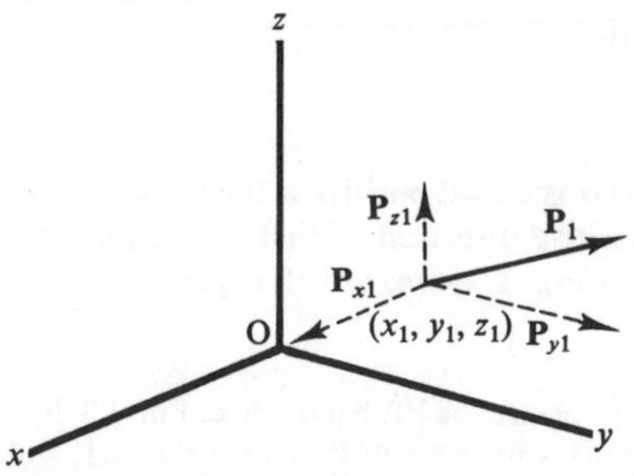

Fig. 1·2–3

moments of its components about the x-axis. Suppose P_1 acts at point (x_1, y_1, z_1), then the component P_{z1} has a moment arm y_1 about the x-axis and hence produces a moment $P_{z1} y_1$ about this axis. Similarly, the component P_{y1} produces a moment of $-P_{y1} z_1$, while the component P_{x1} is parallel to the x-axis and produces no moment about it.

Hence

$$\left.\begin{aligned}&\text{moment of } \mathbf{P}_1 \text{ about } x\text{-axis} = P_{z1}y_1 - P_{y1}z_1\\&\text{Similarly, it can be shown that}\\&\text{moment of } \mathbf{P}_1 \text{ about } y\text{-axis} = P_{x1}z_1 - P_{z1}x_1\\&\text{moment of } \mathbf{P}_1 \text{ about } z\text{-axis} = P_{y1}x_1 - P_{x1}y_1\end{aligned}\right\} \quad (1.2\text{–}8)$$

Hence, if the system of forces $\mathbf{P}_1, \mathbf{P}_2 \ldots \mathbf{P}_n$ produces no moments about any of the coordinate axes, then

$$\begin{bmatrix} \sum (P_z y - P_y z) \\ \sum (P_x z - P_z x) \\ \sum (P_y x - P_x y) \end{bmatrix} = 0 \quad (1.2\text{–}9)\dagger$$

In Eqn 1.2–9, the moment about the x-axis has been expressed as $\sum (P_z y - P_y z)$ instead of $\sum (P_y z - P_z y)$ because of a **sign convention for moments**, which will be adhered to throughout this book.

Suppose a moment **M**, of magnitude M, acts in a plane as shown in Fig. 1.2–4(a); the moment is represented by a vector with double arrow-heads whose length represents the magnitude M and whose direction is normal to the plane in which **M** acts. The direction of the arrow is that of the advance of a right-handed screw turned in the same sense as the moment **M**. Thus, if the sense of **M** is reversed, as shown in Fig. 1.2–4(b), the arrow points in the reversed direction.

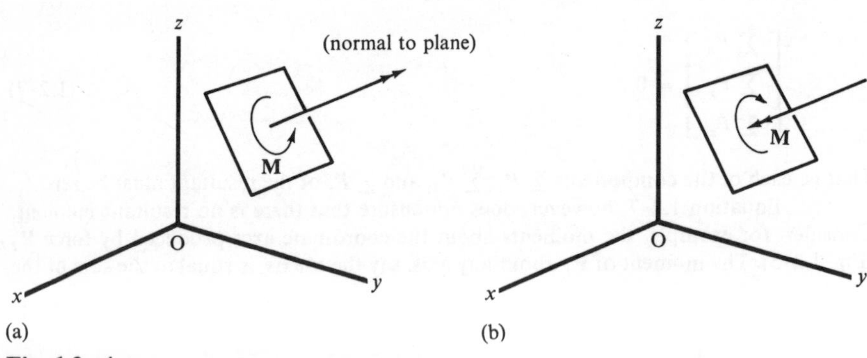

Fig. 1·2–4

A moment acting about a coordinate axis is called **positive** if the arrow of the vector representing the moment points in the **positive** direction of that axis. Thus, the moments in Fig. 1.2–5(a) are all positive; those in Fig. 1.2–5(b) are all negative.

† Readers familiar with vector algebra will recognize the left-hand side of Eqn 1.2–9 as the summation of the vector products $\sum (\mathbf{P}_i \times \mathbf{r}_i)$, where the force vector $\mathbf{P}_i = [P_{xi}\ P_{yi}\ P_{zi}]^T$ and the position vector $\mathbf{r}_i = [(0 - x_i)\ (0 - y_i)\ (0 - z_i)]^T = [-x_i\ \ -y_i\ \ -z_i]^T$. Hence Eqn 1.2–9 may be expressed as a determinantal equation:

$$\sum \begin{vmatrix} \mathbf{i} & \mathbf{j} & \mathbf{k} \\ P_x & P_y & P_z \\ -x & -y & -z \end{vmatrix} = 0$$

where **i**, **j**, and **k** are unit vectors in the x-, y-, and z-directions respectively.

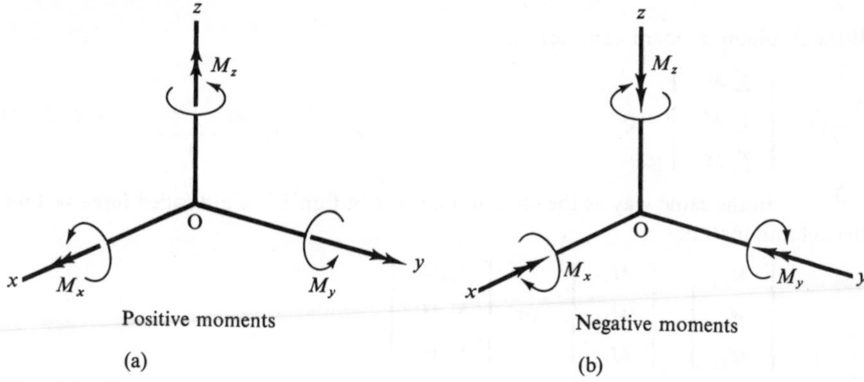

Fig. 1·2–5

If there are a number of moments $\mathbf{M}_1, \mathbf{M}_2 \ldots \mathbf{M}_n$, as represented by the vectors in Fig. 1.2–6, then these moments can be specified by their components about the *x*-, *y*-, and *z*-axes, in the same way as the forces in Fig. 1.2–2 were specified by Eqn 1.2–5. Thus

$$\mathbf{M}_1 = \begin{bmatrix} M_{x1} \\ M_{y1} \\ M_{z1} \end{bmatrix}; \quad \mathbf{M}_2 = \begin{bmatrix} M_{x2} \\ M_{y2} \\ M_{z2} \end{bmatrix}; \quad \ldots; \quad \mathbf{M}_n = \begin{bmatrix} M_{xn} \\ M_{yn} \\ M_{zn} \end{bmatrix} \tag{1.2–10}$$

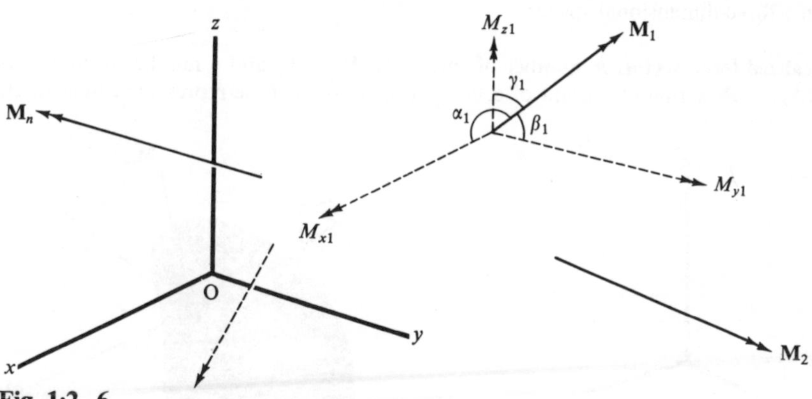

Fig. 1·2–6

The resultant of the *n* moments will be the vector sum:

$$\mathbf{M}_1 + \mathbf{M}_2 + \cdots + \mathbf{M}_n = \begin{bmatrix} M_{x1} \\ M_{y1} \\ M_{z1} \end{bmatrix} + \begin{bmatrix} M_{x2} \\ M_{y2} \\ M_{z2} \end{bmatrix} + \cdots + \begin{bmatrix} M_{xn} \\ M_{yn} \\ M_{zn} \end{bmatrix}$$

$$= \begin{bmatrix} M_{x1} + M_{x2} + \cdots + M_{xn} \\ M_{y1} + M_{y2} + \cdots + M_{yn} \\ M_{z1} + M_{z2} + \cdots + M_{zn} \end{bmatrix}$$

$$= \begin{bmatrix} \sum M_x \\ \sum M_y \\ \sum M_z \end{bmatrix} \tag{1.2–11}$$

If the resultant moment vanishes, then

$$\begin{bmatrix} \Sigma M_x \\ \Sigma M_y \\ \Sigma M_z \end{bmatrix} = 0 \tag{1.2–12}$$

In the same way as the column matrices in Eqn 1.2–5 are called **force vectors**, the column matrices

$$\begin{bmatrix} M_{x1} \\ M_{y1} \\ M_{z1} \end{bmatrix}, \begin{bmatrix} M_{x2} \\ M_{y2} \\ M_{z2} \end{bmatrix}, \quad \text{or} \quad \begin{bmatrix} \Sigma M_x \\ \Sigma M_y \\ \Sigma M_z \end{bmatrix}$$

the elements of which are components of the moments, are often called **moment vectors**. We have seen in Figs. 1.2–1 and 1.2–4 that force vectors and moment vectors can be represented by straight lines in a three-dimensional space. Vectors which can be so represented are physical quantities having both magnitude and direction; other examples of such physical vectors include velocity vectors and acceleration vectors which are already well known to the reader. It should be pointed out, however, that the traditional definition of a vector as 'a quantity having magnitude and direction' is often found to be unnecessarily restricted, and later in this book* we shall freely refer to matrices of order $n \times 1$, i.e. n-component column matrices, as vectors. Of course such a vector (i.e. an $n \times 1$ matrix) represents a set of n quantities, and cannot be represented by a straight line in a three-dimensional space.

Generalized force vector. A number of forces $\mathbf{P}_1, \mathbf{P}_2 \ldots \mathbf{P}_n$ and a number of moments $\mathbf{M}_1, \mathbf{M}_2 \ldots \mathbf{M}_n$ acting at a point of a body (Fig. 1.2–7) can be represented by a single

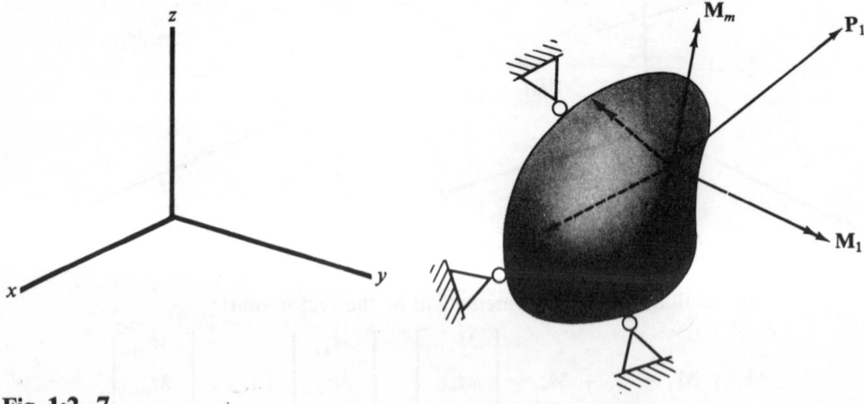

Fig. 1·2–7

force $\mathbf{P}_R$ and a single moment $\mathbf{M}_R$. $\mathbf{P}_R$ is the resultant of all the forces and $\mathbf{M}_R$ is the resultant of all the moments. From Eqns 1.2–6 and 1.2–11,

$$\mathbf{P}_R = \begin{bmatrix} \Sigma P_x \\ \Sigma P_y \\ \Sigma P_z \end{bmatrix} \quad \text{and} \quad \mathbf{M}_R = \begin{bmatrix} \Sigma M_x \\ \Sigma M_y \\ \Sigma M_z \end{bmatrix}$$

* See, for example, Chapter 5, Eqn 5.2–2 and Chapter 7, Eqn 7.9–2.

Sometimes it is convenient to use a single letter, say **P**, to denote $[P_R \ M_R]^T$, i.e.

$$\mathbf{P} = \begin{bmatrix} P_R \\ \\ M_R \end{bmatrix} = \begin{bmatrix} \sum P_x \\ \sum P_y \\ \sum P_z \\ \sum M_x \\ \sum M_y \\ \sum M_z \end{bmatrix} \tag{1.2–13}$$

If only one force and one moment act at the point, then Eqn 1.2–13 becomes

$$\mathbf{P} = \begin{bmatrix} P_x \\ P_y \\ P_z \\ M_x \\ M_y \\ M_z \end{bmatrix} \tag{1.2–14}$$

The quantity **P** in the left-hand side of Eqn 1.2–13 or 1.2–14 will be referred to as a **generalized force vector** or simply as a **generalized force**. It should be noted that the generalized force vector in fact represents two vectors—a force vector and a moment vector, and that these two vectors cannot be represented by a *single* line in the three-dimensional space.

Later in the text (subsequent to Chapter 3) the same symbol **P** may be used to represent a **generalized force system** acting on a body, but it will be so described (as in Section 4.8(a)) and there should be no confusion.

1.3 Equilibrium of a body

Figure 1.3–1 shows a body in equilibrium under the action of a system of *n* forces and *m* moments which are arbitrarily directed in space. From Newton's second law of motion, we must have

(a) the resultant force vanishes, and

(b) the resultant moment (due to the *n* forces and the *m* moments) vanishes.

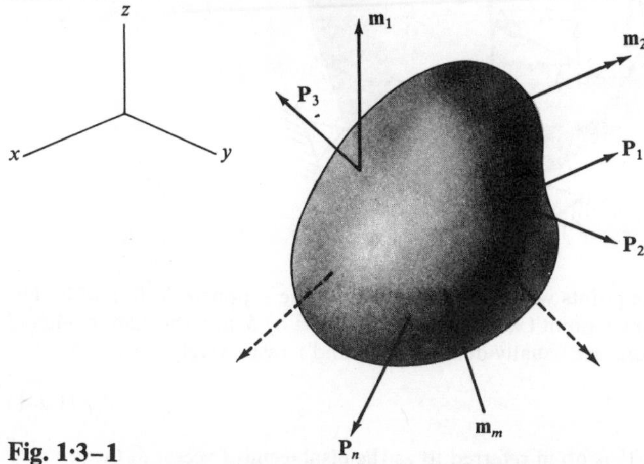

Fig. 1·3–1

From Eqns 1.2–7, 1.2–9, and 1.2–12, the necessary and sufficient conditions of equilibrium are:

$$\begin{bmatrix} \sum P_x \\ \sum P_y \\ \sum P_z \end{bmatrix} = 0 \tag{1.3–1(a)}$$

and

$$\begin{bmatrix} \sum (P_z y - P_y z) \\ \sum (P_x z - P_z x) \\ \sum (P_y x - P_x y) \end{bmatrix} + \begin{bmatrix} \sum m_x \\ \sum m_y \\ \sum m_z \end{bmatrix} = 0 \tag{1.3–1(b)}$$

Equation 1.3–1(b) is often written simply as

$$\sum M_x = 0; \qquad \sum M_y = 0; \qquad \sum M_z = 0 \tag{1.3–1(c)}$$

where $\sum M_x$ represents the quantity $\sum (P_z y - P_y z) + \sum m_x$, and so on. Note carefully that Eqns 1.3–1 are independent of the choice of coordinate axes; x, y, z can be any three mutually perpendicular directions. If the body is in equilibrium, then Eqns 1.3–1 must be satisfied, irrespective of the orientation of the coordinate axes and the position of the origin of the coordinate system.

In the particular case of a two-dimensional body which lies in, say the xy-plane and which is under the action of forces and moments lying in that plane, the necessary and sufficient conditions of equilibrium reduce to

$$\sum P_x = 0; \qquad \sum P_y = 0; \qquad \sum M = 0 \tag{1.3–2}$$

where $\sum M$ is understood to mean $\sum M_z$.

1.4 Displacements and rotations

Figure 1.4–1 shows a two-dimensional body, such as a uniform lamina, subjected to an in-plane force **P** applied at a point b. In general, **P** will cause a displacement at all points

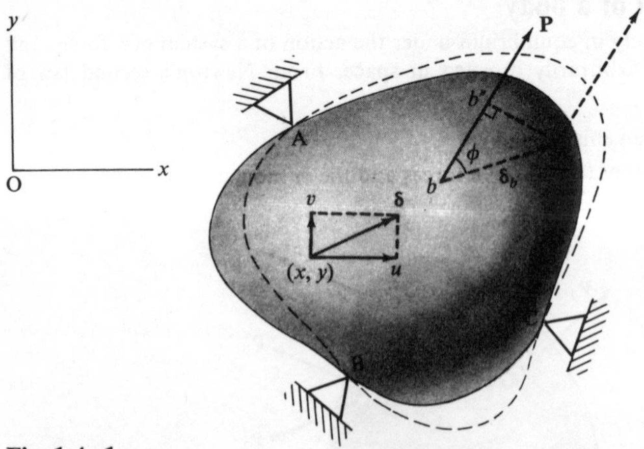

Fig. 1·4–1

of the body, except those points which are restrained by the supports A, B, and C. The displacement at an arbitrary point (x, y) is usually denoted by $\boldsymbol{\delta}$, and the components of $\boldsymbol{\delta}$ in the x- and y-directions are usually denoted by u and v respectively.

$$\boldsymbol{\delta} = \begin{bmatrix} u \\ v \end{bmatrix} = [u \;\; v]^T \tag{1.4–1}$$

The column matrix $[u \;\; v]^T$ is often referred to as the **displacement vector** at (x, y).

Note that the displacement vector $\boldsymbol{\delta}_b$ at b is in general not in the same direction as **P**. The component of $\boldsymbol{\delta}_b$ in the direction of **P** is sometimes called the **corresponding displacement** or corresponding deflection; i.e.

$$\text{corresponding displacement} = bb'' = \delta_b \cos \phi \tag{1.4–2}$$

At any point (x, y), in addition to the linear displacement $\boldsymbol{\delta}$, there is in general also a rotation $\boldsymbol{\theta}$. For the two-dimensional body in Fig. 1.4–1, the rotation is about the z-axis only. In a three-dimensional body, $\boldsymbol{\theta}$ will have components about each of the three coordinate axes:

$$\boldsymbol{\theta} = [\theta_x \ \theta_y \ \theta_z]^T \tag{1.4–3}$$

The column matrix $[\theta_x \ \theta_y \ \theta_z]^T$ is the **rotation vector** at the point (x, y, z). Similarly, the linear displacement vector $\boldsymbol{\delta}$ will have three components in a three-dimensional body:

$$\boldsymbol{\delta} = [u \ v \ w]^T \tag{1.4–4}$$

Sometimes, the single letter $\boldsymbol{\delta}$ is used to represent both $\boldsymbol{\delta}$ of Eqn 1.4–4 and $\boldsymbol{\theta}$ of Eqn 1.4–3:

$$\boldsymbol{\delta} = [u \ v \ w \ \theta_x \ \theta_y \ \theta_z]^T \tag{1.4–5}$$

The six-component column matrix in Eqn 1.4–5 is sometimes called a **generalized displacement vector**, or simply a **generalized displacement**. Note that a generalized displacement vector in fact represents two vectors—a linear displacement vector $[u \ v \ w]^T$ and a rotation vector $[\theta_x \ \theta_y \ \theta_z]^T$, and these two vectors cannot be represented by a *single* line in the three-dimensional space.

The previous discussions on corresponding displacements for the two-dimensional case in Fig. 1.4–1 can also be extended to the three-dimensional case. That is, if at a point in a body the generalized force vector is $[P_x \ P_y \ P_z \ M_x \ M_y \ M_z]^T$ and the generalized displacement vector is $[u \ v \ w \ \theta_x \ \theta_y \ \theta_z]^T$, then the *components* of the generalized displacement vector are the corresponding displacements of the respective *components* of the generalized force vector, even though in general the displacement vector $[u \ v \ w]^T$ is not in the same direction as the force vector $[P_x \ P_y \ P_z]^T$ and $[\theta_x \ \theta_y \ \theta_z]^T$ is not in the same direction as $[M_x \ M_y \ M_z]^T$.

1.5 Stresses—notation and sign convention

In Fig. 1.5–1(a), a **prismatic** bar, i.e. a bar of uniform cross-section, is subjected to an axial force **P**. Imagine that the bar is cut at section x–x perpendicular to the bar axis, so that the lower portion is isolated as a **free body** (Fig. 1.5–1(b)). By considering the equi-

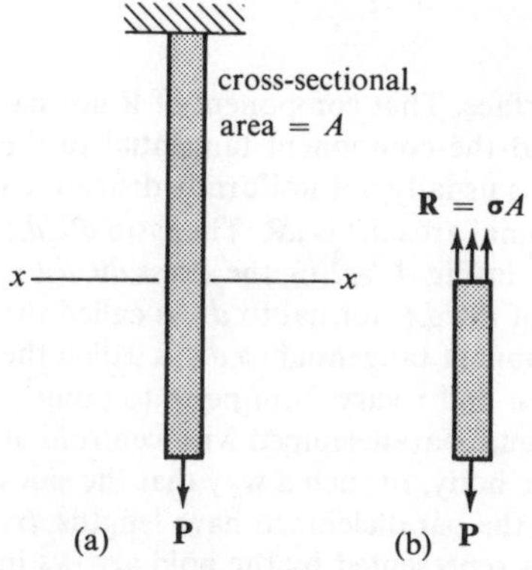

Fig. 1·5–1

librium of the free body, it is seen that there must be forces $\boldsymbol{\sigma}$ per unit area acting at the upper end of the free body such that their resultant $\mathbf{R}(=\boldsymbol{\sigma}A)$ balances the force **P**. Hence

$$\sigma = \frac{P}{A} \tag{1.5–1}$$

The quantity $\boldsymbol{\sigma}$, which is the force per unit area, is called the **stress** in the bar, and **R** is the **internal force** in the bar and is sometimes referred to as the **stress resultant** of $\boldsymbol{\sigma}$.

If a three-dimensional body (Fig. 1.5–2(a)) is in equilibrium under a system of forces and moments, then the stresses in the body can be studied similarly. Imagine the body to be cut by a plane, so that the free body to the right side of the imaginary plane is as shown in Fig. 1.5–2(b). Since the free body is in equilibrium, there must be an internal force **R** on the cut face having such magnitude and direction and point of application that Eqns 1.3–1(a) and (b) are satisfied.

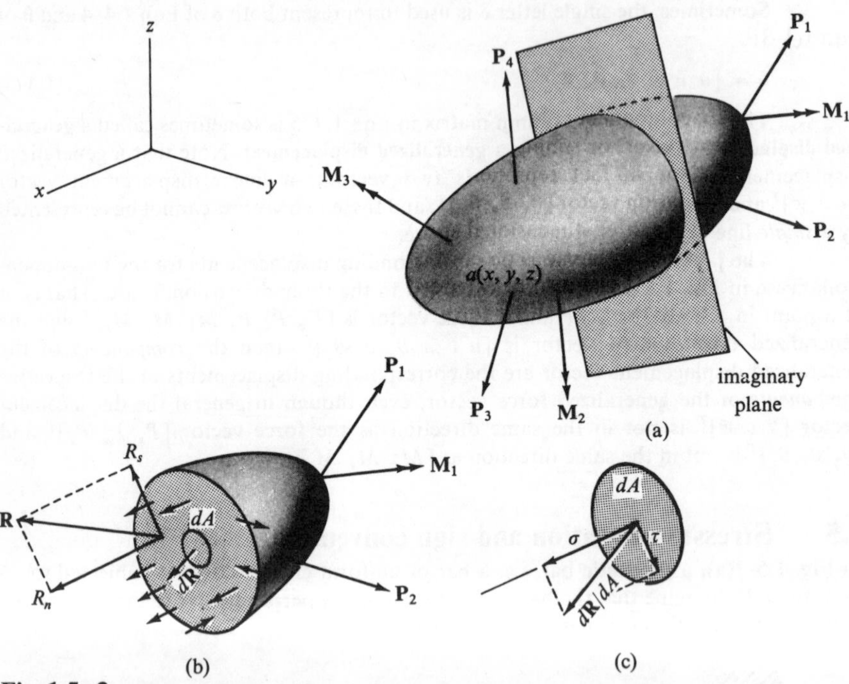

Fig. 1·5–2

In general, the force **R** is oblique to the surface. That component of **R** normal to the surface, $\mathbf{R}_n$, is the internal **normal force**, and the component tangential to the surface, $\mathbf{R}_s$, is the internal **shear force**. Note that **R** is usually not uniformly distributed across the surface. Suppose the force on an infinitesimal area dA is $d\mathbf{R}$. The ratio $d\mathbf{R}/dA$ is the stress at the point where dA is taken. As shown in Fig. 1.5–2(c), the stress $d\mathbf{R}/dA$ is in general oblique to the area dA. The component of $d\mathbf{R}/dA$ normal to dA is called the **normal stress** and is usually denoted by $\boldsymbol{\sigma}$; the component tangential to dA is called the **shear stress** and is usually denoted by $\boldsymbol{\tau}$. In general, $\boldsymbol{\sigma}$ and $\boldsymbol{\tau}$ vary from point to point.

Suppose we now imagine that an infinitesimal parallelepiped with centroid at point a (x, y, z) in Fig. 1.5–2(a) is cut away from the body, in such a way that the sides are parallel to the coordinate axes. Let the sides of the parallelepiped have lengths dx, dy, and dz and let the stresses on the six faces be as represented by the bold arrows in Fig. 1.5–3.

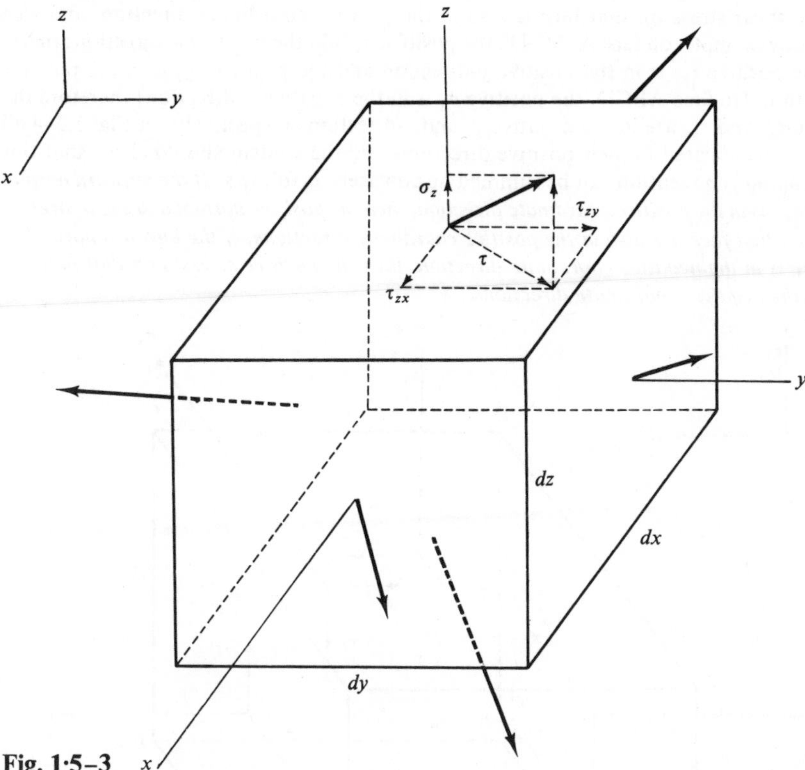

Fig. 1·5–3

As the magnitudes and directions of these stresses are not yet known, arbitrary magnitudes and directions have been assumed. Consider the stress on a face normal to the z-axis.

As explained previously, the stress on the face can be resolved into two components: a normal stress σ_z in the z-direction and a shear stress τ. The shear τ itself can be further resolved into two components—τ_{zx} in the x-direction and τ_{zy} in the y-direction. Each of the stresses on the other five faces can be similarly resolved into components in the coordinate directions, so that the stress components for the parallelepiped is as shown in Fig. 1.5–4.

The stress notation and sign convention in Fig. 1.5–4 are those which have been practically universally adopted in structural mechanics. The **stress notation** is as follows: Normal stresses are denoted by σ. The suffix attached to σ indicates the coordinate axis which is normal to the face on which σ acts. Thus, σ_x acts on a face normal to the x-axis, and σ_y and σ_z on faces normal to the y- and z-axes respectively. Shear stresses are denoted by τ, followed by two suffixes. The first suffix indicates the face on which the stress acts and the second suffix indicates the direction of the shear stress. Thus, τ_{xy} acts on a face normal to the x-axis and acts in the y-direction. τ_{yx} acts on a face normal to the y-axis and acts in the x-direction.

The **sign convention for stresses** is as follows: a normal stress is defined as positive if it is a **tensile** stress, i.e. if it is directed away from the surface on which it acts; a **compressive** stress, i.e. one directed towards the surface, is defined as negative. Thus, on a face such as A′B′C′D′ (Fig. 1.5–4) the positive normal stress is in the positive x-direction, and on face ABCD it is in the negative x-direction. The positive direction of a shear stress depends on the positive direction of the normal stress on that face. The rule is: If the positive normal stress on a face is in the positive coordinate direction then the positive shear stress on that face is also in the positive coordinate direction, and vice versa. For example, on face A′B′C′D′, the positive σ_x is in the positive x-direction; therefore the positive τ_{xy} is in the positive y-direction and the positive τ_{xz} is in the positive z-direction. On face ABCD, the positive σ_x is in the negative x-direction; therefore the positive τ_{xy} and τ_{xz} are in the negative y- and z-directions respectively. In Fig. 1.5–4 all stresses are indicated in their positive directions and the student should check that this is so. The sign convention can be summed up concisely as follows: *If the outward normal to the face is in the positive coordinate direction, then the positive normal and shear stresses acting on that face are also in the positive coordinate directions; if the outward normal to the face is in the negative coordinate direction, then the positive stresses on that face are also in the negative coordinate directions.*

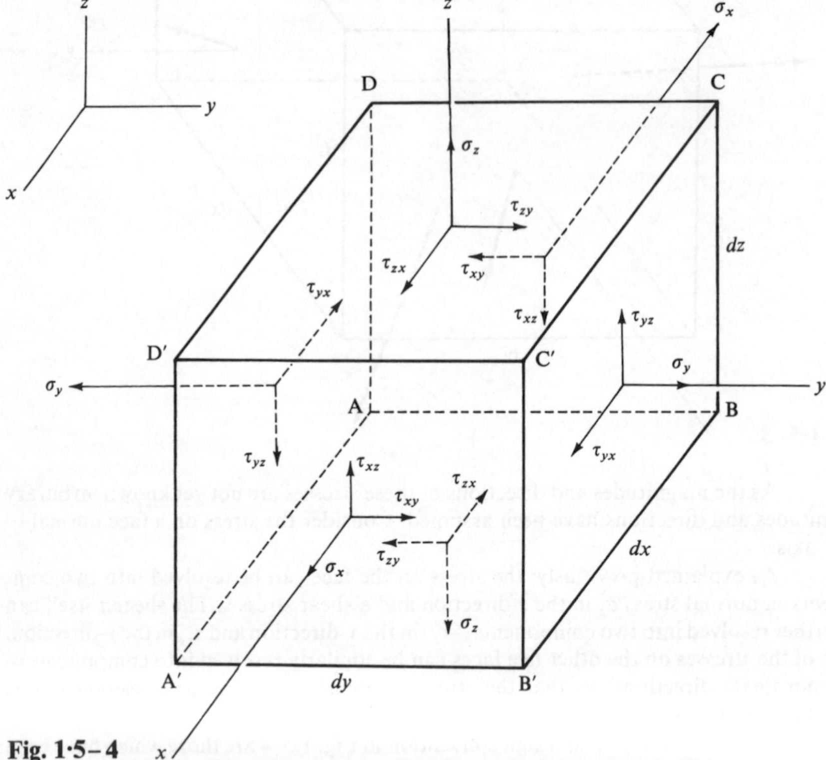

Fig. 1·5–4

Next consider the equilibrium of the parallelpiped in Fig. 1.5–4. Moments about the z-axis are caused by the shear forces $\tau_{xy}\,dy\,dz$ on faces A′B′C′D′ and ABCD, and by the shear forces $\tau_{yx}\,dx\,dz$ on faces BB′C′C and AA′D′D. Since no rotation occurs about the z-axis, summation of moments about this axis must be zero. Then

$$(\tau_{xy}\,dy\,dz)\,dx = (\tau_{yx}\,dx\,dz)\,dy$$

i.e.

$$(\text{force}) \times \text{lever arm} = (\text{force}) \times \text{lever arm}$$

i.e.

$$\tau_{xy} = \tau_{yx}$$

Similarly,

$$\tau_{yz} = \tau_{zy} \qquad (1.5\text{–}2)$$

$$\tau_{zx} = \tau_{xz}$$

Thus the nine symbols representing the states of stress on the faces of the parallelepiped may be reduced to six, and the column matrix of these stresses represented by

$$\boldsymbol{\sigma} = [\sigma_x \ \sigma_y \ \sigma_z \ \tau_{xy} \ \tau_{yz} \ \tau_{zx}]^T \tag{1.5–3}$$

The column matrix in Eqn 1.5–3 is often referred to as the **stress vector** at point (x, y, z).

The shear stresses τ_{xy} and τ_{yx} are equal in magnitude and are referred to as **complementary shear stresses**, while the same complementary relationships exist between τ_{yz} and τ_{zy} and between τ_{zx} and τ_{xz}.

In Fig. 1.5–4, the edges of the elementary parallelepiped are parallel to the respective axes of the cartesian system Oxyz, and its centroid is the point a (x, y, z) of the body in Fig. 1.5–2(a). Consider another elementary parallelepiped whose centroid is at the same point a, but whose edges are parallel to the axes of a new cartesian system OXYZ. Let

$$\boldsymbol{\sigma}\,(\mathrm{O}XYZ) = [\sigma_X \ \sigma_Y \ \sigma_Z \ \tau_{XY} \ \tau_{YZ} \ \tau_{ZX}]^T \tag{1.5–4}$$

be the stress vector for this second parallelepiped. The components of the stress vector in Eqn 1.5–4 would be different from those in Eqn 1.5–3. However, given the components of the stress vector $\boldsymbol{\sigma}$ (Oxyz), those of $\boldsymbol{\sigma}$ (OXYZ) can be readily determined, as will be shown in Chapter 11, Section 11.2. The same complementary relationships will exist and

$$\begin{aligned} \tau_{XY} &= \tau_{YX} \\ \tau_{YZ} &= \tau_{ZY} \\ \tau_{ZX} &= \tau_{XZ} \end{aligned} \tag{1.5–5}$$

1.6 Strains

Displacements of a general body were discussed in Section 1.4. The displacements that occur in most engineering structures under service conditions are small compared with the dimensions of the structures, and in this section we shall define the terms **normal strain** and **shear strain** and derive expressions for them. *Both the definitions and the resulting expressions are only applicable when displacements are small.* General definitions of these strains, which are applicable to both large and small displacements, will be given in Chapter 11, Section 11.4. However, this book is mainly concerned with small displacements of structures, so that the results obtained in this section are applicable unless otherwise stated.

Normal strain. Consider first the one-dimensional case of the prismatic bar in Fig. 1.6–1. Suppose for some reason, such as the application of an axial force or an increase in temperature, the bar extends from its initial length of l to the final length l'. Let the displacement at a point A be u, where the variable u is a function of x, being equal to

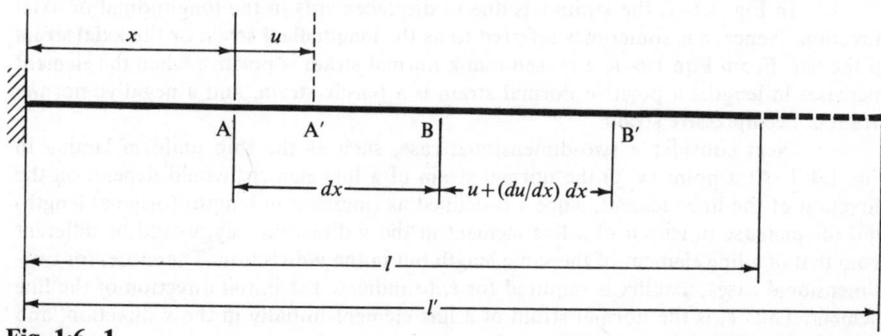

Fig. 1·6–1

zero at the fixed end and $l' - l$ at the free end. Consider another point B at dx from A before displacement occurs. The displacement at B will be $u + (du/dx)dx$, in which the second term arises from the stretching of the member between A and B.

As a result of the extension of the bar, A moves to A′ (i.e. AA′ = u) and B moves to B′ [i.e. BB′ = $u + (du/dx)dx$]. Hence

$$\text{coordinate of A}' = x + u$$
$$\begin{aligned}\text{coordinate of B}' &= \text{coordinate of B} + u + (du/dx)dx \\ &= x + dx + u + (du/dx)dx\end{aligned}$$

Then

$$\begin{aligned}\text{length A}'\text{B}' &= [x + dx + u + (du/dx)dx] - (x + u) \\ &= dx + (du/dx)dx\end{aligned}$$

The *normal* strain or *direct* strain of the bar at point A is defined as

$$\varepsilon = \frac{\text{increase in length of AB}}{\text{original length of AB}}$$

and is hence dimensionless.

From Fig. 1.6–1,

$$\varepsilon = \frac{\text{A}'\text{B}' - \text{AB}}{\text{AB}} = \frac{[dx + (du/dx)dx] - dx}{dx} = \frac{du}{dx} \tag{1.6–1}$$

If du/dx is constant over the length of the bar, then

$$\begin{aligned}\frac{du}{dx} &= \frac{u \text{ of free end} - u \text{ of fixed end}}{l} \\ &= \frac{(l' - l) - 0}{l} = \frac{l' - l}{l}\end{aligned}$$

i.e.

$$\varepsilon = \frac{l' - l}{l} \tag{1.6–2}$$

where du/dx is constant.

As will be seen in subsequent chapters, many engineering structures are built up of a number of straight bars of uniform cross-sections. If such a bar is subjected to axial forces, the quantity du/dx is very nearly constant throughout the length of the bar, and Eqn 1.6–2 is often used to determine the normal strain in the bar.

In Fig. 1.6–1, the strain ε is due to displacements in the longitudinal or axial direction. Hence, ε is sometimes referred to as the **longitudinal strain** or the **axial strain** of the bar. From Eqn 1.6–1, it is seen that a normal strain is positive when the element increases in length; a positive normal strain is a **tensile strain**, and a negative normal strain is a **compressive strain**.

Next consider a two-dimensional case, such as the thin uniform lamina in Fig. 1.4–1. At a point (x, y) the normal strain of a line element would depend on the direction of the line element, since ε is defined as (increase in length)/(original length) and the increase in length of a line element in the x-direction, say, would be different from that of a line element of the same length but in the y-direction. Therefore, for two-dimensional cases, a suffix is required for ε, to indicate the initial direction of the line element. Thus ε_x is the normal strain of a line element initially in the x-direction, and ε_y is that of a line element initially in the y-direction.

Let $m_1(x, y)$ and $m_2(x + dx, y)$ be two points initially in the x-direction as shown in Fig. 1.6–2.

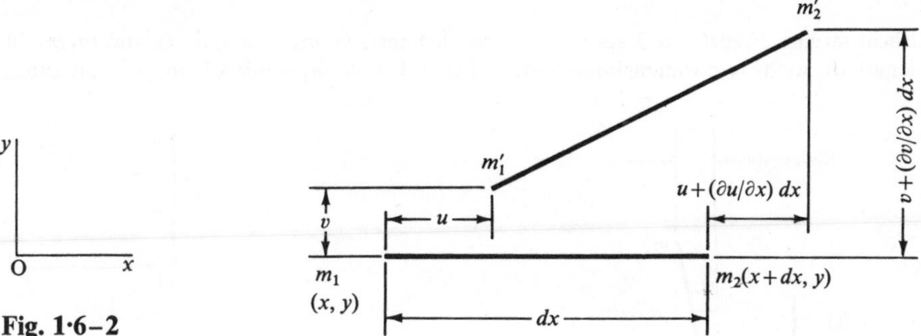

Fig. 1·6–2

Let m_1' and m_2' be the respective positions of m_1 and m_2 after displacement. Then, by definition,

$$\varepsilon_x = \frac{m_1' m_2' - m_1 m_2}{m_1 m_2} \tag{1.6–3}$$

Note that the displacement of points m_1 and m_2 to m_1' and m_2' as shown in Fig. 1.6–2 means that the line element $m_1 m_2$ undergoes both a change in length and a rotation about the z-axis, which is perpendicular to the plane of the paper. However, this rotation does not affect the definition of the normal strain ε_x, which is still defined as (increase in length)/(original length). If the displacements at m_1 are u, v then:

$$\begin{aligned} x\text{-coordinate of } m_1' &= x + u \\ x\text{-coordinate of } m_2' &= x\text{-coordinate of } m_2 + [u + (\partial u/\partial x)dx] \\ &= (x + dx) + [u + (\partial u/\partial x)dx] \end{aligned}$$

then

$$\begin{aligned} x\text{-projection of } m_1' m_2' &= [x + dx + u + (\partial u/\partial x)dx] - (x + u) \\ &= dx + (\partial u/\partial x)dx \end{aligned}$$

Similarly,

$$\begin{aligned} y\text{-projection of } m_1' m_2' &= [y + v + (\partial v/\partial x)dx] - (y + v) = (\partial v/\partial x)dx \\ \text{Length } m_1' m_2' &= \{[dx + (\partial u/\partial x)dx]^2 + [(\partial v/\partial x)dx]^2\}^{1/2} \\ &= [(dx)^2 + 2(\partial u/\partial x)(dx)^2 + (\partial u/\partial x)^2(dx)^2 + (\partial v/\partial x)^2(dx)^2]^{1/2} \\ &= [(dx)^2 + 2(\partial u/\partial x)(dx)^2]^{1/2} \end{aligned}$$

by neglecting infinitesimal quantities of the fourth order

$$= [1 + 2(\partial u/\partial x)]^{1/2}\, dx = [1 + (\partial u/\partial x)]\, dx$$

when $\partial u/\ x$ is small.

From Eqn 1.6–3,

$$\varepsilon_x = \frac{m_1' m_2' - m_1 m_2}{m_1 m_2} = \frac{[1 + (\partial u/\partial x)]dx - dx}{dx} = \frac{\partial u}{\partial x} \tag{1.6–4}$$

Similarly, it can be shown that the normal strain of a line element initially in the y-direction is $\varepsilon_y = \partial v/\partial y$. In a three-dimensional body, there is a displacement w in the z-direction, and we have $\varepsilon_z = \partial w/\partial z$. That is, for a general body

$$\varepsilon_x = \frac{\partial u}{\partial x}; \qquad \varepsilon_y = \frac{\partial v}{\partial y}; \qquad \varepsilon_z = \frac{\partial w}{\partial z} \tag{1.6–5}$$

where u, v, w are respectively the x-, y-, and z-components of the displacement at the point where ε_x, ε_y, and ε_z occur.

Shear strains. Figure 1.6–3 shows two line elements, $m_1 m_2$ of length dx and $m_1 m_3$ of length dy, in the two-dimensional body in Fig. 1.4–1. $m_1 m_2$ is initially in the x-direction

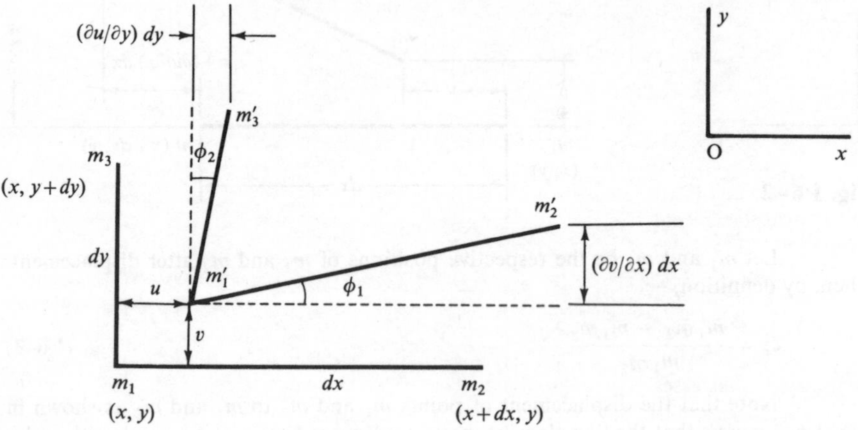

Fig. 1·6–3

and $m_1 m_3$ is initially in the y-direction. After deformation of the body, m_1, m_2, m_3 occupy positions m_1', m_2', and m_3'. The **shear strain** γ_{xy} is defined as the change in value of the right angle between two line elements initially in the x- and y-directions, i.e.

$$\begin{aligned}\gamma_{xy} &= \angle m_3 m_1 m_2 - \angle m_3' m_1' m_2' \\ &= \phi_1 + \phi_2 \end{aligned} \tag{1.6–6}$$

where angles are measured in radians.

In Fig. 1.6–3, the y-projection of $m_1' m_2'$ is $(\partial v/\partial x)dx$ as the student can easily verify, and since the angles are small

$$\begin{aligned}\phi_1 = \sin \phi_1 &= \frac{(\partial v/\partial x)dx}{m_1' m_2'} = \frac{(\partial v/\partial x)dx}{m_1 m_2(1 + \varepsilon_x)} \quad \text{(from Eqn 1.6–4)} \\ &= \frac{(\partial v/\partial x)dx}{m_1 m_2} \quad \text{since } \varepsilon_x \text{ is small compared with unity} \\ &= \frac{\partial v}{\partial x} \quad \text{since } m_1 m_2 = dx\end{aligned}$$

Similarly, the student should verify that

$$\phi_2 = \frac{\partial u}{\partial y}$$

then

$$\gamma_{xy} = \phi_1 + \phi_2 = \frac{\partial v}{\partial x} + \frac{\partial u}{\partial y}$$

Note that γ_{yx} is synonymous with γ_{xy} (but that the shear stresses τ_{yx} and τ_{xy} act on mutually perpendicular planes and are only numerically equal).

For a three-dimensional body, there will also be the shear strain γ_{yz} (or γ_{zy}), which is defined as the change in value of the right angle between two line elements initially in the y- and z-directions, and the shear strain γ_{zx} (or γ_{xz}) which is similarly defined. The reader should derive the expressions for γ_{yz} and γ_{zx} and verify that,

$$\gamma_{xy} = \frac{\partial v}{\partial x} + \frac{\partial u}{\partial y}$$

$$\gamma_{yz} = \frac{\partial w}{\partial y} + \frac{\partial v}{\partial z}$$

$$\gamma_{zx} = \frac{\partial u}{\partial z} + \frac{\partial w}{\partial x} \qquad (1.6\text{–}7)$$

where u, v, w are respectively the components of the displacement at the point where these strains occur. Note that Eqns 1.6–5 and 1.6–7 have been derived on the assumption that displacements are small so that strains are small compared with unity. The meaning of **small displacement** will be further explained in Chapter 4, Section 4.11 and Chapter 11, Section 11.10.

In Eqns 1.6–5 and 1.6–7 the strains are expressed in terms of the three displacement components u, v, w, which are taken as continuous functions of the coordinates x, y, z and considered to have continuous derivatives with respect to x, y, z. These strain-displacement relations are geometric, i.e. they are the same whether the displacements are caused by stresses or by change of temperature or by other factors.

These equations show that for **rigid-body displacement**, i.e. where u, v, w remain constant throughout the body, all strains are zero. They also show that at any point (x, y, z) of a body, there are only six independent strains. The column matrix whose components are these six strains is called the **strain vector** at point (x, y, z) and is usually denoted by

$$\boldsymbol{\varepsilon} = [\varepsilon_x \ \varepsilon_y \ \varepsilon_z \ \gamma_{xy} \ \gamma_{yz} \ \gamma_{zx}]^T \qquad (1.6\text{–}8)$$

At the end of Section 1.5, it was stated that, given the stress vector $\boldsymbol{\sigma}(\mathrm{O}xyz)$ with respect to cartesian system Oxyz, the stress vector $\boldsymbol{\sigma}(\mathrm{O}XYZ)$ could be determined. Similarly, given $\boldsymbol{\varepsilon}(\mathrm{O}xyz)$, $\boldsymbol{\varepsilon}(\mathrm{O}XYZ)$ can be determined as will be shown in Chapter 11, Section 11.4.

1.7 Stress–strain relations

Stresses and strains were dealt with separately in the previous two sections, and it was pointed out that strains could be caused by factors other than stresses. In this section, we shall discuss the relation between stresses and the strains which are caused by these stresses.

From Robert Hooke's experimental observations it is now known that if a rectangular block (Fig. 1.7–1) is subjected to uniformly distributed normal stress σ_x,

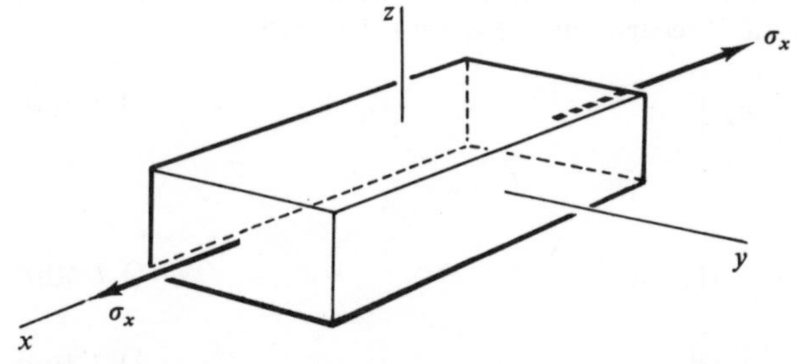

Fig. 1·7–1

then the normal strain ε_x which occurs as a result of the application of σ_x is proportional to σ_x, i.e.

$$\frac{\sigma_x}{\varepsilon_x} = E \quad \text{or} \quad \varepsilon_x = \frac{\sigma_x}{E} \tag{1.7–1}$$

where E is an experimentally determined proportionality constant called the **modulus of elasticity** or **Young's modulus** of the material of the block; since strain is defined as the ratio (change of length)/(original length) and is hence dimensionless, it is clear from Eqn 1.7–1 that E has the same dimension as stress. The proportionality of stress and strain in Eqn 1.7–1 is known as Hooke's law. For an **isotropic** material, i.e. one having the same properties in all directions, E would be the same in all directions, so that if instead of applying σ_x, uniformly distributed stress σ_y is applied to the face normal to the y-axis, then

$$\frac{\sigma_y}{\varepsilon_y} = E \quad \text{or} \quad \varepsilon_y = \frac{\sigma_y}{E} \tag{1.7–2}$$

Similarly, if σ_z alone is applied

$$\frac{\sigma_z}{\varepsilon_z} = E \quad \text{or} \quad \varepsilon_z = \frac{\sigma_z}{E} \tag{1.7–3}$$

Referring to Fig. 1.7–1, if the stress σ_x is removed, the strain ε_x will tend to disappear. If the strain disappears completely upon removal of stresses, the material is said to be **perfectly elastic**. Most materials used in structural engineering can be regarded as being perfectly elastic for stresses below a certain value known as the **elastic limit**, which varies with the material. Also, for a given material, Hooke's law is obeyed only for stresses below a certain value known as the **proportional limit** of the material. For most structural materials, it is usual to assume that the proportional limit is coincident with the elastic limit. Therefore, when stresses are within the elastic limit, the material is perfectly elastic and Hooke's law is also obeyed; a material in such a condition is said to be **linearly elastic**.

Also, in considering the block in Fig. 1.7–1, attention was focused on the strain ε_x which was in the same direction as σ_x. Actually, ε_x is always accompanied by lateral strains ε_y and ε_z which are proportional to ε_x but are of opposite sign, and which are given by

$$\varepsilon_y = \varepsilon_z = -\nu\,\varepsilon_x = -\nu\frac{\sigma_x}{E} \tag{1.7–4}$$

when σ_x only is acting. The constant ν is called **Poisson's ratio** of the material. In structural analysis, it is usually adequate to assume that E and ν each has the same magnitude both when the material is in tension and when it is in compression.

Suppose the block in Fig. 1.7–1 is subjected simultaneously to stresses σ_x, σ_y, and σ_z. Now that Poisson's ratio is recognized, the strain ε_x becomes

$$\varepsilon_x = \frac{1}{E}[\sigma_x - \nu(\sigma_y + \sigma_z)] \tag{1.7–5(a)}$$

Similarly,

$$\varepsilon_y = \frac{1}{E}[\sigma_y - \nu(\sigma_z + \sigma_x)] \tag{1.7–5(b)}$$

$$\varepsilon_z = \frac{1}{E}[\sigma_z - \nu(\sigma_x + \sigma_y)] \tag{1.7–5(c)}$$

These equations completely define the deformation of the block under the normal stresses σ_x, σ_y, and σ_z. Note that the normal stresses will produce only normal strains; there would be no shear strains, i.e. each face of the block will remain a rectangle. In general, it can be proved that for isotropic and linearly elastic materials,

(a) the normal strains ε_x, ε_y, and ε_z are functions of the normal stresses σ_x, σ_y, and σ_z only and are independent of any shear stresses that may be present; and

(b) the shear strain γ_{xy} is a function of the shear stress τ_{xy} only and is independent of normal stresses and also independent of the shear stresses τ_{yz} and τ_{zx}; similarly γ_{yz} is a function of τ_{yz} only, and γ_{zx} is a function of τ_{zx} only.

Hence in the general case of a block subjected simultaneously to normal and shear stresses (Fig. 1.5–4), the normal strains are still given by Eqn 1.7–5, while the shear strains are given by

$$\gamma_{xy} = \frac{\tau_{xy}}{G}; \qquad \gamma_{yz} = \frac{\tau_{yz}}{G}; \qquad \gamma_{zx} = \frac{\tau_{zx}}{G} \tag{1.7–6}$$

The constant G is called the **modulus of rigidity** or the **shear modulus** of the material. It can be shown that the elastic constants E, G, and ν are related by the equation

$$G = \frac{E}{2(1 + \nu)} \tag{1.7–7}$$

Equations 1.7–5 and 1.7–6 can be summed up in matrix form as

$$\begin{bmatrix} \varepsilon_x \\ \varepsilon_y \\ \varepsilon_z \\ \gamma_{xy} \\ \gamma_{yz} \\ \gamma_{zx} \end{bmatrix} = \frac{1}{E} \begin{bmatrix} 1 & -\nu & -\nu & 0 & 0 & 0 \\ -\nu & 1 & -\nu & 0 & 0 & 0 \\ -\nu & -\nu & 1 & 0 & 0 & 0 \\ 0 & 0 & 0 & 2(1+\nu) & 0 & 0 \\ 0 & 0 & 0 & 0 & 2(1+\nu) & 0 \\ 0 & 0 & 0 & 0 & 0 & 2(1+\nu) \end{bmatrix} \begin{bmatrix} \sigma_x \\ \sigma_y \\ \sigma_z \\ \tau_{xy} \\ \tau_{yz} \\ \tau_{zx} \end{bmatrix}$$

i.e.

$$\boldsymbol{\varepsilon} = \mathbf{N}\boldsymbol{\sigma} \tag{1.7–8}$$

where the constant G has been replaced by $E/[2(1 + \nu)]$.

Also, from Eqns 1.7–5 and 1.7–6 the stresses can be expressed in terms of the strains, and the reader should verify that

$$\begin{bmatrix} \sigma_x \\ \sigma_y \\ \sigma_z \\ \tau_{xy} \\ \tau_{yz} \\ \tau_{zx} \end{bmatrix} = \frac{E}{(1+\nu)(1-2\nu)} \begin{bmatrix} 1-\nu & \nu & \nu & 0 & 0 & 0 \\ \nu & 1-\nu & \nu & 0 & 0 & 0 \\ \nu & \nu & 1-\nu & 0 & 0 & 0 \\ 0 & 0 & 0 & \frac{1-2\nu}{2} & 0 & 0 \\ 0 & 0 & 0 & 0 & \frac{1-2\nu}{2} & 0 \\ 0 & 0 & 0 & 0 & 0 & \frac{1-2\nu}{2} \end{bmatrix} \begin{bmatrix} \varepsilon_x \\ \varepsilon_y \\ \varepsilon_z \\ \gamma_{xy} \\ \gamma_{yz} \\ \gamma_{zx} \end{bmatrix}$$

i.e.

$$\boldsymbol{\sigma} = \mathbf{D}\boldsymbol{\varepsilon} \tag{1.7–9}$$

Note that matrix **D** in Eqn 1.7–9 is the inverse of the matrix **N** in Eqn 1.7–8, but that it is easier to obtain **D** from Eqns 1.7–5 and 1.7–6 than by inverting **N**.

Plane stress. By **plane stress** is meant a state of stress in which $\sigma_z = \tau_{yz} = \tau_{zx} = 0$. Therefore, in a state of plane stress, the non-zero stresses are only σ_x, σ_y, and τ_{xy}.

Equation 1.7–8 then reduces to

$$\begin{bmatrix} \varepsilon_x \\ \varepsilon_y \\ \gamma_{xy} \end{bmatrix} = \frac{1}{E}\begin{bmatrix} 1 & -\nu & 0 \\ -\nu & 1 & 0 \\ 0 & 0 & 2(1+\nu) \end{bmatrix}\begin{bmatrix} \sigma_x \\ \sigma_y \\ \tau_{xy} \end{bmatrix} \tag{1.7–10}$$

From Eqn 1.7–10, the stresses can be expressed in terms of the strains, namely,

$$\begin{bmatrix} \sigma_x \\ \sigma_y \\ \tau_{xy} \end{bmatrix} = \frac{E}{1-\nu^2}\begin{bmatrix} 1 & \nu & 0 \\ \nu & 1 & 0 \\ 0 & 0 & \dfrac{1-\nu}{2} \end{bmatrix}\begin{bmatrix} \varepsilon_x \\ \varepsilon_y \\ \gamma_{xy} \end{bmatrix} \tag{1.7–11}$$

It should be pointed out that in a state of plane stress, ε_z is not equal to zero, but is given by Eqn 1.7–5(c) with $\sigma_z = 0$, i.e.

$$\varepsilon_z = -\frac{\nu(\sigma_x + \sigma_y)}{E} \tag{1.7–12}$$

Plane strain. By **plane strain** is meant a state of strain in which $\varepsilon_z = \gamma_{yz} = \gamma_{zx} = 0$. Therefore, in a state of plane strain, only the strains ε_x, ε_y, and γ_{xy} exist.

Equation 1.7–9 then reduces to

$$\begin{bmatrix} \sigma_x \\ \sigma_y \\ \tau_{xy} \end{bmatrix} = \frac{E}{(1+\nu)(1-2\nu)}\begin{bmatrix} 1-\nu & \nu & 0 \\ \nu & 1-\nu & 0 \\ 0 & 0 & \dfrac{1-2\nu}{2} \end{bmatrix}\begin{bmatrix} \varepsilon_x \\ \varepsilon_y \\ \gamma_{xy} \end{bmatrix} \tag{1.7–13}$$

from which the strains can be expressed in terms of stresses, namely,

$$\begin{bmatrix} \varepsilon_x \\ \varepsilon_y \\ \gamma_{xy} \end{bmatrix} = \frac{1+\nu}{E}\begin{bmatrix} 1-\nu & -\nu & 0 \\ -\nu & 1-\nu & 0 \\ 0 & 0 & 2 \end{bmatrix}\begin{bmatrix} \sigma_x \\ \sigma_y \\ \tau_{xy} \end{bmatrix} \tag{1.7–14}$$

Note that in a state of plane strain, σ_z is not zero but is given by Eqn 1.7–5(c):

$$0 = \frac{1}{E}[\sigma_z - \nu(\sigma_x + \sigma_y)]$$

hence

$$\sigma_z = \nu(\sigma_x + \sigma_y) \tag{1.7–15}$$

Chapter 2
Structural mechanics—statically determinate plane frames

2.1 Concurrent forces in a plane

A number of forces lying in one plane are described as **co-planar**—if they intersect at one point they are **concurrent**.

If a number of concurrent co-planar force $\mathbf{P}_1$, $\mathbf{P}_2 \ldots \mathbf{P}_5$ intersect at A, as shown in Fig. 2.1–1(a), their resultant **R** will be given by Eqn 1.2–6 and 7.

$$\mathbf{R} = [\Sigma P_x \quad \Sigma P_y]^{\mathrm{T}} = 0 \quad \text{for equilibrium} \qquad (2.1\text{–}1)$$

In general, *the projection of the resultant on any axis is equal to the algebraic sum of the projections of the individual forces on that axis*, and that summation can be carried out graphically as shown in Fig. 2.1–1(b), in which the forces $\mathbf{P}_1 \ldots \mathbf{P}_5$ are represented by the vectors ab, bc. . .ef, and the resultant, **R**, is represented in magnitude and direction by the vector af. The polygon a′b′c′d′e′f′ is a **force polygon** or polygon of forces, and may be drawn from any origin, and take the forces in any order, but the closing vector a′f′ will always represent the RESULTANT—see Fig. 2.1–1(c). If a′ and f′ coincide, the resultant is zero, and the concurrent system is in equilibrium—if they did not, a force applied at A, of magnitude a′f′ but in the direction f′a′, would restore equilibrium and is sometimes called an EQUILIBRANT.

If it is known that a system of concurrent forces is in equilibrium, the fact that the force polygon must close permits the determination of any two unknowns, which may be either the magnitude or direction of particular forces. This follows from Eqn 1.2–7, which, in this co-planar case, reduces to

$$\Sigma P_x = \Sigma P_y = 0$$

It follows, also, *that if a body is in equilibrium under the actions of three non-parallel and co-planar forces, they must be concurrent*, for the equilibrant must pass through the point of intersection of the other two as seen in Fig. 2.1–1(d).

Example 2.1–1. Determine, graphically and analytically, the resultant of the forces $\mathbf{P}_1 \ldots \mathbf{P}_5$, shown in Fig. 2.1–2(a), which are defined in the table below.

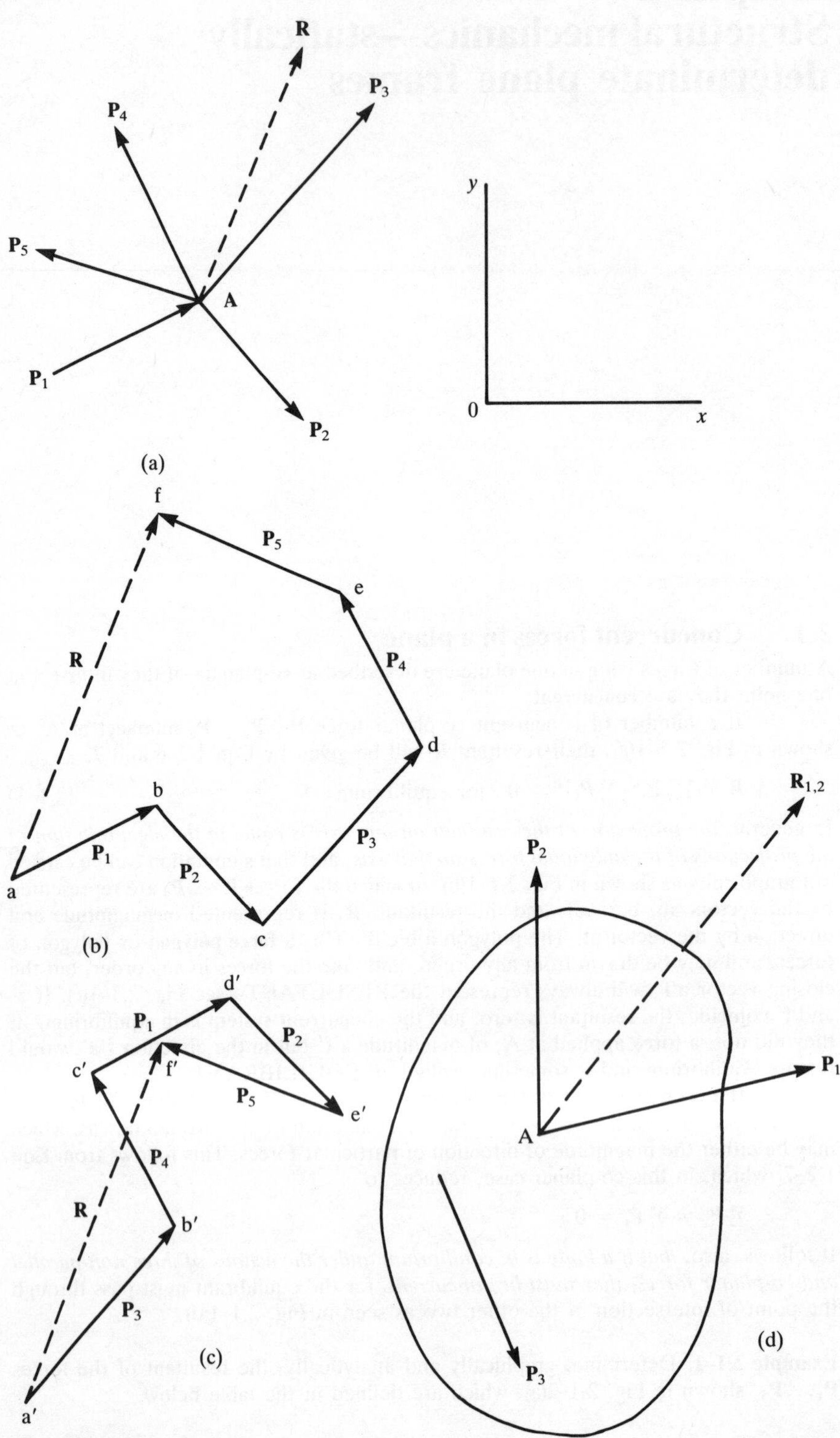
R
P3
P4
P5
A
P1
P2
(a)
y
0
x
f
P5
e
R
P4
d
b
P1
P3
P2
a
c
(b)
R1,2
P2
d′
P1
P2
c′
f′
P5
e′
P1
A
P4
R
b′
P3
(c)
(d)
a′
P3

Fig. 2·1–1

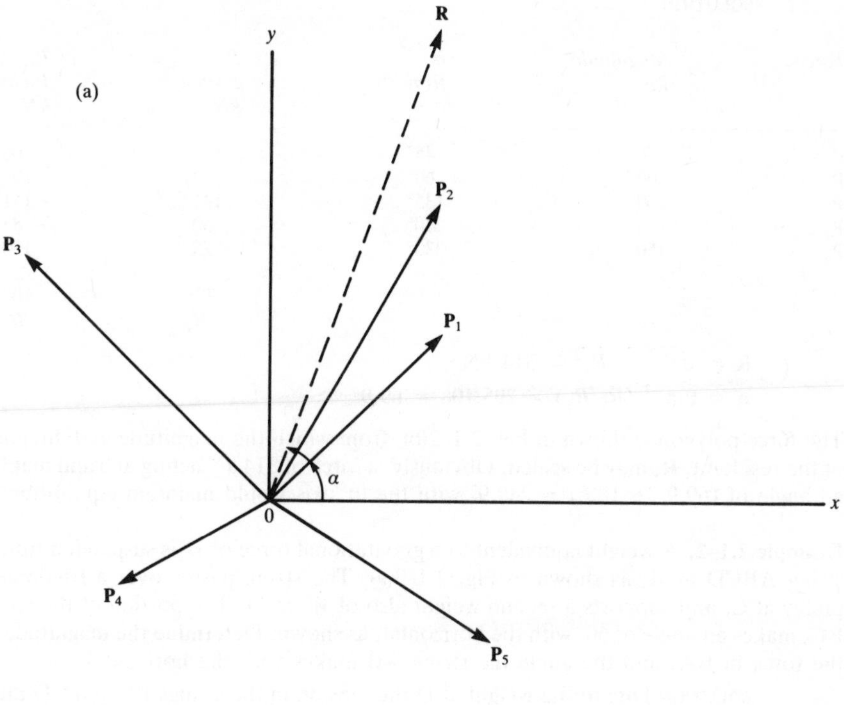

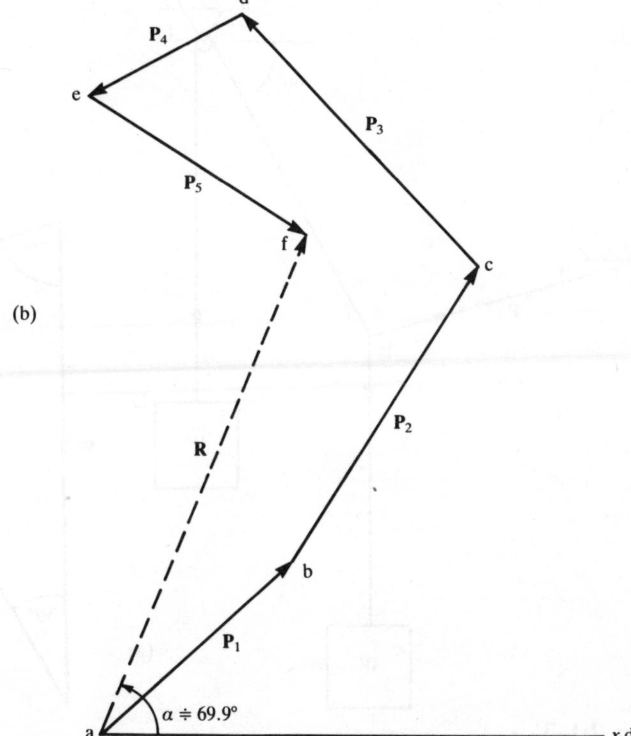

Fig. 2·1–2

SOLUTION

Force	*Magnitude* *kN*	α *from 0x*	P_y $P \sin\alpha$ *kN*	P_x $P\cos\alpha$ *kN*
$\mathbf{P}_1$	150	45°	106	106
$\mathbf{P}_2$	200	60°	173	100
$\mathbf{P}_3$	200	135°	141	−141
$\mathbf{P}_4$	100	210°	− 50	− 87
$\mathbf{P}_5$	150	330°	− 75	130
			295 R_y	108 R_x

$$\mathbf{R} = \sqrt{R_y^2 + R_x^2} = 314 \text{ kN}$$
$$\alpha = \tan^{-1}(R_y/R_x) = 295/108 = 69.9°$$

The force polygon is shown in Fig. 2.1–2(b), from which the magnitude and direction of the resultant, **R**, may be scaled. Obviously, a force of 314 kN acting at 0 and making an angle of (69.9° + 180°) = 249.9° with the 0x axis would maintain equilibrium.

Example 2.1–2. A weight equivalent to a gravitational force of W is suspended from a string ABCD at B, as shown in Fig. 2.1–3(a). The string passes over a frictionless pulley at C, and supports a second weight also of W, at D. The portion of the string BC, makes an angle of 60° with the horizontal, as shown. Determine the magnitude of the force in BA, and the angle the string AB makes with the horizontal.

SOLUTION Due to the weight at D the tension in the strings BC and CD must be W, since the pulley is frictionless. The triangle of forces for point B is shown in Fig.

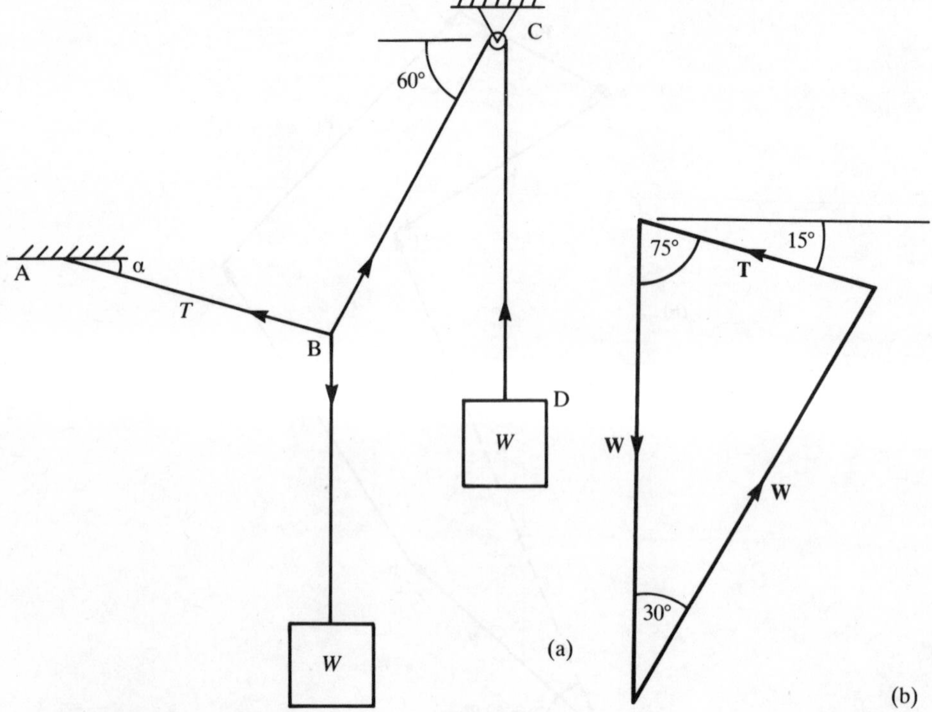

Fig. 2·1–3

2.1–3(b), in which those equal forces of W subtend equal angles of 75°. The force, T, in the string AB = $2W \cos 75° = 0.52W$—it will be tensile.

Example 2.1–3. A weightless bar AB (shown in Fig. 2.1–4(a)), of length L, is hinged to a rigid wall at A, and supported at B by a string which is attached to the wall at C, vertically above A. The bar is subjected to a vertical downwards force of W at the third point of its length. What are the force in the tie, T, and the magnitude and direction of the reaction, R, from the wall?

SOLUTION The three forces acting on the bar, and shown in Fig. 2.1–4(a), are the tension in the tie, the reaction at A, and the force W. Since the bar is in equilibrium the three forces must be concurrent, and the reaction, **R**, must pass through the point of intersection of W with the tie. The triangle of forces for the bar is shown in Fig. 2.1–4(b), and it should be emphasized that the angle between the line of action of **R**, and the tie, is not 90°. Calculations show that R makes an angle of 59.2° with the horizontal, and that **R** = $0.78W$ and $T = 0.52W$.

2.2 Parallel forces in a plane

If a system of forces acting in a plane is not concurrent the conditions defined by Eqn 2.1–1 will not be sufficient to ensure equilibrium. When two parallel forces of equal magnitude but opposite sense act as shown in Fig. 2.2–1, it can be seen that if an x axis parallel to these forces is chosen both the equations $\sum P_x = 0$ and $\sum P_y = 0$ will be satisfied. The two forces are not in equilibrium, however, and the total moment about some point m, distant d from one of the forces, will be

$$P(a + d) - Pd = Pa$$

The two forces constitute a **couple**, and the moment of the couple about any point in the plane in which it acts will be the product of the magnitude of one of the forces and

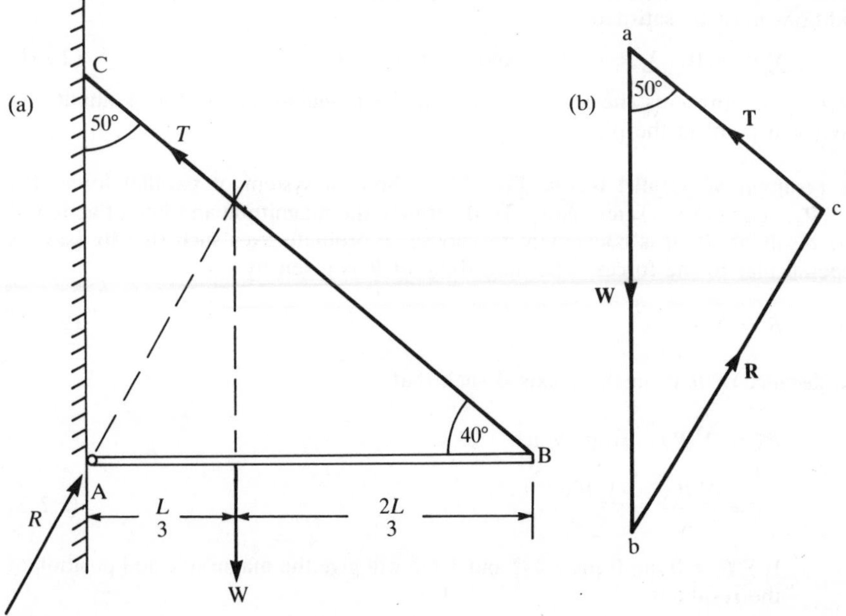

Fig. 2·1–4

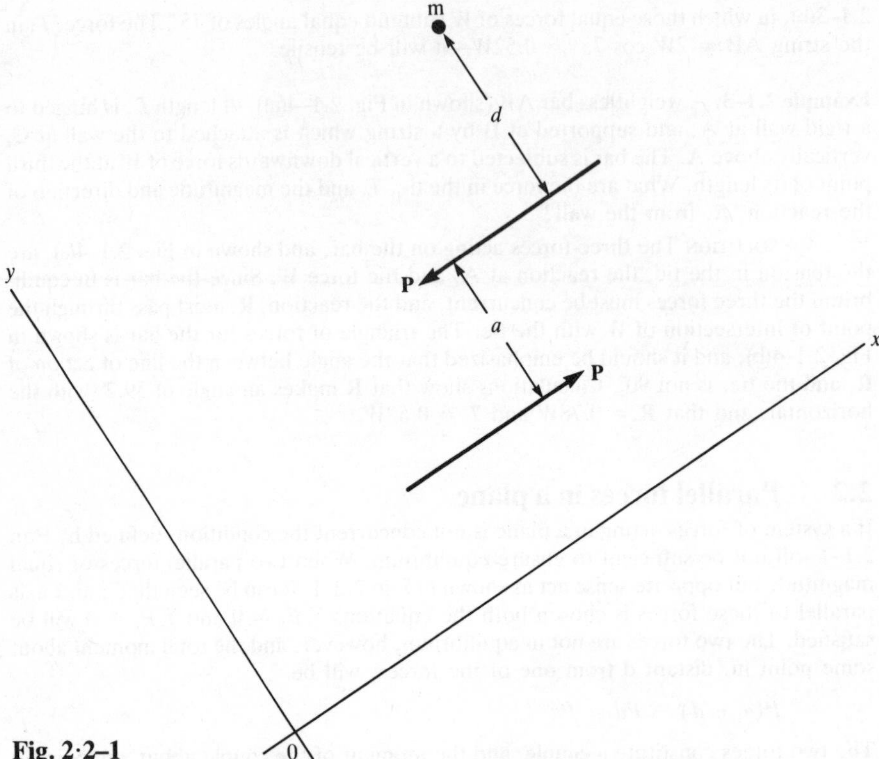

Fig. 2·2–1

the perpendicular distance between them—in Fig. 2.2–1 the sense of the couple is counter-clockwise. Hence, for a system of parallel forces to be in equilibrium, three conditions must be satisfied

$$\sum P_x = 0, \quad \sum P_y = 0 \quad \text{and} \quad \sum M = 0 \qquad (2.2\text{–}1)$$

where $\sum M$ represents the algebraic sum of the moments of the forces about any convenient point in the plane.

The resultant of parallel forces. Fig. 2.2–2 shows a system of parallel forces $\mathbf{P}_1$, $\mathbf{P}_2 \ldots \mathbf{P}_n$, acting in the same plane. To determine the magnitude and line of action of their resultant, $\mathbf{R}$, it is convenient to choose co-ordinate axes such that $0x$ (say) is perpendicular to the forces. The magnitude of $\mathbf{R}$ is given by

$$R = \sum_{i}^{n} P_i \qquad (2.2\text{–}2)$$

The distance of $\mathbf{R}$ from the y axis is such that

$$R\bar{x} = \sum_{1}^{n} P_i x_i \quad \text{from which}$$

$$\bar{x} = \frac{\sum P_i x_i}{R} = \frac{\sum P_i x_i}{\sum P_i} \qquad (2.2\text{–}3)$$

(a) If $\sum P_i \neq 0$ the Eqns 2.2–2 and 2.2–3 will give the magnitude and position of the resultant.

(b) If $\sum P_i = 0$ but $\sum P_i x_i \neq 0$, the resultant of all the forces will be a couple of magnitude $\sum P_i x_i$.

(c) If $\sum P_i = \sum P_i x_i = 0$ the resultant of the system will be zero, i.e. the system will be in equilibrium.

Example 2.2–1. Fig. 2.2–3 shows a beam of length L and negligible weight, supported on smooth rollers at A and B. It carries four forces of equal magnitudes, but different sense, equally spaced. Determine the reactions $\mathbf{R}_A$ and $\mathbf{R}_B$.

SOLUTION If the horizontal direction is regarded as the x direction, it will be

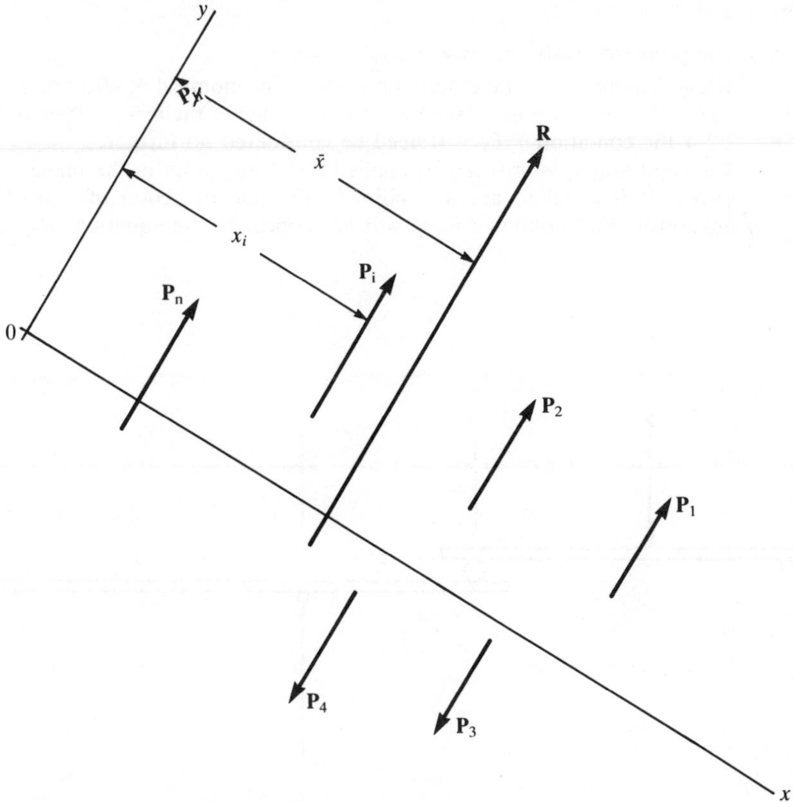

Fig. 2·2–2

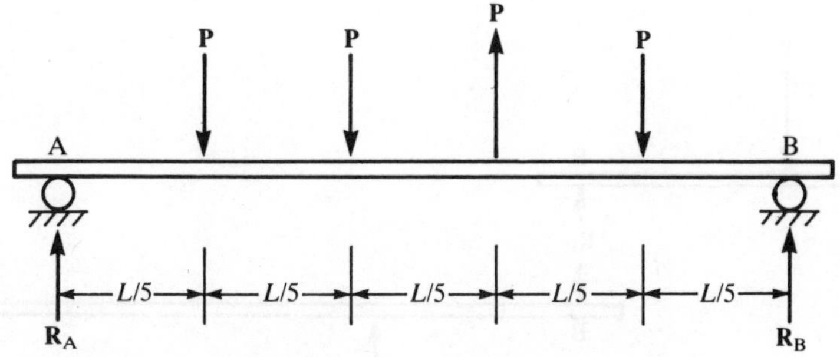

Fig. 2·2–3

seen that the first part of Eqn 2.2–1 is satisfied—$\sum P_x = 0$ and there are no horizontal forces.

since $\quad \sum P_y = 0, \quad R_A + R_B = 3P - P = 2P$ (a)

The equation $\sum M = 0$ may be applied to any convenient point—if to point A, then

$$R_B L = P(L/5) + P(2L/5) - P(3L/5) + P(4L/5) = (4/5)PL$$
$$R_B = 0.8P \qquad \text{(b)}$$

From (a) and (b) $R_A = 1.2P$

The problem illustrates some valuable points.

(i) When dealing with parallel forces it is always economical of effort to choose one of the reference axes (say the *y* axis) parallel to the forces. Then in Eqn 2.2–1 the condition $\sum P_x = 0$ need be considered no further.

(ii) The condition $\sum M = 0$ can be applied about any point in the plane of the forces. If it is taken about a point in the line of action of one of the unknowns, that unknown force will not appear in the equation. R_A could

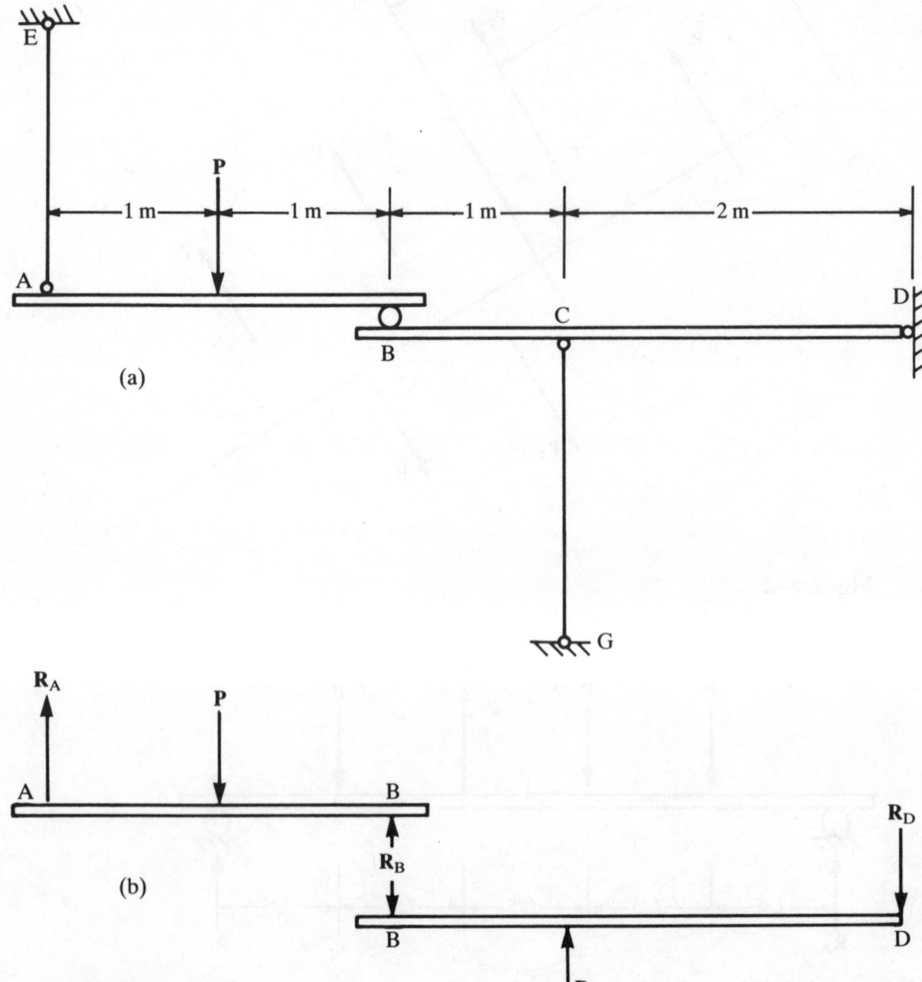

Fig. 2·2–4

have been obtained explicitly by applying the equation $\sum M = 0$ about B, and R_B by applying the equation $\sum M = 0$ about A.

Example 2.2–2. Fig. 2.2–4(a) shows a beam AB, carrying a load P at its mid-point, suspended at A from a vertical string and supported at B on a roller which lies on a beam BCD. The latter is pinned at D to a rigid wall, and supported at C on a pin-ended vertical column. Find the reactions and inter-actions on beams and column.

SOLUTION Fig. 2.2–4(b) shows the **free body diagrams** for the beams. It will be seen by inspection that $R_A = R_B = 0.5P$ (symmetry). Applying the equation $\sum M = 0$ to the point D of the beam BCD

$$R_B \times 3 = R_C \times 2 \quad \text{from which} \quad R_C = 0.75P$$

Consideration of the vertical equilibrium ($\sum P_y = 0$) of beam BCD will then show that

$$R_D = 0.25P$$

The tensile force in AE will then be $0.5P$ and the compressive force in the column CG will be $0.75P$.

2.3 The general case of forces in a plane

Fig. 2.3–1 shows a general system of forces, $\mathbf{P}_1 \ldots \mathbf{P}_n$, lying in the x–y plane. Their resultant, $\mathbf{R}$, will be in the same plane, and from Eqn 1.2–6.

$$\mathbf{R} = [\sum P_x \quad \sum P_y]^T \tag{2.3–1}$$

and its magnitude will be $\sqrt{(\sum P_x)^2 + (\sum P_y)^2}$

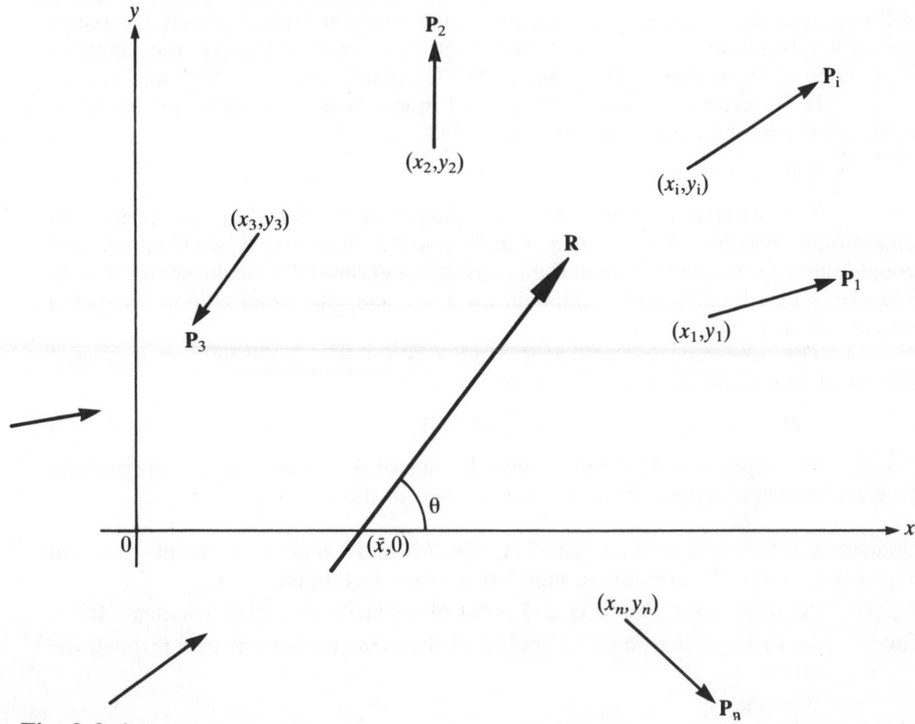

Fig. 2·3–1

If the line of action of the resultant makes an angle θ with the x axis, as shown, then

$$\theta = \tan^{-1} (\Sigma P_y/\Sigma P_x) \tag{2.3–2}$$

The algebraic sum of the moments of all the forces about the z axis (normal to the plane of the paper) is given by Eqn 1.2–9, viz,

$$\Sigma (P_y x - P_x y) \tag{2.3–3}$$

and if R intersects the x axis at the point $(\bar{x}, \text{o})$ the moment of **R** about the z axis is

$$R_y \bar{x} = (\Sigma P_y)\bar{x} \tag{2.3–4(a)}$$

(R_y = component of **R** in the 'y' direction $= \Sigma P_y$)

Equating the two expressions

$$\bar{x} = (\Sigma P_y x - \Sigma P_x y)/\Sigma P_y \tag{2.3–4(b)}$$

Similarly if we assume the resultant intersects the y axis at the point $(0, \bar{y})$, then the moment of the resultant about the z axis can be expressed as

$$-R_x \bar{y} = (-\Sigma P_x)\bar{y} \tag{2.3–5(a)}$$

where the negative sign is used because the moment is clockwise—see Fig. 1.2–5. Then

$$\bar{y} = -\Sigma (P_y x - P_x y)/\Sigma P_x \tag{2.3–5(b)}$$

If moments $\mathbf{m}_1, \mathbf{m}_2 \ldots \mathbf{m}_n$ are acting in the x–y plane, as shown in Fig. 2.3–2, Eqn 2.3–4 and 2.3–5 will need modification to take their effects into account, and

$$\bar{x} = (\Sigma P_y x - \Sigma P_x y + \Sigma m)/\Sigma P_y \tag{2.3–6}$$

where Σm includes all moments with vectors in the $0z$ direction.

It should be noted that the existence of applied moments in the x–y plane will not affect the magnitude or direction of the resultant force, but only the points where that resultant force intersects the co-ordinate axes. Moreover, the points of application of the moments $\mathbf{m}_1 \ldots \mathbf{m}_n$, in the x–y plane, have no effect on $\bar{x}$ and $\bar{y}$.

If the general system of forces and moments is in equilibrium, the force resultant **R** must be zero, and from Eqn 2.3–1

$$\Sigma P_x = \Sigma P_y = 0$$

The two requirements are not sufficient alone, however, to ensure that equilibrium exists for, as was explained in Section 2.2, there may be some unbalanced couple acting in the plane. Equilibrium can only be ensured if it can be shown that no resultant moment exists about points in the x–y plane—the point chosen for such a specific check is immaterial.

The necessary and sufficient conditions for the equilibrium of a general system of forces and moments in a plane are that

$$\Sigma P_x = 0, \quad \Sigma P_y = 0 \quad \text{and} \quad \Sigma M = 0 \tag{2.3–7}$$

The expression ΣM will include the sum of the moments due to forces and to given moments, taken about any axis normal to the x–y plane.

Example 2.3–1. Forces of magnitude 5 N, 6 N, 7 N and 8 N act at the points shown in Fig. 2.3–3, where the coordinate units are metres. Determine:

(a) the magnitude, direction and point of application of their resultant, **R**;

(b) the moment that must be applied in the plane to make **R** pass through the origin.

SOLUTION

(a) $P_x = 5 \cos 330° + 6 \cos 60° + 7 \cos 30° + 8 \cos 315° = 19.05$ N

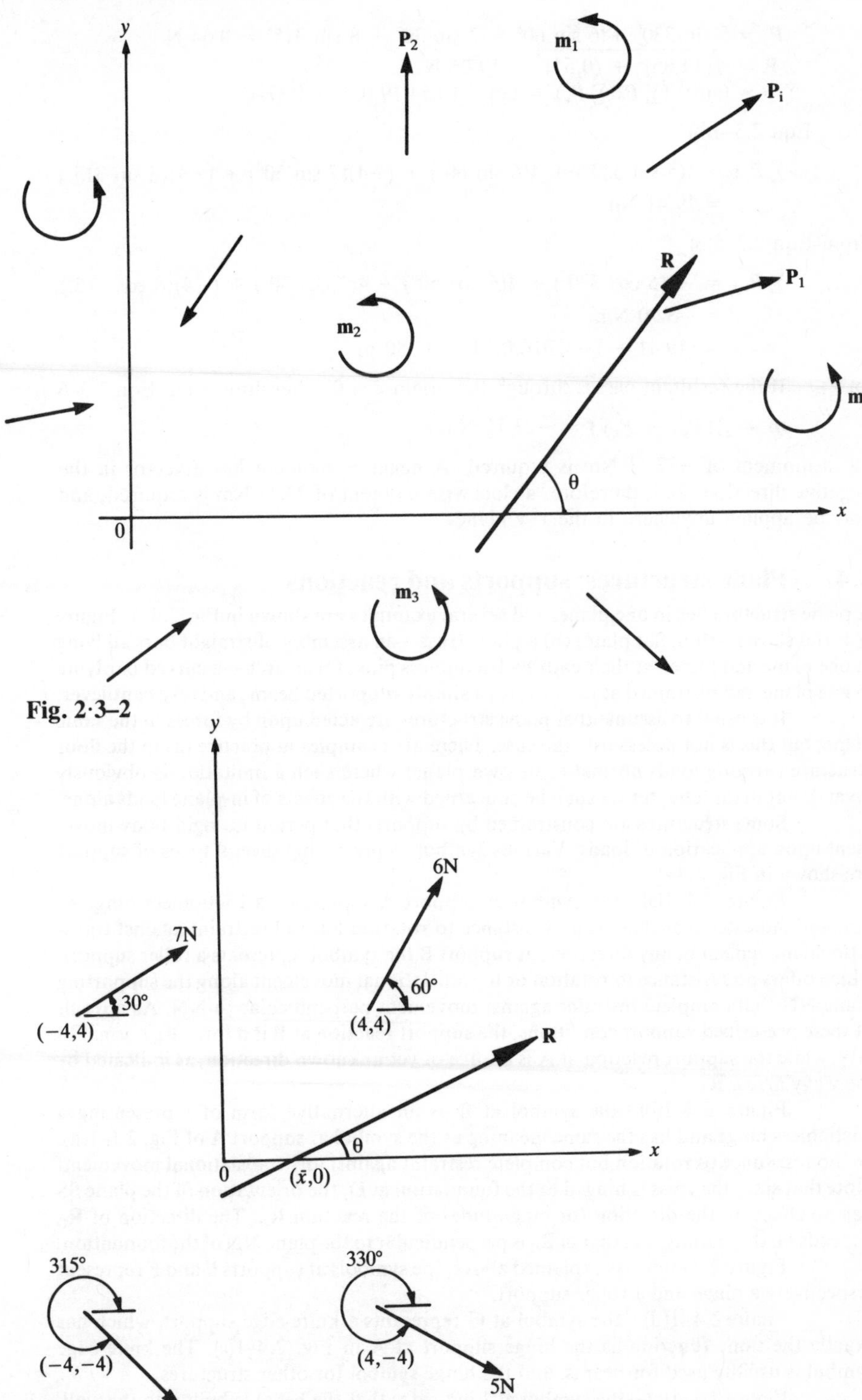

Fig. 2·3–2

Fig. 2·3–3

$$P_y = 5 \sin 330° + 6 \sin 60° + 7 \sin 30° + 8 \sin 315° = 0.54 \text{ N}$$
$$\mathbf{R} = \sqrt{(19.05)^2 + (0.54)^2} = 19.06 \text{ N}$$
$$\theta = \tan^{-1} (\Sigma P_y/\Sigma P_x) = \tan^{-1} (0.54/19.05) = 1°37.4'$$

From Eqn 2.3–4(b)

$$\Sigma P_y x = 4(5 \sin 330°) + 4(6 \sin 60°) + (-4)(7 \sin 30°) + (-4)(8 \sin 315°)$$
$$= 19.41 \text{ Nm}$$

From Eqn 2.3–5(b)

$$\Sigma P_x y = -4(5 \cos 330°) + 4(6 \cos 60°) + 4(7 \cos 30°) + (-4)(8 \cos 315°)$$
$$= -3.70 \text{ Nm.}$$
$$\bar{x} = (19.41 - (-3.70))/0.54 = 42.80 \text{ m.}$$

(b) If the resultant passes through the origin $\bar{x} = 0$. Therefore, from Eqn 2.3–6

$$m = \Sigma(P_y x - P_x y) = -23.11 \text{ Nm}$$

i.e. a moment of -23.11 Nm is required. A negative moment has a vector in the negative direction, and, therefore, a clockwise moment of 23.11 Nm is required, and may be applied anywhere in the x–y plane.

2.4 Plane structures: supports and reactions

A plane structure lies in one plane, and several examples are shown in Fig. 2.4–1. Figure 2.4–1(a) shows a thin, flat plate; (b) a plane truss—an assembly of straight bars all lying in one plane and joined at their ends by frictionless pins; (c) an arch—a curved bar lying in one plane and restrained at its ends; (d) a simply supported beam; and (e) a cantilever.

It is usual to assume that plane structures are acted upon by forces in the same plane, but this is not necessarily the case. There are examples in practice (as in the floor structure carrying loads normal to its own plane) where such a limitation is obviously invalid, but in this chapter we shall be concerned with the effects of in-plane loads alone.

Some structures are constrained by supports that permit no rigid-body movement upon application of loads. Various symbols representing several types of support are shown in Fig. 2.4–1.

Figure 2.4–1(a)—the symbol at support A represents a frictionless hinge or pin, and indicates that there is no resistance to rotation but full restraint against translational movement in any direction; at support B the symbol represents a roller support, which offers no resistance to rotation or to translational movement along the supporting plane NN, but complete restraint against movement perpendicular to NN. As a result of these prescribed support conditions, the support reaction at B is a force $\mathbf{R}_B$ normal to NN, while the support reaction at A is a force of yet unknown direction, as indicated by the wavy arrow $\mathbf{R}_A$.

Figure 2.4–1(b)—the symbol at D is an alternative form of representing a frictionless hinge and has the same meaning as the symbol at support A of Fig. 2.1–1(a), i.e. no resistance to rotation but complete restraint against any translational movement. Note that since the truss is hinged to the foundation at D, the orientation of the plane SS has no effect on the direction (or magnitude) of the reaction $\mathbf{R}_D$. The direction of $\mathbf{R}_D$ depends on the loading, but that of $\mathbf{R}_C$ is perpendicular to the plane NN of the foundation.

Figure 2.4–1(c)—as explained above, the symbols at supports E and F represent respectively a hinge and a roller support.

Figure 2.4–1(d)—the symbol at G represents a knife-edge support, which has exactly the same function as the hinge support at A in Fig. 2.4–1(a). The knife-edge symbol is usually used for beams, and the hinge symbol for other structures.

Figure 2.4–1(e)—the symbol at J indicates that the beam is built into the wall. Such a built-in support offers full restraint against rotation and translational movement.

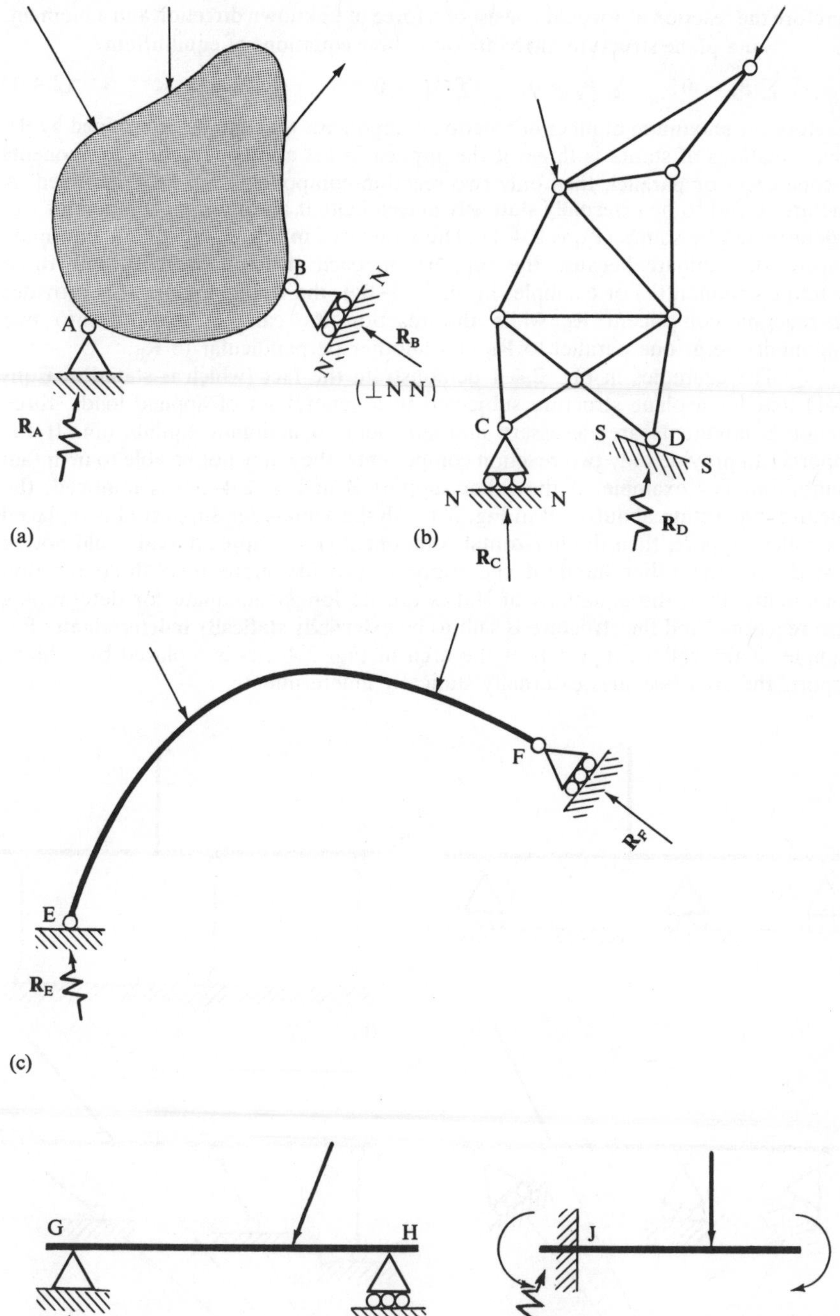

Fig. 2·4–1

Therefore the reaction at J would consist of a force of unknown direction and a moment.

For a plane structure, there are only three equations of equilibrium:

$$\sum P_x = 0; \qquad \sum P_y = 0; \qquad \sum M = 0 \tag{2.4-1}$$

Therefore the maximum number of reaction components that can be determined by the above equations of statics is three; if the applied forces and the reaction components are concurrent or parallel, then only two reaction components can be determined. A structure is said to be **externally statically determinate** if all the support reactions can be determined by statics (Eqns 2.4–1). The structures in Fig. 2.4–1 are all externally statically determinate because the supports in each structure provide only three reaction components. For example, in Fig. 2.4–1(a), the roller support at B provides one reaction component $\mathbf{R}_B$ while the reaction $\mathbf{R}_A$ can be resolved into two components—e.g. one parallel to $\mathbf{R}_B$ and another perpendicular to $\mathbf{R}_B$.

The examples in Fig. 2.4–1 demonstrate the fact (which is stated in Eqns 2.4–1) that for a plane structure subjected to a general set of applied loads, three reaction components are necessary and sufficient to maintain equilibrium. If the supports can provide only two reaction components, they may not be able to maintain equilibrium. For example, if the roller support B in Fig. 2.4–1(a) is removed, the structure may rotate about A; if in Fig. 2.4–1(d) the knife-edge support G is replaced by a roller support, then the horizontal component of the applied load could not be resisted. On the other hand, if the supports provide more than three reaction components, then the equations of statics are no longer adequate for determining these reactions, and the structure is said to be **externally statically indeterminate**. For example, if the roller support F of the arch in Fig. 2.4–1(c) is replaced by a hinge support, the arch becomes externally statically indeterminate.

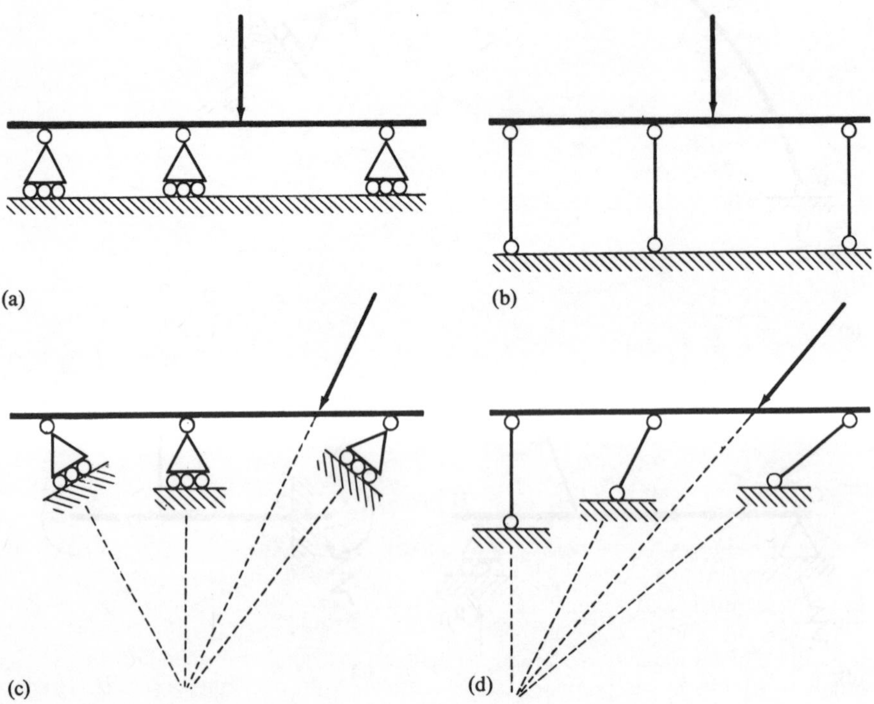

Fig. 2·4–2

Having said that three reaction components are sufficient for maintaining equilibrium in the general case of the plane structure, it is necessary to point out that these reaction components must not be parallel nor should they intersect at one point. For example, the beams in Figs 2.4–2(a) and (b) can be in equilibrium only under a vertical load; similarly the beams in Figs 2.4–2(c) and (d) are in equilibirum only of the load passes through the point of intersection of the support reactions. Arrangements of supports such as shown in Fig. 2.4–2 should be avoided.

2.5 Plane trusses

As explained in Section 2.4 with reference to Fig. 2.4–1(b), a plane truss is an assemblage of skeletal line members all lying in one plane, each member being joined at its ends to other members by means of frictionless pins through especially prepared holes on the axes of the several members meeting at the joints. Such a truss with ideal frictionless joints† or hinges is often referred to as a pin-jointed truss. In practice, trusses are often made of steel, and the members meeting at a joint are generally bolted or riveted to a common gusset plate or else are welded together; hence the ideal case of pin-jointed trusses hardly ever occurs in practice. However, a study of pin-jointed trusses is immensely valuable for developing a sound grounding in the principles of mechanics.

Note that the analysis of pin-jointed trusses has practical application as well; in the design office, calculations of the forces in members based on an assumption of pinned joints are frequently made. Axial forces so calculated are a good indication of those in the actual truss, and are often used for the design of less important trusses, for which the cost of more exact analysis based on more realistic assumptions of joint fixity may not be justified.

In order to form a stable (or rigid) truss, a sufficient number of members have to be used and these have to be arranged in a suitable manner. For example, a rectangular configuration (Fig. 2.5–1) is not stable because its shape can be changed without changing the length of any of the four members or bars. A triangular configuration (Fig. 2.5–2), however, is stable, because its shape cannot be changed without changing

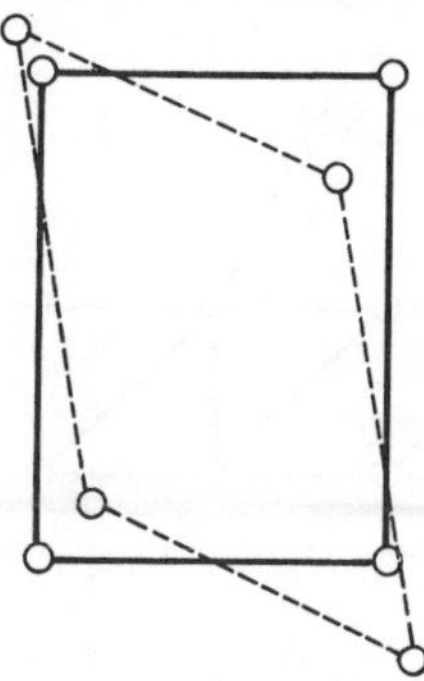

Fig. 2·5–1

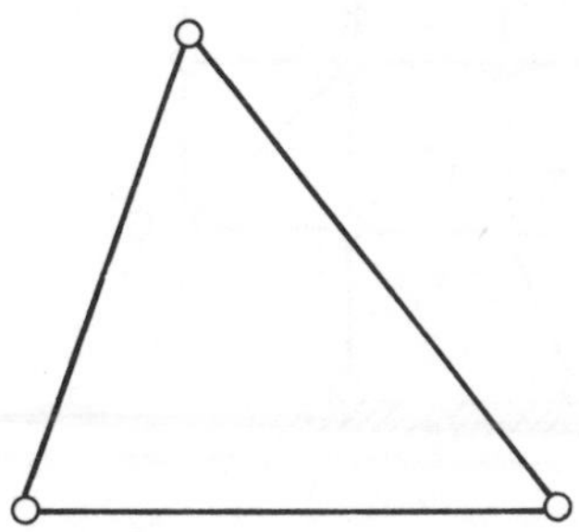

Fig. 2·5–2

the length of any of the members. By beginning with three members arranged to form a pin-jointed triangle, a stable truss can be constructed by successive addition of two new members and a joint; examples of stable trusses so constructed are shown in Fig. 2.5–3. A stable truss so constructed is often referred to as a **simple truss**; for such a truss there exists a definite relationship between the number of members m and the number of joints j. The original triangle consists of three members and three joints.

† In this book the symbol for a (frictionless) pin-joint is a small circle. For example, each of the joints in the truss in Fig. 2.4–1(b) is a pin-joint.

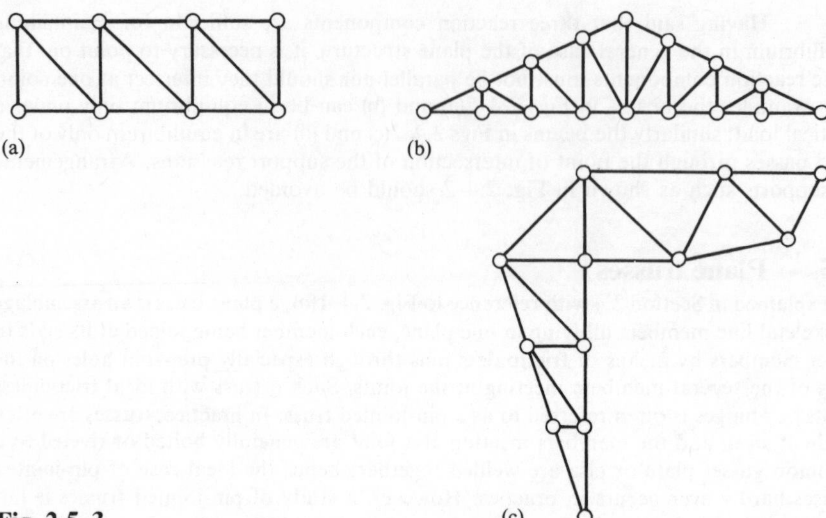

Fig. 2·5–3

Since for each additional joint there are two additional members, the relationship must be

$$m - 3 = 2(j - 3)$$

i.e.

$$m = 2j - 3 \qquad (2.5\text{–}1)$$

Equation 2.5–1 relates the number of members to the number of joints in a simple truss that is not connected to the foundation. It was pointed out in the previous section that three support reaction components were necessary and sufficient for the complete restraint of a plane structure. Hence a plane truss can be rigidly connected to

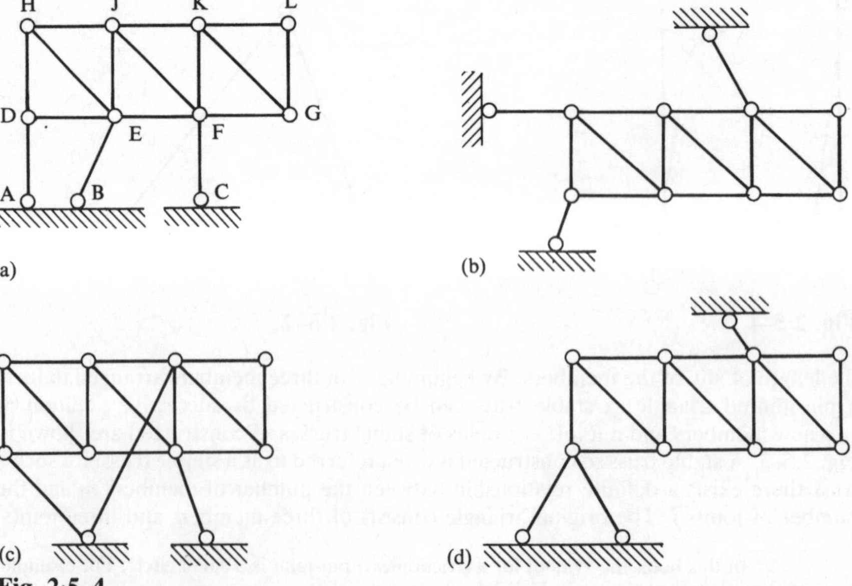

Fig. 2·5–4

the foundation by three members; these three members can be arranged in any manner (Fig. 2.5–4) provided that they are not all parallel or all intersecting at one point (see also Fig. 2.4–2). For a truss so connected to the founcation, Eqn 2.5–1 is modified to

$$m = 2j \tag{2.5–2}$$

where m is now the total number of members *including* those that connect the truss to the foundation, and j is the total number of joints *excluding* those at the foundation. (After the reader has studied the following chapters, he will appreciate that Eqns 2.5–1 and 2.5–2 merely express the condition that the total number of unknown member forces must be equal to the total number of 'degrees of freedom'.)

In Fig. 2.5–4, a truss rigidly connected to the foundation is constructed by first building a self-contained stable truss (Fig. 2.5–3(a)) and then restraining this stable truss to the foundation by three additional members. Figure 2.5–5 shows

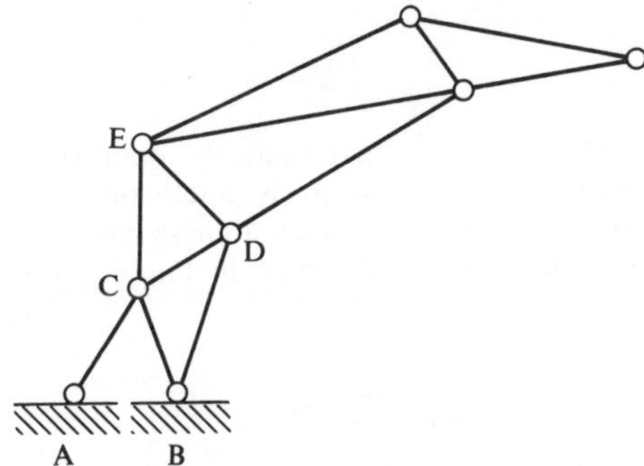

Fig. 2·5–5

another way of constructing a truss which is rigidly held to the foundation. Starting at two convenient points, A and B, on the foundation, joint C is established by two members AC and BC. Joint D is then established by two additional members CD and BD; similarly joint E is established by CE and DE, and so on. The manner of construction of the truss in Fig. 2.5–5 makes it obvious that the member–joint relationship in Eqn 2.5–2 is satisfied, provided it is remembered that m includes the members AC, BC, and BD, but j excludes the jonts A and B at the foundation.

When it was stated above, with reference to Fig. 2.5–4, that a self-contained stable truss could be rigidly held to the foundation by three members, it was implied that the forces in these three members were axial. External forces applied to the joints of a pin-jointed truss can only cause axial (i.e. tensile or compressive) forces in the members. Figure 2.5–6(a) shows a typical member, ab, of a truss loaded at the joints only. Since the joints a and b are frictionless pin joints, no moment can be transmitted to the member ab at the joints. Let $\mathbf{F}_a$ and $\mathbf{F}_b$ be the forces transmitted to member ab through the pins at a and b respectively. For ab to be in equilibrium, $\mathbf{F}_a$ and $\mathbf{F}_b$ must be equal in magnitude and opposite in sense; they must also act along the same straight line, i.e. along the axis of member ab. Figure 2.5–6(b) shows however, that if the effect of self-weight of the member is included then the forces transmitted to the member ab through the pins at joints a and b will not act along the member axis. The lines of action of $\mathbf{F}_a$ and $\mathbf{F}_b$ must be such that the three forces, $\mathbf{F}_a$, $\mathbf{F}_b$, and $\mathbf{W}$ (which represents the weight of the member ab), intersect at one point: the three forces must, of course, also form a triangle of forces that closes. In many trusses used in practice, the weights of the trusses are neglibible compared with the forces applied to them; in this chapter and in that following the weights of the individual members will be neglected unless a contrary statement is made.

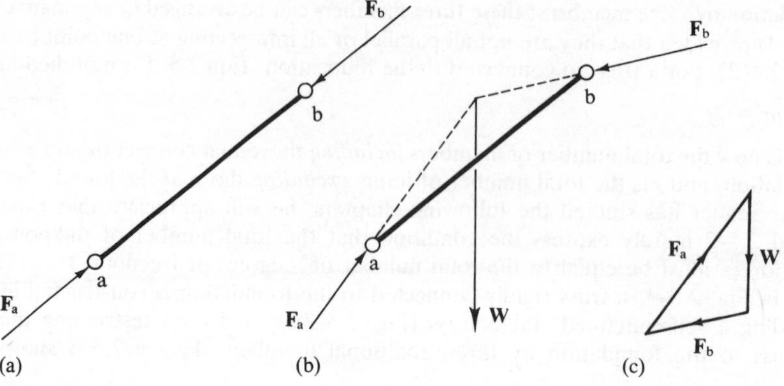

Fig. 2·5–6

Referring again to Eqns 2.5–1 and 2.5–2, note that while the number of members satisfying these equations is sufficient to build, respectively, a self-contained stable truss and a truss rigidly held to the foundation,it is essential that these members should be arranged in a suitable manner. For example, the truss in Fig. 2.5–7(a)

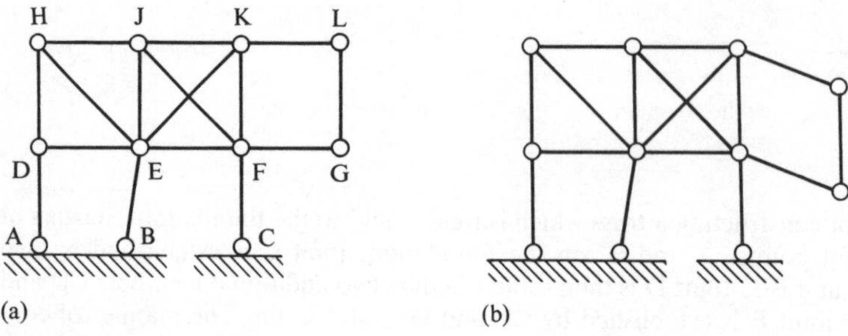

Fig. 2·5–7

satisfies the member–joint relationship of Eqn 2.5–2; in fact this truss is a modification of that in Fig. 2.5–4(a), formed by removing member KG and adding a member KE. Figure 2.5–7(b) shows clearly that the shape of the truss can be changed without changing the length of any member; it can be seen that the truss is not stable. Equations 2.5–1 and 2.5–2 are significant in that they give the minimum number of members required to build the stablę trusses. If a truss has less than this number of members, it cannot be stable; if it has more than this number of members then, as will be explained in Section 2.6, the equations of statics are no longer sufficient for the determination of the forces in the members of the truss under the action of applied forces, and the truss is said to be **internally statically indeterminate**. A structure which is either externally statically indeterminate (as discussed in Section 2.4) or internally statically indeterminate, or both, can be referred to as being statically indeterminate; a structure is said to be **statically determinate** when it is both internally statically determinate and externally statically determinate. Some of the more important differences between statically determinate structures and statically indeterminate structres are summarized in Table 2.5–1; the reader will appreciate the significance of these differences after he has studied the next few chapters.

Table 2.5–1

Statically determinate structures	*Statically indeterminate structures*
Cross-sectional dimensions and values of modulus of elasticity do not affect internal forces.	Internal forces depend on relative cross-sectional dimensions of members and on relative values of modulus of elasticity of members.
Slight† support displacement or slight† lack of fit of a member does not affect internal forces.	Internal forces can be significantly affected even by slight† support displacement or slight† lack of fit of a member.

† Not causing gross change of geometry of structure.

Example 2.5–1. Explain whether the plane trusses in Fig. 2.5–8 are (a) statically determinate, and (b) stable.

SOLUTION

(a) *Truss in Fig. 2.5–8(a)*

No. of members $m = 15$ (including members 1, 2, 3).

No. of joints $j = 7$ (excluding joints H and J at the foundation).

Then $m > 2j$ and Eqn 2.5–2 is not satisfied.

Hence the truss is internally statically indeterminate. Since the supports H and J provide only three reaction components (one at H in direction of member 1, and two at J), the truss is externally statically determinate.

To examine the stability of the truss, consider it as consisting of a truss ABCDEFG rigidly held to the foundation by members 1, 2, and 3. Truss ABCDEFG can be regarded as having been built by beginning with ABC, and then establishing each additional joint with two additional members (e.g. joint D with BD and CD) until joint G is fixed by members FG and CG. Subsequently an extra member AG is added. Hence the truss is stable; in fact it is stable even without the extra bar AG.

Conclusion: The truss in Fig. 2.5–8(a) is stable but statically indeterminate.

(b) *Truss in Fig. 2.5–8(b)*

The truss can be considered as a truss ABCDEFG rigidly held to the foundation by three bars, 1, 2, 3. Hence it is externally statically determinate. The truss ABCDEFG has 11 members and 7 joints. Since $m = 2j - 3$, Eqn 2.2–1 is satisfied, and hence the truss ABCDEFG is internally statically determinate. As explained in (a) above, it is also stable.

Conclusion: The truss in Fig. 2.5–8(b) is stable and statically determinate.

(c) *Truss in Fig. 2.5–8(c)*

As explained in (a) and (b) above, the truss ABCDEFG is stable and internally statically determinate. However, it is connected to the foundation by hinges A and G, each of which is capable of providing two reaction components. Therefore the truss is externally statically indeterminate.

Note that the structure can be considered as consisting of a stable and internally statically determinate truss BDEFC which is attached to the foundation by four bars—AB, AC, GC, and GF.

Conclusion: The structure in Fig. 2.5–8(c) is stable but statically indeterminate.

(d) *Truss in Fig. 2.5–8(d)*

This truss itself is identical to that in Fig. 2.5–8(c), except that the one here is connected to the foundation by a pin-joint at A (providing two reaction components) and a roller-bearing at G (providing one reaction component).

Conclusion: The truss in Fig. 2.5–8(d) is stable and statically determinate.

(e) *Truss in Fig. 2.5–8(e)*

No. of members $m = 11$.

No. of joints $j = 7$.

Hence $m = 2j - 3$ and Eqn 2.5–1 is satisfied.

However, the members are so arranged that the truss is not stable. For example, distortion to the shape indicated by the dotted lines is possible without changing the length of any member.

The supports provide two reaction components at A and one at G; hence the structure is externally statically determinate. However, statical determinancy is irrelevant in this case, because the structure is not stable.

Conclusion: The truss in Fig. 2.5–8(e) is not stable.

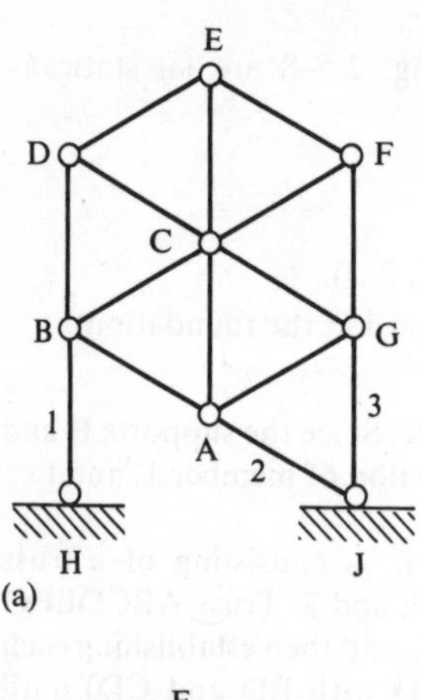

(a)

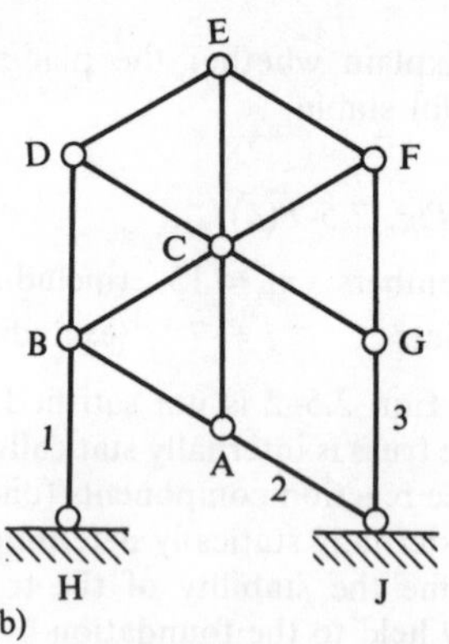

(b)

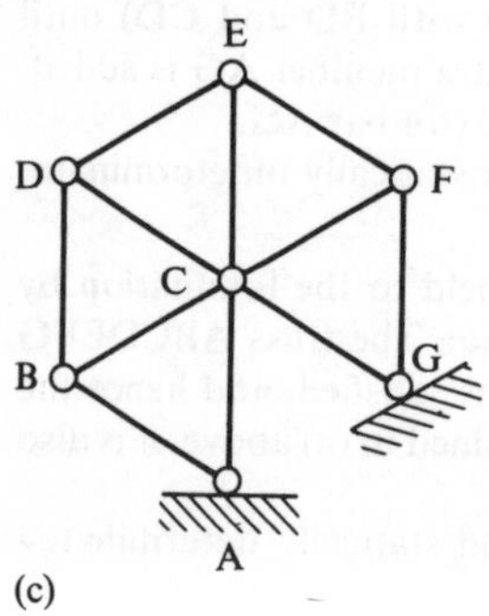

(c)

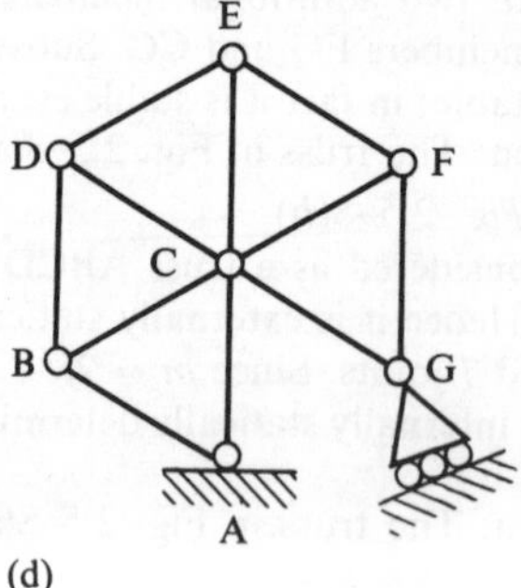

(d)

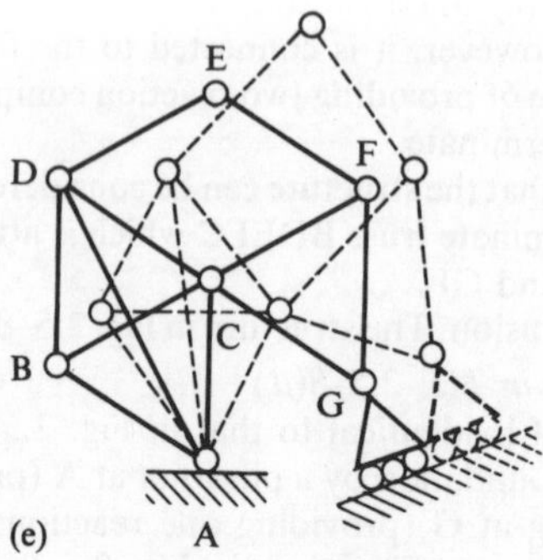

(e)

Fig. 2·5–8

Example 2.5–2. Explain whether the trusses in Fig. 2.5–9 are (a) stable, (b) statically determinate.

SOLUTION

(a) *Truss in Fig. 2.5–9(a)*

No. of members $m = 29$.

No. of joints $j = 16$.

Hence $m = 2j - 3$ and Eqn 2.5–1 is satisfied.

However, an examination of Fig. 2.5–9(a) shows that the truss does not satisfy the definition of a 'simple truss', which is constructed by beginning with a triangle and then successively establishing each additional joint with two additional members. But the truss can be considered as made up of trusses ABC and DEF which are connected together by bars 1, 2, and 3. In the figure, the trusses ABC and DEF are shaded, for clarity. The reader should verify that each of these shaded trusses is a simple truss, being both stable and statically determinate. Since these two rigid trusses are connected together by three non-parallel, non-intersecting bars, it can be concluded that the 'compound-truss' ABDFEC is stable and also statically determinate. In fact, a truss such as that in Fig. 2.5–9(a), which is made up of two (or more) simple trusses rigidly connected together, is called a **compound truss**.

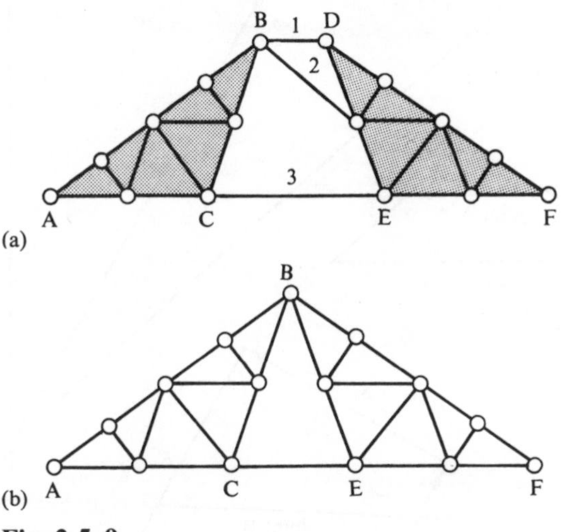

Fig. 2·5–9

(b) *Truss in Fig. 2.5–9(b)*

The only difference between this truss and the one in Fig. 2.5–9(a) is that here the two simple trusses are rigidly held together, not by three bars, but by a hinge (or pin-joint) at B and a bar CE. The hinge B and the bar CE together provide three forces.

Hence it can be concluded that the compound truss in Fig. 2.5–9(b) is stable and statically determinate.

2.6 Analysis of plane trusses

Trusses can be efficiently analysed by the matrix computer method given later in Chapter 7. However, a study of the various methods of analysis by hand is an invaluable means of developing a sound understanding of the equilibrium of coplanar forces. Plane trusses can be analysed by hand using (a) the method of joints, (b) the method of sections, or (c) a combination of the two methods.

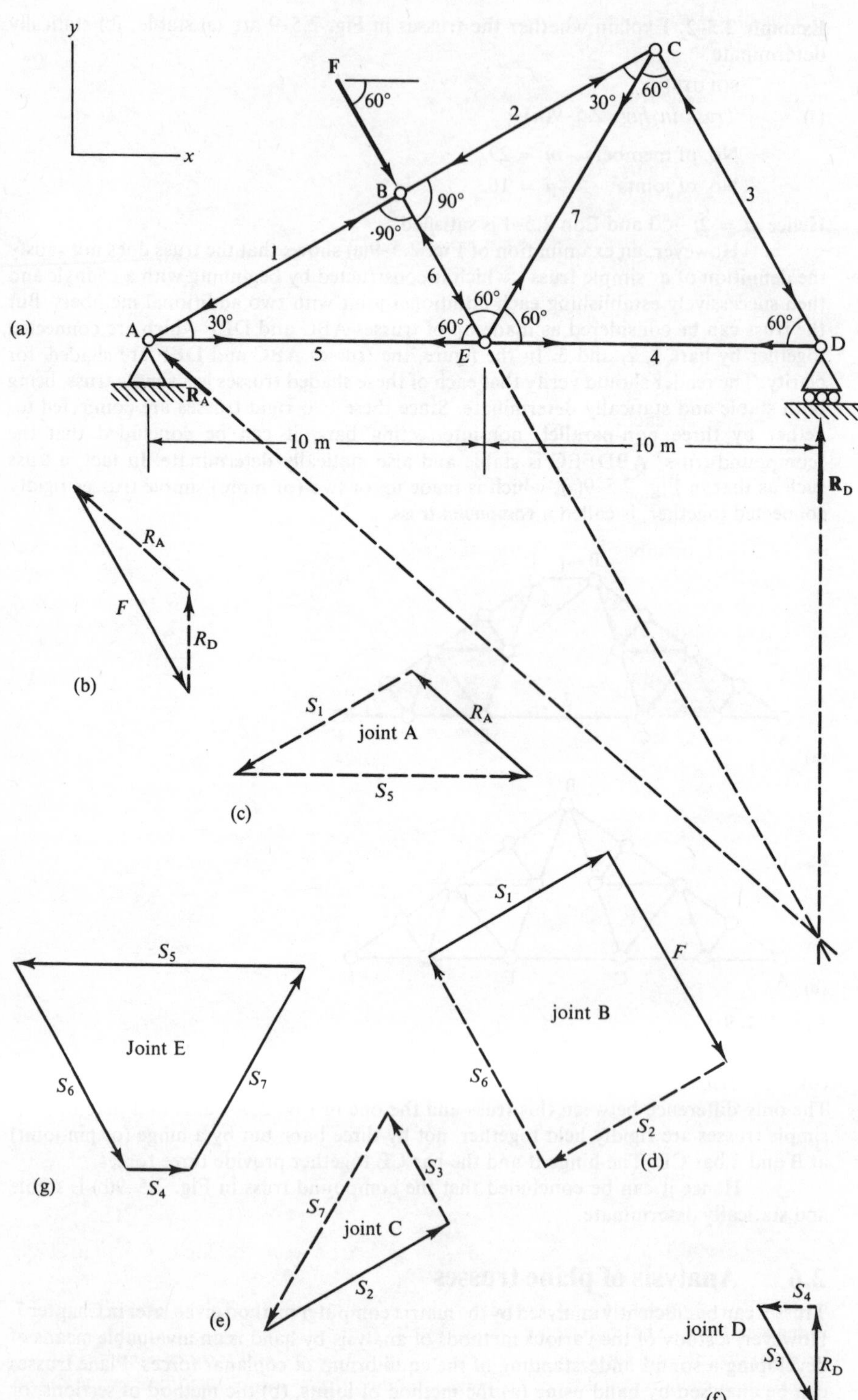
y
x
F
60°
C
30°
60°
2
B
90°
·90°
3
7
1
6
60°
(a)
A
30°
60°
60°
60°
D
5
E
4
R_A
10 m
10 m
R_D
R_A
F
R_D
(b)
S_1
R_A
joint A
S_5
(c)
S_1
F
S_5
joint B
Joint E
S_6
S_7
S_6
S_2
(d)
(g)
S_4
S_3
S_7
joint C
S_2
(e)
S_4
joint D
S_3
R_D
(f)

Fig. 2·6–1

Method of joints. The **method of joints** consists of the successive application of

$$\sum P_x = 0, \qquad \sum P_y = 0$$

to the joints of the truss, thereby determining the unknown axial forces of the members. The method is illustrated by the following worked example.

Example 2.6–1. Figure 2.6–1(a) shows a statically determinate truss acted on by a force **F** applied at joint B. Using the method of joints, determine graphically the axial forces in all the members and the support reactions at A and D.

SOLUTION Since each joint is in equilibrium under the action of the member axial forces (and the applied forces and support reactions if any), then these forces acting on any individual joint must form a closed polygon. If, at any joint, only two force components are unknown then these can be determined by drawing a force polygon. An examination of Fig. 2.6–1(a) shows that there are at least three unknowns at every joint. However, since the entire truss is in equilibrium under the action of the applied force and the support reactions, the latter can be determined by a force triangle, noting that the direction of the reaction $\mathbf{R}_A$ must be such that the three forces **F**, $\mathbf{R}_A$, and $\mathbf{R}_D$ intersect at one point. With $\mathbf{R}_A$ and $\mathbf{R}_D$ determined from a force triangle (Fig. 2.6–1(b)), a force triangle can be drawn for joint A (Fig. 2.6–1(c)), thereby determining the axial forces S_1 and S_5 in members 1 and 5 respectively. Figure 2.6–1(c) shows that S_1 acts towards joint A, that is member 1 is in compression; likewise, S_5 acts away from joint A, showing that member 5 is in tension. Arrow-heads are now added in members 1 and 5 to indicate that (a) member 1 is in compression, so that its axial force acts towards both joints A and B; (b) member 2 is in tension so that its axial force acts away from both joints A and E.

Now that S_1 is known, a force polygon can be constructed for joint B (Fig. 2.6–1(d)), determining S_2 and S_6. Arrow-heads are now added on to members 2 and 6 (Fig. 2.6–1(a)) to indicate compression in both members.

In a similar way, force polygons are drawn successively for joints C and D (Figs. 2.6–1(e) and (f)). The force polygon for joint E (Fig. 2.6–1(g)) is drawn from the member forces acting on that joint which are now all known. The closing of this polygon is an indication that the previous force polygons have been constructed correctly.

In each of the force polygons in Figs. 2.6–1(b) to (f) inclusive, the unknown forces determined from the polygon have been shown by dotted lines.

Attention is drawn to two significant points: (a) the layout dimensions of the truss are immaterial as far as the member forces are concerned—such forces depend only on the shape of the truss; (b) the cross-sectional sizes of the members and their modulus of elasticity have no effect on member forces.

In Example 2.6–1, the graphical analysis has been carried out by constructing a force polygon for the truss as a whole, together with a separate polygon for each joint. The result is that the force vector for each member appears in two force polygons, e.g. vector for S_1 appears in Figs. 2.6–1(c) and (d). It is possible to avoid this duplication by combining Figs. 2.6–1(b) to (g) into a single force diagram called a **Maxwell diagram**, in which no force vector is duplicated. To construct a Maxwell diagram, it is necessary to adhere to a system of notation; a commonly used system, called **Bow's notation**, is explained below with reference to the truss in Fig. 2.6–1(a), which for clarity is redrawn in Fig. 2.6–2(a).

According to Bow's notation the spaces between the lines of action of the external forces and internal (member) forces acting on the joints of the truss are designated with letters a, b, c, etc. Thus, in Fig. 2.6–2(a), the entire space between $\mathbf{R}_A$ and **F** is denoted by 'a', that between **F** and $\mathbf{R}_D$ by 'b', that between $\mathbf{R}_D$ and $\mathbf{R}_A$ by 'c', and that between the internal forces in members AB and BE by 'd', and so on. Using this system of notation, each external force or internal member force can be identified by the spaces on each side of that force. For example the force **F** acting at joint B lies between the spaces 'a' and 'b' and is referred to as *ab*. Note that movement from one space to the next space should be *consistently clockwise* (or consistently anticlockwise) both when considering the forces acting on a joint or the forces acting in a member. Thus, when forces acting on joint B are considered, the force in member 2 is denoted by *be*, but when the forces acting on joint C are considered the force in the same member 2 will be denoted by *eb*.

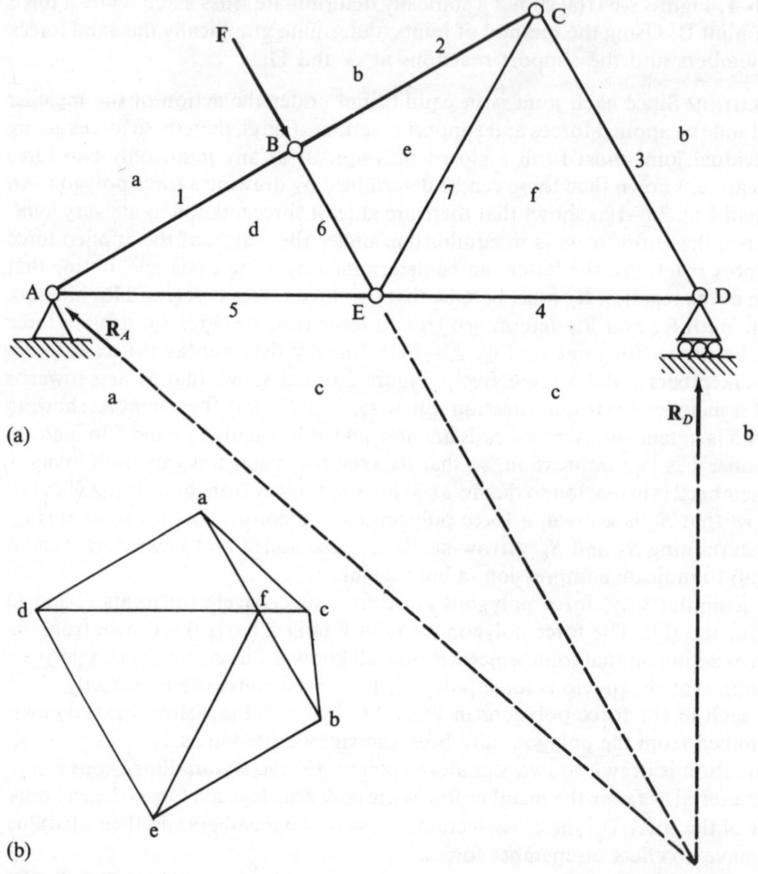

Fig. 2·6–2

The Maxwell diagram for the truss in Fig. 2.6–2(a) is constructed as follows:

STEP 1: Construct the force polygon for the forces **F**, $\mathbf{R}_A$, and $\mathbf{R}_D$. This is triangle abc in Fig. 2.6–2(b), where *ab* represents **F** which lies between spaces 'a' and 'b, *bc* represents $\mathbf{R}_D$ which lies between spaces 'b' and 'c', and *ca* represents $\mathbf{R}_A$ which lies between spaces 'c' and 'a'.

STEP 2: Construct the force polygon for a joint at which only two unknown forces act. Either joint A or joint D can be chosen. Let us choose joint A. Going round joint A clockwise, the force $\mathbf{R}_A$ is denoted as *ca*, the force in member 1 as *ad*, and the force in member 5 as *dc*. The force polygon, cad, is as shown in Fig. 2.6–2(b), which shows that the force in member 1, *ad*, acts towards joint A and is therefore a compression; similarly, *dc* acts away from joint A and hence member 5 is in tension.

STEP 3: Construct force polygon for joint B, say, which is now acted on by only two unknown forces—*be* in member 2 and *ed* in member 6. The force polygon, dabed, is as shown in Fig. 2.6–2(b).

STEP 4: Force polygons ebfe and fbcf are similarly constructed for joints C and D, respectively.

STEP 5: The closure of force polygon defcd for joint E provides a check of the correctness of the Maxwell diagram.

Note that in a Maxwell diagram the sense of direction of the vectors representing the various member forces need not be indicated by arrow-heads. This is because each force vector is represented by the line joining two letters and the order of the letters gives the direction of the force vector. For example, suppose it is required to know the magnitude and sense of the force in member 7. The magnitude is immediately determined from the length of the line joining e and f in Fig. 2.6–2(b). To determine whether member 7 is in tension or compression, we can consider either the forces acting on joint E or those on joint C. If joint E is considered, then the force in member 7 is represented by *ef* in Fig. 2.6–2(b); since *ef* is directed away from E, member 7 is in tension. If joint C is considered, the force in member 7 is represented by *fe* (since we always move clockwise round a joint); *fe* is directed away from C and again member 7 is in tension.

Example 2.6–2. Briefly explain the basis of the method of joints for statically determinate trusses. Using the method of joints, determine analytically the reactions and member forces in the truss in Fig. 2.6–1(a).

SOLUTION If there are j joints in a truss, then since there are two equations ($\sum P_x = 0$; $\sum P_y = 0$) to each joint, there will be a total of $2j$ equations. If the truss is supported on a hinge and a roller, such as the one in Fig. 2.6–1(a), the unknowns are the m member forces and the three reaction components. Since Eqn 2.5–1 states that

$$m + 3 = 2j$$

the number of equations will be exactly equal to the number of unknowns. If the truss is connected to the foundation by three bars, as in Figs. 2.5–4 and 2.5–5, then the total number of unknowns is m, where m now includes the three bars which hold the truss to the foundation. In this case, the member–joint relationship (Eqn 2.5–2) is

$$m = 2j$$

so that there are again exactly as many equations as there are unknowns.

Note, however, that while it is always possible to set up the $2j$ equations and solve them formally, it is rarely desirable to do so. Instead, we should first try to determine those member forces which can be found by inspection or by very little calculation. With as many as possible of such readily determined forces calculated, the number of simultaneous equations that have to be solved is greatly reduced. Indeed, in many cases, it is not even necessary to solve more than two simultaneous equations. This point will be made clear by the following calculations for the truss in Fig. 2.6–1(a), which, for clarity, has been redrawn again as Fig. 2.6–3.

Let S_1 to S_7 inclusive be the (assumed) *tensile* forces in members 1 to 7 respectively. Let X_A and Y_A be the x- and y-components of the support reaction at A, i.e. $\mathbf{R}_A = [X_A \ Y_A]^T$; let $\mathbf{R}_D$ be the support reaction at D which is known to be in the y-direction.

STEP 1: Consider equilibrium of whole truss (Fig. 2.6–3). $\sum M = 0$ at joint A gives

$$R_D \times 20 = F \times 10 \cos 30°$$

then

$$R_D = 0{\cdot}433F$$

$\sum P_y = 0$ gives

$$Y_A + R_D = F \sin 60°$$

where R_D is already known, then

$$Y_A = 0{\cdot}433F$$

$\sum P_x = 0$ gives

$$X_A + F\cos 60° = 0$$

hence

$$X_A = -0{\cdot}5F$$

Note: It could have been written down by inspection that $Y_A = R_D = 0{\cdot}433F$ and that $X_A = -0{\cdot}5F$. The line of action of the force **F** passes through B and E; hence **F** at B can be replaced by a force **F** at E, for the purpose (only) of determining the support reactions. A force **F** at E inclined at 60° to the horizontal, is equivalent to a horizontal component of $F\cos 60°$ and a vertical component of $F\sin 60°$, both applied at joint E. The horizontal component $F\cos 60°$ is resisted entirely by support A; hence $X_A = -0.5F$. The vertical component is resisted equally by supports A and D; hence $Y_A = R_D = 0.5F\sin 60° = 0.433F$.

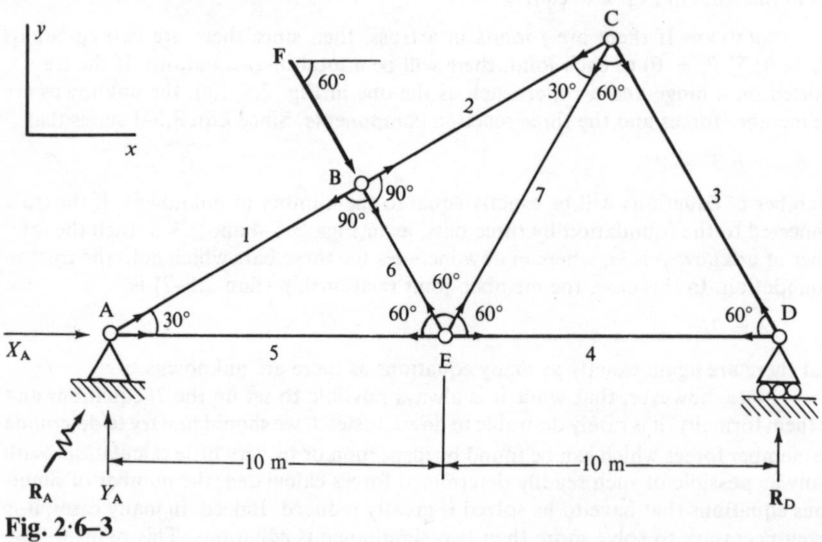

Fig. 2·6–3

STEP 2: Equilibrium of joint A. Condition $\sum P_y = 0$ gives

$$S_1 \sin 30° + Y_A = 0$$

where Y_A is $0.433F$ (from Step 1). Then

$$\underline{S_1 = -0.866F}$$

(The negative sign for S_1 indicates that member 1 is actually in compression, and not in tension as assumed.) Condition $\sum P_x = 0$ gives

$$S_5 + X_A + S_1\cos 30° = 0$$

where $X_A = -0.5F$ and $S_1 = -0.866F$. Then

$$\underline{S_5 = 1.25F}$$

STEP 3: Equilibrium of joint B. For equilibrium in direction BC, since $S_2 = S_1$,

$$\underline{S_2 = -0.866F}$$

For equilibrium in direction BE, since $S_6 + F = 0$,

$$\underline{S_6 = -F}$$

STEP 4: Equilibrium of joint D. Condition $\sum P_y = 0$ gives

$$S_3 \sin 60° + R_D = 0$$

where $R_D = 0.433F$ (from Step 1). Then

$$\underline{S_3 = -0.5F}$$

Condition $\sum P_x = 0$ gives

$$S_4 + S_3 \cos 60° = 0$$

$$S_4 = \underline{0.25F}, \quad \text{since} \quad S_3 = -0.5F$$

STEP 5: Equilibrium of joint C. For equilibrium in direction CB, we must have

$$S_2 + S_7 \cos 30° = 0$$

where $S_2 = -0.866F$ (from Step 3). Then

$$S_7 = \underline{F}$$

Forces S_1 to S_7 inclusive are now all determined. As a check, consider equilibrium of joint E.

Condition $\sum P_x = 0$ demands that

$$S_7 \cos 60° + S_4 - S_6 \cos 60° - S_5 = 0$$

which is identically satisfied for the values of S_7, S_4, S_6, and S_5 calculated above. Condition $\sum P_y = 0$ demands that

$$S_6 \sin 60° + S_7 \sin 60° = 0$$

which is again identically satisfied.

Example 2.6–2 above indicates that while in general a truss of j joints can give rise to $2j$ simultaneous equations, in fact one usually does not have to solve many simultaneous equations when using the method of joints analytically. In the particular case of Example 2.6–2 it was never necessary even to solve two simultaneous equations. Actually it can be said that for a 'simple truss' it will not be necessary to solve more than two simultaneous equations at a time. By definition, in any simple truss there is at least one joint (and the last joint established in accordance with the rule of construction is one such joint) at which only two members meet. Hence, if we start with this joint, then consider the last but one joint established in accordance with the rule, and then work backwards joint by joint, it would be possible to determine all member forces without solving more than two simultaneous equations at a time. For the more complicated case of a 'compound truss' it may be necessary sometimes to solve more than two simultaneous equations at a time. For some trusses, called **complex trusses,**† it may be necessary to solve many simultaneous equations at one time; such trusses should in general be analysed by the computer method presented in Chapter 7.

† A complex truss is defined as one which does not satisfy either the definition of a 'simple truss' or the definition of a 'compound truss'.

Method of sections. The **method of sections** is particularly useful when it is required only to determine the forces in a few members of the truss. In this method, a section is passed through the truss to cut the members whose forces are to be determined. This imaginary section divides the truss into two parts, each of which is in equilibrium under the action of the externally applied forces and the internal forces of the cut members. In the general case, these external and internal† forces are not concurrent; hence from the three conditions

$$\sum P_x = 0, \qquad \sum P_y = 0, \qquad \sum M = 0$$

up to three member forces can be determined. The application of the method is illustrated in the following worked examples.

Example 2.6–3. Determine the forces in members CD, CM, LM, and DM of the truss in Fig. 2.6–4(a), caused by the applied forces shown.

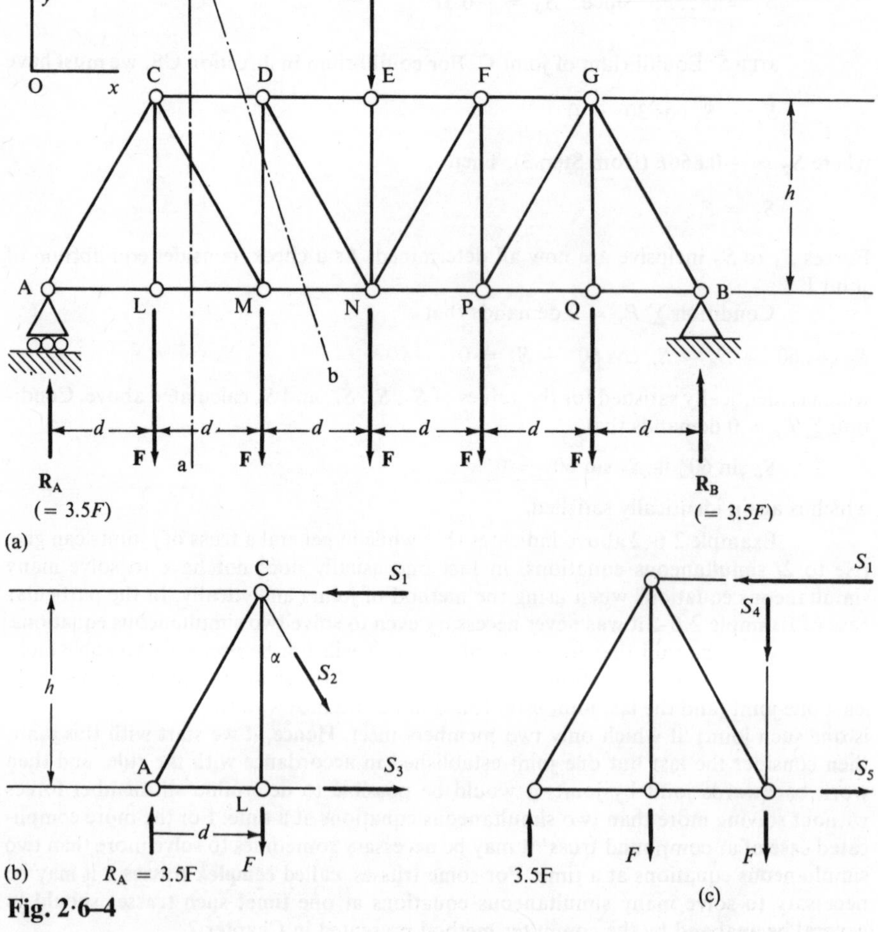

Fig. 2·6–4

† Actually the internal forces of the cut members now become 'external' forces acting on each part of the cut truss.

SOLUTION First consider the equilibrium of the portion of the truss to the left of section a–a (Fig. 2.6–4(b)). Applying condition $\sum P_y = 0$,

$$S_2 \cos \alpha + F = R_A$$

where R_A is $3.5F$ by inspection. Hence

$$S_2 = \frac{2.5F}{\cos \alpha} = 2.5 \frac{\sqrt{(h^2 + d^2)}}{h} F$$

Applying condition $\sum M = 0$ to point C, since $S_3 \times h = 3.5F \times d$,

$$S_3 = 3.5 \frac{d}{h} F$$

(Both S_1 and S_2 pass through point C and hence do not enter into the moment equation.)

Applying condition $\sum P_x = 0$,

$$S_1 = S_3 + S_2 \sin \alpha$$

$$= 3.5\left(\frac{d}{h}\right) F + \left[\frac{2.5\sqrt{(h^2 + d^2)}}{h}\right] F \left[\frac{d}{\sqrt{(h^2 + d^2)}}\right]$$

$$= 6\left(\frac{d}{h}\right) F$$

To determine the axial force in member DM, consider the equilibrium of the portion to the left of section b–b (Fig. 2.6–4(c)).

Condition $\sum P_y = 0$ gives

$$S_4 = R_A - F - F = 1.5F$$

Note that in the free-body diagrams (Fig. 2.6–4(b) and (c)), the forces S_2, S_3, and S_5 have been assumed to be tensile while S_1 and S_4 have been assumed to be compressive. These assumptions were based only on intelligent guesses. We could have assumed, of course, all axial forces to be tensile and left it to the calculations to show whether they are actually tensile or compressive. However, the reader will gradually discover that except for relatively complicated trusses, it is usually possible to determine by inspection whether the force in a member is tensile or compressive. In any case, even if we guess wrongly, our error will be revealed in the subsequent calculations by a negative sign. For example, if we had assumed S_4 to be tensile, the calculations would have shown S_4 to be $-1.5F$.

Example 2.6–4. Determine the force in member HE of the truss in Fig. 2.6–5(a).

SOLUTION Consider the equilibrium of that part of the truss to the left of section a–a (Fig. 2.6–5(b)).

Since both the unknown forces S_{HC} and S_{DE} pass through joint A, it will be most convenient to apply the condition $\sum M = 0$ to this joint.

$$S_{HE} \times d = F \times \frac{l}{4}$$

Thus

$$S_{HE} = \frac{Fl}{4d} \quad \text{(compression)}$$

The lever-arm distance, d, can be calculated or else measured from a drawing.

The above calculations show that the force in member HE depends only on the force **F** and not on forces **P** and **Q**. The explanation for this is simple: Consideration of vertical equilibrium for joint D shows that the member HD is unstressed. Hence joint H is under the action of four forces only (Fig. 2.6–5(c)); of these, the forces in members HA and HC are collinear. For equilibrium of joint H in a direction perpendicular to HC, we must have

$$S_{HE} \sin \beta = F \sin \alpha$$

That is, S_{HE} depends on **F** only. As an exercise the reader should show that the force in member EJ depends only on force **Q**; that is, if **Q** is zero, then member EJ is unstressed. Similarly, show that the force in member CE depends only on **F** and **Q** and is independent of **P**. This means that, if the truss in Fig. 2.6–5(a) is acted on by force **P** only, then members CE, HE, JE, HD, and JG are all unstressed. The ability to determine by inspection which members are unstressed often greatly simplifies the analysis.

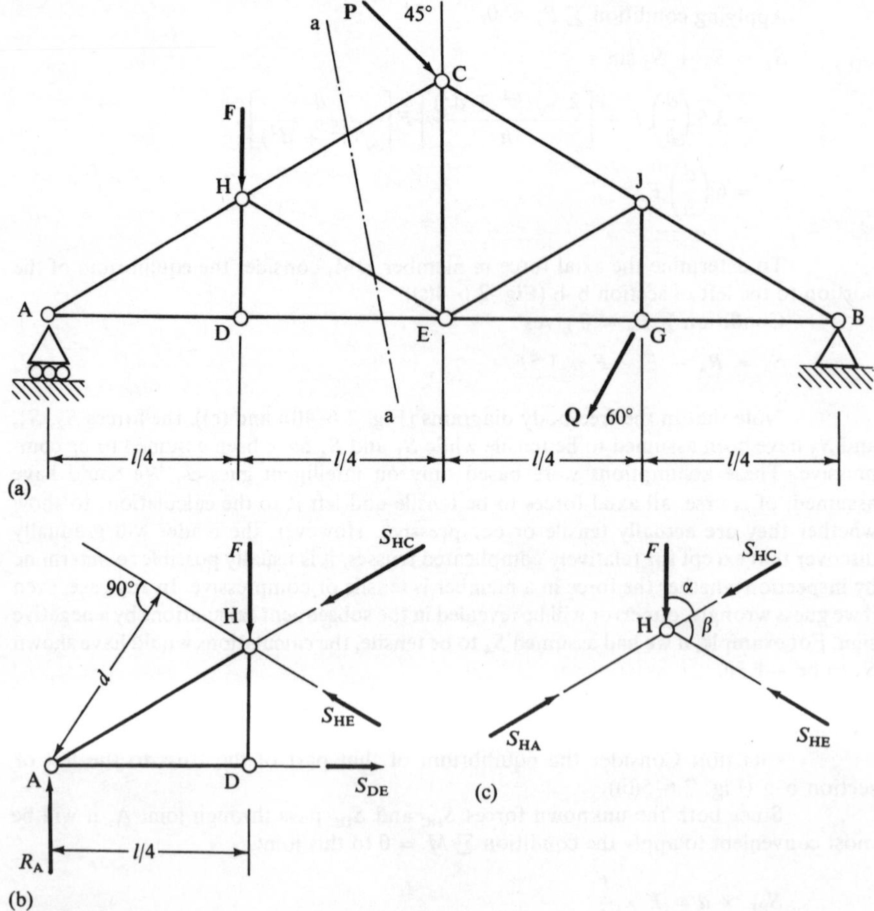

Fig. 2·6–5

Combined use of method of joints and method of sections. It is sometimes advantageous to use both the method of joints and the method of sections in the same problem. For example, suppose it is required to determine the forces in all the members of the statically determinate truss in Fig. 2.6–6. Pass a section to cut members 1, 2, and 3, and consider the equilibrium of the portion DCE, which is under the action of the applied force **P** and the forces S_1, S_2, and S_3 in members 1, 2, and 3 respectively. By inspection,

$$S_2 = 0, \quad \text{since} \quad \sum \text{horizontal forces} = 0$$

and

$$S_3 = \tfrac{3}{4}P \ \text{(compression)}; \qquad S_1 = \tfrac{1}{4}P \ \text{(compression)}$$

(If the reader is unable to tell by inspection that $S_3 = \frac{3}{4}P$ and $S_1 = \frac{1}{4}P$, he should apply the condition $\sum M = 0$ to joints C and D successively.) Once the forces in members 1, 2, and 3 have been determined by the method of sections, the forces in the other members can easily be determined by the method of joints. Actually, since S_2 is zero, we know by inspection that members FA and FB must be unstressed, and consequently, member AB must also be unstressed.

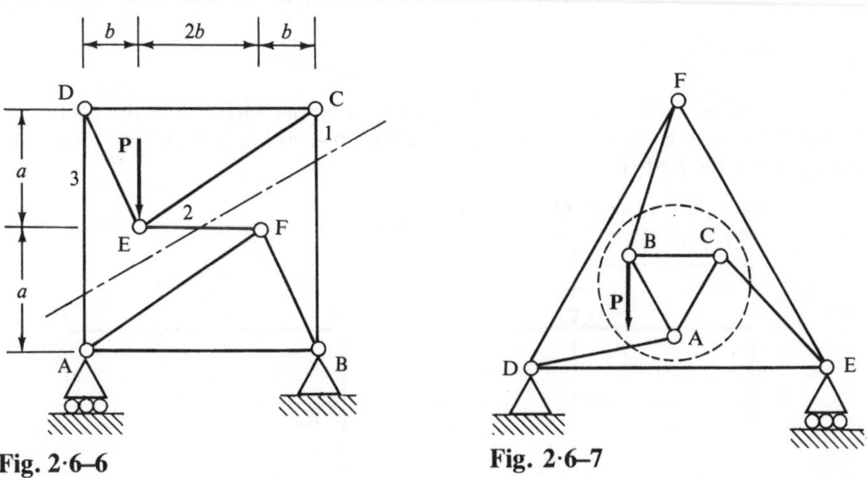

Fig. 2·6–6

Fig. 2·6–7

Figure 2.6–7 also shows a statically determinate truss which cannot be solved by the method of joints alone. However, if a circular section (a section need not be straight!) is passed through members AD, BF, and CE to isolate the portion ABC, the equilibrium of this subtruss ABC would give three equations

$$\sum P_x = 0; \qquad \sum P_y = 0; \qquad \sum M = 0$$

from which the forces in the three cut members can be determined. When these three forces are known the forces in the other members are readily found by the method of joints.

2.7 Influence lines—the effects of moving loads

So far in this book we have considered the effects on a structure caused by forces which act at particular fixed points. This is not always the case; a bridge, for example, is built in order that loads may move across it—the loads remain constant, but their point of application varies across the span. If a force changes its position on a structure the effects it causes will also vary, and it is frequently necessary to know how the changing positions of a particular load may vary the support reactions and the internal stresses in the structure. The diagram which shows how some effect varies as a single load rolls across the span is known as an **influence line**.

2.7–1 Influence lines for reaction in beams

Consider the simply supported beam of length L, shown in Fig. 2.7–1(a) subjected to a single rolling force of magnitude P units. It is necessary to determine the values of the reactions R_A and R_B for all possible positions of the load between A and C.

If the load is x from A ($0 < x < L$) then, taking moments about A

$$R_B = \frac{x}{a} P \text{ units}$$

that is, R_B varies from 0 when $x = 0$ (the load is at A) to $((a + b)/a)P$ when the load is at C, and having the value P when the load is at B. The figure shows, at (b), the variation of the coefficient f in the equation $R_B = fP$, where f is an **influence coefficient**—the value of R_B due to a load in a particular position will be the product of the magnitude of the load and the influence coefficient for that position. The effect of a 'train' of loads a fixed distance apart rolling across the span will be given by summation

$$R_B = \sum_i^n f_i P_i$$

Similarly, the influence line for R_A (showing the variation of the coefficient by which a load must be multiplied to give R_A) is shown in Fig. 2.7–1(c) and represents the equation for f when

$$R_A = fP = ((a - x)/a)P$$

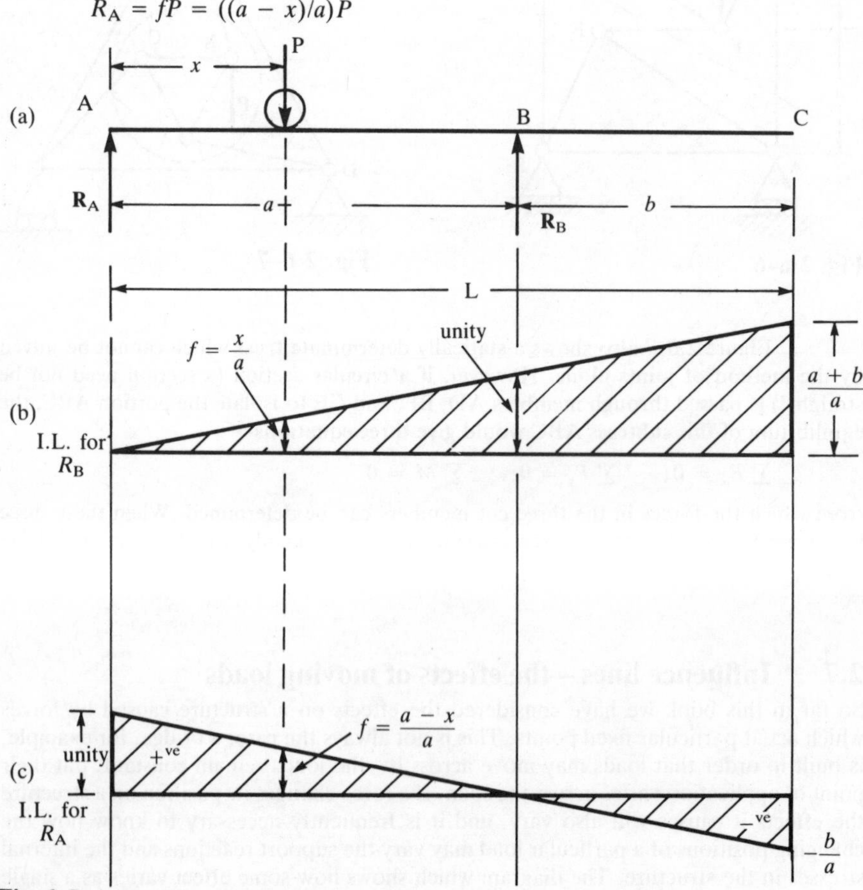

Fig. 2·7–1

It will be noted that when $x > a$ the coefficient becomes negative, implying that when $x > a$ the reaction at A must be downwards.

If a load rolling across a span is uniformly distributed with an intensity of w per unit length, its effect may readily be calculated using the appropriate influence line. An elementary portion of the load, of length dx, will apply an equivalent point load of $w\,dx$, and its effect on any particular variable will be $f(w\,dx)$ where f is the ordinate of the appropriate influence line, assumed constant over that elementary length. The total effect of the load extending from D to E in Fig. 2.7–2(a) will be

$$\int_{x\,\mathrm{at\,D}}^{x\,\mathrm{at\,E}} fw dx = w \int_{x\,\mathrm{at\,D}}^{x\,\mathrm{at\,E}} f dx$$

= $w \times$ (the area of the influence line above the load).

Inspection of the influence line for R_B shows that a short load would cause a maximum value of R_B if placed as shown in Fig. 2.7–2(b), and of R_A if as shown in Fig. 2.7–2(c). If the load was centred about B (Fig. 2.7–2(d)) it would cover equal positive and negative areas of the influence line for R_A, which would be zero, therefore, and the whole of the load would be balanced on R_B.

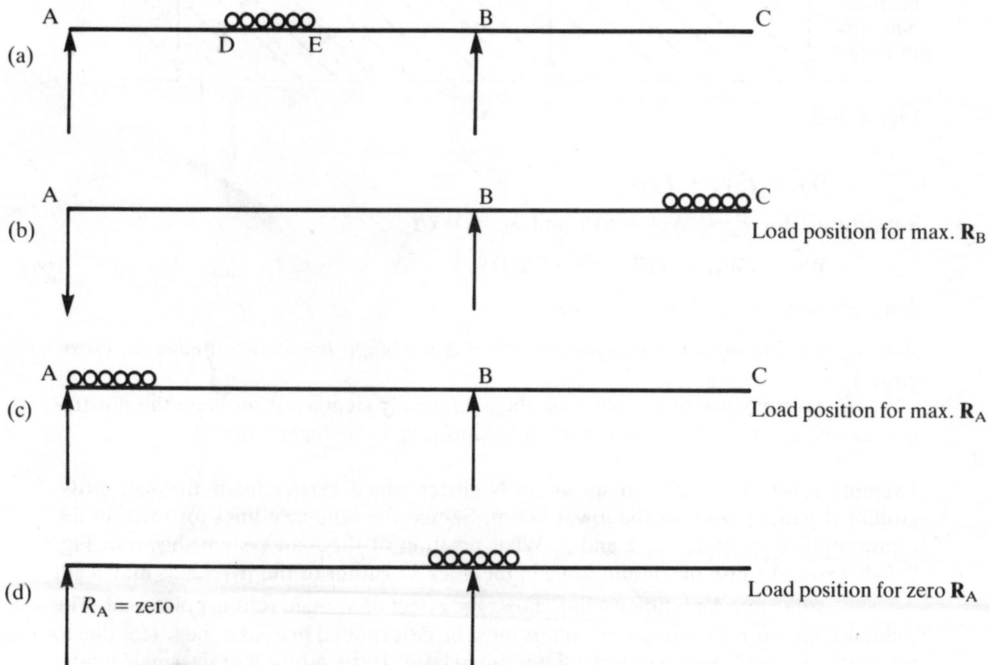

Fig. 2·7–2

2.7–2 Influence lines for force in the members of a plane pin-jointed truss

Large bridges frequently consist of two (or more) longitudinal trusses (e.g. Fig. 2.7–4(a)) supporting a roadway which transmits a load to the trusses through cross-girders at the panel points (points such as C or D). The road or railway is supported on longitudinal beams or a slab of span l, simply supported on adjacent cross-girders as shown in Fig. 2.7–3. A load, W, distant x from the left-hand end, will produce interactions between the longitudinal and cross-girders as shown thereon. Those effects can be calculated, and the total effect of W

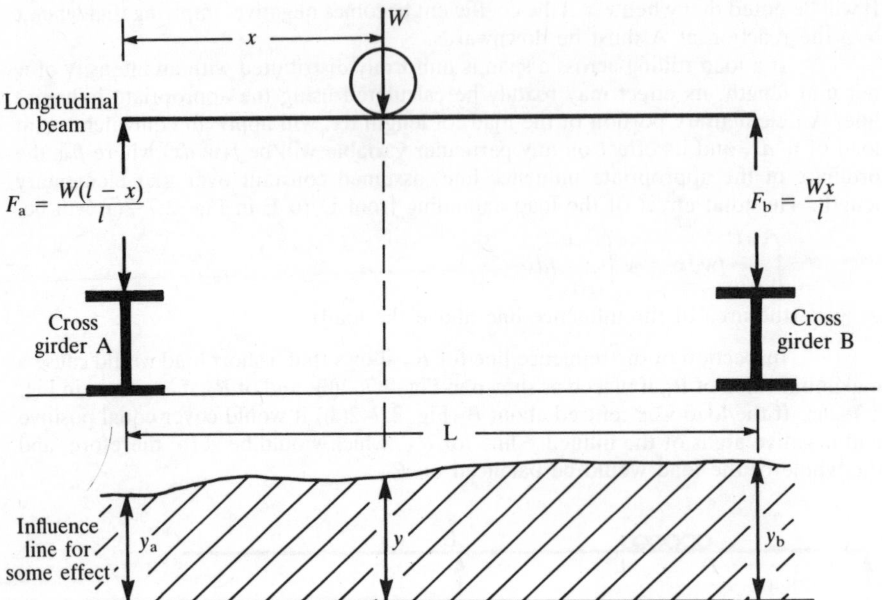

Fig. 2·7–3

$$Wy = F_a y_a + F_b y_b$$

Substituting for $F_a = W(l - x)/l$ and $F_b = Wx/l$

$$Wy = (W(l - x)/l)y_a + (Wx/l)y_b$$

from which $y = y_a + x(y_b - y_a)/l$

showing that *the influence line for any effect is a straight line between adjacent cross-girders.*

To concentrate attention on the analytically significant problem the illustrative examples given below are written as applying to a single truss.

Example 2.7–1. Fig. 2.7–4(a) shows an N-girder which carries loads through cross-girders at panel points on the lower boom. Sketch the influence lines for force in the representative members 1, 2 and 3. What position of the load system shown in Fig. 2.7–4(b) would cause maximum force in member 2? Either of the two loads may lead.

SOLUTION An influence line shows the effect of a single rolling point load. The influence lines for force in the members must be determined first, and the forces due to the particular load system must be determined later. If the position of the single load is defined by x, the distance of the load from A, it is plain that there will only be gradual changes in the force in members as x increases in the range $0 < x < 15$ (i.e. as the load moves from A to C). Similarly, changes will only be gradual whilst load is in the range $18 < x < 24$. If the equations connecting the forces in the members with the positions of a single load in these two ranges can be determined, the equation in the range $15 < x < 18$ is known to be linear—see Fig. 2.7–3.

$0 < x < 15$ (load between A and C)

Taking moments about the intersection of members 2 and 3, and considering the equilibrium of that part of the truss to the right

$$F_1 4 + 9R_B = 0 \qquad F_1 = -2.25R_B$$

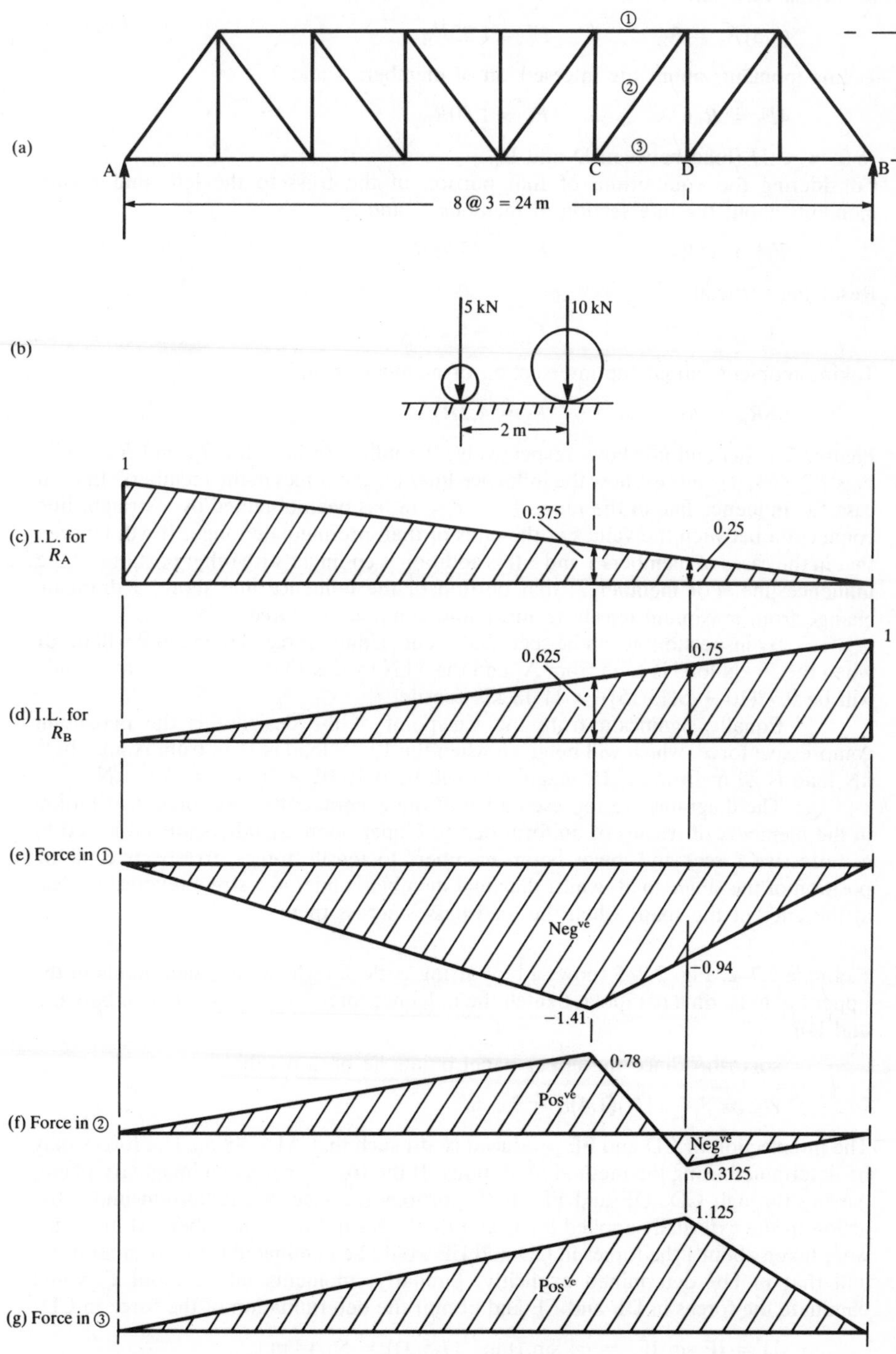
①
②
③
4 m
(a)
A
C
D
B
8 @ 3 = 24 m
5 kN
10 kN
(b)
2 m
1
0.375
0.25
(c) I.L. for R_A
0.625
0.75
1
(d) I.L. for R_B
(e) Force in ①
Neg^ve
−0.94
−1.41
0.78
Pos^ve
(f) Force in ②
Neg^ve
−0.3125
1.125
Pos^ve
(g) Force in ③

Fig. 2·7–4

Resolving vertically

$$(4/5)F_2 = R_B \qquad F_2 = 1.25R_B$$

Taking moments about the intersection of members 1 and 2

$$4F_3 = R_B 6 \qquad F_3 = 1.50R_B$$

18 < x < 24 (load between D and B)
Considering the equilibrium of that portion of the truss to the left, and taking moments about the intersection of members 2 and 3.

$$F_1 4 + 15R_A = 0 \qquad F_1 = -3.75R_A$$

Resolving vertically

$$(4/5)F_2 + R_A = 0 \qquad F_2 = -1.25R_A$$

Taking moments about the intersection of members 1 and 2

$$18R_A - 4F_3 = 0 \qquad F_3 = 4.5R_A$$

Figures 2.7–4(c) and (d) show, respectively, the influence lines for R_A and R_B, while Figs 2.7–4(e), (f) and (g) show the influence lines for the forces in the members. In each case the influence line in the range $15 < x < 18$ has been obtained by a straight line connection between the values at the ends of that intermediate range. It will be seen that in the cases of members 1 and 3 this addition is co-linear with other portions of the influence line. For member 2, that portion of the influence line shows a dramatic change from maximum tensile to maximum compressive force.

By inspection, it can be seen that the maximum force in member 2 will occur when the 10 kN load is 15 m from A, and the 5 kN load is 13 m from A. Its magnitude will be $0.78(10 + 5(13/15)) = 11.18$ kN tensile.

Equally significant from the viewpoint of the designer is the maximum compressive force, which will be given when the 10 kN load is 18 m from A and the 5 kN load is 20 m from A. Its magnitude will be $0.31(10 + 5(4/6)) = 4.17$ kN.

The diagrams are representative of the general pattern of force distribution in the members of trusses of uniform depth. Upper boom members are subjected to compressive forces and lower boom members to tensile forces, irrespective of the position of the single load, whilst diagonal members undergo a stress reversal (unless at the ends of the span) when the load passes across them.

Example 2,7–2. Fig. 2.7–5 shows a bow-string girder in which the panel points in the upper boom lie on a parabola. Sketch the influence lines for force in the members DC and DF.

SOLUTION Since the upper panel points lie on a parabola

$$DE = 8 - (1/4)(1/4)8 = 7.5 \text{ m}$$

The intersection of CD and FE produced is at I such that AI = 48 m. The forces may be determined using the method of sections. If the truss is cut by an imaginary plane, passing through CD, DF and FE, both portions must be in equilibrium under the action of the externally applied loads and the forces in the cut members. If moments were taken about I the forces in CD and EF would be eliminated from consideration, and that in DF determined explicitly. Similarly, moments taken about F would eliminate the forces in DF and EF and permit the determination of the force in CD.

$$IJ = IF \sin I\hat{F}J = 64 \sin(\tan^{-1}(7.5/4)) = 56.47 \text{ m}$$

$$FG = 2\left(\frac{\text{Area } \triangle DCF}{DC}\right) = \frac{32}{\sqrt{4^2 + (0.5)^2}} = 7.94 \text{ m}$$

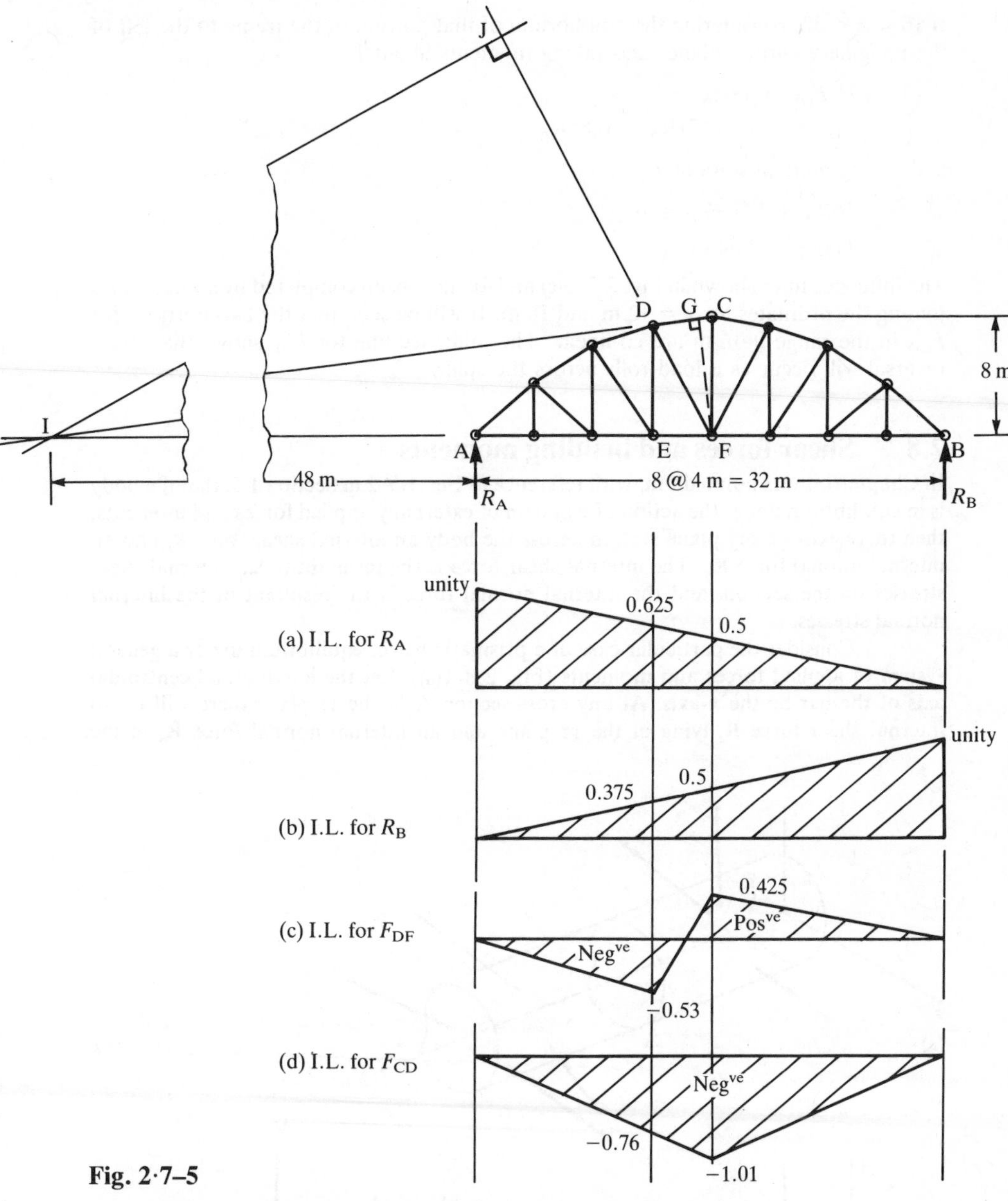

Fig. 2·7–5

Now, defining the position of the load by its distance x, from A.

If $0 < x < 12$—considering the equilibrium of that portion of the frame to the right of the imaginary cutting plane, and taking moments about I

$$\text{IJ } F_{DF} + 80R_B = 0$$
$$F_{DF} = -(80/56.47)R_B = -1.417R_B$$

and taking moments about F

$$F_{DC}\, GF + 16R_B = 0$$
$$F_x = -(16/7.94)R_B = -2.02R_B$$

If $16 < x < 32$, considering the equilibrium of that portion of the frame to the left of the imaginary cutting plane, and taking moments about I

$$IJ\ F_{DF} = 48R_A$$
$$F_{DF} = (48/56.47)R_A = 0.85R_A$$

and taking moments about F

$$16R_A + FG\ F_{CD} = 0$$
$$F_{CD} = -2.02R_A$$

The influence lines shown in Fig. 2.7–5(c) and (d) have been completed by straight lines joining the ordinates for $x = 12$ m and 16 m. It will be seen that the two portions for F_{CD} in the range 0–16 m are co-linear. The influence line for F_{DF} shows that stress reversal will occur as a load rolls across the span.

2.8 Shear forces and bending moments

In Chapter 1 it was pointed out, with reference to Fig. 1.5–2 in Section 1.5, that if a body is in equilibrium under the action of a system of externally applied forces and moments, then there exist at any plane section across the body an internal shear force $\mathbf{R}_s$ and an internal normal force $\mathbf{R}_n$. The internal shear force is the resultant of the internal shear stresses on the section, and the internal normal force is the resultant of the internal normal stresses.

Consider the particular case of a prismatic bar in equilibrium under a general system of applied forces and moments (Fig. 2.8–1(a)). Let the longitudinal centroidal axis of the bar be the x-axis. At any cross-section A in the yz plane there will be an internal shear force $\mathbf{R}_s$ lying in the yz plane and an internal normal force $\mathbf{R}_n$ in the

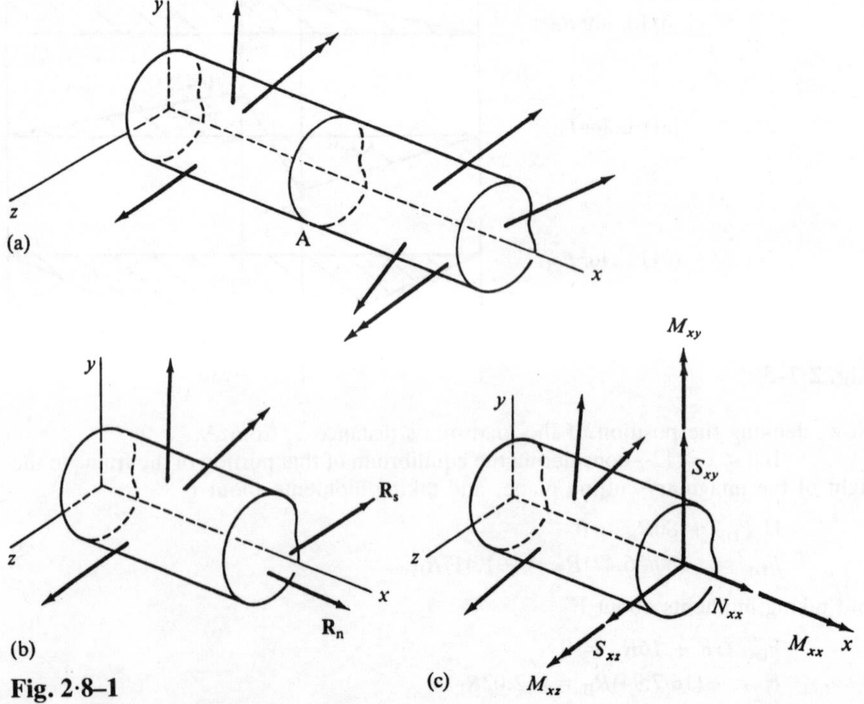

Fig. 2·8–1

x-direction. In general neither $\mathbf{R}_s$ nor $\mathbf{R}_n$ need pass through the centroid of the cross-section (Fig. 2.8–1(b)). However, the shear force $\mathbf{R}_s$ can be replaced by a shear force (of the same magnitude R_s) passing through the centroid and a **twisting moment** (or torque) M_{xx} about the x-axis; and this shear force through the centroid can be resolved into components S_{xy} and S_{xz} in the y- and z-directions respectively (Fig. 2.8–1(c)). Similarly, the normal force $\mathbf{R}_n$ can be replaced by a normal force N_{xx} passing through the centroid, together with moments M_{xy} and M_{xz}. In Fig. 2.8–1(c), the meanings of the suffixes xx, xy, and xz are the same as those for stresses in Fig. 1.5–4, namely, the first suffix denotes the direction of the normal to the surface, and the second suffix denotes the direction of the force or moment. For example:

S_{xy} is a force acting on a section whose outward normal is in the positive x-direction (as denoted by the first suffix) and the force itself acts in the positive y-direction (as denoted by the second suffix).

M_{xz} is a moment acting on a section whose outward normal is in the positive x-direction, and the moment itself acts about the z-axis in the positive sense (see also Fig. 1.2–5 for meaning of 'positive sense' of a moment).

The forces S_{xy} and S_{xz} are called the **shear forces** in the section, and the moments M_{xy} and M_{xz} the **bending moments** in the section. The force N_{xx} is the **normal force** and the moment M_{xx} is the **twisting moment** or torque.

In the two-dimensional case of a beam subjected to forces and moments in the xy plane only (Fig. 2.8–2) the only shear force is S_{xy} and the only bending moment is M_{xz}. In this two-dimensional case, it is usual to denote the shear force by S (instead of S_{xy}) and the bending moment by M (instead of M_{xz}); suffixes are not required as no ambiguity exists, and for this reason the normal force N_{xx} may simply be denoted by N.

The free-body diagram in Fig. 2.8–2(b) shows that:

(a) The magnitude of the shear force at any section a–a of a beam is equal to the algebraic sum of the transverse components of the forces acting on either side of the section.

(b) The magnitude of the bending moment at any section of a beam is equal to the algebraic sum of the moments about that section of the forces acting on either side of the section.

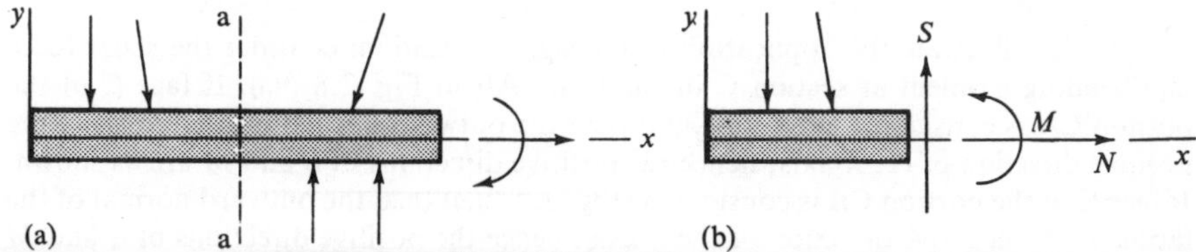

Fig. 2·8–2

To define the direction, or sense, of shear force and bending moment a sign convention is required, and the authors have chosen in this text to adopt the following **sign convention for shear forces and bending moments** (Fig. 2.8–3):

> The x-axis is chosen as the longitudinal centroidal axis of the member. If the outward normal of the face of a section is in the *positive* direction of the x-axis, then the positive shear force vector is in the *positive* y-direction and the positive bending moment vector is also in the *positive* z-direction; if the outward normal of the face of the section is in the *negative* direction of the x-axis, then the positive shear force vector and the positive bending moment vector are in the *negative* y- and *negative* z-directions respectively.

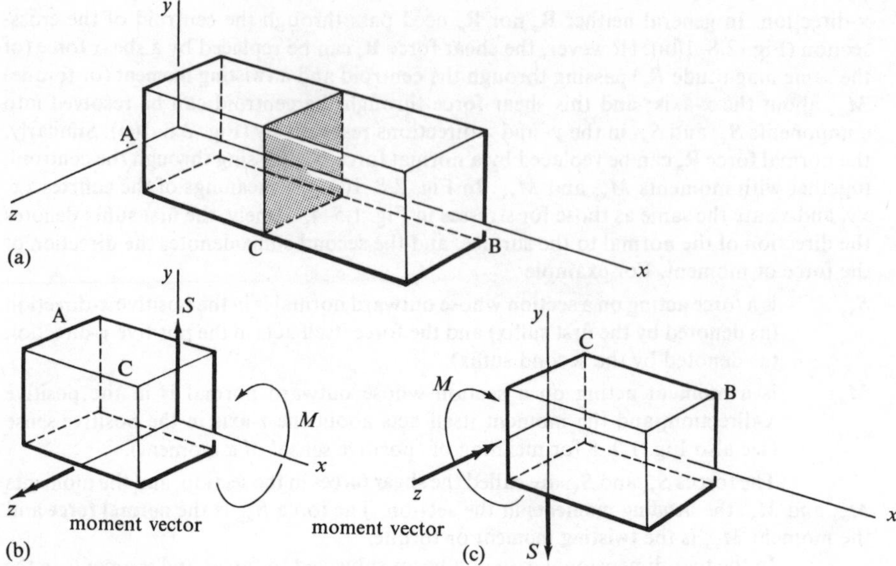

Fig. 2·8–3

The above sign convention has several important advantages:

(a) It is consistent with the sign convention for stresses (Section 1.5, Fig. 1.5–4) which has been almost universally adopted in structural mechanics. Such consistency enables the sign convention to be conveniently applied to the analysis of continua by the finite difference and finite element methods (Chapters 11 and 12; Figs. 11.13–3 and 12.7–2).

(b) It is based on a system of coordinate axes for a member; this is compatible with the matrix computer methods of analysis of skeletal frameworks (Chapters 7 and 8).

(c) It leads to unambiguous interpretations for the general structure (see Example 2.8–2).

To illustrate the application of the sign convention, consider the shear force and bending moment at section C of the beam AB in Fig. 2.8–3(a). If face C of the portion CA is considered (Fig. 2.8–3(b)) then the outward normal of the face is in the positive direction of the x-axis; hence the positive directions of S and M are as shown. If face C of the portion CB is considered (Fig. 2.8–3(c)) then the outward normal of the face is in the negative direction of the x-axis; hence the positive directions of S and M are as shown in Fig. 2.8–3(c). From this example, it is clear that the positive directions for the shear force and bending moment at a section depend on which direction is chosen as the positive x-direction. If in Fig. 2.8–3(a) the positive x-direction is reversed† then the directions for positive S (but not M) in Figs. 2.8–3(b) and (c) will be reversed. In the simple case of a horizontal beam, if the positive x-direction is always to the right and the positive y-direction is always chosen as upwards, then the above sign convention can be easily interpreted as follows:

(a) If the resultant of the forces to the left of a section acts downwards, then the shear force at that section is positive (Fig. 2.8–3(b) 2.8–4(a)).

(b) A 'sagging' bending moment is positive (Fig. 2.8–4(b)).

† If the positive x-direction is reversed while the positive y-direction is retained, then the positive z-direction will be reversed (see Section 1.1 on the right-handed systems).

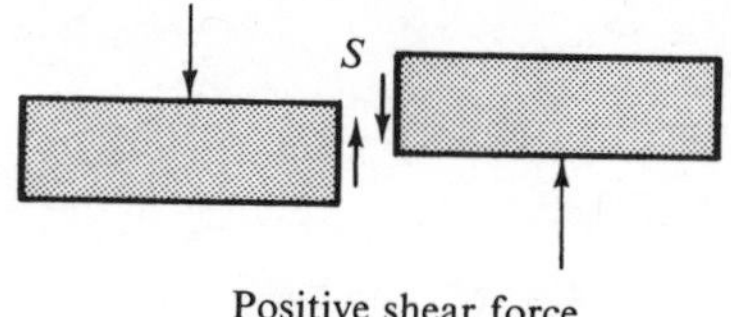

Positive shear force

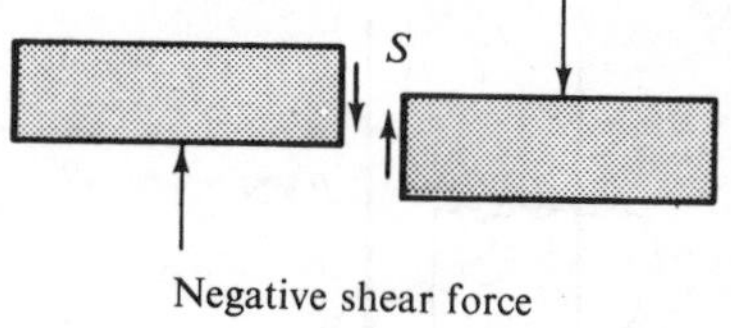

Negative shear force

(a)

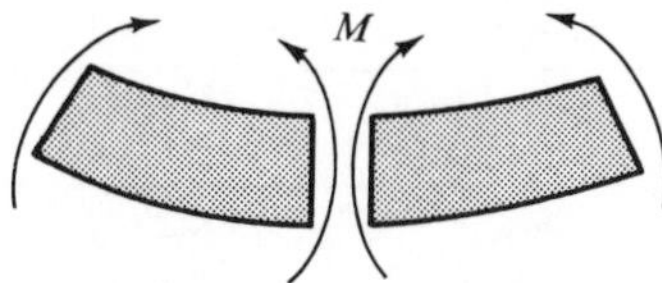

Positive bending moment ('sagging')

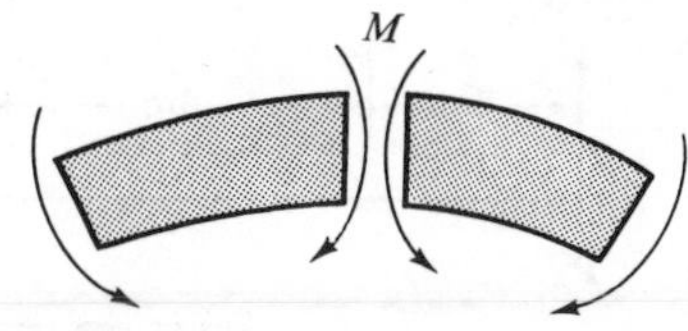

Negative bending moment ('hogging')

(b)

Fig. 2·8–4

Example 2.8–1. Determine the shear force and bending moment at any section of the beam AB in Fig. 2.8–5(a).

SOLUTION By inspection (or successive application of the condition $\sum M = 0$ to B and A):

$$R_A = \tfrac{4}{6} P = 0.67P \text{ kN}; \qquad R_B = \tfrac{2}{6}P = 0.33P \text{ kN}$$

Consider a section a–a at distance x from support A.

For $0 < x < 2$ m, the only force to the left of section a–a is $\mathbf{R}_A$ (0.67 P kN) and it acts upwards. Therefore according to the sign convention in Fig. 2.8–4(a) (which is a particular case of the general sign convention in Fig. 2.8–3) the shear force at section a–a is -0.67 P kN.

For 2 m $< x <$ 6 m, the resultant of the forces to the left of section a–a acts downwards and has a magnitude of $(P - 0.67P)$ kN $= 0.33P$ kN. Hence the shear force at section a–a is $+$ $0.33P$ kN.

The shear force at any section of the beam is as shown in the **shear force diagram** (often referred to as the SF diagram or the SFD) in Fig. 2.8–5(b). Note that, according to the above calculations, the shear forces at sections at A, C, and B (where x exactly equals 0, 2 m and 6 m respectively) are indeterminate. This is because of the theoretical assumption that the forces $\mathbf{R}_A$, $\mathbf{P}$, and $\mathbf{R}_C$ are each concentrated at one point. Such concentrated forces cannot occur in practice because they would immediately cut through the beam, since a concentrated force produces an infinitely large pressure at its point of application. In fact, so-called concentrated forces are distributed over a small length of the beam, so that at section C, for example, the shear-force distribution would appear somewhat as shown by the dotted line in Fig. 2.8–5(b).

Next consider the bending moment at section a–a at x from A:

For $0 < x < 2$ m, the moment is a 'sagging' moment of magnitude $0.67Px$ kN m produced by the reaction $\mathbf{R}_A$. Hence the bending moment at x from A is $+0.67Px$ kN m, i.e. the bending moment varies linearly from 0 for $x = 0$ at A to $+1.33P$ kN m for $x = 2$ m at C.

For 2 m $< x <$ 6 m, the resultant moment at the section is made up of a 'sagging' moment of $0.67Px$ kN m due to $\mathbf{R}_A$ and a 'hogging' moment of $P(x - 2)$ kN m, i.e. the resultant sagging moment is

$$0.67Px - P(x - 2) = (2 - 0.33x)P \text{ kN}$$

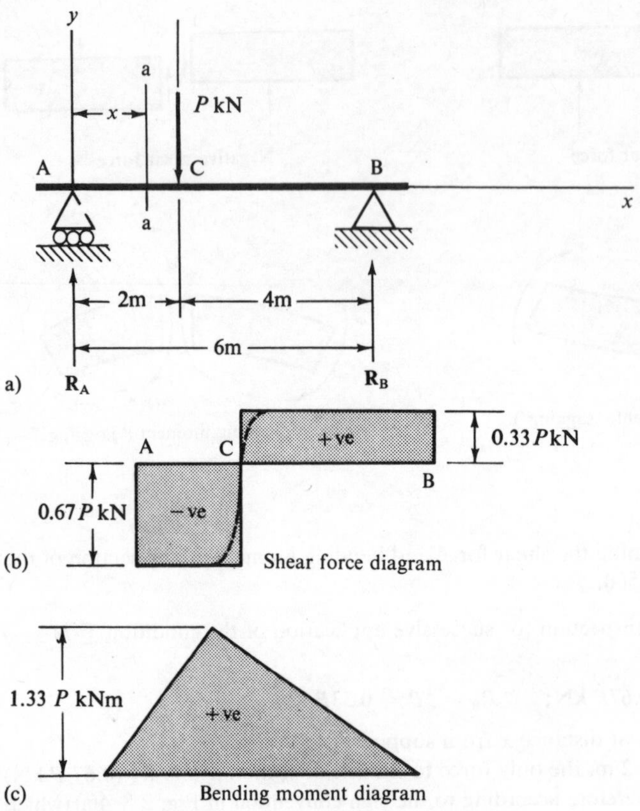

Fig. 2·8–5

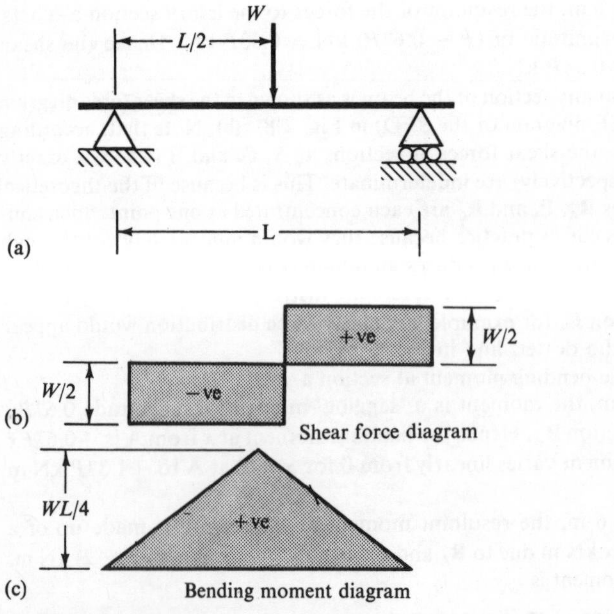

Fig. 2·8–6

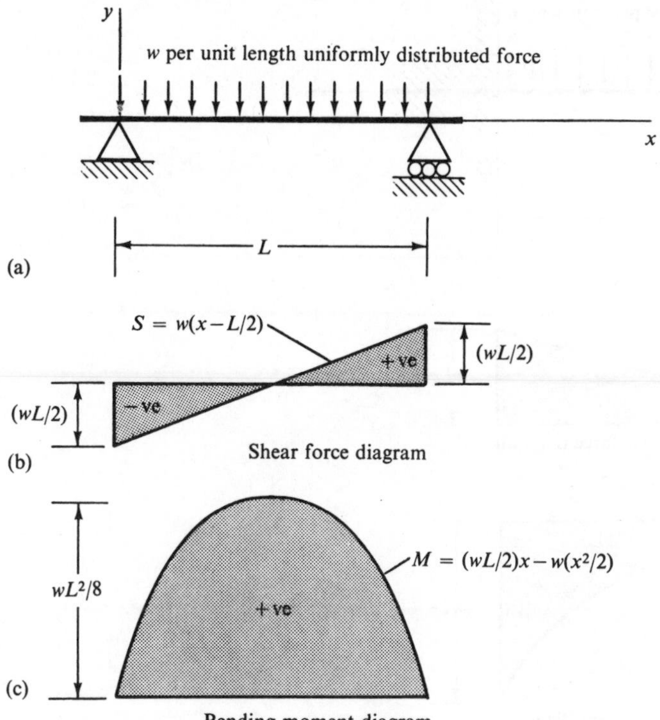
y
w per unit length uniformly distributed force
x
L
(a)
S = w(x−L/2)
+ve
(wL/2)
−ve
(wL/2)
(b)
Shear force diagram
wL²/8
M = (wL/2)x − w(x²/2)
+ve
(c)
Bending moment diagram

Fig. 2·8–7

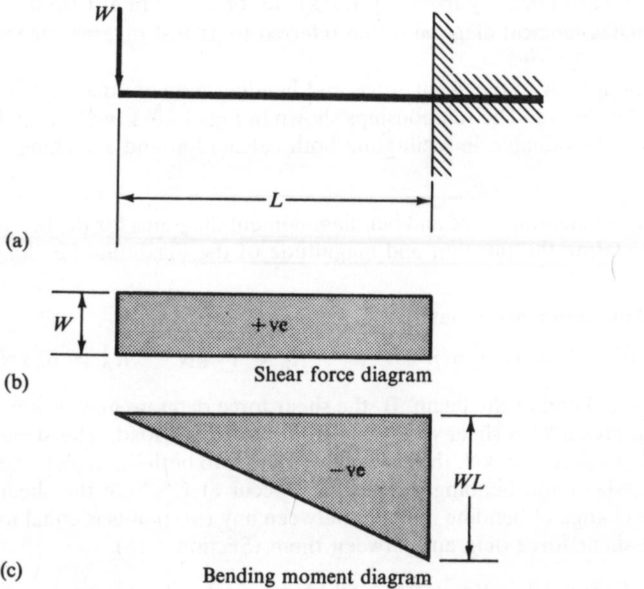
W
L
(a)
W
+ve
(b)
Shear force diagram
−ve
WL
(c)
Bending moment diagram

Fig. 2·8–8

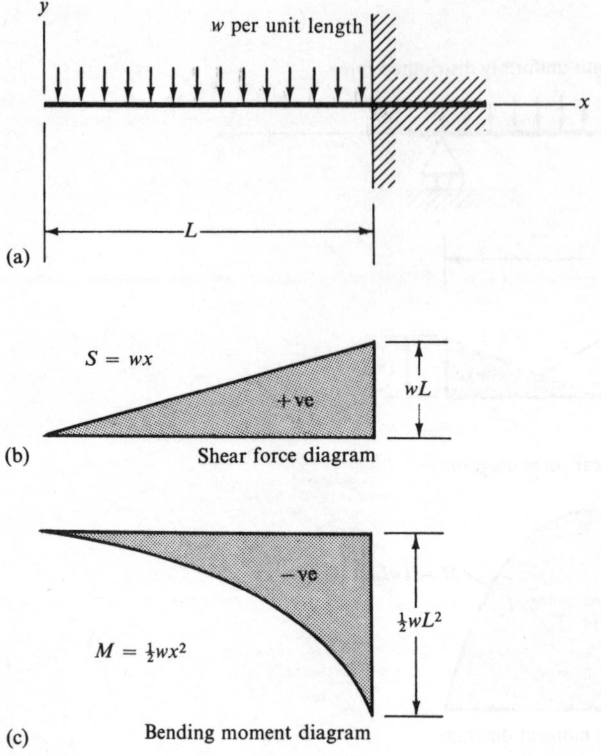

Fig. 2·8–9

i.e. the bending moment varies linearly from $+1.33P$ kN m for $x = 2$ m at C to 0 for $x = 6$ m at B. The **bending moment diagram** (often referred to as BM diagram or the BMD) is as shown in Fig. 2.8–5(c).

The reader should verify the shear force and bending moment diagrams in Figs 2.8–6 to Fig. 2.8–9 inclusive. The relationships shown in Fig. 4.18–1 in Chapter 4 will be found exceptionally valuable in facilitating both calculation and sketching.

Example 2.8–2. Sketch the shearing force and bending moment diagrams for the beam shown in Fig. 2.8–10, stating the position and magnitude of the maximum bending moment.

SOLUTION Taking moments about A

$$15R_B = 5 \times 6 + 12 \times 11 + 6 \times 1 \times 8 \qquad R_B = 14 \text{ kN} \qquad R_A = 10 \text{ kN}$$

Beginning at the right-hand end of the beam, B, the shear force diagram may now be drawn (Fig. 2.8–10(b)), giving zero shear at C, 2 m from the 12 kN load. The shear immediately to the right of A is − 10 kN, thus providing a check to both the sketch and the calculation above. Maximum bending moment will occur at C where the shear force is zero. Since the change of bending moment between any two points is equal to minus the area of the shear force diagram between them (Section 4.18),

$$M_C = M_A - (-5 \times 10 - \tfrac{1}{2}(4 \times 4)) = 58 \text{ kNm}$$

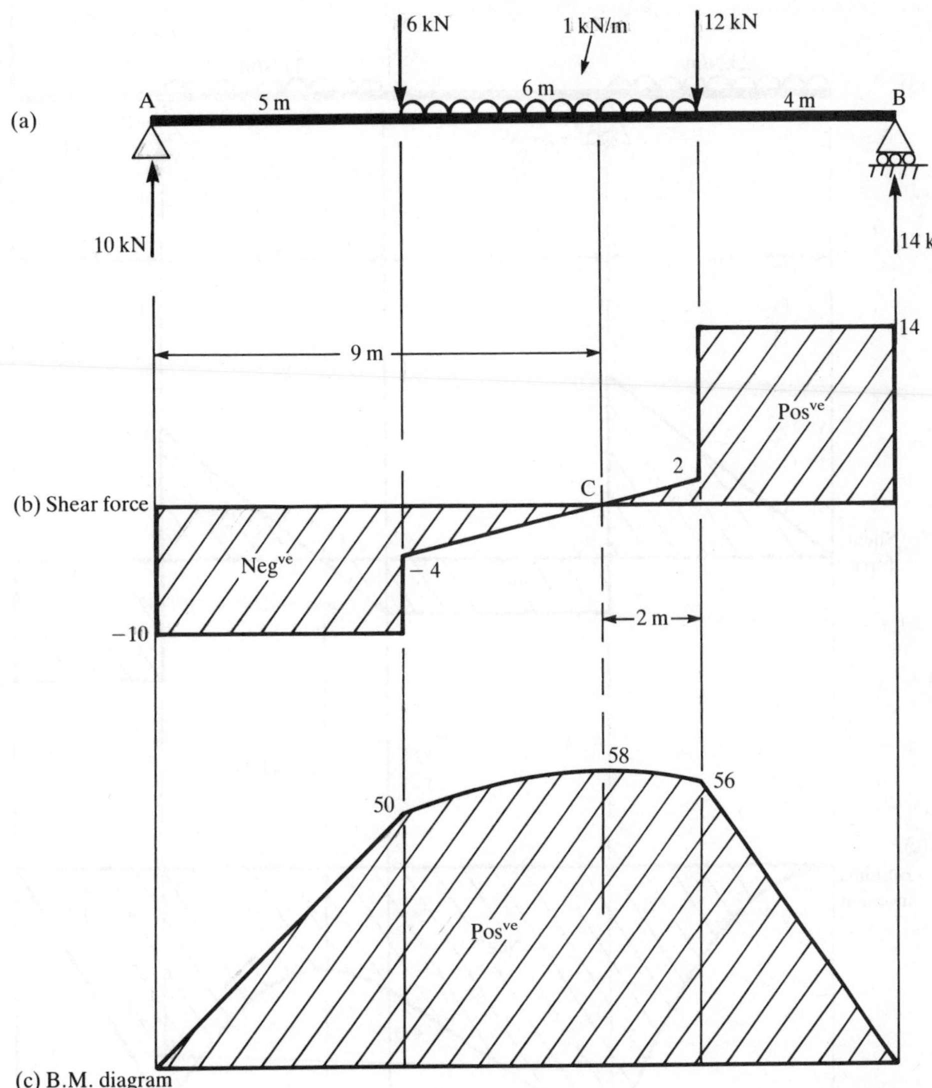

Fig. 2·8–10

Example 2.8–3. Sketch the bending moment and shear force diagrams for the beam AB shown in Fig. 2.8–11. Determine the general shape of the bent beam.

SOLUTION Taking moments about A

$$6R_B + 2 \times 3 \times 1.5 = 2 \times 3 \times 4.5 + 4 \times 8 \qquad R_B = 8.33 \text{ kN}$$

from which $R_A = 16 - 8.33 = 7.67$ kN

Again, following the directions of the applied loads and the reactions, the shear force diagram may be completed, and the bending moment diagram drawn from the fact that the slope of the latter is equal to minus the ordinate hof the shear force diagram (Section 4.18).

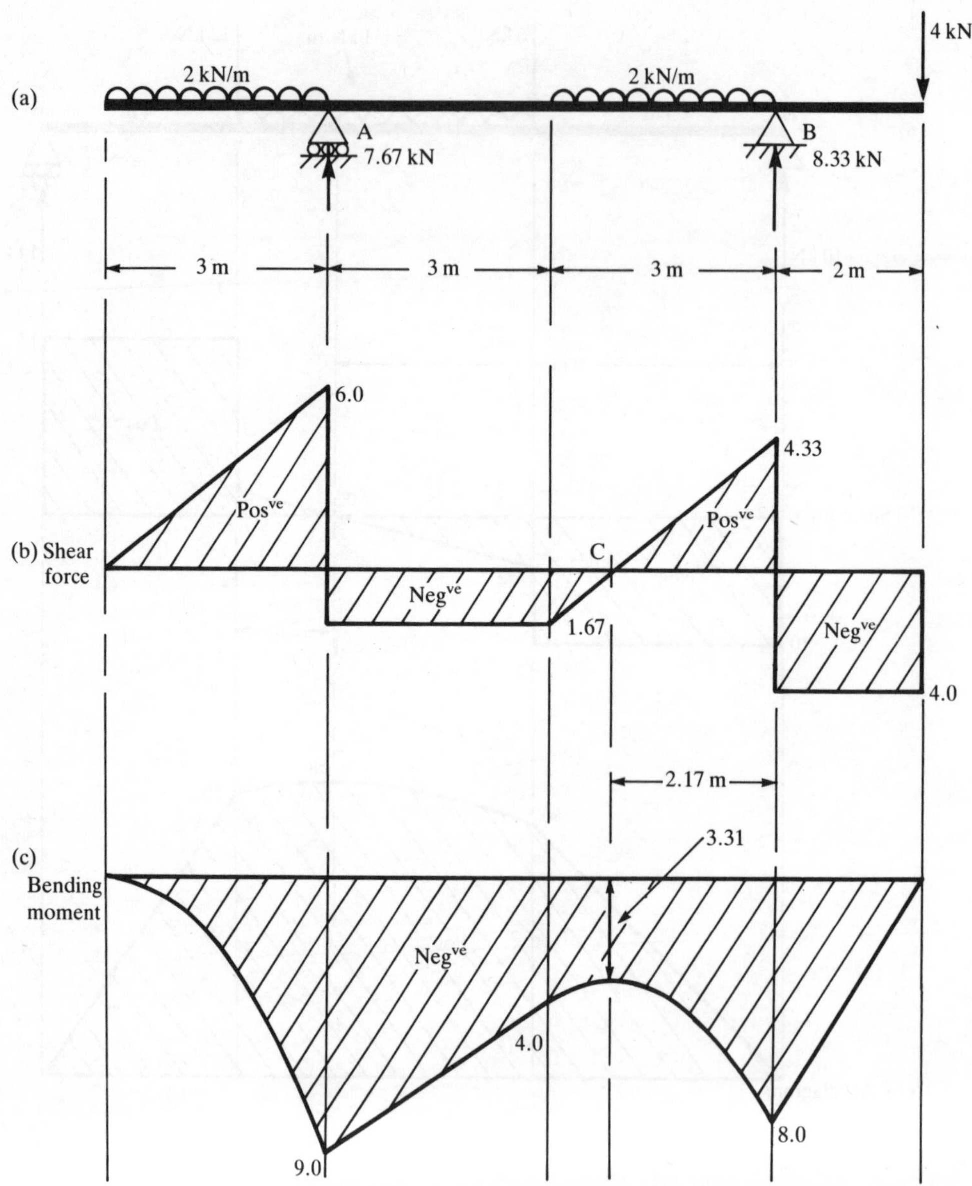

Fig. 2·8–11

An algebraic maximum occurs at C, 2.17 m to the left of B, and has the value of 3.31 kNm.

Throughout the beam the bending moment is wholly negative and, therefore, the curvature is negative also. The beam is supported at A and B, and its bent shape is wholly convex upwards.

Example 2.8–4. A beam of span L, of uniform flexural rigidity, carries a distributed load which varies linearly in intensity from zero at the left-hand support to w per unit

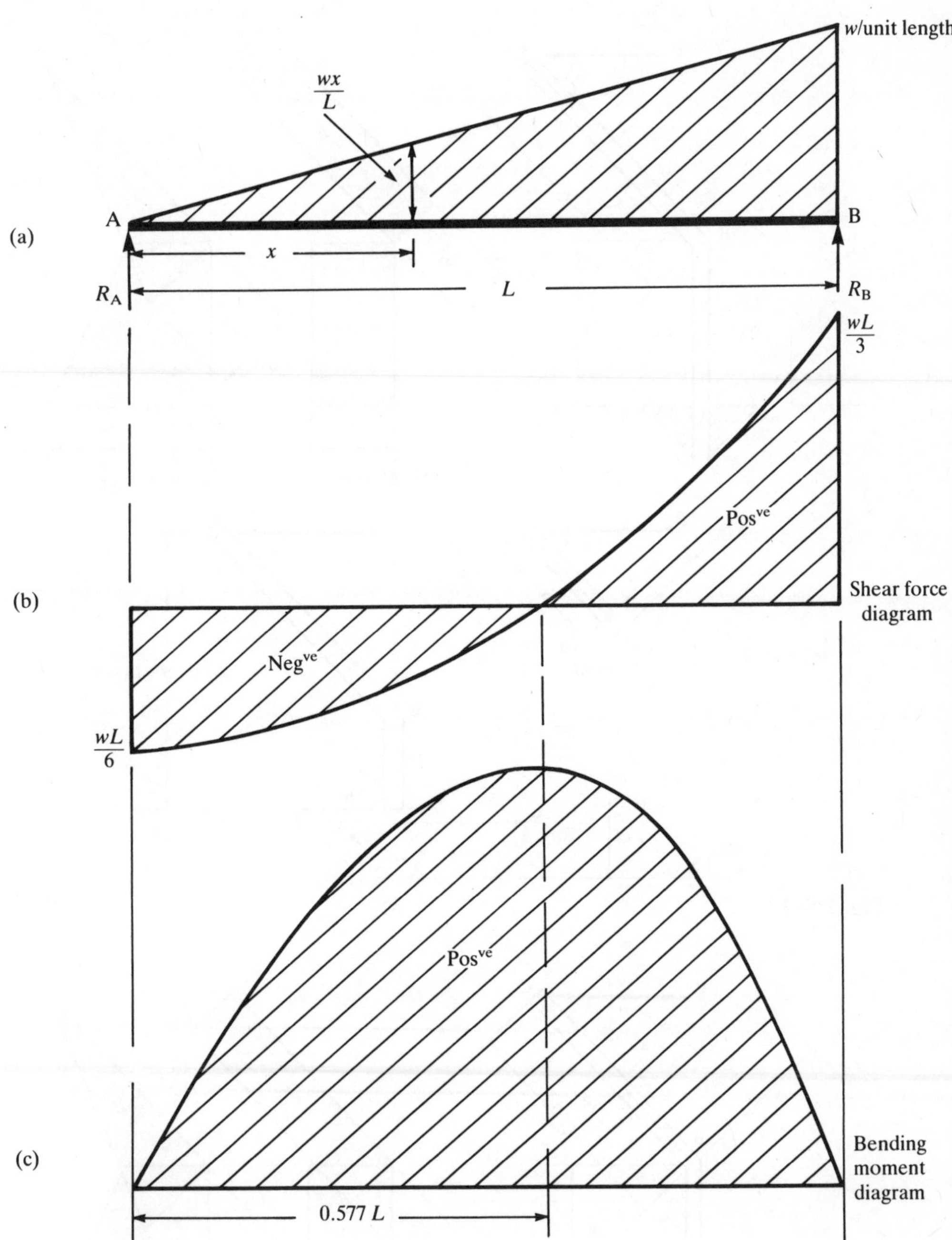

Fig. 2·8–12

length at the right, as shown in Fig. 2.8–12. Find the distance from the left-hand support of the point of maximum bending moment, sketch the shear force and bending moment diagrams, and give numerical values if $L = 10$ m and $w = 2$ kN/m.

SOLUTION The total load on the beam is $(wL/2)$, and taking moments about A will show that $R_B = (wL/3)$ and $R_A = (wL/6)$.

(a)

(d) SF diagram

(e) BM diagram

(b)

(c)

Fig. 2·8–13

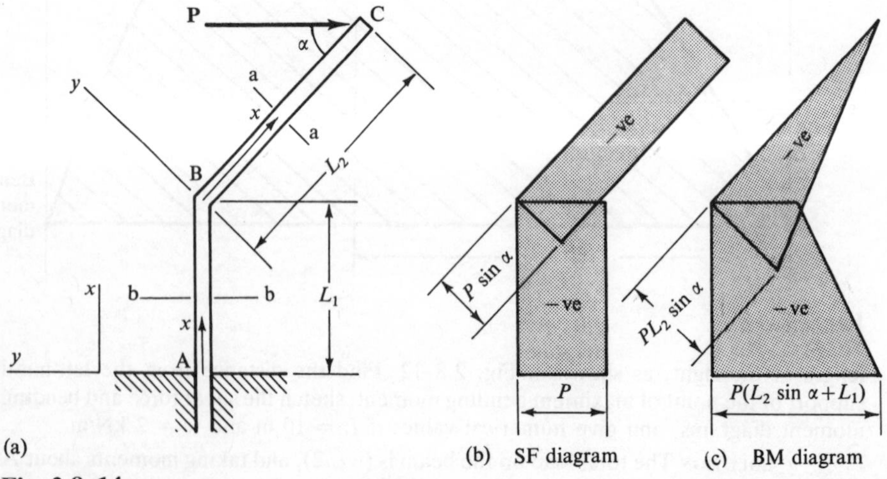

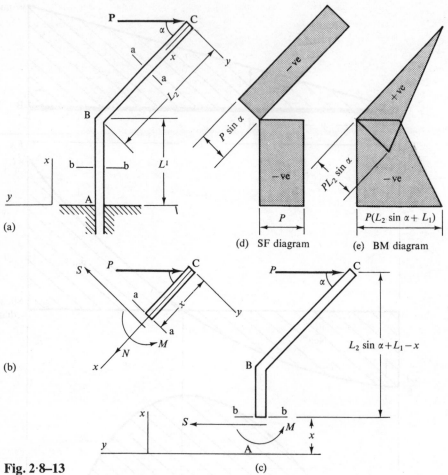

(a) (b) SF diagram (c) BM diagram

Fig. 2·8–14

The bending moment at a point x from A $= (wL/6)x - (wx/L)(x/2)(x/3)$
$= (wLx/6) - (wx^3/6L)$

Differentiating with respect to x, and equating to zero, shows that a maximum will occur at $x = 0.577L$ where $M = \dfrac{wL^2}{9\sqrt{3}} = 0.064wL^2$.

The shapes of the shear force and bending moment diagrams are shown at (b) and (c).

Example 2.8–5. Draw the shearing force and bending moment diagrams for the bent member ABC shown in Fig. 2.8–13(a).

SOLUTION Since the given structure is not a horizontal beam, the simplified sign convention in Fig. 2.8–4 is not suitable; its use in this case is confusing since phrases such as 'forces to the left of the section act downwards' or 'sagging moment' are ambiguous. Hence we should refer to Fig. 2.8–3 for the sign convention.

Suppose coordinate axes for member AB and those for member BC are chosen as shown in Fig. 2.8–13(a). Consider a section a–a (Fig. 2.8–13(b)). It is easy to verify that the directions of the shear force S and the bending moment M are as indicated. Since the outward normal to the face of the section is in the positive x-direction, it is seen that S is negative (because the shear force vector is in the negative y-direction) and that M is positive (because the bending moment vector is in the positive z-direction, i.e. perpendicular to the plane of the paper and pointing towards the reader). Also, from equilibrium consideration, the magnitudes of S and M are

$S = P \sin \alpha$ (negative, according to sign convention)
$M = Px \sin \alpha$ (positive, according to sign convention)

M therefore varies linearly from 0 at C to $PL_2 \sin \alpha$ at B, while S remains constant at $-P \sin \alpha$ throughout length CB.

At a section b–b in member AB at distance x from A the directions of S and M can be shown to be as in Fig. 2.8–13(c). In this case the outward normal to the section is in the negative x-direction while the shear force vector and the bending moment vector are in the positive y- and z-directions respectively. Hence the sign convention makes both S and M negative. For equilibrium of the free body in Fig. 2.8–13(c),

$S = P$ (negative)
$M = P(L_2 \sin \alpha + L_1 - x)$ (negative)

M therefore varies linearly from $-PL_2 \sin \alpha$ at B to $-P(L_2 \sin \alpha + L_1)$ at A.

The shear force diagram and the bending moment diagram are shown in Figs. 2.8–13(d) and (e) respectively.

Suppose, instead of choosing member coordinate axes as shown in Fig. 2.8–13(a), we choose axes as shown in Fig. 2.8–14(a). The reader should verify that the shear force diagram and the bending moment diagram become as shown in Figs. 2.8–14(b) and (c) respectively.

This example has demonstrated that the sign of the shear force or bending moment at a section depends on the system of coordinate axes chosen for the member. The necessity to specify a system of coordinate axes for each member appears cumbersome, but actually it is a highly efficient and systematic way of handling structures, and is in fact a standard procedure in matrix computer structural analysis (see Chapter 7).

Example 2.8–6. Draw the shearing force and bending moment diagrams for the members of the rectangular rigid-jointed frame shown in Fig. 2.8–15.

SOLUTION In such a problem it is helpful to separate the frame into its

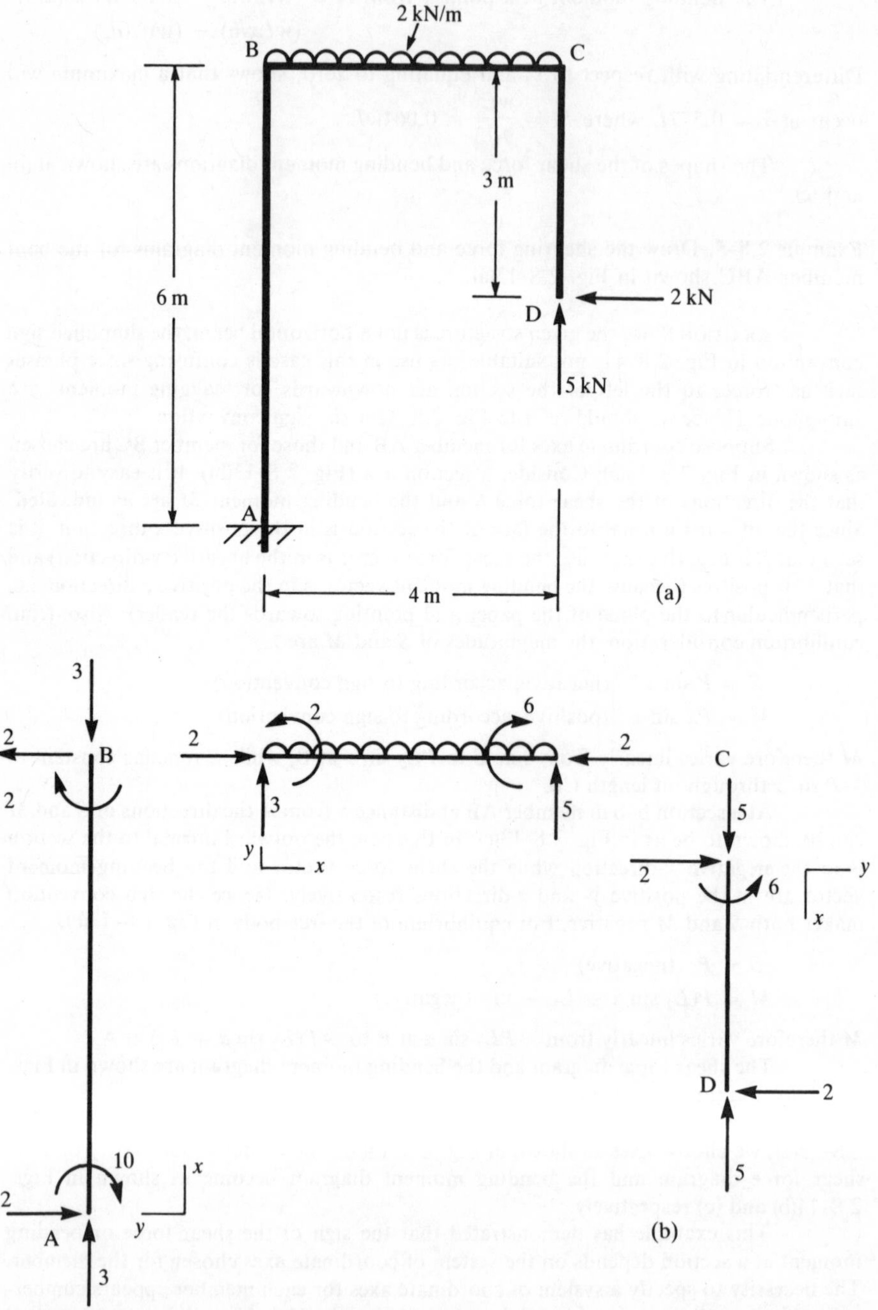

constituent members, as shown in Fig. 2.8–15(b) and to consider the equilibrium of the parts separately, but in a systematic manner. Starting from a point of application of known forces (in this case D) it is readily possible to evaluate the forces and moments which must act at C to ensure equilibrium. Equal and opposite forces will act, therefore, on the end C of member BC and, in turn, the equilibrium of BC will require

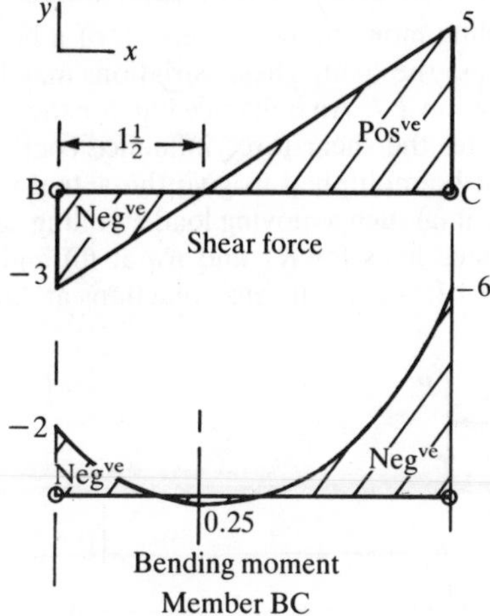

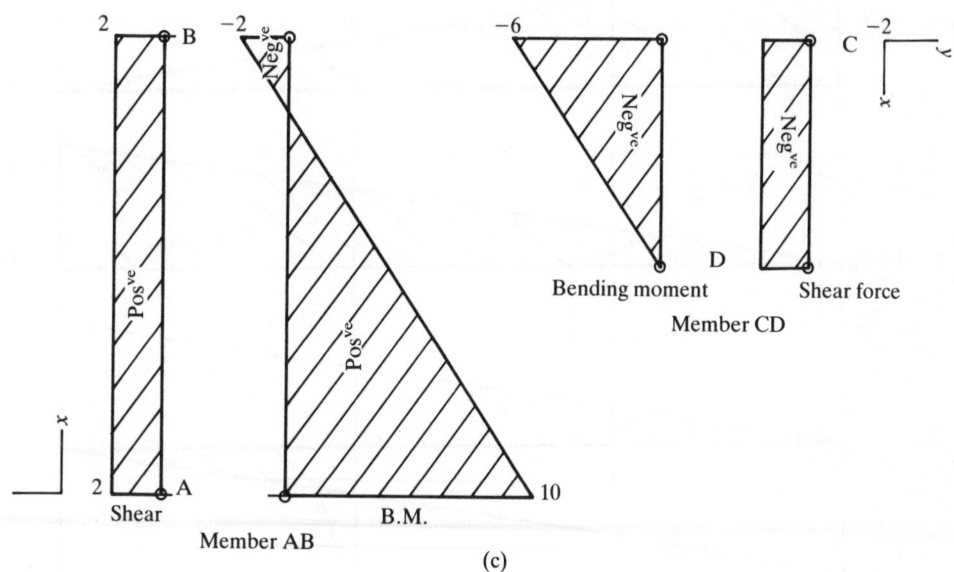

Fig. 2·8–15

certain moments and forces at end B. These moments and forces will act in the reverse direction at end B of member AB.

Having obtained the forces required at A it is always advisable to check the equilibrium of the whole frame, to ensure that the forces required at A to ensure the equilibrium of AB do, together with the forces at D, ensure the equilibrium of the whole frame. The shear force and bending moment diagrams for the constituent parts are shown in Fig. 2.8–15(c).

2.9 Influence lines for shear force and bending moment

The shear force and bending moment at any section of a beam subjected to moving loads will vary as loads cross the span. These variations may be represented helpfully by influence lines—see Section 2.7. An influence line for shear force at *some particular point* will show, as ordinate, the shear force influence coefficient by which a rolling load of magnitude P must be multiplied to give the actual shear force *at that point.*

Fig. 2.9–1 shows at (a) such a moving load P rolling across a simply supported span of L, and the influence lines for R_A and R_B at (b) and (c).

A load P, distant x from A will cause reactions at A and B of Pf_{RA} and Pf_{RB} respectively.

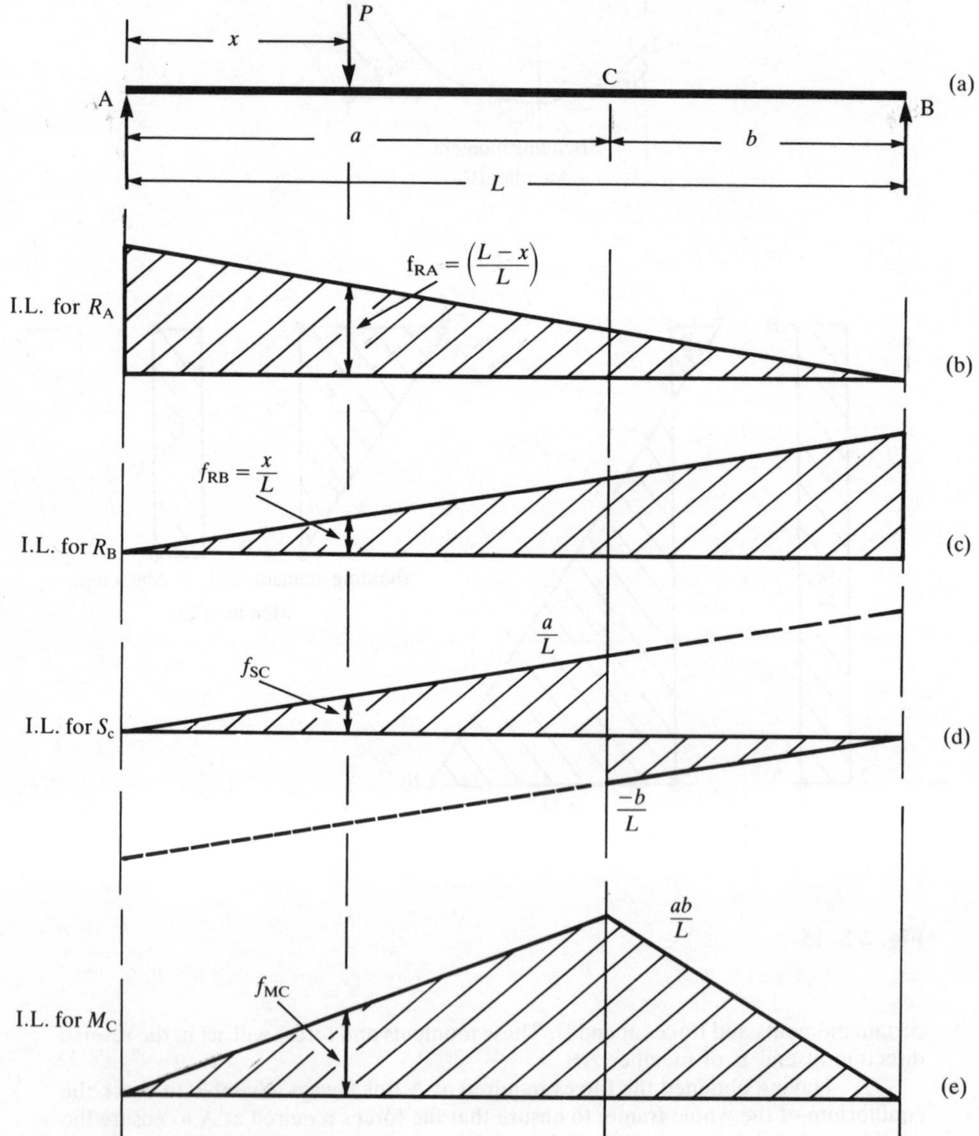

Fig. 2·9–1

It should be noted that, dimensionally, a force P is multiplied by f to give a reaction R, which is itself, a force. The coefficient in this case is dimensionless—a number only. Taking moments

$$R_B = Pf_{RB} = Px/L \quad \text{and} \quad f_{RB} = x/L$$

Similarly, $f_{RA} = (L - x)/L$

Now, considering a point distant x from A—if $x < a$, that is, if the load is to the left of C.

Shear at C $= Pf_{SC} = R_B = Pf_{RB} = Px/L$ hence $f_{SC} = x/L$

Bending moment at C $= Pf_{MC} = bR_B = bPf_{RB}$ $f_{MC} = bx/L$

Similarly, if $a < x < L$, that is, if the load is to the right at C.

Shear at C $= -R_A$ $f_{SC} = -(L - x)/L$

Bending moment at C $= Pf_{MC} = Paf_{RA}$ $f_{MC} = a(L - x)/L$

The completed influence lines for shear and bending moment at C may now be drawn, and are shown in Fig. 2.9–1(d) and (c). In the former, the ordinate is dimensionless, since shear is itself a force—the bending moment influence line, however, has dimensions of length. If will be noted that as a load rolls across the span the bending moment at C (and at every other point in the span) is consistently positive. As the load rolls across C, however, the shear there changes sign from a maximum positive to a maximum negative value. Maximum bending moment due to a single point load occurs at C when the load is directly over C.

Example 2.9–1. A simply supported beam AB of span 20 m is crossed by travelling loads. Determine the greatest shear and bending moment experienced at a point C, distant 8 m from the left-hand support due to passage across the span of:

(a) A uniformly distributed load of 5 kN/m extending over a length of 6 m.

(b) Two point loads of 3 kN and 2 kN, 6 m apart. They may cross in either direction, and either may lead.

SOLUTION The influence lines for shear and bending moment at C are shown in Fig. 2.9–2(a) and (b).

(a) The greatest positive shear at C will occur when the front of the load extends to the left of C, and appears as shown in Fig. 2.9–2(c). Numerically that shear will be (the load intensity) × (the area of the influence line covered) and will be

$$5 \times (6/2)(0.4 + (2/8)0.4) = 15 \times 0.4 \times 1.25 = 7.5 \text{ kN}$$

The greatest negative shear will occur when the load is just to the right of C, as in (d), and the magnitude will be

$$5 \times (6/2)(0.6 + 0.3) = 15 \times 0.9 = 13.5 \text{ kN}$$

which is the larger of the two possibilities.

The load will cause the greatest bending moment at C when it is so arranged over C that the ordinates cd and ef, at the two ends of the load, are equal—as shown in Fig. 2.9–2(e). Thus the load is placed in such a way that C divides the span and the load in the same ratio. The statement is obvious enough from that figure—if the load is moved a small distance to the right (as shown), or left, the area of the influence line thus uncovered would be greater than the additional amount covered at the other end.

$$(M_c)_{max} = (5 \times 6/2)(3.36 + 4.80) = 122.40 \text{ kNm}$$

It will be noted that the influence line technique may be used to position the load to

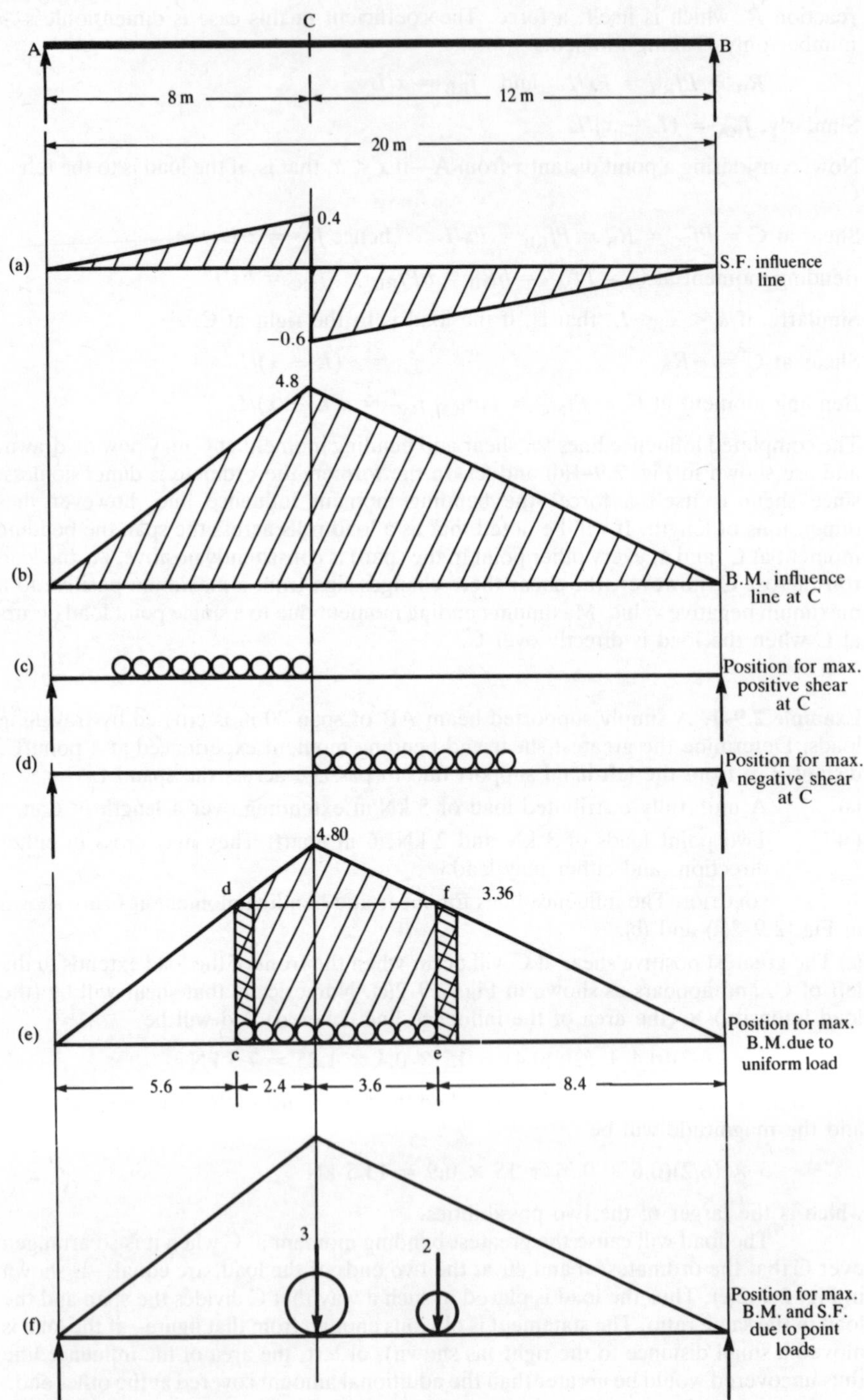
A
C
B
8 m
12 m
20 m
0.4
(a)
S.F. influence line
−0.6
4.8
(b)
B.M. influence line at C
(c)
Position for max. positive shear at C
(d)
Position for max. negative shear at C
4.80
d
f
3.36
(e)
c
e
Position for max. B.M.due to uniform load
5.6
2.4
3.6
8.4
3
2
(f)
Position for max. B.M. and S.F. due to point loads

Fig. 2·9–2

give a desired effect and, thereafter, the magnitude of the effect may be calculated directly by simple statics.

(b) Consideration of the shape of shear force and bending moment diagrams resulting from point loads shows that maximum values occur under one of the loads, when the second is so arranged as to contribute most effectively towards the maximum. The juxtaposition shown in Fig. 2.9–2(f) will obviously cause maximum negative shear at C of $-(3 \times 0.6 + 2 \times (6/12) \times 0.6) = -2.4$ kN—positioning the 2 kN load to the left of C obviously results in a reduction of this maximum.

Maximum bending moment at C will occur when one of the loads is at C. The shape of the influence diagram shows that the second load will have more effect if placed to the right of C. The distribution shown in the figure (at f) produces the greatest value of the bending moment at C, which is

$$(3 \times 4.80 + 2 \times (6/12) \times 4.80) = 19.20 \text{ kNm}$$

Problems

2.1. (a) Explain, without calculations, why the forces in bars 1 and 2 are of equal magnitude. (b) Determine *by inspection* all member forces.

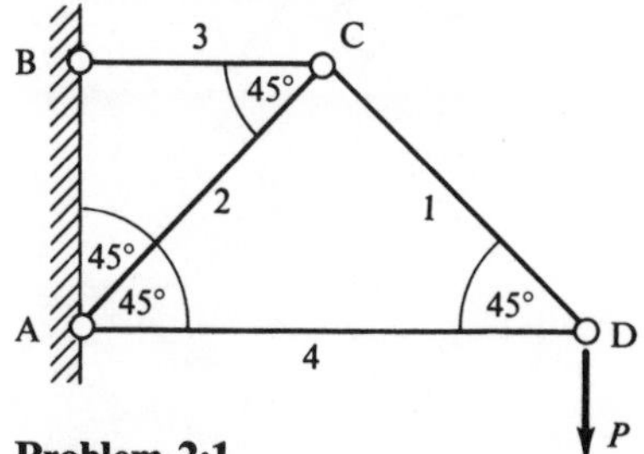

Problem 2·1

Ans. (a) $\sum$ vertical forces $= 0$ at C; (b) $S_1 = P\sqrt{2}$; $S_2 = -P\sqrt{2}$; $S_3 = 2P$; $S_4 = -P$.

2.2. Determine the forces in all the members.

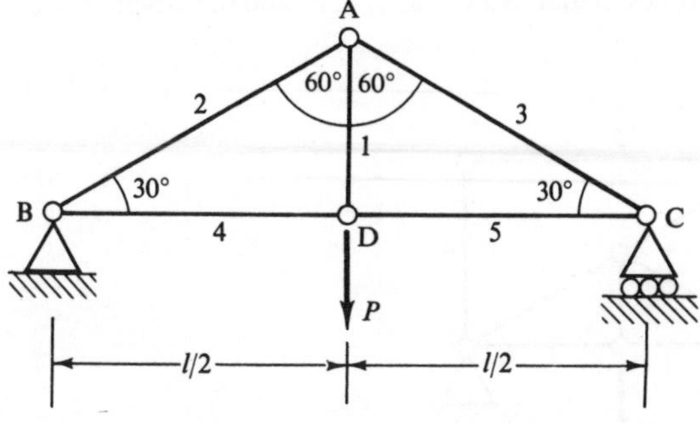

Problem 2·2

Ans. $S_1 = P$; $S_2 = S_3 = -P$; $S_4 = S_5 = P\sqrt{(3)}/2$.

2.3. The truss in this problem is that of Problem 2.2 with additional members 6 to 11 inclusive. Using the results of Problem 2.2, determine *by inspection* all member forces in this new truss.

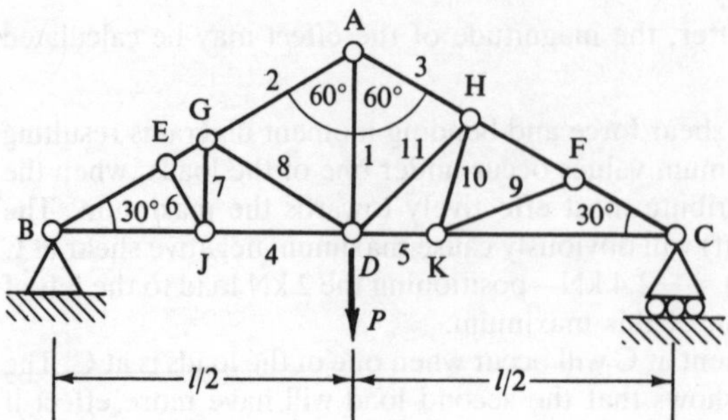

Problem 2·3

Ans. $S_6 = 0$ (Why?); $S_7 = S_8 = S_9 = S_{10} = S_{11} = 0$; S_1 to S_5 as in Problem 2·2 (Why?); $S_{GE} = S_{EB} = S_2$; $S_{HF} = S_{FC} = S_3$.

2.4. Determine all member forces *by inspection*.

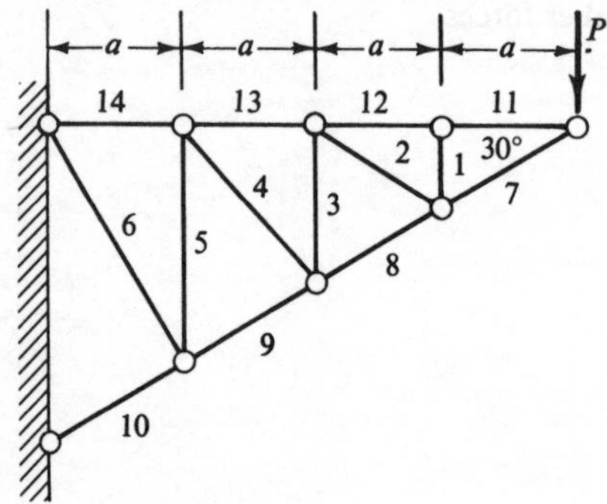

Problem 2·4

Ans. S_1 to $S_6 = 0$; hence S_7 to $S_{10} = -P \operatorname{cosec} 30°$; S_{11} to $S_{14} = P \cot 30°$.

2.5. Determine *by inspection* the forces in members 1, 2, 3, 4, 5, and 6. (Given $\alpha = \beta$.)

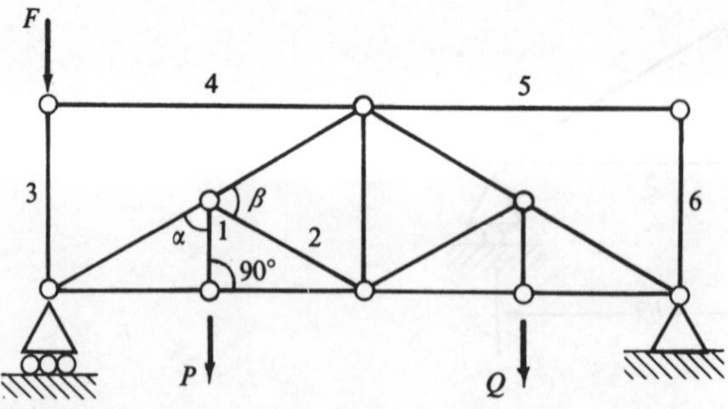

Problem 2·5

Ans. $S_1 = P$; $S_2 = -P$; $S_3 = -F$; $S_4 = 0$; $S_5 = 0$; $S_6 = 0$.

2.6. Using the method of sections, (a) determine the forces in members 1 and 2; (b) write down the equations from which the force in member 3 can be determined.

Ans. (a) $S_1 = -P\sqrt{2}$; $S_2 = -P\sqrt{2}$.
(b) $S_3 \sin 30° - S_4 \sin 30° = P$; $S_3 = -S_4$.

2.7. Explain how the method of sections can be used to determine the forces in members CE and DG.

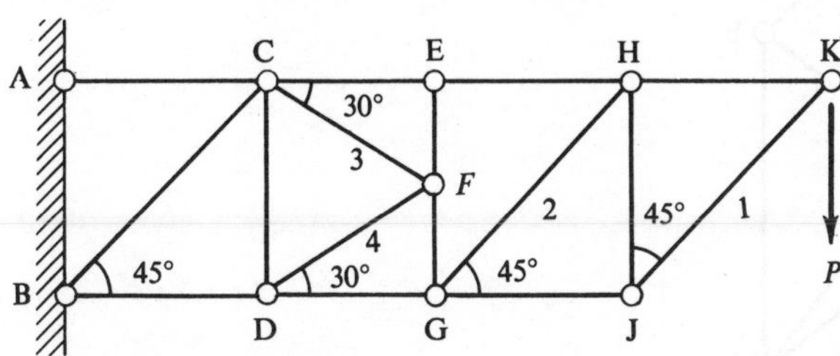

Problems 2·6 and 2·7

Ans. Pass curved section through members CE, EF, FG, and DG; then consider equilibrium of portion to the right. S_{CE} and S_{DG} are given by two conditions: (a) $\sum$ horizontal forces = 0; (b) $\sum$ moments about F = 0.

2.8. (a) Determine *by inspection* the forces in members 1, 2, and 3. (b) Write down an equation from which the force in member 4 can be found.

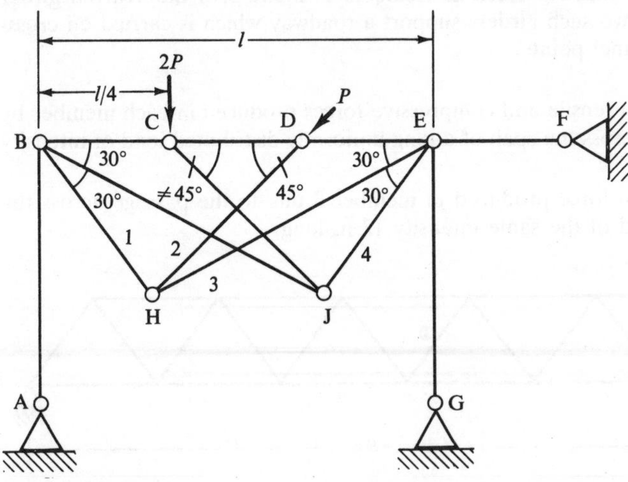

Problem 2·8

Ans. (a) $S_2 = -P$; $S_3 = -S_2 \cos 15° = P \cos 15°$; $S_1 = P \sin 15°$.
(b) For equilibrium of subtruss BCJ, $S_4 \times l \cos 30° = 2P \times l/4$.

2.9. (a) Determine *by inspection* the forces in members 1 and 2. (b) Determine the support reactions. (c) Determine the force in member 3.

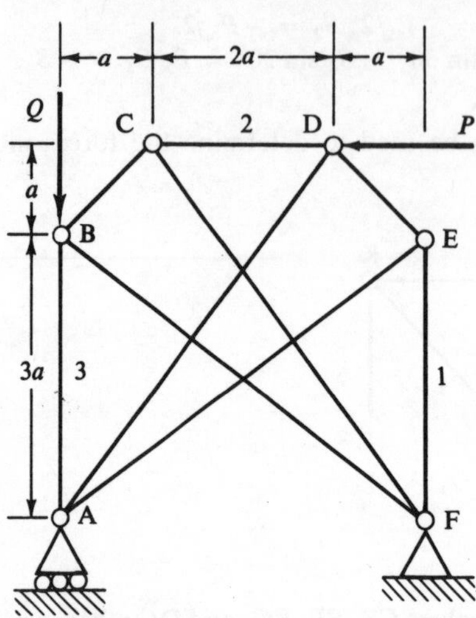

Problem 2·9

Ans. **(a) For equilibrium of subtruss ADE: $S_1 = 0$, $S_2 = -P$. (b) $H_A = 0$; $V_A = Q + P$ (up); $H_F = P$; $V_F = P$ (down). (c) $S_3 = -(Q + P)$.**

2.10. Sketch influence lines for force in members 1, 2 and 3 of the Warren girder shown in the figure. Two such girders support a roadway which is carried on cross-girders at the lower-panel points.

Determine:

(a) The maximum tensile and compressive forces produced in each member by the passage across the span of a long uniformly distributed load of intensity 4 kN/m.

(b) The maximum force produced in member 3 due to the passage across the span of a load of the same intensity 12 m long.

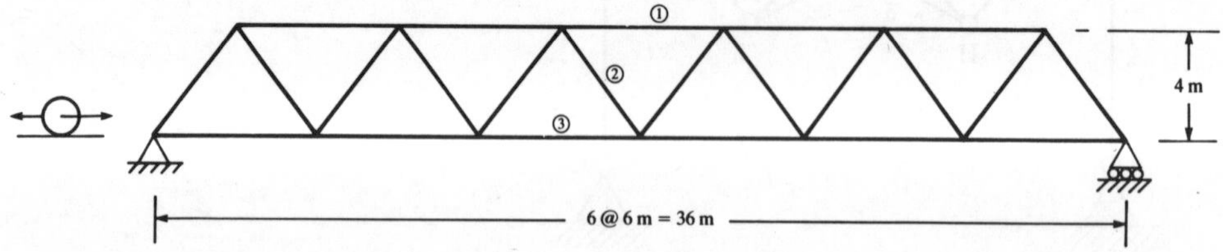

Problem 2·10

Ans. (a) 1) 81.0; 2) + 13.5. – 6.0; 3) 76.50 kN. (b) 34.84 kN.

2.11. The dead load over the whole span of the bridge truss is equivalent to a uniformly distributed load of 10 kN/m. What is the maximum intensity of a long live load which may cross the bridge without causing a change of sign of the forces in the diagonals of the two centre panels?

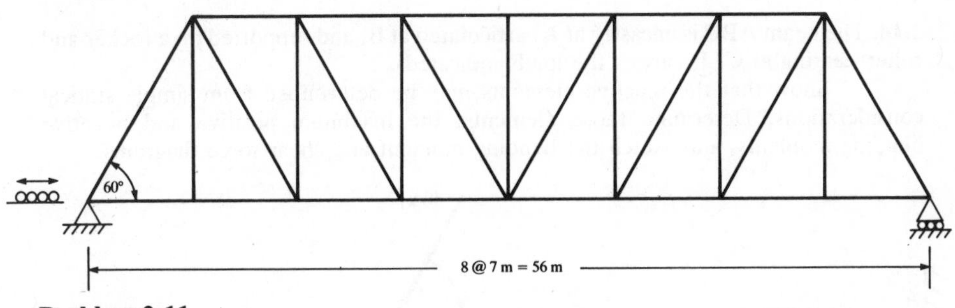

Problem 2·11

Ans. 7.77 kN/m.

2.12. Sketch the shearing force and bending moment diagrams for the beam ABCDEF shown.

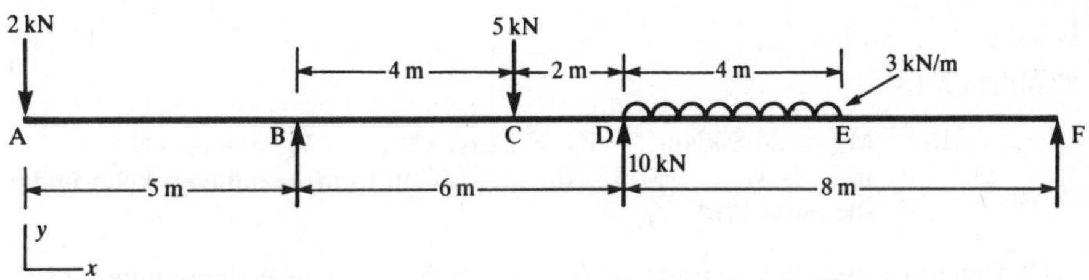

Problem 2·12

Ans.

		B		C		D		E
Shear	2.0		−3.71	−3.71	1.29	1.29	−8.71	3.29 kN
B.M.		−10.0		4.84		2.26		13.16 kNm.

Max. B.M. is 14.96 kNm 1.097 m to the left of E.

2.13. Sketch the shearing force and bending moment diagrams for the beam shown in the figure, indicating the magnitudes of any maxima, and the positions of any points of contraflexure.

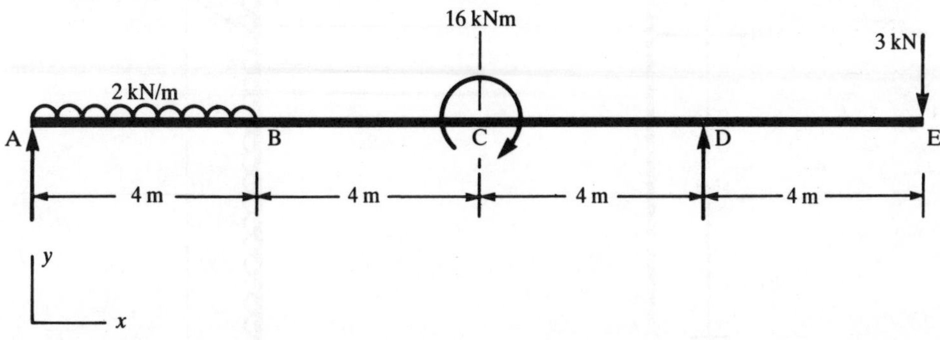

Problem 2·13

Ans. 4.70 kNm at 2.17 m from A, 2.67 kNm and 13.33 kNm at C and 12.00 kNm at D. Points of contraflexure at 11.63 and 7.27 m from E.

2.14. The beam ABC is encastré at A, articulated at B, and supported on a rocker and roller bearing at C. It carries the loads indicated.

Show that the reactive elements may be determined from simple statical considerations. Determine those elements, the maximum positive and negative bending moments, and sketch the bending moment and shear force diagrams.

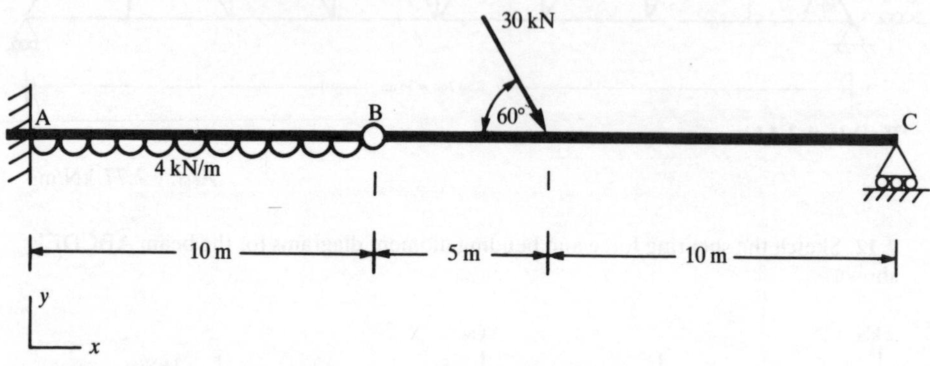

Problem 2·14

Ans. $M_A = 26.8$ kNm $V_A = 22.68$ kN ↓ $V_C = 8.66$ kN ↑
max. B.M. = −37.5 kNm, 5.67 m from A and 86.6 kNm under the point load.

2.15. Determine the reactive forces at A and D in the rectangular rigid-jointed frame shown and the bending moments at B, C and E. Where, and what, is the maximum bending moment?

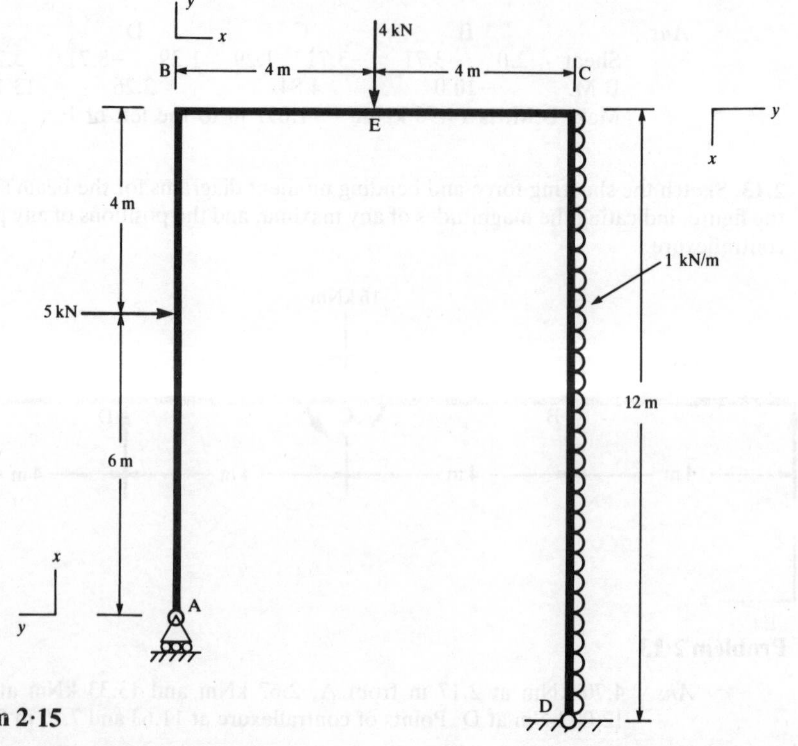

Problem 2·15

Ans. $V_A = 6$ kN ↑ $V_D = 2$ kN ↓ $H_D = 7$ kN→
$M_B = -20$ $M_E = 4$ $M_C = 12$ kNm
Max. moment is 24.5 kNm, 7 m from D.

2.16. The rectangular rigid jointed frame ABCDE is hinged at A, C and E and supports the loads shown. Determine the reactions at A and E and sketch the shear force and bending moment diagrams.

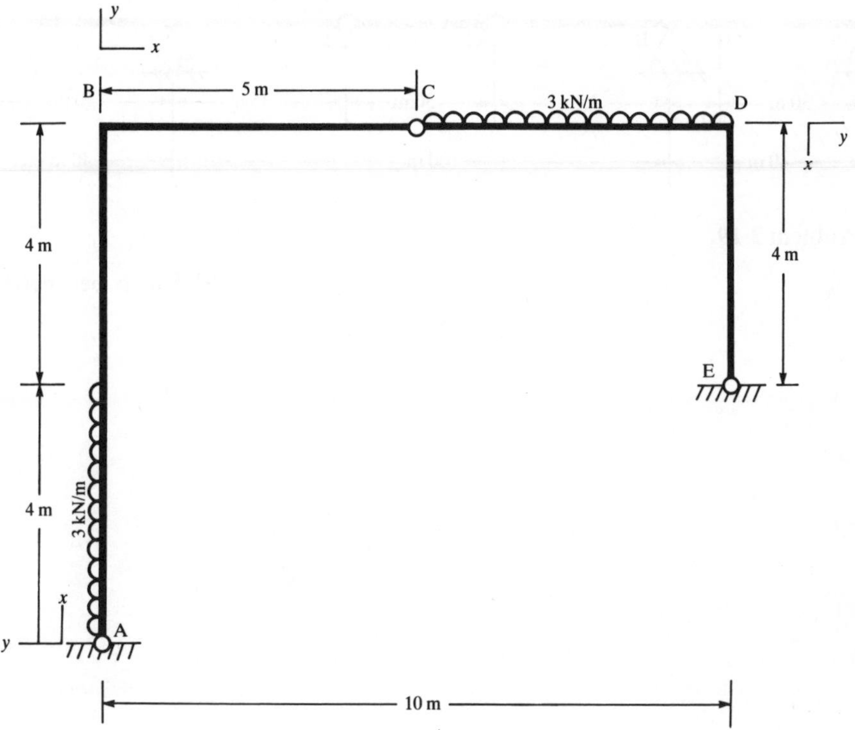

Problem 2·16

Ans. $V_A = 3.4$ ↑ $H_A = 6.88$← $V_E = 11.6$ ↑ $H_E = 5.12$← kN.
Interactive forces at C are 5.12 and 3.4 kN
Maximum B.M. in AB is 7.89 kNm, 2.29 m from A.

2.17. A horizontal beam AB, of length 20 m, carries a uniformly distributed load of 4 kN/m and point loads of 12 and 18 kN, respectively, at its ends A and B. It is to be supported on two equal reactions, 8 m apart. Where should these be placed? What, and where, is the maximum bending moment?

Ans. 5.6 m and 13.6 m from A.
133.1 kNm at the support nearer to B.

2.18. A horizontal beam of length L is supported on two symmetrically placed reactions and carries a uniformly distributed load over its entire length. Where should the reactions be placed to ensure that the maximum bending moment in the beam has a minimum value? Where are the points of contra-flexure?

Ans. Reactions at $0.207L$ from each end. Points of contraflexure at $0.293L$ from each end.

2.19. The 'cantilever and suspended central span' form, used in many major bridges, is shown schematically in the figure. Sketch influence lines for the reaction at B, and for the bending moment at point G in the span AB.

If the dead load on the structure is equivalent to a uniformly distributed load of 2000 units per metre of total span of 210 metres, what is the least intensity of a long travelling load which will cause a change of sign in the bending moment at G?

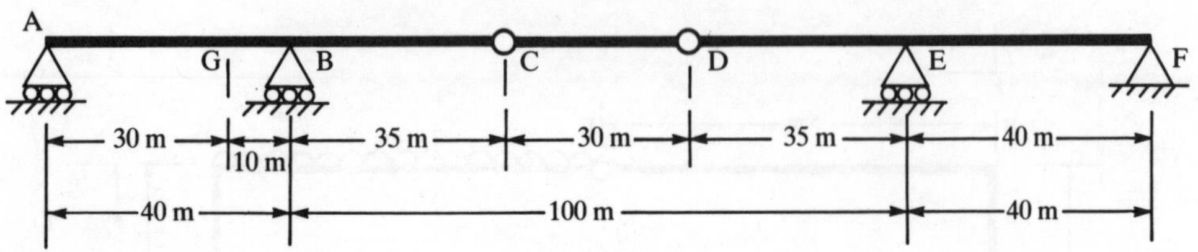

Problem 2·19

Ans. 9375 units per metre.

Chapter 3
Space statics and determinate space structures

3.1 Space statics

Space statics is concerned with the equilibrium of forces in a three-dimensional space. The basic concepts are as given in Sections 1.2 and 1.3. That is, in the general case of a number of forces $\mathbf{P}_1, \mathbf{P}_2 \ldots \mathbf{P}_n$ and a number of moments $\mathbf{m}_1, \mathbf{m}_2 \ldots \mathbf{m}_m$ acting in space, as shown in Fig. 1.3–1, the resultant force $\mathbf{R}$ is given in magnitude and direction by

$$\mathbf{R} = [\sum P_x \ \sum P_y \ \sum P_z]^T \tag{3.1–1}$$

i.e.

$$R = \sqrt{[(\sum P_x)^2 + (\sum P_y)^2 + (\sum P_z)^2]} \tag{3.1–2}$$

and

$$\cos \bar{\alpha} = \frac{\sum P_x}{R}; \qquad \cos \bar{\beta} = \frac{\sum P_y}{R}; \qquad \cos \bar{\gamma} = \frac{\sum P_z}{R} \tag{3.1–3}$$

where $\bar{\alpha}$, $\bar{\beta}$ and $\bar{\gamma}$ (Fig. 3.1–1) are the angles which $\mathbf{R}$ makes with the respective coordinate directions.

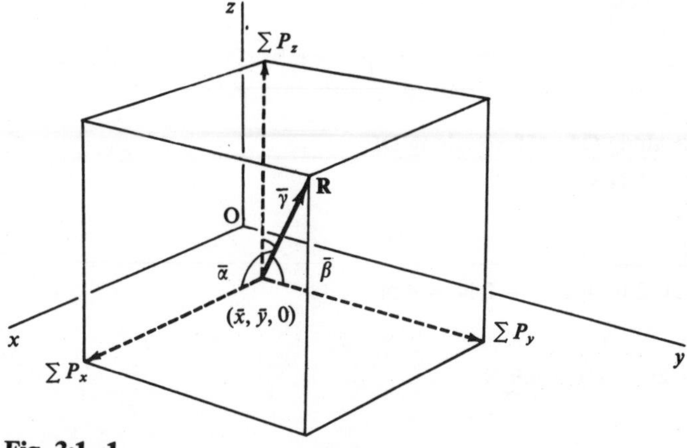

Fig. 3·1–1

The above equations make it clear that neither the magnitude nor the direction of the resultant force is dependent on the applied moments $\mathbf{m}_1, \mathbf{m}_2 \ldots \mathbf{m}_m$, which affect only the position of the line of action of **R**. Suppose **R** intersects the xy plane at $(\bar{x}, \bar{y}, 0)$. From the footnote on page 4,

$$\text{moment of } \mathbf{R} \text{ about origin } 0 = \mathbf{R} \times \mathbf{r} \tag{3.1–4}$$

where **r** is now the vector $(-\bar{x}\mathbf{i} - \bar{y}\mathbf{j})$. Example 3.1–1 explains how to determine $\bar{x}$ and $\bar{y}$.

If a body is in equilibrium under the action of a system of forces and moments in space (Fig. 1.3–1) then we have, from Eqns 1.3–1:

$$\sum P_x = 0; \qquad \sum P_y = 0; \qquad \sum P_z = 0 \tag{3.1–5(a)}$$

$$\sum M_x = 0; \qquad \sum M_y = 0; \qquad \sum M_z = 0 \tag{3.1–5(b)}$$

where $\sum M_x$ represents the sum of the moments about the x-axis, due to both the given forces and the given moments, and similarly for $\sum M_y$ and $\sum M_z$.

†**Example 3.1–1.** Figure 3.1–2 shows a cube of sides 1 m acted on by eight forces and a moment as shown. Determine the magnitude and direction of the resultant force **R** and the coordinates $(\bar{x}, \bar{y}, \bar{z})$ of the point at which **R** intersects the face OACB. Is there a resultant moment in addition to the resultant force **R**?

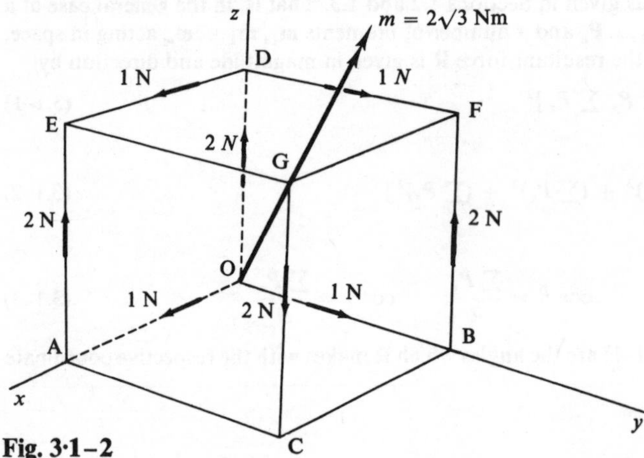

Fig. 3·1–2

SOLUTION

$$\sum P_x = \text{force along OA} + \text{force along DE}$$
$$= 1\text{ N} + 1\text{ N} = 2\text{ N}$$

similarly,

$$\sum P_y = 2\text{ N}$$
$$\sum P_z = 2\text{ N} + 2\text{ N} + 2\text{ N} - 2\text{ N} = 4\text{ N}$$

Then

$$R = \sqrt{(2^2 + 2^2 + 4^2)} = \underline{4.9\text{ N}}$$

† See Appendix 1 of Ref. 1 for a more detailed discussion of this type of problem.

The angles $\bar{\alpha}$, $\bar{\beta}$, and $\bar{\gamma}$ which **R** makes with the *x*-, *y*-, and *z*-axes are:

$$\bar{\alpha} = \cos^{-1}\frac{\sum P_x}{R} = \cos^{-1}\frac{2}{4.9} = \underline{66°}$$

$$\bar{\beta} = 66°; \qquad \bar{\gamma} = \cos^{-1}\frac{4}{4.9} = \underline{35.3°}$$

Thus the unit vector in the direction of **R** is

$$\mathbf{e}_R = 0.408\mathbf{i} + 0.408\mathbf{j} + 0.816\mathbf{k} \text{ where } 0.408 = \cos\bar{\alpha} \text{ etc.}$$

Let $\mathbf{M} = [M_x\ M_y\ M_z]^T$ be the resultant moment due to the given forces and moment. Taking moments about the point 0,

$$\begin{aligned} M_x &= -(\text{Force along DF}) \times 1 + (\text{Force along BF}) \times 1 \\ &\quad -(\text{Force along GC}) \times 1 + m_x \\ &= -1 \times 1 + 2 \times 1 - 2 \times 1 + 2\sqrt{3} \times (1/\sqrt{3})\ \text{Nm} = 1\ \text{Nm} \end{aligned}$$

Similarly,

$$M_y = 1 \times 1 - 2 \times 1 + 2 \times 1 + 2\sqrt{3} \times (1/\sqrt{3})\ \text{Nm} = 3\ \text{Nm}$$

$$M_z = 2\sqrt{3} \times (1/\sqrt{3}) = 2\ \text{Nm}$$

That is, $\mathbf{M} = [M_x\ M_y\ M_z]^T = \mathbf{i} + 3\mathbf{j} + 2\mathbf{k}$ Nm

M can be considered as made up of two components: $\mathbf{M}_R$ parallel to **R** and $\mathbf{M}_N$ normal to **R**. $\mathbf{M}_R$ is of course simply the projection of **M** on **R**. That is,

$$\begin{aligned} \mathbf{M}_R &= (\mathbf{M} \cdot \mathbf{e}_R)\,\mathbf{e}_R = (\mathbf{i} + 3\mathbf{j} + 2\mathbf{k}) \cdot (0.408\mathbf{i} + 0.408\mathbf{j} + 0.816\mathbf{k})\,\mathbf{e}_R \\ &= 1.33\mathbf{i} + 1.33\mathbf{j} + 2.66\mathbf{k}\ \text{Nm} \end{aligned}$$

$$\mathbf{M}_N = \mathbf{M} - \mathbf{M}_R = -0.33\mathbf{i} + 1.67\mathbf{j} - 0.66\mathbf{k}\ \text{Nm}$$

And $\mathbf{M}_N$ must be equal to $\mathbf{R} \times \mathbf{r}$ as given by Eqn 3.1–4. That is

$$-0.33\mathbf{i} + 1.67\mathbf{j} - 0.66\mathbf{k} = \begin{vmatrix} \mathbf{i} & \mathbf{j} & \mathbf{k} \\ 2 & 2 & 4 \\ -\bar{x} & -\bar{y} & 0 \end{vmatrix}$$

giving $\bar{x} = -0.418\text{m}$, $\bar{y} = -0.083\text{m}$

Thus the given system of forces and moment is reduced to the single force **R** acting through the point (−0.418, −0.083, 0) plus the moment $\mathbf{M}_R$. Note that $\mathbf{M}_R$ cannot be made to vanish.

3.2 Constraint of space structures

A space structure is a structure in a three-dimensional space. Equations 3.1–5 show that the maximum number of support reaction components that can be determined from the conditions of static equilibrium is six. A space structure is said to be externally statically determinate if all the support reactions can be determined by statics. In general, a space structure constrained by six suitable support reaction components is externally statically determinate; if the reaction components are not suitably arranged, then even a large number of such components cannot maintain equilibrium of the structure. If the structure is rigidly constrained by more than six reaction components, it is externally indeterminate.

To study the constraint of a structure in space, consider the rectangular body in Fig. 3.2–1. First, any point A in the body can be completely restrained against linear displacements in space by three smoothly jointed bars such as A1, A2, and A3 which do not all lie in one plane. For space structures, a small circle such as the one at A, or the ones at 1, 2, and 3, represents a frictionless ball-and-socket joint which offers no restraint against rotation of the joint about any axis through that joint. To understand why the

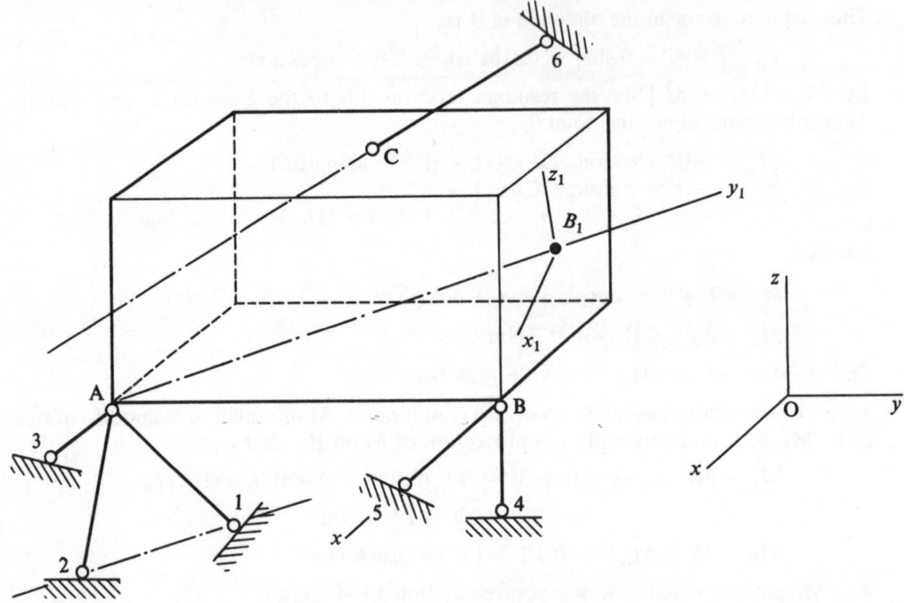

Fig. 3·2–1

three bars must not lie in one plane, we note from Fig. 3.2–1 that bars A1 and A2 restrain displacement of A in the plane A–1–2, but rotation of this plane about axis 1–2 is possible. Such rotation is prevented by a bar A3 where joint 3 does not lie on axis 1–2. Coming back to the rectangular body, we now see that any other point on it, such as B, can only move on the surface of a sphere of centre A and radius AB. Such movement at B will be completely restrained by the two bars B4 and B5. The bar B4 is shown to be in the z-direction and B5 in the x-direction; in fact, of course, any two non-collinear bars in a plane normal to the y-axis will do. Indeed, they do not even have to lie in that plane; all that is required is that they are capable of providing reaction components in (any) two perpendicular directions on a plane normal to the y-direction at B. (Similarly, if instead of restraining point B, we wish to restrain another point B_1, we require two bars which are capable of providing restraints in two perpendicular directions in plane $x_1 B_1 z_1$, which is normal to the y_1-direction in the figure.) With points A and B thus restrained, the body can only rotate about axis AB and such rotation is prevented by a bar C6, the direction of which does not intersect axis AB. From the above discussion, it is clear that a space structure can always be completely restrained against rigid body movements by six bars or reaction components provided in the following successive steps:

(a) restraint of any point A by three non-coplanar bars;

(b) restraint of any other point B by two non-collinear bars which lie in a plane not passing through point A,

(c) restraint of any other point C, not on AB, by a bar which does not intersect the line AB.

While six reaction components provided in the above manner will guarantee complete restraint of a structure in space, it should be noted that these six components can also be arranged in some other suitable manner to fulfil the same purpose. The ultimate test is that the six conditions,

$$\sum P_x = 0; \quad \sum P_y = 0; \quad \sum P_z = 0;$$
$$\sum M_x = 0; \quad \sum M_y = 0; \quad \sum M_z = 0$$

can be satisfied under any system of applied forces. For a plane structure it is easy to see whether the three reaction components actually provided are suitably arranged or not; for a space structure restrained by six reaction components, it is sometimes difficult to tell whether they can maintain equilibrium. However, a practical test exists: after the reader has studied Chapter 7, he will realize that if the support reactions do not provide adequate restraint, then the stiffness matrix $\mathbf{K}_s$, modified to allow for support restraints, will be singular (see Section 7.10).

3.3 Determinate space structures

In this section, we shall analyse the support reactions of externally statically determinate space structures. There are six reaction components in each determinate space structure, and these can always be determined by solving the six simultaneous equations (Eqns 3.1–5) which represent the six conditions of equilibrium. However, in many cases it is possible to determine one or more of these reaction components by inspection or by some very simple calculations. With as many as possible of such readily determined components calculated, the number of simultaneous equations that have to be solved is appreciably reduced. This method of solution is illustrated in the following examples, which deal with structures specially chosen to help the reader develop his skill in the application of the principles of statics.

Example 3.3–1. A circular bar AB, of radius r metres, lies on the horizontal plane (Fig. 3.3–1(a)) and is rigidly attached to two vertical walls at A and B. A uniformly distributed twisting moment of intensity T newton-metres per metre of bar length acts on the bar. Determine the reactions at A and B.

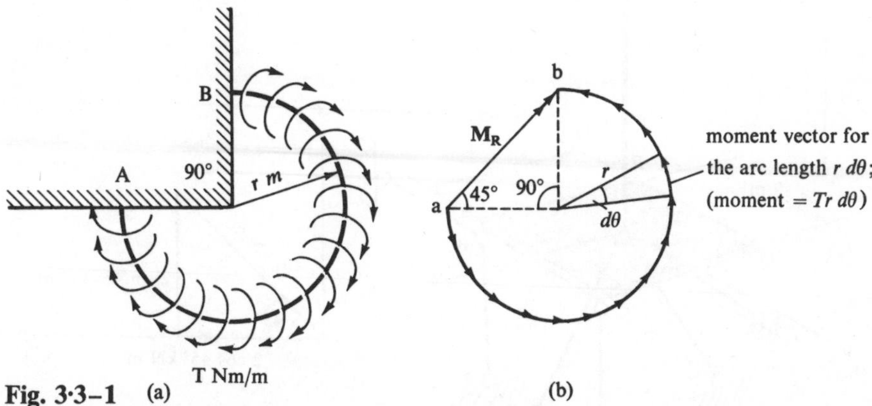

Fig. 3·3–1 (a) (b)

SOLUTION Consider an infinitesimal arc length of the bar subtending an angle $d\theta$ radians at the centre. The twisting moment acting on this length $r\,d\theta$ is $Tr\,d\theta$ newton-metres. As explained in Section 1.2, this moment can be represented by the vector shown in Fig. 3.3–1(b). Thus the distributed moment acting on the bar is represented by an infinite number of moment vectors following one another head to tail along the axis of the bar. The moment resultant $\mathbf{M}_R$ of these moment vectors is given by the line ab.

$$M_R = \text{length ab} = Tr/\cos 45° \text{ Nm}$$

By symmetry the reactions at A and B are each $-\mathbf{M}_R/2$, where the negative sign shows that the reactions are equal and opposite to the resultant $\mathbf{M}_R$. Specifically,

$$\text{Bending moment at A} = \tfrac{1}{2}M_R \cos 45° = \tfrac{1}{2}Tr \text{ Nm}$$

$$\text{Twisting moment at A} = \tfrac{1}{2}M_R \sin 45° = \tfrac{1}{2}Tr \text{ Nm}$$

and similarly at B.

Example 3.3–2. A Y-shaped frame is in the *yz* plane and is rigidly attached to the central point of a thin square plate in the *xy* plane (Fig. 3.3–2(a)). The underside of the plate at A is hinged to three non-coplanar bars which are anchored to the foundation at points 1, 2, and 3. The plate is further attached to the foundation by bar B4 in the *z*-direction, bar B5 in the *x*-direction, and bar C6 in the *z*-direction. The frame is acted on by a twisting moment of 2 kN m at F and a horizontal force of 1 kN in the *x*-direction at H. Determine the forces in bars B4, B5, and C6.

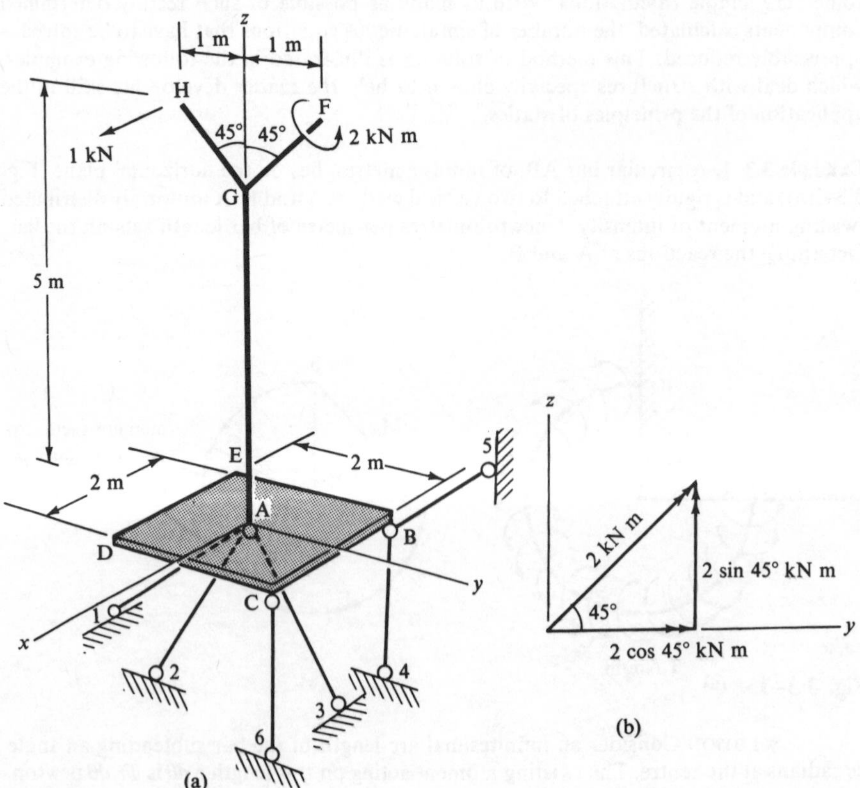

Fig. 3·3–2

SOLUTION Figure 3.3–2(b) shows that the 2 kN m twisting moment at F has a component $m_y(\text{F})$ about the y-axis and a component $m_z(\text{F})$ about the z-axis:

$$m_y(\text{F}) = (2 \text{ kN m}) \cos 45^\circ = 1.41 \text{ kN m}$$

$$m_z(\text{F}) = (2 \text{ kN m}) \sin 45^\circ = 1.41 \text{ kN m}$$

For the purpose of calculating support reactions, the 1 kN force at H is equivalent to the following force and moments acting at A:

a force $P_x(\text{H})$ of 1 kN in the x-direction, plus a moment $m_y(\text{H})$ of 1 kN × 5 m = 5 kN m, about the y-axis, plus

a moment $m_z(\text{H})$ of 1 kN × 1 m = 1 kN m, about the z-axis.

In other words, the square plate is acted on by the following force and moments at point A:

(a) A force P_x of 1 kN, which is entirely resisted by the three bars at A.

(b) A moment about the y-axis of magnitude

$$\sum M_y = m_y(\text{F}) + m_y(\text{H}) = 1.41 + 5 = 6.41 \text{ kN m}$$

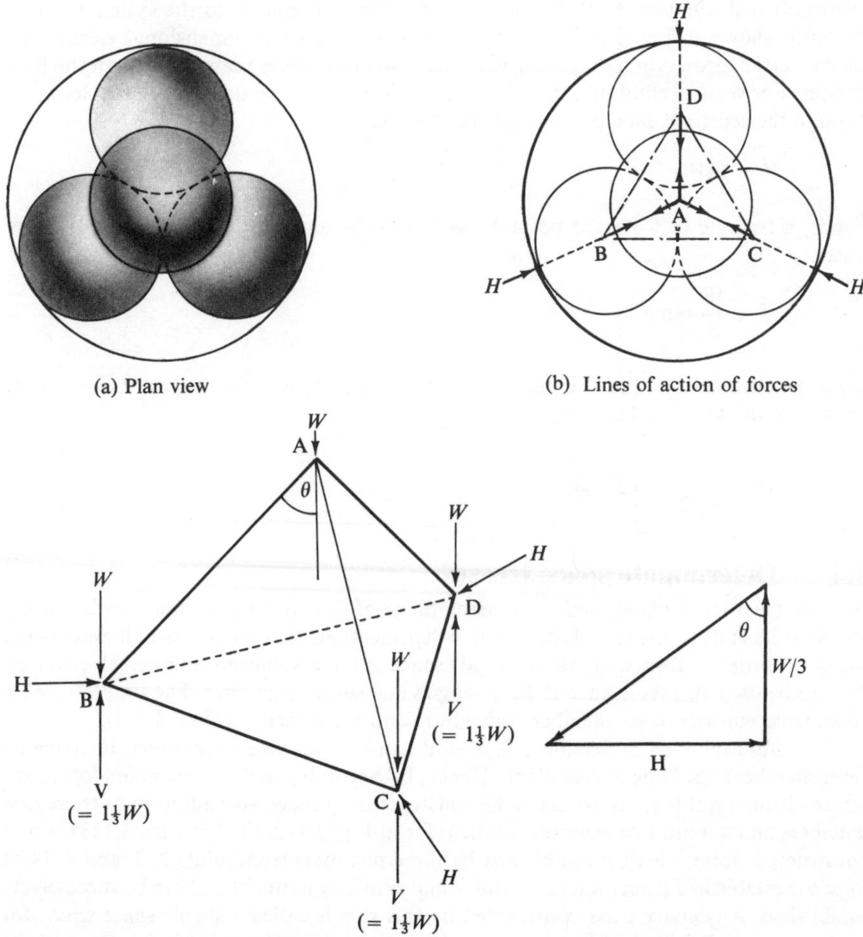

(a) Plan view

(b) Lines of action of forces

Fig. 3·3–3 (c) Force system (d) Force triangle at B

By inspection, this moment is resisted entirely by a couple constituted by a compressive force in C6 and a tensile force in B4.

Compressive force in C6 = tensile force in B4

$$= \sum M_y/(\text{length BC})$$
$$= 6.41/2 = 3.2 \text{ kN}$$

(c) A moment about the z-axis of magnitude

$$\sum M_z = m_z(\text{F}) + m_z(\text{H}) = 1.41 + 1 = 2.41 \text{ kN m}$$

By inspection, this moment produces a compressive force in bar B5, such that:

compressive force in B5 × 1 m = 2.41 kN m

Hence compressive force in B5 = $\underline{2.41 \text{ kN}}$

Example 3.3–3. Four smooth spheres of equal radius form a pyramid on a smooth table (Fig. 3.3–3(a)). The bottom three spheres are constrained by a cylinder of such diameter that these three spheres are just touching each other and the cylinder. If each sphere weighs W, determine the contact force between each bottom sphere and the cylinder.

SOLUTION The contact forces H between the bottom spheres and the cylinder are due to the weight W of the top sphere, which is transmitted from the centre A of the top sphere through the centres B, C, and D of the bottom spheres to the cylinder, along the paths shown in Fig. 3.3–3(b). Fig. 3.3–3(c) shows a three-dimensional view of the regular tetrahedron ABCD together with the system of forces that act through the four vertices. For vertical equilibrium of the system, the vertical reaction V of the table acting through the centre of each bottom sphere must be

$$V = 1\tfrac{1}{3}W$$

Hence, a triangle of force for point B, say, must be as shown in Fig. 3.3–3(d), from which

$$H = \frac{W}{3} \tan \theta$$

Since the tetrahedron ABCD is regular, i.e. the edges are equal in length, it can readily be shown that $\tan \theta = 1/\sqrt{2}$. Thus

$$H = \frac{W}{3\sqrt{2}} = 0.236W$$

3.4 Determinate space trusses

A space truss is a three-dimensional assemblage of line members, each member being joined at its ends to the foundation or to other members by frictionless ball-and-socket joints. In order to form a stable (or rigid) space truss, a sufficient number of members have to be used and these have to be arranged in a suitable manner. The simplest stable space truss consists of six members joined to form a tetrahedron (Fig. 3.4–1).

As explained in Section 3.2, a joint can be completely restrained in space by three members not lying in one plane. Hence, by beginning with six members forming a tetrahedron, a stable space truss can be constructed by successive addition of three new members and a joint. For example, in the truss in Fig. 3.4–2, the basic truss 1234 is first constructed. Joint 5 is then established by three members from joints 2, 3, and 4. Next joint 6 is established from joints 2, 3, and 5, and similarly joints 7 to 12 can be successively established. Any space truss constructed by this rule is called a **simple space truss**; for such a truss a definite relationship exists between the number of members m and the number of points j. The original tetrahedron consists of six members and four joints. Since for each additional joint there are three additional members, the relationship must be

$$m - 6 = 3(j - 4)$$

i.e.

$$m = 3j - 6 \tag{3.4–1}$$

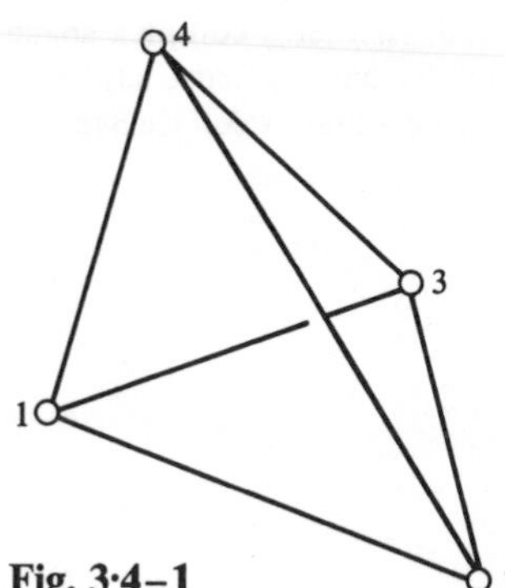

Fig. 3·4–1

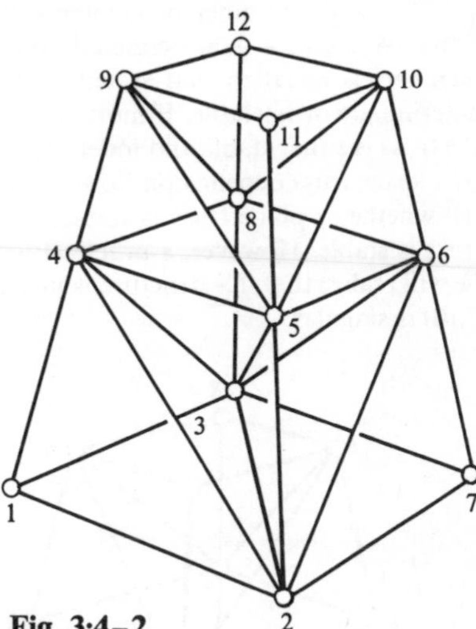

Fig. 3·4–2

Equation 3.4–1 relates the number of members to the number of joints in a simple space truss that is not connected to the foundation. It was pointed out in Section 3.2 that six reaction components were necessary and sufficient for the complete restraint of a space structure. Hence a space truss can be rigidly connected to the foundation by six members arranged in a suitable way. For a space truss so connected to the foundation Eqn 3.4–1 is modified to

$$m = 3j \tag{3.4–2}$$

where m is now the total number of members including those that connect the truss to the foundation, and j is the number of joints excluding those at the foundation. (After the reader has studied the next several chapters, he will appreciate that the total number of unknown member forces must be equal to the total number of 'degrees of freedom'.)

Figure 3.4–3 shows another method of constructing a simple space truss that is rigidly attached to the foundation. Starting at convenient points A, B, C on the foundation, joint 1 is established by three members; similarly joint 2 is established by three members from B, C, and D. Next, joints 3 to 8 are successively established by three additional members to each joint. Obviously, the member–joint relationship for such a truss is that of Eqn 3.4–2.

Equations 3.4–1 and 3.4–2 are significant in that they give the minimum number of members required to build the stable trusses. If a truss has less than this number of members, it cannot be stable; if it has more, then, as will be explained in Section 3.5, the equations of statics are no longer sufficient for the determination of the forces in the members under the action of applied forces, and the truss becomes internally statically indeterminate. As for plane structures, a space structure is said to be statically indeterminate if it is either externally or internally statically indeterminate. Similarly, a space structure is said to be statically determinate only when it is both internally and externally statically determinate.

The reader should verify that Eqns 3.4–1 and 3.4–2 apply equally to stable and statically determinate **compound space trusses**, each of which can be defined as a space truss made up of two (or more) simple space trusses rigidly connected together by six suitably arranged members. In fact these equations apply also to **complex space trusses**, which are space trusses not satisfying either the definition of simple space trusses or of compound space trusses. Hence these member–joint relationships are general.

If the number of members in a given space truss is less than that stated by Eqns 3.4–1 or 3.4–2, we conclude immediately that it is not stable. If the number of members is equal to that stated by the equation, then the truss is either stable and determinate or unstable. If more members are used than required by the equation then the truss is either stable and indeterminate or unstable. As for plane trusses, the stability of a space truss depends on how the members are arranged: while it is usually easy to tell whether a plane truss is stable, it is often difficult to tell whether a complex space truss is stable. However, a practical test exists: after the reader has studied Chapter 7, he will realize that if a structure is unstable, the modified stiffness matrix $\mathbf{K}_s$ (see Section 7.10) is singular.

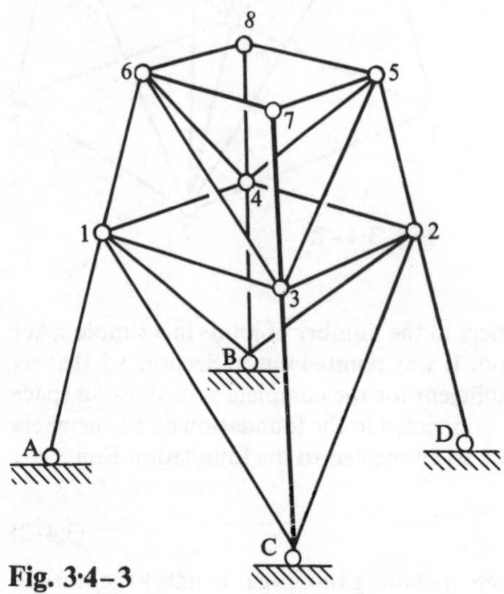

Fig. 3·4–3

3.5 Analysis of determinate space trusses

Space trusses of all types, simple, compound, or complex, can be efficiently analysed by the matrix computer methods to be given later in this book. Internal or external indeterminancy presents no problem, and unsuitable arrangement of members that lead to instability can be detected. Hence, methods of analysis of space trusses by hand have little value as practical tools. However, analysis by hand of suitably chosen determinate space trusses inculcates a mastery of the principles of statics and an ability to apply them. It is for this reason that this section is included, and the trusses studied here are special cases chosen with the above aim in mind.

All determinate space trusses can be analysed by the method of joints. If there are j joints, then since there are three equations ($\sum P_x = 0$; $\sum P_y = 0$; $\sum P_z = 0$) to each joint, there will be a total of $3j$ equations. If the truss is rigidly attached to the foundation, then the unknowns are the m member forces. Since Eqn 3.4–2 states that

$$m = 3j$$

the number of equations will be exactly equal to the number of unknowns. Note, however, that while it is always possible to set up the $3j$ equations and solve them formally, it is rarely desirable to do so. In any case, for a 'simple space truss' it will not be necessary to solve more than three simultaneous equations at a time. By definition, in any simple space truss there is at least one joint (and the last joint to be established in accordance with the rule of construction is one such joint) at which only three members meet. Hence by starting with this joint, then considering the last but one joint established in accordance with the rule, and then working backwards joint by joint, it would be possible to determine all member forces without solving more than three simultaneous equations at a time. For the more complicated case of compound space trusses it may be necessary to solve more than three simultaneous equations at a time; for complex space trusses it may be necessary to solve many simultaneous equations. Compound or complex space trusses should in general be solved by computer methods.

When analysing a space truss by hand, those member forces which can be obtained by inspection or by very little calculation should first be determined. With as many as possible of such readily calculated forces determined, the solution for the rest of the member forces is greatly simplified. For this purpose, the following simple rules are very useful:

RULE 1: If all except one of the members meeting at a joint lie in one plane, then the force in this odd member is zero if no external force acts at the joint. If an external force acts at the joint, then the force in the odd member must be such that its component normal to the plane of the other members is equal to the corresponding component of the external force.

RULE 2: If three non-coplanar members meet at a joint not acted on by an external force, then the force in each member must be zero.

RULE 3: If two of four non-coplanar members at a joint are collinear, then the forces in each of the two non-collinear members can be determined easily by considering force components normal to the plane containing the other three members.

RULE 4: If all except two non-collinear members at a joint have zero force, then these two members must also have zero force if no external force acts at the joint.

The reader should verify the correctness of the above four rules.

Example 3.5–1. A space truss ABCDEF is hinged to a vertical wall in the zx plane at A, B, C, and D (Fig. 3.5–1(a)). Joints A, B, F, and E lie in the horizontal xy plane, with BA along the x-axis and BF along the y-axis. Joints B, C, and D lie on the vertical z-axis. The horizontal angles between bars 2, 3, and 5, and the angles which bars 1 and 6 make with the vertical are as shown. Determine all the bar forces due to a vertical force **P** at joint E.

SOLUTION At joint F, all bars except bar 1 lie in the horizontal plane. Hence, by Rule 1, the force in bar 1 is zero, i.e.

$$S_1 = \underline{0}$$

At joint E, S_6, the force in bar 6 is immediately determined from the condition $\sum P_z = 0$:

$$S_6 \cos 60^\circ = P$$

Then

$$S_6 = \underline{2P} \text{ (compressive)}$$

The horizontal component of S_6 is

$$H_6 = S_6 \sin 60^\circ = 1.73P$$

From Fig. 3.5–1(b), it is clear that

$$S_4 = H_6 \cos 60^\circ = \underline{0.87P} \quad \text{(tensile)}$$

$$S_5 = H_6 \sin 60^\circ = \underline{1.5P} \quad \text{(tensile)}$$

By symmetry (Fig. 3.5–1(b)),

$$S_3 = H_6 = \underline{1.73P} \quad \text{(compressive)}$$

$$S_2 = S_4 = \underline{0.87P} \quad \text{(tensile)}$$

In this example, we do not even have to solve more than one equation at a time.

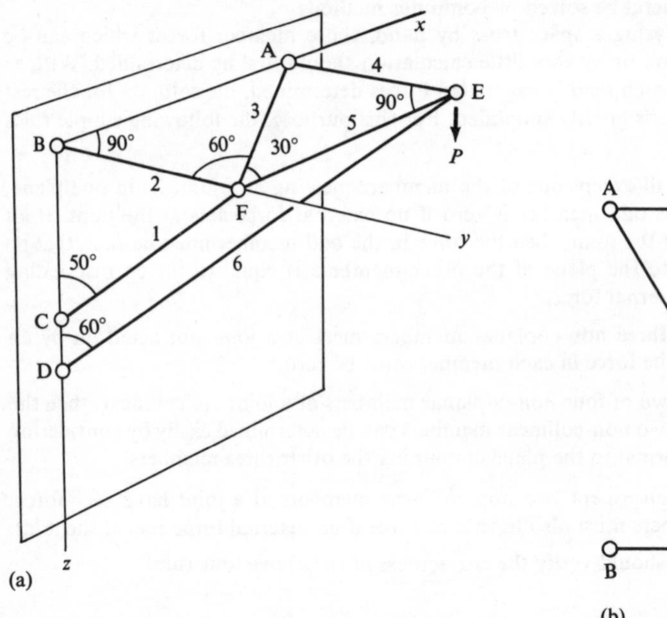

Fig. 3·5–1

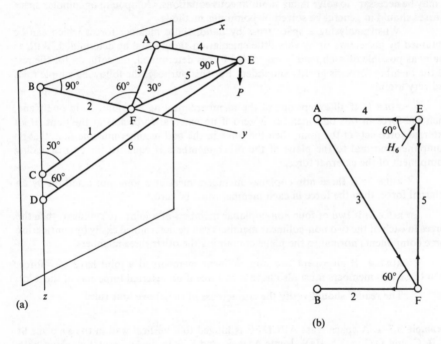

Example 3.5–2. If, in Example 3.5–1 (Fig. 3.5–1), the force **P** is applied at joint F instead of joint E, determine the forces in all members.

SOLUTION By Rule 2, members 4, 5, and 6 are all unstressed (dotted lines, Fig. 3.5–2).

At joint F, where **P** acts, member 1 is the only member not lying in the horizontal *xy* plane. Therefore (Rule 1), considering vertical equilibrium of the joint,

$$S_1 \cos 50^\circ = P$$

Hence

$$S_1 = \underline{1.55P} \quad \text{(compressive)}$$

Again, the force in member 3 is the only force at joint F not lying in the *yz* plane. Hence (Rule 1), considering force components normal to that plane, we must have

$$S_3 \sin 60^\circ = 0 \quad \text{i.e.} \quad S_3 = \underline{0}$$

whence, by inspection,

$$S_2 = S_1 \sin 50° = \underline{1.19P} \quad \text{(tensile)}$$

$$S_4 = S_5 = S_6 = \underline{0} \quad \text{as stated above}$$

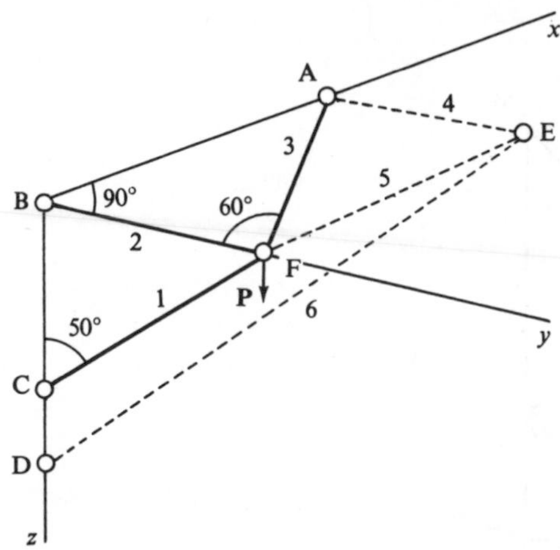

Fig. 3·5–2

Example 3.5–3. Figure 3.5–3(a) shows a determinate space truss made up of horizontal, vertical, and inclined members. The horizontal members and the vertical members are all of equal length. Member GJ is a diagonal of a cube, while all other inclined members are diagonals of equal squares. The truss is acted on by a vertical force **Q** at joint L; at joint E it is acted on by a horizontal force **P** which is inclined at 60° to member EH. Determine all member forces.

SOLUTION By Rule 1, member 1 is unstressed; similarly member 2 is unstressed. By Rule 2 (with $S_1 = 0$), members 3, 4, and 5 are unstressed. By Rule 2 (with $S_2 = 0$), members 6, 7, and 8 are unstressed. By Rule 2 (with $S_5 = S_6 = 0$), members 9, 10, and 11 are unstressed.

At joint G, only the forces in four members (12, 13, LG, and GC) are unknown. Whence, by Rule 3, $S_{12} = S_{13} = 0$.

At joint E, with S_{10} known to be zero, then, for vertical equilibrium of the joint, member 14 must be unstressed.

At joint F, with $S_{13} = 0$, member 16 is the only one not lying in the plane of the others. Hence by Rule 1, member 16 is unstressed, i.e. $S_{16} = 0$.

By similar reasoning at joint H, member 15 is unstressed, i.e. $S_{15} = 0$.

With members 1 to 16 inclusive being unstressed, the only other members that might carry forces are as shown in Fig. 3.5–3(b).

By inspection,

$$S_{LG} = S_{GC} = \underline{Q} \quad \text{(compression)}$$

$$S_{FE} = P \sin 60° = \underline{0.87P} \quad \text{(compression)}$$

$$S_{EH} = P \cos 60° = \underline{0.50P} \quad \text{(tension)}$$

$$S_{FA} = S_{FE} \sec 45° = 0.87P \times 1.41 = \underline{1.23P} \quad \text{(tension)}$$

$$S_{FB} = S_{FA} \cos 45° = \underline{0.87P} \quad \text{(compression)}$$

$$S_{HA} = S_{EH} \sec 45° = 0.50P \times 1.41 = \underline{0.70P} \quad \text{(compression)}$$

$$S_{HD} = S_{HA} \cos 45° = \underline{0.5P} \quad \text{(tension)}$$

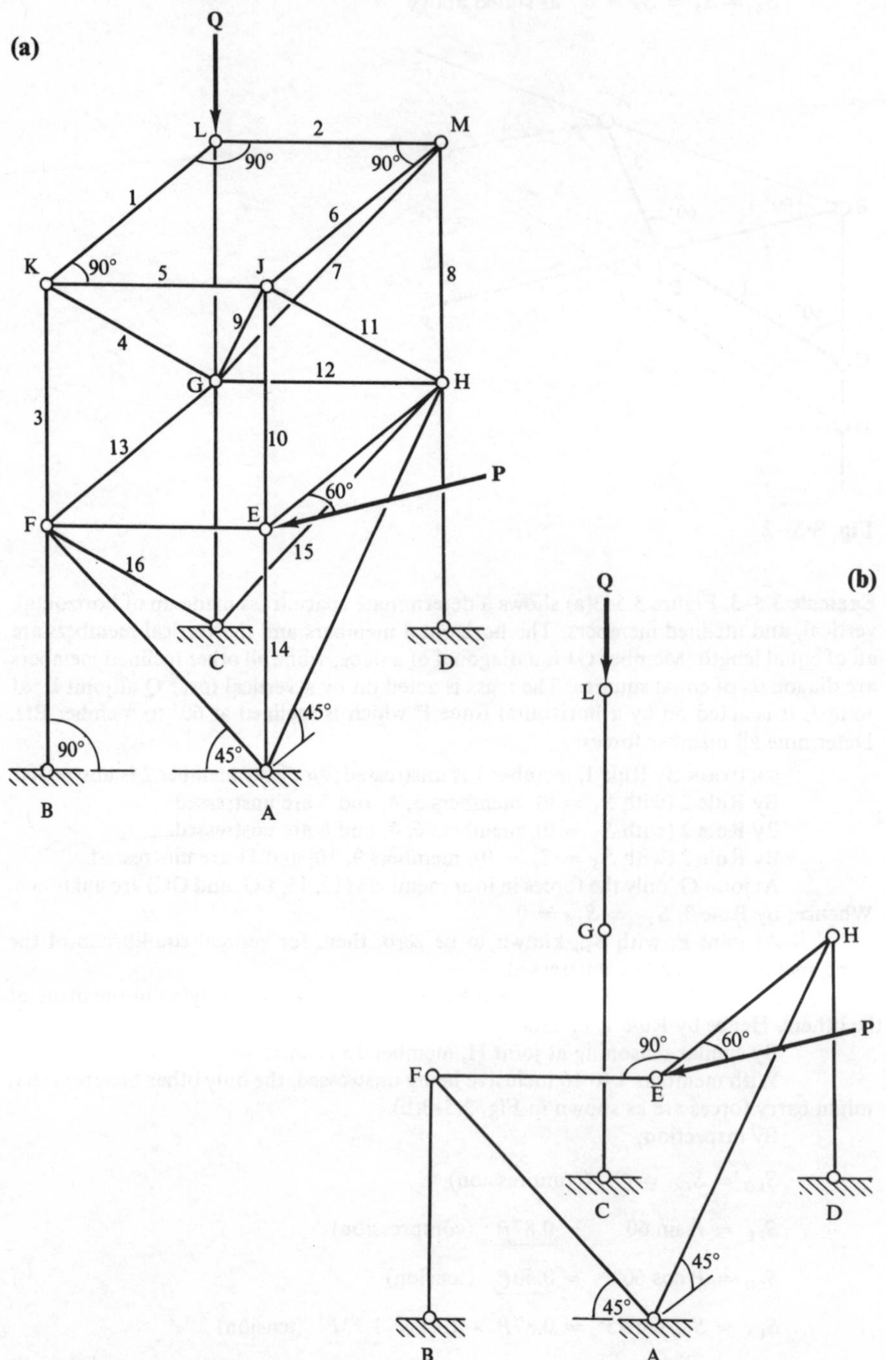

Fig. 3·5–3

Example 3.5–4. The space truss ABCDEFGH in Fig. 3.5–4(a) is in the form of a cube.

(a) Show that it is stable and determinate.

(b) Determine the force in each member due to the forces **P**, which are applied at joints C and E and the lines of action of which are along CE.

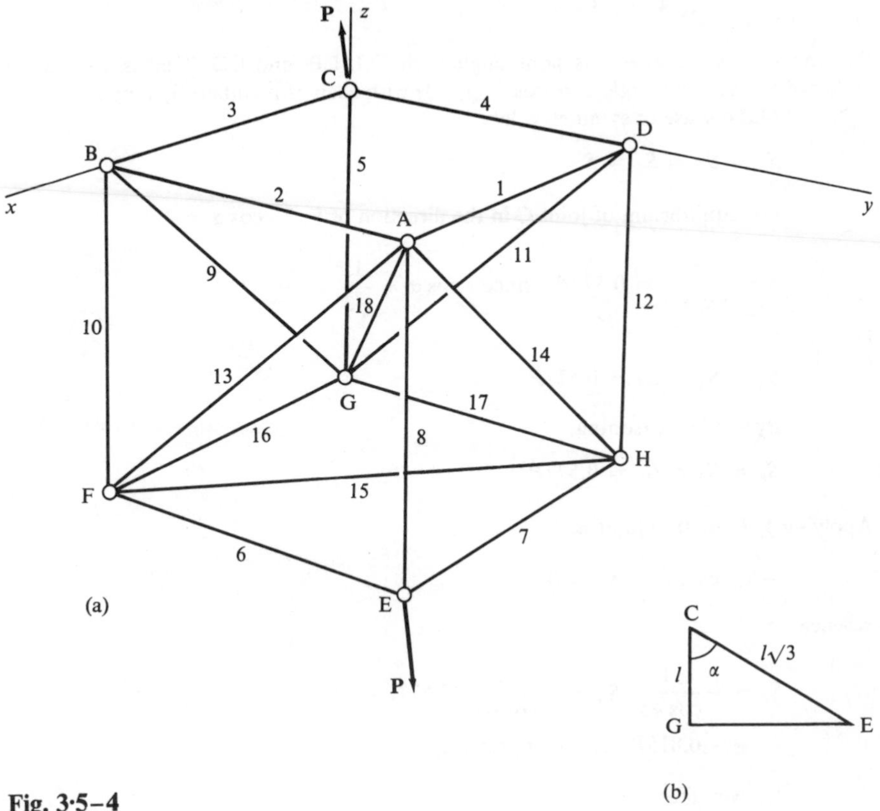

Fig. 3·5–4

SOLUTION

(a) The truss can be constructed in accordance with the rule for the construction of simple space trusses: starting with the tetrahedron AFHE, we can establish the other joints successively, as follows:

joint G from joints A, F, and H;
joint D from joints A, H, and G;
joint B from joints A, F, and G;
joint C from joints G, B, and D.

Hence, the truss ABCDEFGH is both stable and determinate.

(b) Let $S_1, S_2, S_3, \ldots, S_{18}$ represent the tensile forces in members 1, 2, . . . , 18 respectively. At joint D, all members except member 1 lie in the *yz* plane. Hence, by Rule 1,

$$S_1 = \underline{0}$$

Similarly,

$$S_2 = \underline{0}$$

To determine the forces in members 3, 4, and 5, we note that the length of the line CE is $l\sqrt{3}$, where l is the length of an edge of the 'cubic' truss. Hence the angle α which CE makes with CG (Fig. 3.5–4(b)) is given by

$$\cos\alpha = \frac{l}{l\sqrt{3}} = \frac{1}{\sqrt{3}}$$

By symmetry, CE makes this same angle with CG, CB, and CD. That is, the line of action of **P** makes an angle $\alpha = \cos^{-1}(1/\sqrt{3})$ with each of members 3, 4, and 5.

Making use of symmetry, let

$$S_3 = S_4 = S_5 = S$$

For equilibrium of joint C in the direction of **P**, $3S\cos\alpha = P$

$$S = \frac{P}{3\cos\alpha} = 0.577P \quad \text{since} \quad \cos\alpha = \frac{1}{\sqrt{3}}$$

i.e.

$$S_3 = S_4 = S_5 = \underline{0.577P}$$

By similar reasoning,

$$S_6 = S_7 = S_8 = \underline{0.577P}$$

Applying $\sum P_x = 0$ to joint B,

$$-S_9\cos 45^\circ - S_3 = 0$$

whence

$$S_9 = -\frac{1}{\cos 45^\circ}S_3 = -\frac{1}{\cos 45^\circ}(0.577P)$$
$$= -\underline{0.815P} \quad \text{(i.e. compressive)}$$

By symmetry,

$$S_{11} = -\underline{0.815P}$$

Applying $\sum P_z = 0$ at joint B,

$$-S_{10} - S_9\cos 45^\circ = 0$$

Then

$$S_{10} = -(-0.815P)\cos 45^\circ = \underline{0.577P}$$

By symmetry,

$$S_{12} = \underline{0.577P}$$

Applying $\sum P_z = 0$ at joint F,

$$S_{13}\cos 45^\circ + S_{10} = 0$$

Hence

$$S_{13} = -\frac{1}{\cos 45^\circ}(0.577P) = \underline{-0.815P}$$

By symmetry,

$$S_{14} = \underline{-0.815P}$$

Applying $\sum P_x = 0$ at joint H

$$S_{15} \cos 45^\circ + S_{14} \cos 45^\circ + S_7 = 0$$

Then

$$S_{15} \cos 45^\circ = -(-0.815P) \cos 45^\circ - 0.577P$$
$$= 0$$

Hence

$$S_{15} = \underline{0}$$

With $S_{15} = 0$, then member 16 is the only member at joint F not lying on a plane normal to the x-axis. Hence (Rule 1)

$$S_{16} = \underline{0}$$

By symmetry,

$$S_{17} = \underline{0}$$

To determine S_{18}, let its components in the coordinate directions be X_{18}, Y_{18}, and Z_{18} respectively.

Consider equilibrium of joint A. $\sum P_x = 0$ requires that

$$X_{18} - S_{14} \cos 45^\circ = 0$$

Then

$$X_{18} = S_{14} \cos 45^\circ = -0.815P \cos 45^\circ$$
$$= -0.577P$$

$\sum P_y = 0$ requires that

$$Y_{18} - S_{13} \cos 45^\circ = 0$$

Then

$$Y_{18} = S_{13} \cos 45^\circ = -0.815P \cos 45^\circ$$
$$= -0.577P$$

$\sum P_z = 0$ requires that

$$Z_{18} - S_8 - S_{13} \cos 45^\circ - S_{14} \cos 45^\circ = 0$$
$$Z_{18} = -0.577P + 0.815 \cos 45^\circ + 0.815P \cos 45^\circ = 0$$
$$Z_{18} = -0.577P$$

the magnitude of S_{18} is

$$\sqrt{(X_{18}^2 + Y_{18}^2 + Z_{18}^2)} = P$$

Since S_{18} has negative components in all coordinate directions, it must be tensile, i.e.

$$S_{18} = \underline{P}$$

This example demonstrates that, even though the general method of joints would mean solving $3j - 6 = 3 \times 8 - 6 = 18$ simultaneous equations in this case, in the actual solution above we did not even have to solve two simultaneous equations at any time.

Problems

3.1. Three forces $\mathbf{P}_1$, $\mathbf{P}_2$, and $\mathbf{P}_3$ meet at origin O of cartesian coordinate system Oxyz. It is known that $\mathbf{P}_1$, of magnitude 30 N, passes through the point (8, 2, 8); $\mathbf{P}_2$, of magnitude 120 N, passes through the point (4, 6, 0); $\mathbf{P}_3$, of magnitude 90 N, passes through the point (3, −6, 12). Determine the magnitude and direction of the resultant.

Ans. 160 N; direction cosines are $l = 0.67$; $m = 0.41$; $n = 0.62$.

3.2. A bar AB is in the form of a circular arc of radius a and it subtends an angle of 120° at the centre. If the bar is rigidly attached to the foundation at A, and a twisting moment of intensity m per unit length of arc is applied throughout its length, determine the support reaction at A.

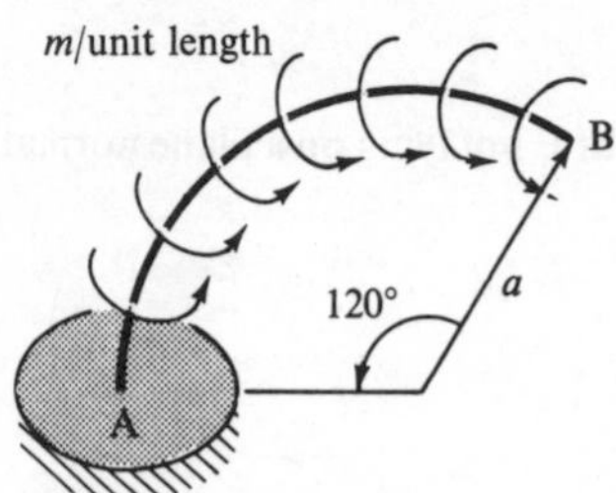

Problem 3·2

Ans. Bending moment at A = $3ma/2$; twisting moment at A = $\sqrt{3}ma/2$.

3.3. Moments M_1 and M_2 act respectively in the planes of the faces AA′C′C and AA′B′B of the prismatic block. Determine *by inspection* the magnitude of the moment M_3 which must be applied in the plane ABC to maintain equilibrium.

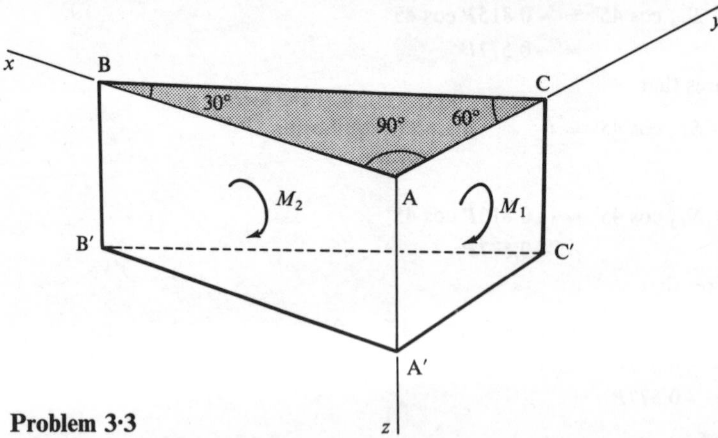

Problem 3·3

Ans. Equilibrium impossible—moments in mutually perpendicular planes.

3.4. If in Problem 3.3, equilibrium is maintained by a moment M_3 acting in the plane BCC′B′, determine (a) the relationship between the magnitudes of M_1, M_2, and M_3; (b) the sense of M_3.

Ans. (a) $M_3 = M_1/\sin 30° = M_2/\sin 60°$; (b) moment vector M_3 acts towards face BCC′B′.

3.5. A prismatic block is obtained by removing a quarter of a cube by cutting along a vertical diagonal plane and along a horizontal plane. Forces P_1, P_2, P_3, and P_4 act normal to the respective faces, while moments M_1, M_2, M_3, and M_4 act in the planes of these faces. If the block is in equilibrium, determine *by inspection* the relative magnitudes and directions of the forces and moments.

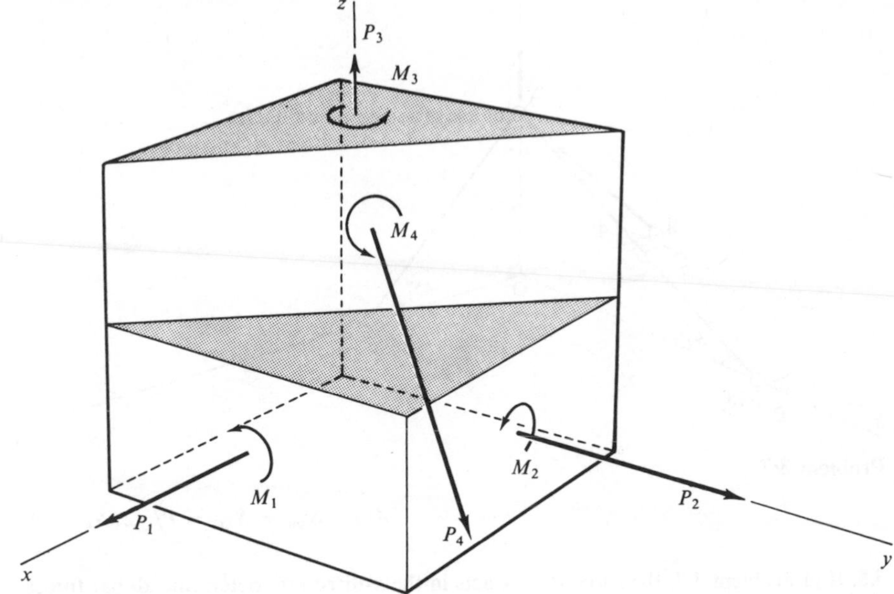

Problem 3·5

Ans. $P_3 = M_3 = 0$ (Why?); $P_4 = -P_1\sqrt{2} = -P_2\sqrt{2}$; $M_4 = -M_1\sqrt{2} = -M_2\sqrt{2}$.

3.6. Forces of magnitude P each act on the corners of a cube of edge length a as shown. Determine *by inspection* the magnitude and direction of the force F and the moment M which must be applied at the origin O to maintain equilibrium. If the point of application of F (alone) is moved from O to another corner of the cube, would equilibrium be disturbed? If that of M (alone) is moved, would equilibrium be disturbed?

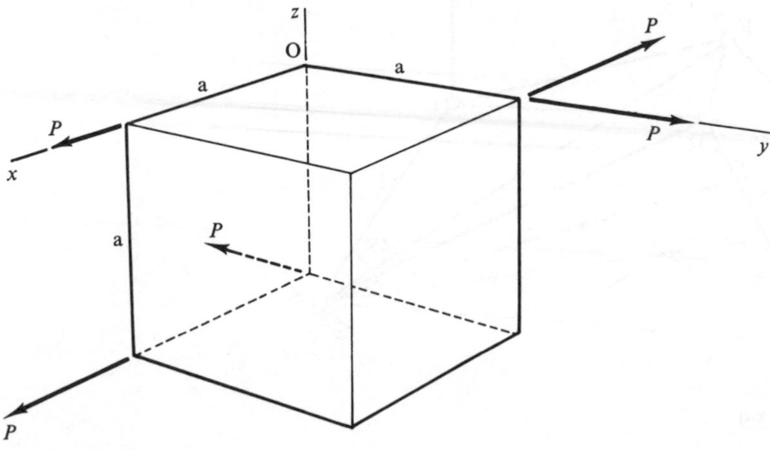

Problem 3·6

Ans. $F = [-P\ 0\ 0]^T$; $M = [Pa\ Pa\ -Pa]^T$; Yes; No.

3.7. A space truss consists of bars AB and AC in the *yz* plane and bar AD in the *zx* plane. A force *P* acts at joint A in the *z*-direction. Determine *by inspection* all bar forces.

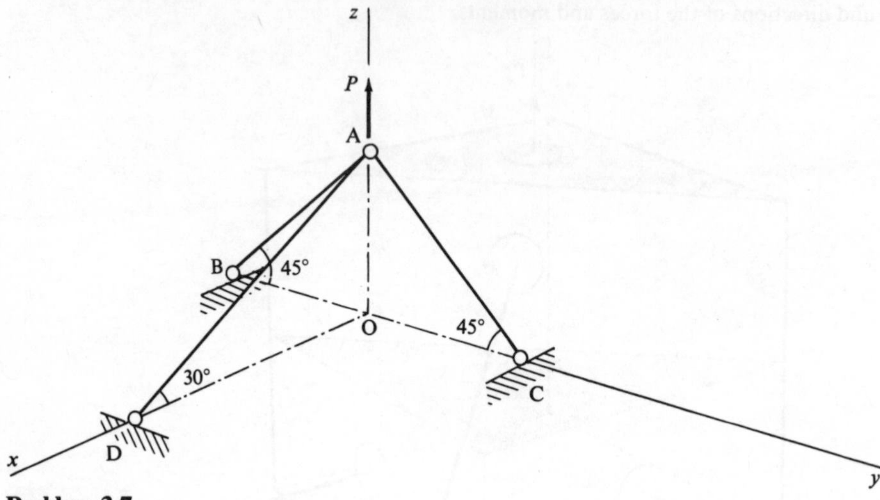

Problem 3·7

Ans. $S_{AB} = S_{AC} = P/\sqrt{2}$; $S_{AD} = 0$.

3.8. If in Problem 3.7, the force *P* at A acts in the *x*-direction, determine all bar forces.

Ans. $S_{AD} = -2P/\sqrt{3}$; $S_{AB} = S_{AC} = P/\sqrt{6}$.

3.9. In the space truss ABCDEF, the equilateral triangles ABC and DEF lie on horizontal planes. △ DEF is symmetrically oriented with respect to △ ABC so that joints D, E, and F are vertically above the respective medians of △ ABC. Each of the members AD, BE, and CF is inclined at 45° to the vertical. Satisfy yourself that joints such as A, B, E, and D lie in one plane. Hence determine *by inspection* the forces in members 1, 2, and 3.

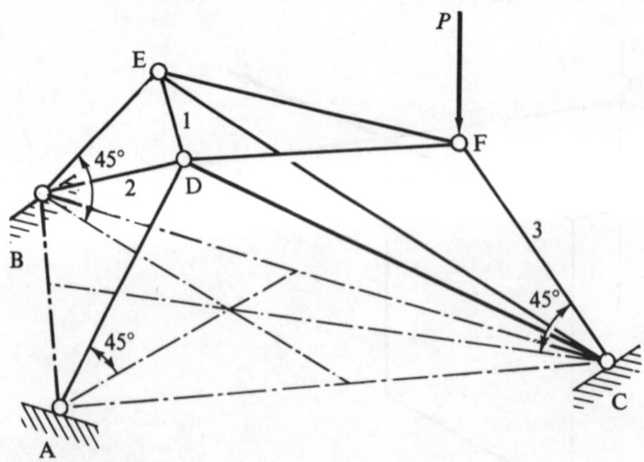

Problem 3·9

Ans. $S_1 = 0$ (Rule 1 at joint E); $S_2 = 0$ (Rule 1 at joint D); $S_3 = -P\sqrt{2}$ (Rule 1 at joint F).

3.10. In the space truss ABCDEFGH, the horizontal plane containing the square EFGH is 10 m above the horizontal plane containing the square ABCD. A vertical force of *P* kN acts at joint F. Determine *by inspection* the forces in (a) bar 1; (b) bar 2; (c) bars 3, 4, 5, 6, 7, and 8.

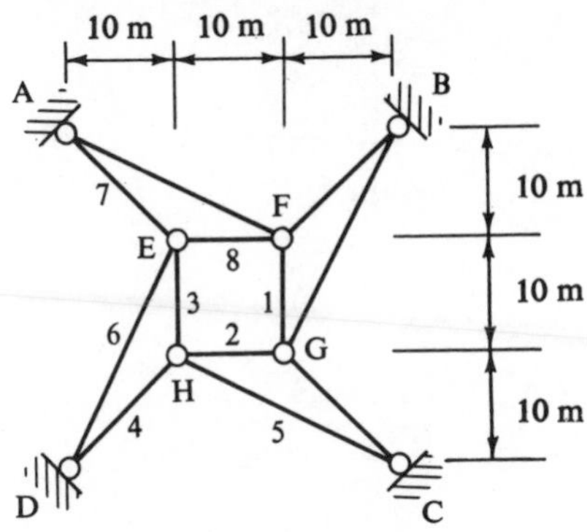

(a) Plan

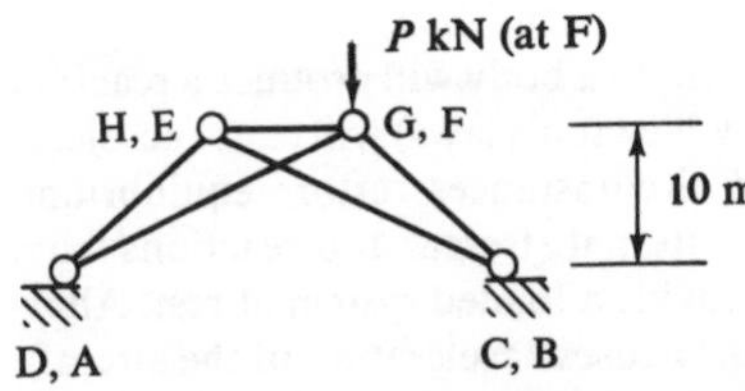

(b) Elevation

Problem 3·10

Ans. $S_1 = -P$ kN (take moments about AB of forces at joint F); $S_2 = 0$ (Rule 1 at joint G); $S_3 = S_4 = S_5 = 0$ (Rule 2 with $S_2 = 0$); similarly $S_6 = S_7 = S_8 = 0$.

References

1 Kong, F.K., Prentis, J.M. and Charlton, T.M., 'Principle of virtual work for a general deformable body—a simple proof', *The Structural Engineer*, 61A, No. 6, June 1983, pp. 173–9.

Chapter 4
Basic structural concepts

4.1 Actions and reactions

The application of some external force or action system to a body will produce a reaction system which either maintains the state of rest or (by Newton's laws) will cause acceleration to take place and to continue until changed circumstances restore equilibrium. Thus the application of load to a bridge produces internal stresses and reactions from the foundations of the abutments which together provide a loaded system at rest. Alternatively the increase of thrust from an aircraft engine causes acceleration of the aircraft, until the increased drag of air resistance and the other combined aerodynamic forces restore equilibrium at some uniform (but increased) velocity. In both cases, the action and reaction systems (which might be very complex) can each be compounded to give six force vectors which will be equal in magnitude and opposite in direction.

The two systems, of action and reaction may be regarded as essentially independent and dependent respectively. For the purposes of structural analysis loads occur in any pattern of distribution or magnitude, while reactions are consequentially produced in the pattern dictated by the particular shape and properties of the structure.

The number of reactive elements is not limited in any way. It was shown in Eqns 1.3–1(a) and (b) that six equations of equilibrium arise in the general case and it follows, therefore, that only six reactive elements can be calculated from statistical considerations alone (only three in the case of a two-dimensional frame, owing to the reduced number of equations).

The simple beam supported by a pin at one end and by a pin and roller connection at the other (shown in Fig. 4.1–1(a)) is such a case, where the three reactive elements due to any load system may be explicitly determined. The encastré arch of Fig. 4.1–1(b), with three reactive elements at each abutment, as shown, cannot be analysed so simply, but the magnitudes of those reactive elements can be determined after consideration of the deformation properties of the arch rib. Moreover, it is not necessary to prevent deflection at the point of application of a reaction; in Fig. 4.1–1(c) a loaded beam is shown supported on the usual pin/pin and roller connection and, in addition on two elastic springs which deform proportionately to the load carried. In this case the forces in the springs are reactions which act on the beam and are produced by the loads. Equally, the forces supporting the springs could be regarded as the reactions in a larger system, and the forces in the springs as part of an internal force system giving equilibrium in the various portions of the structure; for example, the upwards reaction exerted by the supporting ground on one spring must equal the reactive force of the spring on the beam and the compressive force in the spring which is thus in equilibrium.

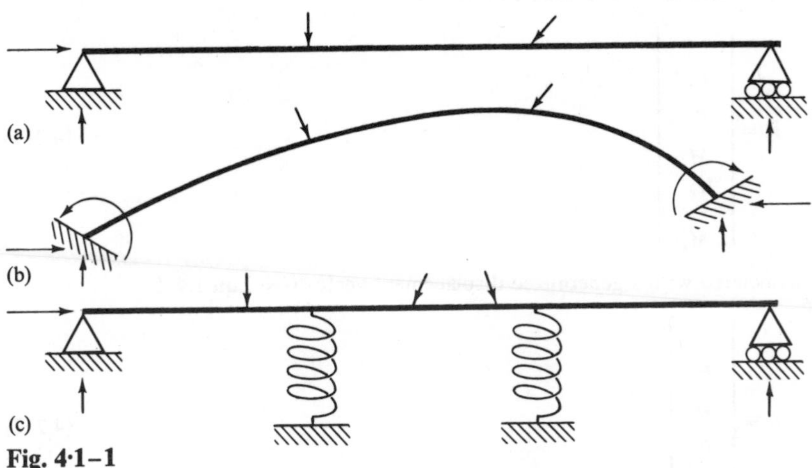

Fig. 4·1–1

4.2 Displacements and corresponding displacements

In any structure the application of some general force system consisting of both independently variable actions and some dependent reactions, will produce both a system of internal actions and a pattern of deformation. All points in the structure will move (unless prevented by some postulated external restraint or by some fortuitous combination of effects) and calculations must involve consideration of the three vectorial systems of actions (applied forces), internal actions, and displacement.

The stresses at any internal point are customarily associated with the system of displacements by the choice of coordinate axes which are selected for ease of computation from considerations of structural shape. For example, in a two-dimensional framed structure one of the axes of reference is frequently chosen as being parallel to the critically important structural member, or to the greatest possible number of members, while the second axis is automatically at right angles. Applied forces are by definition both independent in possible magnitude and arbitrary in direction, however, and to reduce these to a conformable state for computational purposes each must be resolved into components (or in the planes of) the coordinate axes.

With each action or component of action, therefore, there is an associated or 'corresponding' displacement, which occurs at the point considered and in the same direction (see Section 1.4).

As an example a point in a loaded structure will be displaced and the component of that displacement which takes place in the line of, and in the direction of application of a linear force acting at that point, is said to 'correspond' with it. Similarly a rotation of that point which took place in the direction of a moment applied at that point would correspond with it—but these two simple cases are hardly representative of the general case. It must be assumed that the external forces applied at any point couid consist of any number of linear loads or moments acting in any direction. These could be reduced (for computational convenience) to a generalized force vector $\mathbf{P}$ at that point (see Eqn 1.2–14) with six components in the six coordinate directions chosen. It should be noted that such a generalized vector could not arise by compounding the effects of a number of point loads alone (since all can be represented by force components of different magnitudes in the same three directions), but will occur from the combined effects of forces and moments. The former will give components in the direction of the coordinate axes and the latter produce couples in the planes containing them. This conventional representation of the applied forces can be used to define the corresponding movements and thus the generalized force vector

$$\mathbf{P} = \begin{bmatrix} P_x \\ P_y \\ P_z \\ M_x \\ M_y \\ M_z \end{bmatrix} \tag{4.2–1}$$

will be associated with a generalized displacement vector (see Eqn 1.4–5)

$$\boldsymbol{\delta} = \begin{bmatrix} u \\ v \\ w \\ \theta_x \\ \theta_y \\ \theta_z \end{bmatrix} \tag{4.2–2}$$

in which there is a geometrical association between each component of force and the component of displacement which corresponds with it. It may be that, in certain special cases, the component(s) of force or displacement in some particular direction(s) is zero, but this would be revealed simply by a zero term in the appropriate column vector and the consistency of representation would be unaffected.

Similarly, at an internal point the stress vector representing internal action (see Eqn 1.5–3)

$$\boldsymbol{\sigma} = \begin{bmatrix} \sigma_x \\ \sigma_y \\ \sigma_z \\ \tau_{xy} \\ \tau_{yz} \\ \tau_{zx} \end{bmatrix} \tag{4.2–3}$$

has an associated or corresponding strain vector (see Eqn 1.6–8)

$$\boldsymbol{\varepsilon} = \begin{bmatrix} \varepsilon_x \\ \varepsilon_y \\ \varepsilon_z \\ \gamma_{xy} \\ \gamma_{yz} \\ \gamma_{zx} \end{bmatrix} \tag{4.2–4}$$

These associated stress and strain vectors can be exceptionally complex and will vary from point to point in the cross-section of a member and from one cross-section to another along its length. The process of analysis can be simplified considerably if the structure is reduced to a **skeletal** or line form and the stress and strain vectors appropriate at a cross-section replaced by the internal action and deformation vectors which represent the summed effects of stress and strain at that cross-section. Thus, if direction x is measured along the axis of a member,

$$F_x = \int \sigma_x \, dA$$

where dA represents a small area of the cross-section and the integration is carried out over the whole of the cross-section. The generalized stress vector $\boldsymbol{\sigma}$ can be replaced by a generalized internal force vector

$$\mathbf{F} = \begin{bmatrix} F_1 \\ \vdots \\ F_6 \end{bmatrix} \tag{4.2–5}$$

with which is an associated corresponding relative deformation vector

$$\mathbf{e} = \begin{bmatrix} e_1 \\ \vdots \\ e_6 \end{bmatrix} \tag{4.2–6}$$

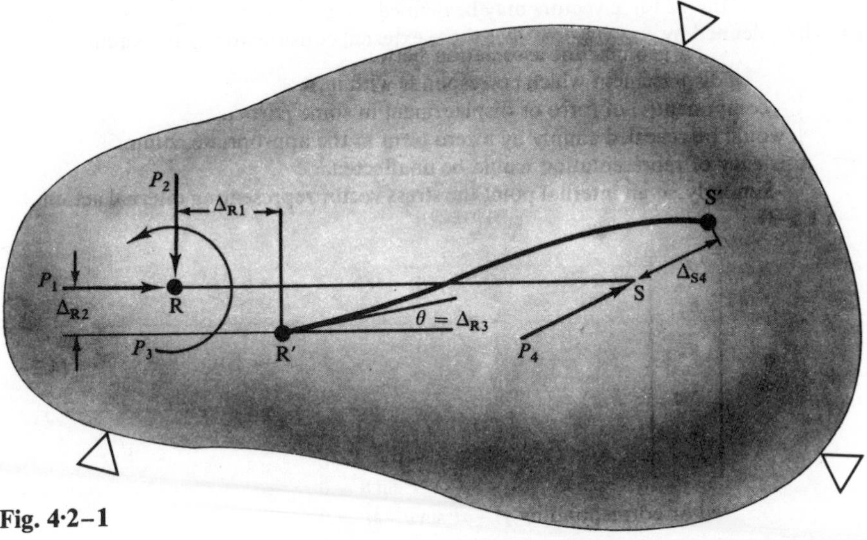

Fig. 4·2–1

There are, therefore, corresponding vectors such as those represented by Eqns 4.2–1 and 4.2–2 for loads and displacements at nodal points, by Eqns 4.2–3 and 4.2–4 for internal stresses and strains at points in the structure, and by Eqns 4.2–5 and 4.2–6 for internal actions at a cross-section. Each of these equations (applying as it does to a single loaded point, or to some particular cross-section) may be regarded as one of a much larger general family, involving external loads at all possible points of application, with their corresponding displacements, or a general representation of the stresses or internal actions in a structure with the corresponding strains or deformations. There will still be, however, a 'paired' relationship between forces (or stresses) and displacements (or strains) and it should be noted that only one of these three pairs can be arbitrarily defined. If the general load (or displacement) vector for any structure is laid down, then the displacement (or load) vector may be calculated, together with the general vectors defining stress or internal action with their corresponding deformations. The form of these vectors will depend on the reaction system, and on the properties of the structural material.

Occasionally it is possible in some particular problem to analyse the structure without resolving the small number of applied forces involved into their formal components. In these cases one may use an independent system of axes at each loaded point, defining the displacement corresponding to an applied force as being that component of the deflection of the loaded point measured in the direction of application of the force. A structure, supported on three reactions is shown in Fig. 4.2–1. Forces P_1, P_2, and P_3 are applied at a point R, and a force P_4 at point S, causing R and S to move to R′ and S′ respectively, and the line RS to deform as shown. The corresponding displacements are indicated on the figure.

4.3 Dependent and independent actions and displacements

Within any section of a structure the necessary conditions of equilibrium and compatibility must be maintained, and these will have a bearing on the interdependence of the appropriate force and displacement vectors.

For example, at each of the two ends of some portion of a member, six force vectors may act, but since the equilibrium equations (see Eqns 1.3–1(a) and (b)) are themselves only six in number, and since the portion must be in equilibrium, it follows that only 6 of the 12 force vectors may be defined by external considerations. The other six will be defined by a combination of those external considerations and equilibrium.

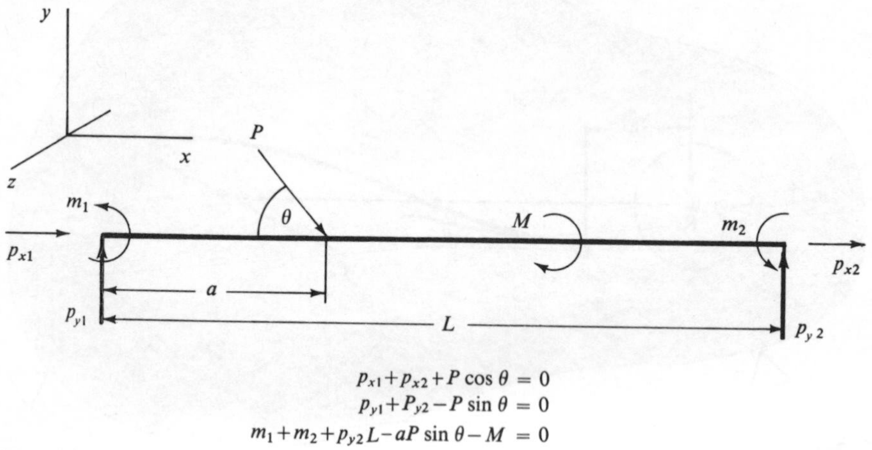

$$p_{x1}+p_{x2}+P\cos\theta = 0$$
$$p_{y1}+p_{y2}-P\sin\theta = 0$$
$$m_1+m_2+p_{y2}L-aP\sin\theta-M = 0$$

Fig. 4·3–1

In the simple two-dimensional skeletal member shown in Fig. 4.3–1 the vectors representing forces at the two ends will each contain three terms, but since relevant equations of equilibrium are three in number, and Eqns 1.3–1(a) and (b) reduce to

$$\sum P_x = 0, \quad \sum P_y = 0, \quad \text{and} \quad \sum M_z = 0$$

only three of the force terms can be specified arbitrarily—others will be implicitly specified by those equations of equilibrium.

Similarly, the displacement vectors defining movement at the two ends of the portion of a member must be associated by considerations of compatibility in the intervening length. If the movement of one end, in the direction of the length of the member, is defined by some external displacement, then considerations of internal compatibility require that the movement at the other, in the same direction, should only differ by an amount equal to the change in length of the member. The relationship between the two movements is dependent on the stiffness of the member between the points considered, and on the forces in the line of the member. Similarly in Fig. 4.1–1(c) the magnitude of the compression of either of the springs depends not only on the compression of the other, but also on the flexibility of the beam and the loads which are applied to it.

4.4 Superposition of actions and displacements—influence coefficients

Influence coefficients. It is usual to assume that a displacement caused at any point in a structure is linearly dependent on the magnitude of the loads applied. Such an assumption depends on the existence of proportionality and reversibility between the stress applied to, and the strain in any small element of material of the structure (which we loosely define as **elasticity**). The converse is not necessarily true and an assumption of elasticity does not imply that deflections will be proportional to load, as will be shown in Section 4.5. If deflections alter the structural geometry in a significant way, and thereby change the manner in which potential energy is stored in the deflected structure, an increase in load will not cause a proportionate increase in the deflection.

(a) In much of structural analysis, however, the proportionality between load (applied at some point, in some particular direction) and displacement (measured at any point in any chosen direction) may be assumed and, from this assumption, the effect of several loads acting independently may be determined. If a single load, acting at a specified point in a specified direction increased from zero to P_n it would produce a rigid body movement of the structure unless that movement was prevented by reactions adequate to ensure equilibrium. In that case deflections would occur throughout the structure and if at some point m, in some particular direction it was Δ_m such that

$$\Delta_m \propto P_n$$

then

$$\Delta_m = f_{mn} P_n \tag{4.4–1}$$

as shown in Fig. 4.4–1.

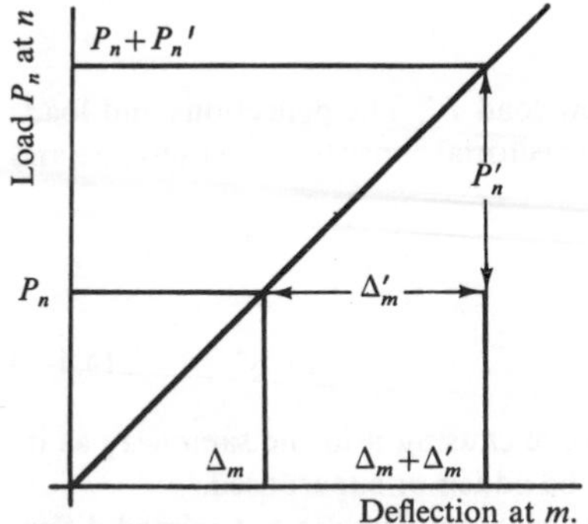

Fig. 4·4–1

In this equation f_{mn} is a **flexibility influence coefficient** defining the flexibility of the structure by the relationship between P and Δ, and using subscripts to define particular directions of measured deflection and applied load as in Fig. 4.4–2. f_{mn} is equal numerically, but not dimensionally to the deflection at some point in a specified direction at m due to unit load applied at some point in a specified direction n.

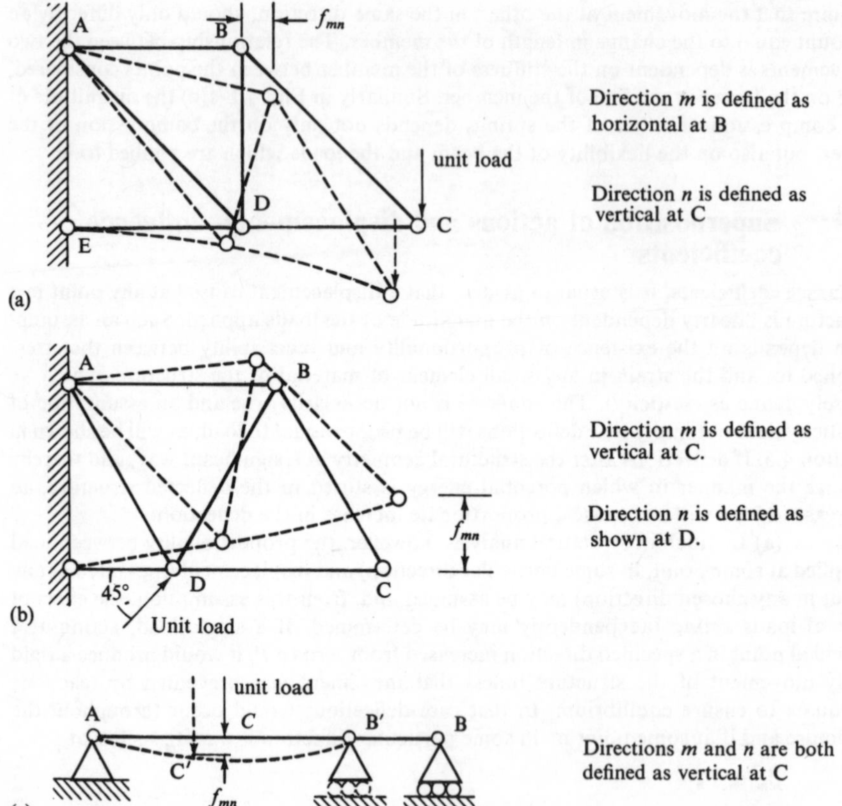

Fig. 4·4–2

If the relationship of Eqn 4.4–1 holds, then equally

$$\Delta'_m = f_{mn} P'_n$$

where Δ' represents the deflection caused by a new load P'. The deflections and loads may be added since care has been taken to maintain vectorial consistency in each case and

$$\Delta_m + \Delta'_m = f_{mn} P_n + f_{mn} P'_n$$

so that

$$\Delta_m + \Delta'_m = f_{mn}(P_n + P'_n) \tag{4.4–2}$$

thus indicating that $(\Delta_m + \Delta'_m)$ is related to the force causing it in the same way as its component parts—the forces and deflections may be added or superposed.

(b) A particular deflection and load may be proportionate but related differently in the form

$$P_n = k_{nm} \Delta_m \tag{4.4–3}$$

where k_{nm} is a **stiffness influence coefficient** equal numerically to the force required at one point in a specified direction *n* to produce unit displacement at another point in a specified direction *m*, all other displacements being prevented at that point. If the relationship exists in this form then the possibility of superposition still exists since, if a force P'_n

causes a displacement Δ'_m, superposing one force on another (which is permissible since the subscripts define the nature of the vectorial quantities involved)

$$\begin{aligned} P_n + P'_n &= k_{nm}\Delta_m + k_{nm}\Delta'_m \\ &= k_{nm}(\Delta_m + \Delta'_m) \end{aligned} \tag{4.4–4}$$

indicating that the total deflection would be equal to the summed effects of the two forces acting separately. Here, as before, it is assumed that the application of the forces P and P' produce reactive systems of forces in equilibrium with them, and the effect of these reactive systems is taken into account when calculating the stiffness influence coefficients k_{nm}.

(c) It will be understood that for any two points there are an infinite number of influence coefficients f_{mn} and k_{nm} depending on the specified directions m and n at those two points, although several different considerations could reduce that number to more manageable proportions. Moreover, it will be seen that any particular value of an influence coefficient is dependent on the form of reactive constraint which is assumed—if that reactive system changes each of the infinite number of values for the coefficients will also change.

The relationship between the flexibility and stiffness of a structure is of considerable importance in structural analysis, and will be developed later in Chapter 5. It will be seen however, from Eqns 4.4–1 and 4.4–3 that in the very simple case of a structure in which load and deflection could only be applied and measured at one point, the two coefficients would be reciprocally related and

$$f_{mn} = \frac{1}{k_{nm}}$$

In the general case, as will be shown, the relationship is much more complex.

4.5 Where superposition may not occur—nonlinear behaviour

The fact that the effects of two loads may be superposed (that the deflections due to two loads applied simultaneously is equal to the sum of the deflections due to the loads applied separately) is dependent solely on the assumption that load and deflection are linearly related, and plainly the principle must not be used where nonlinear conditions prevail. It must not be assumed, however (as was pointed out in Section 4.4), that linearity between stress and strain necessarily implies a linear relationship between loads and deflections and, therefore, that the principle of superposition may be applied with safety. Even when a proportionate relationship between stress and strain exists, changes in the position of structural nodal points—that is, changes in the structural geometry due to deformation—can ensure that load and deflection are not linearly related. This may arise from the change of geometry of the structure alone, or because that change alters the method by which potential energy is stored in the body.

As examples of these two possibilities we may consider Figs. 4.5–1 and 4.5–2.

(a) Figure 4.5–1(a) shows a light, extensible string of length $2L$, tautly fastened between two fixed points. If a load of $2W$ be applied at the centre (as in Fig. 4.5–1(b)), the string will stretch, and the point of application of the load will deflect. Symmetry demands that the deflection will be vertically downward and (parenthetically) will correspond with the direction of the applied load at that point. Consideration of the vertical equilibrium of the load shows that

$$2(\text{tension in the string}) \cos\theta = 2W$$

and

$$\text{tension} = \frac{\sqrt{(L^2 + \delta^2)}}{\delta} W$$

Now if the elasticity of string is such that it extends by unity for a tensile force of k in it, then

$$\sqrt{(L^2 + \delta^2)} - L = \frac{\sqrt{(L^2 + \delta^2)}}{k\delta} W$$

$$\delta\left(1 - \frac{L}{\sqrt{(L^2 + \delta^2)}}\right) = \frac{W}{k}$$

It will be seen that, when δ is very small the rate of increase of δ with respect to W is very large, but that the rate decreases with increasing value of W, as shown in Fig. 4.5–1(c).

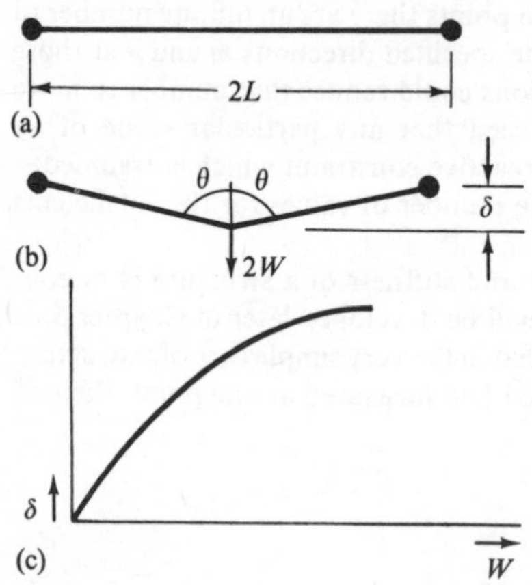

Fig. 4·5–1

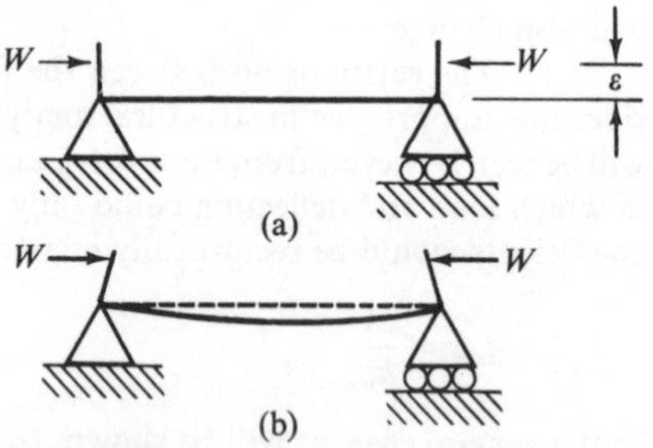

Fig. 4·5–2

(b) A second example is shown in Fig. 4.5–2(a), where an elastic beam, simply supported on pin/pin and roller connections, is acted upon by a load parallel to the axis, but at a slight eccentricity, ε. Potential energy will be stored in the loaded structure due to direct compression of the beam, but, as shown in Fig. 4.5–2(b) loading will occasion bending, causing the increased potential energy to be stored disproportionately in bending, as well as compression. In this case, if W be the applied load, A the cross-sectional area of the strut, and Z its section modulus then the maximum stress in the strut is

$$\frac{W}{A} + \frac{W(\varepsilon + y)}{Z}$$

and since y is obviously dependent on W (although not linearly so), the maximum stress will not vary linearly with W, even if there is a linear relationship between stress and strain. This example of behaviour in compression is of great structural importance, and will be dealt with more extensively in Chapter 9.

For the principle of superposition to be valid, therefore, it is necessary to ensure that all deformations are so small as to make no significant change in the geometry of the structure, and the discerning engineer must be alive to the possibility that even small changes can have a considerable effect on structural behaviour.

4.6 Compatibility

The deformation of any structure as a result of loads applied to it arises from the deformation of the elements of which it is composed. The bent shape of the beam in Fig. 4.5–2(b) arises from the combined effects of shortening (an overall compressive strain) and

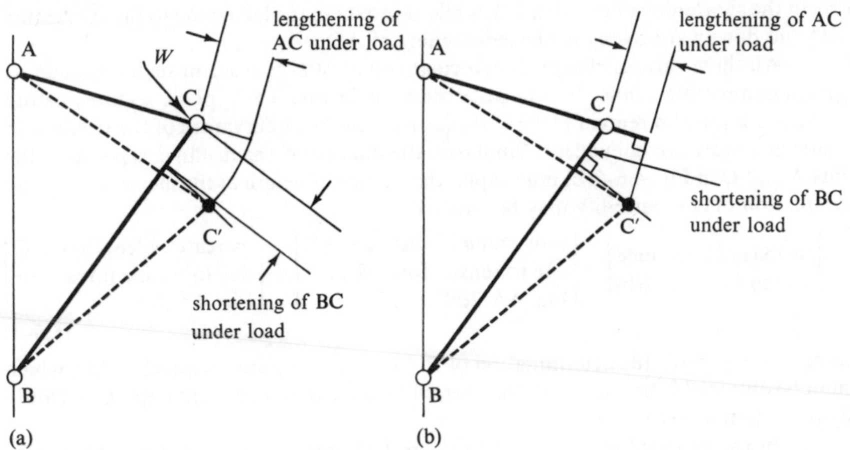

Fig. 4·6–1

bending which causes a compressive strain in the concave portions of the beam with an appropriate tensile strain on the convex face.

The elementary unloaded pin-jointed frame ABC, in Fig. 4.6–1(a) shows a node C connected to a rigid wall by members AC and BC. When load is applied in the direction of W, member AC stretches and BC shortens, and the new position of C (C′) may be found by striking two arcs, with centres A and B and radii equal respectively to the loaded lengths of the two members. Since it is usual to assume that changes in the lengths of members are small in relation to the members, the new position of C may be obtained with great accuracy by drawing lines perpendicular to the original direction of the members as shown in Fig. 4.6–1(b). This principle is used in the graphical determination of deflections, known as the 'Williot diagram'. It will not be dealt with here.

Similarly in the pin-jointed structure ABCDE of Fig. 4.6–2(a) the deflection of point B is to the right, arising from the extension of member AB, and downward

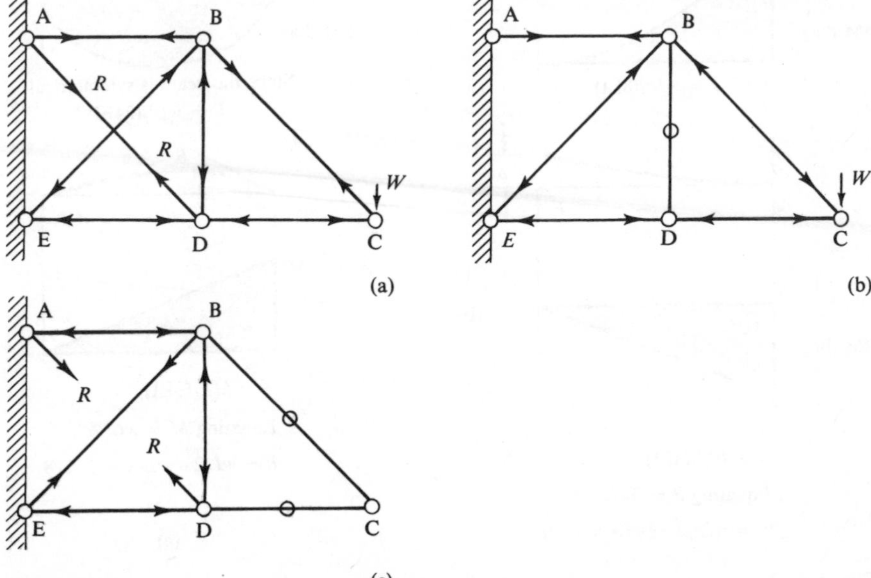

Fig. 4·6–2

owing to the shortening of member EB, while D moves to the left owing to the shortening of ED and downward owing to the lengthening of AD.

All these various elemental deformations must take place in such a manner as to give a compatible whole. In a flexible beam, as in Fig. 4.5–2, plane sections before bending are known to remain plane and thus the centres of curvature of the tensile and compressive faces are coincident. Similarly, the change of the distance separating the points A and D in Fig. 4.6–2(b) must equal the change of length of the member AD, and and equation of compatibility may be written:

$$\begin{bmatrix}\text{increase of distance}\\ \text{AD in Fig. 4.6–2(b)}\end{bmatrix} - \begin{bmatrix}\text{shortening of distance AD}\\ \text{due to tensile force } R \text{ in}\\ \text{Fig. 4.6–2(c)}\end{bmatrix} = \begin{bmatrix}\text{increase in length of AD}\\ \text{due to tensile force } R \text{ in}\\ \text{Fig. 4.6–2(a)}\end{bmatrix} \quad (5.6–1)$$

This equation permits the determination of R, the unknown force in member AD, which cannot be obtained from the equations of equilibrium alone. The point is discussed more fully in Section 4.16(b).

In the propped cantilever of Fig. 4.6–3(a) there are four reactive constraints (three at the encastré end, and one at the prop). In this case, however, of two dimensions only, there are only three equations of equilibrium, and one unknown is effectively

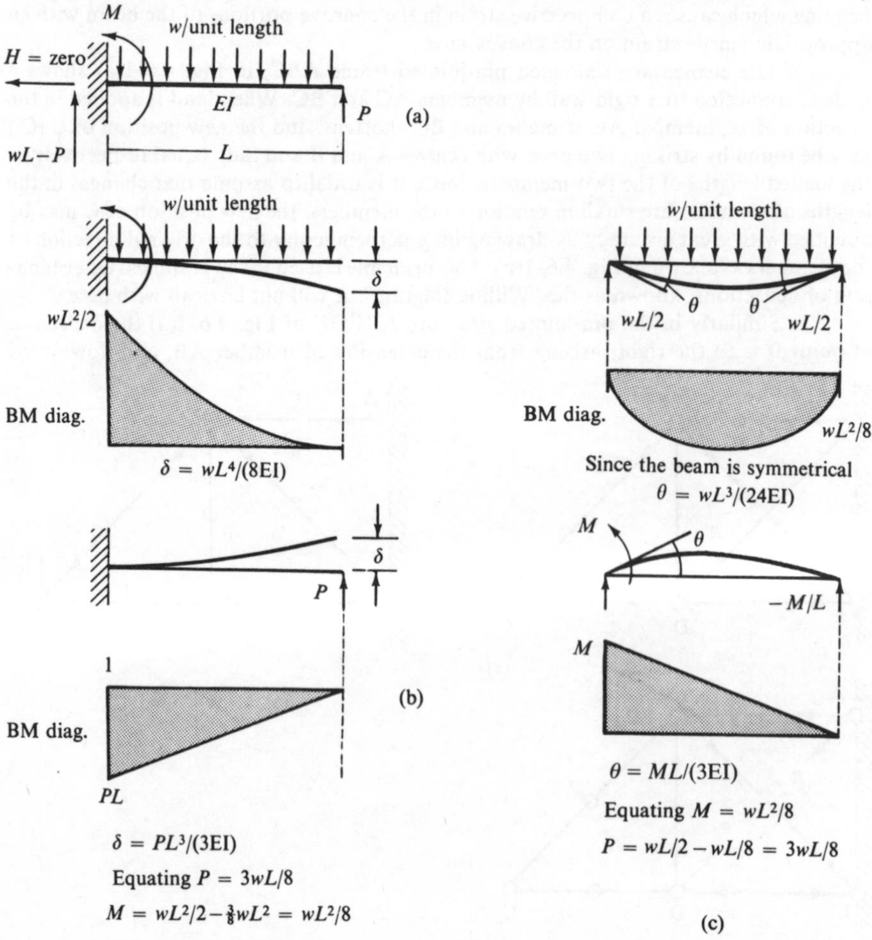

Fig. 4·6–3

implied by the elasticity of the structure. An equation of compatibility must be used to effect the solution. Several methods are possible, but two which rely on flexibility methods (the explicit statement of displacement as a function of applied loads) are given below.

In Fig. 4.6–3(b) the propping force P is treated as the unknown element, and the downward deflections due to applied load, and upward deflections due to the prop, at the free end of the cantilever, are equated. The calculation of the individual quantities is developed more fully later in this chapter. In Fig. 4.6–3(c) the unknown is regarded as the fixing moment M, and the solution equates the clockwise rotation caused at A by the applied load acting on the equivalent simply supported beam with the anti-clockwise rotation caused by M. Both methods are seen to give the same values for the propping force and the moment at the encastré end, but care must be taken to ensure that, when superposing two partial solutions to give a desired whole, the superposition involves all the factors necessary to give equilibrium. In Fig. 4.6–3(b), the loaded free cantilever has a fixing moment as shown, while in Fig. 4.6–3(c) the simply supported beam has equal vertical reactions at its ends, and these must be taken into account when completing the solution.

4.7 Work and complementary work

If a member AB as shown in Fig. 4.7–1(a) is subjected to a tensile force F an extension of the member, e, will be produced and, in the general (nonlinear) case a graph of load against extension might appear as shown in Fig. 4.7–1(b). The area below the curve represents the work done by the force in moving its point of application B to B′, and potential energy stored in the bar in its strained condition (and known as strain energy) will be equal to it, so that an equation exists,

$$U = \text{strain energy stored} = \text{work done} = \int_0^e F\,de \qquad (4.7\text{–}1)$$

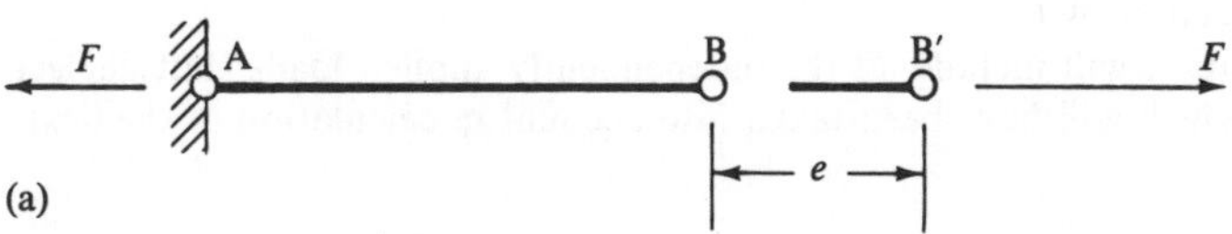

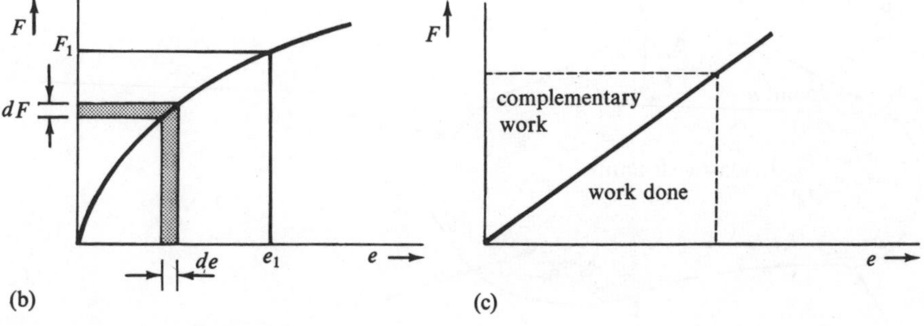

Fig. 4·7–1

in which the integration is carried through the entire range of deformation. The area above the curve is complementary, and clearly is equal to

$$\int_0^{F_1} e\,dF$$

so that a further equation exists,

$$C = \text{complementary energy} = \text{complementary work done}$$

$$= \int_0^{F_1} e\,dF \tag{4.7–2}$$

These two equations are of considerable importance in structural analysis, but conceptual difficulties do arise because no physical significance can be given to complementary energy and complementary work. If, however, the relationship between extension of the member and the force producing it is linear, as shown in Fig. 4.7–1(c), then obviously, the two items will be equal, as may be seen from the figure.

4.8 Linearity of load-deflection relationship

(a) Linearity of load and deflection. When an elastic structure is acted upon by a system of loads which is in equilibrium (or which is put in equilibrium by the reactive forces called into effect) deformation of the structure will take place, and a loaded point will move, not only in directions corresponding to loads applied at that point, but also in others as shown in Fig. 4.8–1. At one point (point 1) in the particular direction corresponding to one of the loads acting there, P_1, there will be a corresponding displacement

$$\Delta_1 = f_{11}P_1 + f_{12}P_2 + f_{13}P_3 \cdots f_{1n}P_n \tag{4.8–1}$$

where f_{1n} is a flexibility influence coefficient as defined in Section 4.4, equal numerically to the displacement at point 1, in the direction of P_1, owing to a unit force applied at point n in the direction of P_n.

The summation will include all the independently applied loads (but neglect the reactive system which will have been taken into account in calculation of the flexi-

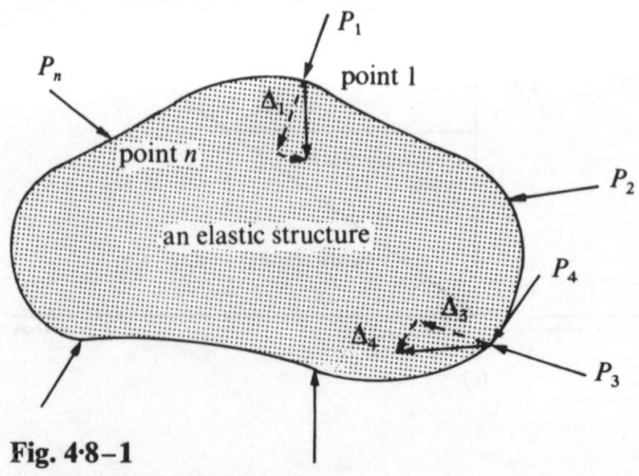

Fig. 4·8–1

bility influence coefficients as in Section 4.4) and there will be similar expressions for the displacements corresponding to all such loads, of the form:

$$\Delta_2 = f_{21}P_1 + f_{22}P_2 + f_{23}P_3 \ldots f_{2n}P_n$$
$$\Delta_3 = f_{31}P_1 + f_{32}P_2 + f_{33}P_3 \ldots f_{3n}P_n$$
$$\vdots$$
$$\Delta_n = f_{n1}P_1 + f_{n2}P_2 + f_{n3}P_3 \ldots f_{nn}P_n$$

These may be written simply as

$$\mathbf{\Delta} = \mathbf{fP} \tag{4.8–2}$$

where $\mathbf{\Delta}$ is the general displacement vector (a column matrix of $n \times 1$ terms) representing the displacements corresponding to all loads, $\mathbf{P}$ represents the general loads system (again a column matrix of $n \times 1$ terms), and $\mathbf{f}$ is the structure flexibility matrix of $n \times n$ terms.

One of those deflections (say Δ_1, of Eqn 4.8–1) is clearly known if the various flexibility coefficients are determined, and the relationship between that displacement and its corresponding load will be

$$\frac{\Delta_1}{P_1} = f_{11} + f_{12}\frac{P_2}{P_1} + f_{13}\frac{P_3}{P_1} + \cdots + f_{1n}\frac{P_n}{P_1} \tag{4.8–3}$$

This relationship will not be linear unless the ratios P_n/P_1 remain constant throughout the range of loading considered (sometimes called a **monotonic** loading system), but, in general, if the values of the loads may increase independently and a particular displacement may increase owing to the action of other loads, without change in the magnitude of the load corresponding to that particular displacement, the relationship between Δ_1 and P_1 could appear as shown in Fig. 4.8–2. The various phases indicated in that diagram are:

O–a a period of general increase in Δ_1, with varying ratios of P/P_1;

a–b a period of monotonic behaviour, during which the ratios P/P_1 remain constant and Δ_1/P_1 is linear;

b–c a period during which Δ_1 increases due to changing values of $P_2 \ldots P_n$, although P_1 itself remains constant;

c–d a second period of general increase in Δ_1 and in forces P.

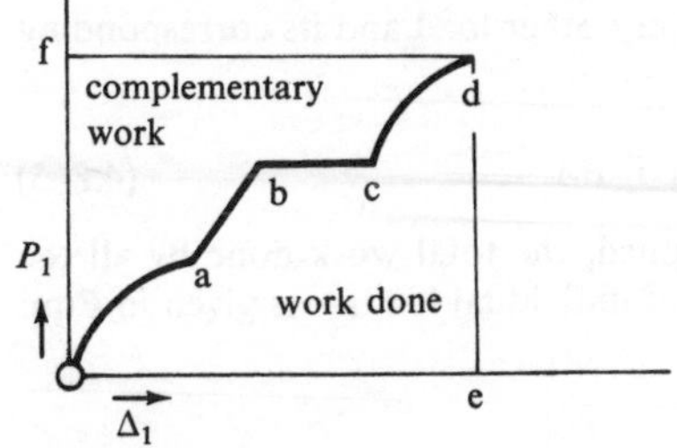

Fig. 4·8–2

(b) **Work done by a general load system.** The work done by the load P_1 of Fig. 4.8–2 will be

$$\int P_1\, d\Delta_1$$

and will be the area Oabcde under the graph between the curve and the axis of displacement, while the complementary work done by the same load will be the area Oabcdf between the curve and the axis of load and defined as $\int \Delta_1 \, dP_1$. It will be seen that during the period of the loading programme defined by the line bc, the force P_1 did work (in that the area under the curve was increasing) while no change occurred in the complementary work done by it. Equally it might be that P_1 should increase with no corresponding work in the movement of its point of application. In such a case P_1 would do no work, but its complementary work would increase.

In general, therefore, the work done by a particular load will be

$$\int P_1 \, d\Delta_1 = \eta_1 P_1 \Delta_1 \tag{4.8–4}$$

in which η_1 is a factor dependent on the shape of the load–displacement relationship and such that

$$0 < \eta_1 < 1$$

If the loading system increased monotonically, maintaining a constant ratio between all the loads, then the relationship between P_1 and Δ_1 would be continuously linear as shown in Fig. 4.8–3 and the work done by P_1 would be

$$\tfrac{1}{2}\text{Oe.de} = \tfrac{1}{2}P_1 \Delta_1$$

and

$$\eta_1 = \tfrac{1}{2}$$

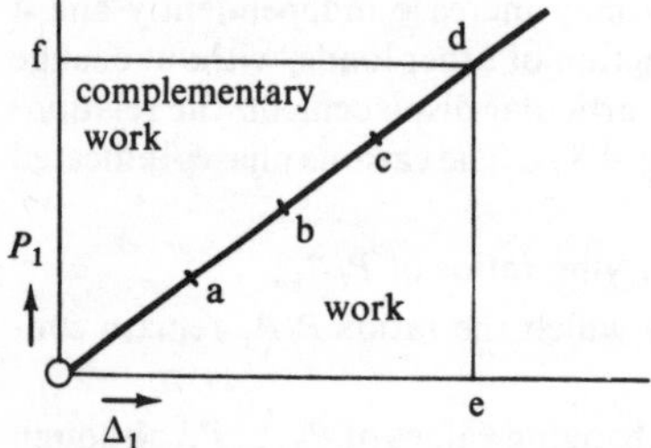

Fig. 4·8–3

A linearity of relationship would exist between every other load and its corresponding displacement, and the total work done

$$\tfrac{1}{2}P_1 \Delta_1 + \tfrac{1}{2}P_2 \Delta_2 + \tfrac{1}{2}P_3 \Delta_3 \ldots \tfrac{1}{2}P_n \Delta_n = \tfrac{1}{2}\sum P\Delta \tag{4.8–5}$$

If the loading system was not monotonically applied, the total work done by all the loads would be obtained by summing the effects of individual loads—as given in Eqn. 4.8–4, and would be

$$\sum_{1}^{n} \eta_1 P_1 \Delta_1$$

It will be shown more fully, however, in Section 4.9 that the work done by all the loads in reaching some final state must be independent of the loading programme by which that final state is achieved and that the **principle of conservation of energy** requires that the total work done is

$$\tfrac{1}{2}\sum_{1}^{n} P_1 \Delta_1 \tag{4.8–6}$$

In such a case the work done by any particular force may vary with the loading programme, but the summed effect of all is a constant and may be conveniently written in matrix terms as

$$\tfrac{1}{2}\mathbf{P}^T\mathbf{\Delta} \tag{4.8–7}$$

where the symbol for transposition is used since work is a scalar quantity.

In a completely general case where loads $P_1 \ldots P_n$ are allowed to vary independently when acting on a nonlinear structure, the total work done will be

$$\eta\mathbf{P}^T\mathbf{\Delta} \tag{4.8–8}$$

where η represents the combined effects of the various $\eta_1 \ldots \eta_n$ in Eqn 4.8–4.

(c) **Effects of distributed loads.** An elastic structure may be subjected to loads—whether linear or in the form of moments—applied at particular points, and, in addition, to loads distributed over a defined area. The latter may be included in Eqn 4.8–7 by regarding elementary areas of application as equivalent to point loads. If **p** represents the intensity of externally applied distributed loads or moments which are assumed to be uniformly distributed over an elementary area dA, defined as in Eqn 1.2–14 by the expression

$$\mathbf{p} = \begin{bmatrix} p_x \\ p_y \\ p_z \\ m_x \\ m_y \\ m_z \end{bmatrix}$$

and if $\boldsymbol{\delta}$ be the corresponding displacement matrix

$$\boldsymbol{\delta} = \begin{bmatrix} u \\ v \\ w \\ \theta_x \\ \theta_y \\ \theta_z \end{bmatrix}$$

in which each term corresponds with the appropriate term in the load matrix then the work done by distributed loads is

$$\tfrac{1}{2}\int \mathbf{p}^T\boldsymbol{\delta}\, dA$$

and we may now write

$$\text{Work done by all applied loads} = \tfrac{1}{2}\mathbf{P}^T\mathbf{\Delta} + \tfrac{1}{2}\int \mathbf{p}^T\boldsymbol{\delta}\, dA \tag{4.8–9}$$

4.9 Energy—strain and complementary

The work done by an applied load system must be stored in the structure in the form of potential energy. This energy is stored as a result of the differing strains of its various elements, and, in the special case of an elastic structure would be recovered when the loads are removed, since we have assumed that the stress–strain relationship is reversible and since the principle of the conservation of energy requires that energy is indestructible. It follows that this stored energy will depend on the final magnitudes of the applied loads with their corresponding displacements, and not on the manner by which that final state is achieved, since it would otherwise be possible to create energy by loading and unloading the structure by different loading routes.

The potential energy stored in the form of strain in the general case will be essentially of the form $\eta \boldsymbol{\sigma}^T \boldsymbol{\varepsilon}$ per unit volume where η has the same significance as in Eqn 4.8–8 and defines the summed effect of varying loading changes (being $\frac{1}{2}$ when the structural response is linear and elastic) and $\boldsymbol{\sigma}$ and $\boldsymbol{\varepsilon}$ are the corresponding stress and strain vectors defined in Section 4.2. As pointed out in that section, however, it is often convenient in a member of a skeletal structure to devote analytical attention to the internal action and deformation vectors which represent the summed effects of stress and strain respectively at particular cross-sections. These internal actions will consist of an axial force and two shear forces, with a torque and two bending moments as represented by the vector $\mathbf{F}$ and will produce corresponding displacements $\mathbf{e}$, so that the strain energy in a general case could be written as $\eta \mathbf{F}^T \mathbf{e}$ (η again having the same significance as in Eqn 4.8–8). There will be an equality between the work done by the applied loads and the strain energy, U, stored, so that

$$\begin{aligned} U &= \text{work done by applied loads} \\ &= \eta \sum_{1}^{n} P_1 \Delta_1 + \eta \int \mathbf{p}^T \boldsymbol{\delta}\, dA \\ &= \text{strain energy stored} \end{aligned} \tag{4.9–1}$$

Similarly there will be an equality between the complementary work done by all the applied loads and the complementary energy stored in the structure. As shown in Section 4.7 complementary work done by a varying load P moving through a distance Δ is $\int \Delta\, dP$ (Section 4.8(b)), but Fig. 4.8–2 shows that

$$\begin{aligned} \int \Delta_1 dP_1 &= P_1 \Delta_1 - \int P_1 d\Delta_1 \\ &= (1 - \eta_1) P_1 \Delta_1 \end{aligned}$$

so that

$$\begin{aligned} C &= \text{complementary work done by the applied loads} \\ &= (1 - \eta) \sum_{1}^{n} P_1 \Delta_1 + (1 - \eta) \int \mathbf{p}^T \boldsymbol{\delta}\, dA \\ &= \text{complementary energy} \end{aligned} \tag{4.9–2}$$

It will be seen from the form of Eqns 4.9–1 and 4.9–2 that the complementary and strain energies become equal when

$$\begin{aligned} 1 - \eta &= \eta \\ \eta &= \tfrac{1}{2} \end{aligned}$$

that is, in the special case of a linear elastic structure in which a linear and reversible relationship exists between any single load and its corresponding displacement. The same equality was shown to exist in a linear case in Section 4.7, when dealing with the extension of a single member.

4.10 Superposition of strain energies

Energy is stored in the structure with each load application, and it might be thought that the energies stored as a result of the separate applications of different loads, being scalar, could readily be added to give the total effect of the loads applied simultaneously. This will not necessarily be so, as can be seen from Fig. 4.8–2 where in section bc the point of application of a load, P_1, has moved as a result of changes in other loads without any change in P_1 itself. As a result P_1 has done work, causing an increase in the strain energy of the structure.

If one system of forces applied to a structure causes another system to do work, then the simple addition of energies resulting from the applications of the separate systems is inadmissible. If a member is subjected to an axial force and a torque, strain energy will be stored in it as a result of the deformations caused by both axial and shear stresses. In this case the former are caused by axial force alone and the latter by torque alone, with the result that the energies caused by the two effects separately can be added. If, however, the same member was acted upon by a torque and a shear causing shear stresses (and therefore shear strains) on the same plane, the superposition of one action upon another would cause work to be done by stresses set up by the first action. In such a case addition of the energies may only be made provided the effect of all loads on the displacement of points of application of others is taken into account. The point will be developed further in the next section.

Example 4.10–1. The pin-jointed truss of Fig. 4.10–1(a) consists of members of equal cross-sectional area and of the same material. Discuss the admissibility of determining the strain energy stored as a result of the action of the loads shown by simple addition of the various effects of each.

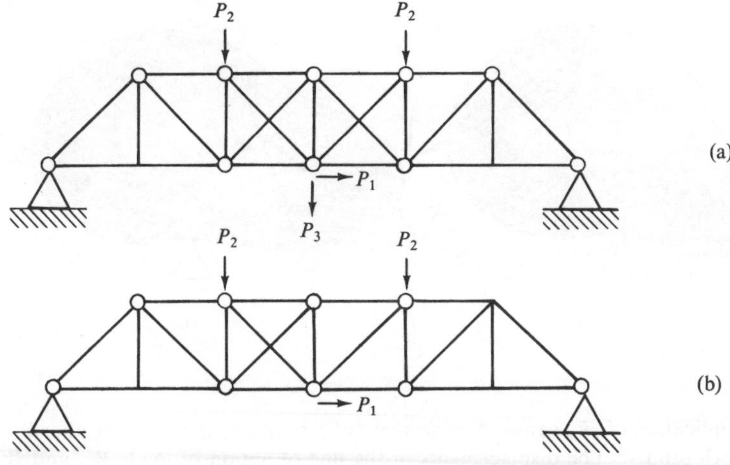

Fig. 4·10–1

SOLUTION The structure is symmetrical about a vertical centre line, as are the load systems defined as P_2 and P_3. Application of these loads will therefore cause no horizontal movement of the point of application of P_1. However, the action of P_2 will cause vertical displacement of the point of application of P_3 (and vice versa), although application of P_1 will not cause displacements corresponding either to P_2 or P_3.

It follows that

$$U_{P2+P1} = U_{P2} + U_{P1}$$
$$U_{P3+P1} = U_{P3} + U_{P1}$$
$$U_{P2+P3} \neq U_{P2} + U_{P3}$$

The above arguments follow from the symmetry of the figure. In Fig. 4.10–1(b) in which no symmetry exists

$$U_{P_1+P_2} \neq U_{P_1} + U_{P_2}$$

since application of the forces P_2 will cause horizontal displacement corresponding to P_1.

4.11 Reciprocality of influence coefficients in elastic structures

In Section 4.4 reference was made to the influence coefficients f_{mn} and k_{nm}, the flexibility and stiffness influence coefficients respectively, where f_{mn} is the displacement caused in a specified direction m by the application of unit load in some specified direction n and k_{nm} the force required in some specified direction n to produce unit displacement in some specified direction m. It will be shown that in a linear elastic structure there is a reciprocal relationship between influence coefficients with similar subscripts and that

$$f_{mn} = f_{nm} \quad \text{and} \quad k_{nm} = k_{mn}$$

These equalities imply that the matrices **f** and **k**, the flexibility and stiffness matrices of the structure, are symmetrical about their leading diagonals.

(a) **Flexibility influence coefficients: Clerk Maxwell's reciprocal theorem.** Consider some linear elastic structure supported by reactive elements which do no work, to which two loads W_m and W_n may be applied in turn as shown in Figs. 4.11–1(a) and 4.11–1(b). In

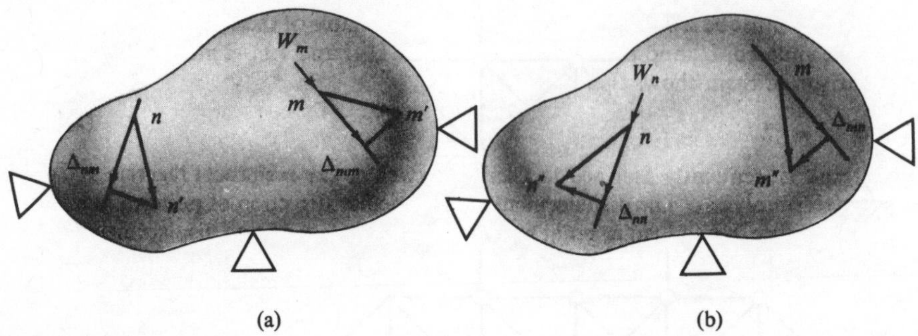

Fig. 4·11–1

the former, application of W_m causes displacement of n to n' and m to m' (together with rotation of both points). The displacements in the line of action of loads W_m and W_n (i.e. corresponding with them) are

$$\Delta_{mm} = f_{mm} W_m \quad \text{and} \quad \Delta_{nm} = f_{nm} W_m$$

Similarly if W_n is applied, m and n will be displaced to m″ and n″ (as in Fig. 4.11–1(b)) and the displacements corresponding to W_m and W_n will be

$$\Delta_{mn} = f_{mn} W_n \quad \text{and} \quad \Delta_{nn} = f_{nn} W_n$$

Now if W_m is applied first to the structure the work done by W_m will be

$$\tfrac{1}{2} W_m \Delta_{mm} = \tfrac{1}{2} W_m (f_{mm} W_m)$$

and if W_n is applied subsequently, W_n will do work of

$$\tfrac{1}{2} W_n \Delta_{nn} = \tfrac{1}{2} W_n (f_{nn} W_n)$$

but the point of application of W_m will be displaced and additional work will be done by W_m of

$$W_m \Delta_{mn} = W_m \ (f_{mn} W_n)$$

(it will be noted that, during this movement there is no change of W_m and the factor of $\frac{1}{2}$ is omitted). The total work done by the two loads, and therefore the strain energy stored in the structure will be

$$U_1 = \tfrac{1}{2} f_{mm} W_m^2 + \tfrac{1}{2} f_{nn} W_n^2 + f_{mn} W_m W_n \tag{4.11–1}$$

Now, if the structure be unloaded, thus restoring it to its original condition and the loads applied singly again, but in the reverse order, i.e. if W_n is applied first, it will do work of

$$\tfrac{1}{2} W_n \Delta_{nn} = \tfrac{1}{2} W_n (f_{nn} W_n)$$

Now if W_m is applied, it will do work of

$$\tfrac{1}{2} W_m \Delta_{mm} = \tfrac{1}{2} W_m (f_{mm} W_m)$$

but application of W_m will cause displacement of n and therefore cause W_n to do work of

$$W_n \Delta_{nm} = W_n \ f_{nm} W_m$$

and the total work done by the two loads, which will be stored in the structure in the form of strain energy is

$$U_2 = \tfrac{1}{2} f_{mm} W_m^2 + \tfrac{1}{2} f_{nn} W_n^2 + f_{nm} W_m W_n \tag{4.11–2}$$

The same final state of the structure, in equilibrium under the two loads W_m and W_n, has thus been achieved by two different routes and, as pointed out in Section 4.9, the same energy must be stored in the structure or the principle of the conservation of energy would be violated. U_1 and U_2 expressed by Eqns 4.11–1 and 4.11–2 must be equal, and a comparison of them shows that

$$f_{mn} = f_{nm} \tag{4.11–3}$$

This result is frequently ascribed to Clerk Maxwell as the **reciprocal theorem** and may be written simply as: 'The displacement in a particular direction of point m in a linear elastic structure due to load at point n will equal the corresponding displacement caused at n by the same corresponding load at m.'

In the proof given it will be seen that application of a second load causes displacement of the point of application of the first load and, therefore, causes that first load to do work. The total strain energy of the structure cannot be obtained simply by the addition of the work done by the two loads acting separately (given by the first two terms in Eqns 4.11–1 and 4.11–2), but includes a third interactive term from which the principle of reciprocality may be established.

(b) **Stiffness influence coefficients.** If we now consider, as before, a linear elastic structure supported in such a way that the reactive elements do no work and devote attention to two points r and s in it as shown in Fig. 4.11–2. If r is displaced a unit distance in the direction R, all other displacements at r (in the direction at right angles R' or in rotation) and all displacements at point s being prevented, then three forces would be required at each of points r and s:

$$k_{RR}, k_{R'R}, k_{\theta_r R}, k_{SR}, k_{S'R}, k_{\theta_s R}$$

these being the appropriate influence coefficients as defined in Section 4.4. It will be noted that if any movement of s or of r other than in the direction R occurred, the forces generated would not be stiffness influence coefficients as previously defined.

Now let r be displaced a distance q_R in the direction R (causing forces $k_{RR} q_R$ and $k_{R'R} q_R$ and $k_{\theta_r R} q_R$ at r and $k_{SR} q_R$, $k_{S'R} q_R$ and $k_{\theta_s R} q_R$ at s). If this is followed by a

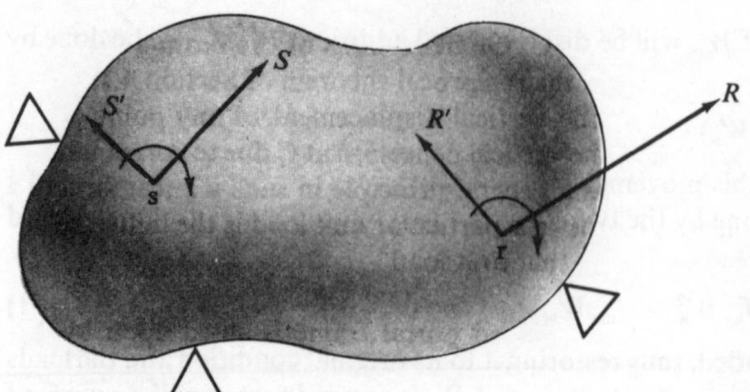

Fig. 4·11–2

displacement q_S at s (all other displacements being prevented) then the total work done, and energy stored

$$U_1 = \tfrac{1}{2}k_{RR}q_R^2 + k_{SR}q_Rq_S + \tfrac{1}{2}k_{SS}q_S^2$$

Similarly, if q_S is imposed first, followed by q_R

$$U_2 = \tfrac{1}{2}k_{RR}q_R^2 + k_{RS}q_Sq_R + \tfrac{1}{2}k_{SS}q_S^2$$

and since the energies must be equal

$$k_{SR} = k_{RS}$$

establishing the reciprocality of the stiffness influence coefficients.

(c) **Application of reciprocality of coefficients.** Structural analysis frequently requires the determination of some quantity which varies with the movement of load upon a structure. The graphical representation of such a variation is termed an **influence line** (see Chapter 2, Section 2.7).

An example might be the vertical displacement of point C in the simply supported elastic beam AB of Fig. 4.11–3(a), caused by the passage across the beam of a load W. Such a variation can adequately be portrayed by an 'influence line' as in Fig. 4.11–3(b) in which the deflection of C due to unit value of W is plotted as ordinate against the appropriate position of W as abscissa.

Let W be at point X, causing vertical deflections δ_x at X and δ_c at C. Then, using flexibility influence coefficients and the normal notation

$$\delta_x = Wf_{xx} \quad \text{and} \quad \delta_c = Wf_{cx}$$

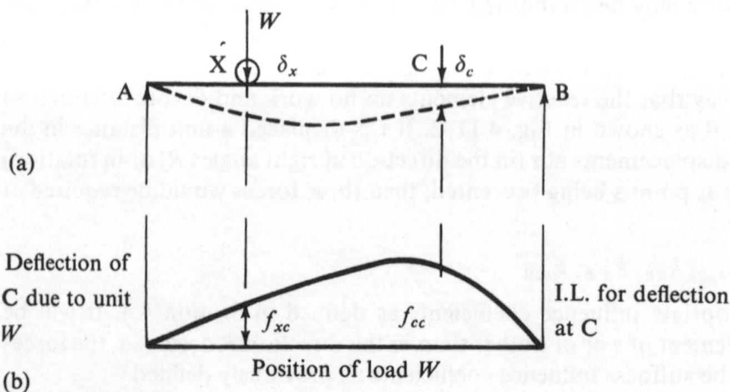

Fig. 4·11–3

Now, as a result of applying a unit load at C, the deflections of the beam at X and C would be f_{cx} and f_{cc} respectively and, from the reciprocal theorem of Section 4.11(a), f_{cx} and f_{xc} would be equal. It follows that the vertical displacement of any point on the beam due to unit vertical load at C gives the vertical deflection at C due to corresponding unit load at the point in question, and, as a general principle in such a linear structure that: 'the deflected shape of a structure due to a particular unit load is the influence line for the effect (deflection) corresponding to that unit load'.

Example 4.11–1. In Fig. 4.11–4(a) a rigid-jointed portal frame is acted upon by two forces, a unit moment at X and a unit linear force at Y. Is the rotation produced at Y by the unit moment equal to the vertical deflection produced at X by the unit force?

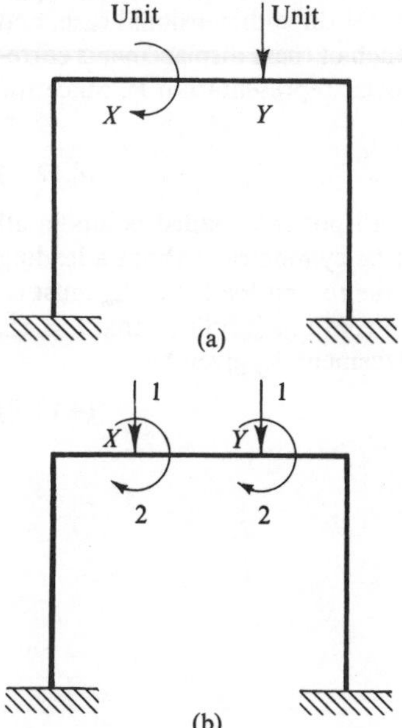

Fig. 4·11–4

SOLUTION Attention must always be concentrated on corresponding forces, and to see clearly what is happening directions 1 and 2 corresponding to vertical and angular forces respectively at X and Y as shown in Fig. 4.11–4(b) should be defined. Using the notation of flexibility influence coefficients, then unit moment at X produces:

angular rotation at X of $f_{x2,\,x2}$;
vertical deflection at X of $f_{x1,\,x2}$;
angular rotation at Y of $f_{y2,\,x2}$;
vertical deflection at Y of $f_{y1,\,x2}$.

Similarly unit vertical force at Y produces:

angular rotation at X of $f_{x2,\,y1}$;
vertical deflection at X of $f_{x1,\,y1}$;
angular rotation at Y of $f_{y2,\,y1}$;
vertical deflection at Y of $f_{y1,\,y1}$.

The two values under discussion are $f_{y2,\,x2}$ and $f_{x1,\,y1}$ and it can be seen there is no reciprocality.

4.12 Generalized reciprocal theorem in an elastic structure. Betti's theorem

The reciprocality of specific influence coefficients can be shown to be a specialized case of a much more general principle ascribed to Betti, although here the findings of Section 4.11 will be used to establish that principle. Stated in its simplest terms, the principle is that: 'If a linear elastic structure is subjected at different times to two different loading systems, the sum of the products of the loads of one system and the corresponding displacements of the second will equal the sum of the products of the loads of the second system and the corresponding displacements of the first.'

A force system **P** acts at points in a linear structure causing displacements $\mathbf{\Delta}_p$. All points in the structure may be displaced and will have coordinate movements equal to the degrees of freedom of each point (six in the general three-dimensional case, and three for two dimensions) and it is prudent to imagine each of these displacements corresponding to a force (which may be zero) in the vectorial representation **P**. Since the structure is elastic,

$$\mathbf{\Delta}_p = \mathbf{fP} \tag{4.12–1}$$

where **f** is a flexibility matrix which, since it concerns all possible loaded points in all possible directions must be square. Moreover, it must be symmetrical about a leading diagonal since (from Section 4.11) the deflection at m due to unit load at n, f_{mn} must be equal to the deflection at n due to unit load at m, f_{nm}. From this it follows that $\mathbf{f} = \mathbf{f}^T$. If a second load system **Q** is applied it will cause displacement $\mathbf{\Delta}_Q$ given by

$$\mathbf{\Delta}_Q = \mathbf{fQ} \tag{4.12–2}$$

and the scalar quantity

$$\begin{aligned}
\underset{1\times n}{\mathbf{P}^T}\,\underset{n\times 1}{\mathbf{\Delta}_Q} &= \underset{1\times n}{\mathbf{P}^T}\,\underset{n\times n}{\mathbf{f}}\,\underset{n\times 1}{\mathbf{Q}} \\
&= [\mathbf{P}^T\,\mathbf{f}\,\mathbf{Q}]^T \\
&= \mathbf{Q}^T\,\mathbf{f}^T\,\mathbf{P} \\
&= \underset{1\times n}{\mathbf{Q}^T}\,\underset{n\times n}{\mathbf{f}}\,\underset{n\times 1}{\mathbf{P}} \\
&= \mathbf{Q}^T\,\mathbf{\Delta}_p
\end{aligned} \tag{4.12–3}$$

which is also scalar. Clearly much depends here on the properties of the flexibility matrix, and it should be emphasized that the generality of the proof given herein does not eliminate the possibility that some forces in the **P** and **Q** matrices may be zero.

A similar result may be derived from considerations of the energy stored in a structure, following the argument used in Section 4.11(a).

If the force system **P** is applied to the structure, causing displacements $\mathbf{\Delta}_p$ then work will be done equal to $\frac{1}{2}\mathbf{P}^T\mathbf{\Delta}_p$. Now if the force system **Q** be added, causing displacements $\mathbf{\Delta}_Q$, work will be done by **Q** of $\frac{1}{2}\mathbf{Q}^T\mathbf{\Delta}_Q$. During this latter load application, however, **P** will also do work of $\mathbf{P}^T\mathbf{\Delta}_Q$ so that the total stored energy is

$$\tfrac{1}{2}\mathbf{P}^T\mathbf{\Delta}_p + \tfrac{1}{2}\mathbf{Q}^T\mathbf{\Delta}_Q + \mathbf{P}^T\mathbf{\Delta}_Q \tag{4.12–4}$$

Similarly if **Q** was applied first, followed by **P** the total strain energy stored, which must be identical with that of Eqn 4.12–4, will be

$$\tfrac{1}{2}\mathbf{P}^T\mathbf{\Delta}_p + \tfrac{1}{2}\mathbf{Q}^T\mathbf{\Delta}_Q + \mathbf{Q}^T\mathbf{\Delta}_p \tag{4.12–5}$$

Comparison of the two equations shows that

$$\mathbf{P}^T\mathbf{\Delta}_Q = \mathbf{Q}^T\mathbf{\Delta}_p \tag{4.12–6}$$

which is the result given previously.

Example 4.12–1. A non-uniform beam AB is found to have an angular stiffness k_{AA} at A when the beam is propped at A and encastré at B. The moment produced at B is $C_{AB}k_{AA}$. If the same beam is encastré at A and propped at B it will have an angular stiffness of k_{BB} at B and the moment produced at A will be $C_{BA}k_{BB}$. Show that $C_{AB}k_{AA} = C_{BA}k_{BB}$.

SOLUTION The two loading cases are shown in Fig. 4.12–1(a) and (b) and, applying the generalized reciprocal theorem

$$k_{AA}(0) + F_{A1}(0) + C_{AB}k_{AA}(1) + F_{B1}(0) = C_{BA}k_{BB}(1) + F_{A2}(0) + k_{BB}(0) + F_{B2}(0)$$

from which

$$C_{AB}k_{AA} = C_{BA}k_{BB}$$

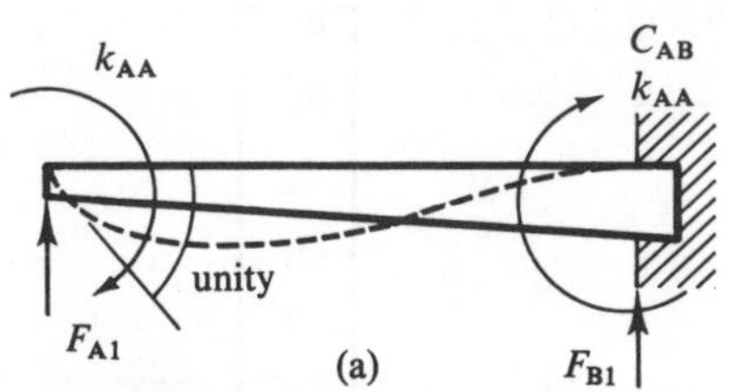

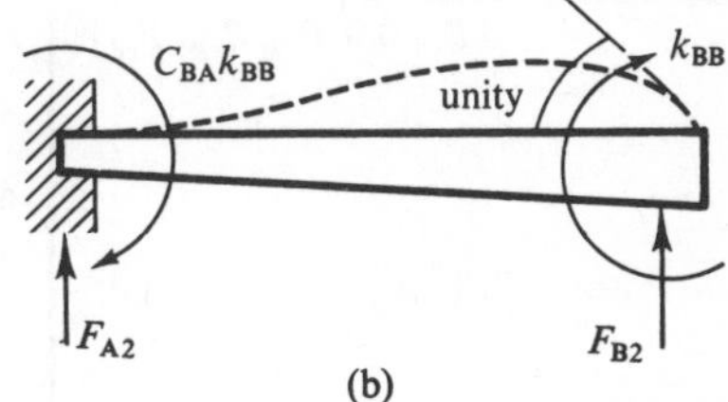

Fig. 4·12–1

4.13 Mueller-Breslau's principle. Model analysis

(a) **Mueller-Breslau's principle.** Consider a linear elastic structure indicated in Fig. 4.13–1 supported by reactions which do no work, and acted upon by two different load systems in turn. In the first, let a load P_{sS} act at s in the direction S, causing displacements δ at that point which will be different in the coordinate directions and generating forces P (different in the coordinate directions) at another point r. Reference is made below to the 12 vectors involved in a two-dimensional case but no generality is lost by this limitation and visualization is easier.

First system

	At s			*At* r		
	sS	sS'	$s\theta$	rR	rR'	$r\theta$
Forces (1)	P_{sS}	zero	zero	P_{rR}	$P_{rR'}$	$Pr\theta$
Displacements (2)	δ_{sS}	$\delta_{sS'}$	$\delta_{s\theta}$	zero	zero	zero

It should be noted that the displacements at r are recorded as zero since, by implication there can be no discontinuity at this point. Moreover, for the same reason, there will be no displacement at other points such as r.

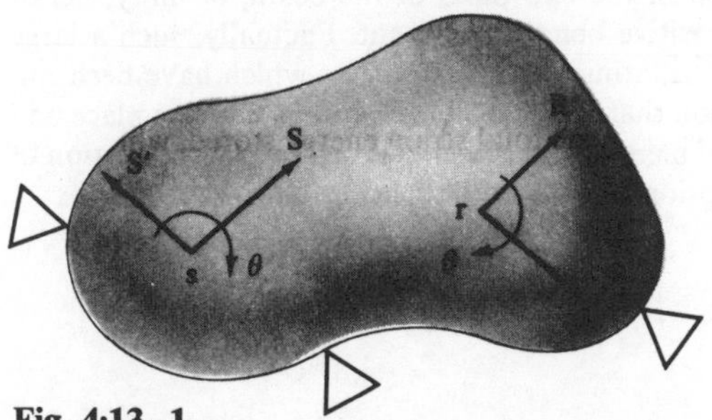

Fig. 4·13–1

Now let a second force system be applied by movement of point r through a distance Δ_{rR} in the direction R. No loads are applied at s, which moves through a distance Δ in the coordinate directions at s and no other movement is to be permitted at r although obviously forces such as Q will be required there.

Second system

	At s			*At* r		
	sS	sS'	$s\theta$	rR	rR'	$r\theta$
Forces (3)	zero	zero	zero	Q_{rR}	$Q_{rR'}$	$Q_{r\theta}$
Displacements (4)	Δ_{sS}	$\Delta_{sS'}$	$\Delta_{s\theta}$	Δ_{rR}	zero	zero

The generalized reciprocal theorem may now be applied by summing the corresponding products of lines 1 and 4 and equating to the summed products from lines 2 and 3:

$$[P_{sS}\;0\;0\;P_{rR}\;P_{rR'}\;P_{r\theta}]\begin{bmatrix}\Delta_{sS}\\ \Delta_{sS'}\\ \Delta_{s\theta}\\ \Delta_{rR}\\ 0\\ 0\end{bmatrix} = [0\;0\;0\;Q_{rR}\;Q_{rR'}\;Q_{r\theta}]\begin{bmatrix}\delta_{sS}\\ \delta_{sS'}\\ \delta_{s\theta}\\ 0\\ 0\\ 0\end{bmatrix}$$

from which

$$P_{rR} = -\frac{\Delta_{sS}}{\Delta_{rR}}P_{sS}$$

The force P_{rR} caused at r by a unit force P_{sS} may be determined separately by the displacement Δ_{sS} of s caused by a unit movement of r which corresponds with P_{rR}. As it may be generally expressed, 'the deflected shape of a structure due to a particular unit distortion represents the influence line for the effect corresponding to that distortion'.

In applying this principle, which is due to Mueller-Breslau, care must be taken to ensure that complete correspondence exists both between the unknown effect whose magnitude is sought and the applied unit distortion, and between the applied load causing the effect and the displacement of its point of application. Moreover the scale of the distortion must be considered small, so that its application causes no change in the essential geometry of the structure. An example can be seen if reference is made to the influence line for bending moment at some point P, distant a and b from the two ends of a simply supported beam of span L (Fig. 4.13–2). The influence line will be a triangle with a maximum ordinate at P of ab/L as in Fig. 4.13–2(b). It could be obtained by giving a unit distortion at P corresponding to the desired effect, as shown in (c), and in this case we require a relative rotation of the two parts of the beam, of unity, i.e. of 1 rad, corresponding to the sign of positive bending moment. Factually, such a large deformation could not occur without vitiating other hypotheses which have been implicitly accepted, such as the assumption that no large deformations will take place and that the projected length of a deformed member remains unaltered by the application of load. It is possible to use these assumptions, however, in calculation so that

$$y_p = a\theta = b\phi$$

from which

$$\phi = \frac{a}{b}\theta$$

But $\theta + \phi = 1 \text{ rad} = \theta + a\theta/b$, hence

$$\theta = \frac{b}{a+b} \quad \text{and} \quad y_p = a\theta = \frac{ab}{a+b} = \frac{ab}{L}$$

The influence line for shear at P may be obtained equally simply by giving the two parts of the beam at P a unit distortion corresponding with the convention of positive shear as shown in Figs. 4.13–2(e) and (f). This distortion must take place in such a way that no other distortion at P occurs, that is, particularly, that there is neither relative rotation nor horizontal displacement of the two ends of the cut beam at P. It follows that the two must move in such a way as to remain parallel, and that the lines shown dotted will contain the influence lines for shear at all intermediate points in the span.

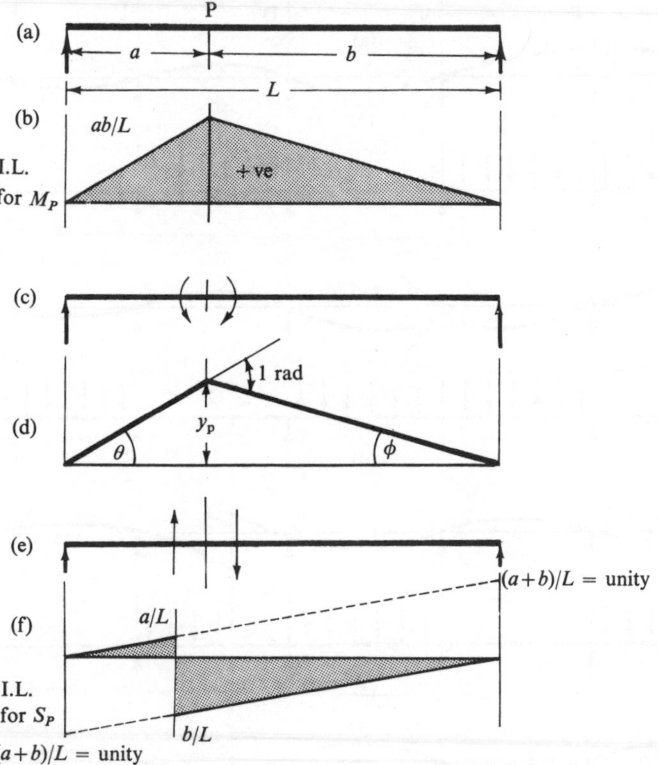

Fig. 4·13–2

These examples are, of course, of an exceptionally simple nature, and it is most unlikely that Mueller-Breslau's principle would be used in the determination of these particular influence lines since both could be obtained more simply by direct means. The principle is of considerable value, however, in more complex cases. It forms the basis of one of the two major subdivisions of model analysis in which a structure which is far too complicated to be analysed by orthodox methods is solved by experimental techniques, as dealt with more fully in Subsection 4.13(b), but it can be of considerable help in reducing the generality of a particular problem to permit of a relatively simple solution.

An example might be the design of the floor for a warehouse, assumed to be a continuous beam, with a large number of spans AB, BC, CD, DE, and EF as shown in

Fig. 4.13–3(a). The beams must support their own weight, together with the floor slab above them, and analysis of the effects of these loads is simple, but it is known that any or all of the various spans may be loaded to perform the function for which the warehouse is intended. Simple visualization of a deformation pattern allows one to sketch—without any idea of numerical magnitudes involved—the shape of the influence line for reaction at B (say) as shown in Fig. 4.13–3(b) and to conclude therefrom that maximum reaction due to the live loads will occur at B when spans AB, BC, and DE are loaded as shown in Fig. 4.13–3(c). Knowing what particular load position causes a maximum effect, a simple 'static' calculation permits the determination of the effect's magnitude. It will be seen that in all cases the influence lines are curved (as always occurs with a

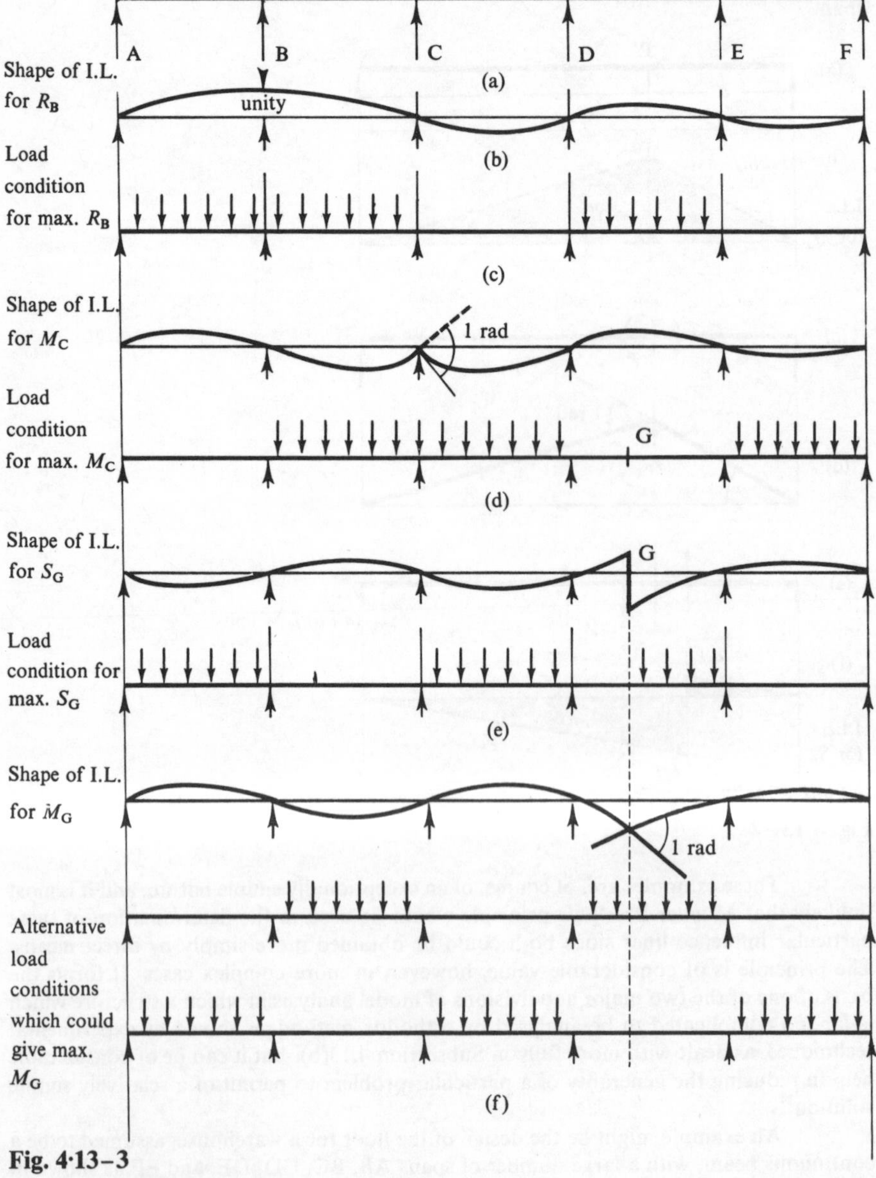

Fig. 4·13–3

continuous statically indeterminate structure), but the shapes of the influence lines for internal points such as E (shown in Figs. 4.13–3(e) and (f)) are of particular interest. The appropriate unit movements of the beam must occur without applying any relative movement in other directions and, for example, vertical separation of the two parts of the beam at G in Fig. 4.13–3(e) must not be accompanied by relative rotation of their ends, while the relative rotation required in Fig. 4.13–3(f) must not be accompanied by vertical separation.

(b) **Model analysis.** Structural analysis must depend on the idealization of structural form and the various aspects of material behaviour, in order that both may be reduced to a form which permits handling by the computational methods available. It is obviously true that the form of computing equipment available influences the type of problem which may be tackled—the existence of the computer allows us to analyse more complicated structures than would be possible with a slide rule alone. It is less obvious, but equally true, that the extended computing facilities now available, and the ease of calculation which they imply, allow computations to be made with several different idealizations to determine which represents most adequately the real behaviour of the structure. It would frequently be helpful, however, to have a model of the structure available in order that its behaviour under load could readily be ascertained and there are cases (even in the structure in which linear behaviour can be assumed), in which the model can provide the most ready answer to a problem of analysis. These models take two forms, depending on the method of use and of interpretation of the results, they are (a) direct methods, and (b) indirect methods.

In a direct method of model analysis a true-to-scale (usually) model is made, and loaded with an appropriately reduced version of the applied loads. The material is usually identical with that of the full-scale structure, or has some conveniently reduced value of the modulus of elasticity and an identical value of Poisson's ratio, and measurements of strain and deflection may be related directly to the behaviour of the prototype. In indirect methods Mueller-Breslau's principle is used and the effect of applied loads (usually in determining the magnitudes of implied reactions) are calculated from influence line techniques. The applied unit distortions can be relatively large, measurable with a rule and hand lens, or so small as to require a measuring microscope. In the latter case considerable refinements are necessary in the instruments and methods for applying the various unit distortions, and, since deformations are small many precautions are necessary to counteract the effects of changes of temperature and, occasionally, of humidity from the respiration of the observer.

In the indirect methods it is possible to ensure similarity between model and prototype in various ways, in order to ensure that the former correctly simulates different aspects of the latter's behaviour. If, for example, strain energy was stored in the structure in a direct form only, as a result of tension or compression in the members, then it would be necessary for the model to represent the prototype consistently in the length of its members, their cross-sectional areas, and their moduli, although the scales need not be the same for each parameter. If bending strain energy only is important (and this assumption has already been made in arches and most rigidly framed civil engineering structures) the model should be consistent in its representation of length, of second moment of area, and the moduli of members, although here again the scales for the various parameters need not be identical. If complete similarity of behaviour is required, in which the model represents its original in behaviour due to extension, bending, torsion, and shear, complete similarity must exist in all dimensions and it would frequently be more simple to adopt a direct method of model analysis, rather than to use Mueller-Breslau's principle. Details for the interested reader will be found in Charlton [1, 2] (see References at the end of this chapter).

4.14 Principle of virtual work

As indicated in Chapter 1 and in Sections 4.1 and 4.2 the application to any structure of a generalized force system which can conveniently be represented by the vector **P** will produce both a reaction system and an internal stress system characterized by the vector $\boldsymbol{\sigma}$. In Section 4.2 it was pointed out that analysis in terms of stresses could be exceptionally complex since the energy stored in the structure as a result of strain, if written in terms of stress and strain, must be expressed as a function of their product for each elementary volume of material, i.e. as

$$\iint \left\{ \int_0^{\varepsilon} \boldsymbol{\sigma}^T \, d\boldsymbol{\varepsilon} \right\} d\,(\text{vol.})$$

The total energy in a skeletal member such as a beam would be derived, therefore, after integration over a particular cross-section (since, in general, stress will vary from point to point in it) and, subsequently, integration over the entire length of a member to allow for varying bending moments, shears, etc. Calculation can be much simplified if externally applied loads are regarded as producing internal actions **F** (a direct force, a shear and a bending moment in the two-dimensional case and additional shear, torque, and bending moment in the case of three dimensions) at particular cross-sections. If required at some particular point the stress could then be calculated relatively simply from the effects of the appropriate internal actions.

Thus, application of a general load system **P** to any structure produces a reactive system and internal actions **F** from which the stresses $\boldsymbol{\sigma}$ may be computed. The structure will change shape, giving displacements of the points of application of the loads, internal deformations associated with the internal actions, and strains, all of which have been represented by the corresponding vectors $\boldsymbol{\Delta}$, **e**, and $\boldsymbol{\varepsilon}$.

Using this notation the principle of virtual work may be developed quite simply. It is of exceptional power and importance and proves to be the basis for many specific approaches in structural analysis each of which is of a limited application only, although the principle is itself applicable widely to structures of all types. The authors are indebted to T. M. Charlton[3,4] for what appears to them to be a wholly satisfying and elegant derivation.

If a system of concurrent forces $P_1 \ldots P_N$, which have a resultant P_R, act upon a particle, and if as a result the particle moves through a distance Δ_R in the direction of P_R, work will be done by the various forces.

If Δ_R is small, so that the directions and magnitudes of the various forces may be regarded as unchanged, and if the component of movement of the particle corresponding to each individual force is $\Delta_1 \ldots \Delta_n$ then the principle of the conservation of energy requires that

$$P_R \Delta_R = P_1 \Delta_1 + P_2 \Delta_2 + P_3 \Delta_3 \ldots P_n \Delta_n$$

Now if the force system is in equilibrium, that is if $P_R = 0$ then

$$P_1 \Delta_1 + P_2 \Delta_2 + P_3 \Delta_3 \ldots P_n \Delta_n = 0 \tag{4.14–1}$$

Essentially Eqn 4.14–1 gives expression to the principle of virtual work. It states that, if a particle in equilibrium is caused to suffer any small arbitrary displacement, the net work done by forces acting on the particle is zero. The only limitation relates to the displacement involved which must be such that the magnitudes and directions of forces do not change. Clearly there is no requirement that forces should be linear only, and the same equation may contain moments, torques, or shear forces, requiring only that the displacement term used with a particular force should correspond with it. Moreover, it must be understood that Eqn 4.14–1 sums to zero because of the relationships between the various P and Δ terms, and not because the displacements necessarily approach zero when considered to be small enough.

When applying the principle to structures we must pay particular attention to the sign of work done by internal actions undergoing a corresponding displacement. Figures 4.14–1(a) and (b) show members in a structure subject to a positive action F taking the form of direct tensile force in (a) and of torque in (b). In the former the two ends of a member AB carrying a tensile force F move through a and b respectively—in the latter the two ends of member AB, carrying equal moments rotate through angles a and b.

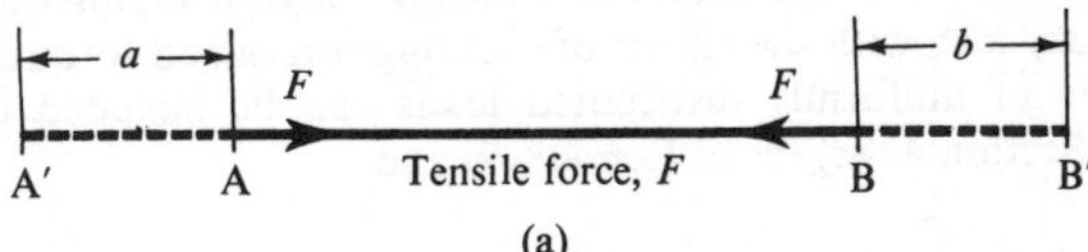

(a)

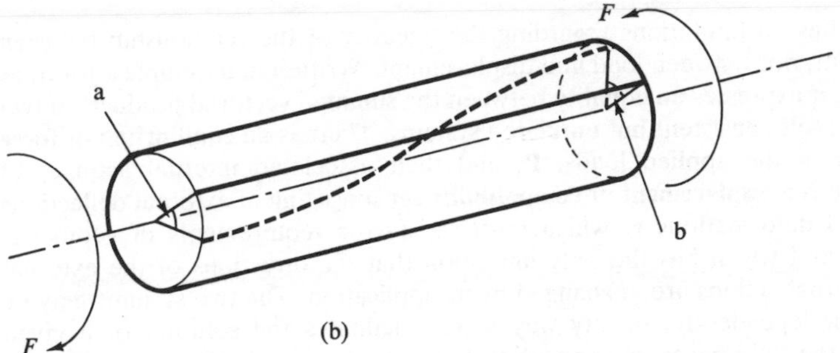

(b)

Fig. 4·14–1

The work done on the member (and therefore stored in it) is

$$Fa + Fb = F(a + b) = Fe$$

where e is the corresponding displacement of the internal action. However, the work done on the points of fixity at the ends of each member will be

$$(-Fa) + (-Fb) = -Fe$$

If now we consider some virtual deformation of the structure defined by the vectors $\overline{\mathbf{\Delta}}$, $\overline{\mathbf{e}}$, and $\overline{\boldsymbol{\varepsilon}}$ all of which are compatible and conform to the requirements of Eqn 4.14–1 (the word 'virtual' is used here to indicate that the deformation has a particular effect rather than a particular origin—it can be quite arbitrary; the dictionary gives one definition of 'virtual' as 'being such in essence or effect though not in name or appearance'), then the work done on all points in the structure will be

$$\mathbf{P}^T\overline{\mathbf{\Delta}} - \mathbf{F}^T\overline{\mathbf{e}} = 0 \tag{4.14–2}$$

The same equation may be arrived at more directly. During the virtual movement $\overline{\mathbf{\Delta}}$ work will be done on the structure of $\mathbf{P}^T\overline{\mathbf{\Delta}}$ (as shown in Section 4.8(b)). If the system remains in equilibrium, this work will be stored in the structure in the form of strain energy, and it may conveniently be written as $\mathbf{F}^T\overline{\mathbf{e}}$ rather than in the more elemental form $\int \boldsymbol{\sigma}^T\overline{\boldsymbol{\varepsilon}}\, d\,(\text{vol.})$, but there will be an equality

$$\mathbf{P}^T\overline{\mathbf{\Delta}} = \mathbf{F}^T\overline{\mathbf{e}} = \int \boldsymbol{\sigma}^T\overline{\boldsymbol{\varepsilon}}\, d\,(\text{vol.}) \tag{4.14–3}$$

It is in this form that the principle of virtual work is normally used in structural analysis, and it must be emphasized that the equation should include all applied loads and all points in the structure as developed earlier. The second term, involving the internal actions with their corresponding deformations must include the varying effects of F

within any one member, as well as the more obvious effects within all the members. If all actions were in the form of direct force as in a pin-jointed structure with members of constant cross-sectional area, then

$$\mathbf{F}^T\bar{\mathbf{e}} = \sum F\bar{e}$$

where summation would take place for all members, but bending effects would be represented by expressions of the form $\sum \int M\, d\bar{\theta}$ where the virtual deformation with its associated action must be integrated along the length of a member before summation takes place, as would be necessary also with the effects of shearing forces and torques.

If required, the effects of uniformly distributed loads can be included in Eqn 4.14–3 as was indicated in Section 4.8(c) (as in Eqn 4.8–9) and

$$\mathbf{P}^T\bar{\Delta} + \int \mathbf{p}^T\bar{\boldsymbol{\delta}}\, dA = \mathbf{F}^T\bar{\mathbf{e}} = \int \boldsymbol{\sigma}^T\bar{\boldsymbol{\varepsilon}}\, d\,(\text{vol.}) \qquad (4.14\text{–}4)$$

This equality has no limitations regarding the linearity of the relationship between stress and strain, nor between load and displacement. Written in its simplest form, as in Eqn 4.14–2, it expresses an equality between the summed vectorial products of two corresponding, self-consistent but unrelated systems. There is an **equilibrium** or **force set**, consisting of the applied loads, **P**, and their associated internal actions, **F**; secondly, there is a **displacement** or **compatibility set** consisting of external deflections Δ and internal deformations **e**, which must satisfy the requirements of geometric compatibility, and which has the only limitation that the directions of the external forces and internal actions are unchanged by its application. The two systems may be chosen quite independently, in any way which facilitates the solution of a given problem, and the only necesary connection between them is that actions and their associated displacements must correspond. The choice of the systems most helpful is a matter for judgement and experience, and students should concentrate on the distinctions between those used in specific problems. In the examples which follow, the equilibrium and compatibility sets will be indentified.

4.15 Applications of the principle of virtual work (1)

(a) **Unit load method for deflections.** In the last section the principle of virtual work was established and represented in the Eqns 4.14–2 and 4.14–3, it being assumed therein that the products of internal actions (or stresses) with their corresponding deformations (or strains) were summed over the entire structure for all members and all cross-sections. In that form the equations are very suitable to determine the deflection at some particular point and in a particular direction, owing to a given load system. If the actual deflection of a particular point is required, which will occur in a direction generally unknown, it may be obtained readily by determining the component movements in two directions at right angles and compounding.

It has been pointed out that each product in Eqn 4.14–3

Equilibrium set

$$\mathbf{P}^T\bar{\Delta} = \mathbf{F}^T\bar{\mathbf{e}} = \int \sigma^T\bar{\varepsilon}\, d(\text{Vol}) \qquad (4.15\text{–}1)$$

Compatibility set

consists of two corresponding parts, a force system in equilibrium and a displacement system which must be compatible; one or the other of these parts must be so small as to ensure that basic geometry is not disturbed. If the displacement system is defined as the deflected shape of a structure under load then $\bar{\boldsymbol{\Delta}}$ becomes the actual (but corresponding) movements of loaded points, $\bar{\mathbf{e}}$ the real deformations of members corresponding to internal actions, and $\bar{\varepsilon}$ the strains occurring at all points. If a load system is chosen in which a unit force acts at the particular point and in the particular direction in which the deflection is to be determined, then **P** represents this unit load alone (since any reactive system called into play will do no work), $\mathbf{F}_U$ the internal actions created by the unit load with its reactive system, and $\boldsymbol{\sigma}_U$ the resulting stresses in the structure.

Equation 4.15–1 thus becomes

$$1.\Delta = \mathbf{F}_U^T \mathbf{e} = \int \boldsymbol{\sigma}_U^T \boldsymbol{\varepsilon} \, d\,(\text{vol.}) \tag{4.15–2}$$

where the bar has been omitted since the displacements are real and finite and Δ is the particular deflection required. It should be emphasized that there are no limits (other than geometrical ones) to the applicability of this formula. It applies equally to the calculations of deflection due to temperature changes and to creep and shrinkage, as well as to the effects of load in linear and non-elastic structures, but it can be written more explicitly if attention is confined to structures with a linear load–deflection relationship and elastic behaviour. In this case the internal actions at any cross-section of a structure in two dimensions are threefold—direct force, moment, and shear, usually characterized by F, M, and Q respectively—and their corresponding deformations

$$\mathbf{e} = \text{force deformation} = \frac{F}{AE}\,ds$$

$$\text{moment deformation} = \frac{M}{EI}\,ds$$

$$\text{shear deformation} = \frac{KQ}{GA}\,ds$$

where A and I represent the geometrical properties of a cross-section;
E and G are the moduli of direct elasticity and shear;
ds is an element of length of one member;
K is a shape factor representing the distribution of shear stress over a cross-section, equal to 1.2 for a rectangular cross-section, and 1.11 for a solid circular cross-section.

The internal actions produced by the unit load may be written in representative form as F_U (force), M_U (moment), and Q_U (shear) and Eqn 4.15–2 now becomes

$$\Delta = \sum \int \frac{F_U F}{AE}\,ds + \sum \int \frac{M_U M}{EI}\,ds + \sum K \int \frac{Q_U Q}{AG}\,ds \tag{4.15–3}$$

where the $\sum$ sign implies summation for all members of the frame and $\int ds$ integration along the length of a member. If each member has a constant cross-sectional area this reduces to

$$\Delta = \sum \frac{F_U F L}{AE} + \sum \frac{1}{EI} \int M_U M\,ds + \sum \frac{K}{AG} \int Q_U Q\,ds \tag{4.15–4}$$

In the general three-dimensional case of an elastic structure there will be six internal actions at a particular cross-section and Eqn 4.15–3 will be extended by the addition of two terms for moment and shear in directions orthogonal to those defined therein, and a third term to show the effect of torque which will be of the form

$$\sum \int \frac{T_U T}{GJ}\,dx$$

where T_U and T are the torques produced by the unit load and the applied loads respectively and where (in a circular cross-section only) J is the polar second moment of area

Section	J_x in formula $\theta = \frac{TL}{GJ_x}$
 Solid circle	$J_x = \frac{\pi r^4}{2}$
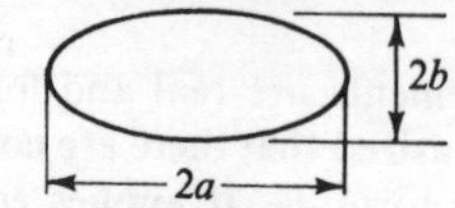 Solid ellipse	$J_x = \frac{\pi a^3 b^3}{a^2 + b^2}$
 Hollow circle	$J_x = \frac{\pi(r_2^4 - r_1^4)}{2}$
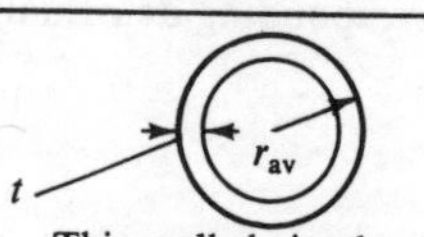 Thin-walled circular tube	$J_x = 8\pi r_{av} t^3$
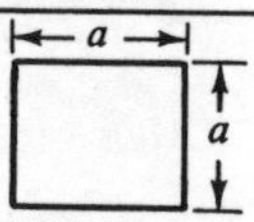 Solid square	$J_x = 0.1406a^4$
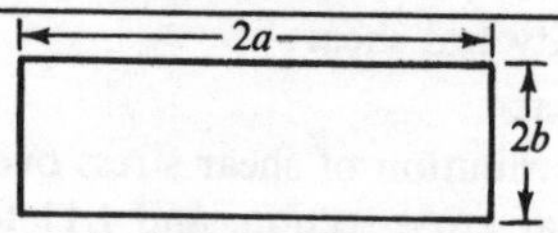 Solid rectangle	$J_x = ab^3\left[\frac{16}{3} - 3.36\frac{b}{a}\left(1 - \frac{b^4}{12a^4}\right)\right]$
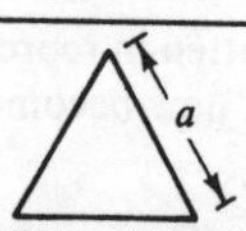 Equilateral triangle	$J_x = \frac{a^4\sqrt{3}}{80}$
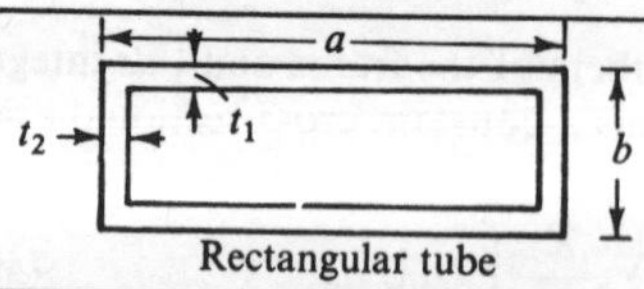 Rectangular tube	$J_x = \frac{2t_1 t_2(a-t_2)^2(b-t_1)^2}{at_2 + bt_1 - t_2^2 - t_1^2}$
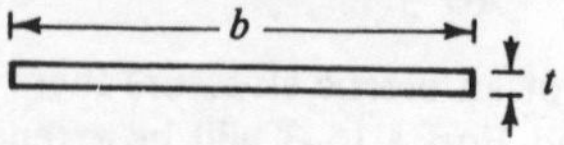 Thin rectangle	$J_x = \frac{1}{3} bt^3$
Any figure built up of thin rectangles	$J_x = \frac{1}{3} \sum bt^3$

Fig. 4·15–1 The shear centre for all the above sections, as for all doubly symmetrical sections coincides with the centroid.

of the cross-section. Equivalent values of J for other cross-sections are given in Fig. 4.15–1.

If the determination of deflection in a linear elastic pin-jointed structure is necessary, then only the first terms of the right-hand sides of Eqns 4.15–3 and 4.15–4 are necessary.

Example 4.15–1. Calculate the deflection at the free end of the pin-jointed cantilever frame shown in Fig. 4.15–2.

SOLUTION The direction of this deflection is unknown, and therefore we must determine the components of deflection in two convenient directions at right angles (in this case horizontal and vertical) and subsequently compound to give both the real magnitude and deflection. It is best to lay out calculations in a tabular form:

Equilibrium set

$$1\Delta = \sum F_u \frac{FL}{AE}$$

Compatibility set

Member	*Area* (1)	*Length* (2)	*Force due to applied loads F* (3)	*Force due to unit horizontal load at C* F_{UH} (4)	*Force due to unit vertical load at C* F_{UV} (5)	$\dfrac{F_{UH}FL}{AE}$ (6)	$\dfrac{F_{UV}FL}{AE}$ (7)
AB	A	L	0	0	1	0	0
BC	A	$L\sqrt{2}$	0	0	$\sqrt{2}$	0	0
CD	A	L	0	-1	-1	0	0
DE	A	L	$-W$	-1	-2	$\dfrac{WL}{AE}$	$\dfrac{2WL}{AE}$
AD	A	$L\sqrt{2}$	$W\sqrt{2}$	0	$\sqrt{2}$	0	$\dfrac{2WL\sqrt{2}}{AE}$
BD	A	L	$-W$	0	-1	0	$\dfrac{WL}{AE}$
						$\Delta_H = \sum = \dfrac{WL}{AE}$	$\Delta_V = \sum = \dfrac{3WL}{AE} + 2\sqrt{2}\dfrac{WL}{AE}$

Columns 1 and 2 define the geometrical properties of the members which must be known, but which need not be identical. Equally, there is no requirement that the modulus shall be the same for all members of the structure and a further column could be introduced here if that complication proved necessary.

Column 3 gives the forces produced in each member by the applied load system, which must be recorded with an appropriate sign; here, for example, by regarding tensile forces as positive and compressive forces as negative. In this case it will be seen that the structure is essentially simple and that only members BD, AD, and ED carry load. Columns 4 and 5 give the forces caused in each member by unit horizontal and vertical forces respectively, applied at the free end C, where a knowledge of deflection is required. The member forces may be of different sign from those in column 3.

Columns 6 and 7 give, respectively, the contribution made by each member to the required movement of the free end. Since, as noted under columns 4 and 5, the unit forces need not have the same sign as the force caused by applied load, terms in 6 and 7 may be either positive or negative. Summing (algebraically) the contribution made by each member gives

$$\Delta_H = \frac{WL}{AE}$$

$$\Delta_V = \frac{WL}{AE}(3 + 2\sqrt{2}) = 5.828\frac{WL}{AE}$$

Both these expressions are positive, and since the convention of direction must coincide with the directions of the unit loads—by implication the directions of deflection on the left-hand side of Eqn 4.15–2 must correspond with the unit loads applied—C moves to the left and downward. The actual deflection of C will be

$$\frac{WL}{AE}\sqrt{[1^2 + (5.828)^2]} = \frac{WL}{AE}\sqrt{34.9} = \frac{5.93WL}{AE}$$

and its direction $\tan^{-1} 1/5.828 = \tan^{-1} 0.172 = 9.77°$ to the left of vertical.

Example 4.15–2. If, in the pin-jointed frame of Fig. 4.15–2(a) the members BC and DC are subjected to temperature increases of t, what will be the resulting deflections of the free end? The coefficient of linear expansion for the two members is α.

SOLUTION In such a case it is wise to revert to the form of the relationship given in Eqn 4.15–2, which may be rewritten as

$$\delta = \sum F_U e$$

The lay-out is again preferably tabular:

Equilibrium set

$$\boxed{1\Delta = \sum F_u e}$$

Compatibility set

Member	*Length* (1)	*e = Lαt* (2)	*Force due to unit horizontal load at C,* F_{UH} (3)	*Force due to unit vertical load at C,* F_{UV} (4)	$F_{UH}e$ (5)	$F_{UV}e$ (6)
AB	L	0	0	1	0	0
BC	$L\sqrt{2}$	$+L\sqrt{(2)}\alpha t$	0	$\sqrt{2}$	0	$2L\alpha t$
CD	L	$+L\alpha t$	-1	-1	$-L\alpha t$	$-L\alpha t$
DE	L	0	-1	-2	0	0
AD	$L\sqrt{2}$	0	0	$\sqrt{2}$	0	0
BD	L	0	0	-1	0	0
					$\Delta_H = -L\alpha t$	$\Delta_V = L\alpha t$

Column 1 gives the geometrical properties of the members as before, while column 2 defines the change of length of each. It is plain that no difficulty is created if members are made of different materials or are subject to different temperatures.

Columns 3 and 4 give the force caused in each member by unit loads acting in the directions shown in Fig. 4.15–2(b) and (c). The final summations in columns 5 and 6 indicate that the component displacements of the free end are vertically downward (corresponding to the unit load), and horizontally to the right (opposed to the unit horizontal load). Compounding the movement in the two directions we see that the resultant deflection is $\sqrt{(2)}L\alpha t$ downwards and to the right in a direction at 45° to the horizontal.

The following example illustrates the method used for beams and frames subject to bending only.

Example 4.15–3. Figure 4.15–3 shows an elastic cantilever AB of length L and uniform cross-section throughout its length, subjected to a uniformly distributed load of w per unit length. Determine the deflection of its free end.

SOLUTION The bending moment caused by the load at some point R distant x from the free end will be $M = -wx^2/2$.

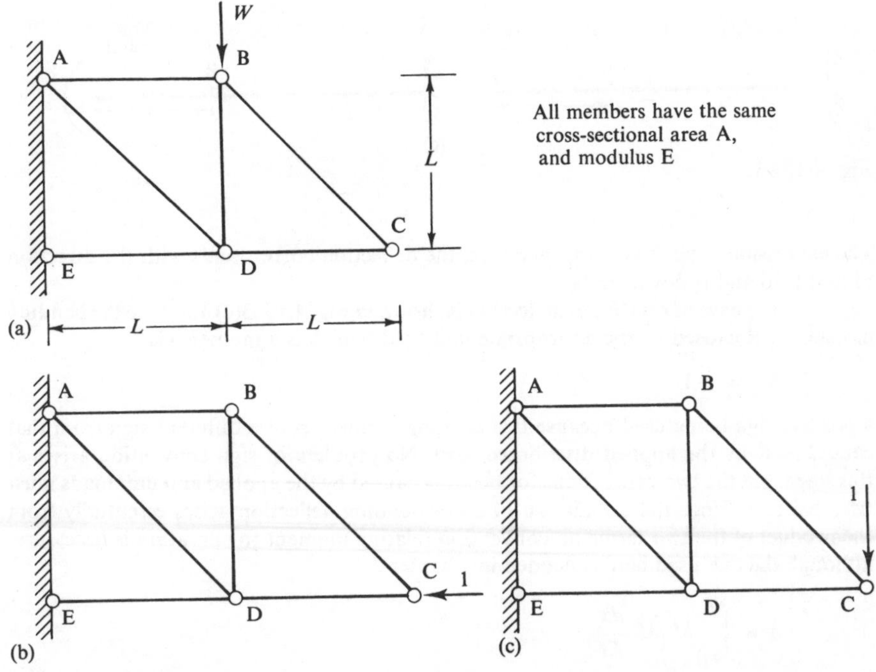

Fig. 4·15–2

Now if the sole contribution to deflection at any point is made by the moments in the elastic structure, and the second term on the right-hand side of Eqns 4.15–3 and 4.15–4 is the only significant one (which implies that the only strain energy stored in the structure is in bending) then the vertical and rotational deflections of the free end may be obtained by applying unit corresponding forces at the free end B, as shown in Figs. 4.15–3(b) and (c). In the case of the former, the bending moment at R caused by unit load at B is

$$M_U = -x$$

Then

$$\Delta = \int_0^L \frac{M_U M\,dx}{EI} = \int_0^L (-x)\left(\frac{-wx^2}{2}\right)\frac{dx}{EI}$$

$$= \frac{1}{EI}\frac{wL^4}{8}$$

Equilibrium set

$$1\Delta = \int_0^L \frac{M_u M\,dx}{EI}$$

Compatibility set

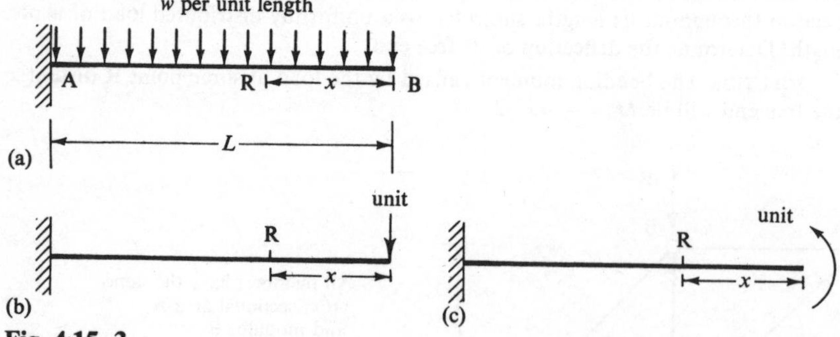

Fig. 4·15–3

The expression is positive and, therefore, the deflection corresponds with the direction of unit load and is downwards.

The case of rotational deflection is shown in Fig. 4.15–3(c) and here the bending moment at R caused by the appropriate unit load (which is a moment) is

$$M_U = +1$$

a positive sign being used because this bending moment is of a different sign from that caused at R by the applied distributed load. No problem of sign convention arises at this stage, but the two expressions for bending caused by the applied and unit loads must be consistent. Since the calculation of corresponding deflection arises essentially from the product of the two terms no convention relating moment to curvature is necessary, although the one used here is defined in Chapter 2.

$$\Delta = \int_0^L M_U M \frac{dx}{EI}$$

$$= \int_0^L (+1)\left(-\frac{wx^2}{2}\right)\frac{dx}{EI}$$

$$= -\frac{1}{EI}\frac{wL^3}{6}$$

showing that the rotation of the free end caused by the applied load does not correspond with that of Fig. 4.15–3(c) but is clockwise.

Example 4.15–4. If a point load is applied at the centre of a uniform elastic beam as shown in Fig. 4.15–4(a) determine the central deflection.

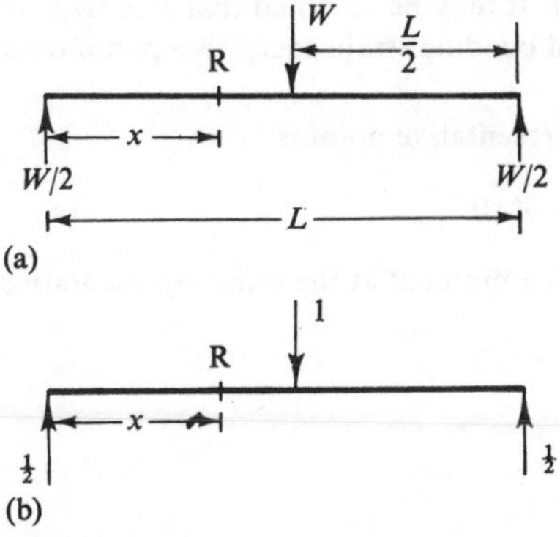

Fig. 4·15–4

SOLUTION

Equilibrium set

$$1\Delta = \int_0^L M_u M \frac{dx}{EI}$$

Compatibility set

The behaviour of the beam is symmetrical about a vertical centre line and in integration we can double the contribution made to deflection by half the beam. In this case the bending moment caused at R, a representative point in the beam,

$$M = \frac{W}{2}x$$

$$M_U = \frac{x}{2} \quad \text{(from Fig. 4.15–4(b))}$$

and

$$\Delta = \int M_U M \frac{dx}{EI}$$

$$= 2\int_0^{L/2} \left(\frac{x}{2}\right)\left(\frac{Wx}{2}\right)\frac{dx}{EI}$$

$$= \frac{2}{EI}\left[\frac{Wx^3}{12}\right]_0^{L/2}$$

$$= \frac{1}{EI}\frac{WL^3}{48}$$

It will be seen that the use of the obvious symmetry of the structure allows us to express the bending moment at R by a single expression, rather than to use a more cumbersome expression between $x = L/2$ and $x = L$ or alternatively, to find the total central deformation by summing the contributions made by the left-hand and right-hand portions separately.

Example 4.15–5. A bar of material of uniform cross-section is bent into an arc of a circle of radius R, subtending an angle at the centre of 90°. It is built into a rigid support at one end and supports a load of W at the other, as shown in Fig. 4.15–5. Determine the vertical and horizontal deflections of the free end. It may be assumed that R is large in relation to the cross-sectional dimensions and that bending strain energy is a paramount part of the total.

SOLUTION The bending moment at a representative point is

$$M = WR(1 - \cos\phi) \quad \text{(from Fig. 4.15–5(a))}$$

From Fig. 4.15–5(b) it can be seen that the bending moment at the same representative point due to unit vertical load is

$$M_{UV} = R(1 - \cos\phi)$$

and the vertical deflection

$$\delta_V = \int M_{UV}\frac{M}{EI}\,ds$$

$$= \int_0^{\pi/2} \frac{R(1 - \cos\phi)WR(1 - \cos\phi)R\,d\phi}{EI}$$

$$= \frac{WR^3}{EI}\int_0^{\pi/2}(1 - \cos\phi)^2\,d\phi$$

$$= \frac{WR^3}{EI}\left(\frac{3\pi}{4} - 2\right)$$

which corresponds to the direction of the unit load and is downwards.

From Fig. 4.15–5(c) the bending moment at a representative point due to unit horizontal load is

$$M_{UH} = -R\sin\phi$$

$$\boxed{1\delta_V = \int M_{UV}M\frac{ds}{EI}}$$

$$\boxed{1\delta_H = \int M_{UH}M\frac{ds}{EI}}$$

The horizontal deflection will be

$$\delta_H = \int M_{UH}M\frac{ds}{EI}$$

$$= \int_0^{\pi/2}(-R\sin\phi)WR(1 - \cos\phi)\frac{R\,d\phi}{EI}$$

$$= \frac{WR^3}{EI}\int_0^{\pi/2}(\sin\phi\cos\phi - \sin\phi)\,d\phi$$

$$= \frac{WR^3}{EI}\int_0^{\pi/2}\left(\frac{\sin 2\phi}{2} - \sin\phi\right)d\phi$$

$$= \frac{WR^3}{EI}\left(-\frac{1}{2}\right)$$

This deflection is opposed to the direction of unit load in Fig. 4.15–5(c) and the free end moves $WR^3/2EI$ to the left.

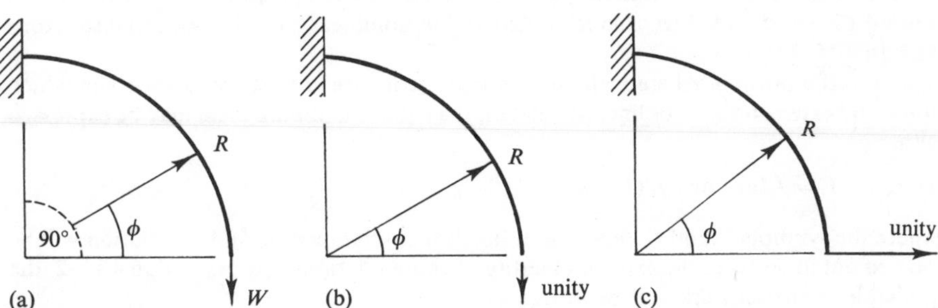

Fig. 4·15–5

(b) **Derivation of Mueller-Breslau's principle using the principle of virtual work.** In Section 4.13(a) it was shown that, in a linear elastic structure the deflected shape due to some particular unit distortion represented the influence line for the effect corresponding to that distortion. Proof of the principle depended on the use of the generalized (or Betti's) reciprocal theorem, but an elegant proof may be given using the principle of virtual work.

In Section 4.14 it has been demonstrated that there is an equality between the internal and external parts of the summed products of two corresponding systems, a force system in equilibrium and an arbitrarily chosen (but geometrically compatible) displacement system. Let the former consist of a unit load applied at any representative point. Obviously a reaction system will be produced thereby, but internal effects **F** will be produced at all points in the structure and would vary in their magnitude depending on the position of that unit load. The manner in which that variation occurs is usually represented by an influence line.

Now, let the displacement system be that caused by a small distortion (considered to be unity) of the structure corresponding to one particular effect, all other displacements at the same point being prevented.

In Eqn. 4.14–3

$$\mathbf{P}^T\overline{\mathbf{\Delta}} = \mathbf{F}^T\overline{\mathbf{e}}$$

we now have

Displacement system caused by a single unit internal distortion

$$\mathbf{P}^T\overline{\mathbf{\Delta}} = \mathbf{F}^T\overline{\mathbf{e}}$$

Force system caused by a single, unit, external load

from which

$$1.\Delta = F.1$$

$$F = \Delta$$

The magnitude of some internal effect, therefore, due to a single applied load of unity, is equal to the corresponding deflection of the point of application of the load, caused by a unit distortion corresponding to the internal effect. The argument is made in completely general terms and may be extended to all points in the structure, so that the deflected shape of the structure due to the unit distortion gives the influence line for the effect which corresponds.

(c) **Virtual work where elasticity may not be assumed.** Equation 4.15–2 (Section 4.15(a)) was completely general and, as was pointed out, is applicable to all structures whether a linear load–deflection relationship exists or not. Developments of that equation in Eqns 4.15–3 and 4.15–4 were so restricted and in nonlinear cases it is essential to return to Eqn 4.15–2.

If a pin-jointed statically determinate structure consists of members in which force and extension are not linearly related, the equation joining them may be expressed either as

$$F = f'(e) \quad \text{or} \quad f''(F) = e$$

where the symbols f' and f'' represent a functional relationship, and calculation will be carried out in terms of the explicit quantity of those relationships. As in Eqn 4.15–2, the principle of virtual work will be written as

Displacement system caused by real loads

$$1.\bar{\Delta} = \mathbf{F}_U^T\bar{\mathbf{e}}$$

Load system due to an applied load of unity where deflection is required

The forces F_U due to applied unit load in the structure may be calculated quite readily, and will be numerical only since the structure is statically determinate. The extensions of each member caused by the real loads may be calculated if

$$e = f''(F)$$

since e may be calculated for each member explicitly in terms of the force caused in it by the applied loads. If, however, the force in any member is expressed as a function of its extension then, since the force in it is known from the various equilibrium conditions in the structure, the extension must be calculated subsequently and this can only be done generally if the relationship is limited to the second degree and is of the form

$$F = Ce + De^2$$

in which C and D are constants since

$$e = -\frac{C}{2D} \pm \left(\frac{C^2}{4D^2} + \frac{F}{D}\right)^{\frac{1}{2}}$$

If the load–extension relationship involves higher powers of extension than the second, no convenient general formula exists and mathematical complexities make solution in this fashion impossible.

4.16 Applications of the principle of virtual work (2)

It has already been pointed out (in Sections 2.4 and 2.5) that structures may be indeterminate as a result of internal redundancy or arising from the support conditions, and it will be convenient to take these cases separately.

(a) **Indeterminacy due to the number of reactive constraints.** It has been shown that a fixed number of equilibrium equations exist, three in the case of a two-dimensional structure and six in the general case of three dimensions. Only this number of reactive constraints can be determined, therefore, from these equations alone, and, if a structure is supported in more complicated fashion, conditions of compatibility must arise when assessing the magnitudes of reactive elements. Examples are shown in Figs. 4.16–1(a) to (f). The first of these, the propped cantilever shown at (a), has three reactive constraints at the encastré (left-hand) end, in general, but, since the load shown in the sketch is

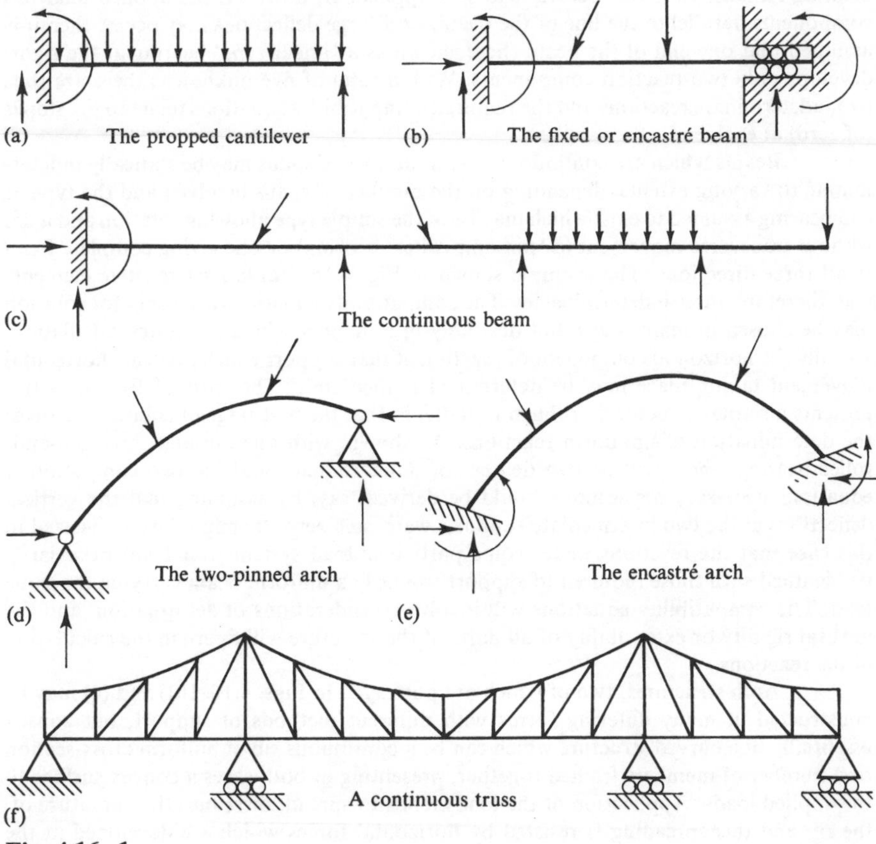

Fig. 4·16–1

purely vertical, no horizontal constraint will be developed at that end. There remain only two equations of equilibrium (vertical equilibrium and moment) to determine the three other effective constraints and the structure shown is statically indeterminate to the first degree (i.e. there is a surplus or redundant element of reaction) as it would be under a more general loading condition. In such a case the third equation of equilibrium (horizontal) would be used to establish immediately any horizontal component of reaction developed at the left-hand end, but the two remaining equilibrium equations would still be inadequate—of themselves—to complete the determination of the reactions. In all cases some recourse must be made to a compatibility equation defining the movement, if any, of the free end.

In Fig. 4.16–1(b) a load system containing both vertical and horizontal components is shown acting on a beam which is built in, encastré, or direction fixed (all three names will be encountered in various texts) at its ends. It is implied that, although loads are applied to the beam, the method of support is such as to prevent any rotation of the ends, which remain parallel to their unloaded conformation (horizontal as shown in the sketch). In addition to the normal transverse reactions (vertical in the case of vertical loads) moments, known as the fixing moments, must be developed at each end. It is usual to assume that any component of reaction parallel to the beam, i.e. horizontal in the diagram, is only developed at one end. This limitation is of no significance when vertical loads alone may be assumed to act on a horizontal beam, since it is usual to assume that any resulting deformations are small and that the projected length of any span after bending remains as it was before load was applied. If, however, the applied load has components parallel to the line of the beam, or if large deflections may occur, then it is assumed that one end of the beam (here shown as a 'linear-type' bearing at the right) develops only two reaction components. With a total of five unknowns there are thus two indeterminate reactions and the required compatibility equations relate to the slopes (of zero) at each end.

Beams which are continuous over a number of spans may be statically indeterminate to varying extents, depending on the number of spans involved and the type of end bearing assumed to exist which may be of the simple type allowing rotation and axial without transverse movement (a 'pin-and-roller' bearing) or one giving complete fixity in all three directions. The example shown in Fig. 4.16–1(c) has six reactive elements and, therefore, three indeterminacies. The compatibility equations necessary for solution may be chosen in many ways, but desirably in a manner which facilitates calculation. Usually the horizontal component of reaction at that support which prevents horizontal movement taking place may be determined immediately. The truss of Fig. 4.16–1(f) presents a similar structural problem initially, in that the first stage of solution involves the determination of unknown reactions. As shown, with one pin and three pin-and-roller bearings there will be two degrees of indeterminacy and the two compatibility equations necessary for solution could be derived, say, by assuming that the vertical deflections at the two intermediate supports were each zero. It should be emphasized in this case that the reactions under some particular load system would not necessarily be identical with those required to support similarly a uniform beam carrying the same load. The compatibility equations will involve considerations of deformation, and the flexural rigidity or extensibility of all parts of the structure will figure in the calculation of the reactions.

Arch structures, two of which are portrayed in Figs. 4.16–1(d) and (e) may be constructed in many differing forms with different methods of support, but consist essentially of a curved structure which can be a continuous rib of uniform cross-section or a number of members framed together, presenting in both cases a convex surface to the applied loads. Application of the loads tends to spread, or reduce the curvature of, the rib and this spreading is resisted by horizontal forces which are developed at the reactions, known in this case as the springings. In the two-pinned arch (Fig. 4.16–1(d)) horizontal and vertical components of reaction are produced at each springing giving one degree of indeterminacy, while the encastré arch, in which horizontal, vertical, and rotational movements are all prevented at each springing has three indeterminacies. To avoid the necessary calculation of these indeterminacies and to avoid the unknown stresses which would be developed in any indeterminate structure by unknown displacements of the foundations occurring as a result of load application, it is common to build arch-type structures with a pin allowing rotation to take place at some intermediate point in the span. Examples are shown in Figs. 4.16–2(a) and (b), where application of load produces both horizontal and vertical components of reaction at A and B, giving four unknowns—one more than can be determined by the usual three equations of equilibrium for the structure as a whole. However, the construction of a pin at C, which allows relative rotations of the two parts of the structure about that point, gives an additional equation of equilibrium, namely that of the bending moment at C,

$$M_C = 0$$

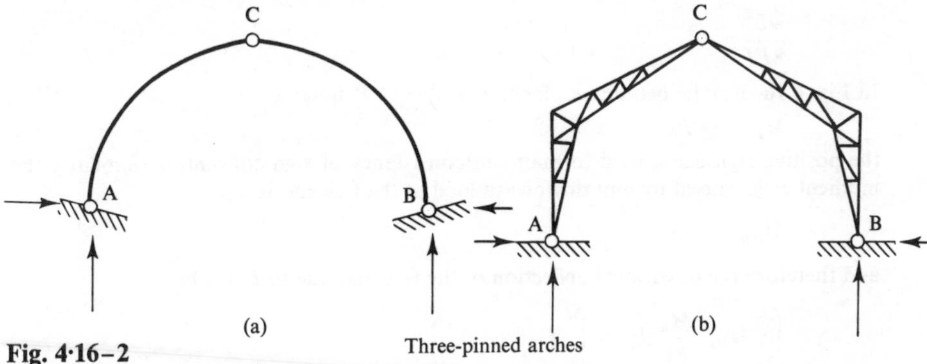

Fig. 4·16–2

thus allowing immediate solution to be made without recourse to compatibility considerations. The intermediate hinge, C, can be constructed at any convenient point and, although it is usual to ensure symmetry in a bridge span (Fig. 4.16–2(a)), the apex of the roof (whether it be symmetrically placed or not) is more usual in the workshop or hangar type of structure shown in Fig. 4.16–2(b).

Example 4.16–1. Propped cantilever under uniform load conditions. A cantilever of length L and constant flexural rigidity EI is propped at its free end to the same level as the built-in support. It carries a load of intensity w per unit length across the entire span as shown in Fig. 4.16–3. Determine the propping force.

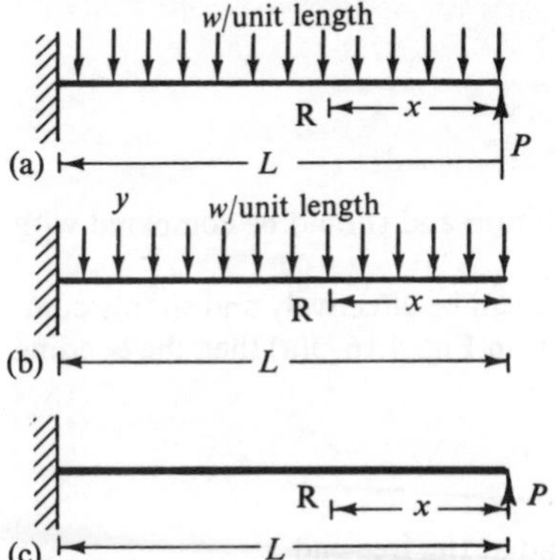

Fig. 4·16–3

SOLUTION

$$1\Delta = \int M_{UR}\frac{M_R}{EI}dx = \text{zero}$$

Solution may be effected by dividing the problem into primary and complementary parts, in which the effects of the load and the prop are treated separately in Figs. 4.16–3(b) and (c) respectively. The former has already been dealt with in Section 4.15 (in Example 4.15–3) and it has been shown that if bending stress predominates and the contribution made to the total strain energy in the structure by stresses other than bending can be neglected, then the downward deflection of the free end is

$$\frac{wL^4}{8EI}$$

In Fig. 4.16–3(c) the bending moment at R due to P alone is

$$M_R = +Px$$

the positive sign being used to maintain consistency of sign convention. Similarly the moment at R caused by unit downward load at the free end is

$$M_{UR} = -x$$

and therefore the downward deflection of the free end due to P will be

$$\int_0^L M_{UR} \frac{M_R}{EI} dx = \int_0^L (-x)(Px) \frac{dx}{EI}$$

$$= -\frac{PL^3}{3EI}$$

If the propped and encastré ends remain at the same level, the deflection of the free end will be zero and

$$\frac{wL^4}{8EI} - \frac{PL^3}{3EI} = 0 \qquad (4.16\text{–}1)$$

from which $P = \frac{3}{8}wL$ and the bending moment at R,

$$M_R = \frac{wx^2}{2} - Px$$

$$= \frac{wx^2}{2} - \frac{3wL}{8}x$$

$$= \frac{wLx}{2}\left(\frac{x}{L} - \frac{3}{4}\right)$$

It will be seen that this is a **flexibility** method of solution and should be compared with Section 4.6 and Fig. 4.6–3.

The primary and complementary solutions can be effectively and simply combined in one single operation. If we consider point R in Fig. 4.16–3(a) then the bending moment at R

$$M_R = -\frac{wx^2}{2} + Px$$

and the moment caused at R by downward unit load at the free end

$$M_{UR} = -x \quad \text{(from Fig. 4.16–3(c))}$$

The deflection corresponding to the unit load will be

$$\int_0^L M_{UR} M_R \frac{dx}{EI} = \int_0^L \left(-\frac{wx^2}{2} + Px\right)(-x)\frac{dx}{EI}$$

$$= \frac{wL^4}{8EI} - \frac{PL^3}{3EI} = 0 \qquad (4.16\text{–}2)$$

No difficulty would have been occasioned if the prop had been itself elastic and had compressed by an amount f due to unit load in it. The extent of this overall compression would be Pf and Eqn 4.16–2 would become

$$\frac{wL^4}{8EI} - \frac{PL^3}{3EI} = Pf$$

still giving explicit solution for P.

It will be noted that in none of these various methods of solution has it been stated explicitly that the method of support at the fixed end of the cantilever is such as to ensure that no vertical or rotational movements take place. Nevertheless, these restrictions have been implicitly made. The conditions of overall equilibrium will require that a vertical force and a couple exist at the point of fixity, but the development of the principle of virtual work in Eqn 4.14–2 has assumed that reactive components do no work. The reactive elements, therefore, can have no corresponding displacement.

Example 4.16–2. Propped cantilever with central point load. The same uniform elastic cantilever is subjected to a point load of W, applied midway between the encastré end and the prop, as shown in Fig. 4.16–4. What is the propping force?

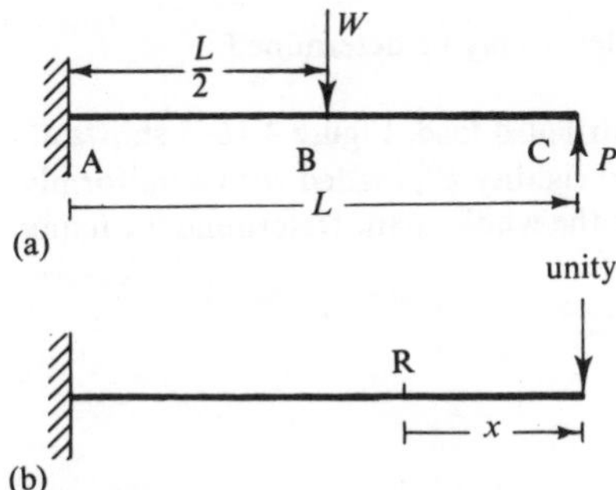

Fig. 4·16–4

SOLUTION

$$1\Delta = \int M_{UR}\frac{M_R}{EI}dx = \text{zero}$$

The expression

$$\int \frac{M_{UR}M_R}{EI}dx$$

is summed for the various parts of the beam in which differing expressions for bending moment exist, and equated to zero.

Between B and C, $0 < x < L/2$

$$M_R = +Px$$

$M_{UR} = -x$ (due to unit downward load at the free end)

Between A and B, $L/2 < x < L$

$$M_R = -W\left(x - \frac{L}{2}\right) + Px$$

$$M_{UR} = -x$$

Then

$$\int M_{UR} M_R \frac{dx}{EI} = \int_0^{L/2} (-x)(+Px)\frac{dx}{EI} + \int_{L/2}^{L} (-x)\left[-W\left(x - \frac{L}{2}\right) + Px\right]\frac{dx}{EI}$$

$$= -\frac{PL^3}{3EI} + \frac{W}{EI}\left(\frac{x^3}{3} - \frac{x^2 L}{4}\right)_{L/2}^{L}$$

$$= -\frac{PL^3}{3EI} + \frac{5WL^3}{48EI}$$

If the prop is incompressible and supports the free end of the cantilever to the same level as the encastré end

$$P = \frac{5W}{16}$$

and the bending moment at all points in the cantilever may be determined.

Example 4.16–3. Encastré beam with uniformly distributed load. Figure 4.16–5 shows an elastic fixed beam of length L and uniform flexural rigidity EI, loaded with a uniformly distributed load of intensity w per unit length over the whole span. Determine the fixing moments at the encastré ends.

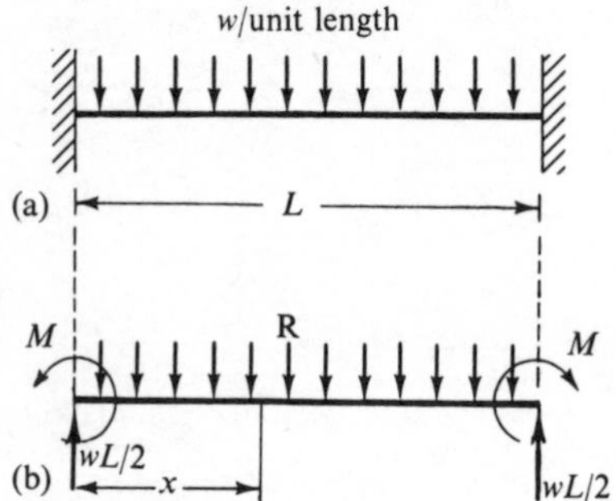

Fig. 4·16–5

SOLUTION

$$1\Delta = \int M_{UR} M_R \frac{dx}{EI} = \text{zero}$$

The general problem, as explained earlier has two statical indeterminacies, but since the structure is symmetrical about a vertical centre line it is obvious that the fixing moment at each end will be equal in magnitude and that both vertical reactions will be $wL/2$. The single unknown then is the fixing moment M, and the bending moment at some representative point R, distant x from the left-hand end

$$M_R = \frac{wL}{2}x - \frac{wx^2}{2} - M$$

The bending moment caused at R by unit couple at the left-hand end, corresponding to M is

$$M_{UR} = -1$$

Hence

$$\int_0^L M_{UR} M_R \frac{dx}{EI} = 2\int_0^{L/2} M_{UR} M_R \frac{dx}{EI}$$

$$= \frac{2}{EI}\int_0^{L/2}(-1)\left(\frac{wL}{2}x - \frac{wx^2}{2} - M\right)dx$$

$$= \frac{2}{EI}\left(\frac{ML}{2} - \frac{wL^3}{24}\right)$$

Since this expression must equal zero for there is no rotation of the end corresponding to the unit couple

$$M = \frac{wL^2}{12} \tag{4.16–3}$$

Example 4.16–4(a). Encastré beam with central point load. Determine the fixing moments caused in the same uniform encastré beam of length L by a central point load of W as shown in Fig. 4.16–6(a).

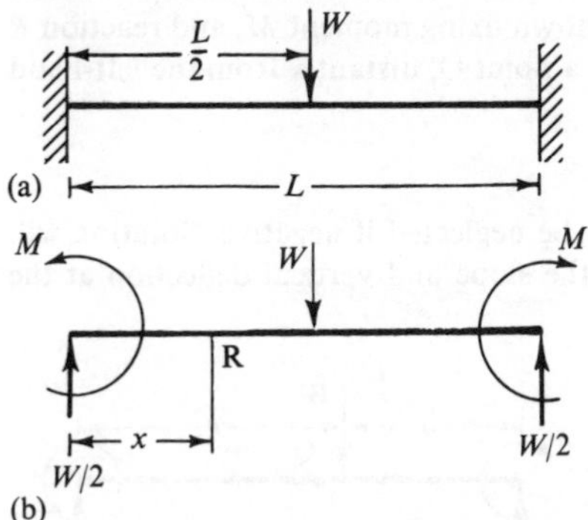

Fig. 4·16–6

SOLUTION

$$1\Delta = \int M_{UR}M_R\frac{dx}{EI} = \text{zero}$$

Again symmetry dictates that the vertical reactions at each end are $W/2$ upwards and that the fixing moments will be equal in magnitude. If we treat the moment at the left-hand end as being the only unknown, as shown in Fig. 4.16–6(b), then the moment at a representative point R will be

$$M_R = \frac{W}{2}x - M \quad \text{if} \quad 0 < x < \frac{L}{2}$$

and the moment at R caused by unit couple at the left-hand end

$$M_{UR} = -1$$

from which

$$\int M_{UR}\ M_R\ \frac{dx}{EI} = 2\int_0^{L/2}(-1)\left(\frac{Wx}{2} - M\right)\frac{dx}{EI}$$

$$= \frac{2}{EI}\int_0^{L/2}\left(M - \frac{Wx}{2}\right)dx$$

$$= \frac{2}{EI}\left(\frac{ML}{2} - \frac{WL^2}{16}\right)$$

and if the end is completely encastré so that no rotation takes place

$$M = \frac{WL}{8} \tag{4.16–4}$$

Example 4.16–4(b). Encastré beam with unsymmetrical point load. Determine the fixing moments when the same uniform encastré beam carries an unsymmetrical load as shown in Fig. 4.16–7(a).

SOLUTION Since the applied load is solely vertical, no horizontal component of reaction will be developed and the beam will be in equilibrium with four components of reaction as shown in Fig. 4.16–7(b). Only two equations of equilibrium remain to allow a solution to be made and, since no considerations of symmetry permit the reduction of the number of two unknowns, compatibility equations, solved simultaneously, must be used. It is convenient to choose these as the unknown fixing moment M, and reaction R at the left-hand end. Concentrating attention on a point Q, distant x from the left-hand end then

$$M_Q = -M + Rx - W[x - a]$$

where the term in square brackets $[x - a]$ may be neglected if negative. Solution will follow from the two expressions which equate the slope and vertical deflection at the left-hand end to zero. Considering these in turn

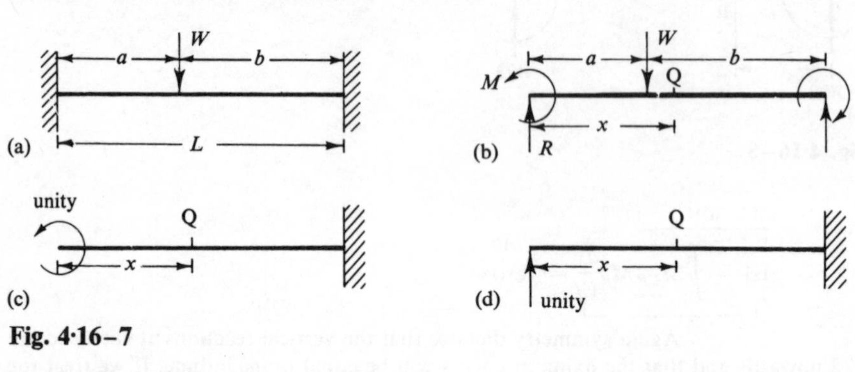

Fig. 4·16–7

$$1\Delta = \int M_U \frac{M_Q dx}{EI} = \text{zero}$$

$$1\theta = \int M_U \frac{M_Q}{EI} dx = \text{zero}$$

Solve simultaneously

Slope at the left-hand end

$$M_U \text{ (from Fig. 4.16–7(c))} = -1$$

$0 < x < a$,

$$\int_0^a M_U M_Q \frac{dx}{EI} = \int_0^a (-1)(-M + Rx)\frac{dx}{EI} = \frac{1}{EI}\left[Mx - \frac{Rx^2}{2}\right]_0^a$$

$$= \frac{1}{EI}\left(Ma - \frac{Ra^2}{2}\right) \tag{4.16–5}$$

$a < x < L,$

$$\int_a^L M_U M_Q \frac{dx}{EI} = \int_a^L (-1)(-M + Rx - W[x - a]) \frac{dx}{EI}$$

$$= \frac{1}{EI}\left(Mx - \frac{Rx^2}{2} + \frac{W}{2}[x - a]^2\right)_a^L$$

$$= \frac{1}{EI}\left[ML - \frac{RL^2}{2} + \frac{W}{2}(L - a)^2 - Ma + \frac{Ra^2}{2}\right] \quad (4.16\text{–}6)$$

Summing Eqns 4.16–5 and 4.16–6 for the whole beam and equating to zero

$$\frac{1}{EI}\left(ML - \frac{RL^2}{2} + \frac{W}{2}[L - a]^2\right) = 0 \quad (4.16\text{–}7)$$

Now considering vertical upwards deflection at the left-hand end

M_U (from Fig. 4.16–7(d)) $= +x$

$0 < x < a,$

$$\int_0^a M_U \, M_Q \frac{dx}{EI} = \int_0^a (-M + Rx)(+x) \frac{dx}{EI}$$

$$= \frac{1}{EI}\left[-\frac{Mx^2}{2} + \frac{Rx^3}{3}\right]_0^a$$

$$= \frac{1}{EI}\left(-\frac{Ma^2}{2} + \frac{Ra^3}{3}\right) \quad (4.16\text{–}8)$$

$a < x < L,$

$$\int_a^L M_U M_Q \frac{dx}{EI} = \int_a^L (-M + Rx - W[x - a])(+x) \frac{dx}{EI}$$

$$= \frac{1}{EI}\left[-\frac{Mx^2}{2} + \frac{Rx^3}{3} - \frac{Wx^3}{3} + \frac{Wax^2}{2}\right]_a^L$$

$$= \frac{1}{EI}\left(-\frac{ML^2}{2} + \frac{RL^3}{3} + \frac{Ma^2}{2} - \frac{Ra^3}{3} - \frac{WL^3}{3} - \frac{Wa^3}{6} + \frac{WaL^2}{2}\right) \quad (4.16\text{–}9)$$

Summing Eqns 4.16–8 and 4.16–9 for the whole beam and again equating to zero

$$\frac{1}{EI}\left(-\frac{ML^2}{2} + \frac{RL^3}{3} - \frac{WL^3}{3} - \frac{Wa^3}{6} + \frac{WaL^2}{2}\right) = 0 \quad (4.16\text{–}10)$$

Equations 4.16–7 and 4.16–10 must now be solved simultaneously. Since both equate to zero, the term involving the flexural rigidity may be cancelled from both and, simplifying

$$2ML - RL^2 + WL^2 - 2WaL + Wa^2 = 0$$

$$-3ML^2 + 2RL^3 - 2WL^3 + 3WaL^2 - Wa^3 = 0$$

and solving

$$\left.\begin{aligned} M &= \frac{Wa(L-a)^2}{L^2} = \frac{Wab^2}{L^2} \\ R &= \frac{Wb^2(b+3a)}{L^3} \end{aligned}\right\} \tag{4.16–11}$$

Use of the normal equations of equilibrium will now show that at the right-hand end the fixing moment and the reaction are, respectively

$$\frac{Wba^2}{L^2} \quad \text{and} \quad \frac{Wa^2(a+3b)}{L^3} \tag{4.16–12}$$

Example 4.16–5. Encastré beam with linear variation of load. The formulae developed in the last subsection can helpfully be used in more general situations of a uniformly distributed load covering part of the span, or of a distributed load which is not uniformly spread, but arranged in some fashion which can be conveniently expressed in mathematical (and integrable) terms.

In Fig. 4.16–8(a), for example, the same uniform encastré beam is shown loaded by a total W, which is distributed with an intensity varying linearly from zero at the left-hand end to $2W/L$ at the right.

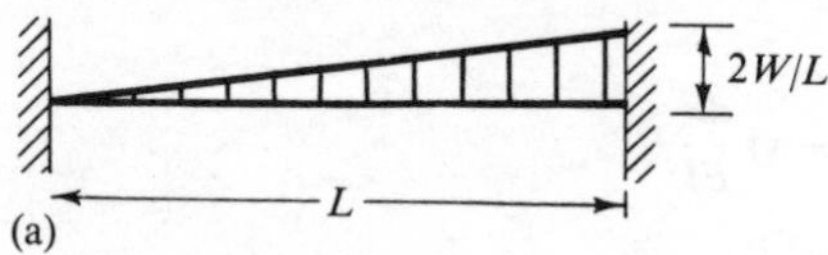

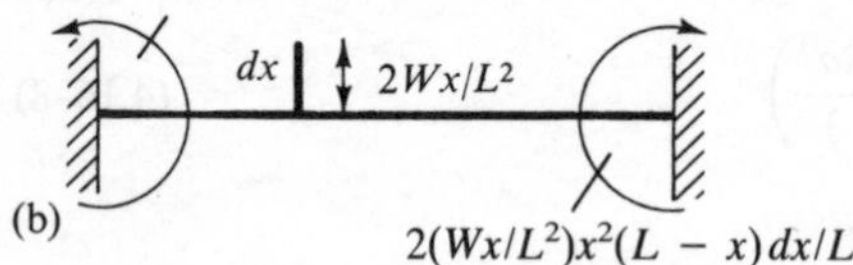

Fig. 4·16–8

SOLUTION At a point distant x from the left-hand end the intensity (Fig. 4.16–8(b)) will be $2Wx/L^2$ and the fixing moments caused by a small portion of the load extending over a length dx (such that the intensity might there be regarded as constant) will be at the left-hand end,

$$\frac{2Wx}{L^2}\cdot\frac{x(L-x)^2}{L^2}\,dx$$

and at the right-hand end,

$$\frac{2Wx}{L^2}(L-x)\frac{x^2}{L^2}\,dx$$

from Eqns 4.16–11 and 4.16–12. The total effects of the load will therefore be at the left-hand end,

$$\int_0^L \frac{2Wx^2(L-x)^2}{L^4}\,dx = \frac{WL}{15}$$

and at the right-hand end, (4.16–13)

$$\int_0^L \frac{2Wx^3(L-x)}{L^4}\,dx = \frac{WL}{10}$$

Clearly the limits of integration would be adjusted to conform to any particular load system.

Example 4.16–6. Uniform encastré beam with relative sinkage of supports. If in the same uniform encastré beam there is a sinkage of the right-hand end relative to the left (i.e. a clockwise relative movement of the two ends) without any rotation at either of the two fixings, as shown in Fig. 4.16–9 there will be a moment and reaction developed at each end Determine these reactive elements.

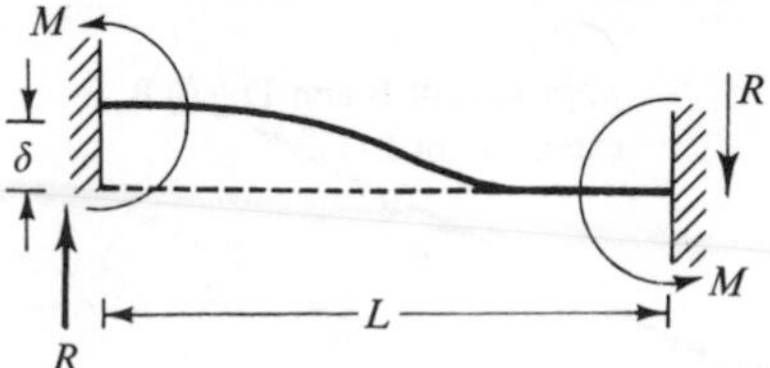

Fig. 4·16–9

$$1\theta = \int M_{UM}\frac{M_R}{EI}dx = \text{zero}$$

$$1\delta = \int M_{UR}\frac{M_R}{EI}dx = \delta$$

Solve simultaneously

SOLUTION From considerations of equilibrium these will be of equal magnitude and opposite sign. The bending moment at some representative point distant x from the left-hand end will be

$$M_R = Rx - M$$

The moment caused by unit couple at the left-hand end will be

$$M_{UM} = -1$$

and that by unit vertical upwards reaction at the left-hand end will be

$$M_{UR} = x$$

so that

$$\int M_R M_{UM}\frac{dx}{EI} = \int_0^L (Rx - M)(-1)\frac{dx}{EI} = 0$$

$$\int M_R M_{UR}\frac{dx}{EI} = \int_0^L (Rx - M)(x)\frac{dx}{EI} = \delta$$

Completing the integration

$$\frac{1}{EI}\left(ML - \frac{RL^2}{2}\right) = 0$$

$$\frac{1}{EI}\left(-\frac{ML^2}{2} + \frac{RL^3}{3}\right) = \delta$$

from which

$$M = \frac{6EI\delta}{L^2} \quad \text{and} \quad R = \frac{12EI\delta}{L^3} \qquad (4.16\text{–}14)$$

Reactive elements of the same magnitude but in the opposite directions will be developed at the right-hand end.

(b) **Indeterminate pin-jointed elastic structure.** If a pin-jointed structure is adequately braced the position of each node is defined in relation to the others and (to take a simple example) in Fig. 4.16–10(a) the distance between nodes B and D is fixed. If, now, an additional member is inserted between these two nodes it must either fit precisely into place or, since this is difficult to ensure, its insertion will cause forces in the member BD

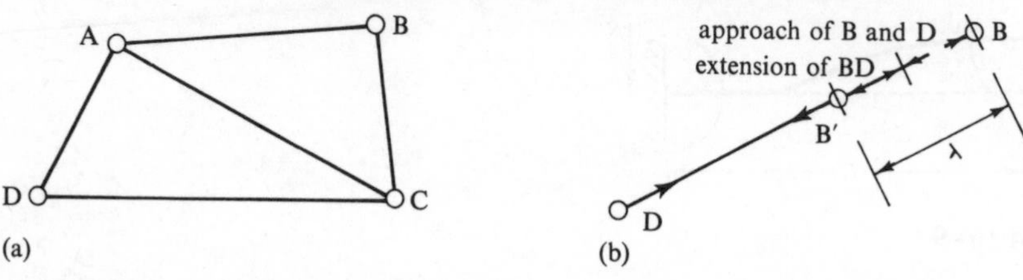

Fig. 4·16–10

and in the other members of the structure to give compatibility. This will be achieved by causing simultaneously an extension (or compression) of the member BD and a reduction (or increase) of the distance between nodes B and D which together equal the initial difference in length, or lack of fit, as shown in Fig. 4.16–10(b). This lack of fit causes forces in the members irrespective of the application of load to the structure, and, plainly, the effect of loads must be compounded with the self-straining occasioned by a redundant member. This can conveniently be done by a single method of solution.

If an indeterminate pin-jointed structure, shown in Fig. 4.16–11(a) is acted upon by external loads W, the method of analysis must take account of the extension of

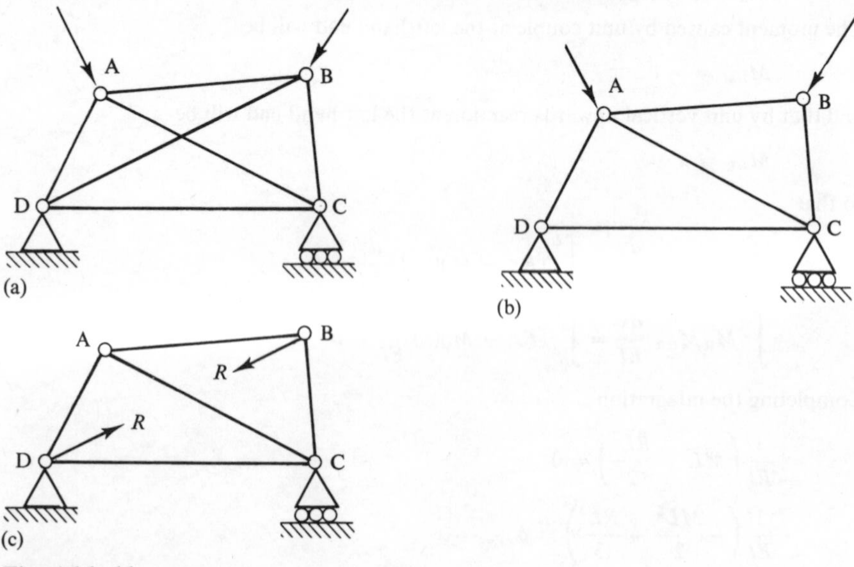

Fig. 4·16–11

the members. There will be some freedom in the choice of that member styled as the indeterminacy (in Fig. 4.16–11(a) any one could be so chosen), but it will be assumed to be BD and that the force caused in it, by the applied loads W, is R. In any other member the force will be

$$F = KW + F_{UR}R \tag{4.16–14}$$

where K is some coefficient dependent only on the geometry of the structure and the position and magnitude of the loads W which can best be determined by assuming BD does not exist and the remaining (determinate) structure alone carries the loads as shown in Fig. 4.16–11(b). Similarly, F_{UR} will also be a coefficient, dependent on the geometry, obtained by neglecting the applied loads and calculating the forces caused in the remaining determinate structure by forces R applied at B and D as in Fig. 4.16–11(c). It will be noted that the expression of Eqn 4.16–14 also applies to the member BD, where K is zero and F_{UR} is unity, giving

$$F_{BD} = R \quad \text{(as assumed)}$$

Now in the original structure of Fig. 4.16–11(a) the principle of virtual work can be applied and

$$\mathbf{P}^T\mathbf{\Delta} = \mathbf{F}^T\mathbf{e}$$

as developed in Eqn 4.14–2. If the deformation system, characterized by $\mathbf{\Delta}$ and $\mathbf{e}$, is chosen to represent the deflected structure, and if the members of the structure are linearly elastic then for each member

$$e = \frac{FL}{AE} = \frac{L}{AE}(KW + F_{UR}R)$$

in which L, A, and E are the length, cross-sectional area and modulus respectively. If the force system, $\mathbf{P}$ and $\mathbf{F}$, is that occasioned by unit tensile force in the member BD then the corresponding deformation will be λ, the initial lack of fit (too short) in that member and the forces in other members will be F_{UR} so that

$$\begin{aligned} 1.\lambda &= \sum F_{UR}\frac{(KW + F_{UR}R)}{AE}L \\ &= \sum F_{UR}\frac{KW}{AE}L + \sum F_{UR}\frac{F_{UR}}{AE}RL \end{aligned} \tag{4.16–15}$$

It will be noted that Eqn 4.16–15 gives representative expression to the compatibility requirement that the initial lack of fit is equal to the sum of the approach of B and D due to the applied loads, the approach of B and D due to forces R, and the extension of member BD, for, in that member, where F_{UR} is unity, the final term of the equation is RL/AE. It should be compared with Fig. 4.6–2 and Eqn 4.6–1.

Example 4.16–7. The truss shown in Fig. 4.16–12 has 14 members each of the same cross-sectional area of 5000 mm², and modulus 200 kN/mm², pinned at the lettered nodes. Vertical loads of 30 kN are applied at F and G as shown. Determine the forces in all members.

SOLUTION The most convenient lay-out is a tabular one, as for deflections in pin-jointed structures. Care must be taken to include all the members—the indeterminate is

$$\boxed{1\lambda} = \sum F_U(KW + F_UR)\frac{L}{AE} = \text{zero}$$

often forgotten! In this case, with eight nodes, there is one such member which could be chosen anywhere in the panel BCGF, in which the redundancy occurs. We have assumed it to be the member BG, which is removed from Figs. 4.16–12(b) and (c). Columns 1 and 2

Member (*1*)	*Length* (m) (*2*)	$KW + F_U R$ (*3*)	F_U (*4*)	$F_U(KW + F_U R)L$ (*5*)	*Force* (kN) (*6*)
AB	3	0			0
BC	3	$30 - \frac{\sqrt{2}}{2}R$	$-\frac{\sqrt{2}}{2}$	$\frac{\sqrt{2}}{2}\left(-90 + 3\frac{\sqrt{2}}{2}R\right)$	36.2
CD	3	0			0
EF	3	-30			-30
FG	3	$-30 - \frac{\sqrt{2}}{2}R$	$-\frac{\sqrt{2}}{2}$	$\frac{\sqrt{2}}{2}\left(90 + 3\frac{\sqrt{2}}{2}R\right)$	-23.8
GH	3	-30			-30
AE	3	-30			-30
EB	$3\sqrt{2}$	$30\sqrt{2}$			42.4
BF	3	$-30 - \frac{\sqrt{2}}{2}R$	$-\frac{\sqrt{2}}{2}$	$\frac{\sqrt{2}}{2}\left(90 + 3\frac{\sqrt{2}}{2}R\right)$	-23.8
FC	$3\sqrt{2}$	$0 + 1.R$	1	$0 + 3\sqrt{(2)}R$	-8.8
CG	3	$-30 - \frac{\sqrt{2}}{2}R$	$-\frac{\sqrt{2}}{2}$	$\frac{\sqrt{2}}{2}\left(90 + 3\frac{\sqrt{2}}{2}R\right)$	-23.8
GB	$3\sqrt{2}$	$+1.R$	1	$3\sqrt{(2)}R$	-8.8
CH	$3\sqrt{2}$	$30\sqrt{2}$			42.4
HD	3	-30			-30
				$\frac{\sqrt{2}}{2}\left(180 + 12\frac{\sqrt{2}}{2}R\right) + 6\sqrt{(2)}R$	

merely record the designation of the member, with its length. In column 3 the force in any member is recorded in two parts, the first (KW) coming from calculations based on Fig. 4.16–12(b) and the second ($F_U R$) from Fig. 4.16–12(c). In the former it will be noted that, by symmetry, the reactions at A and D are both vertical and equal to 30 kN and that there is zero force in member *FC* owing to the fact that no shear is carried in that panel. In the calculations of ($F_U R$) it may be seen that the effects of the redundancy are contained within the panel BCGF. Summation of $F_U(KW + F_U R)L$ in column 5 only involves six members, but that sum may be equated to zero, since there is no initial lack of fit, thus implying that the numerical value of the modulus does not affect the answer.

$$\frac{\sqrt{2}}{2}\left(180 + 12\frac{\sqrt{2}}{2}R\right) + 6\sqrt{(2)}R = 0$$

from which $R = -8.8$ kN.

It will be noted that the units of force are dictated by the choice of units used for KW in column 3 and hence R is in kilonewtons. Knowing this value the final column (6) giving the force in each member may be obtained by substitution in column 3.

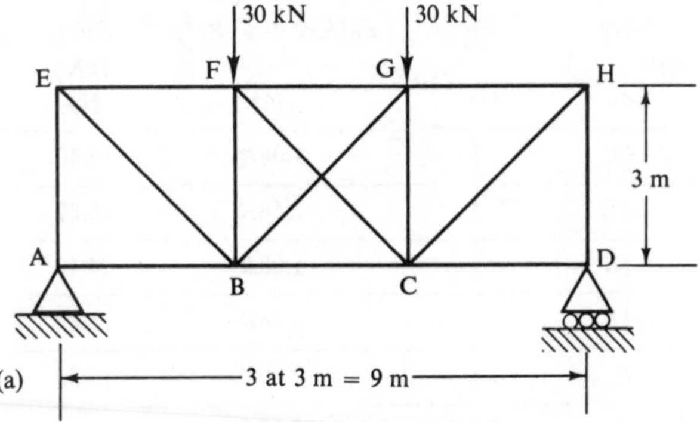

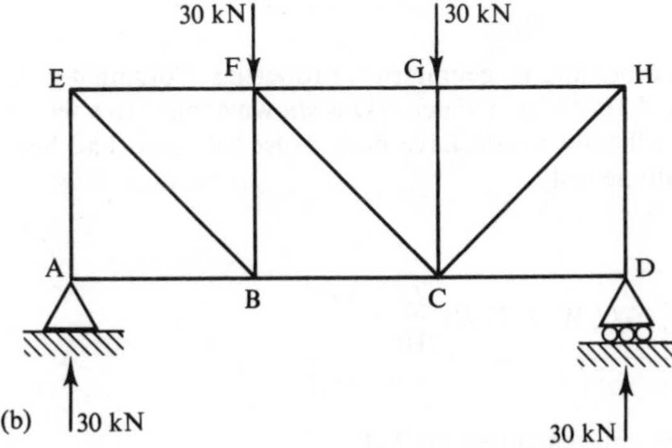

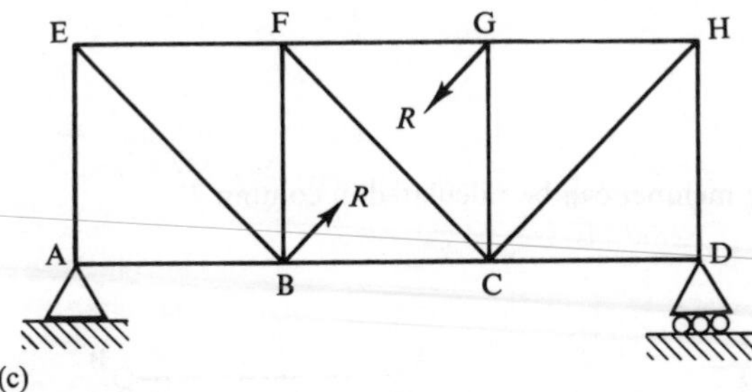

Fig. 4·16–12

Example 4.16–8. The frame of Fig. 4.16–13 is made of steel of modulus 200 kN/mm². Diagonals are of 500 mm² cross-sectional area, but other members of area 1 000 mm². Member AD is 1 mm too short but is forced into place. Find the force in each member.

SOLUTION In this case the chosen redundancy must be that with the known lack of fit, and the direction of force in it arranged to correspond with the lack of fit. AD was too short and must therefore extend.

Member	*Area* (mm²)	*Length* (mm)	*Force* $KW + F_U R$	F_U	$F_U(KW + F_U R)\frac{L}{A}$	*Force* (kN)
(1)	(2)	(3)	(4)	(5)	(6)	(7)
AB	1000	3000	$-\frac{3}{5}R$	$-\frac{3}{5}$	$1.08R$	−4.40
BD	1000	4000	$-\frac{4}{5}R$	$-\frac{4}{5}$	$2.56R$	−5.87
DC	1000	3000	$-\frac{3}{5}R$	$-\frac{3}{5}$	$1.08R$	−4.40
CA	1000	4000	$-\frac{4}{5}R$	$-\frac{4}{5}$	$2.56R$	−5.87
CB	500	5000	$+R$	$+1$	$10.00R$	7.33
AD	500	5000	$+R$	$+1$	$10.00R$	7.33
					$27.28R$	

Columns 1, 2, and 3 define the member and its geometrical properties. Column 4 gives the force in it, calculated from Fig. 4.16–13(b) in which AD is shown subject to a tensile force, R. It will be seen that no difficulty would have been caused if loads had been applied (causing forces KW) simultaneously.

$$1\lambda = \sum F_U(KW + F_U R)\frac{L}{AE} = \lambda$$

Summing from column 6, and remembering that

$$\frac{F_U(KW + F_U R)L}{AE} = \lambda$$

$$\frac{27.28\,R}{200} = 1$$

$$R = 7.33 \text{ kN}$$

From this the actual force in each member can be calculated in column 7.

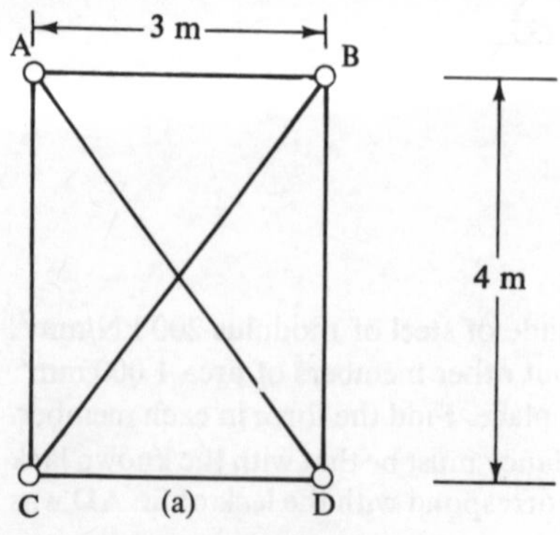

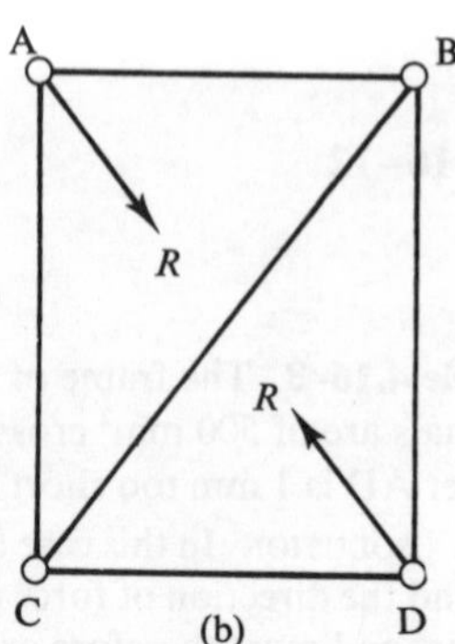

Fig. 4·16–13

The frame in this example is quite arbitrarily chosen for numerical convenience rather than for any practical significance, but the student should note the large forces involved as a result of a small lack of fit.

If there is more than one degree of indeterminacy

If the pin-jointed structure has two more members than are necessary to brace the structure adequately there will be two members which could each produce self-straining due to an initial lack of fit, and two compatibility equations will determine the magnitudes of the forces involved.

The frame of Fig. 4.16–14(a) will be found indeterminate to the second degree and, although any member in each of the panels ABEF and BCDE could be chosen as indeterminate it has been assumed to be AE and CE in which there are forces R_1 and R_2. Forces for all other members may be obtained by summing the effects of the loads (Fig. 4.16–14(b)), of R_1 (Fig. 4.16–14(c)) and of R_2 from Fig. 4.16–14(d) to give

$$F = KW + F_{UR_1} R_1 + F_{UR_2} R_2$$

If λ_1, and λ_2 are the amounts by which AE and CE respectively are too short to fit between their nodal points without straining, then successive application of the virtual work principle will show that

$$\lambda_1 = \sum F_{UR_1} \frac{(KW + F_{UR_1} R_1 + F_{UR_2} R_2)L}{AE}$$

and

$$\lambda_2 = \sum F_{UR_2} \frac{(KW + F_{UR_1} R_1 + F_{UR_2} R_2)L}{AE}$$

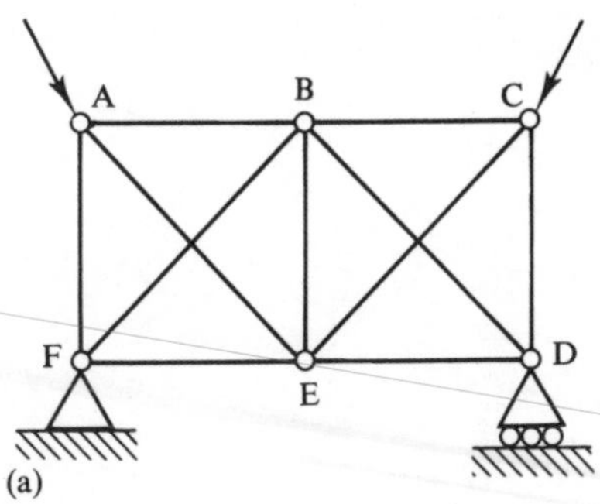

(a)

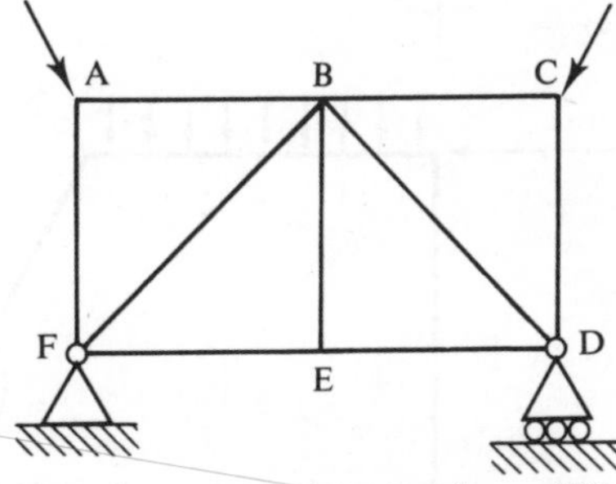

Force in a representative member $= KW$

(b)

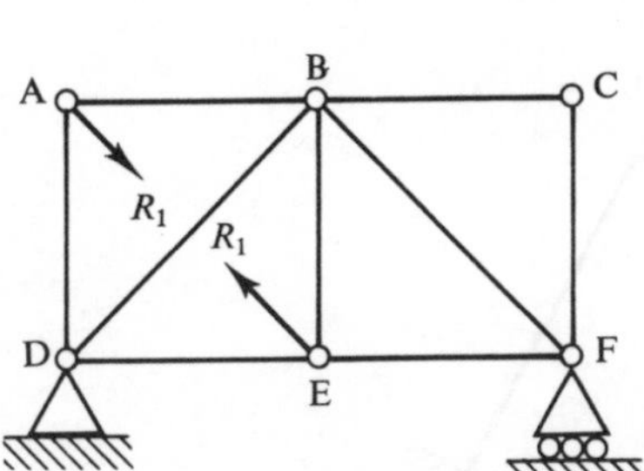

Force in a representative member $= F_{UR_1} R_1$

(c)

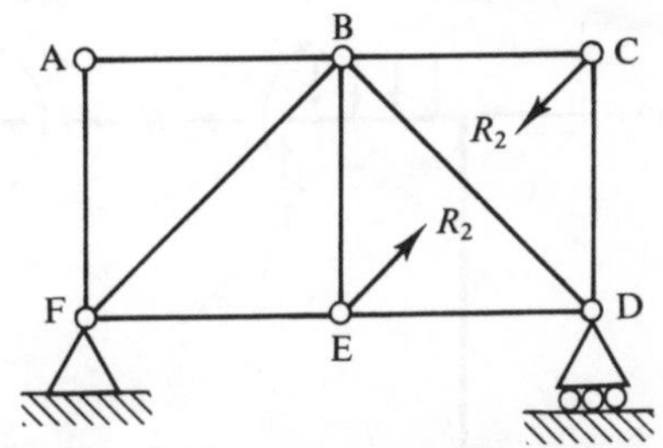

Force in a representative member $= F_{UR_2} R_2$

(d)

Fig. 4·16–14

Both these expressions contain the two unknowns, and solution must necessarily involve simultaneous consideration of the two equations.

(c) **Rigid-jointed frame.** Increasingly, simple and more complex frames are manufactured in steel or concrete with rigid joints capable of transmitting moments from members on one side to the other. This can lead to a stiffer structure for a given weight of material, but increases the complexity of analysis and the need to ensure both a high quality of workmanship (see Example 4.16–8) and well-prepared foundations (see Example 4.16–6).

The frame shown in Fig. 4.16–15(a) having six reactive elements is indeterminate to the third degree, and there is considerable freedom in the choice of the unknowns. Usually in such a case these will be taken as occurring at one of the points of support, as in Fig. 4.16–15(b), or at some convenient intermediate point as in Fig. 4.16–15(c). In both cases there will be no movement of the points of application of the unknowns, relative to the foundation (Fig. 4.16–15(b)) or other parts of the structure, and solution may follow the application of three equations similar to Eqn 4.15–4 in

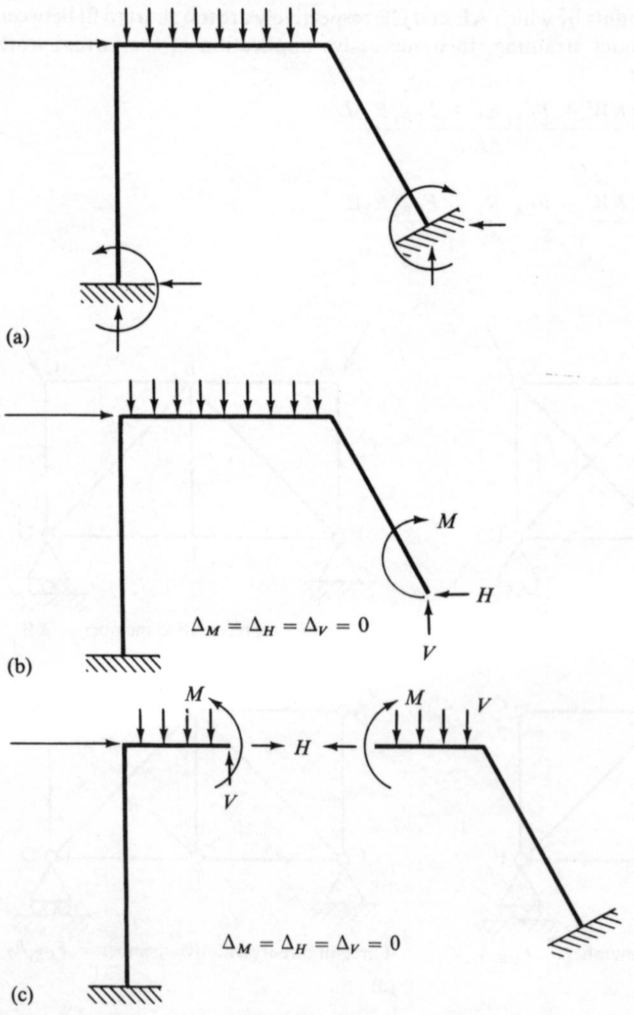

Fig. 4·16–15

which the left-hand side (the corresponding deflection) equals zero. Usually in civil engineering structures it will be found that the contributions made to the right-hand sides by terms involving direct force and shear are small in relation to that by bending moment, and it is quite common to use no other. The assumption is warranted when members are long and slender, but as these become increasingly stiff and stocky, possible inaccuracies become progressively greater.

Example 4.16–9. The single-bay portal frame ABCDE, of Fig. 4.16–16 is of constant cross-section throughout, and pinned to rigid foundations at A and E. It is loaded along the beam with 10 kN centrally placed and a uniformly distributed load of 10 kN/m. Determine the bending moments at B and D.

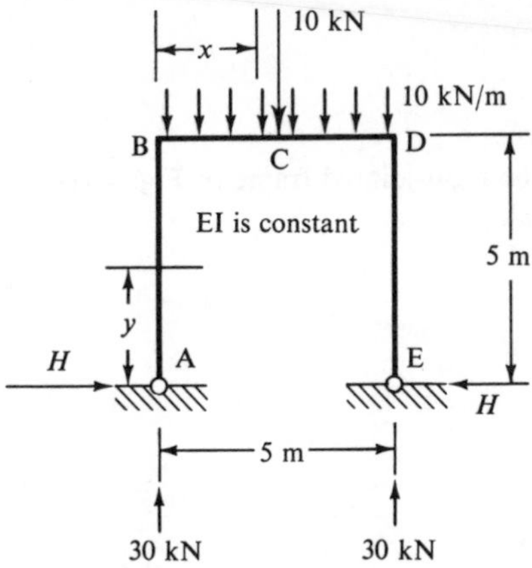

Fig. 4·16–16

SOLUTION Since the frame and load are both symmetrical the vertical components of reaction at each support will be 30 kN. The horizontal components will be equal in magnitude and opposite in direction since no other horizontal forces act on the frame, which is thus indeterminate to the first degree. H may conveniently be found, using bending terms only, from Eqn. 4.15–4, which may be written

$$\Sigma \frac{1}{EI} \int M M_U \, ds = 0$$

Component effects from the various members may be summed.

Member AB

Considering a representative point distance y above A

$$M = -Hy \qquad M_U = -y$$

$$\frac{1}{EI} \int M M_U \, ds = \frac{1}{EI} \int_0^5 H y^2 \, dy = \frac{125H}{3EI}$$

Member BC

Again, considering a representative point distance x from B, $0 < x < 2\frac{1}{2}$,

$$M = 30.x - 5H - 10\frac{x^2}{2} \quad M_U = -5$$

$$\frac{1}{EI} \int M M_U \, ds = \frac{1}{EI} \int_0^{2\frac{1}{2}} \left(-150x + 25H + 50\frac{x^2}{2} \right) dx$$

$$= \frac{1}{EI}\left[-150\frac{x^2}{2} + 25Hx + 50\frac{x^3}{6}\right]_0^{2\frac{1}{2}}$$

$$= \frac{1}{EI}(-338.8 + 62.5H)$$

Summing the components from AB and BC and equating to zero (strictly the sum should be doubled since the frame is symmetrical, but we are equating to zero),

$$0 = -338.8H + H(62.5 + 41.7)$$

$$H = \frac{338.8}{104.2} = 3.25 \text{ kN}$$

from which the bending moments at B and D are

$$5H = 16.25 \text{ kN m}$$

Example 4.16–10. Determine the reactions in the rigid-jointed frame of Fig. 4.16–17 The flexural rigidities of all members are equal.

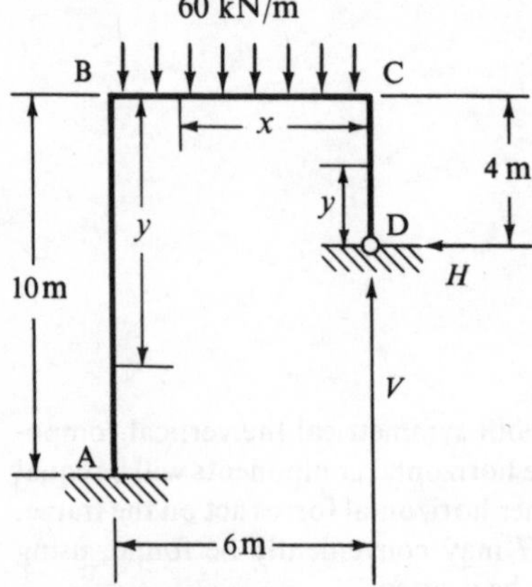

Fig. 4·16–17

SOLUTION With three possible components of reaction at A and two at D, the frame is indeterminate to the second degree. If the two unknowns are assumed to be H and V at the pin D, then solution may be obtained from the simultaneous equations:

$$1\delta_H = \int M_{UH}\frac{M}{EI}ds = 0$$

$$1\delta_v = \int M_{UV}\frac{M}{EI}ds = 0$$

Solve simultaneously

$$\Sigma\frac{1}{EI}\int MM_{UV}\,ds = 0$$

$$\Sigma\frac{1}{EI}\int MM_{UH}\,ds = 0$$

Taking the members in turn:

Member DC

$$M = Hy$$

$$M_{UH} = y \qquad M_{UV} = 0$$

$$\frac{1}{EI}\int MM_{UH}\, ds = \frac{1}{EI}\int_0^4 Hy^2\, dy = \frac{H}{EI}\frac{64}{3}$$

$$\frac{1}{EI}\int MM_{UV}\, ds = 0$$

Member CB

$$M = 4H - Vx + \frac{wx^2}{2}$$

$$M_{UH} = 4 \qquad M_{UV} = -x$$

$$\frac{1}{EI}\int MM_{UH}\, ds = \frac{1}{EI}\int_0^6 \left(16H - 4Vx + \frac{240x^2}{2}\right) dx$$

$$= \frac{1}{EI}(96H - 72V + 8640)$$

$$\frac{1}{EI}\int MM_{UV}\, ds = \frac{1}{EI}\int_0^6 \left(-4xH + Vx^2 - \frac{wx^3}{2}\right) dx$$

$$= \frac{1}{EI}(-72H + 72V - 9720)$$

Member BA

$$M = H(4 - y) - 6V + \frac{60.6^2}{2}$$

$$M_{UH} = (4 - y) \qquad M_{UV} = -6$$

$$\frac{1}{EI}\int MM_{UH}\, ds = \frac{1}{EI}\int_0^{10} [H(4 - y)^2 - 6V(4 - y) + 1080(4 - y)]\, dy$$

$$= \frac{1}{EI}(93.3H + 60V - 10{,}800)$$

$$\frac{1}{EI}\int MM_{UV}\, ds = \frac{1}{EI}\int_0^{10} [-6H(4 - y) + 36V - 6480]\, dy$$

$$= \frac{1}{EI}(60H + 360V - 64{,}800)$$

Summing all the component parts and equating to zero:

$$\Delta_H = 210.67H - 12V - 2160 = 0$$

$$\Delta_V = -12H + 432V - 74{,}520 = 0$$

from which $H = 20.11$ kN and $V = 173.06$ kN.

Using these values and the equations of equilibrium, then at A:

the horizontal reaction = 20.11 kN (inwards);
the vertical reaction = 186.94 kN (upwards);
the moment = 79.03 kN m clockwise.

Example 4.16–11. The completed analysis of the frame of Example 4.16–10 is shown in Fig. 4.16–18. If the magnitudes of the horizontal deflexion at B and the rotation of B were required it would be necessary to evaluate these separately by applying, successively, corresponding unit forces at B, as shown in Fig. 4.16–19(a) and (b). The frame is still indeterminate to the second degree, however, and it might be thought that the determination of the appropriate moments throughout the frame presented major problems in themselves. They need not do so, however. In our development of the calculations up to this stage we have used a **Displacement Set** which conforms to the real deflected shape of the frame. We may choose, therefore, **Equilibrium Sets** which are in equilibrium with the applied loads and facilitate our calculations, as shown in Fig. 4.16–19(c) and (d). From these, calculations are confined to the member AB only.

Using the axes shown in Fig. 4.16–18:

Bending moment in member AB: $M = 20.11x - 79.03$

$$\theta_B = \int_A^B M_u \frac{M}{EI} dx \quad \text{where } M_u = 1 \text{ (see Fig. 4.16–19(c))}$$

$$= \int_0^{10} (1) \frac{(20.11x - 79.03)}{EI} dx$$

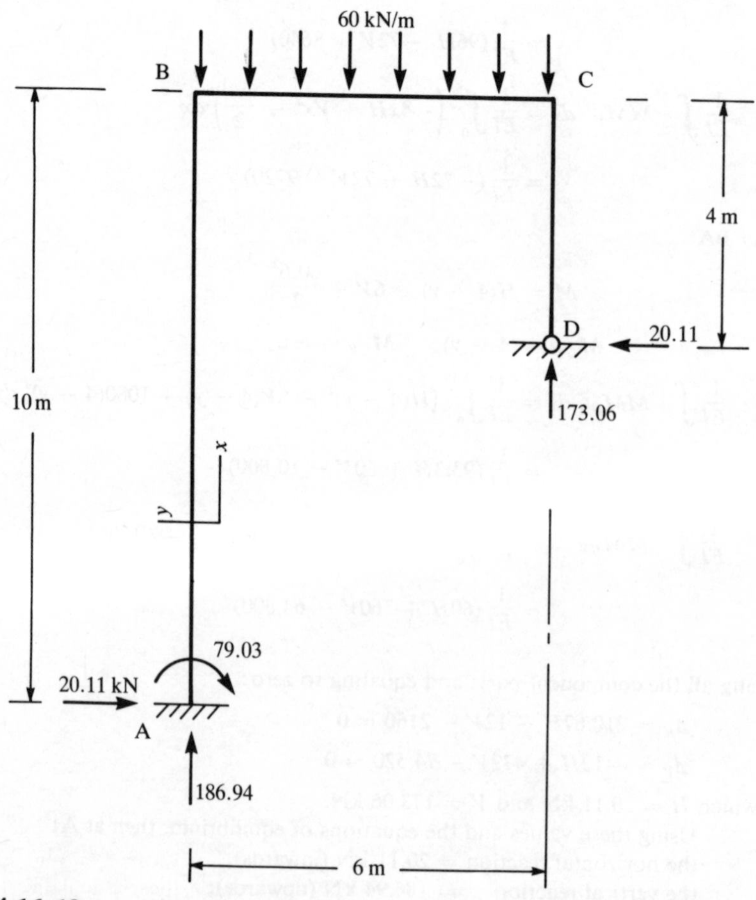

Fig. 4·16–18

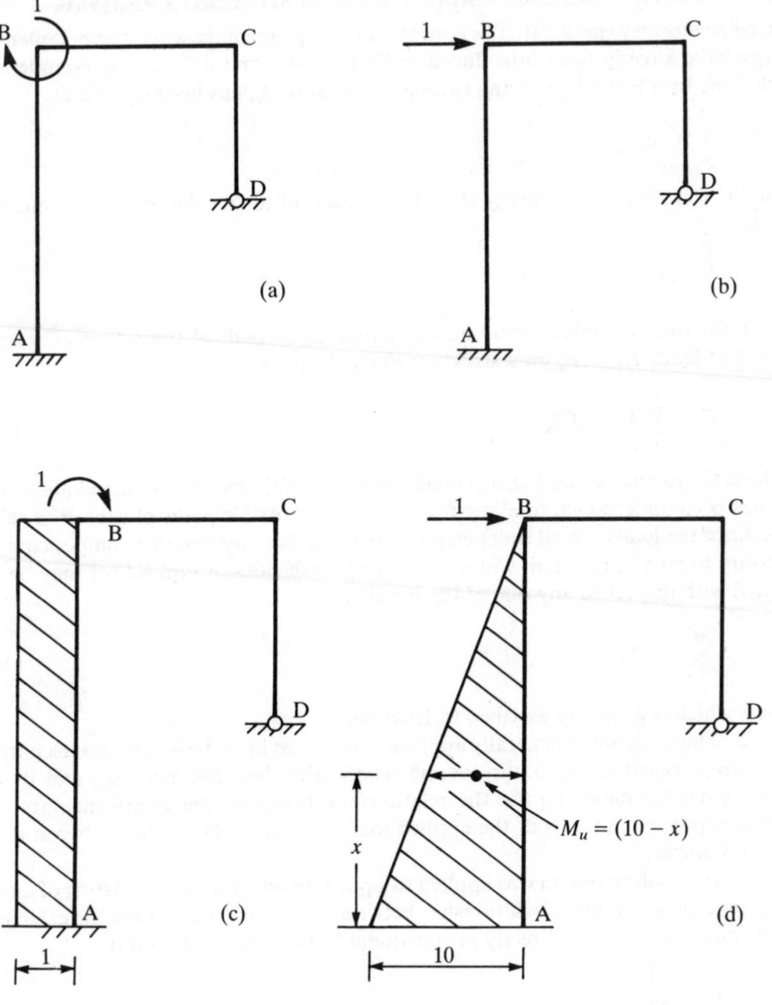

Fig. 4·16–19

$$= \frac{215.2}{EI} \text{ radians}$$ (clockwise, i.e. in the same sense as the unit moment applied at B)

$$\delta_{BH} = \int_A^B M_U \frac{M}{EI} dx \quad \text{where } M_u = (10 - x) \text{ (see Fig. 4.16–19(d))}$$

$$= \int_1^{10} (10 - x)\frac{(20.11x - 79.03)}{EI} dx$$

$$= -\frac{599.8}{EI}$$ (to the left, i.e. the −ve sign indicates that δ_{BH} is in the opposite direction to that of the unit force applied at B).

The discerning student will appreciate the similarity between these calculations and those possible from Fig. 4.16–18 using Moment Area Methods.

4.17 Energy methods—application in structural analysis

(a) **Complementary energy.** (i) The concepts of complementary work and complementary energy have already been introduced in Sections 4.7 and 4.9 and the complementary work done by a force P_m moving through a distance Δ_m has been defined as

$$\int \Delta_m \, dP_m$$

while the complementary energy stored as a result of the displacement of some internal action is

$$\int e \, dF$$

Clearly the total complementary energy stored as a result of the action of several independent loads $P_1 \ldots P_n$ on a structure in equilibrium will be

$$C = \sum_1^n \int \Delta_m \, dP_m \tag{4.17–1}$$

in which the various contributions made by the several loads are summed after integration has been undertaken, to allow for the various possible paths of each P–Δ relationship. Since the loads are all independent variables (for any reactive components found necessary to give equilibrium will do no work), both sides of Eqn 4.17–1 may be differentiated with respect to any one of the loads and

$$\frac{\partial C}{\partial P_m} = \Delta_m \tag{4.17–2}$$

a result which is generally ascribed to Engesser.

This equation is generally applicable for it has been derived without a particular assumption regarding the nature of the relationship between any load and its corresponding displacement. To use the relationship, however, the complementary energy must be expressed in terms of the applied loads, and the relationship between Δ and P must be known.

If a loading system was applied to a pin-jointed structure consisting of members with a reversible and linear relationship between stress and strain, then the extension, e, of any member would be directly proportional to the force F in it, and

$$e = \frac{FL}{AE}$$

where L, A, and E represent the length, cross-sectional area and modulus of the member. The complementary energy will involve summation extended over all members

$$C = \Sigma \int e \, dF$$

$$= \Sigma \int \frac{FL}{AE} \, dF$$

and

$$\Delta_n = \frac{\partial C}{\partial P_n} = \frac{\partial}{\partial P_n} \Sigma \int \frac{FL}{AE} \, dF$$

$$= \frac{\partial}{\partial P_n} \Sigma \frac{F^2 L}{2AE}$$

$$= \Sigma \frac{\partial F}{\partial P_n} \cdot \frac{FL}{AE}$$

Now $\partial F/\partial P_n$ for a particular member is numerically equal to the force caused in that member by unit value of P_n. Then

$$\Delta_n = \Sigma \frac{F_U FL}{AE} \tag{4.17–3}$$

which should be compared with Eqn 4.15–4. It will be seen that the first term on the right-hand side of that equation is identical with Eqn 4.17–3, which could be extended in a precisely similar way to include those more general (but still linear elastic) structures in which energy may be stored in bending, shear, and torsion, as well as a result of axial force alone.

(ii) In Section 4.16(b) it was pointed out that in an indeterminate structure self-straining may be caused by forcing into place redundant members which are not initially of the right size and shape. If, in a structure subjected to forces $P_1 \ldots P_n$, causing corresponding deflections $\Delta_1 \ldots \Delta_n$ there are j redundant elements, and if the internal action F_k in some member has an initial lack of fit λ_k then the complementary energy

$$C = \sum_1^n \int \Delta\, dP + \sum_1^j \int \lambda_k\, dF_k$$

The forces $P_1 \ldots P_n$ are, by definition, independent and

$$\frac{\partial C}{\partial F_k} = \lambda_k \tag{4.17–4}$$

If there is no lack of fit of the member then

$$\frac{\partial C}{\partial F_k} = 0 \tag{4.17–5}$$

and in its application this equation would frequently become indistinguishable from Eqn 4.17–2. This point becomes immediately clear if we consider the indeterminate reactive constraints in a structure which is intrinsically statically determinate, as in Fig. 4.17–1. The two-pinned arch develops four reactive constraints, one of which (see Section 4.16(a)) will be indeterminate unless account is taken of structural compatibility.

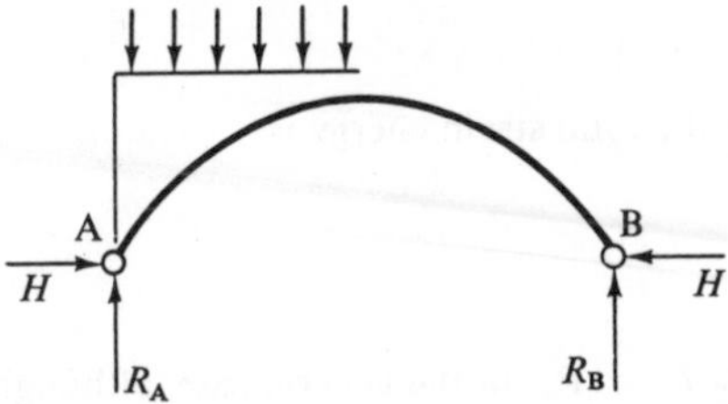

Fig. 4·17–1

It is usual to regard this as the horizontal component of reaction, which must be developed at each pin and which must be equal in magnitude, H, but of opposite sign at each, as shown, since the applied load is only vertical. Analysis could follow from either Eqn 4.17–5, since there is no lack of fit of the arch rib in the horizontal direction, or from Eqn 4.17–2 since, owing to the unyielding nature of the abutments there is no deflection of the point of application of H in its line of action.

As quoted in Eqn 4.17–5, the equation has complete generality, but may be developed simply for use with a linear elastic pin-jointed structure (as shown in Fig. 4.16–11, in which member BD has been taken as the indeterminacy). Combining the forces caused in all members of the statically determinate primary structure by the applied loads, Fig. 4.16–11(b) with the forces caused by the redundancy Fig. 4.16–11(c), the force in any member is

$$F = KW + F_{UR}R$$

where K is a coefficient dependent on the shape of the structure and the magnitude of the applied loads and F_{UR} is numerically equal to the force caused in each member by unit R. Then the complementary energy stored in the structure

$$C = \sum \int e\,dF = \sum \int \frac{FL}{AE}\,dF$$

$$= \sum \frac{F^2 L}{2AE}$$

λ, the initial lack of fit of member BD is given by

$$\lambda = \frac{\partial C}{\partial R} = \frac{\partial}{\partial R} \sum \frac{F^2 L}{2AE}$$

$$= \sum \frac{\partial F}{\partial R} \frac{FL}{AE}$$

$$= \sum \frac{F_{UR}(KW + F_{UR}R)L}{AE} \qquad (4.17\text{–}6)$$

which should be compared with Eqn 4.16–15 obtained by application of virtual work methods.

(b) Strain energy. It was pointed out in Sections 4.7 and 4.9 that the work done by applied forces must be stored in the distorted structure as a form of potential energy, the energy of strain. If

$$\int P\,d\Delta$$

represents the work done by a single load then the total strain energy is

$$U = \sum_{1}^{n} \int P\,d\Delta$$

when a structure is acted upon by several loads $P_1 \ldots P_n$. In the general case, although this strain energy may be readily comprehended as the sum of the work done by external loads, it can only be used as a starting-point for structural calculations in some special cases.

In a linear elastic structure subjected to loads $P_1 \ldots P_n$ the corresponding deflections will be of the form

$$\Delta_1 = f_{11}P_1 + f_{12}P_2 + f_{13}P_3 + f_{14}P_4 \cdots f_{1n}P_n \qquad (4.17\text{–}7)$$

where f_{mn} is a flexibility influence coefficient as defined in Section 4.6. In such a structure it has already been demonstrated in Section 4.11 that

$$f_{mn} = f_{nm}$$

Now the work done by each applied load $= \frac{1}{2}P\Delta$ and therefore

$$U = \tfrac{1}{2}\mathbf{P}^T\mathbf{\Delta} = \sum_1^n \tfrac{1}{2}P\Delta$$

$$= \tfrac{1}{2}P_1(f_{11}P_1 + f_{12}P_2 + f_{13}P_3 \ldots f_{1n}P_n)$$
$$+ \tfrac{1}{2}P_2(f_{21}P_1 + f_{22}P_2 + f_{23}P_3 \ldots f_{2n}P_n)$$
$$+ \tfrac{1}{2}P_3(f_{31}P_1 + f_{32}P_2 + f_{33}P_3 \ldots f_{3n}P_n)$$
$$+ \tfrac{1}{2}P_n(f_{n1}P_1 + f_{n2}P_2 + f_{n3}P_3 \ldots f_{nn}P_n)$$
$$= \tfrac{1}{2}(f_{11}P_1^2 + f_{22}P_2^2 \ldots f_{nn}P_n^2) + (f_{12}P_1P_2 \ldots$$
$$f_{1n}P_1P_n) + (f_{23}P_2P_3 \ldots f_{2n}P_2P_n) \ldots \text{etc.} \qquad (4.17\text{–}8)$$

in which the first bracket term arises from the leading diagonal of the flexibility matrix, the second from reciprocal terms containing P_1, and the third from reciprocal terms which do not contain P_1.

The rate at which U increases with P_1 is given by differentiating Eqn 4.17–8 with respect to P_1 and, since the loads are independent variables (any reactions have done no work), and since the flexibility coefficients are properties of the structure alone,

$$\frac{\partial U}{\partial P_1} = f_{11}P_1 + f_{12}P_2 + f_{13}P_3 \ldots f_{1n}P_n$$
$$= \Delta_1$$

Similarly, $\partial U/\partial P_2 = \Delta_2$, etc., and

$$\frac{\partial U}{\partial P_n} = \Delta_n \qquad (4.17\text{–}9)$$

a result which should be compared with Eqn 4.17–2. The similarity is not surprising since, although the latter is absolutely general it has been shown that in the special case of a linear elastic structure the strain and complementary energies are equal as in Fig. 4.8–3. Equation 4.17–9 might more readily be established by the simple statement

$$\frac{\partial C}{\partial P_n} = \Delta_n \quad \text{(from Eqn 4.17–2)}$$

and since, in a linear elastic structure

$$U = C$$

in such a structure $\partial U/\partial P_n = \Delta_n$. By such a reasoning process we may readily accept that in such a structure with a redundancy having a lack of fit λ

$$\frac{\partial U}{\partial R} = \lambda \qquad (4.17\text{–}10)$$

(c) **Use of strain-energy theorems.** It has been shown that the principle of virtual work and methods of complementary energy give methods of structural solution which are indistinguishable in the general case, and that strain-energy methods are identical when applied to certain clearly defined structures and loading cases. This fact has led to some confusion in the past since strain energy, which can be so readily equated to the summed effects of the work done by various loads can be easily comprehended; it has also been assumed that equations of the form of 4.17–9 and 4.17–10 (usually ascribed to Castigliano) are generally applicable.

Strain-energy methods in the forms here outlined may only be used when the structure is composed of members obeying a linear (and reversible) stress–strain relationship. It is true that this convenient assumption is frequently made and that strain-energy methods are therefore frequently applicable but the limitations on their use remain.

Example 4.17–1. Determine the reactive forces in the two-pinned parabolic arch rib in Fig 4.17–2, carrying symmetrical point loads as shown. The flexural rigidity (EI) of the arch rib is proportional to the secant of the angle of slope of the arch rib, and has the value EI_c at the crown of the arch. It may be assumed that the arch rib is linearly elastic.

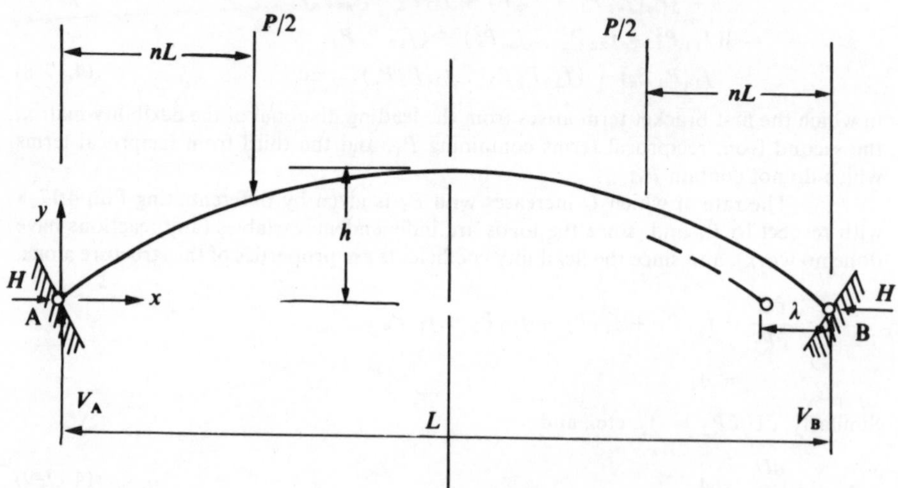

Fig. 4·17–2

SOLUTION With two possible components of reaction at A and two at B, the arch is indeterminate to the first degree. By symmetry, and by taking moments about either hinge

$$V_A = V_B = \frac{P}{2}$$

The remaining unknown horizontal reaction forces (both equal in magnitude to H) will be determined using strain energy methods.

If it is now assumed that a horizontal displacement λ is allowed to occur at support B in the directions corresponding to H then the value of λ is given (since the structure is linearly elastic) by Eqn 4.17–10:

$$\lambda = \frac{\partial U}{\partial H}$$

U is here the strain energy due to bending, as it is found that axial and shear forces make a negligible contribution to the total energy stored in an arch rib of conventional proportions.

Hence

$$\lambda = \frac{\partial U}{\partial H} = \frac{\partial}{\partial H}\left[\int_0^l \frac{M^2\, ds}{2EI}\right] \tag{4.17–11}$$

In this expression, s is measured along the arch rib, and l is its total length.

The bending moment M is the only term which varies with H, and Eqn 4.17–11 may be written

$$\lambda = \int_0^l \frac{\frac{\partial M}{\partial H} M ds}{EI} \tag{4.17–12}$$

The expressions for M are

$$M = V_A x - Hy \quad (0 < x < nL)$$

$$\text{and} \quad M = PnL/2 - Hy \quad (nL < x < L/2) \tag{4.17–13}$$

Therefore $\dfrac{\partial M}{\partial H} = -y$

which is the bending moment due to a single unit horizontal force corresponding to the deflection λ.

Hence Eqn 4.17–12 may be written

$$\lambda = \int_0^l \frac{M_u M ds}{EI} \tag{4.17–14}$$

and this is identical to the expression developed from virtual work considerations, as for example in Eqn 4.15–3.

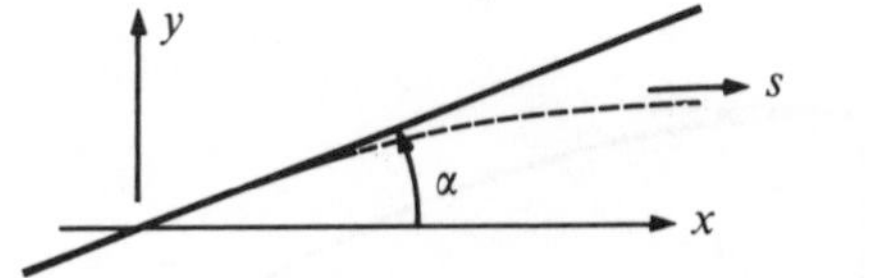

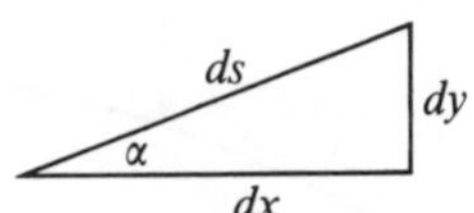

Fig. 4·17–3

From Fig. 4.17–3 it is seen that

$$ds = dx \sec \alpha$$

and if the arch rigidity varies in the same manner then Eqn. 4.17–14 becomes

$$\lambda = \frac{1}{EI_c} \int_0^L M_u \, M dx$$

Insertion of the bending moment expression and use of symmetry gives

$$\lambda = \frac{2}{EI_c} \int_0^{nL} (Hy^2 - Pxy/2) dx + \frac{2}{EI_c} \int_{nL}^{L/2} (Hy^2 - PnLy/2) dx$$

$$= \frac{2H}{EI_c} \int_0^{L/2} y^2 \, dx - \frac{P}{EI_c} \int_0^{nL} xy \, dx - \frac{PnL}{EI_c} \int_{nL}^{L/2} y dx \tag{4.17–15}$$

The solution is therefore reduced to the evaluation of three integrals, and if the equation of the parabola is taken as

$$y = \frac{4hx}{L^2} (L - x) \tag{4.17–16}$$

then $$\int_0^{L/2} y^2 dx = \frac{4h^2L}{15} \qquad \int_0^{nL} xydx = \frac{n^3hL^2}{3}(4 - 3n)$$

$$\int_{nL}^{L/2} ydx = \frac{hL}{3}(1 - 6n + 4n^3) \tag{4.17–17}$$

In reality, the value of λ is of course zero if there is no settlement of the foundations, and Eqn 4.17–15 gives

$$H = \frac{5PL}{8h}(n - 2n^3 + n^4) \tag{4.17–18}$$

The assumption regarding the secant variation of the rib rigidity is frequently made in the analysis of parabolic arches and is introduced to allow the integrals to be solved readily. As the slope of arch ribs is generally small no great error is introduced.

Example 4.17–2. Determine the reaction forces in the two-pinned parabolic arch rib shown in Fig 4.17–4, carrying the single point load P indicated. The arch rib has the same properties as that of Example 4.17–1.

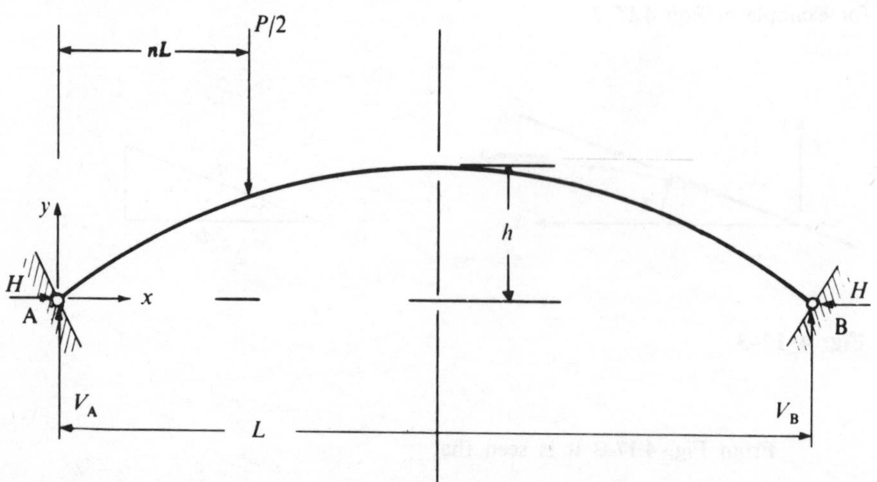

Fig. 4·17–4

This example is very similar to the preceding one, and the same solution method could be used. Hence

$$V_A = \frac{P(1 - n)L}{L} = P(1 - n) \qquad V_B = Pn$$

These reaction forces could now be used to set up bending moment expressions similar to those in Eqn 4.17–13, and these inserted in Eqn 4.17–14. The integration intervals would have to be 0 to nL and nL to L. The whole calculation would follow very closely that already performed.

An alternative approach is to employ the earlier result directly. The present unsymmetrical loading case may be considered as the superposition of two other cases —one symmetric and the other anti-symmetric (Fig. 4.17–5).

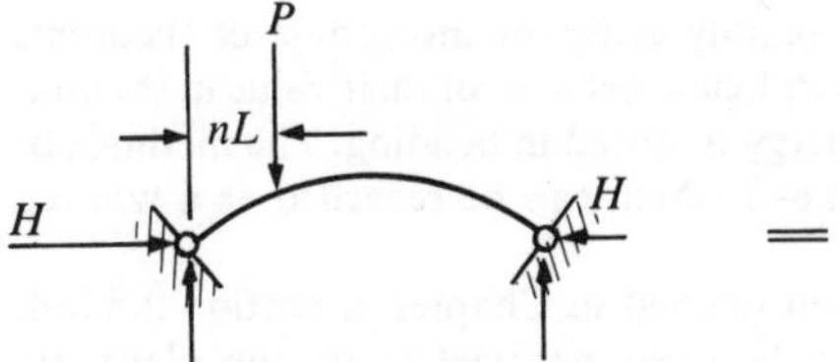

Fig 4·17–5

The answer to the symmetric case is already known.

The anti-symmetric case can best be understood by examining the effects of the two loads individually. For if the left hand (downwards) load is acting on its own it causes equal inwards reaction forces (say) H_3 at the two pinned supports. If the right hand (upwards) load is now considered on its own, and the arch examined from the rear, it can be seen that the reaction force is again H_3, but this time outwards. The net reaction H_2 is therefore zero.

The horizontal reaction force for the unsymmetrically loaded case of Example 4.17–2 is therefore the same as that for the symmetric case of Example 4.17–1, carrying the same total load P. This will come as no surprise to readers who have noted the symmetric nature of Eqn 4.17–18 (this may be checked by substituting $(1 - n)$ for n).

This technique, of examining symmetric and anti-symmetric cases in the solution of a problem is of wide applicability in linear structural analysis, and is in no way confined to arches. It can frequently happen that the two individual cases can be quite trivial, although the unsymmetric problem may appear formidable.

In the case of encastré arches, with three components of reaction at each support, the use of this device changes the problem from one involving the solution of three equations in three unknowns to the solution of two distinct problems, each involving two unknowns. Without some form of computer, the latter is less labour. The reader may verify for himself that in a symmetrical encastré arch subjected to symmetrical loads, the vertical components of reaction are equal and known, whilst under anti-symmetrical loads the vertical components of reaction will be equal in magnitude to the load in each half-span but in opposite directions.

4.18 Moment-area methods

The principle of virtual work is a powerful tool of general applicability in the calculation of displacements in any structure, and has been used in Section 4.15(a) for the determination of deflections in linear elastic beams and frames. In the former case, however, deflections can frequently be evaluated more readily using the moment–area theorems which are ascribed to Mohr, and which are given below because of their value in the analysis of those particular structures in which energy is stored in bending. The methods of this section have already been used in Fig. 4.6–3 which can be regarded as a worked example of their application.

The sign convention has already been defined in Chapter 2, Section 2.8 but, for convenience, the important relationships between parameters in one plane are brought together in Fig. 4.18–1.

(a) y = deflection +

$\frac{dy}{dx}$ = slope +

$\frac{d^2y}{dx^2}$ = curvature +

$EI\frac{d^2y}{dx^2}$ = bending moment M +

$-EI\frac{d^3y}{dx^3}$ = shear $S = -\frac{dM}{dx}$ +

$EI\frac{d^4y}{dx^4}$ = load $w = -\frac{dS}{dx}$ +

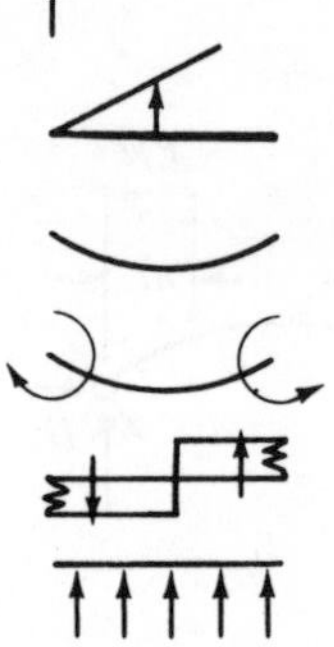

(b) w = load

S = Shear $= -\int w\,dx$

M = bending moment $= -\int S\,dx = \iint w\,dx\,dx$

curvature $= \frac{M}{EI}$

θ = slope $= \int \frac{M}{EI}dx = \iiint \frac{w}{EI}dx\,dx\,dx$

y = deflection $= \iint \frac{M}{EI}dx\,dx = \iiiint \frac{w}{EI}dx\,dx\,dx\,dx.$

Fig. 4·18–1

(a) **First moment–area theorem.** If S and T, two points on an elastic beam subject to applied loads are distant x_1 and x_2 respectively from some origin of coordinates, as shown in Fig. 4.18–2, then since

$$\frac{d^2y}{dx^2} = \frac{M}{EI}$$

where EI is the appropriate flexural rigidity, and

$$\frac{dy}{dx} = \int \frac{M}{EI}dx$$

inserting the appropriate limits of integration,

$$\left[\frac{dy}{dx}\right]_{x_1}^{x_2} = \int_{x_1}^{x_2} \frac{M}{EI}\,dx$$

$$\theta_2 - \theta_1 = \int_{x_1}^{x_2} \frac{M}{EI}\,dx \tag{4.18–1}$$

or, in words, the change of slope between two points on a bent beam is equal to the area of the M/EI diagram between those points.

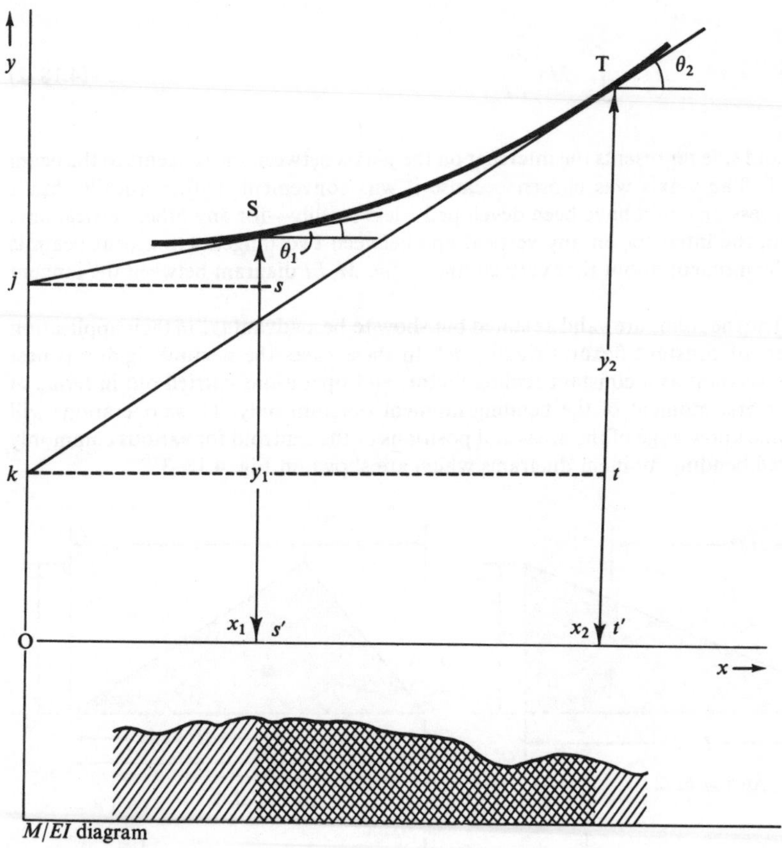

Fig. 4·18–2

(b) **Second moment–area theorem.** Tangents to the beam at S and T will obviously make angles θ_1 and θ_2 respectively with the axis of x and will cut the y-axis at j and k (say).

Now, as shown in Fig. 4.18–1,

$$y = \int\int \frac{M}{EI}\,dx\,dx$$

which, integrating by parts, gives

$$y = x\int \frac{M}{EI}\,dx - \int \frac{Mx\,dx}{EI} = x\frac{dy}{dx} - \int \frac{Mx}{EI}\,dx$$

and inserting the appropriate limits of integration

$$y_2 - y_1 = x_2\theta_2 - x_1\theta_1 - \int_{x_1}^{x_2} \frac{Mx}{EI}\,dx$$

But $y_2 - x_2\theta_2 = tt' =$ Ok, and $y_1 - x_1\theta_1 = ss' =$ Oj so that

$$tt' = ss' - \int_{x_1}^{x_2} \frac{Mx}{EI}\,dx$$

and

$$\text{jk} = ss' - tt' = \int_{x_1}^{x_2} Mx\,\frac{dx}{EI} \tag{4.18–2}$$

The left-hand side represents the intercept on the y-axis between the tangents to the beam at S and T. The y-axis was chosen because it was convenient mathematically, but a similar expression could have been developed—less readily—for any other vertical line. To sum up, the intercept on any vertical line between two tangents to a bent beam is equal to the moment about that vertical line of the M/EI diagram between the tangent points.

The theorems are valid as stated but show to best advantage in their application to members of constant flexural rigidity EI. In these cases the flexural rigidity is best taken into account as a constant scaling factor, and operations carried out in terms of the area or first moment of the bending moment diagram only. These operations will involve some knowledge of the areas and positions of the centroid for various commonly encountered bending moment diagrams which are shown in Fig. 4.18–3.

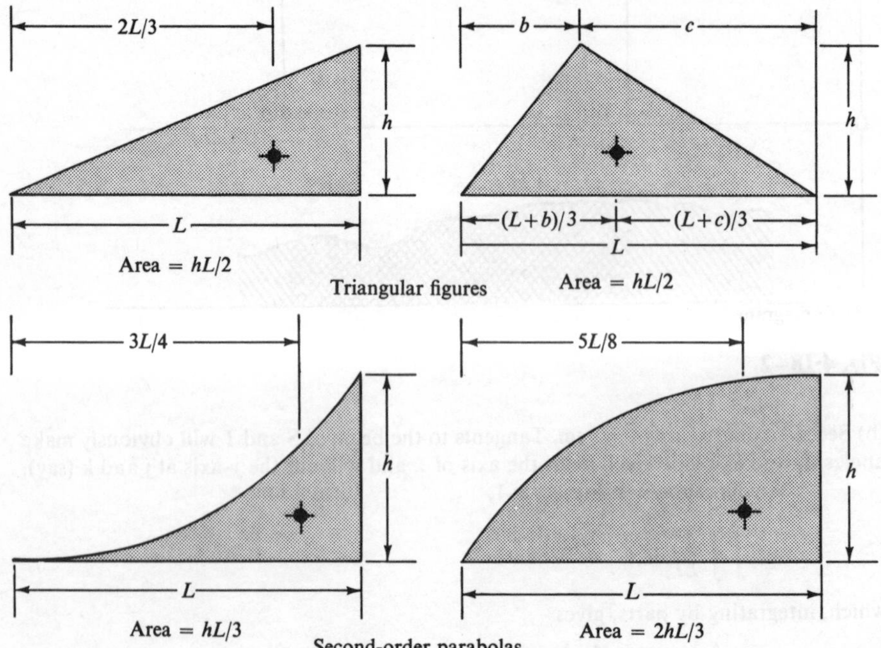

Fig. 4·18–3

(c) **Applications of the method.** The following examples illustrate application of the method in practical problems.

Example 4.18–1. A uniform cantilever of length L and flexural rigidity EI is loaded with
(a) a point load of W at its free end;
(b) a uniformly distributed load of intensity w per unit length, as shown in Figs. 4.18–4(a) and (b). Determine the slope and vertical displacement of the free end in each case.

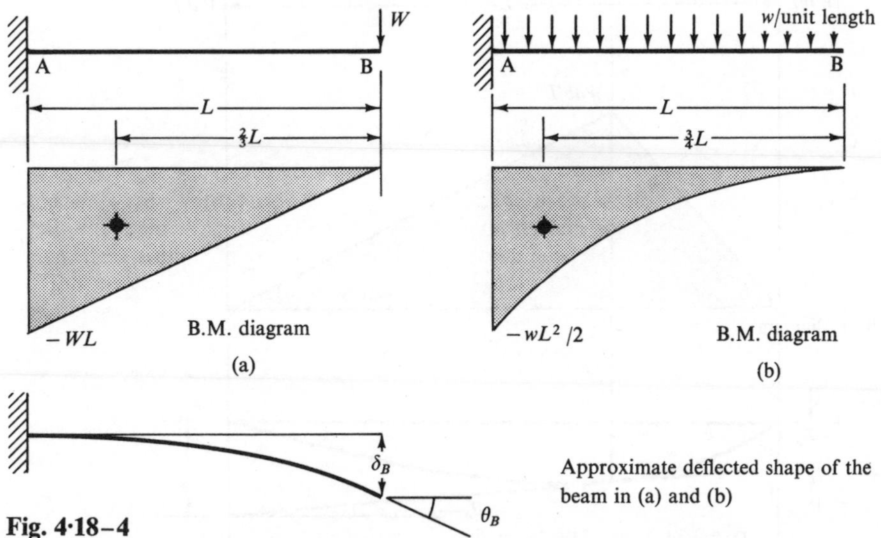

Fig. 4·18–4

SOLUTION (a) In the case of the point load the bending moment diagram will be linear as shown, increasing from zero at the free end to $-WL$ at the haunch, A.

The change of slope between any two points will be (the area of the bending moment diagram between those two points)/EI. The slope at the haunch, A is zero, and therefore the slope at B will be

$$\frac{1}{EI}\left[\tfrac{1}{2}.L(-WL)\right] = -\frac{WL^2}{2EI}$$

The tangent to the beam at A will be a horizontal line and the intercept made on a vertical line through B, by the tangents at A and B will give the vertical displacement of B directly. This will be equal to (the moment of the bending moment diagram between A and B about the vertical through B)/EI and will be

$$\frac{1}{EI}\left[\tfrac{1}{2}.L(-WL)\tfrac{2}{3}L\right] = -\frac{WL^3}{3EI}$$

the negative sign indicating that the beam at B is below the tangent at A.

(b) Applying similar arguments to the uniformly loaded case

$$\theta_B = \frac{1}{EI}\left[\tfrac{1}{3}.L\left(-\frac{wL^2}{2}\right)\right] = -\frac{wL^3}{6EI}$$

and

$$\delta_B = \frac{1}{EI}\left[\tfrac{1}{3}.L\left(-\frac{wL^2}{2}\right)\frac{3L}{4}\right] = -\frac{wL^4}{8EI}$$

Example 4.18–2. A simply supported beam AB, of length L and uniform flexural rigidity EI is subjected to a point load of W at C, distant a from A and b from B, as shown in Fig. 4.18–5. Determine the slope and deflection of the beam at C.

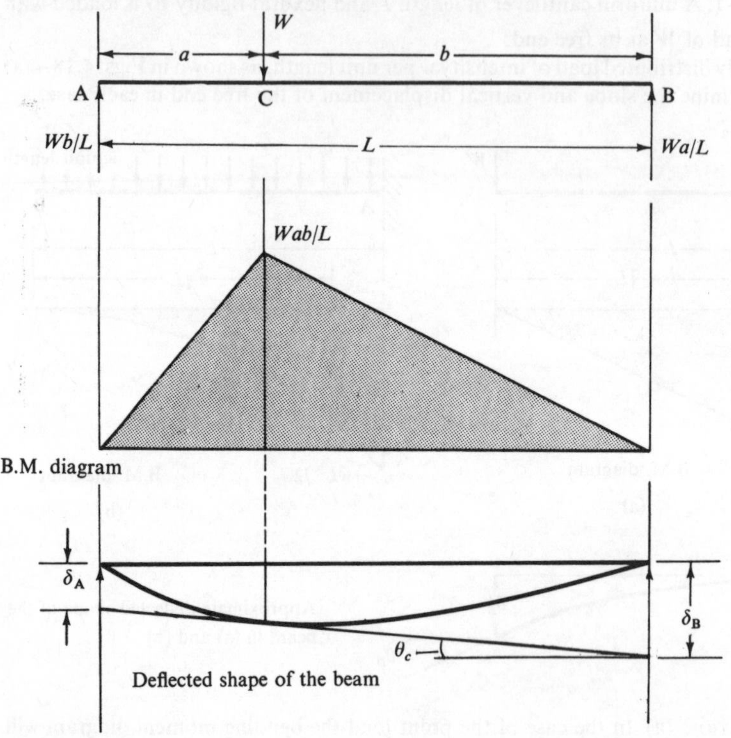

Fig. 4·18–5

SOLUTION Taking moments about A, $R_B = Wa/L$ from which the bending moment at C = Wab/L and the bending moment diagram will appear as shown in the sketch.

Now considering the deflected shape of the beam, on which the tangent to the beam at C is drawn, it may be seen that if δ_A and δ_B respectively, represent the amounts by which the beam at A and B is above the tangent, then

$$\delta_B - \delta_A = L\theta_c$$

Now

$$\delta_A = \frac{1}{EI}\cdot\frac{Wab}{L}\cdot\frac{a}{2}\cdot\frac{2a}{3} = \frac{Wa^3b}{3EIL}$$

Similarly,

$$\delta_B = \frac{1}{EI}\cdot\frac{Wab}{L}\cdot\frac{b}{2}\cdot\frac{2b}{3} = \frac{Wab^3}{3EIL}$$

Hence

$$L\theta_c = \frac{Wab^3}{3EIL} - \frac{Wa^3b}{3EIL}$$

$$= \frac{Wab}{3EIL}(b^2 - a^2)$$

Then

$$\theta_c = \frac{Wab}{3EIL^2}(b^2 - a^2)$$

Now the deflection of the beam at C is

$$\delta_A + \frac{a}{L}(\delta_B - \delta_A) = \delta_A \frac{L - a}{L} + \frac{a}{L}\delta_B$$

$$= \frac{1}{L}(\delta_A b + \delta_B a)$$

$$= \frac{1}{L}\left(b \cdot \frac{Wa^3b}{3EIL} + a \cdot \frac{Wab^3}{3EIL}\right)$$

$$= \frac{Wa^2b^2}{3EIL^2}(a + b)$$

$$= \frac{Wa^2b^2}{3EIL}$$

References

1 Charlton, T. M. *Model Analysis of Structures.* Spon, 1954.
2 Charlton, T. M. *Model Analysis of Plane Structures.* Pergamon Press, 1966.
3 Charlton, T. M. *Energy Principles in Applied Statics.* Blackie & Son Ltd, 1959.
4 Charlton, T. M. *Principles of Structural Analysis.* Longman Group, 1969.

Problems

4.1. (a) A rectangular parallelepiped of dimensions *dx*, *dy*, and *dz* is acted on by stresses σ_x, σ_y, and σ_z causing corresponding strains ε_x, ε_y, and ε_z. Show that the strain energy stored in the volume is

$$\tfrac{1}{2}[\varepsilon_x \ \varepsilon_y \ \varepsilon_z][\sigma_x \ \sigma_y \ \sigma_z]^T \, dx \, dy \, dz$$

(b) If, in addition to the above normal stresses, shear stresses τ_{xy}, τ_{yz}, and τ_{zx} are also acting, show that the strain energy is

$$\tfrac{1}{2}\boldsymbol{\varepsilon}^T\boldsymbol{\sigma} \, dV$$

where $\boldsymbol{\varepsilon}$ is the strain vector $[\varepsilon_x \ \varepsilon_y \ \varepsilon_z \ \gamma_{xy} \ \gamma_{yz} \ \gamma_{zx}]^T$; $\boldsymbol{\sigma}$ is the stress vector $[\sigma_x \ \sigma_y \ \sigma_z \ \tau_{xy} \ \tau_{yz} \ \tau_{zx}]^T$; and $dV = dx \, dy \, dz$.

4.2. The uniform elastic cantilever of length *L*, cross-sectional area *A* and flexural rigidity *EI*, shown in the figure is subjected to force in varying directions at its free end, n. Determine the flexibility influence coefficients, f_{mn}, which define the magnitudes of the deflections *Δ*, at a point m distant *x* from n.

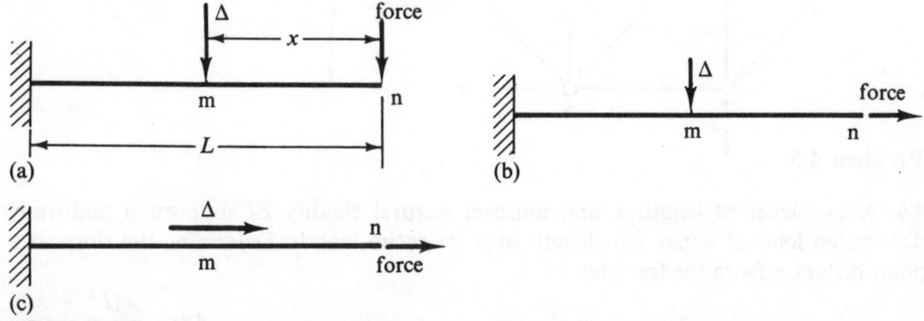

Problem 4·2 *Ans.* (a) $[(L - x)^2(x + 2L)]/6$; (b) Zero; (c) $(L - x)/AE$.

4.3. A beam of non-uniform section ABC, symmetrical about its centre B, is required to bridge two equal spans AB and BC each of length 50 m. A model of the beam is made of length 100 cm and supported at A and C. A load applied at B produces deflections in the model given by the table below, in which x is the distance from and A and y the vertical deflection.

x cm	0	10	20	30	40	50
y cm	0	0.36	0.64	0.84	0.96	1.00

What would be the maximum bending moments in the prototype beam if AC was covered by a uniformly distributed load of 10 kN/m?

Ans. 4167 kN m; 1389 kN m.

4.4. In the truss shown the ratio of length to cross-sectional area, f, is constant for all members. Calculate the horizontal and vertical displacements of C due to a horizontal force of F applied at that point.

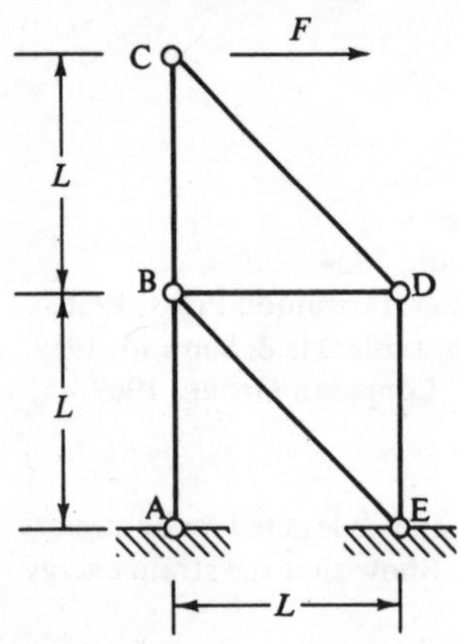

Problem 4·4

Ans. $\dfrac{11Ff}{E}$; $\dfrac{-3Ff}{E}$.

4.5. In the truss shown the ratio of length to cross-sectional area, f, is constant for all members. Show that the horizontal displacement at S due to unit vertical load at T (which equals the vertical displacement at T due to horizontal load at S) is $-5f/9E$.

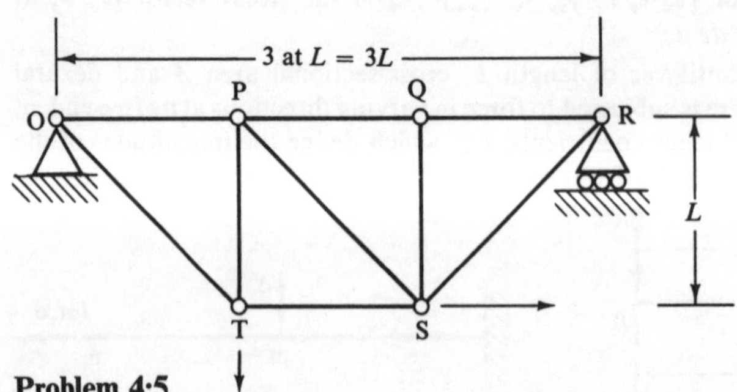

Problem 4·5

4.6. A cantilever of length L and uniform flexural rigidity EI supports a uniformly distributed load of w per unit length over its entire length. Determine the slope at a point distant x from the free end.

Ans. $\dfrac{w(L^3 - x^3)}{6EI}$.

4.7. A beam ABC of length $2L$ and uniform flexural rigidity EI is simply supported at the same level at the end A, and the mid-point B. A force P acts transversely on the beam at the free end C. Determine the slope and deflection at C.

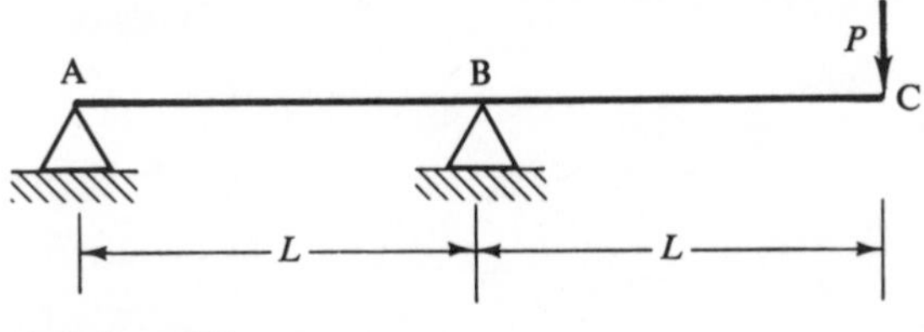

Problem 4·7

Ans. $\dfrac{5PL^2}{6EI}$; $\dfrac{2PL^3}{3EI}$.

4.8. What would be the deflection of point C in Problem 4.7 if the beam was supported at A by a spring, which extended by unit distance due to a force k in it?

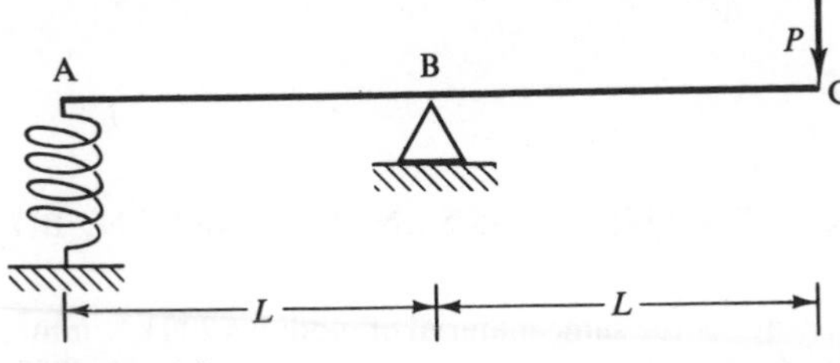

Problem 4·8

Ans. $P\left(\dfrac{2L^3}{3EI} + \dfrac{1}{k}\right)$

4.9. A cantilever ABC of length $2L$ is encastré at A. The portion AB, nearer to the haunch, of length L is of flexural rigidity $2EI$ and the portion nearer the tip, BC, is of length L and flexural rigidity EI. A uniformly distributed load of intensity w per unit length covers the portion BC. Show that the vertical deflection of the tip C is $11wL^4/12EI$.

4.10. A bent bar ABC is encastré at its lower end A and consists of a vertical portion AB of length $2L$ and a horizontal portion BC of length L. The flexural rigidity EI is constant throughout. A uniformly distributed load of intensity w per unit length covers the portion BC. Show that the vertical and horizontal deflections of the tip C are $9wL^4/8EI$, and wL^4/EI respectively.

4.11. In the truss shown in the figure all members are of the same material of modulus 200 kN/mm² and a cross-sectional area of 2000 mm². Determine the horizontal movement of D and the force in all members due to the applied force of 50 kN at B.

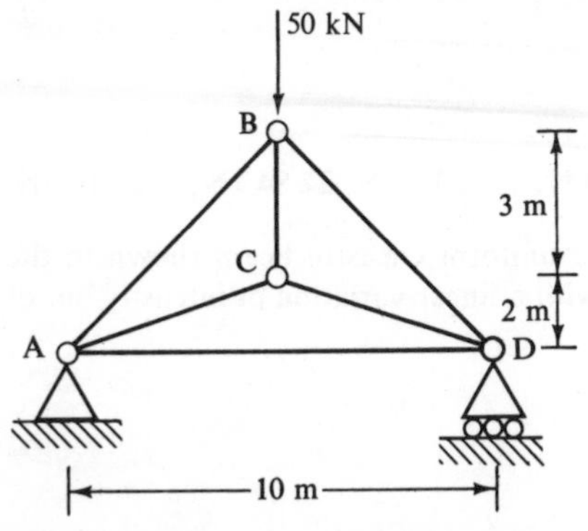

Problem 4·11

Ans. 0.74 mm; AB −30.9 kN; AC −8.5 kN, BC −6.3 kN; AD 29.7 kN, BD −30.9 kN, CD −8.5 kN.

4.12. In the frame shown the cross-sectional areas of AC and BD are 12×10^3 mm^2, and of all other members 20×10^3 mm^2. The members are of steel, of modulus 200 kN/mm^2. Determine the force in the members due to the given loading.

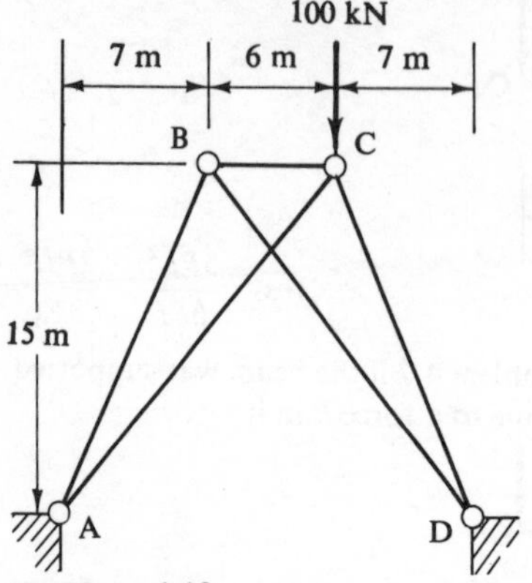

Problem 4·12

Ans. AB −4.8 kN; AC −40.6 kN; BC −5.5 kN; CD −76.5 kN; BD 5.7 kN.

4.13. In the frame shown all members are of the same material of modulus 200 kN/mm^2. Horizontal, diagonal, and vertical members have cross-sectional areas of 2000, 4000, and 8000 mm^2 respectively. Determine the forces in members OS, PT, PR, and QS.

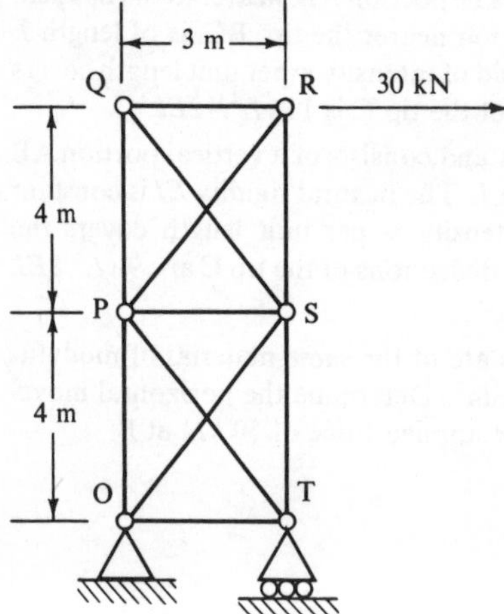

Problem 4·13 *Ans.* 27.84 kN; −22.16 kN; 27.84 kN; −22.16 kN.

4.14. Determine the fixed-end moments in the uniform encastré beam shown in the figure, when subjected to a symmetrical load, with a linear variation of intensity but of total magnitude $wL/2$.

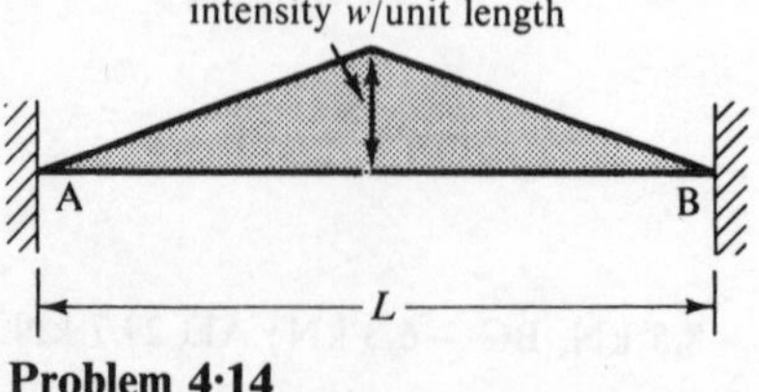

Problem 4·14 *Ans.* $\dfrac{5wL^2}{96}$.

4.15. Determine the fixed-end moments in the uniform encastré beam shown in the figure, when subjected to a uniformly distributed load of intensity w per unit length over half the span.

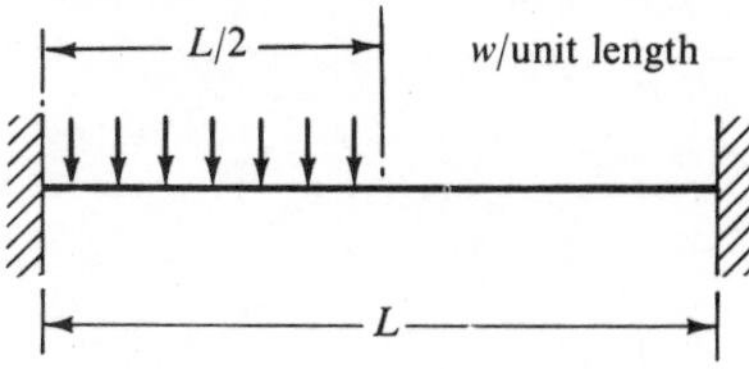

Problem 4·15

Ans. $\dfrac{11wL^2}{192}$; $\dfrac{5wL^2}{192}$.

4.16. A beam of variable depth but of a material with elastic properties is used as a fixed beam. When a unit vertical load is applied at an intermediate point C, the equation for the deflected shape of the beam is $y = kx(L - x)^2$, where x is the distance measured from one end, k a constant, L the span and y the vertical deflection. Show that the vertical deflection at C due to a uniform load of intensity w per unit length covering the whole span is $kwL^4/12$.

4.17. A uniform beam of length $3L$ is subjected to a uniformly distributed load of w per unit length. It is supported over three equal spans as shown. Determine the vertical components of reaction and the bending moments at each support.

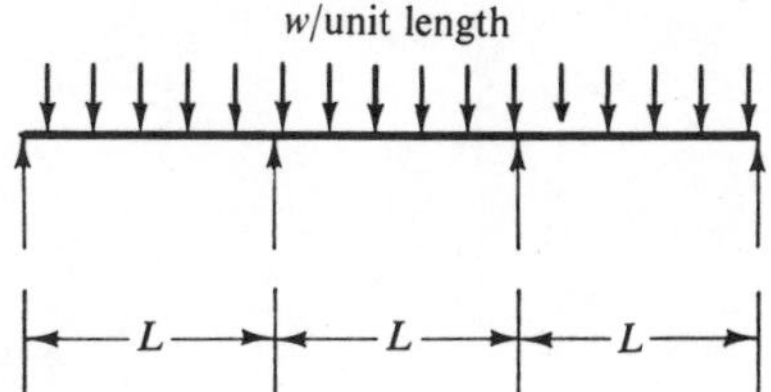

Problem 4·17

Ans. $\frac{4}{10}wL$; $\frac{11}{10}wL$; $\frac{11}{10}wL$; $\frac{4}{10}wL$; 0; $\frac{1}{10}wL^2$; $\frac{1}{10}wL^2$; 0.

4.18. A uniform beam of length $L(1 + \alpha)$ is simply supported at its extremities A and C, and at an intermediate point B, L from end A. A uniformly distributed load of intensity w per unit length covers the entire beam. If the supports A, B, and C remain on the same level, determine the bending moment at the intermediate support B.

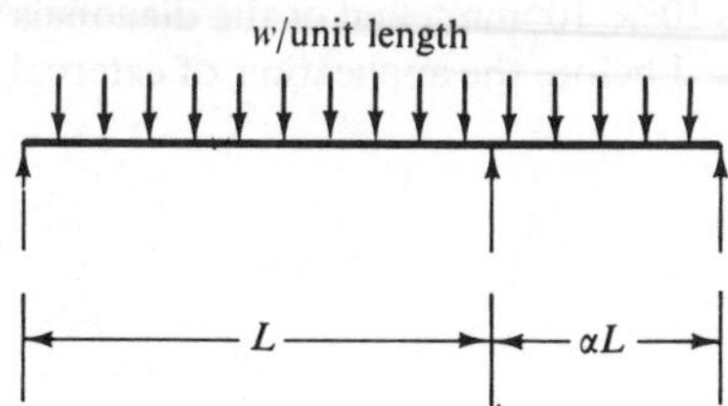

Problem 4·18

Ans. $\dfrac{wL^2}{8}(1 - \alpha + \alpha^2)$.

4.19. A rectangular rigid-jointed portal frame is of uniform cross-section throughout, and is pinned to rigid foundations. The horizontal beam is of length L and flexural rigidity EI_B and the stanchions of height h and flexural rigidity βEI_B. If the portal is subjected to a uniformly distributed beam load of w per unit length, show that the horizontal thrust (neglecting effects other than that of bending) is $wL^3\beta/[4h(2h + 3L\beta)]$.

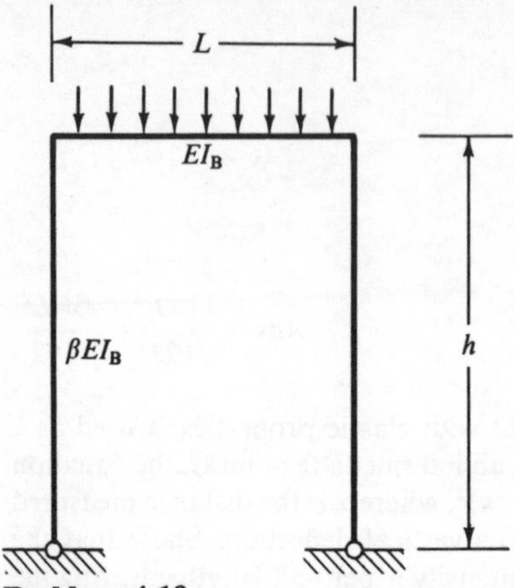

Problem 4·19

4.20. The frame ABC is of constant cross-section throughout and is built in to rigid foundations at A and C. Determine the components of reaction at the foundations due to the loads shown.

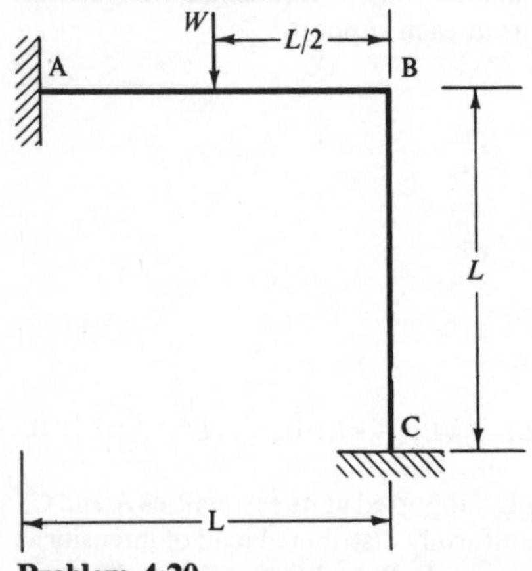

Problem 4·20

Ans. A $\dfrac{5WL}{32}$, $\frac{5}{16}W$; C $\dfrac{WL}{32}$, $\frac{11}{16}W$.

4.21. In the pin-jointed framework shown the cross-sectional area of the booms (horizontals) is 20×10^3 mm², of the posts (verticals) is 10×10^3 mm², and of the diagonals 15×10^3 mm². Assuming the frame to be unstressed before the application of external load, determine the force in each member.

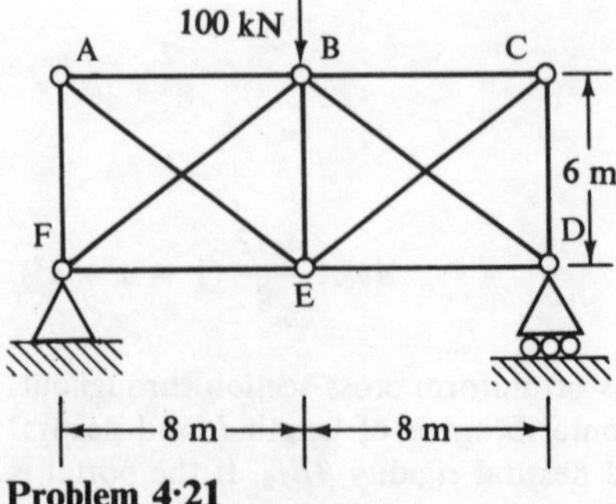

Problem 4·21

Ans. $F_{BE} = 37.0$ kN compressive.

4.22. A circular ring of mean radius R and flexural rigidity EI is subjected to two radial forces each of magnitude W acting outwards from the centre at opposite ends of a diameter. Show that, considering the effects of bending only, the extension of the ring in the direction of the forces is

$$\frac{WR^3}{EI}\left(\frac{\pi}{4} - \frac{2}{\pi}\right).$$

4.23. The elastic cantilever AB shown in the figure is braced by a wire BC which is slack in the unloaded condition. It is found that a force X in direction 1, or a force Y in direction 2 will cause the wire to become taut. When the tie is slack a unit force in direction 1 produces deflections δ_{11} and δ_{21} in directions 1 and 2 respectively; when the tie is taut the same unit force produces deflections δ'_{11} and δ'_{21} respectively.

Show that

$$\frac{\delta_{11}}{\delta_{21}} = \frac{\delta'_{11}}{\delta'_{21}}$$

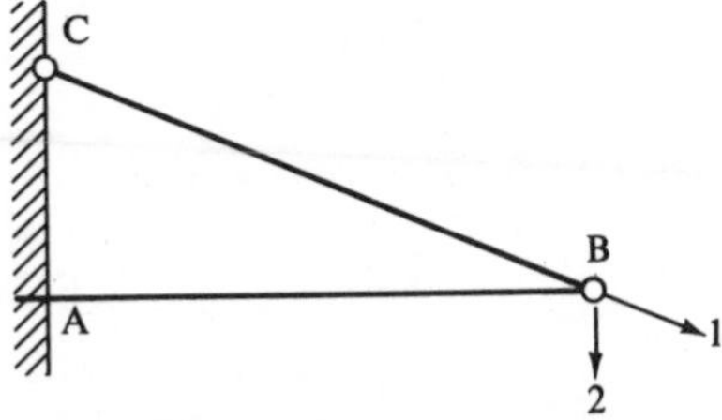

Problem 4·23

4.24. The pin-jointed frame shown is pinned to a rigid foundation at A and F and is subjected to a horizontal sway force of magnitude 10 kN applied at D. All diagonal members lie at 45° to the vertical and the materials and cross-sectional areas of all members are identical. Show that the force produced in member CD is approximately 4.69 kN.

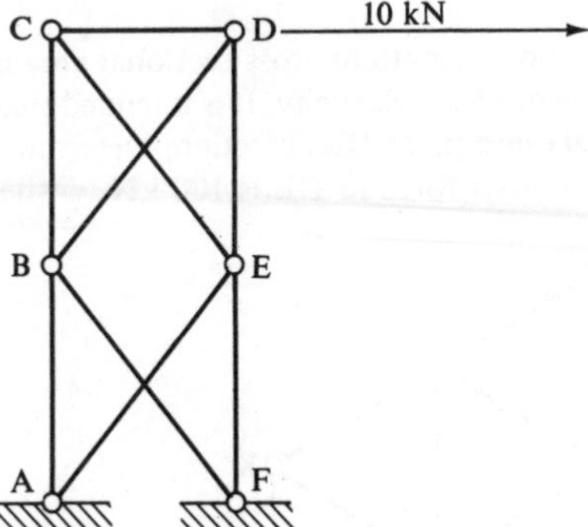

Problem 4·24

4.25. In the frame shown all horizontal and vertical members are of aluminium alloy (modulus of direct elasticity 66.7 kN/mm^2) and of 25 $\times$ 10^3 mm^2 cross-sectional area, while the diagonals are of steel (modulus of direct elasticity 200 kN/mm^2) and of 6.25 $\times$ 10^3 mm^2 cross-sectional area. The diagonal which would carry compressive force under the external load shown is pre-tensioned to prevent the occurrence. Determine the magnitude of the necessary pre-tension.

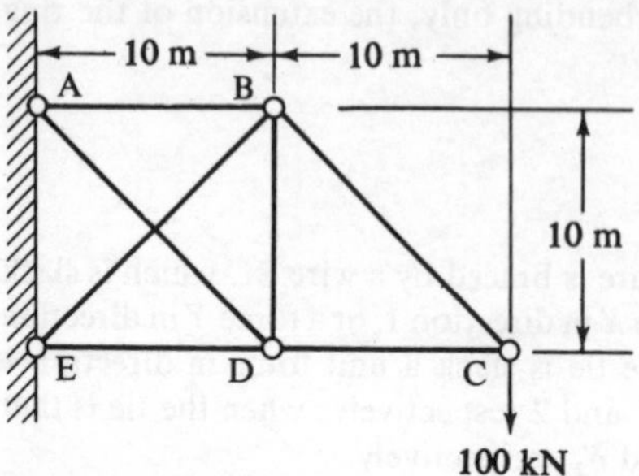

Problem 4·25

Ans. $F_{EB} = 77.5$ kN.

4.26. The flexural rigidities of the three members of the portal frame ABCD (shown in the figure) are equal. The frame is pinned at A and D while the joints B and C are rigid. Show that, under the given loading, the horizontal thrust developed is

$$\frac{3Wab}{2c(3L + 2c)}$$

a b W B C c A D L

Problem 4·26

4.27. In all members of the structure shown the ratio of length to cross-sectional area is constant and all members are of the same modulus of direct elasticity. It is intended that the members should be pre-stressed by forcing DH into place after erection. Determine the necessary lack of fit of this member if the required force in DE is 100 kN tensile.

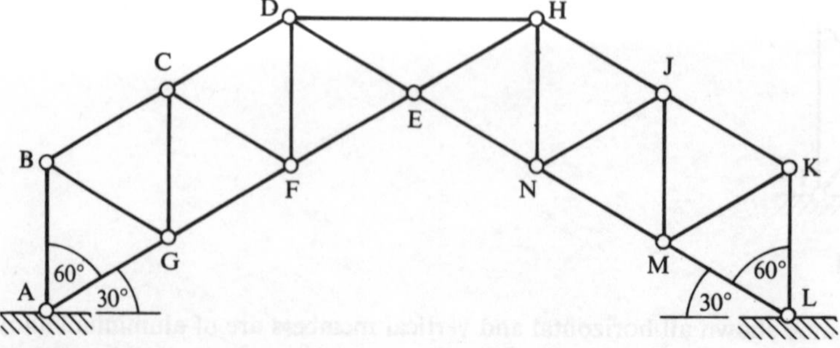

Problem 4·27

Ans. $381\sqrt{(3)}L/AE$.

4.28. A cantilever AB, encastré at A is formed into a quadrant of a circle of radius R. The free end, B, is acted on by a moment (in a plane at right angles to the cantilever section) and a torque, both of the same magnitude M. The cantilever is of circular cross-section of diameter d ($d \ll R$) and $G = \frac{2}{5}E$, where G and E are the shear and direct modulus of elasticity respectively. Neglecting all contributions to the deflection other than those due to bending and torsion, show that the twist and change of slope at the free end B are equal and approximately of magnitude 33.45 (MR/Ed^4).

4.29. A uniform beam is continuous over a large number of spans each of length L. Application of moment M at one end produces unit rotation at that end.

Show that a beam of the same cross-section if simply supported over a span of $0.865L$ has the same stiffness.

4.30. A two pinned parabolic arch rib having a secant variation of flexural rigidity has a span of 40 m and rise of 4 m. If the arch carries a uniformly distributed load of 15 kN/m between points 25 m and 35 m from one end determine the values of the reaction forces.

Ans. Horizontal 203 kN
Vertical 37.5 kN, 112.5 kN

Chapter 5
Stiffness and flexibility

5.1 Stiffness influence coefficients—prismatic member

The concept of a stiffness influence coefficient has been introduced in general terms in Section 4.4, and in this section actual values of the coefficients will be calculated for straight members of constant cross-sectional area (prismatic members). Such members comprise the bulk of those employed in structural engineering. It will be assumed that displacements are small, so that secondary effects, such as the shortening of a beam due to bending, may be ignored, and that axial forces are much less than the Euler load.

A typical member is shown in Fig. 5.1–1. Associated with the member is a set of member axes, which will not, in general, coincide with the axes for other members of the structure. The x-axis is defined as lying along the centroidal line for the member, the positive direction being from end 1 to end 2. The y- and z-axes complete a right-handed system, and those are chosen to be principal axes for the member cross-section. It is assumed that a force applied in any one principal plane causes displacements in that plane only, and this implies that the shear centre of the section coincides with the centroid. For a discussion on shear centre see S. Timoshenko[1] in the References at the end of this chapter.

The notation used in this and later chapters for member displacements and forces has already been used in Chapter 4, but is repeated here for completeness.

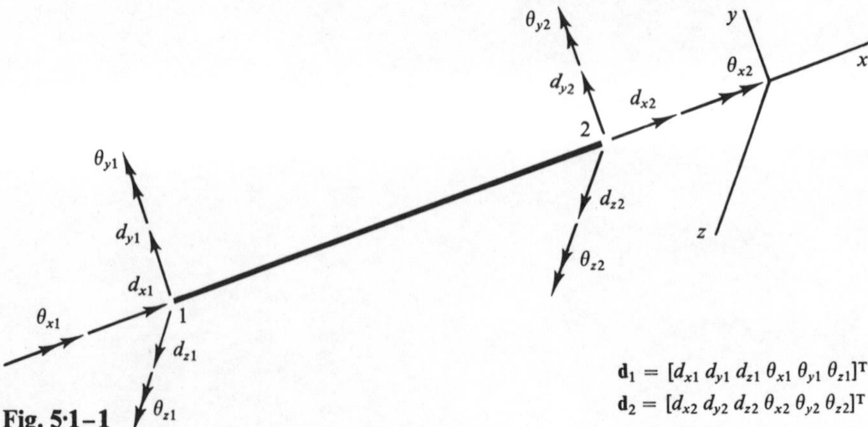

Fig. 5·1–1

There is the possibility of three linear displacements and three rotations at each end of the member, and these are illustrated in Fig. 5.1–1.

The letter d denotes linear displacements, and θ denotes rotations. The first suffix denotes the displacement direction, or the axis about which a rotation takes place, while the second suffix denotes the member end concerned. There are thus 12 possible displacement components for the member, or 12 degrees of freedom.

Associated with each displacement there is a corresponding force or moment, and these are shown in Fig. 5.1–2: p denotes direct forces and m denotes moments; p_x is an axial thrust; p_y and p_z are shears; m_x is a torsional moment; and m_y and m_z are bending moments.†

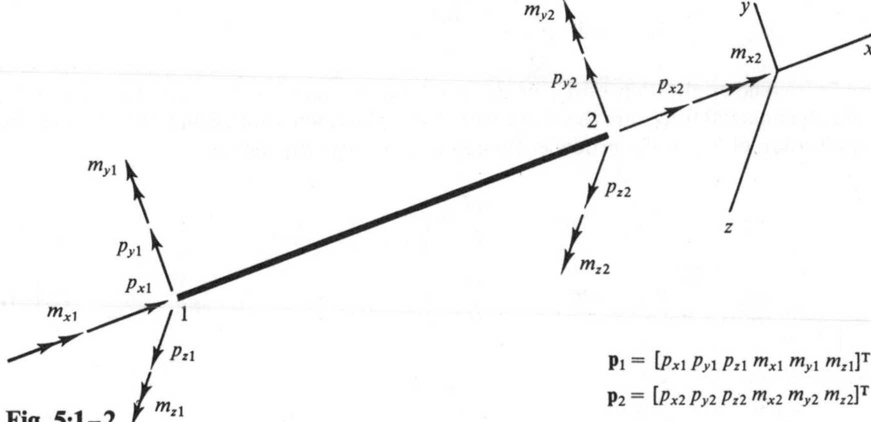

Fig. 5·1–2

The physical properties of the member are designated in the conventional manner—E, G, L, and A denote Young's modulus, shear modulus, length, and cross-sectional area respectively. The principal second moments of area for bending are I_y and I_z, the subscripts indicating the axes about which the second moments are taken. The polar second moment of area, which should logically be denoted by I_x is denoted by J which is the conventional symbol in torsion studies. As previously the word 'force' will frequently be used to imply either a direct force or a bending moment. Similarly, a 'displacement' may be either a linear displacement or a rotation.

The stiffness influence coefficients of the member are the actions imposed by the supporting medium when unit displacements occur in isolation at each end of the member in turn. These unit displacements are assumed to occur one at a time: while all other displacements are held at zero. They are indicated in Figs. 5.1–3 to 5.1–6. The resulting forces are always in equilibrium, and therefore six general equations may be drawn up:

$$\left.\begin{aligned} p_{x1} + p_{x2} &= 0 \\ p_{y1} + p_{y2} &= 0 \\ p_{z1} + p_{z2} &= 0 \end{aligned}\right\} \qquad (5.1\text{–}1)$$

$$\left.\begin{aligned} m_{x1} + m_{x2} &= 0 \\ m_{y1} + m_{y2} &= -p_{z1}L \\ m_{z1} + m_{z2} &= p_{y1}L \end{aligned}\right\} \qquad (5.1\text{–}2)$$

† Strictly speaking these forces and moments should be denoted as p_{xx1}, p_{xy1}, p_{xz1}, m_{xx1}, m_{xy1}, and m_{xz1} etc., since they are acting on a face normal to the x-axis. However, in the case of a line member (such as we have here) the first suffix x can be omitted without ambiguity.

Considering first displacement d_{x1} in Fig. 5.1–3(a):

$$p_{x1} = \frac{EA}{L} d_{x1} \tag{5.1–3}$$

Hence

$$p_{x2} = -\frac{EA}{L} d_{x1} \quad \text{(by Eqn 5.1–1)}$$

Displacement d_{x2} is treated in a similar manner—Fig. 5.1–5(a). Then

$$p_{x2} = \frac{EA}{L} d_{x2} \quad \text{and} \quad p_{x1} = -\frac{EA}{L} d_{x2}$$

The application of d_{x1} or d_{x2} produces restraints in the axial direction only, and all non-axial forces are therefore zero. As displacements are assumed to be small, the application of θ_{x1} or θ_{x2} produces torsional restraints only, hence

$$m_{x1} = \frac{GJ}{L}\theta_{x1} \qquad m_{x2} = -\frac{GJ}{L}\theta_{x1}$$

$$m_{x2} = \frac{GJ}{L}\theta_{x2} \qquad m_{x1} = -\frac{GJ}{L}\theta_{x2} \tag{5.1–4}$$

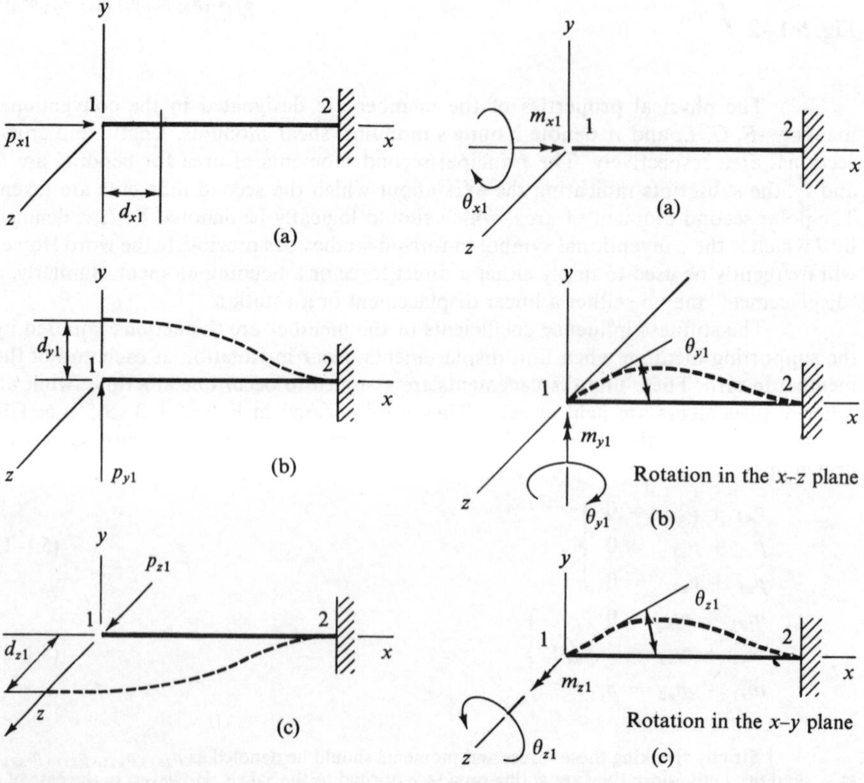

Fig. 5·1–3 Fig. 5·1–4

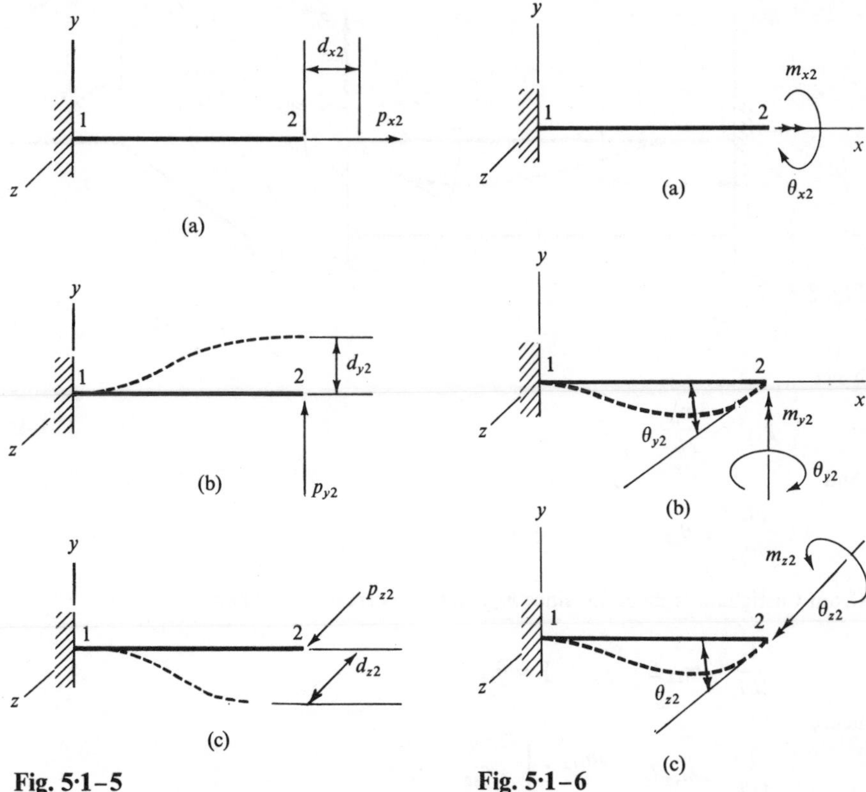

Fig. 5·1–5

Fig. 5·1–6

The influence coefficients involving θ_z, d_y, θ_y, and d_z will be determined using strain-energy methods (Castigliano's theorem, Section 4.17(b)) as an example of their use. The principle of virtual work could be used instead, and would lead to exactly the same expressions.

Rotation θ_z. The member 1–2 is initially straight, and is given an end rotation θ_{z2}. The bending moment at a section a distance from x from end 1 (M) is given by:

$$M = -m_{z1} + p_{y1}x \tag{5.1–5}$$

(see bending moment sign convention in Section 2.8). But

$$m_{z1} + m_{z2} - p_{y1}L = 0 \tag{5.1–6}$$

then

$$M = -p_{y1}(L - x) + m_{z2}$$

The strain energy (U) is

$$U = \int_0^L \frac{M^2\,dx}{2EI_z} = \frac{1}{2EI_z}\left[p_{y1}^2\frac{L^3}{3} + m_{z2}^2 L - p_{y1}m_{z1}L^2\right] \tag{5.1–7}$$

$$\frac{\partial U}{\partial p_{y1}} = d_{y1} = 0$$

(from Castigliano's theorem Eqn 4.17–9, since p_{y1} and d_{y1} correspond). Thus

$$\frac{2L^3}{3}p_{y1} - m_{z2}L^2 = 0$$

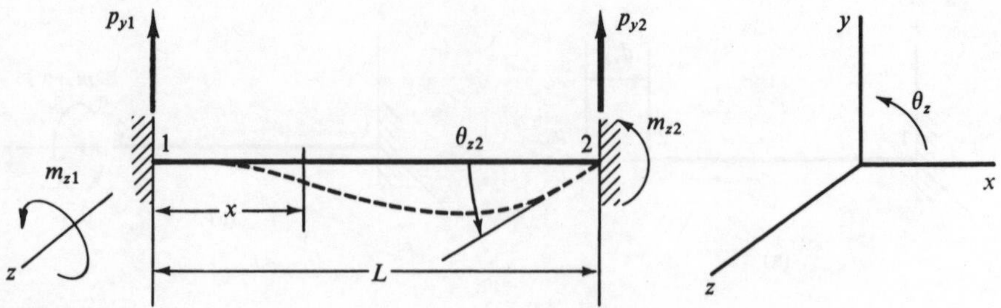

Fig. 5·1–7

hence

$$p_{y1} = \frac{3m_{z2}}{2L} \tag{5.1–8}$$

And

$$\frac{\partial U}{\partial m_{z2}} = \theta_{z2}$$

(from Castigliano's theorem, since m_{z2} and θ_{z2} correspond). Then

$$\frac{1}{2EI_z}[2m_{z2}L - p_{y1}L^2] = \theta_{z2}$$

hence

$$\frac{1}{2EI_z}\left[2m_{z2}L - \frac{3m_{z2}}{2}L\right] = \theta_{z2}$$

giving

$$m_{z2} = \frac{4EI_z}{L}\theta_{z2} \tag{5.1–9}$$

and

$$p_{y1} = \frac{6EI_z}{L^2}\theta_{z2} \tag{5.1–10}$$

From Eqn 5.1–6:

$$m_{z1} = p_{y1}L - m_{z2} = \frac{3m_{z2}}{2} - m_{z2} = \frac{m_{z2}}{2}$$

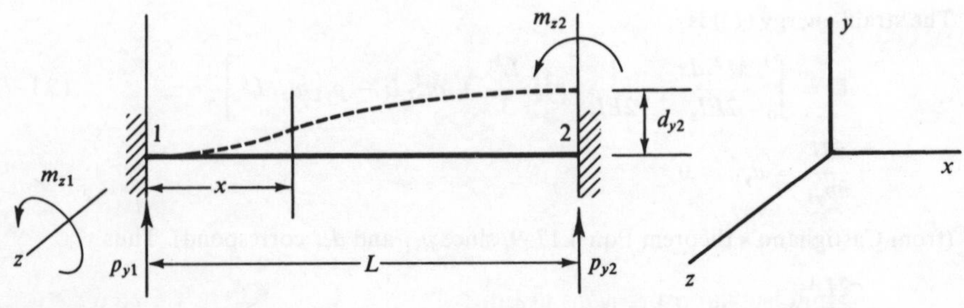

Fig. 5·1–8

Hence

$$m_{z1} = \frac{2EI_z}{L}\theta_{z2} \tag{5.1–11}$$

Similar expressions to Eqns 5.1–9 to 5.1–11 can be set up for rotation θ_{z1} by careful transposition of the suffices.

Displacement d_y. The initially straight member 1–2 is given a displacement d_{y2}. The bending moment at x is:

$$\begin{aligned} M &= -m_{z1} + p_{y1}x \\ &= -m_{z1} - p_{y2}x \end{aligned} \tag{5.1–12}$$

Then

$$U = \frac{1}{2EI_z}\left[m_{z1}^2 L + p_{y2}^2 \frac{L^3}{3} + m_{z1}p_{y2}L^2\right] \tag{5.1–13}$$

$$\frac{\partial U}{\partial m_{z1}} = 0$$

(from Castigliano's theorem, since θ_{z1} is zero). Thus

$$2m_{z1}L + p_{y2}L^2 = 0$$

hence

$$p_{y2} = -\frac{2m_{z1}}{L} \tag{5.1–14}$$

And

$$\frac{\partial U}{\partial p_{y2}} = d_{y2}$$

(from Castigliano's theorem, since p_{y2} and d_{y2} correspond). Then

$$\frac{1}{2EI_z}\left[2p_{y2}\frac{L^3}{3} + m_{z1}L^2\right] = d_{y2}$$

Hence

$$m_{z1} = -\frac{6EI_z}{L^2}d_{y2} \tag{5.1–15}$$

and from Eqn 5.1–6,

$$m_{z2} = -p_{y2}L - m_{z1} = 2m_{z1} - m_{z1} = -\frac{6EI_z}{L^2}d_{y2} \tag{5.1–16}$$

From Eqn 5.1–14

$$p_{y2} = \frac{12EI_z}{L^3}d_{y2} \tag{5.1–17}$$

Again, similar expressions to Eqns 5.1–15 to 5.1–17 can be set up for a displacement d_{y1} by transposition of the suffixes.

The terms still to be considered are those involving θ_y and d_z. These may be deduced from those in θ_z and d_y respectively (Eqns 5.1–9 to 5.1–11 and Eqns 5.1–15 to 5.1–17) with the aid of Fig. 5.1–9.

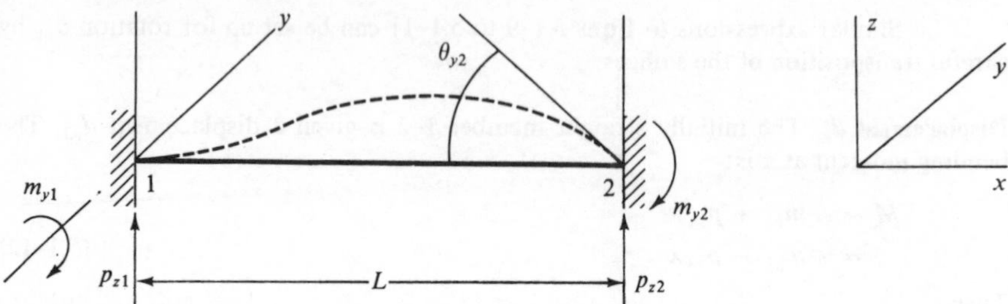

Fig. 5·1–9

5.2 Member stiffness and flexibility equations

The results of the previous section can be summarized in a single matrix equation for member stiffness shown in Eqn 5.2–1.

To allow Young's modulus E to be taken as a common factor the shear modulus G has been replaced by $E/2(1 + \nu)$ (ν is Poisson's ratio).

Equation 5.2–1 may be written in matrix form as:

$$\mathbf{p} = \mathbf{Kd} \tag{5.2–2}$$

This is the **member stiffness equation**. **p** and **d** are 12-term vectors of member force and displacement respectively, and **K** is a 12×12 member stiffness matrix. This is the stiffness matrix for the most general case of a prismatic member in space (neglecting shear deformation), and with the implicit condition that the deformations are so small as to leave the basic geometry unchanged.

Many structural members require less than the full number of 12 degrees of freedom to express their deformations. Since a member in a space truss can have no moments transmitted to it through its hinged ends, its deformation depends only on the three linear displacements at each end, giving it a total of six degrees of freedom. The stiffness matrices in such cases may be obtained by selecting the relevant terms from the full matrix of Eqn 5.2–1. Several examples appear in Chapter 7.

It is important to note the symmetry of the member stiffness matrix **K**, already discussed in Section 4.11.

It is not possible to solve the Eqns 5.2–2 as matrix **K** is essentially singular. This is because the member may be given an arbitrary rigid body movement (a movement involving no deformation) without affecting the end forces. There is thus an infinite number of vectors **d** in Eqn 5.2–2 for any given vector **p**. For example, the member BE of the plane pin-jointed truss in Fig. 5.2–1(a) is normal to the members DE and EC. So long as no external load is applied at E the force in BE is always zero, although loads at B and C produce deflections at the member ends B and E. Similarly, if the force F in Fig. 5.2–1(b) is collinear with BE, then the force in BE is always F, whatever dis-

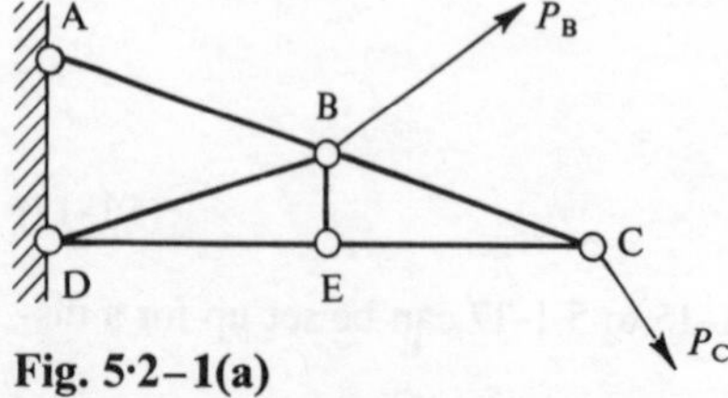

Fig. 5·2–1(a)

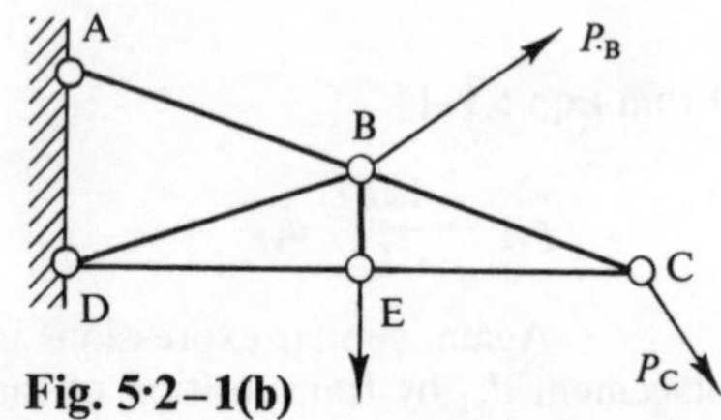

Fig. 5·2–1(b)

$$
\begin{array}{c}
1\\2\\3\\4\\5\\6\\7\\8\\9\\10\\11\\12
\end{array}
\begin{bmatrix}
p_{x1}\\ p_{y1}\\ p_{z1}\\ m_{x1}\\ m_{y1}\\ m_{z1}\\ p_{x2}\\ p_{y2}\\ p_{z2}\\ m_{x2}\\ m_{y2}\\ m_{z2}
\end{bmatrix}
= E
\begin{array}{c}
\begin{array}{cccccccccccc}
1 & 2 & 3 & 4 & 5 & 6 & 7 & 8 & 9 & 10 & 11 & 12
\end{array}\\
\begin{bmatrix}
A/L & 0 & 0 & 0 & 0 & 0 & -A/L & 0 & 0 & 0 & 0 & 0\\
0 & \frac{12I_z}{L^3} & 0 & 0 & 0 & \frac{6I_z}{L^2} & 0 & \frac{-12I_z}{L^3} & 0 & 0 & 0 & \frac{6I_z}{L^2}\\
0 & 0 & \frac{12I_y}{L^3} & 0 & \frac{-6I_y}{L^2} & 0 & 0 & 0 & \frac{-12I_y}{L^3} & 0 & \frac{-6I_y}{L^2} & 0\\
0 & 0 & 0 & \frac{J}{2L(1+v)} & 0 & 0 & 0 & 0 & 0 & \frac{-J}{2L(1+v)} & 0 & 0\\
0 & 0 & \frac{-6I_y}{L^2} & 0 & \frac{4I_y}{L} & 0 & 0 & 0 & \frac{6I_y}{L^2} & 0 & \frac{2I_y}{L} & 0\\
0 & \frac{6I_z}{L^2} & 0 & 0 & 0 & \frac{4I_z}{L} & 0 & \frac{-6I_z}{L^2} & 0 & 0 & 0 & \frac{2I_z}{L}\\
-A/L & 0 & 0 & 0 & 0 & 0 & A/L & 0 & 0 & 0 & 0 & 0\\
0 & \frac{-12I_z}{L^3} & 0 & 0 & 0 & \frac{-6I_z}{L^2} & 0 & \frac{12I_z}{L^3} & 0 & 0 & 0 & \frac{-6I_z}{L^2}\\
0 & 0 & \frac{-12I_y}{L^3} & 0 & \frac{6I_y}{L^2} & 0 & 0 & 0 & \frac{12I_y}{L^3} & 0 & \frac{6I_y}{L^2} & 0\\
0 & 0 & 0 & \frac{-J}{2L(1+v)} & 0 & 0 & 0 & 0 & 0 & \frac{J}{2L(1+v)} & 0 & 0\\
0 & 0 & \frac{-6I_y}{L^2} & 0 & \frac{2I_y}{L} & 0 & 0 & 0 & \frac{6I_y}{L^2} & 0 & \frac{4I_y}{L} & 0\\
0 & \frac{6I_z}{L^2} & 0 & 0 & 0 & \frac{2I_z}{L} & 0 & \frac{-6I_z}{L^2} & 0 & 0 & 0 & \frac{4I_z}{L}
\end{bmatrix}
\end{array}
\begin{bmatrix}
d_{x1}\\ d_{y1}\\ d_{z1}\\ \theta_{x1}\\ \theta_{y1}\\ \theta_{z1}\\ d_{x2}\\ d_{y2}\\ d_{z2}\\ \theta_{x2}\\ \theta_{y2}\\ \theta_{z2}
\end{bmatrix}
\quad (5.2\text{–}1)
$$

placements are produced at B and E by loads P_B and P_C. This argument assumes throughout that the displacements are small.

There is thus no equation:

$$\mathbf{d} = \mathbf{K}^{-1}\mathbf{p} \tag{5.2–3}$$

($\mathbf{K}^{-1}$ would be a member flexibility matrix).

An equation similar to Eqn 5.2–3 can be written, however, in terms of the member's distortions, where distortion is defined as the displacement of end 2 of a member relative to end 1. In Fig. 5.2–2 a plane frame member 1–2 is shown as having

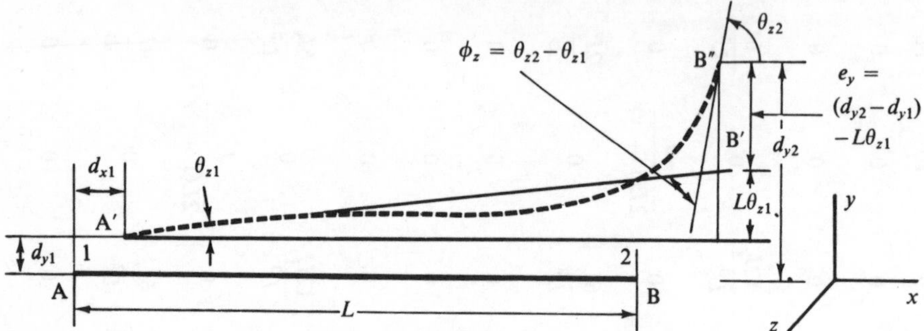

Fig. 5·2–2

been displaced from its initial position before loading (A–B) to its final position (A′–B″) after loading. The complete displacement can be thought of as having occurred in two parts:

(a) A rigid body movement from A–B to A′–B′. The position is defined by the final values of the displacements at end 1 (by $\mathbf{d}_1$).

(b) A deformation of 1–2 into the final position A′–B″.

As end 1 is already in its final position and orientation additional displacements are required at end 2 only. These additional displacements are the member distortions denoted by **e**, and are uniquely related to the forces on the member.

In this example of a member in a plane frame **e** is a three-term vector, but for a space frame member **e** will have six terms. The forces associated with the member deformations are those at end 2 ($\mathbf{p}_2$).

The modified form of Eqn 5.2–1 can now be deduced by setting displacement $\mathbf{d}_1$ equal to zero. Displacements $\mathbf{d}_2$ are then the member distortions. (Distortions in the x-, y-, and z-directions are denoted by e_x, e_y, and e_z respectively, and distortions about the x-, y-, and z-axes are denoted by ϕ_x, ϕ_y, and ϕ_z respectively.)

$$\begin{bmatrix} p_{x2} \\ p_{y2} \\ p_{z2} \\ m_{x2} \\ m_{y2} \\ m_{z2} \end{bmatrix} = E \begin{bmatrix} A/L & 0 & 0 & 0 & 0 & 0 \\ 0 & \dfrac{12I_z}{L^3} & 0 & 0 & 0 & -\dfrac{6I_z}{L^2} \\ 0 & 0 & \dfrac{12I_y}{L^3} & 0 & \dfrac{6I_y}{L^2} & 0 \\ 0 & 0 & 0 & \dfrac{J}{2L(1+\nu)} & 0 & 0 \\ 0 & 0 & \dfrac{6I_y}{L^2} & 0 & \dfrac{4I_y}{L} & 0 \\ 0 & -\dfrac{6I_z}{L^2} & 0 & 0 & 0 & \dfrac{4I_z}{L} \end{bmatrix} \begin{bmatrix} e_x \\ e_y \\ e_z \\ \phi_x \\ \phi_y \\ \phi_z \end{bmatrix} \tag{5.2–4}$$

The stiffness matrix here is the lower right-hand 6 × 6 block of **K**. The equivalent equation to Eqn 5.2–3 is:

$$\mathbf{e} = \mathbf{F}\mathbf{p}_2 \tag{5.2–5}$$

where $\mathbf{p}_2$ is the six-term vector of forces at end 2, **e** is a six-term vector of member distortions, and **F** is the inverse of the stiffness matrix of Eqn 5.2–4.

$$\mathbf{F} = \frac{1}{E}\begin{bmatrix} \frac{L}{A} & 0 & 0 & 0 & 0 & 0 \\ 0 & \frac{L^3}{3I_z} & 0 & 0 & 0 & \frac{L^2}{2I_z} \\ 0 & 0 & \frac{L^3}{3I_y} & 0 & -\frac{L^2}{2I_y} & 0 \\ 0 & 0 & 0 & \frac{2L(1+\nu)}{J} & 0 & 0 \\ 0 & 0 & -\frac{L^2}{2I_y} & 0 & \frac{L}{I_y} & 0 \\ 0 & \frac{L^2}{2I_z} & 0 & 0 & 0 & \frac{L}{I_z} \end{bmatrix} \tag{5.2–6}$$

The terms of **F** are a set of flexibility influence coefficients (see Section 4.4). They are the displacements which occur at end 2 of a member having end 1 rigidly held when unit loads are applied in turn at end 2. For example, the value $L^2/2I_z$ in the sixth column of the second row indicates that a member having a unit moment about the z-axis at end 2 has a relative displacement in the y-direction between ends 1 and 2 (e_y) of $L^2/2I_z$.

F is symmetric as expected (see Section 4.12).

Where there are fewer than 12 degrees of freedom, and fewer than six terms in the vector **e** are required to state the distortion of a member, a correspondingly smaller form of **F** can be set up. An example is given in Eqn 7.20–1, for a member in a plane frame.

5.3 Transformation of axes

The system of axes for a prismatic member used in earlier sections of this chapter was defined in Section 5.1. The x-axis was defined as coinciding with the centroidal line of the member. In a structure of many members there would thus be as many systems of axes. Before the internal actions in the members of the structure can be related, all forces and deflections must be stated in terms of one single system of axes common to all—the **structure axes**. The member axes are a right-handed system x, y, and z and the structure axes are also a right-handed system, x', y', and z' (Fig. 5.3–1).

The prime used to indicate structure axes is also used with forces and deflections. Thus, the force vector acting on end 1 of a member in structure axes is $\mathbf{p}'_1$, and

$$\mathbf{p}'_1 = [p'_{x1}\ p'_{y1}\ p'_{z1}\ m'_{x1}\ m'_{y1}\ m'_{z1}]^T \tag{5.3–1}$$

$\mathbf{p}'_2$, $\mathbf{d}'_1$, and $\mathbf{d}'_2$ are similarly defined vectors for force at end 2 and for deflection respectively.

The axis x makes angles θ'_{xx}, θ'_{xy}, and θ'_{xz} with three axes x', y', and z', and the cosines of these three angles are known as the 'direction cosines' of x with respect to

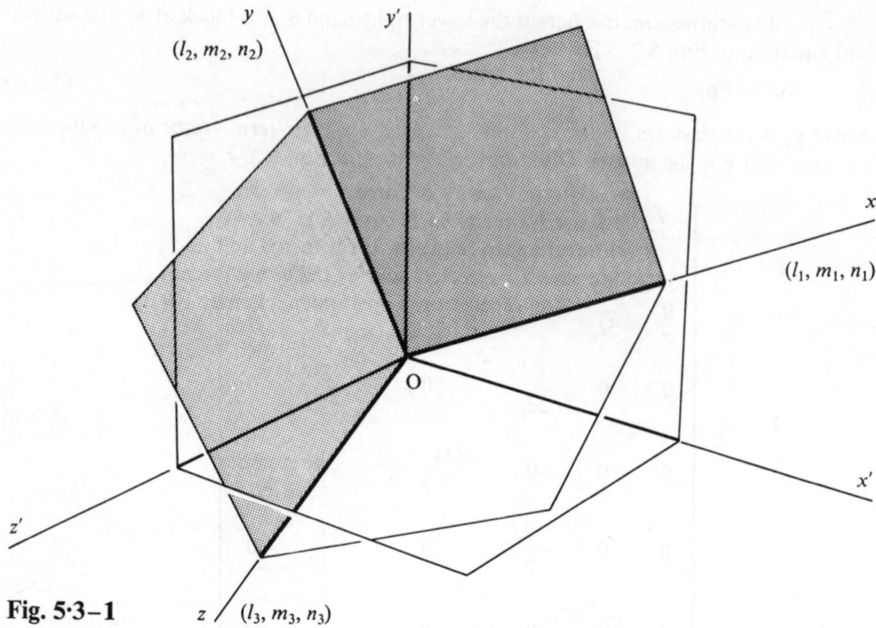

Fig. 5·3–1

x', y', and z'. They are denoted by l_1, m_1, n_1 respectively. Similarly, y and z have direction cosines with respect to x', y', and z' and these are denoted by l_2, m_2, n_2 and l_3, m_3, n_3.

If some vector OP is defined as in Fig. 5.3–2, then it has projections OA, OB, and OC on axes x', y', and z' respectively, as shown.

Then

component of OP in direction x

= projection of vector on x-axis

= sum of the projections of OA, OB, and OC on x

$= \mathrm{OA}l_1 + \mathrm{OB}m_1 + \mathrm{OC}n_1$

$= [l_1 \; m_1 \; n_1][\mathrm{OA} \; \mathrm{OB} \; \mathrm{OC}]^T$

i.e., $x = [l_1 \; m_1 \; n_1][x' \; y' \; z']^T$

Similar expressions can be developed for the components in the y- and z-directions (see Jeffrey[2]). These can be summarized in matrix form as:

$$\begin{bmatrix} x \\ y \\ z \end{bmatrix} = \begin{bmatrix} l_1 & m_1 & n_1 \\ l_2 & m_2 & n_2 \\ l_3 & m_3 & n_3 \end{bmatrix} \begin{bmatrix} x' \\ y' \\ z' \end{bmatrix} \tag{5.3–2}$$

or

$$\begin{bmatrix} x \\ y \\ z \end{bmatrix} = [\mathbf{R}_0] \begin{bmatrix} x' \\ y' \\ z' \end{bmatrix} \tag{5.3–3}$$

x, y, and z can be thought of as having been initially coincident with x', y', and z' and having been rotated through appropriate angles to bring them to their final positions. The direction cosine matrix of Eqn 5.3–2 can therefore be thought of as the 3×3 rotation matrix $\mathbf{R}_0$. Any quantity defined in terms of axes x', y', and z' can be redefined in terms of axes x, y, z by premultiplying by the rotation matrix. When used to redefine member forces and deflections in structure axes, this process is conventionally referred to as **transformation of axes**, and the symbol **T** is used. Hence

$$\mathbf{p}_1 = \mathbf{T}\mathbf{p}_1' \qquad \mathbf{p}_2 = \mathbf{T}\mathbf{p}_2'$$

$$\mathbf{d}_1 = \mathbf{T}\mathbf{d}_1' \qquad \mathbf{d}_2 = \mathbf{T}\mathbf{d}_2' \tag{5.3–4}$$

$\mathbf{p}_1$, $\mathbf{p}_2$, and $\mathbf{d}_1$, $\mathbf{d}_2$ are the six-term vectors of force and displacement at ends 1 and 2 respectively of member 1–2 used already in Section 5.1. Matrix **T** is thus 6 × 6. On transformation there is no interference between direct forces and moments. That is to say, direct forces in structure axes are affected only by the direct forces in member axes, and moments in structure axes are affected only by moments in member axes. The form of **T** used here is:

$$\mathbf{T} = \begin{bmatrix} \mathbf{R}_0 & 0 \\ 0 & \mathbf{R}_0 \end{bmatrix} \tag{5.3–5}$$

The terms and order of **T** are varied to suit the number and type of terms in associated vectors **p** and **d**. Several examples occur in Chapter 7.

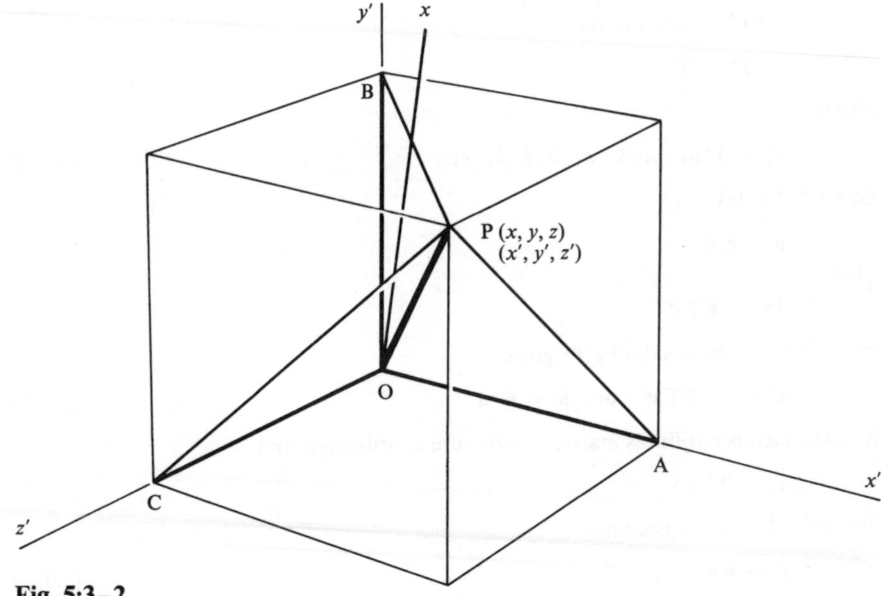

Fig. 5·3–2

One of the properties of the transformation matrix is that it is orthogonal, and its inverse is equal to its transpose. This may be demonstrated by carrying out the matrix multiplication $\mathbf{T}\mathbf{T}^T$. If the product equals unit matrix then $\mathbf{T}^T = \mathbf{T}^{-1}$. Then

$$\mathbf{T}\mathbf{T}^T = \begin{bmatrix} \mathbf{R}_0 & 0 \\ 0 & \mathbf{R}_0 \end{bmatrix} \begin{bmatrix} \mathbf{R}_0^T & 0 \\ 0 & \mathbf{R}_0^T \end{bmatrix}$$

$$= \begin{bmatrix} \mathbf{R}_0\mathbf{R}_0^T & 0 \\ 0 & \mathbf{R}_0\mathbf{R}_0^T \end{bmatrix}$$

The next stage is to examine the product $\mathbf{R}_0\mathbf{R}_0^T$.

$$\mathbf{R}_0\mathbf{R}_0^T = \begin{bmatrix} l_1 & m_1 & n_1 \\ l_2 & m_2 & n_2 \\ l_3 & m_3 & n_3 \end{bmatrix} \begin{bmatrix} l_1 & l_2 & l_3 \\ m_1 & m_2 & m_3 \\ n_1 & n_2 & n_3 \end{bmatrix}$$

From this multiplication two types of terms result:

(a) On the leading diagonal of the product matrix are terms such as $l_1^2 + m_1^2 + n_1^2$. This is the sum of the squares of the direction cosines between axes x and x', y and y', and z and z' and equals unity, i.e.

$$l_1^2 + m_1^2 + n_1^2 = 1$$

(b) In the off-diagonal positions the terms are of the type $l_1 l_2 + m_1 m_2 + n_1 n_2$, and all such sums equal zero (for a discussion of space vectors see Jeffrey[3]), i.e.

$$l_1 l_2 + m_1 m_2 + n_1 n_2 = 0$$

hence

$$\mathbf{R}_0\mathbf{R}_0^T = \text{unit matrix}$$

$$\mathbf{T}\mathbf{T}^T = \text{unit matrix}$$

$$\mathbf{T}^T = \mathbf{T}^{-1}$$

Hence

$$\mathbf{p}_1' = \mathbf{T}^T\mathbf{p}_1 \quad \text{and} \quad \mathbf{d}_1' = \mathbf{T}^T\mathbf{d}_1\text{, etc.} \tag{5.3–6}$$

Eqn 5.2–2 states

$$\mathbf{p} = \mathbf{K}\,\mathbf{d}$$

Then

$$\mathbf{T}\mathbf{p}' = \mathbf{K}\mathbf{T}\,\mathbf{d}'$$

Premultiplying both sides by $\mathbf{T}^T$ gives:

$$\mathbf{p}' = (\mathbf{T}^T\mathbf{K}\mathbf{T})\mathbf{d}' \quad \text{or} \quad \mathbf{p}' = \mathbf{K}'\,\mathbf{d}' \tag{5.3–7}$$

$\mathbf{K}'$ is the member stiffness matrix in structure coordinates, and

$$\mathbf{K}' = \mathbf{T}^T\mathbf{K}\mathbf{T} \tag{5.3–8}$$

Similarly, Eqn 5.2–4 becomes

$$\mathbf{e}' = \mathbf{F}'\mathbf{p}_2' \tag{5.3–9}$$

where $\mathbf{F}' = \mathbf{T}^T\mathbf{F}\mathbf{T}$. The orders of the two matrices $\mathbf{T}$ in Eqns 5.3–8 and 5.3–9 will, of course, not be the same. For Eqn 5.3–8:

$$\mathbf{T} = \begin{bmatrix} \mathbf{R}_0 & & & \\ & \mathbf{R}_0 & & \\ & & \mathbf{R}_0 & \\ & & & \mathbf{R}_0 \end{bmatrix} (12 \times 12)$$

For Eqn 5.3–9:

$$\mathbf{T} = \begin{bmatrix} \mathbf{R}_0 & \\ & \mathbf{R}_0 \end{bmatrix} \quad (6 \times 6)$$

Although in theory the direction cosine matrices for each member of a structure may be set up from the orientation of the members with respect to the structure axes, in practice this can cause some difficulty. It is convenient, therefore, to restate the matrix $\mathbf{R}_0$ in terms of the projections of the members on the structure axes. This can most easily be done by imagining the members as initially lying in the x'-direction with their y- and z-axes coinciding with y' and z', and then moving by a series of three rotations to their final positions. The rotations are (1) a rotation α about the y-axis; (2) a rotation β about the z-axis; and (3) a rotation γ about the x-axis. (Although there is a number of ways in which a member might be moved from its initial to its final position it is essential to this derivation that the order indicated is preserved.)

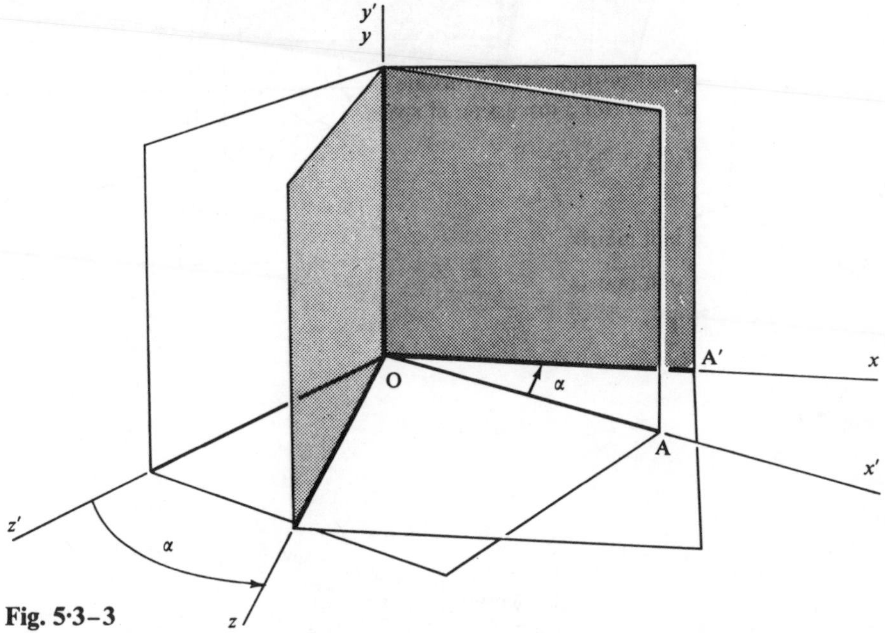

Fig. 5·3–3

Rotation α—about y-axis. Figure 5.3–3 shows the rotation α of the member OA to OA′. It is a rotation in the xz plane. The rotation matrix $\mathbf{R}_\alpha$ after the style of Eqn 5.3–2 is

$$\mathbf{R}_\alpha = \begin{bmatrix} \cos\alpha & 0 & -\sin\alpha \\ 0 & 1 & 0 \\ \sin\alpha & 0 & \cos\alpha \end{bmatrix} \tag{5.3–10}$$

At this stage,

$$\begin{bmatrix} x \\ y \\ z \end{bmatrix} = \mathbf{R}_\alpha \begin{bmatrix} x' \\ y' \\ z' \end{bmatrix} \tag{5.3–11}$$

Rotation β—about z-axis. Figure 5.3–4 shows the rotation β of the member from position OA′ to OA″. It is a rotation in the xy plane. The rotation matrix $\mathbf{R}_\beta$ is

$$\mathbf{R}_\beta = \begin{bmatrix} \cos\beta & \sin\beta & 0 \\ -\sin\beta & \cos\beta & 0 \\ 0 & 0 & 1 \end{bmatrix} \tag{5.3–12}$$

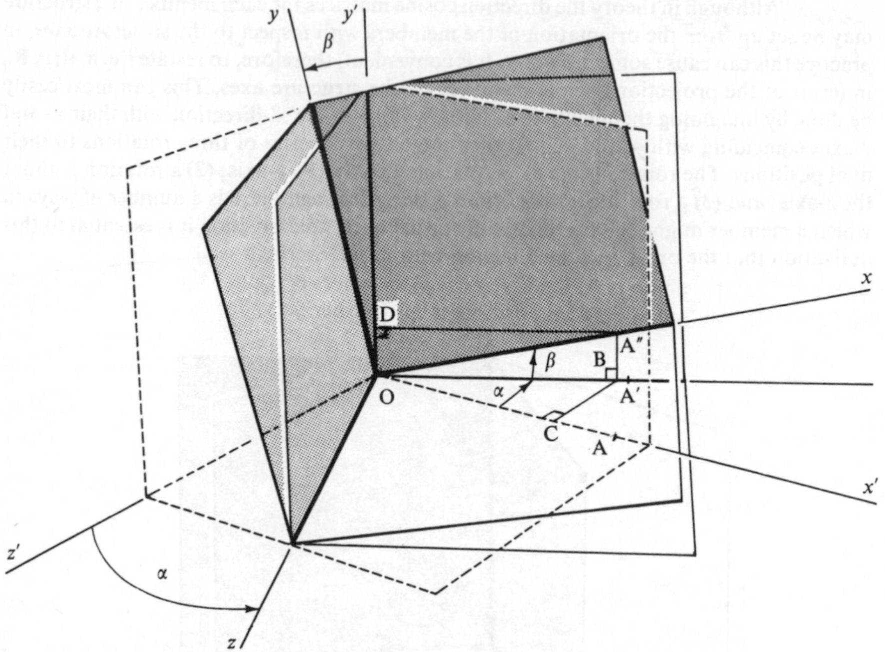

Fig. 5·3–4

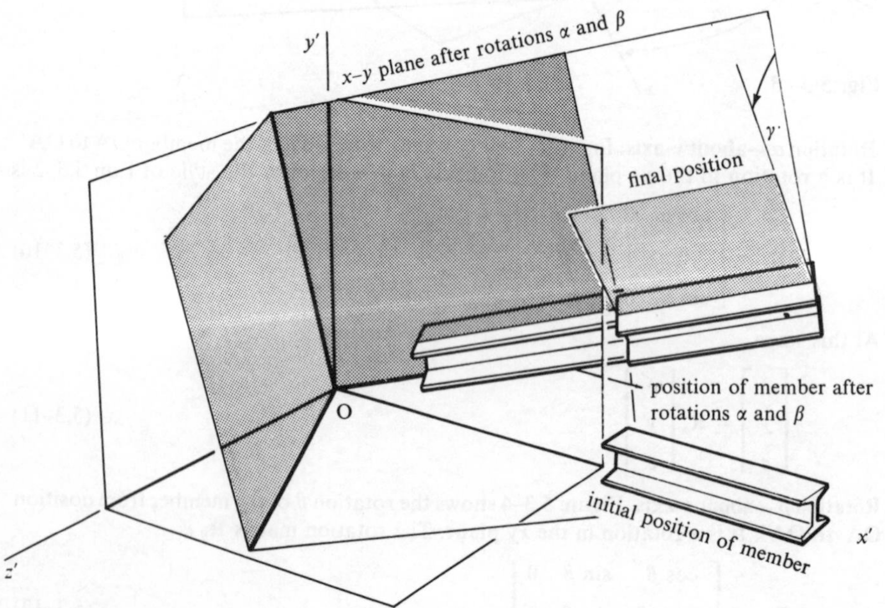

Fig. 5·3–5

The effect of this second rotation is obtained by premultiplying the result of Eqn 5.3–11 by $\mathbf{R}_\beta$. At this stage,

$$\begin{bmatrix} x \\ y \\ z \end{bmatrix} = \mathbf{R}_\beta \mathbf{R}_\alpha \begin{bmatrix} x' \\ y' \\ z' \end{bmatrix} \tag{5.3–13}$$

Rotation γ—about x-axis. Rotations α and β bring the member to its final position, but its z-axis need not be in the $x'z'$ plane. If, for instance, the $x'y'$ plane is a vertical plane, and the member in question is an I section with its web vertical when first placed along the x'-axis, then the web is still in a vertical plane after rotations α and β. If the web is inclined to the vertical in the final position, then a further rotation γ is required as shown in Fig. 5.3–5.

Figure 5.3–6 shows the final rotation γ in the yz plane about the member's own centroidal line.

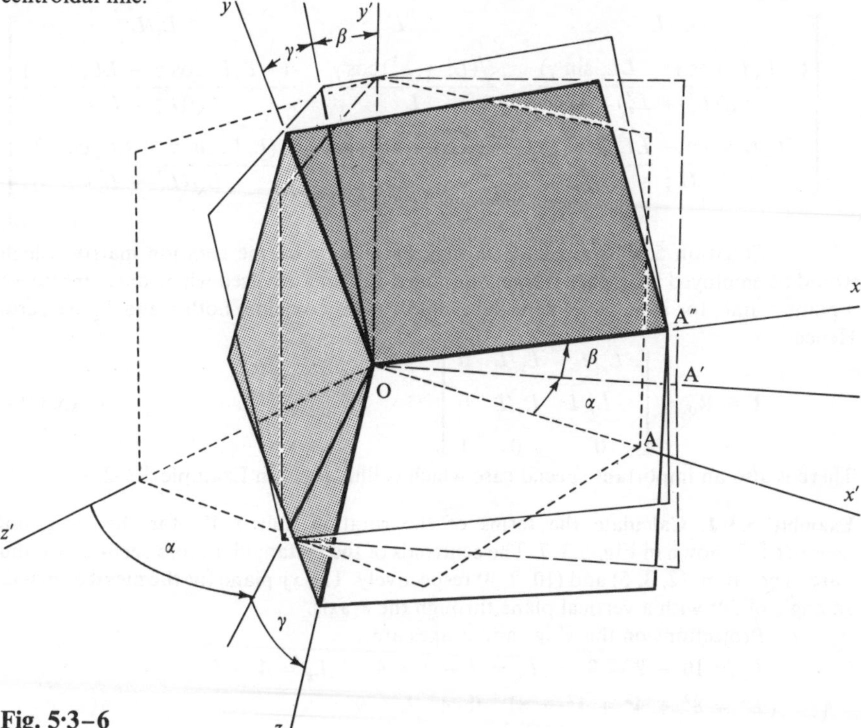

Fig. 5·3–6

The rotation matrix $\mathbf{R}_\gamma$ is

$$\mathbf{R}_\gamma = \begin{bmatrix} 1 & 0 & 0 \\ 0 & \cos\gamma & \sin\gamma \\ 0 & -\sin\gamma & \cos\gamma \end{bmatrix} \tag{5.3–14}$$

The effect of this final rotation is obtained by premultiplying the result of Eqn 5.3–13 by $\mathbf{R}_\gamma$. The final state is

$$\begin{bmatrix} x \\ y \\ z \end{bmatrix} = \mathbf{R}_\gamma \mathbf{R}_\beta \mathbf{R}_\alpha \begin{bmatrix} x' \\ y' \\ z' \end{bmatrix} \tag{5.3–15}$$

Then

$$\mathbf{R}_0 = \mathbf{R}_\gamma \mathbf{R}_\beta \mathbf{R}_\alpha$$

If the member OA is of length L, and its projections in its final position OA″ on the x'-, y'-, and z'-axes are L_x, L_y, and L_z respectively, then it can be seen from Fig. 5.3–4 that

$$\mathrm{OC} = L_x \qquad \mathrm{OD} = \mathrm{BA''} = L_y \qquad \mathrm{BC} = -L_z$$

$$\mathrm{OB} = \sqrt{(L^2 - L_y^2)} = \sqrt{(L_x^2 + L_z^2)}$$

Hence

$$\cos\alpha = \mathrm{OC}/\mathrm{OB} = L_x/\sqrt{(L_x^2 + L_z^2)}$$

$$\sin\alpha = -L_z/\sqrt{(L_x^2 + L_z^2)}$$

$$\cos\beta = \sqrt{(L_x^2 + L_z^2)}/L$$

$$\sin\beta = L_y/L$$

$$\mathbf{R}_0 = \mathbf{R}_\gamma \mathbf{R}_\beta \mathbf{R}_\alpha$$

$$= \begin{bmatrix} L_x/L & L_y/L & L_z/L \\ \dfrac{(-L_xL_y\cos\gamma - LL_z\sin\gamma)}{L\sqrt{(L_x^2 + L_z^2)}} & \dfrac{\sqrt{(L_x^2 + L_z^2)}\cos\gamma}{L} & \dfrac{(-L_yL_z\cos\gamma + LL_x\sin\gamma)}{L\sqrt{(L_x^2 + L_z^2)}} \\ \dfrac{(L_xL_y\sin\gamma - LL_z\cos\gamma)}{L\sqrt{(L_x^2 + L_z^2)}} & -\dfrac{\sqrt{(L_x^2 + L_z^2)}\sin\gamma}{L} & \dfrac{(L_yL_z\sin\gamma + LL_x\cos\gamma)}{L\sqrt{(L_x^2 + L_z^2)}} \end{bmatrix} \tag{5.3–16}$$

Equation 5.3–16 gives the most general form of the rotation matrix, which would be employed in a space frame. Simplified versions are used where appropriate. In a plane frame, for instance, where the z- and z'-axes coincide, both γ and L_z are zero. Hence

$$\mathbf{T} = \mathbf{R}_0 = \begin{bmatrix} L_x/L & L_y/L & 0 \\ -L_y/L & L_x/L & 0 \\ 0 & 0 & 1 \end{bmatrix} \tag{5.3–17}$$

There is also an important special case which is illustrated in Example 5.3–2.

Example 5.3–1. Calculate the terms of the rotation matrix $\mathbf{R}_0$ for the structural member 1–2 shown in Fig. 5.3–7. The centroids of the rectangular cross-section at 1 and 2 are in positions (2, 3, 5) and (10, 7, 4) respectively. The xy plane for the member makes an angle of 30° with a vertical plane through the x-axis.

Projections on the x', y', and z' axes are:

$$L_x = 10 - 2 = 8 \qquad L_y = 7 - 3 = 4 \qquad L_z = 4 - 5 = -1$$

$$L^2 = 8^2 + 4^2 + 1^2 = 81 \quad \text{then} \quad L = 9$$

Angle $\gamma = 30°$, $\cos\gamma = \sqrt{(3)}/2$, and $\sin\gamma = \frac{1}{2}$. Hence

$$\mathbf{R}_0 = \begin{bmatrix} 8/9 & 4/9 & -1/9 \\ \dfrac{-16\sqrt{3} + 4.5}{9\sqrt{65}} & \dfrac{\sqrt{3}\sqrt{65}}{18} & \dfrac{2\sqrt{3} + 36}{9\sqrt{65}} \\ \dfrac{16 + 4.5\sqrt{3}}{9\sqrt{65}} & -\dfrac{\sqrt{65}}{18} & \dfrac{-2 + 36\sqrt{3}}{9\sqrt{65}} \end{bmatrix}$$

$$= \begin{bmatrix} 0.889 & 0.445 & -0.111 \\ -0.320 & 0.775 & 0.544 \\ 0.328 & -0.449 & 0.831 \end{bmatrix}$$

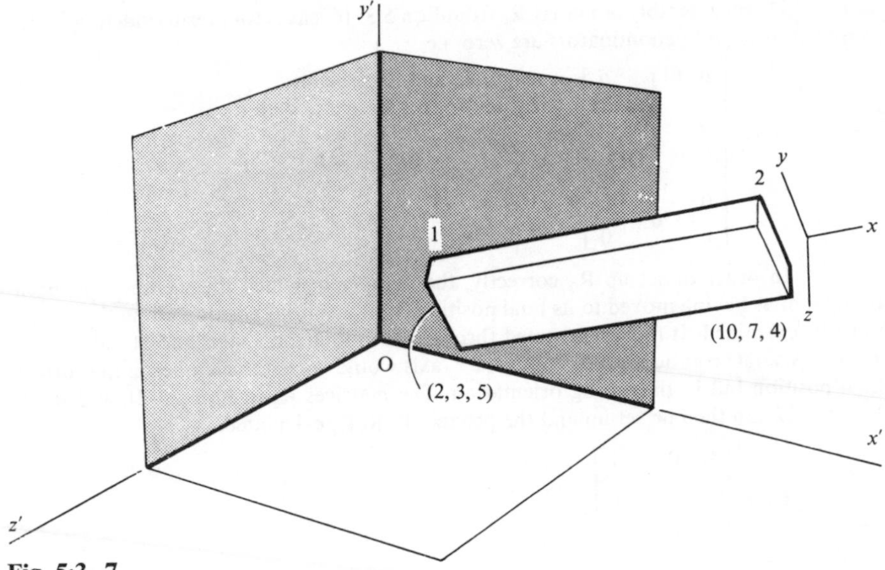

Fig. 5·3–7

(The reader should check the orthogonality of matrix $\mathbf{R}_0$ by calculating the product $\mathbf{R}_0\mathbf{R}_0^T$.)

Example 5.3–2. Calculate the terms of the rotation matrix $\mathbf{R}_0$ for the structural member 1–2 shown in Fig. 5.3–8. The centroidal line of 1–2 is coincident with the y'-axis, and the z-axis makes an angle γ with the x'-axis.

$$L_x = L_z = 0 \qquad L_y = L$$

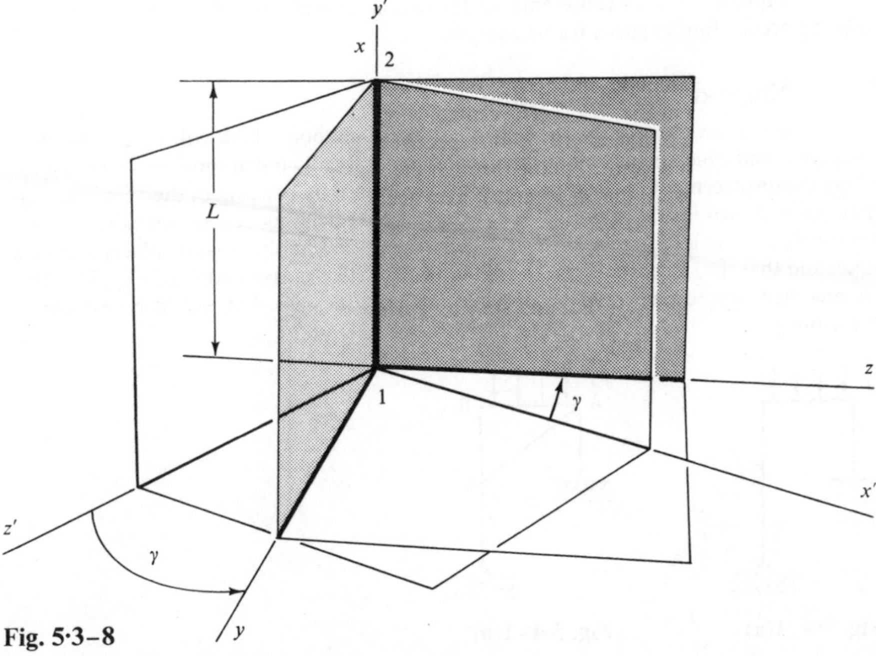

Fig. 5·3–8

Direct assembly of matrix $\mathbf{R}_0$ from Eqn 5.3–16 leaves four terms indeterminate, as numerators and denominators are zero, i.e.

$$\mathbf{R}_0 = \begin{bmatrix} 0 & 1 & 0 \\ \frac{0}{0} & 0 & \frac{0}{0} \\ \frac{0}{0} & 0 & \frac{0}{0} \end{bmatrix}$$

In order to set up $\mathbf{R}_0$ correctly, the manner in which the member has been thought of as having moved to its final position from its initial position along the x'-axis must be examined. It must be rotated through 90° about the y'-axis and 90° about the x'-axis. A single rotation of 90° about the z'-axis would leave the member in the correct final position but in the wrong orientation. The matrices $\mathbf{R}_\alpha$ in Eqn 5.3–10 and $\mathbf{R}_\beta$ in Eqn 5.3–12 can then be set up and the product $\mathbf{R}_\gamma\mathbf{R}_\beta\mathbf{R}_\alpha$ calculated.

$$\mathbf{R}_\alpha = \begin{bmatrix} 0 & 0 & -1 \\ 0 & 1 & 0 \\ 1 & 0 & 0 \end{bmatrix}$$

$$\mathbf{R}_\beta = \begin{bmatrix} 0 & 1 & 0 \\ -1 & 0 & 0 \\ 0 & 0 & 1 \end{bmatrix}$$

Then

$$\mathbf{R}_0 = \mathbf{R}_\gamma\mathbf{R}_\beta\mathbf{R}_\alpha = \begin{bmatrix} 0 & 1 & 0 \\ \sin\gamma & 0 & \cos\gamma \\ \cos\gamma & 0 & -\sin\gamma \end{bmatrix} \tag{5.3–18}$$

Equation 5.3–18 represents an important special case of matrix $\mathbf{R}_0$ which is likely to occur often in space frame analysis.

5.4 Slope-deflection method

This method is an example of the general stiffness method which will be discussed in Chapter 7, and which is suitable for the analysis of statically indeterminate *plane* frames, where the displacement of the joints is primarily caused by bending of the members. An example is shown in Fig. 5.4–1(a). It is assumed that bending can be considered quite separately from axial forces and shears. This assumption implies that displacements are small and that the structure is not triangulated, as with members ABC of Fig. 5.4–1(b). (A modified procedure would be possible In the latter case, but would be extremely laborious.)

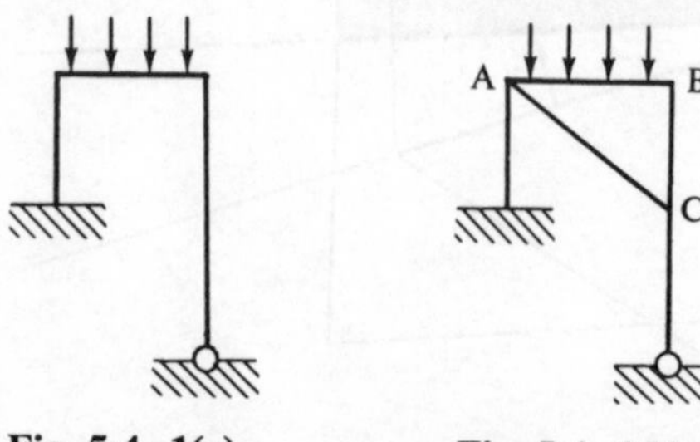

Fig. 5·4–1(a) Fig. 5·4–1(b)

Before proceeding with a description of the method, the slope-deflection equations will be derived. Figure 5.4–2 shows an initially straight prismatic member 1–2 of length L, second moment of area I, and having end rotations and lateral displacements. The suffix z has been omitted as unnecessary.

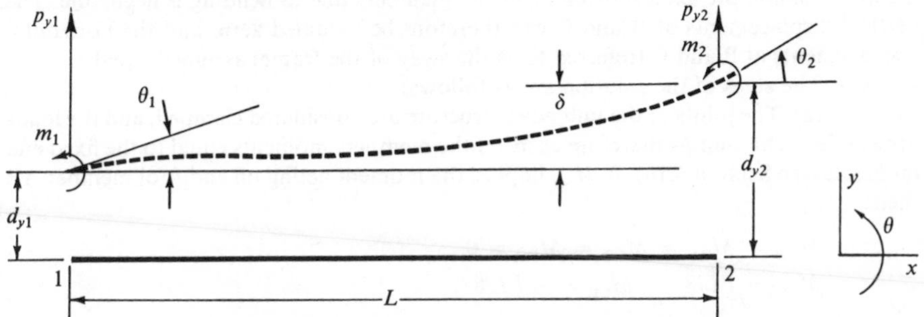

Fig. 5·4–2

The relations between the moments and displacements may be extracted from Eqn 5.2–1. These relations involve the difference between d_{y2} and d_{y1} and it is convenient to define δ as the relative lateral displacement.

$$\delta = d_{y2} - d_{y1}$$

Hence

$$m_1 = \frac{6EI}{L^2} d_{y1} + \frac{4EI}{L} \theta_1 - \frac{6EI}{L^2} d_{y2} + \frac{2EI}{L} \theta_2 \tag{5.4–1}$$

A similar equation may be set up for m_2. In matrix form these may be stated as:

$$\begin{bmatrix} m_1 \\ m_2 \end{bmatrix} = \frac{EI}{L} \begin{bmatrix} 4 & 2 & -6/L \\ 2 & 4 & -6/L \end{bmatrix} \begin{bmatrix} \theta_1 \\ \theta_2 \\ \delta \end{bmatrix} \tag{5.4–2}$$

These are the slope-deflection equations.

The use of these equations in analysing simple frames will now be discussed using the structure of Fig. 5.4–3 as an example.

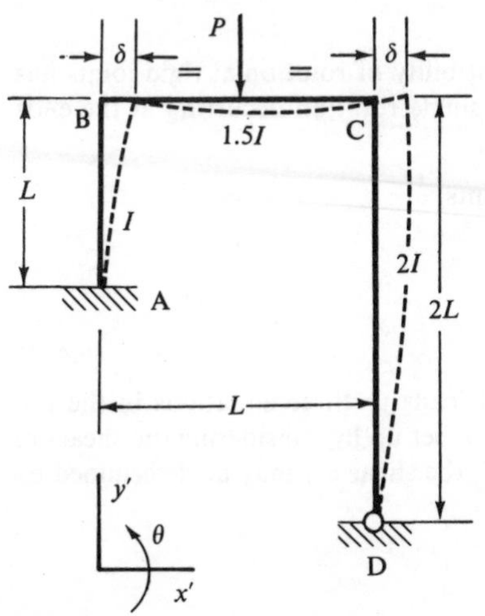

Fig. 5·4–3

The member AB is encastré at A, B, and C are rigid joints, and there is a pinned support at D. The section properties and lengths are as indicated. All members have the same Young's modulus. The loading consists of a central load P on BC. A possible deflection pattern for the structure is indicated by dotted lines. As the bending displacements are small, the axial shortening of the members due to bending is negligible. The vertical displacements at B and C can therefore be assumed zero, and the horizontal displacements at B and C (referred to as the sway of the frame) assumed equal.

The steps of the solution are as follows:

(a) The joints of the unloaded structure are considered clamped, and the loads are applied. The clamps therefore exert on the members moments equal to the fixed end moments (see Section 4.16). If M_{AB} implies the moment acting on end A of member AB then:

$$M_{AB} = M_{BA} = M_{CD} = M_{DC} = 0$$
$$M_{BC} = PL/8 \qquad M_{CB} = -PL/8$$

(b) The structure, with its joints free, is now analysed for a set of loads equal in value, but opposite in sense to the fixed end moments. The slope-deflection equations (5.4–1 and 5.4–2) may now be used to calculate the total moment acting on the ends of each member:

$$\begin{bmatrix} M_{AB} \\ M_{BA} \end{bmatrix} = \frac{EI}{L}\begin{bmatrix} 2 & 6 \\ 4 & 6 \end{bmatrix}\begin{bmatrix} \theta_B \\ \frac{\delta}{L} \end{bmatrix}$$

$$\begin{bmatrix} M_{BC} \\ M_{CB} \end{bmatrix} = \frac{1.5EI}{L}\begin{bmatrix} 4 & 2 \\ 2 & 4 \end{bmatrix}\begin{bmatrix} \theta_B \\ \theta_C \end{bmatrix}$$

$$\begin{bmatrix} M_{CD} \\ \\ M_{DC} \end{bmatrix} = \frac{2.0EI}{2L}\begin{bmatrix} 4 & 2 & \frac{6}{2} \\ \\ 2 & 4 & \frac{6}{2} \end{bmatrix}\begin{bmatrix} \theta_C \\ \theta_D \\ \frac{\delta}{L} \end{bmatrix} \tag{5.4–3}$$

In writing these six equations the compatibility of rotation at rigid joints has been recognized. For instance, θ_B represents the single rotation occurring at the ends B of AB and BC.

(c) Equilibrium is now applied at the joints.

$$M_{BA} + M_{BC} = -PL/8$$
$$M_{CB} + M_{CD} = +PL/8$$
$$M_{DC} = 0 \quad \text{(pin at D)} \tag{5.4–4}$$

(d) Substitution of Eqns 5.4–3 in 5.4–4 leads to three equations in the unknowns θ_B, θ_C, θ_D, and δ. A further equation can be set up by considering the shears in the columns AB and CD. Returning to Fig. 5.4–2, the shear p_{y1} may be determined by taking moments about the right-hand end:

$$p_{y1}.L - m_1 - m_2 = 0$$

or

$$p_{y1} = \frac{m_1 + m_2}{L} \tag{5.4–5}$$

The total shear on the structure is zero (or if there is an applied horizontal load, it sums to that load). In this example:

$$\frac{M_{AB} + M_{BA}}{L} + \frac{M_{CD} + M_{DC}}{2L} = 0 \tag{5.4–6}$$

(e) The four equations now obtained are:

$$\begin{bmatrix} -PL/8 \\ +PL/8 \\ 0 \\ 0 \end{bmatrix} = \frac{EI}{L} \begin{bmatrix} 10 & 3 & 0 & 6 \\ 3 & 10 & 2 & 3 \\ 0 & 2 & 4 & 3 \\ 6 & 3 & 3 & 15 \end{bmatrix} \begin{bmatrix} \theta_B \\ \theta_C \\ \theta_D \\ \dfrac{\delta}{L} \end{bmatrix}$$

These four equations may now be solved for the four unknowns θ_B, θ_C, θ_D, and δ.

$$\theta_B = -2.40 \times 10^{-2} \frac{PL^2}{EI} \qquad \theta_C = 2.04 \times 10^{-2} \frac{PL^2}{EI}$$

$$\theta_D = -1.68 \times 10^{-2} \frac{PL^2}{EI} \qquad \delta = 8.86 \times 10^{-3} \frac{PL^3}{EI} \tag{5.4–7}$$

(f) The values of the rotations and the sway may now be substituted back into Eqns 5.4–3 to obtain the member moments. The final bending moments in the members are obtained by adding these latter values to the moments of Step 1. The shears in the members are obtained from Eqn 5.4–5.

The bending moment diagram is shown in Fig. 5.4–4. It may be drawn by noting on which sides of the members the moment vectors produce tensile stresses.

It can be seen from this example that this method of analysis when handled manually is suitable only where the number of unknown displacements is small. Even this simple structure involved the solution of four simultaneous equations. The stiffness

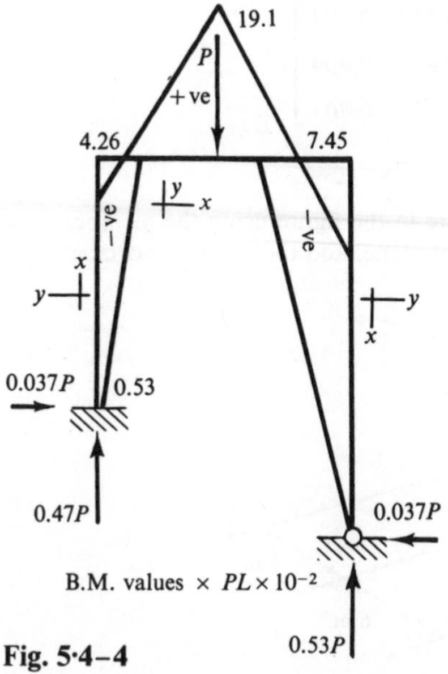

Fig. 5·4–4

method of analysis to be discussed in Chapter 7 is an adaptation of this process designed for computer handling.

Problems

5.1. Set up the rotation matrices for the members of the space frame in the figure using the member axes indicated. In each case the y-axis lies in a vertical plane, with the positive direction being upwards (except for member AC). The z-axis is always horizontal.

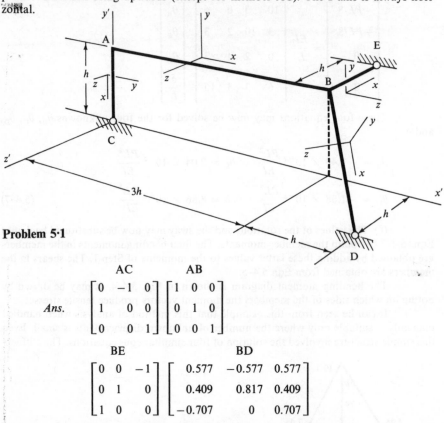

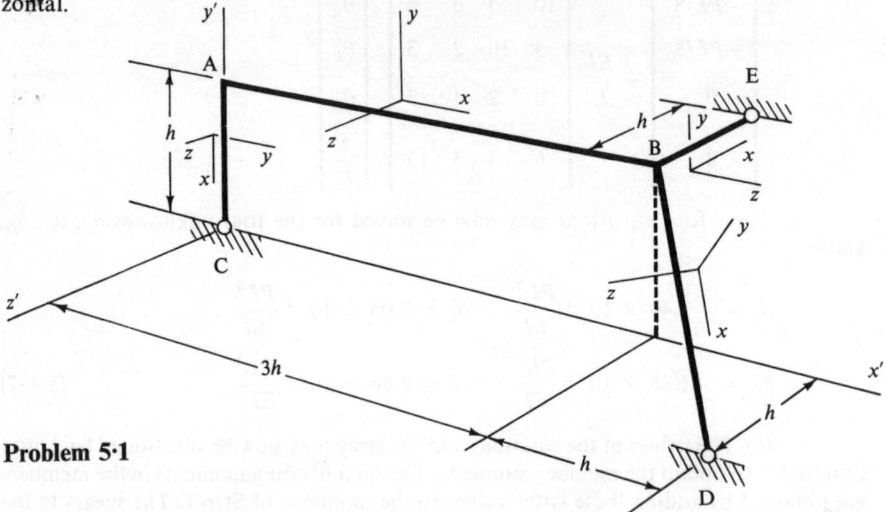

Problem 5·1

Ans.

AC, AB:

$$\begin{bmatrix} 0 & -1 & 0 \\ 1 & 0 & 0 \\ 0 & 0 & 1 \end{bmatrix} \quad \begin{bmatrix} 1 & 0 & 0 \\ 0 & 1 & 0 \\ 0 & 0 & 1 \end{bmatrix}$$

BE, BD:

$$\begin{bmatrix} 0 & 0 & -1 \\ 0 & 1 & 0 \\ 1 & 0 & 0 \end{bmatrix} \quad \begin{bmatrix} 0.577 & -0.577 & 0.577 \\ 0.409 & 0.817 & 0.409 \\ -0.707 & 0 & 0.707 \end{bmatrix}$$

AC is a special case similar to Example 5.3–2.

5.2. Set up the rotation matrices for the structure in the figure using the member axes indicated. In each case the y-axis lies in a vertical plane, and the z-axis is horizontal.

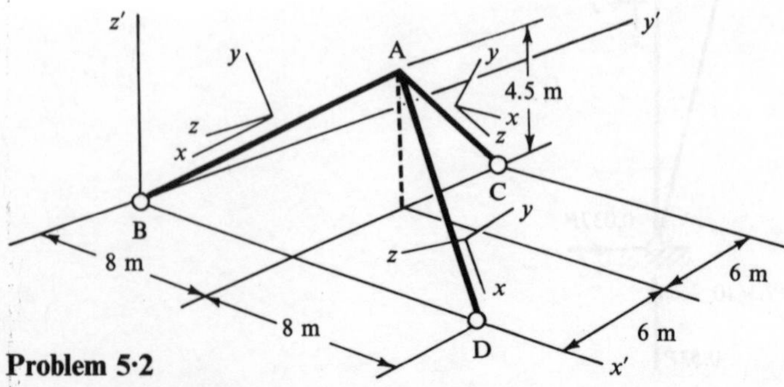

Problem 5·2

Ans.

$$\mathbf{R}_0 = \overset{AB}{\begin{bmatrix} -0.73 & -0.55 & -0.41 \\ -0.35 & -0.22 & 0.91 \\ -0.59 & 0.81 & 0 \end{bmatrix}} \overset{AC}{\begin{bmatrix} 0 & 0.80 & -0.60 \\ 0 & 0.60 & 0.80 \\ 1 & 0 & 0 \end{bmatrix}} \overset{AD}{\begin{bmatrix} 0.73 & -0.55 & -0.41 \\ 0.35 & -0.22 & 0.91 \\ -0.59 & -0.81 & 0 \end{bmatrix}}$$

5.3. The universal-beam purlins and rafters of the symmetrical warehouse structure in the figure are arranged such that the flanges are in parallel planes, as shown in the enlarged detail. Using the axes shown set up the rotation matrix for the member A–B.

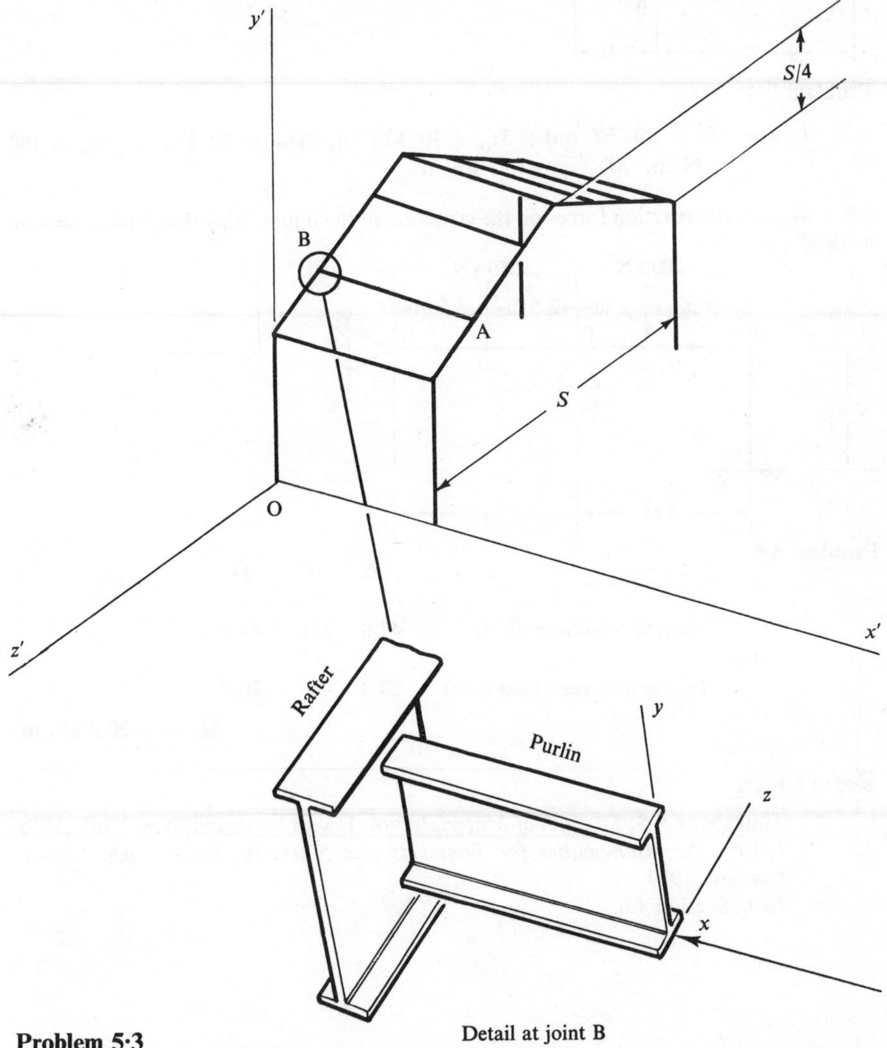

Problem 5·3

Detail at joint B

Ans. $\mathbf{R}_0 = \begin{bmatrix} -1 & 0 & 0 \\ 0 & 0.894 & 0.447 \\ 0 & 0.447 & -0.894 \end{bmatrix}$

5.4 Calculate the bending moments at A, B and D of the plane rigid-jointed structure in the figure using the slope-deflection method when the dimension L is 10m.

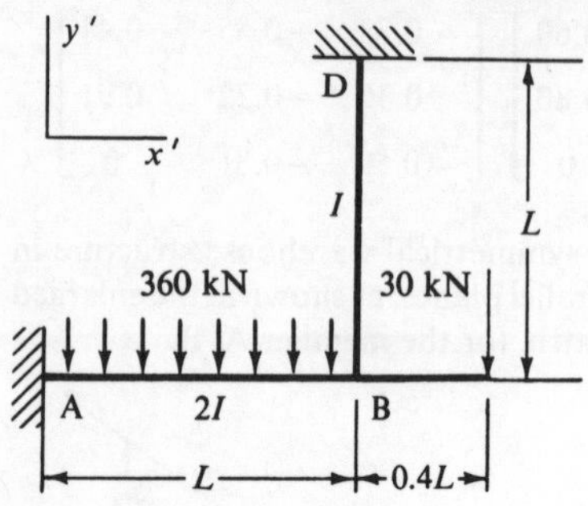

Problem 5·4

Ans. $\theta_B = 150/EI$ rad.; $M_D = 30$ kN m; $M_{BD} = 60$ kN m; $M_A = 360$ kN m; $M_{BA} = -180$ kN m.

5.5. Calculate the reaction forces on the structure in the figure using the slope-deflection method.

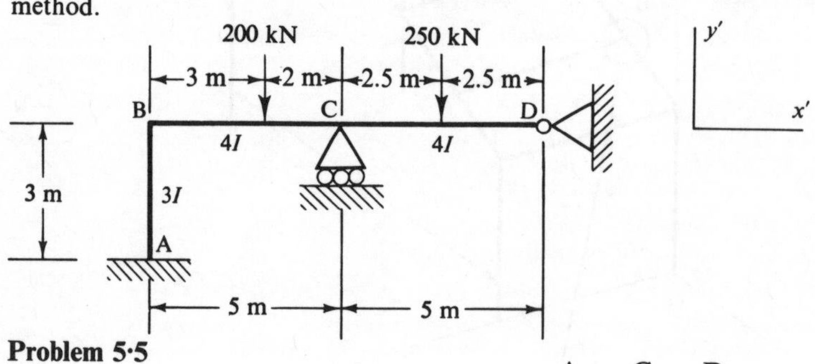

Problem 5·5

Ans.

	A	C	D
Vertical reactions (kN)	47.8	318	84.4
Horizontal reactions (kN)	20.8	—	−20.8

$M_A = -20.8$ kN m.

References

1 Timoshenko, S. *Strength of Materials*, vol. 1. Van Nostrand, New York, 1955.
2 Jeffrey, A. *Mathematics for Engineers and Scientists*, Section 9.9. Nelson, London, 1979.
3 *Ibid.*, Section 4.6.

Chapter 6
Moment distribution

6.1 Moment-distribution method

Prior to the development and ready availability of computers the difficulty of solution of large numbers of simultaneous equations had led engineers and scientists towards the use of methods of successive approximation for the solution of their problems. These methods rely on the continuing application of a relatively simple (but limited) procedure, which gives at any instant an approximate solution to the problem investigated. Repetition of the process can increasingly refine the accuracy of calculation, for the solution converges towards the numerically correct, and indeed it is one of the merits of the process that it can be stopped at any point where further refinement of numerical accuracy is not warranted by the premises.

For structural engineers the difficulties of analysis of a hyperstatic problem were considerably enhanced when the increasing use of welded steelwork or of monolithic concrete construction gave joints which obviously transmitted moment from an end of one member to others framing into it. The assumption that members might safely be regarded as pin-ended was plainly at variance with the facts and structures were erected in increasing numbers which relied for their stability and pleasing form on the rigidity of the joints. An effort to allow for this in analysis leads immediately to increased difficulty in calculation owing to the greater number of terms in the force vector of the matrix stiffness equation for each member and the limitations of classical methods become very speedily apparent. The method of solution known as **moment distribution** was first described in 1930 by Hardy Cross and dominated the field of analysis of such structures for the next 30 years. It remains a most powerful tool for the analyst without computing equipment and provides both a convenient conceptual mechanism and much of the terminology in everyday usage.

6.2 Fundamental concepts—terminology

As originally described, and as it will be dealt with in this text, moment-distribution processes are limited to planar structures, although this need not necessarily be the case; moreover, it is assumed that axial forces have sufficiently little effect on the behaviour of a member as to be of negligible importance. The first limitation will be understood when it is realized that extension to three dimensions increases the complexity of numerical solution, and, in rigid-jointed structures, requires as an added complication a knowledge of the torsional stiffnesses of all members, which are obviously unknown at the initial stages of design and frequently difficult to assess later. The second limitation, which implies the assumption that only the strain energy stored in bending is significant and the neglect of the effect of direct and shear forces and any associated instability is not as important as might appear at first sight. In Chapter 4 Section 4.16(c) it has already been explained that strain energy stored in bending is the greater part of the total in most civil engineering structures. Moreover, the designer does not want large deflections and will proportion members in such a way as to avoid them—he will wish to ensure that the projected length of a deformed member remains essentially as it was before deflection took place, and obviously no great inaccuracy will be introduced by such assumptions in an orthodox structure.

The relationship between forces acting on the ends of a member, and the corresponding deformation is given by the equation

$$\mathbf{p} = \mathbf{K}\mathbf{d} \quad \text{(see Eqn 5.2–2)}$$

in which **K** is the member stiffness matrix and **p** and **d** the force and displacement vectors respectively.

If attention is confined to members of constant cross-section, in which forces act and deformations occur in one plane only then, neglecting the effect of axial and transverse forces, and assuming that no relative sinkage of supports takes place the only remaining forces are m_{z1} and m_{z2} so that, from equation 5.2–1

$$\begin{bmatrix} m_{z1} \\ m_{z2} \end{bmatrix} = \begin{bmatrix} K_{6,6} & K_{6,12} \\ K_{12,6} & K_{12,12} \end{bmatrix} \begin{bmatrix} \theta_{z1} \\ \theta_{z2} \end{bmatrix}$$

$$= \frac{2EI_z}{L} \begin{bmatrix} 2 & 1 \\ 1 & 2 \end{bmatrix} \begin{bmatrix} \theta_{z1} \\ \theta_{z2} \end{bmatrix}$$

where EI_z is the appropriate flexural rigidity and L the length of a member.

The displacements here referred to are the rotations of the ends corresponding to the moments applied to the ends of the member and it is convenient (and in keeping with the established practice of most earlier texts) to use in this chapter only the sign convention of Fig. 6.2–1, in which a 'clockwise moments positive' rule is adopted. With

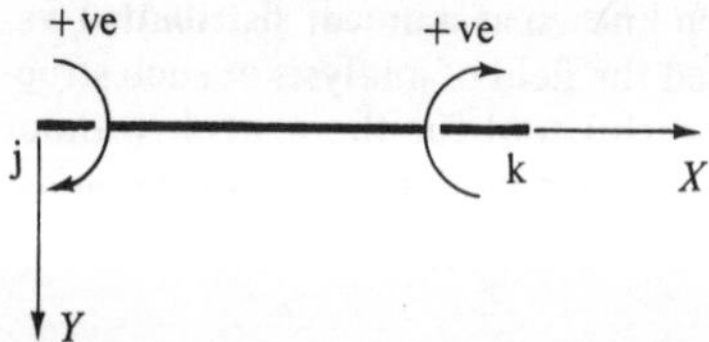

Fig. 6·2–1

the conventions and terminology there indicated the special matrix above may be rewritten to give

$$\begin{bmatrix} M_{jk} \\ M_{kj} \end{bmatrix} = \frac{2EI}{L} \begin{bmatrix} 2 & 1 \\ 1 & 2 \end{bmatrix} \begin{bmatrix} \theta_j \\ \theta_k \end{bmatrix} \qquad (6.2\text{–}1)$$

In the beam shown in Fig. 6.2–2, in which rotation θ_k is prevented it will be seen from this special matrix that

$$\begin{aligned} M_{jk} &= \frac{4EI}{L}\cdot\theta_j \\ M_{kj} &= \frac{2EI}{L}\cdot\theta_j = \tfrac{1}{2}M_{jk} \end{aligned} \qquad (6.2\text{–}2)$$

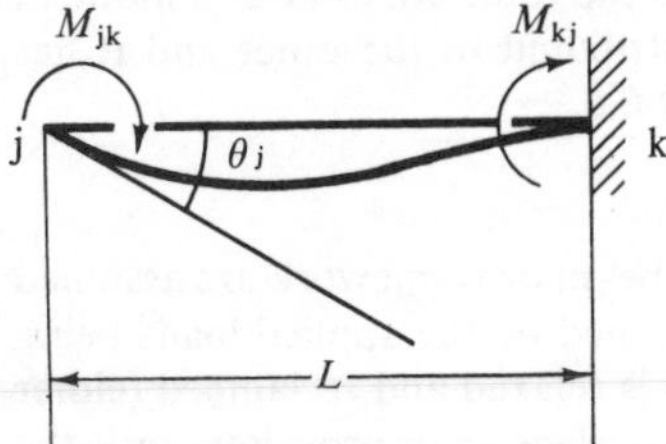

Fig. 6·2–2

The 'stiffness' of the beam jk (defined only as the moment required to cause unit rotation of one of its ends) is $4EI/L$, while application of such a moment causes a 'carry-over' of moment to the other (fixed) end of half its magnitude.

At an intermediate rigid joint B, capable of rotation without translation as shown in Fig. 6.2–3 the application of moment M causes a rotation θ_B and, since compatibility requires that the same rotation occurs in each of the adjoining spans, it follows that

$$M = M_{BA} + M_{BC} = \left(\frac{4(EI)_1}{L_1} + \frac{4(EI)_2}{L_2}\right)\theta_B \qquad (6.2\text{–}3)$$

and that

$$\left.\begin{aligned} M_{AB} &= \frac{2(EI)_1}{L_1}\,\theta_B = \frac{M_{BA}}{2} \\ M_{CB} &= \frac{2(EI)_2}{L_2}\,\theta_B = \frac{M_{BC}}{2} \end{aligned}\right\} \qquad (6.2\text{–}4)$$

It will be seen that a moment applied at an intermediate support (such as B) divides itself between the members meeting at that point, in proportion to their stiffnesses for, from Eqns 6.2–3

$$M_{BA} = \left(\frac{4(EI)_1}{L_1}\right)\theta_B = \left(\frac{4(EI)_1/L_1}{4(EI)_1/L_1 + 4(EI)_2/L_2}\right)M \qquad (6.2\text{–}5)$$

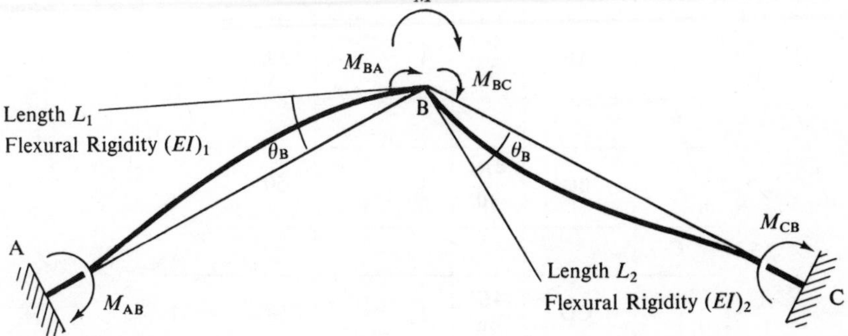

Fig. 6·2–3

Since, in most structures, the material of construction is the same throughout, Eqn 6.2–5 can be simplified to read

$$M_{BA} = \left(\frac{I_1/L_1}{I_1/L_1 + I_2/L_2}\right) M \tag{6.2–6}$$

in which the expression in the brackets is referred to as the **distribution factor** appropriate to a particular member, and the ratio I/L its **stiffness factor**.

At the encastré ends of members remote from the rotation, as at C, a moment is developed equal in magnitude to one-half that distributed to the other and it has become conventional to speak of a **carry-over factor** of 0.5.

6.3 Application to continuous beams

Initially, all the individual members of the continuous beam or framework are assumed to be encastré at their ends, the fixing moments occasioned by the applied loads being provided by external clamps at each joint. Each clamp is relaxed and reclamped (either singly or in groups) allowing rotation of joints to take place as appropriate until the relaxation procedure makes little sensible variation to the deflected shape, at which point the clamps are no longer required and the problem may be regarded as solved. It should be emphasized that only rotation of the joints takes place during the process. If sinkage of a support, or translation of a node occurs during any loading, special arrangements must be made (as will be shown later in Section 6.6) to calculate the effects of that relative sinkage.

The basic steps in a process of distribution are:

(a) To evaluate the distribution factors at each joint using Eqns 6.2–5 or 6.2–6.

(b) To determine the fixed-end moments in each span. These are normally represented in the convention implied in Figs. 6.2–1 and 6.2–2—clockwise positive. The table of fixed-end moments for standard load conditions, Fig. 6.3–1, will be helpful here.

(c) Release the joints in turn, distributing out-of-balance moments to the members meeting at a joint, and inserting carry-overs as appropriate.

(d) Continue until the effect of relaxation at a joint is negligible.

Example 6.3–1 A beam ABCD of three equal spans of uniform flexural rigidity is shown in Fig. 6.3–2. It supports a central point force in span BC and a uniformly distributed force as shown in CD. Determine the support moments.

SOLUTION The steps in the normal method of solution as just outlined are:

(a) the calculation of distribution factors:

Joint	*Member*	$\frac{4EI}{L}$	*Relative stiffnesses*	*Distribution factors*
A				
	AB	$\frac{4EI}{10}$	1	50%
B				
	BC	$\frac{4EI}{10}$	1	50%
C				
	CD	$\frac{4EI}{10}$	1	50%
D				

Fixed-end moments for uniform beams

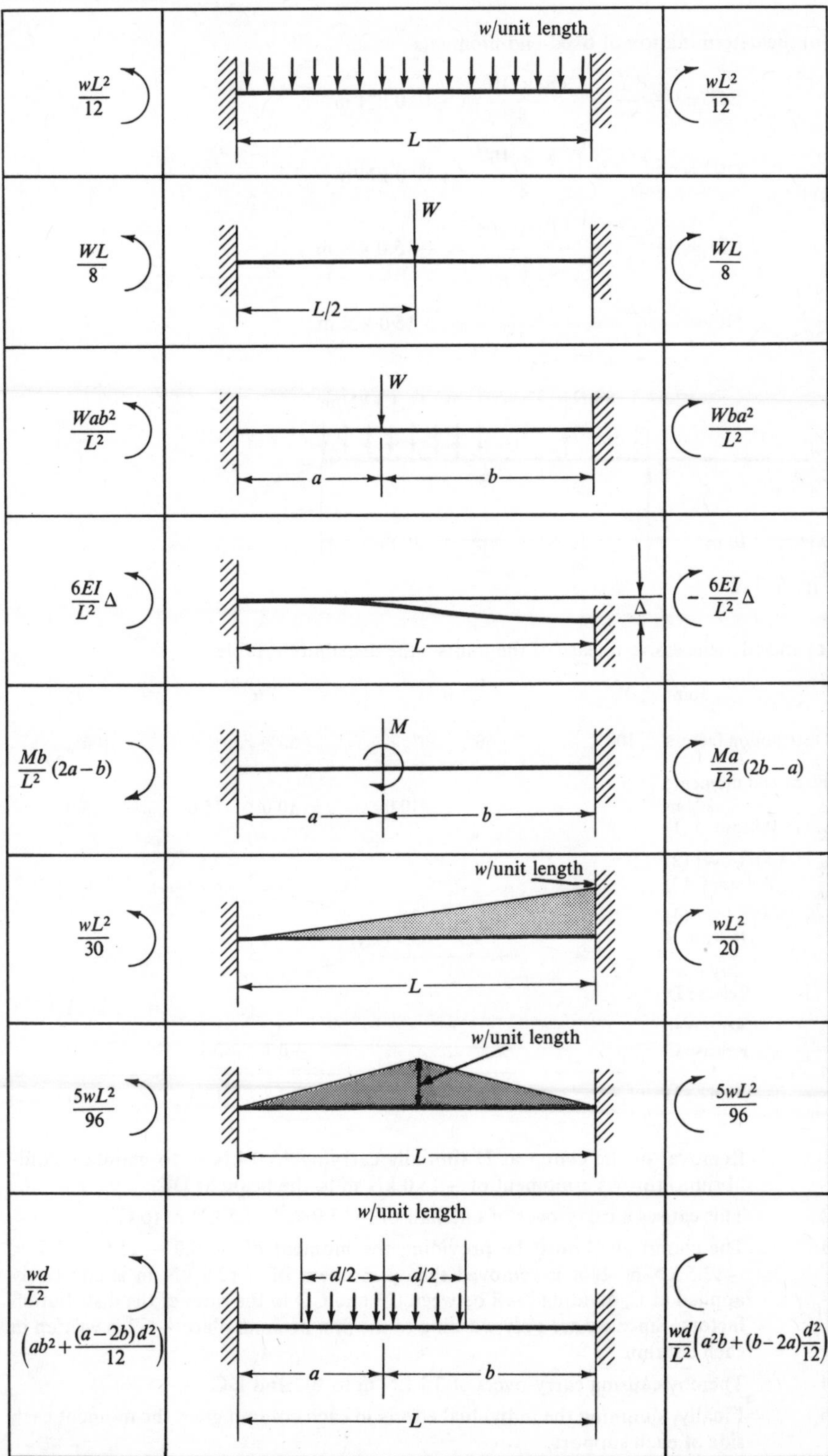

$\frac{wL^2}{12}$
w/unit length
L
$\frac{wL^2}{12}$
$\frac{WL}{8}$
W
L/2
$\frac{WL}{8}$
$\frac{Wab^2}{L^2}$
W
a
b
$\frac{Wba^2}{L^2}$
$\frac{6EI}{L^2}\Delta$
L
Δ
$-\frac{6EI}{L^2}\Delta$
$\frac{Mb}{L^2}(2a-b)$
M
a
b
$\frac{Ma}{L^2}(2b-a)$
$\frac{wL^2}{30}$
w/unit length
L
$\frac{wL^2}{20}$
$\frac{5wL^2}{96}$
w/unit length
L
$\frac{5wL^2}{96}$
$\frac{wd}{L^2}\left(ab^2+\frac{(a-2b)d^2}{12}\right)$
w/unit length
d/2
d/2
a
b
L
$\frac{wd}{L^2}\left(a^2b+(b-2a)\frac{d^2}{12}\right)$

Fig. 6·3–1

(b) the determination of fixed-end moments

$$BC = -\frac{WL}{8} = -\frac{8 \times 10}{8} = -10.0 \text{ kN m}$$

$$CB = +\frac{WL}{8} = +\frac{8 \times 10}{8} = +10.0 \text{ kN m}$$

$$CD = -\frac{wL^2}{12} = -\frac{1.8 \times 10^2}{12} = -15.0 \text{ kN m}$$

$$DC = +\frac{wL^2}{12} = +\frac{1.8 \times 10^2}{12} = +15{\cdot}0 \text{ kN m}$$

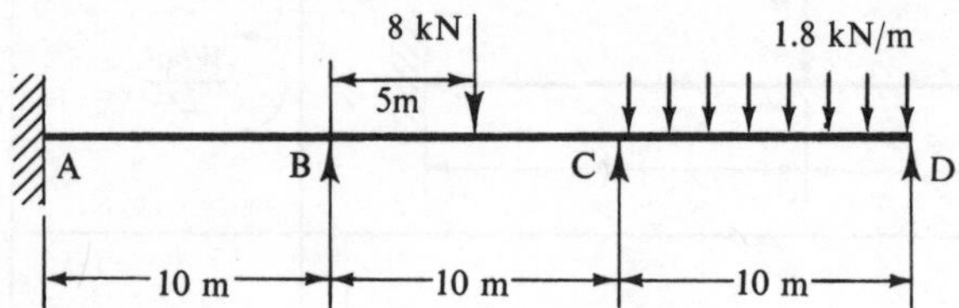

Fig. 6·3-2

(c) and (d) successive release of the joints—the distribution table.

Joint	A	B		C		D
Distribution factors	100%	50%	50%	50%	50%	100%
Fixed-end moments kN m			−10.0	10.0	−15.0	15.0
Release D (1)						−15.0
Carry-over (2)					−7.5	
Release C (3)				6.2	6.2 (say)	
Carry-over (4)			3.1			3.1
Release B		3.5	3.4 (say)			
Carry-over	1.7			1.7		
Release D						−3.1
Carry-over					−1.5	
Release C				−0.1	−0.1	
Sum (5)	1.7	3.5	−3.5	17.8	−17.9	0

1 Removal of the clamp at D (initially carrying 15.0 kN m to maintain equilibrium) throws a moment of −15.0 kN m to the beam at DC.

2 This causes a carry-over of one-half of −15.0 = −7.5 kN m to C.

3 The clamp at C must be providing the moment of +10.0 − 15.0 − 7.5 = −12.5 kN m. If it is removed then a moment of +12·5 kN m is effectively applied at C, dividing itself between CB and CD in the ratio of the distribution factors. Since we are only working to the first decimal place +6.2 is written in each column.

4 Thereby causing carry-overs of 3.1 kN m to BC and DC.

5 Finally, summing the individual effects in each column gives the moment each side of each support.

ALTERNATIVE SOLUTION A more compact lay-out (which does not show to great advantage here, but is most helpful in complex cases) is obtained if all processes of release and of carry-over are carried out simultaneously and the structural significance of this step merits some special thought.

Applying this process to the example of Fig. 6.3–2 will give

Joint	A	B		C		D
Distribution factors (1)	100%	50%	50%	50%	50%	100%
Fixed-end moments (1)			−10.0	10.0	−15.0	15.0
Release (2)		5.0	5.0	2.5	2.5	−15.0
Carry-over (3)	2.5		1.2	2.5	−7.5	1.2
		−0.6	−0.6	2.5	2.5	−1.2
	−0.3		1.2	−0.3	−0.6	1.2
		−0.6	−0.6	0.4	0.5	−1.2
	−0.3		0.2	−0.3	−0.6	0.2
		−0.1	−0.1	0.5	0.4	−0.2
	0.0		0.2	0.0	−0.1	0.2
		−0.1	−0.1	0.0	0.1	−0.2
	0.0		0.0	0.0	−0.1	0.0
		0.0	0.0	0.0	0.1	0.0
	0.0		0.0	0.0	0.0	0.0
Sum (4)	1.8	3.6	−3.6	17.8	−17.8	0.0

1 Distribution factors and fixed-end moments are here unchanged from the earlier solution.

2 Joints B, C, and D are released simultaneously, any out-of-balance moment being provided (by removal of the clamp at each joint) and divided between members at a node in proportion to their distribution factors. After this operation a horizontal line is drawn in each column to indicate that equilibrium is achieved and the clamp restored.

3 Carry-overs of one-half are used in each member, as indicated by arrows. Approximations have been used when necessary to limit figures to the first decimal place.

4 As before, an algebraic sum of the various components of each column gives the total moment at the end of a member.

Several features of this distribution table deserve further comment:

(a) End A is encastré in the given problem and is therefore never released.

(b) End D is a pinned end. During the process it is successively considered encastré, then freed, then reclamped and moments are carried over to it. This can be avoided and will be explained in Section 6.5.

(c) In numerical working, the first decimal place only has been used.

(d) Since end A is encastré, since there is no relative sinkage of A and B, and since no load is applied in the span AB, the final moments M_{AB} and M_{BA} should be in the ratio 1:2.

(e) The solution must give equilibrium at intermediate joints and the sum of all moments at joint B (say) should be zero.

(f) The analyst can only decide that the process has been carried sufficiently far after considering the changes occurring as a result of a complete cycle of operations. Generally it is sufficient to concentrate attention on the magnitudes of carry-over moments, but it is occasionally possible for the carry-over moments reaching all joints to be in the ratio of the distribution factors at those joints. A cycle of operations, therefore, makes no change in the summed value of the end moments.

6.4 Two-dimensional structures

Although the problem dealt with as Example 6.3–1 is representative of a skeletal line structure, there is no inherent limitation which prevents the application of the basic concepts of Section 6.3 to a two-dimensional structure—although occasionally there can be difficulties in the lay-out of the distribution table. It is necessary only to be certain (at this stage—the matter is explored more fully in Section 6.5) that no movement of the nodal points (usually referred to as side-sway) takes place during the loading process.

Example 6.4–1. A rigid-jointed single-bay portal ABCD is shown in Fig. 6.4–1, relative values of the second moment of area being shown alongside each member. It is subjected to a uniformly distributed force along BC and prevented from any sideways movement by a horizontal force *P*, at B. Determine the nodal moments.

SOLUTION A convenient lay-out of the distribution table is achieved in such a case as shown below

Joint	A	B		C		D
Distribution factors	100%	25%	75%	67%	33%	100%
Fixed-end moments kN m			−12.5	12.5		
Release		3.1	9.4	−8.3	−4.2	
Carry-over	1.5		−4.2	4.7		−2.1
Release		1.0	3.2	−3.1	−1.6	
Carry-over	0.5		−1.6	1.6		−0.8
Release		0.4	1.2	−1.1	−0.5	
Carry-over	0.2		−0.6	0.6		−0.2
Release		0.1	0.5	−0.4	−0.2	
Carry-over	0.0		−0.2	0.3		−0.1
Release			0.2	−0.2	−0.1	
Carry-over			−0.1	0.1		
Release			0.1	−0.1		
Sum	2.2	4.6	−4.6	6.6	−6.6	−3.2 kN m

As before, it should be noted that since B and C cannot translate relative to the encastré joints A and D (i.e. the frame has not swayed) and there are no lateral loads, the moments induced at these encastré feet of stanchions AB and DC should be half the moments induced at eaves level.

The propping force required to prevent sway may be readily obtained from consideration of the horizontal equilibrium of beam BC, as shown in the 'exploded' view (or 'free-body' diagram as it is sometimes called) of Fig. 6.4–1(b). Taking moments about the end A of forces acting on the stanchion AB shows that

$$H_A = \frac{M_{AB} + M_{BA}}{L_{AB}} = \frac{2.2 + 4.6}{5} = \frac{6.8}{5}\,\text{kN}$$

and similarly for CD that

$$H_D = \frac{M_{CD} + M_{DC}}{L_{CD}} = \frac{-6.6 - 3.2}{3.33} = \frac{-9.8}{3.33}\,\text{kN}$$

Equating horizontal forces on the stationary beam BC (or on the frame as a whole) shows that

$$P + H_A + H_D = 0$$

i.e.

$$P + \frac{M_{AB} + M_{BA}}{L} + \frac{M_{CD} + M_{DC}}{L_{CD}} = 0 \tag{6.4–1}$$

$$P = \frac{9.8}{3.33} - \frac{6.8}{5} = 1.59 \text{ kN}$$

The determination of this propping force (and of its subsequent elimination in appropriate cases) is of considerable importance in the analysis of more complex frames, and will be discussed further in Sections 6.7 and 6.8.

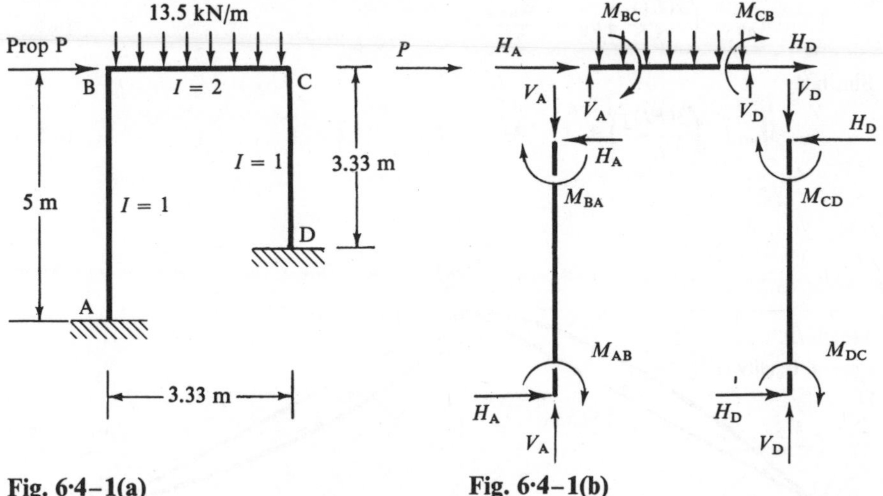

Fig. 6·4–1(a) Fig. 6·4–1(b)

6.5 Pinned and overhanging ends

As was indicated in note (b) following the solution of the problem of Example 6.3–1 the basic method explained earlier involved a complex process by which a pinned end was alternately assumed to be encastré, then freed, causing moments to be carried to it and subsequently balanced out. This labour can be largely eliminated by calculating special distribution factors applicable at a support next to a pinned end. Then, if necessary, that particular joint may be considered as encastré for the calculation of fixed-end moments only, then freed, and thereafter left undisturbed.

If moment M is applied at the junction, B, of two or more beams, such as AB and BC in Fig. 6.2–3, of lengths L_1 and L_2 and flexural rigidities $(EI)_1$ and $(EI)_2$, the general equations are:

$$\begin{bmatrix} M_{AB} \\ M_{BA} \\ M_{BC} \\ M_{CB} \end{bmatrix} = \begin{bmatrix} \frac{4(EI)_1}{L_1} & \frac{2(EI)_1}{L_1} & 0 \\ \frac{2(EI)_1}{L_1} & \frac{4(EI)_1}{L_1} & 0 \\ 0 & \frac{4(EI)_2}{L_2} & \frac{2(EI)_2}{L_2} \\ 0 & \frac{2(EI)_2}{L_2} & \frac{4(EI)_2}{L_2} \end{bmatrix} \begin{bmatrix} \theta_A \\ \theta_B \\ \theta_C \end{bmatrix} \tag{6.5–1}$$

Equilibrium will obviously require that

$$M = M_{BA} + M_{BC}$$

and some simplification would occur if θ_A and θ_C were zero, as shown, or might otherwise be determined.

In Fig. 6.5–1 in which the end C is pinned, while end A is encastré ($\theta_A = 0$), the equilibrium equation that $M_{CB} = 0$ will ensure that $\theta_C = -\theta_B/2$ and that

$$M = (M_{BA} + M_{BC}) = \left(\frac{4(EI)_1}{L_1} + \frac{3(EI)_2}{L_2}\right)\theta_B \tag{6.5–2}$$

while

$$M_{AB} = \left(\frac{2(EI)_1}{L_1}\right)\theta_B = \frac{M_{BA}}{2} \tag{6.5–3}$$

Similarly,

$$M_{BC} = \left(\frac{3(EI)_2}{L_2}\right)\theta_B, \qquad M_{CB} = 0 \tag{6.5–4}$$

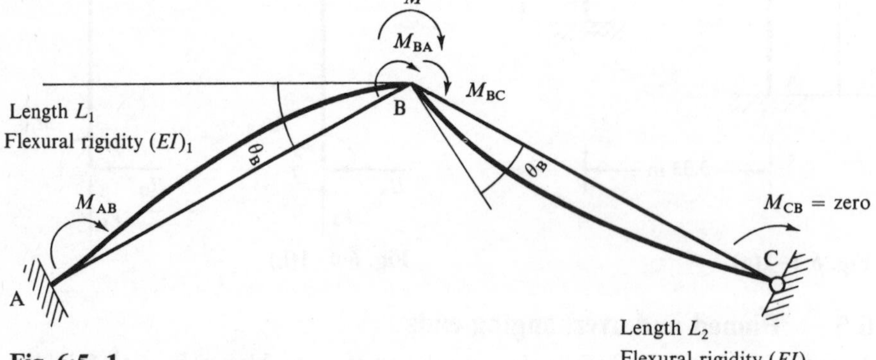

Fig. 6·5–1

If a moment M is applied at the node B in Fig. 6.5–1 it divides itself between the members meeting at B, in proportion to their stiffnesses, and

$$M_{BA} = \left(\frac{4(EI)_1/L_1}{4(EI)_1/L_1 + 3(EI)_2/L_2}\right)M \tag{6.5–5}$$

$$M_{BC} = \left(\frac{3(EI)_2/L_2}{4(EI)_1/L_1 + 3(EI)_2/L_2}\right)M \tag{6.5–6}$$

Assuming, as in Eqn. 6.2–6 that the structure is built of one single material,

$$\left.\begin{aligned} M_{BA} &= \left(\frac{I_1/L_1}{I_1/L_1 + \frac{3}{4}I_2/L_2}\right)M \qquad M_{AB} = \frac{M_{BA}}{2} \\ \text{and} \\ M_{BC} &= \left(\frac{\frac{3}{4}I_2/L_2}{I_1/L_1 + \frac{3}{4}I_2/L_2}\right)M \qquad M_{CB} = 0 \end{aligned}\right\} \tag{6.5–7}$$

from which it will be seen that the stiffness factor of a member with a remote pinned end is $\frac{3}{4}(I/L)$, whilst the appropriate carry-over is zero.

Example 6.5–1. Rework the problem shown in Fig. 6.3–2 using the simplification made possible by the above analysis.

SOLUTION

Support	A	B		C		D	
Stiffness factor		$\frac{I}{10}$		$\frac{I}{10}$		$\frac{3}{4}\frac{I}{10}$	
Distribution factor		50%	50%	57.2%	42.8%		
Fixed-end moments (1)			−10.0	10.0	−15.0	15.0	kN m
Release D (2)						−15.0	
Carry-over (3)					−7.5		
Release B and C (4)		5.0	5.0	7.2	5.3		
Carry-over (5)	2.5		3.6	2.5			
Release B and C		−1.8	−1.8	−1.4	−1.1		
Carry-over (6)	−0.9		−0.7	−0.9			
Release B and C		0.4	0.3	0.5	0.4		
Carry-over	0.2		0.2	0.2			
		−0.1	−0.1	−0.1	−0.1		
Sum	1.8	3.5	−3.5	18.0	−18.0	0	kN m

1 Fixed-end moments are calculated as before, on the assumption that ends B and C in BC and C and D in CD are all encastré, but distribution factors calculated using the stiffness factor appropriate to a member with a pinned end.

2 and 3 The pinned end, D, is then released, once and for all, giving the required equilibrium condition—in this case of zero moment. The appropriate carry-over is entered in the table.

4 and 5 All other joints are released, causing the appropriate carry-overs. It will be noted that no moment is carried to end D, which was released (i.e. put into the pinned-end condition) in 2 and has not been reclamped.

If comparison is made with the basic steps of Section 6.2 listed in Section 6.3, it will be seen that:

(a) distribution factors are evaluated for each joint using Eqns 6.2–6 or 6.5–7 as appropriate;

(b) fixed-end moments are calculated in an identical way in all cases;

(c) a pinned joint is released as the first step, the appropriate carry-over is performed, and the state of the joint is not changed subsequently—i.e. there is no carry-over to this end, and therefore no occasion to effect a subsequent release;

(d) successive iterations continue until out-of-balance moments and carry-overs are sufficiently small to be neglected.

Example 6.5–2. Determine the support moments in the beam of Fig. 6.5–2 in which relative values of the second moment of area are shown alongside each member.

SOLUTION A beam in which some loaded overhanging end occurs, as at joint D, may be treated in a manner very similar to the method used directly with the pinned end at D, Example 6.5–1 (Fig. 6.3–2) for, in both cases the bending moment at D is known. In Fig. 6.3–2 the requisite condition is that the bending moment there is zero, whereas in Fig. 6.5–2 the moment must be that caused by the cantilevered overhang of section DE. The process of distribution remains identical, but the stiffness of section DE (its resistance to rotation) must be zero, and once equilibrium at D is assured there need be no variation in the moment there—D may be treated as pinned.

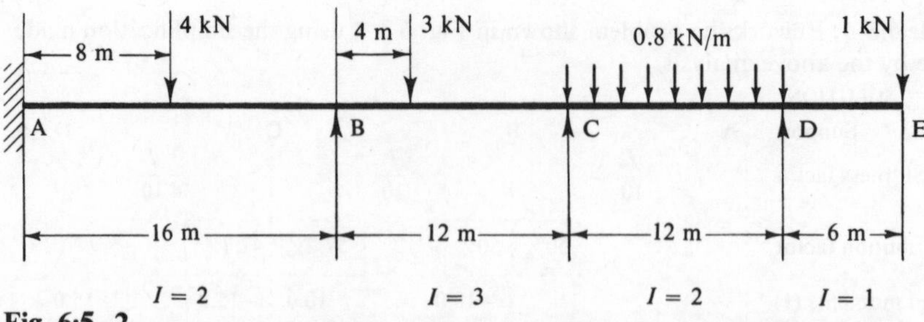

Fig. 6·5–2

In the treatment of this example (and hereafter) the explanatory column detailing the various steps of the process will be omitted for brevity. After the distribution factors (listed in a 'box' at each node) the first figures in any column refer to fixed-end moments. After joint D is balanced all joints are released simultaneously (balancing moments being underlined at each joint) and carry-overs are listed in the line below.

	A	B	B	C	C	D	D
		$\frac{2}{16}$		$\frac{3}{12}$		$\frac{3}{4}\frac{2}{12}$	ZERO
	100%	33%	67%	67%	33%	100%	0%
kN m	−8.0	8.0	−5.3	2.7	−9.6	9.6	−6.0
						−3.6	
					−1.8		
		−0.9	−1.8	5.8	2.9		
	−0.5		2.9	−0.9			
		−1.0	−1.9	0.6	0.3		
	−0.5		0.3	−0.9			
		−0.1	−0.2	0.6	0.3		
	0		0.3	−0.1			
		−0.1	−0.2	0.1			
	0		0	−0.1			
	0		0	0.1			
kN m	−9.0	5.9	−5.9	7.9	−7.9	6.0	−6.0

Particular attention should be paid to the sign of the fixed-end moment originally recorded as −6.0 in DE. In the distribution table, attention is concentrated on the moments acting *on the parts of the beam* and, therefore, although the force of 1 kN acting at E causes a clockwise moment of 6 kN m about D, the clamp at D must react on the beam with a moment of −6.0 kN m.

Parenthetically, it may be noted that since A is encastré, and there is no relative vertical movement of A and B, the change of moment at B will be double the change at A. Any departure from this rule must arise only as a result of rounding-off errors in the distribution process.

6.6 Settlement of supports

During the loading of a structure supports may settle by small amounts and, since the reactive forces are most unlikely to be equal, the settlements of the various supports will differ, causing relative movements of the ends of a particular span. If the relative movements (or the actual movements for all supports) can be estimated, their effects can be included in the general analysis by increasing the apparent fixed-end moments caused by the loads at the first stage of the distribution process. If movements occur in the positive direction of y, at both ends A and B of a prismatic member, as well as rotations θ_A and θ_B,

$$\begin{bmatrix} M_{AB} \\ M_{BA} \end{bmatrix} = \frac{2EI}{L} \begin{bmatrix} 2 & 1 & 3 & -3 \\ 1 & 2 & 3 & -3 \end{bmatrix} \begin{bmatrix} \theta_A \\ \theta_B \\ d_{yA} \\ d_{yB} \end{bmatrix}$$

and the increase of fixed-end moment caused by a relative clockwise sinkage of $\delta(\delta = d_{yB} - d_{yA})$ is given by writing $\theta_A = \theta_B = 0$, i.e.

$$M_{AB} = -\frac{6EI\delta}{L^2} = M_{BA} \qquad (6.6\text{–}1)$$

It is necessary, therefore, to know the modulus of direct elasticity, E, and the actual (as distinct from the relative) values of the appropriate second moments of area, before solution can be achieved. As a general point arising from this particular issue, provided no numerical values of deformation are known or required, the solution of any hyperstatic problem can be based on relative stiffnesses, but if a deformation is to be calculated, or must be used in establishing the basic equations, complete information is necessary.

Example 6.6–1. If the material of the beam in Fig. 6.5–2 has a modulus of 16 units, and the second moment of area of DE is 2 units and if, as a result of the application of the loads, B and C sink by 1 and 2 units respectively, calculate the support moments.

SOLUTION The relative clockwise sinkages of AB : BC : CD will be 1 : 1 : −2, and the additional fixed-end moments $(-6EI\delta/L^2)$ may be written under the fixed-end moments caused by load in the second line of the distribution table, marked with an asterisk below.

	A		B			C			D			E
		$\frac{2}{16}$			$\frac{3}{12}$			$\frac{3}{4}\frac{2}{12}$			ZERO	
	100%		33%	67%		67%	33%		100%	0%		
kN m	−8.0		8.0	−5.3		2.7	−9.6		9.6	−6.0		
*	−1.5		−1.5	−4.0		−4.0	5.3		5.3			
									−8.9			
							−4.4					
			0.9	1.9		6.7	3.3					
	0.5			3.3		0.9						
			−1.1	−2.2		−0.6	−0.3					
	−0.6			−0.3		−1.1						
			0.1	0.2		0.7	0.4					
				0.3		0.1						
			−0.1	−0.2		−0.1						
	0			0		−0.1						
kN m	−9.6		6.3	−6.3		5.3	−5.3		6.0	−6.0		

6.7 Frames in which sway occurs

Attention was drawn in the last section to the fact that, should relative movement of the ends of members occur a knowledge of the numerical value of that movement was needed before analysis might be completed. In many frames, however, loading will imply some sideways movement which is unknown beforehand, and which could not be estimated since any error would imply some departure from the obvious equations of equilibrium. Examples are shown in Fig. 6.7–1. In Fig. 6.7–1(a), the application of any load to a single-bay portal frame (save only the particular case of the application of a symmetrical vertical load to a symmetrical frame) will occasion some sideways movement, and it will be expected that, in the figure, B and C will move to the right.

A vector diagram of displacements may be used to determine the behaviour of the frame. In diagrams of this type displacements are represented in direction, and to an appropriate scale, by lines showing movement from a fixed point in space. In Fig. 6.7–1(b)

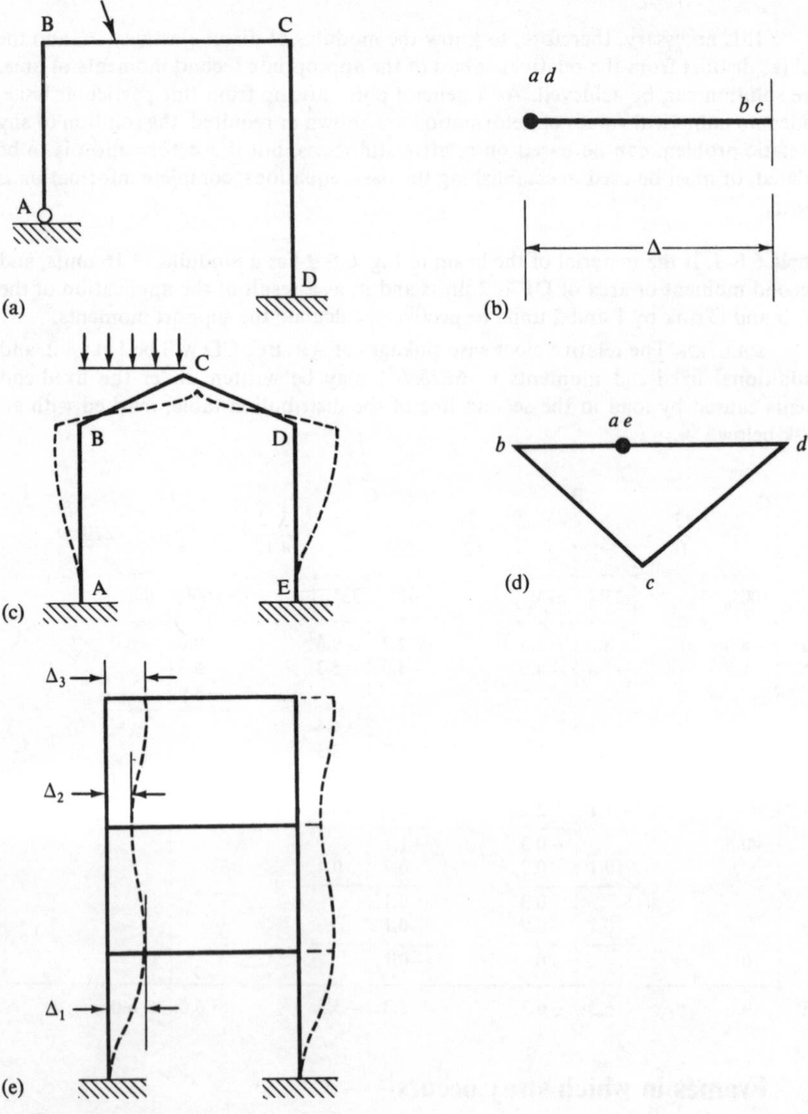

Fig. 6·7–1

the displacement diagram for the frame is shown. A single point representing zero movement indicates the position of nodes A and D which (by definition) are incapable of translational movement within the plane of the frame, although rotation at A could certainly occur. Relative to A, point B must move in an arc of centre A and radius AB and, since movements must be small when compared with the lengths of members, B will effectively move along a line at right angles to AB. If the extent of this movement, Δ say, is assumed, the vector *ab* could be drawn to scale, thereby defining the position of B. But C must move at right angles to DC and, similarly, at right angles to BC (since shortening of members has been neglected), from which it follows that the vectorial positions of *b* and *c* are coincident. The side-sways of nodes B and C are identical and, moreover, there is no relative sinkage of C with respect to B. The frame may be characterized as possessing a single degree of translational freedom since the single measurement Δ defines the position of all nodes. Clearly, there may well be some rotational movements (save at D where the possibility is expressly eliminated), but this rotational movement may be readily taken into account by a moment-distribution process.

The vector diagram of displacements for the ridge-roofed portal, ABCDE of Fig. 6.7–1(c) can only be completed if two of the movements (for example *ab* and *ed*) are known, thus characterizing a frame with two degrees of freedom, while a multi-storey, single-bay frame as in Fig. 6.7–1(e) of *n* storeys will have *n* degrees of freedom, shown in the diagram as Δ_1, Δ_2, and Δ_3, the horizontal deflections at each beam level of the frame.

Using moment distribution the essential method of attack involves $n + 1$ distributions, in one of which the effect of the loads alone (with all possible nodal movements suppressed) is determined, and one further distribution for each of the possible 'sways' indicating the 'degree of freedom' of the structure.

The solution for a single-bay, single-storey frame in which sway might occur, is indicated in Fig. 6.7–2. It consists of one distribution (as in Fig. 6.7–2(b)) in which the

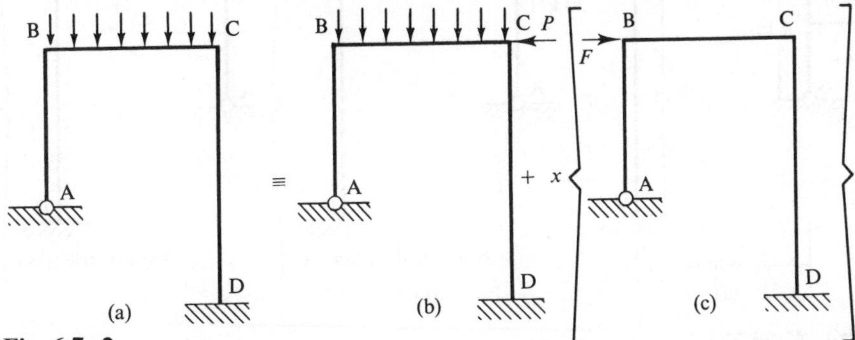

Fig. 6·7–2

existence of a propping force *P*, of unknown magnitude, prevents horizontal movement of the beam BC. The application of moment-distribution methods implies that no translational movement (as distinct from rotational movement) exists, and use of the sway equation (Eqn 6.4–1) allows the calculation of the appropriate propping force *P*. The effect of this propping force is calculated separately (Fig. 6.7–2(c)) and added (with proper regard to the convention of sign employed), to give the required result, and, obviously, the condition connecting the two sections in the simple case of no lateral load as in Fig. 6.7–2 is that

$$P = xF \tag{6.7–1}$$

The effect of the prop *P* is most readily determined by permitting an arbitrary sway to take place, unaccompanied initially be rotation at the joints. Since there is no relative sinkage of B and C (see Fig. 6.7–1(b)), the magnitude of the sway will be equal in the two columns and zero in the beam, thus allowing fixed-end moments proportional to the appropriate displacements to be readily chosen. Distribution of these 'sway moments' allows joints to rotate to their equilibrium positions and, subsequently, the force F necessary to maintain the frame in its swayed position may be determined. The actual amount of side-sway necessary to eliminate the propping force, and therefore the moments caused by the side-sway, are determined from Eqn. 6.7–1.

It will be noted that, although the force needed to cause sway without rotation (or the moments caused at any node by a given force, provided no rotation occurs) can be readily determined; these will inevitably be affected by rotation of the joints.

The use of the sway equation (Eqn. 6.4–1), with the correcting modification implied in Eqn 6.7–1, is a fruitful source of errors. Most of these arise from some misconception of the sign convention used, and it should be realized that this can be implied either by some specific statement or by the unthinking application of a diagrammatic instruction implying direction of a force.

If the loading system postulates some lateral load, as in Fig. 6.7–3(a), there are no changes in the fundamental application, although the sway equation used must make appropriate allowances for the horizontal force which equilibrium requires shall be considered to act at beam level.

Example 6.7–1. Analyse the frame of Fig. 6.7–3. The relative flexural rigidities of AB:BC:CD are 4:2:4.

SOLUTION The no-sway distribution of Fig. 6.7–3(b) is performed in the usual way, using the stiffness factors appropriate to the end-conditions at the foot of each

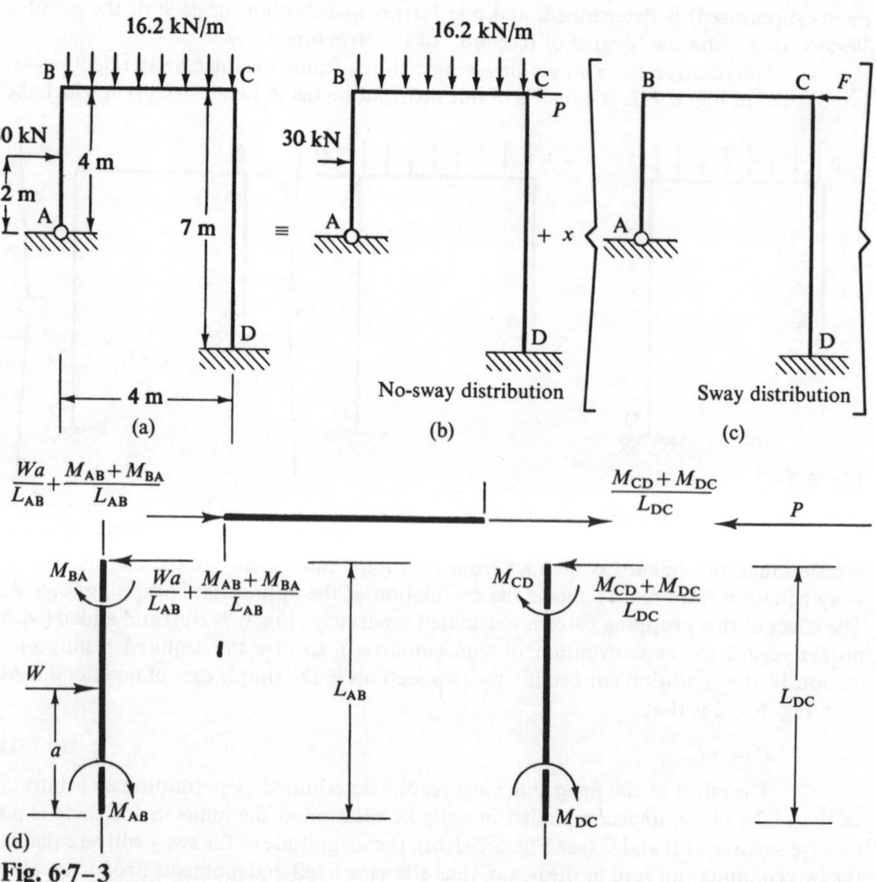

Fig. 6·7–3

column, and fixed-end moments calculated on the initial assumption that both ends of all members are direction-fixed.

	A	B		C		D
	$\frac{3}{4}$		$\frac{2}{4}$		$\frac{4}{7}$	
	100%	60%	40%	47%	53%	100%
kN m	−15.0	15.0	−21.6	21.6		
	15.0	7.5				
		−0.5	−0.4	−10.1	−11.5	
			−5.0	−0.2		−5.7
		3.0	2.0	0.1	0.1	
			0.0	1.0		0.0
				−0.5	−0.5	
			−0.2	0.0		−0.2
		0.1	0.1			
kN m	0.0	25.1	−25.1	11.9	−11.9	−5.9

The method of calculation of the required propping force, P, is detailed in Fig. 6.7–3(d) in the exploded view of the portal frame.

From considerations of equilibrium of the beam it follows that P (of the sign indicated in Fig. 6.7–3(b)) is given by

$$-P + \frac{30 \times 2}{4} + \frac{0.0 + 25.1}{4} + \left(\frac{-11.9 - 5.9}{7}\right) = 0$$

$$P = 15 + 6.27 - 2.54 = 18.73 \text{ kN}$$

The effect of this propping force is best calculated from a separate distribution in which the member ends (at B and C) are given a quite arbitrary (but equal) sway—the force, F, required to give equilibrium after rotation of the joints has occurred, being calculated from a separate application of the sway equation.

It has been established in Eqn 6.6–1 that the moments caused at the ends of some encastré span by clockwise relative sinkage δ are equal to $-(6EI\delta/L^2)$ and it follows, therefore, that for columns swaying by equal amounts (and made, as usual, of the same material) the fixed-end moments will be in the ratio of I/L^2. Quite arbitrary values in this ratio can therefore be applied to the top and bottom of each column, as in the distribution table below, and a separate calculation will give the magnitude of the force F in Fig. 6.7–3(c).

A	B		C		D
100%	60%	40%	47%	53%	100%
441	441			144	144
−441					
	−220				
	−133	−88	−68	−76	
		−34	−44		−38
	20	14	21	23	
		10	7		11
	−6	−4	−3	−4	
		−2	−2		
	1	1	1	1	−2
0	103	−103	−88	88	115

The units and the actual magnitudes of the initial fixed-end moments should be chosen to be of a similar order of magnitude to the final moments from the no-sway distribution. As was done there, it will be noted that the first step is to restore end A to its desired condition as a pinned end, but that distribution is normal thereafter. Although starting with a larger number for the initial out-of-balance moment, it has not been thought necessary to work to the same order of accuracy when completing the distribution table. F can be calculated from the principles of Fig. 6.7–3(d) and it will follow that

$$\frac{0+103}{4}+\frac{88+115}{7}-F=0$$

from which $F = 54.75$ kN

The value of the coefficient x must clearly be such that

$$P + xF = 0$$

and

$$x = -\frac{18.73}{54.75} = -0.342$$

The required answer is obtained by tabular addition of the no-sway and sway moments.

	A	B		C		D	
No-sway moments	0.0	25.1	−25.1	11.9	−11.9	−5.9	kN m
−0.342 (sway moments)	0.0	−35.1	35.1	30.0	−30.0	−39.2	
Required solution	0.0	−10.0	10.0	41.9	−41.9	45.1	kN m

Example 6.7–2. A three-member frame ABCD of constant flexural rigidity is shown in Fig. 6.7–4(a). A and D are encastré, and loads are applied to AB and BC as shown. Determine the nodal moments.

SOLUTION Solution will follow an appropriate combination of the sway and no-sway distributions as shown in Fig. 6.7–4(b). Taking the latter first, we know that the fixed-end moments will be:

$$\text{in AB} \qquad \mp\frac{WL}{8} = \mp 3.75 \text{ kN m}$$

$$\text{in BC} \qquad \mp\frac{wL^2}{12} = \mp 3.75 \text{ kN m}$$

The distribution table will appear as follows:

A	B		C		D	
$\frac{I}{3}$		$\frac{I}{3}$		$\frac{I}{3.605}$		
	50%	50%	54.6%	45.4%		
−3.75	3.75	−3.75	3.75			
	0	0	−2.04	−1.71		
		−1.02	0		−0.86	
	0.51	0.51	0	0		
0.25		0	0.25			
	0	0	−1.14	−0.11		
		−0.07	0		−0.05	
	0.04	0.03	0	0		
0.02		0	0.02	0		
	0	0	−0.01	−0.01		
−3.48	4.30	−4.30	1.83	−1.83	−0.91	kN m

The propping force preventing sway taking place in Fig. 6.7–4(b) may be calculated from the exploded view of Fig. 6.7–4(c). From the horizontal equilibrium of the beam, BC,

$$P = 5 + \frac{M_{AB} + M_{BA}}{3} + \frac{1}{3}\left(M_{DC} + M_{CD} - 2\left[7\tfrac{1}{2} + \frac{M_{BC} + M_{CB}}{3}\right]\right) \quad (6.7\text{–}2)$$

$$= 5 + \frac{0.82}{3} + \frac{1}{3}\left(-2.74 - 2\left[7\tfrac{1}{2} - \frac{2.47}{3}\right]\right)$$

$$= 5.27 + \tfrac{1}{3}(-2.74 - 13.36) = -0.095 \text{ kN}$$

The sway distribution may be effected after determining the relative sinkages of the ends of each member due to some arbitrary sway to the right. If B moves through some distance Δ, at right angles to AB, as shown vectorially in Fig. 6.7–4(d), then C must move at right angles to BC (i.e. vertically) and at right angles to CD. The relative clockwise sinkages of members AB : BC : CD are $\Delta : -\Delta \cot\phi : \Delta \operatorname{cosec}\phi$. The relative fixing moments due to this sway only will be, respectively, in the ratio

$$\frac{\Delta}{3^2} : -\frac{\Delta . 2}{3.3^2} : \frac{\Delta\sqrt{13}}{3.13} = \frac{\Delta}{9} : -\frac{\Delta}{13.5} : \frac{\Delta}{10.81}$$

$$= -1.11 : 0.74 : -0.92$$

These form the fixing moments at the initial stage of a distribution with the same distribution factors.

A	B		C		D
	50%	50%	54.6%	45.4%	
−1.11	−1.11	0.74	0.74	−0.92	−0.92
	0.18	0.19	0.09	0.09	
0.09		0.04	0.09		0.04
	−0.02	−0.02	−0.05	−0.04	
−0.01		−0.02	−0.01		−0.02
	0.01	0.01	0.01		
−1.03	−0.94	0.94	0.87	−0.87	−0.90

The force F (in Fig. 6.7–4(b)) causing the moments of this distribution can be calculated from Fig. 6.7–4(c) if the effects of any loads applied between the nodes are omitted.

$$F + \frac{M_{AB} + M_{BA}}{3} + \frac{1}{3}\left(M_{DC} + M_{CD} - 2\left[\frac{M_{BC} + M_{CB}}{3}\right]\right) = 0 \quad (6.7\text{–}3)$$

$$F - \frac{1.97}{3} + \frac{1}{3}\left(-1.77 - \frac{2}{3}1.81\right) = 0$$

$$F = 0.657 + 0.993 = 1.65 \text{ kN}$$

Now the equilibrium sum of Fig. 6.7–4(b) requires that

$$P = xF \quad (6.7\text{–}4)$$

from which

$$x = -\frac{0.091}{1.65} = -0.055$$

The actual moments may then be obtained by summation:

	A	B		C		D
No-sway moments kN m	−3.48	4.30	−4.30	1.83	−1.83	−0.91
x (sway moments)	0.06	0.05	−0.05	−0.05	0.05	0.05
kN m	−3.42	4.35	−4.35	1.78	−1.78	−0.86

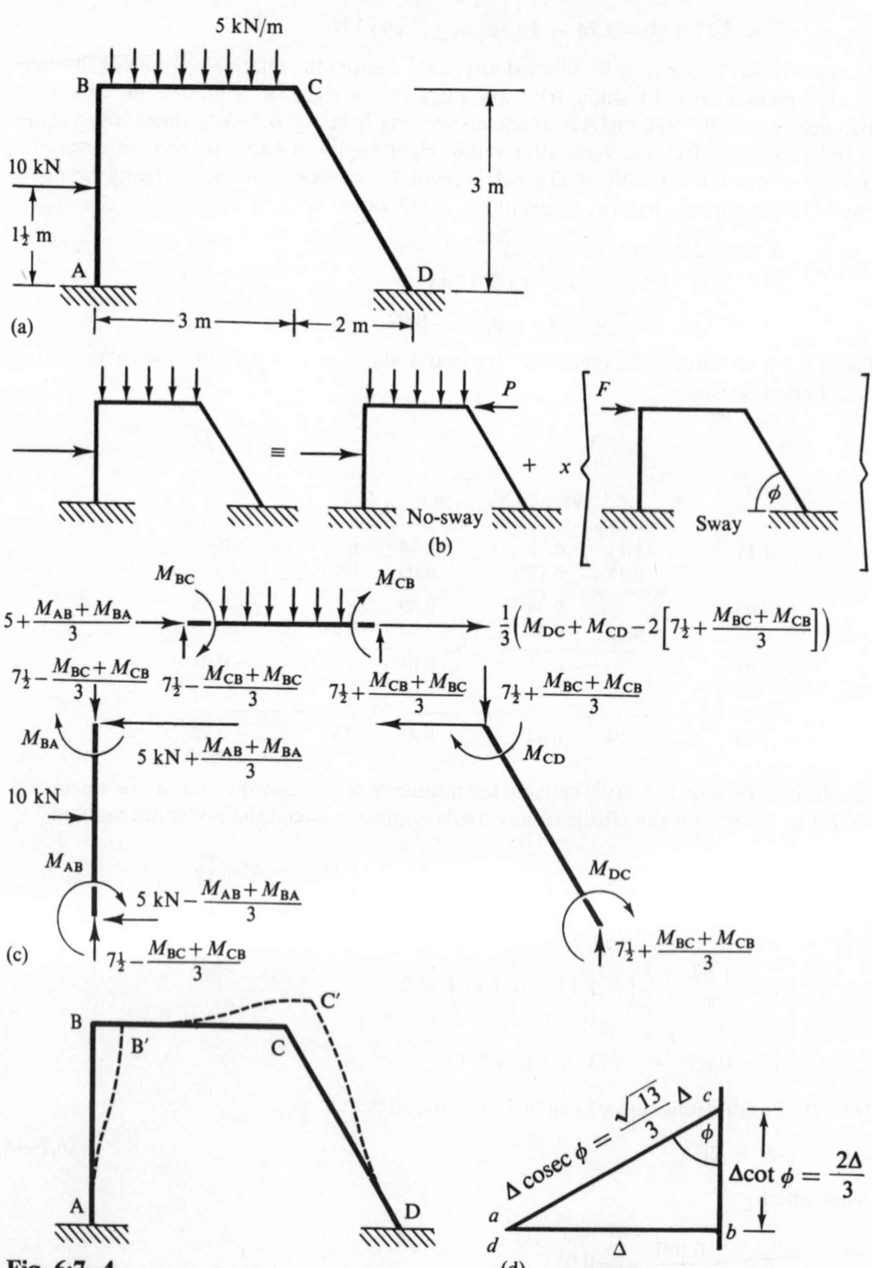

Fig. 6·7–4

6.8 Use of the instantaneous centre of rotation and the principle of virtual work

The principle of virtual work has already been used in Sections 4.14, 4.15 and 4.16. It defines an equality between the summed products of the external and internal parts of a force system in equilibrium and an independently chosen compatible deformation system in the form

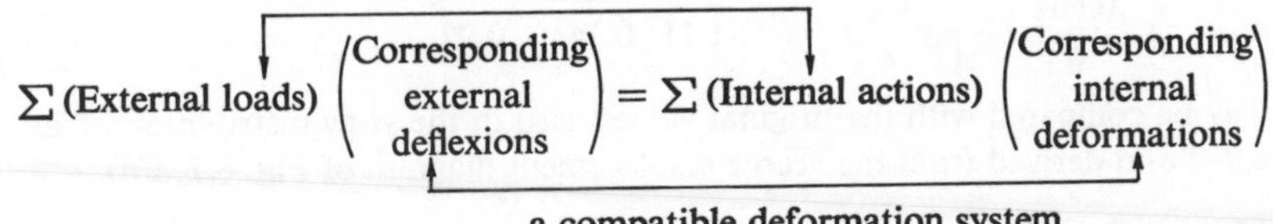

It may be used with advantage, particularly when combined with a judicious use of the instantaneous centre of rotation, to facilitate the calculation of fixed end moments for sway distribution and the derivation of x as defined in Fig. 6.7–4(b).

(a) Use of the instantaneous centre

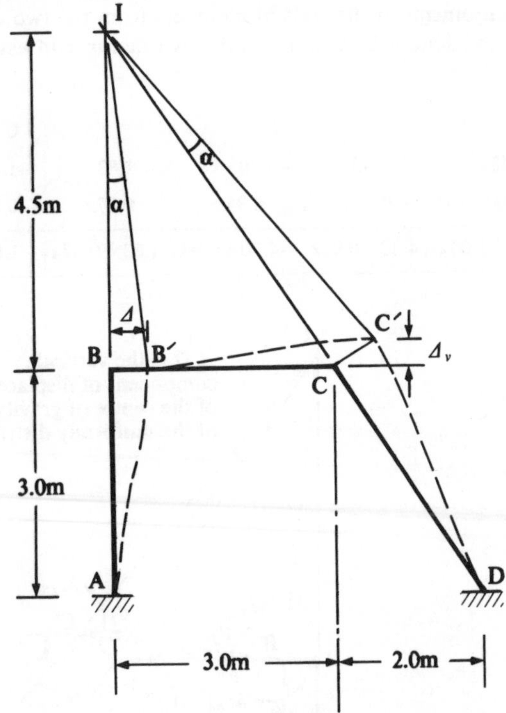

Fig. 6·8–1

In Fig. 6.8–1 the joints of the frame shown originally in Fig. 6.7–4(a) are restrained against rotation. Joint B is given a displacement to B′ of Δ, which must be horizontal since the member AB is assumed not to change in length, i.e. the effect of axial distortions is neglected. C moves to C′, and the member BC thus rotates about its instantaneous centre I, located at the intersection of AB and DC produced. Simple proportion indicates that I will be 4.5 m vertically above B. Member BC rotates through an angle $\alpha = \Delta/4.5$ and since $IC = \sqrt{3^2 + 4.5^2} = 5.4$ m then $CC' = (IC)\alpha = 1.2\Delta$ and $\Delta_v = 3\alpha = 0.667\Delta$. The relative clockwise sinkages of AB:BC:CD are thus Δ: -0.667Δ:1.2Δ giving fixing moments due to this sway only (of the form $\frac{-6EI\Delta}{L^2}$) in the ratio

$$-\frac{1}{3^2}:\frac{0.667}{3^2}:-\frac{1.2}{(3^2+2^2)} = -1.11:0.74:-0.92$$

These may be compared with the original values used in the sway distribution of Example 6.7–2 and derived from the vector displacement diagram of Fig. 6.7–4(d).

(b) Use of the principle of virtual work

Earlier, in Example 6.7–2, it was shown that a proportion of the sway moments obtained from the sway distribution must be added to the no-sway distribution, to ensure compliance with the equilibrium equation shown pictorially in Fig. 6.7–4(b) and notationally in Eqn 6.7–4. The principle of virtual work provides a convenient way of determining the unknown, x, in that equation without the necessity of evaluating the propping and sway forces.

The moments at the ends of members from the two distributions above, are listed below and summed, using the unknown factor x in association with the sway moments.

	A		B		C	D
No sway	−3.48	4.30	−4.30	1.83	−1.83	−0.91
Sway	$-1.03x$	$-0.94x$	$0.94x$	$0.87x$	$-0.87x$	$-0.90x$
	$-3.48-1.03x$	$4.30-0.94x$	$-4.30+0.94x$	$1.83+0.87x$	$-1.83-0.87x$	$-0.91-0.90x$

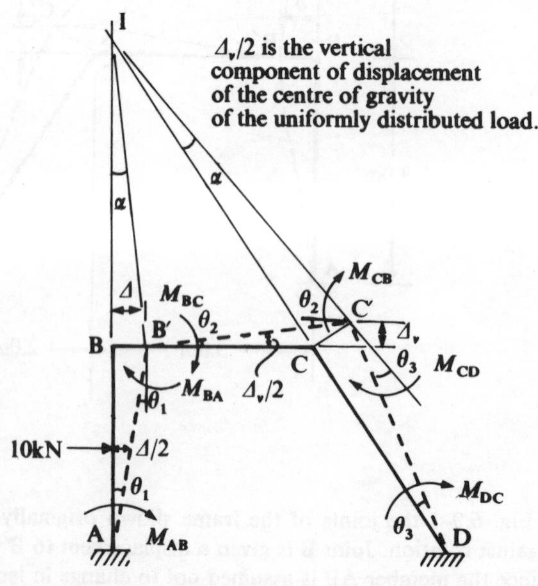

Fig. 6.8–2

It will be remembered that these moments all use a convention of 'clockwise positive' and, when determining the corresponding displacements the same sign convention must be used. If we assume a compatible displacement system as shown in Fig. 6.8–2, in which all members rotate through clockwise angles and the point B moves by a distance Δ to B′, as in Fig. 6.8–1, then the table of external forces and internal actions with their corresponding displacements will appear as in Table 6.8–1.

It will be seen that the sum of the virtual work done by external forces is fortuitously zero but, if it was not, the arbitrarily assumed value of Δ would disappear after equating internal and external effects. In this particular case

$$-0.091 - 1.649x = 0 \text{ from which}$$

$$x = -0.055$$

which should be compared with the solution of Eqn 6.7–4. The result of the applied loads and the sway combined will now follow as at the end of Ex 6.7–2.

Table. 6·8–1

$$\theta_1 = \frac{\Delta}{3} \qquad \theta_2 = -\frac{\Delta_v}{3} = -\frac{2\Delta}{9} \qquad \theta_3 = \frac{CC'}{\sqrt{13}} = \frac{\Delta}{3}$$

Internal Action	Value	Corresponding Displacement	Product
M_{AB}	$-3.48 - 1.03x$	$\frac{\Delta}{3}$	$\Delta(-1.160 - 0.343x)$
M_{BA}	$4.30 - 0.94x$	$\frac{\Delta}{3}$	$\Delta(1.433 - 0.313x)$
M_{BC}	$-4.30 + 0.94x$	$\frac{-2\Delta}{9}$	$\Delta(0.956 - 0.210x)$
M_{CB}	$1.83 + 0.87x$	$\frac{-2\Delta}{9}$	$\Delta(-0.407 - 0.193x)$
M_{CD}	$-1.83 - 0.87x$	$\frac{\Delta}{3}$	$\Delta(-0.610 - 0.290x)$
M_{DC}	$-0.91 - 0.90x$	$\frac{\Delta}{3}$	$\Delta(-0.303 - 0.300x)$
		Sum	$\Delta(-0.091 - 1.649x)$

External Force	Value	Corresponding Displacement	Product
10kN force on AB	10	$\frac{\Delta}{2}$	5Δ
5kN/m on BC	15	$-\frac{\Delta_v}{2} = -\frac{\Delta}{3}$	-5Δ
		Sum	Zero

6.9 Frames with two or more modes of sway

If the frame has two degrees of freedom, as in a two-storey single-bay frame (Fig. 6.9–1) or in a four-member single-bay frame (Fig. 6.7–1(c)), the procedure is similar, but it must be realized that the no-sway distribution using the applied loads will imply in general the existence of two propping forces, the effects of which may only be determined by simultaneous solution.

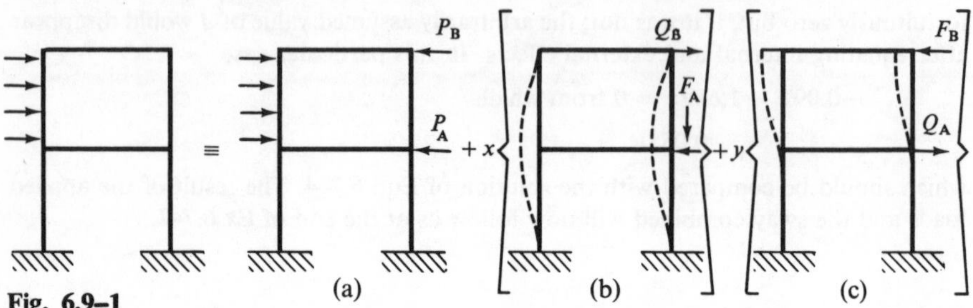

Fig. 6.9–1

In Fig. 6.9–1 the two implied propping forces are P_A and P_B in (a). Separate distributions are made for each of the independent sway modes shown dotted in (b) and (c), each implying the existence of forces F and Q as shown. The amounts of each of the individual sways necessary to give equilibrium may then be calculated from the equations:

$$P_A + F_A x - Q_A y = 0$$

$$P_B - Q_B x + F_B y = 0$$

Similar procedures can be adopted for frames of more than two degrees of freedom, the total number of distributions necessary being $(n + 1)$ for n degrees. Clearly, solution of n simultaneous equations for n unknowns would also be involved, and the labour required for calculation increases disproportionately with increasing complexity of the frame.

6.10 Symmetry and anti-symmetry

Using moment-distribution, as with other methods of structural analysis, it is possible to take advantage of any symmetry of a structure if it is subjected to load systems which are themselves symmetrical or anti-symmetrical about the same geometric centre line. On such a symmetrical structure, symmetrical load systems produce equal and opposite slopes about the centre line, whilst anti-symmetrical loads cause deflexions of equal magnitude and different sign at corresponding points about the centre.

(a) **Symmetrical loads.** Since these cause deformations which are mirrored about the structural centre-line, the slope at that centre line must be zero. If this occurs at a nodal point (that is, if the structure has an even number of spans) we may regard the structure as being effectively encastré at that point, and complete the distribution for one-half of the structure only. Fig. 6.10–1(a) shows a symmetrical four-span continuous beam PQRST, subjected to equal loads at the centres of the outer spans PQ and ST. The final deflected shape is shown dotted and it will be seen that there is zero slope over the support R and that the final form is mirrored about the vertical through R. Analyses could be confined, therefore, to the structure of Fig. 6.10–1(b) in which the release of P (as a pinned end in Section 6.5) followed by the release of Q will give an exact answer, with no repeated cycles of distribution.

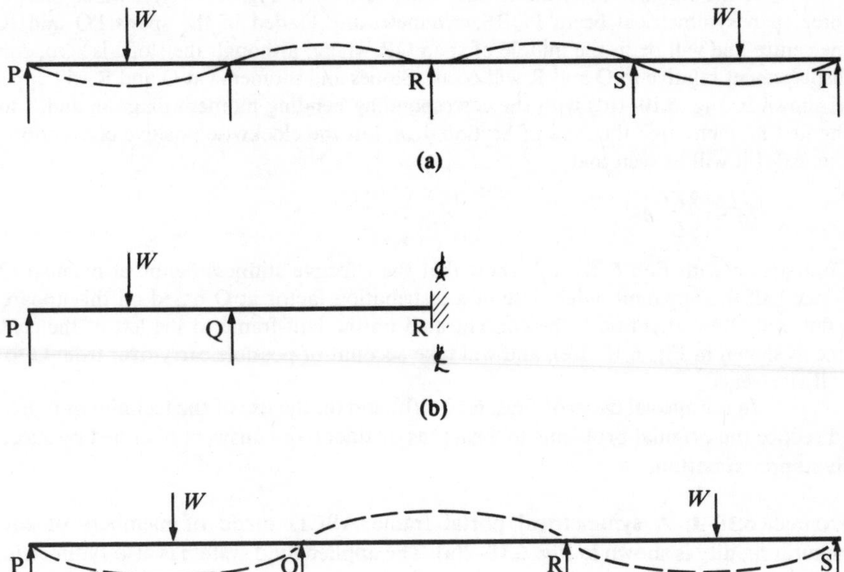
W
W
P
R
S
T
(a)
W
P
Q
R
(b)
W
W
P
Q
R
S
(c)

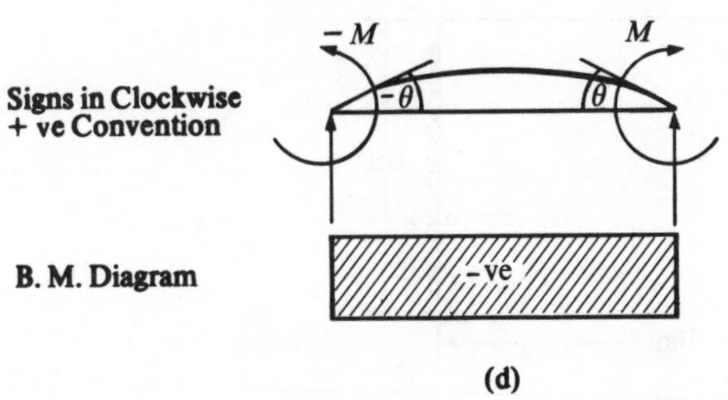
−M
M
−θ
θ
Signs in Clockwise
+ ve Convention
B. M. Diagram
−ve
(d)

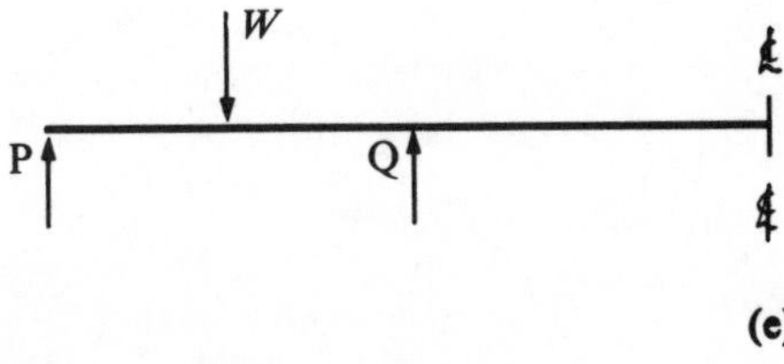
W
P
Q
(e)

Fig. 6·10–1

If the number of spans is odd, however, as in Fig. 6.10–1(c), which shows a three span symmetrical beam PQRS, symmetrically loaded in the spans PQ and RS, the centre line will be in the middle of span QR where, although the slope is zero, some displacement relative to Q and R will occur. Slopes and moments at Q and R will appear as shown in Fig. 6.10–1(d) with the corresponding bending moment diagram and, using the first moment-area theorem of Section 4.18, but the clockwise positive convention of Fig. 6.2–1 it will be seen that

$$M = \frac{2EI}{L}\,\theta$$

Comparison with Eqn 6.2–2 will show that the effective stiffness factor of member QR is one-half the apparent value. Use of a distribution factor at Q based on this apparent value will allow attention to be concentrated on the half-frame to the left of the centre line as shown in Fig. 6.10–1(e), and will take account of possible carry-over from Q to R and vice-versa.

In the special cases of Figs. 6.10–1(b) and (e) the use of the techniques indicated reduce the original problems to 'exact' as distinct from answers obtained by successive approximation.

Example 6.10–1. A symmetrical portal frame ABCD made of members of equal flexural rigidity is shown in Fig. 6.10–2(a). The applied load system is also symmetrical, and the frame will not sway, but deform to give zero slope at the vertical centre line.

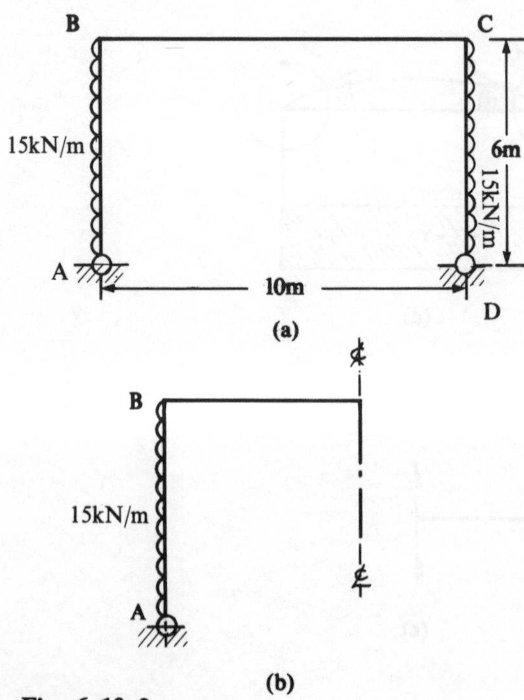

Fig. 6·10–2

We may use the half-frame of Fig. 6.10–2(b), and the distribution table will appear as below.

	A		B		
Modified stiffness factors		$\frac{3}{4}\frac{EI}{L}=\frac{EI}{8}$		$\frac{1}{2}\frac{EI}{L}=\frac{EI}{20}$	
Distribution factors		71.5%	28.5%		
	−45.0	45.0			
Release A	+45.0	22.5			
Release B		−48.2	−19.3		
	0	19.3	−19.3		kN m

(b) **Anti-symmetrical loads.** If a symmetrical structure is subjected to loads which are anti-symmetrical about the centre line as in Figs. 6.10–3(a) and 6.10–4(a), solution of the problem will vary depending on the number of spans. Whether this be even, as in Fig. 6.10–3(a), or odd, as in Fig. 6.10–4(a), the structure will behave as if pinned on the centre line, the pin occurring at a support in the former case and at the centre of the central span in the latter. The structures requiring analysis may thus be modified as shown in Fig. 6.10–3(b), in which the stiffness factor of span QR is $\frac{3}{4}(I/L)$ as in Section 6.5, and Fig. 6.10–4(b) in which the central span has a stiffness factor of $\frac{3}{4}(2I/L) = \frac{3}{2}(I/L)$.

It should be noted that if sway may occur, as in a portal frame, anti-symmetric loading must inevitably cause it, and there is little virtue in the application of an anti symmetric treatment since other difficulties may arise.

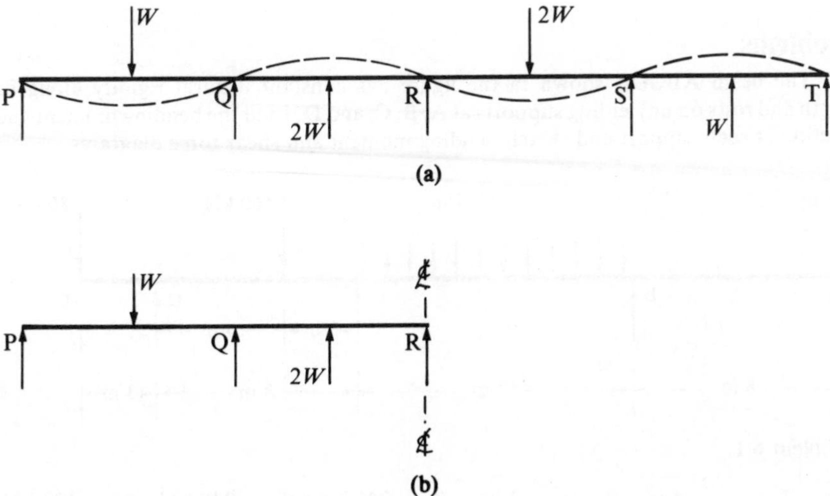

Fig. 6·10–3

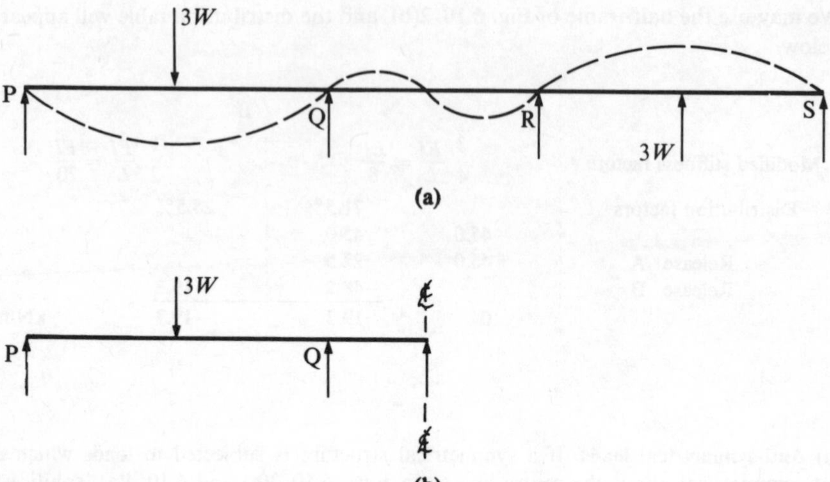

Fig. 6·10–4

There may be occasions, however, particularly in continuous beam structures, where the device of Section 4.17 in which a particular load system is replaced by a combination of a symmetric and anti-symmetric distribution, can be used to advantage.

Problems

6.1. The beam ABCDE shown in the figure has constant flexural rigidity along its length and rests on unyielding supports at A, B, C, and D. Find the bending moment and reaction at each support and sketch bending moment and shear force diagrams.

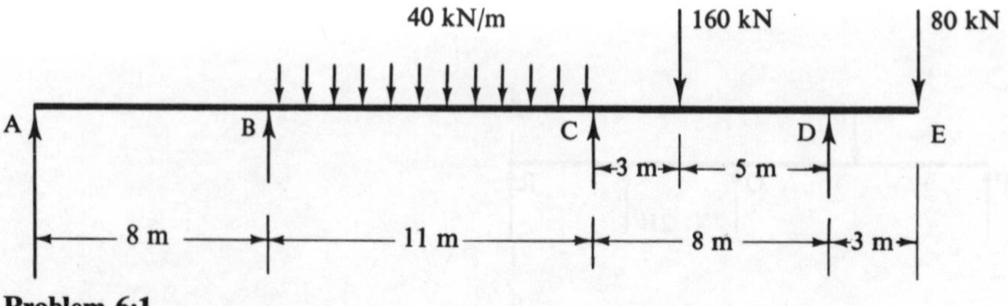

Problem 6·1

Ans. $R_A = -32$ kN; $R_B = 245$ kN; $R_C = 338$ kN; $R_D = 129$ kN.

6.2. The frame shown in the figure is encastré at A and D and rigidly jointed at B and C, the flexural rigidity being constant throughout. Find the point of application on BC of a vertical force which does not produce horizontal movement at B and C.

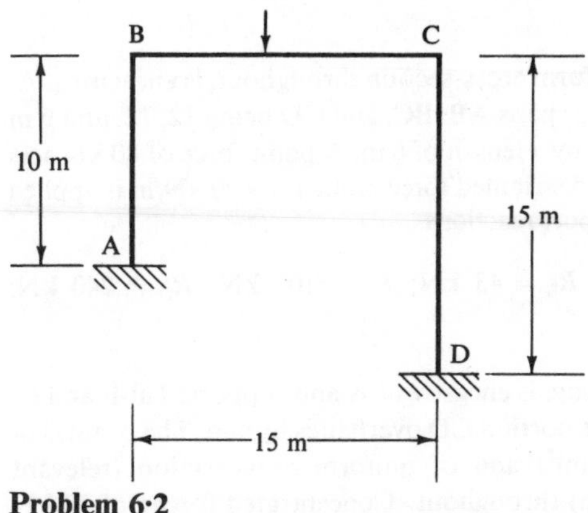

Problem 6·2

Ans. 11.18 m from B.

6.3. A beam of flexural rigidity *EI* is continuous over five equal spans of length *L* metres. Distributed live loading of intensity *w* kilonewtons per metre may be applied to the beam in each or any of the spans. Find the maximum reaction due to live loading which may occur at the end support.

Ans. 0.447*wL* kN m

6.4. The rigid steel frame shown in the figure is supported on rollers at A and C and encastré at P. The relevant second moment of area of both AB and BC is 8×10^8 mm^4 and of BP is 12×10^8 mm^4. Find the magnitude and direction of horizontal deflection at A when a uniformly distributed force of intensity 6 kN/m acts over ABC. State the values of bending moment at B. The modulus of elasticity for steel may be taken as 200 kN/mm^2.

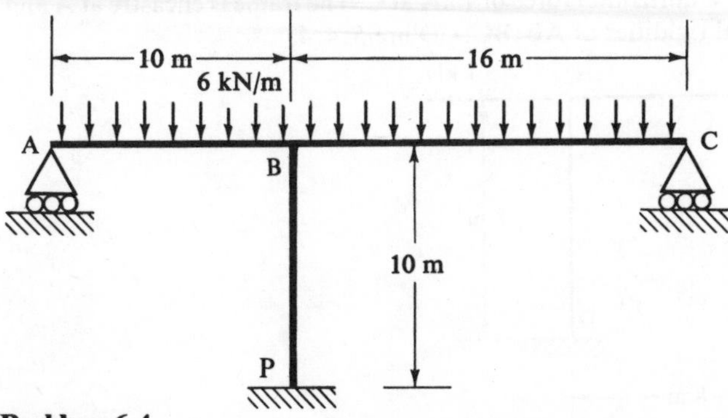

Problem 6·4

Ans. 0.57×10^{-2} m towards C; 130, 28, −158 kN m.

6.5 A rectangular rigid-jointed frame ABCD is encastré at A and D. The vertical stanchions AB and DC are each of length 10 m, and the beam BC also 10 m, but the flexural rigidities of the stanchions are twice that of the beam. A point force of 4 kN acts vertically downward at a point in the beam distant 6 m from B. Determine the support moments.

Ans. A 2.16 kN m; D −1.68 kN m.

6.6. A continuous beam ABCDE of uniform cross-section throughout, is encastré at A, and simply supported at B, C, and D, the spans AB, BC, and CD being 12, 12, and 9 m respectively. The portion DE overhangs by a length of 6 m. A point force of 80 kN acts at the centre of span BC and a uniformly distributed force of intensity 20 kN/m is applied to spans CD and DE. Determine the support reactions.

Ans. $R_A = -6$ kN; $R_B = 43$ kN; $R_C = 103$ kN; $R_D = 240$ kN.

6.7. A continuous beam ABCD, 20 m long, is encastré at A and supported at B and C. Spans AB and BC are each 8 m, while the portion CD overhangs by 4 m. The beam is of steel (modulus of elasticity 200 kN/mm²) and of uniform cross-section (relevant second moment of area 1×10^8 mm⁴ m) throughout. Concentrated forces of 80 kN and 40 kN are applied 2 m from B in the span BC and at D respectively. Application of these loads causes a settlement of the support B of 15 mm. What are the support moments?

Ans. $M_{AB} = -17.0$; $M_{BC} = 5.8$; $M_{CB} = 160.0$ kN m.

6.8. A square, single-bay, rigid-jointed portal ABCD is encastré at A and D (AB and DC being vertical), the three members of the frame being of the same length and flexural rigidity. B is held by a link so that it cannot move horizontally.

A vertical force of magnitude P acts on the beam at a point $0.7L$ from B. Determine the force produced in the link at B and the magnitudes and directions of the axial forces in the members AB, BC, and CD.

Ans. $0.0495P$; $0.267P$; $0.125P$; $0.753P$ all compressive.

6.9. Determine the bending moments at A and D for the rigid-jointed frame shown in the figure, under the action of a horizontal force of 1 kN at C. The frame is encastré at A and D and the relative flexural rigidities of AB:BC:CD are 5:4:4.

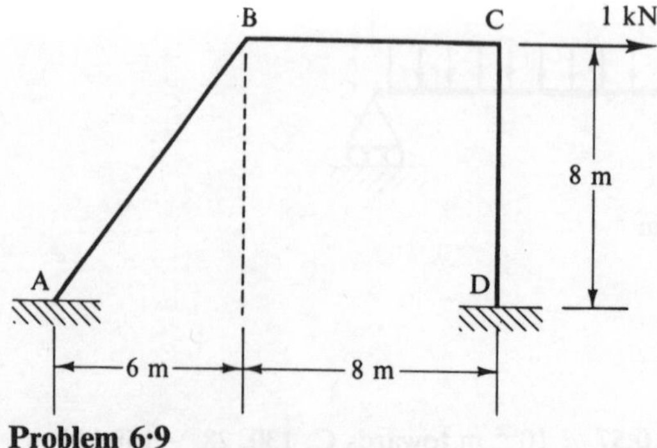

Problem 6·9

Ans. Both −1.50 kN m.

6.10. The frame shown is of uniform modulus throughout, but the second moment of area of the stanchions is twice that of the beam BC. Calculate the bending moments at B and C.

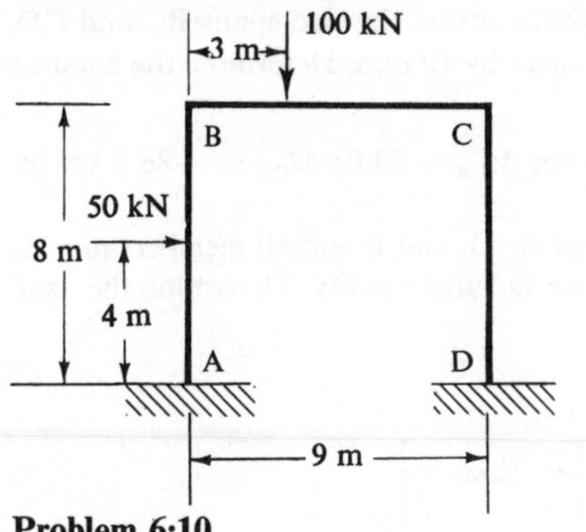

Problem 6·10

Ans. $M_{BC} = -76.4$; $M_{CB} = 95.5$ kN m.

6.11. The built-in beam shown in the figure has a solid rectangular cross-section of uniform width but varying depth. It is to be replaced by a beam of the same width, but of uniform depth $2d$. By what factor may the load P be increased if the maximum stresses are to be the same as in the original beam?

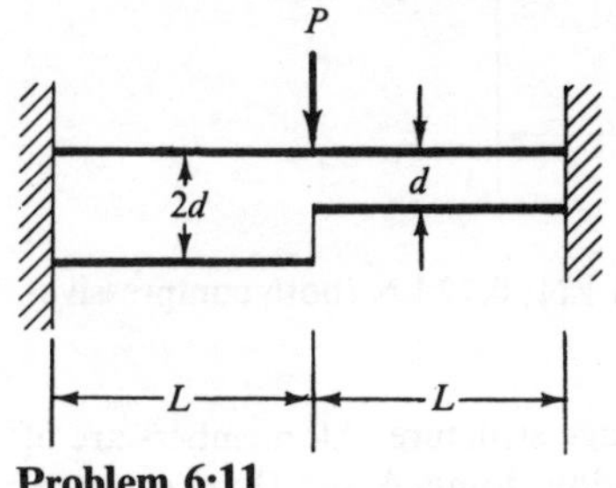

Problem 6·11

Ans. 2.88.

6.12. In the continuous beam shown, the figures in circles in each span show the relative ratios of relevant second moment of area : length. What is the reaction at B?

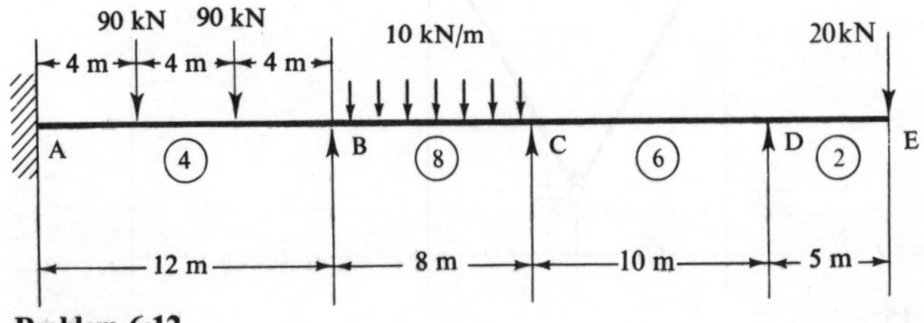

Problem 6·12

Ans. 150 kN (upward).

6.13. A continuous beam ABCD of constant flexural rigidity $EI = 30 \times 10^9$ kN mm^2 is simply supported at A and encastré at D, the three spans being AB = 4 m, BC = 5 m, and CD = 6 m. A concentrated force of 100 kN acts in the centre of span AB, and a uniformly distributed force of intensity 10 kN/m covers the two spans BC and CD. Under the action of these loads the support C sinks by 10 mm. Determine the bending moments in the beam.

Ans. $M_{BA} = -79.90$; $M_{CB} = 33.0$; $M_{DC} = -86.5$ kN m.

6.14. The rigid-jointed frame shown is encastré at A, D, and E and all members may be assumed to be inextensible and to have the same flexural rigidity. Determine the axial forces in BC and CD.

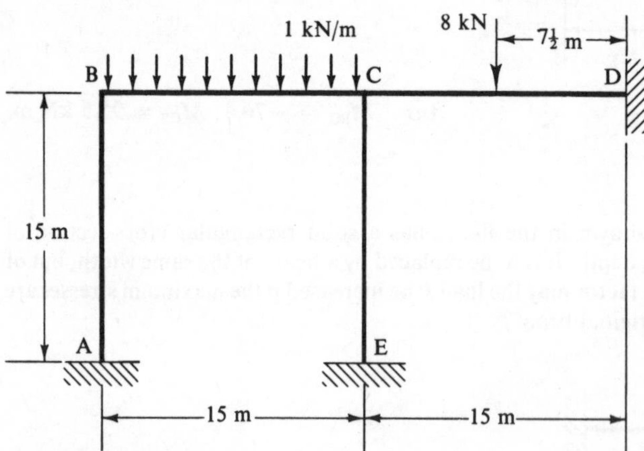

Problem 6·14

Ans. 1.01 kN; 0.72 kN (both compressive).

6.15. The figure shows an idealized portion of a bridge structure. All members are of uniform cross-section and have the same flexural rigidity. Joints A and D are encastré, B, C, and E are rigid, but E is free to rotate relative to its support. Determine the vertical components of reaction at A, D, and E.

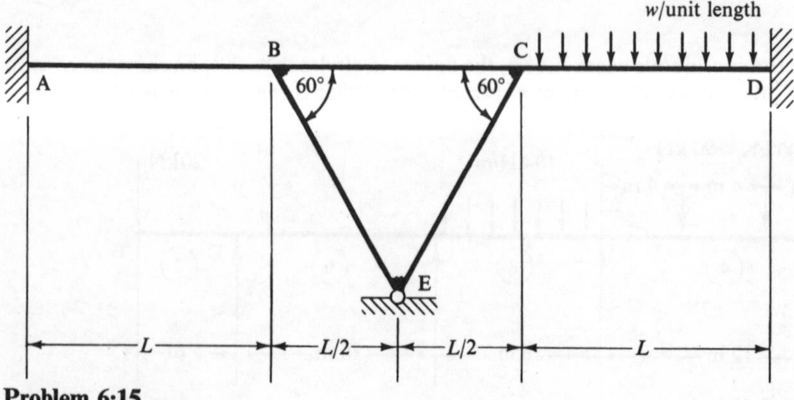

Problem 6·15

Ans. $0.00576wL$; $0.5444wL$; $0.4498wL$.

6.16. What force H is required to prevent horizontal movement of the beam ABC in the elastic rigid-jointed frame shown, which is built into a rigid foundation at D and F and pinned to it at E?

The relevant flexural rigidity of the beam ABC is twice that of the stanchions.

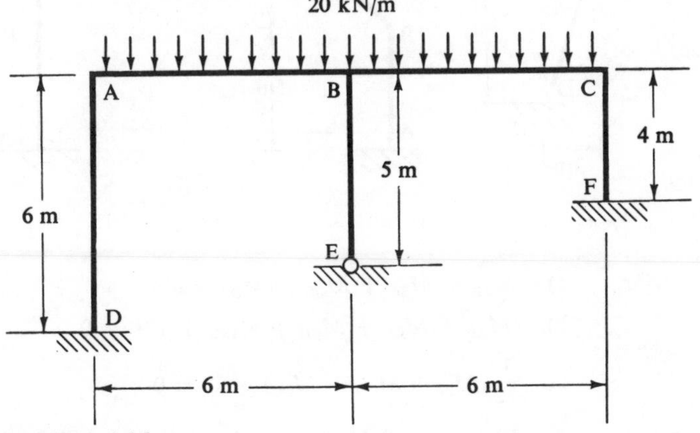

Problem 6·16

Ans. 4.29 kN acting from left to right.

6.17. Obtain the sway equations for each of the frames shown below.

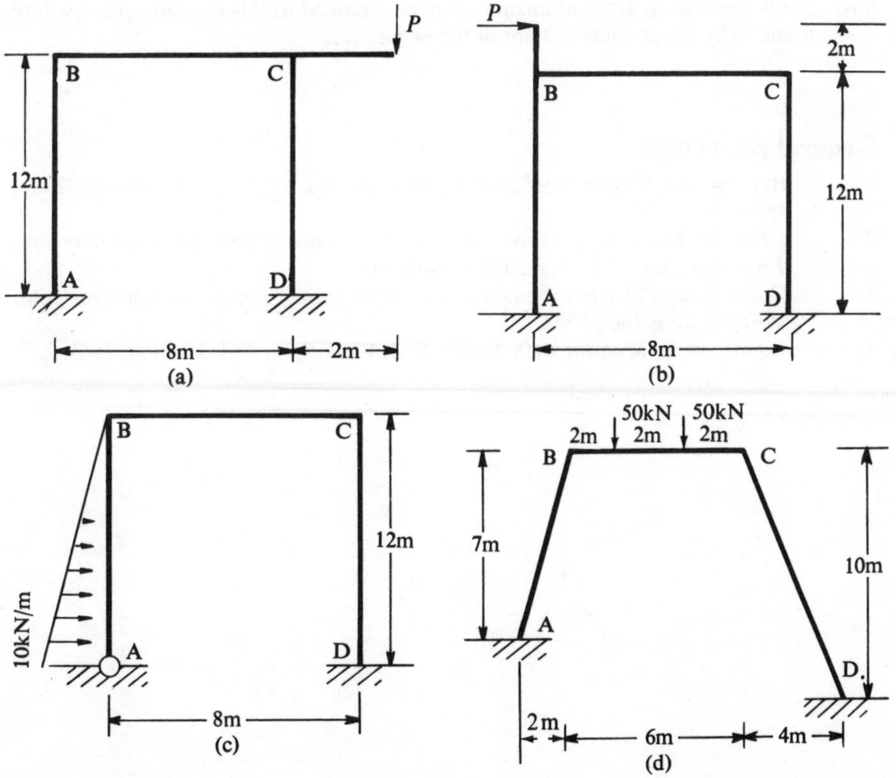

(a)

(b)

(c)

(d)

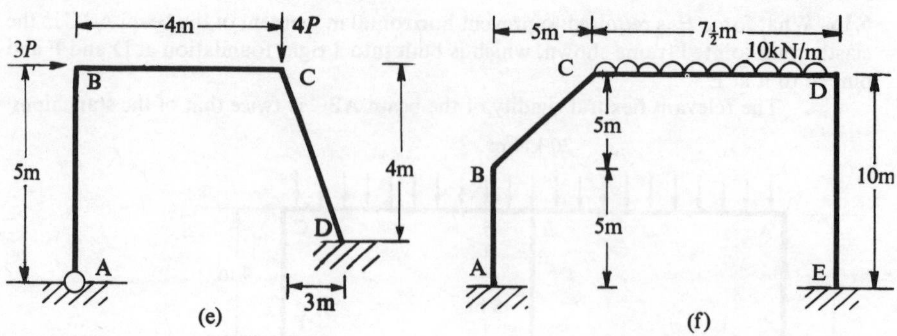

(e) (f)

Problem 6·17

Ans.
a) $M_{AB} + M_{BA} + M_{CD} + M_{DC} = 0$

b) $(M_{AB} + M_{BA} + M_{CD} + M_{DC}) + 12P = 0$

c) $\frac{1}{12}(M_{BA} + M_{CD} + M_{DC}) + 20 = 0$

d) $10\,M_{AB} + 18\,M_{BA} + 15\,M_{CD} + 7\,M_{DC} = 400$

e) $31\,M_{BA} + 35\,M_{CD} - 20\,M_{DC} = 0$

f) i) $6\,M_{AB} + 12\,M_{BA} - 10\,M_{CB} + 4\,M_{DC} = 1125$

ii) $2\,M_{AB} + 2\,M_{BA} + M_{ED} + M_{DE} = 0$

NOTE: Many apparently different answers may be obtained to this question. They will be found to be linear combinations of the two above.

General references

1 Butterworth, S. *Structural Analysis by Moment Distribution*. Longman Group, 1949.

2 Cross, H. The analysis of continuous frames by distributing fixed-end moments. *Proc. Am. Soc. C.E.*, Sept. 1930–April 1932.

3 Cross, H. and Morgan, N. D. *Continuous Frames of Reinforced Concrete*. John Wiley & Sons Inc., 1932.

4 Lightfoot, E. *Moment Distribution*. E. & F. N. Spon Ltd, London, 1961.

Chapter 7
Matrix stiffness method

7.1 Introduction

Until recent years engineers had available only the methods already discussed, and variations of them, to analyse statistically indeterminate structures. For any structure of more than one degree of redundancy, the result was always obtained as a solution to a set of linear equations, the only exception being moment-distribution solutions involving no sway. The engineer often went to considerable lengths to make the set small.

The advent of the electronic computer has removed the problem of solving large sets of equations. But if the computer is to be used to solve the equations, why not use it to set them up in the first place? The hand-based methods of analysis are unsuitable for this purpose, and this chapter will discuss a method which lends itself to straightforward computer application. Although the method is not a new one, its use by engineers is only practicable now that computers are readily available. It is assumed that only comparatively trivial examples (structurally) would be attempted by hand. The aim should at all times be to reduce the amount of input data to a minimum, and variations of the basic method which involve some preliminary hand manipulation have purposely been excluded.

7.2 Simple example—stiffness method

Before discussing a completely general form of the stiffness method of analysis in matrix terms, it is worth spending a little time on a straightforward example involving no matrix algebra. The structure could be analysed more easily by other methods, but it provides a useful illustration of the application of the stiffness approach.

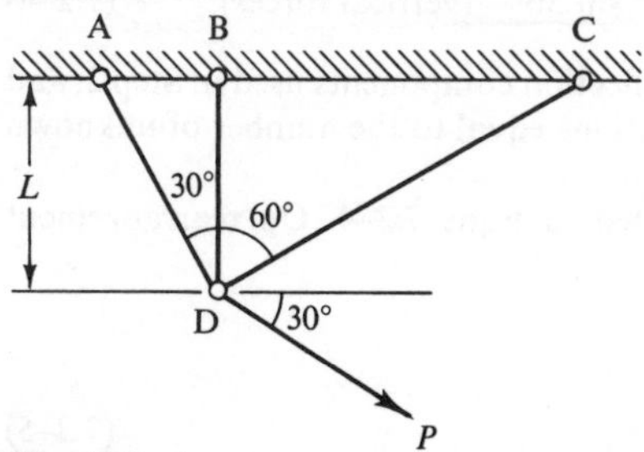

Fig. 7·2–1

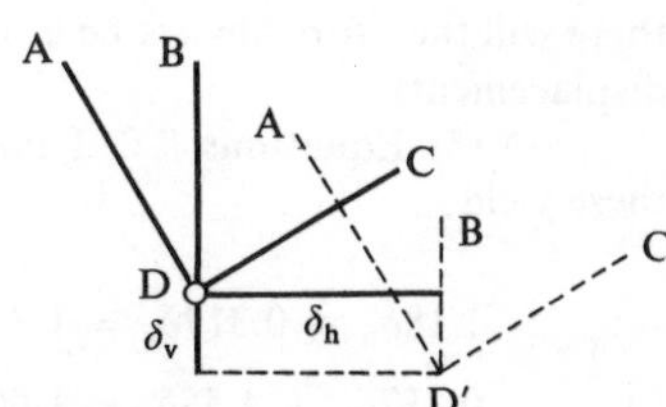

Fig. 7·2–2

The structure shown in Fig. 7.2–1 has three elastic rods pinned at the supports A, B, and C and pinned together at D. For simplicity the rods are of the same material, have the same cross-sectional area A, and are stiff enough to prevent instability. The joint D is acted on by a load P inclined at 30° to the horizontal. The joint D will deflect a small amount to position D′ (Fig. 7.2–2) and this can be resolved into horizontal and vertical components δ_h and δ_v. The engineer wishes to know the values of the member forces, but the solution is directed towards the determination of the displacements. The steps are:

(1) Force/distortion relations are established for each member. In this case the member only carries an axial force F and has an axial extension e.

$$F_{AD} = \frac{EA}{L_{AD}} \cdot e_{AD} = \frac{EA}{L \sec 30^\circ} \cdot e_{AD}$$

$$F_{BD} = \frac{EA}{L} \cdot e_{BD}$$

$$F_{CD} = \frac{EA}{L \sec 60^\circ} \cdot e_{CD} \qquad (7.2\text{–}1)$$

(2) This step involves a recognition of compatibility. The distortions of the members must match the deflections of the structure.

$$e_{AD} = (\delta_h \sin 30^\circ + \delta_v \cos 30^\circ)$$
$$e_{BD} = \delta_v$$
$$e_{CD} = (-\delta_h \sin 60^\circ + \delta_v \cos 60^\circ) \qquad (7.2\text{–}2)$$

(3) Equations 7.2–2 may be substituted in Eqns 7.2–1;

$$F_{AD} = \frac{EA}{L \sec 30^\circ}(\delta_h \sin 30^\circ + \delta_v \cos 30^\circ) = \frac{EA}{L}\left(\frac{\sqrt{3}}{4}\delta_h + \frac{3}{4}\delta_v\right)$$

$$F_{BD} = \frac{EA}{L} \cdot \delta_v \qquad F_{CD} = \frac{EA}{L}\left(-\frac{\sqrt{3}}{4}\delta_h + \frac{\delta_v}{4}\right) \qquad (7.2\text{–}3)$$

(4) Equilibrium is now applied at the joint D. The forces acting on each member, together with the applied external load P must satisfy equilibrium. Hence

$$F_{AD} \sin 30^\circ - F_{CD} \sin 60^\circ = P \cos 30^\circ \quad \text{(horizontal forces)}$$

and

$$F_{AD} \cos 30^\circ + F_{BD} + F_{DC} \cos 60^\circ = P \sin 30^\circ \quad \text{(vertical forces)} \qquad (7.2\text{–}4)$$

Each equation here corresponds to one of the deflection components used in Step 2, and there will therefore always be a number of equations equal to the number of unknown displacements.

(5) Equations 7.2–3 may be substituted in Eqns 7.2–4. On rearrangement these yield:

$$1.18\delta_h + 0.317\delta_v = 1.73LP/EA$$
$$0.317\delta_h + 4.55\delta_v = 1.00LP/EA \qquad (7.2\text{–}5)$$

(6) The pair of linear equations (7.2–5) may now be solved:

$$\delta_h = 1.42PL/AE \qquad \delta_v = 0.16PL/AE$$

(7) Substitution of these displacements into Eqns 7.2–3 gives the required member forces:

$$F_{AD} = \frac{EA}{L}\left(\frac{\sqrt{3}}{4} \times 1.42 + \frac{3}{4} \times 0.16\right)\frac{PL}{AE} = 0.74P$$

$$F_{BD} = 0.16P \qquad F_{CD} = -0.58P$$

The structure analysed here is statistically indeterminate, but no account of the degree of indeterminacy has been taken in the solution. The process would work for any number of redundants. This problem has two unknown displacement components, or two degrees of freedom. Exactly the same series of seven steps could be used to solve a very much larger structure with many more degrees of freedom, and the only important difference would lie in the number of equations of type 7.2–5 to be solved. Even with as few as five or six degrees of freedom, only a computer solution would be reasonable. The remainder of the chapter is concerned with the adaptation of the whole method to computer handling.

7.3 Assumptions

This discussion of the stiffness method will be based on plane, rigid-jointed frames, with encastré supports. It will be assumed that members are straight beams and columns with constant properties between joints or nodes—points at which two or more members meet. All members will behave in a linear, elastic manner and no yielding will occur. Deflections will be sufficiently small for changes of geometry to be ignored. The axial forces in members will be very much less than the respective Euler buckling loads (see Chapter 9). For each member a principal axis of bending lies in the plane of the structure, and the shear centre coincides with the centroid of the section. All applied loads act in the structure plane, and consequently all displacements are also in this plane.

This list of assumptions is not intended to suggest that only structures satisfying these conditions can be analysed by the stiffness method. The removal of these requirements, and the generalization of the method will be discussed in later sections.

7.4 Notation and axes

The right-handed member axis system to be used, and the member forces and displacements have already been discussed in Chapter 5. It is assumed that the x–y plane for each member coincides with the plane of the structure, and the y and z axes are principal axes for the cross-section. As two-dimensional structures are being considered here, only three degrees of freedom occur at each end of each member, and to each there is a corresponding force component. The vectors of member force and displacement $\mathbf{p}_1$, $\mathbf{p}_2$ and $\mathbf{d}_1$, $\mathbf{d}_2$ which have already been introduced in Section 5.1 are:

$$\mathbf{p}_1 = \begin{bmatrix} p_{x1} \\ p_{y1} \\ m_1 \end{bmatrix} \qquad \mathbf{p}_2 = \begin{bmatrix} p_{x2} \\ p_{y2} \\ m_2 \end{bmatrix} \tag{7.4–1}$$

$$\mathbf{d}_1 = \begin{bmatrix} d_{x1} \\ d_{y1} \\ \theta_1 \end{bmatrix} \qquad \mathbf{d}_2 = \begin{bmatrix} d_{x2} \\ d_{y2} \\ \theta_2 \end{bmatrix} \tag{7.4–2}$$

These vectors are illustrated in Figs. 7.4–1 and 7.4–2. The moments and rotations should really have been shown here as m_{z1}, m_{z2} and θ_{z1}, θ_{z2}, but in a plane structure where moments and rotations can occur in one plane only, z can safely be omitted.

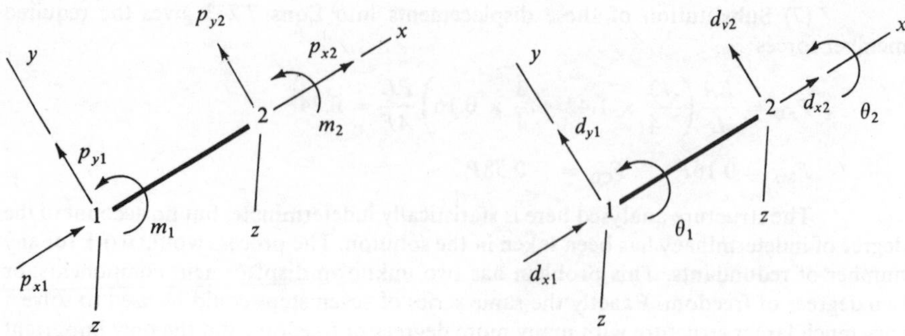

Fig. 7·4–1 **Fig. 7·4–2**

It is important to remember that these are the forces acting *on* the ends of the members, and the displacements *of* the ends of the members. The use of six displacement terms allows the member to experience rigid-body movement as well as elastic deformations.

In order to allow a completely general method of analysis to be established it is necessary to number the joints of a structure in a consistent manner. The pattern to be used follows Fig. 7.4–3.

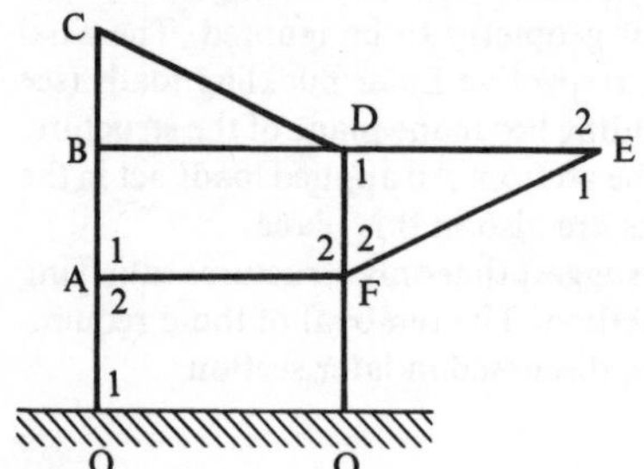

Fig. 7·4–3

All encastré supports are lettered O. All joints are lettered A, B, C, etc., not necessarily in any special order, but without any letter being omitted. The end 1 of any member is at the joint with letter nearer the beginning of the alphabet, as shown in Fig. 7.4–3 with O being assumed to come before A. Each member is known by the letters of the two joints it connects (OA, BD, etc.). (Letters are used here for clarity, but in a computer program numbers are in fact much more convenient.)

In addition to the member-oriented axes, there will be an overall structure-oriented system, denoted by x', y', and z'. The loads applied to the joints of a structure, and the resulting deflections, are not related to any particular member, but are concerned with the structure as a whole and these will therefore be expressed in terms of this overall system.

7.5 Member stiffness

The analysis steps which follow are very similar to those of the simple example in Section 7.2. The first step involves setting up force/deflection relations for each member. (As beams are being discussed here, the terms 'force' and 'deflection' are taken to be general expressions signifying direct forces and moments, and linear deflections and rotations respectively.) Such relations for a space frame member have already been given in Eqn 5.2–1, and the reduced form for a plane frame member is:

$$\begin{bmatrix} p_{x1} \\ p_{y1} \\ m_1 \\ \hdashline p_{x2} \\ p_{y2} \\ m_2 \end{bmatrix} = \left[\begin{array}{ccc:ccc} EA/L & 0 & 0 & -EA/L & 0 & 0 \\ 0 & 12EI/L^3 & 6EI/L^2 & 0 & -12EI/L^3 & 6EI/L^2 \\ 0 & 6EI/L^2 & 4EI/L & 0 & -6EI/L^2 & 2EI/L \\ \hdashline -EA/L & 0 & 0 & EA/L & 0 & 0 \\ 0 & -12EI/L^3 & -6EI/L^2 & 0 & 12EI/L^3 & -6EI/L^2 \\ 0 & 6EI/L^2 & 2EI/L & 0 & -6EI/L^2 & 4EI/L \end{array}\right] \begin{bmatrix} d_{x1} \\ d_{y1} \\ \theta_1 \\ \hdashline d_{x2} \\ d_{y2} \\ \theta_2 \end{bmatrix} \tag{7.5–1}$$

It is convenient to separate the terms connected with the ends 1 and 2 of the member by partitioning, as indicated by the dotted lines.

$$\mathbf{p}_1 = \mathbf{K}_{11}\mathbf{d}_1 + \mathbf{K}_{12}\mathbf{d}_2$$

$$\mathbf{p}_2 = \mathbf{K}_{21}\mathbf{d}_1 + \mathbf{K}_{22}\mathbf{d}_2 \tag{7.5–2}$$

$\mathbf{p}_1$, $\mathbf{p}_2$ and $\mathbf{d}_1$, $\mathbf{d}_2$ are defined in Eqns 7.4–1 and 7.4–2 respectively, and the **K** terms are the 3 × 3 stiffness matrices indicated in Eqn 7.5–1. It is helpful to note:

(1) $\mathbf{K}_{11}$ and $\mathbf{K}_{22}$ are symmetric.

(2) $\mathbf{K}_{12} = \mathbf{K}_{21}^T$.

(3) $\mathbf{K}_{22} = \mathbf{K}_{11}$, with the signs of the off-diagonal elements $6EI/L^2$ reversed.

7.6 Coordinate transformation

Equations 7.5–2 have been expressed in terms of the coordinate systems of the individual members. Step 2 of the simple example involves the application of compatibility to the deflections of the ends of members, and the joints in which they meet. This could be done easily, because simple trigonometrical relations existed between the defined member extensions and joint displacements δ_h and δ_v. For more complex problems a more systematic approach must be made, and this question has already been discussed in Chapter 5. New end force and displacement vectors $\mathbf{p}'$ and $\mathbf{d}'$, similar to Eqns 7.4–1 and 7.4–2 are defined, the primes indicating that they are in terms of structure, not member, axes. (They might be applied to either end 1 or 2 as $\mathbf{p}'_1$, or $\mathbf{p}'_2$.) The unprimed and primed terms are related by the equation:

$$\mathbf{p} = \mathbf{T}\mathbf{p}' \tag{7.6–1}$$

T is a 3 × 3 transformation matrix, and its terms have been given already in Eqn 5.3–17. They are:

$$\mathbf{T} = \begin{bmatrix} L_x/L & L_y/L & 0 \\ -L_y/L & L_x/L & 0 \\ 0 & 0 & 1 \end{bmatrix} = \begin{bmatrix} \cos\beta & \sin\beta & 0 \\ -\sin\beta & \cos\beta & 0 \\ 0 & 0 & 1 \end{bmatrix} \tag{7.6–2}$$

where L_x and L_y are the projections of the member of length L on the x'- and y'-axes, and β is the angle between the x'- and x-axes measured in the direction of the positive rotation vector (Fig. 7.6–1).

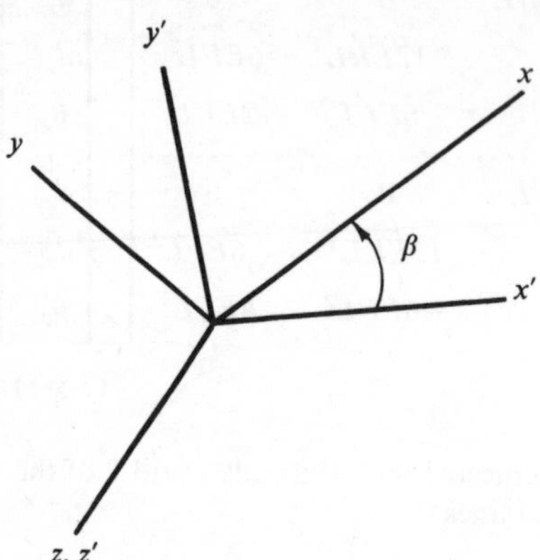

Fig. 7·6–1

As matrix **T** is orthogonal its inverse is equal to its transpose, and Eqn 7.6–1 may be rewritten:

$$\mathbf{p}' = \mathbf{T}^T\mathbf{p} \tag{7.6–3}$$

A similar relation exists for displacements, and while it may be demonstrated geometrically, it is done more neatly by using a virtual work argument.

A member carrying end loads **p** experiences virtual end displacements $\bar{\mathbf{d}}$. These same forces and displacements may be expressed in a structure coordinate system $-\mathbf{p}'$ and $\bar{\mathbf{d}}'$. The work done by the forces is independent of the coordinate system used to express them. Hence

$$\mathbf{p}^T\bar{\mathbf{d}} = \mathbf{p}'^T\bar{\mathbf{d}}'$$

$$(\mathbf{T}\mathbf{p}')^T\bar{\mathbf{d}} = \mathbf{p}'^T\bar{\mathbf{d}}' \quad \text{(by Eqn 7.6–1)}$$

$$\mathbf{p}'^T\mathbf{T}^T\bar{\mathbf{d}} = \mathbf{p}'^T\bar{\mathbf{d}}'$$

But this relation must be independent of the magnitude of $\mathbf{p}'$ as its choice is quite arbitrary. Then

$$\bar{\mathbf{d}}' = \mathbf{T}^T\bar{\mathbf{d}}$$

If it is now assumed that the actual displacements of the structure are small, they may replace the virtual displacements. Hence

$$\mathbf{d}' = \mathbf{T}^T\mathbf{d} \tag{7.6–4}$$

or

$$\mathbf{d} = \mathbf{T}\mathbf{d}' \quad (\mathbf{T} \text{ orthogonal})$$

Equations 7.6–1 and 7.6–4 show the correspondence between force and displacement relations known as **contragredience**. Other examples will occur in later sections.

The member stiffness relations in Eqns 7.5–2 may now be rewritten in terms of the structure coordinate axes:

$$\mathbf{p}_1 = \mathbf{K}_{11}\mathbf{d}_1 + \mathbf{K}_{12}\mathbf{d}_2$$

Hence

$$(\mathbf{T}\mathbf{p}_1') = \mathbf{K}_{11}(\mathbf{T}\mathbf{d}_1') + \mathbf{K}_{12}(\mathbf{T}\mathbf{d}_2')$$

(premultiply both sides by $\mathbf{T}^T$); then

$$\mathbf{p}'_1 = (\mathbf{T}^T\mathbf{K}_{11}\mathbf{T})\mathbf{d}'_1 + (\mathbf{T}^T\mathbf{K}_{12}\mathbf{T})\mathbf{d}'_2$$

or

$$\mathbf{p}'_1 = \mathbf{K}'_{11}\mathbf{d}'_1 + \mathbf{K}'_{12}\mathbf{d}'_2 \tag{7.6–5}$$

where $\mathbf{K}'_{11} = \mathbf{T}^T\mathbf{K}_{11}\mathbf{T}$, etc. Similarly,

$$\mathbf{p}'_2 = \mathbf{K}'_{21}\mathbf{d}'_1 + \mathbf{K}'_{22}\mathbf{d}'_2 \tag{7.6–6}$$

The full form of matrix $\mathbf{K}'_{11}$ is:

$$\mathbf{K}'_{11} = \begin{bmatrix} \frac{C^2EA}{L} + \frac{S^2 12EI}{L^3} & SC\left(\frac{EA}{L} - \frac{12EI}{L^3}\right) & -S\frac{6EI}{L^2} \\ SC\left(\frac{EA}{L} - \frac{12EI}{L^3}\right) & S^2\frac{EA}{L} + C^2\frac{12EI}{L^3} & C\frac{6EI}{L^2} \\ -S\frac{6EI}{L^2} & C\frac{6EI}{L^2} & 4\frac{EI}{L} \end{bmatrix} \tag{7.6–7}$$

($C = \cos\beta$ and $S = \sin\beta$).

The $\mathbf{K}'$ matrices here have the same properties as the unprimed forms.

(1) $\mathbf{K}'_{11}$ and $\mathbf{K}'_{22}$ are symmetric.

(2) $\mathbf{K}'_{12} = \mathbf{K}'^T_{21}$.

(3) $\mathbf{K}'_{22} = \mathbf{K}'_{11}$, with the signs of the off-diagonal terms $6EI/L^2$ reversed.

It is helpful in hand calculations to remember these relations, so that once the $\mathbf{K}'_{11}$ matrix has been set up, the other three may be obtained very easily.

7.7 Compatibility

This and later steps will be illustrated on an actual structure, and that of Fig. 7.7–1 will be used. According to the assumption of Section 7.3 loads may be applied at A or B.

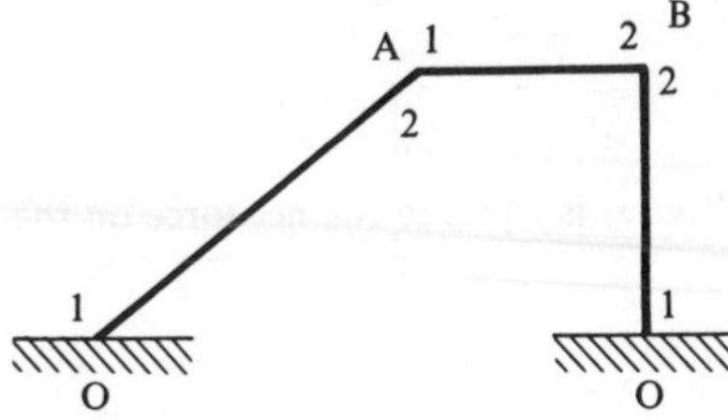

Fig. 7·7–1

The joints have been lettered according to the rules of Section 7.4, and the ends 1 and 2 of each member have been indicated. As in Step 2 of the simple example, there must be compatibility between the end deflections of members, and the joints into which they run.

For OA $\mathbf{d}'_1 = 0 \qquad \mathbf{d}'_2 = \mathbf{\Delta}_A$

For AB $\mathbf{d}'_1 = \mathbf{\Delta}_A \qquad \mathbf{d}'_2 = \mathbf{\Delta}_B$

For OB $\mathbf{d}'_1 = 0 \qquad \mathbf{d}'_2 = \mathbf{\Delta}_B$ (7.7–1)

(As $\mathbf{\Delta}_A$ and $\mathbf{\Delta}_B$, the displacements of joints A and B respectively, must obviously be measured in structure coordinates, the primes have been omitted as superfluous.)

Equations 7.7–1 may be substituted in Eqns 7.6–6 for each member (Step 3 of simple example) and this yields:

For OA $\mathbf{p}'_1 = \mathbf{K}'_{12}\mathbf{\Delta}_A$

$\mathbf{p}'_2 = \mathbf{K}'_{22}\mathbf{\Delta}_A$

For AB $\mathbf{p}'_1 = \mathbf{K}'_{11}\mathbf{\Delta}_A + \mathbf{K}'_{12}\mathbf{\Delta}_B$

$\mathbf{p}'_2 = \mathbf{K}'_{21}\mathbf{\Delta}_A + \mathbf{K}'_{22}\mathbf{\Delta}_B$

For OB $\mathbf{p}'_1 = \mathbf{K}'_{12}\mathbf{\Delta}_B$

$\mathbf{p}'_2 = \mathbf{K}'_{22}\mathbf{\Delta}_B$ (7.7–2)

7.8 Equilibrium

Where several members in a structure meet at a joint there must be equilibrium between the forces on the members and any external loads (Step 4). Figure 7.8–1 shows the forces acting at joint A of the structure in Fig. 7.7–1. The external loads acting are shown on the left (P_{xA}, P_{yA} and m_A), and on the right are shown the loads which must act on the ends of the two members OA and AB.

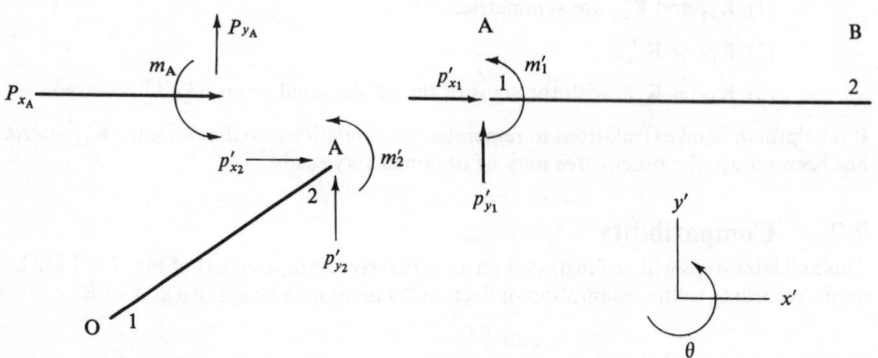

Fig. 7·8–1

The external load $\mathbf{P}_A$ (*in structure coordinates*) is equal to the net force on the joining members. Then

$$P_{xA} = (p'_{x1})_{AB} + (p'_{x2})_{OA}$$

$$P_{yA} = (p'_{y1})_{AB} + (p'_{y2})_{OA}$$

$$m_A = (m'_1)_{AB} + (m'_2)_{OA}$$

or

$$\mathbf{P}_A = (\mathbf{p}'_1)_{AB} + (\mathbf{p}'_2)_{OA}$$

Similarly,

$$\mathbf{P}_B = (\mathbf{p}'_2)_{AB} + (\mathbf{p}'_2)_{OB} \qquad (7.8\text{–}1)$$

7.9 Structure stiffness matrix

Equations 7.7–2 may now be substituted in Eqns 7.8–1. Written in matrix form they become:

$$\begin{bmatrix} \mathbf{P}_A \\ \mathbf{P}_B \end{bmatrix} = \begin{matrix} A \\ B \end{matrix} \overset{\quad A \qquad\qquad\qquad\qquad B}{\begin{bmatrix} (\mathbf{K}'_{22})_{OA} + (\mathbf{K}'_{11})_{AB} & (\mathbf{K}'_{12})_{AB} \\ (\mathbf{K}'_{21})_{AB} & (\mathbf{K}'_{22})_{AB} + (\mathbf{K}'_{22})_{OB} \end{bmatrix}} \begin{bmatrix} \boldsymbol{\Delta}_A \\ \boldsymbol{\Delta}_B \end{bmatrix} \tag{7.9–1}$$

or

$$\mathbf{P} = \mathbf{K}_s \boldsymbol{\Delta} \tag{7.9–2}$$

P and **Δ** are vectors of external loads applied at, and deflections of, the joints of the structure. $\mathbf{K}_s$ is termed the structure stiffness matrix.

$\mathbf{P}_A$ and $\mathbf{P}_B$ are both three-term load vectors, and $\boldsymbol{\Delta}_A$ and $\boldsymbol{\Delta}_B$ are similarly three-term displacement vectors. Equation 7.9–2 is therefore a set of six linear equations (the structure of Fig. 7.7–1 has six degrees of freedom), and if the matrix $\mathbf{K}_s$ is non-singular they can be solved for a given set of loads **P** for the six unknown displacement components in **Δ**. This can be written:

$$\boldsymbol{\Delta} = \mathbf{K}_s^{-1} \mathbf{P} \tag{7.9–3}$$

Once the deflections of the structure are known, appropriate terms may be selected, and substituted in the earlier equations (7.7–2) and (7.6–1) to give member forces first in structure coordinates, and then in member coordinates. The problem is now solved.

Before proceeding, it is worth while examining $\mathbf{K}_s$ carefully. Several points may be noted:

(1) It is square, of an order equal to the number of degrees of freedom, where this number is equal to the total number of displacement components in the complete structure. For a rigidly jointed plane frame the order is thus equal to three times the number of joints, and the total storage required for a structure of j joints is $9j^2$. For a space frame it is $36j^2$.

(2) It is symmetric. This can be seen by noting that the submatrices occurring on the leading diagonal are all $\mathbf{K}'_{11}$ or $\mathbf{K}'_{22}$ matrices, which are themselves symmetric (Section 7.5). The only submatrices occurring off diagonal are $\mathbf{K}'_{12}$ and $\mathbf{K}'_{21}$, and the one is the transpose of the other.

(3) It is non-singular. If the structure is not a mechanism, and is stable, there must be a unique relationship between applied loads and the resulting displacements. Equation 7.9–3 must exist, and therefore $\mathbf{K}_s$ must be non-singular.

(4) It is positive-definite. A real symmetric matrix such as $\mathbf{K}_s$ is said to be positive-definite if the quadratic form $\mathbf{x}^T\mathbf{K}_s\mathbf{x}$ is always positive for any real and non-zero vector **x**. This can be shown quite simply. If a set of forces **P** acts on the joints of a structure of stiffness matrix $\mathbf{K}_s$, causing displacements **x**, the work done (U) is given by:

$$U = \mathbf{x}^T\mathbf{P} = \mathbf{x}^T\mathbf{K}_s\mathbf{x}$$

U is necessarily positive, and therefore $\mathbf{K}_s$ is positive-definite.

The full implication of this conclusion cannot be discussed here, but positive-definiteness is a valuable asset in many equation-solving processes.

(5) It can be built up in a logical way, without going through the steps of Sections 7.7 and 7.8. The top left-hand block of $\mathbf{K}_s$ (Eqn 7.9–1)—the block of the $\mathbf{p}_A$ equation which multiplies $\mathbf{d}_A$—contains the sum of the $\mathbf{K}'_{11}$ and $\mathbf{K}'_{22}$ matrices of all members meeting at A, the $\mathbf{K}'_{11}$ term occurring if the end 1 is joined to A, and $\mathbf{K}'_{22}$ occurring if end 2 is joined to A. The same is true of the lower right-hand block, where $\mathbf{K}'_{11}$ and $\mathbf{K}'_{22}$ for members meeting at B are summed. The off-diagonal positions contain only $\mathbf{K}'_{12}$

or $\mathbf{K}'_{21}$ matrices. $(\mathbf{K}'_{12})_{AB}$ occurs in the A row (the first one), in the B column (the second one). Similarly $\mathbf{K}'_{21}$ occurs in the B row and the A column. Because only one member, AB, joins A and B, there is no summation here. If the member LR occurs in a structure, its stiffness matrices are placed in the overall structure stiffness matrix according to the arrangement in Fig. 7.9–1. Block LL will contain the sum of $\mathbf{K}'_{11}$ or $\mathbf{K}'_{22}$ matrices for all members meeting at L, and RR will contain a similar sum for joint R. Block LR will contain the $\mathbf{K}'_{12}$ matrix for member LR (end 1 of LR being at L) and block RL will contain the $\mathbf{K}'_{21}$ matrix for member LR (end 2 of LR being at R). It can be seen that $\mathbf{K}_s$ can be built up by posting the relevant $\mathbf{K}'$ matrices for each member of a structure in turn to the appropriate position.

(6) It is sparse and banded. It can be seen from Fig. 7.9–1 that all the leading diagonal positions of $\mathbf{K}_s$ will be filled, but off-diagonal positions will only be filled if the joints corresponding to the relevant row and column are joined. In general a joint in a structure is only joined to three or four other joints, and unless $\mathbf{K}_s$ is very small, it tends, therefore, to be sparse.

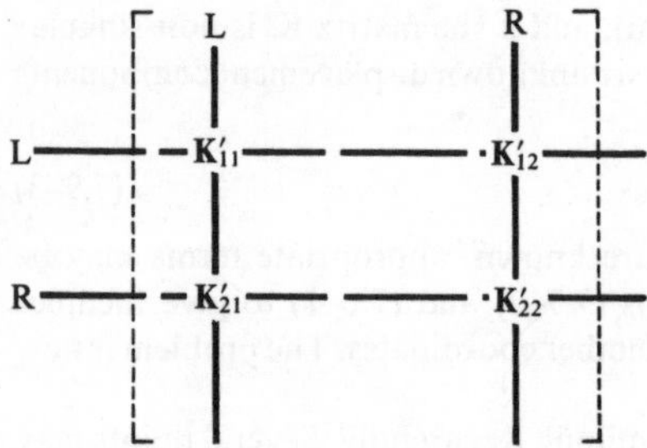

Fig. 7·9–1

The banding is a consequence of the lettering system adopted. If care is taken to keep the 'difference' between the letters at the ends of each member small, all off-diagonal terms are close to the leading diagonal. This can be useful in solving the Eqns 7.9–2 if the order of $\mathbf{K}_s$ is large.

An example of the form of a stiffness matrix for the structure of Fig. 7.9–2 is given in Fig. 7.9–3. A worked example is performed in Section 7.12.

The basic form of the stiffness method has now been discussed. The rest of the chapter will be devoted to its application to structures not satisfying the assumptions in Section 7.3.

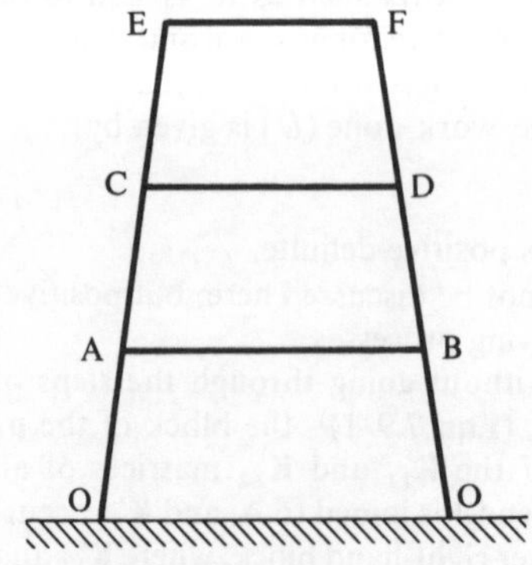

Fig. 7·9–2

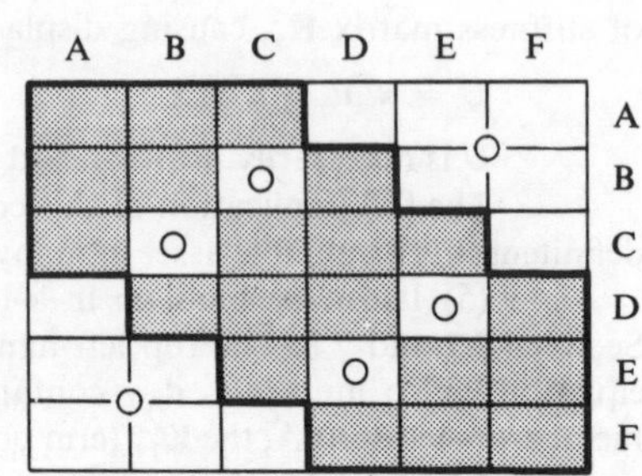

Fig. 7·9–3

7.10 Restrained joints and symmetry

It sometimes occurs in a structure that at a joint where movement in some directions may occur, movement in others is prevented. This happens at a pinned support, for instance, where rotations occur, but linear movement cannot.

Suppose that in the structure of Fig. 7.7–1 horizontal movement is prevented at B (Fig. 7.10–1).

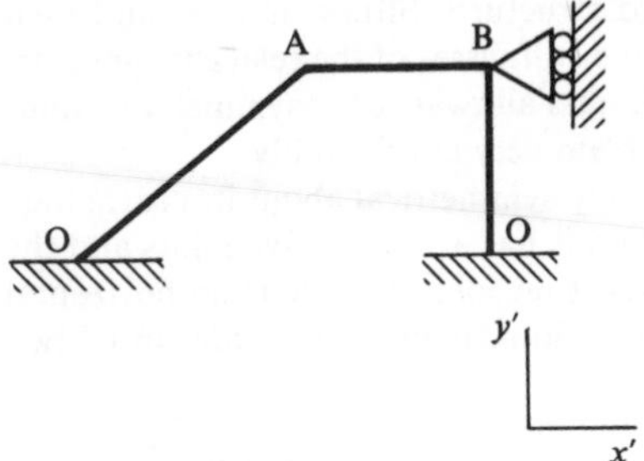

Fig. 7·10–1

The set of Eqns 7.9–1 can be assembled exactly as before, but an additional one must be added.

$$\begin{bmatrix} \mathbf{P}_A \\ \mathbf{P}_B \end{bmatrix} = \begin{bmatrix} (\mathbf{K}'_{22})_{OA} + (\mathbf{K}'_{11})_{AB} & (\mathbf{K}'_{12})_{AB} \\ (\mathbf{K}'_{21})_{AB} & (\mathbf{K}'_{22})_{AB} + (\mathbf{K}'_{22})_{OB} \end{bmatrix} \begin{bmatrix} \mathbf{\Delta}_A \\ \mathbf{\Delta}_B \end{bmatrix} \tag{7.10–1}$$

$$0 = \Delta_{xB} \tag{7.10–2}$$

Substitution of Eqn 7.10–2 in 7.10–1 is equivalent to the elimination of one column from $\mathbf{K}_s$, leaving one more equation (six) in the set than there are unknowns to be determined. Any one of the equations might be omitted, as they form a consistent set, but the load corresponding to the zero deflection is not in fact known. Any additional external load P_{xB} acting in the direction of the zero deflection goes straight to the support, and has no influence on the rest of the structure, but the support force is not known and this equation must therefore be omitted. This is equivalent to setting $P_{xB} = 0$ in the load vector and eliminating the corresponding row from $\mathbf{K}_s$. The five remaining equations may now be solved.

The above process is quite suitable if equations are being solved by hand (any reduction in the number is most valuable), but is not so suitable in computer solution. Removal of rows and columns from $\mathbf{K}_s$ involves its repacking in the computer store, and this is time-consuming. A more economical technique is to

(a) set up the stiffness matrix as though all joints were entirely free;

(b) set rows and columns of $\mathbf{K}_s$ corresponding to zero deflections to zero (thus rendering $\mathbf{K}_s$ singular) and set the corresponding terms in the load vector to zero;

(c) insert a 1 on the leading diagonal of $\mathbf{K}_s$ in zeroed rows (making $\mathbf{K}_s$ non-singular again); this is equivalent to inserting Eqn 7.10–2 in the set.

Pinned or roller supports can be allowed for in this way. Indeed, some computer bureaux programs require that *all* supports are treated like this. If a joint is encastré, it is numbered like the rest and its three deflection components are all set to zero as has been described.

An example using this technique will be given in Section 7.12.

The joint of Fig. 7.10–1 may not be fully restrained, but may be restrained elastically—the reaction may be $\lambda\Delta_{xB}$, for example, where λ is the support stiffness. The effect of this is to cause the external load equation of Eqn 7.8–1 for equilibrium balance in the x-direction at B to be:

$$P_{xB} - \lambda \Delta_{xB} = (p'_{x2})_{AB} + (p'_{x2})_{OB} \tag{7.10–3}$$

Transference of the term in λ to the right-hand side leaves the load vector as in Eqn 7.9–1, and changes $\mathbf{K}_s$ only through the addition of λ to the leading diagonal term in the P_{xB} row. The solution of structures having joints of this kind is therefore merely a question of setting up a normal load vector and structure stiffness matrix, and then modifying the leading diagonal terms by adding the stiffnesses of the relevant supports.

This is a convenient point at which to discuss allowance for symmetry in solution. Such allowance can reduce the size of a problem very considerably.

The pitched portal of Fig. 7.10–2 is entirely symmetrical about its centre line, and is loaded symmetrically. As it stands, however, it has a total of five joints and the stiffness matrix would have to be 15 × 15. However, at joint A there is no horizontal deflection and no rotation—the structure could be looked upon as made up of two structures of the type shown in Fig. 7.10–3.

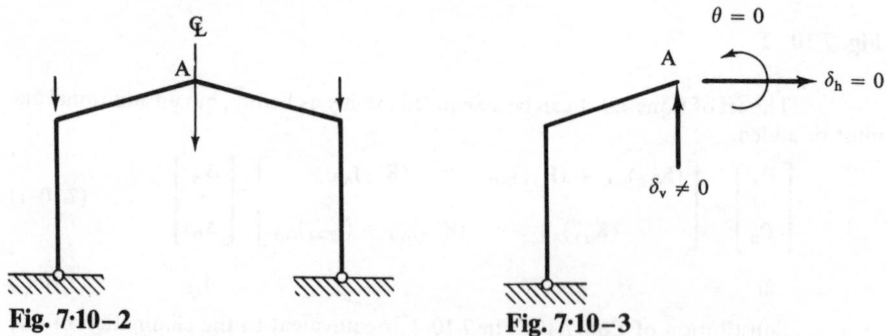

Fig. 7·10–2 **Fig. 7·10–3**

The structure may be analysed by setting up a stiffness matrix for one side only (three joints) and suppressing the rows and columns corresponding to horizontal and rotational displacement at A. A vertical load equal to half that on the original structure will be used at A.

Similar techniques to this may be used in other cases. There is no virtue in using symmetry in small problems, but it can be a great help in large ones, where computer storage is a problem. Each case must be studied on its merits, as additional joints on members cut by the plane of symmetry may sometimes have to be introduced. This could nullify any advantage gained.

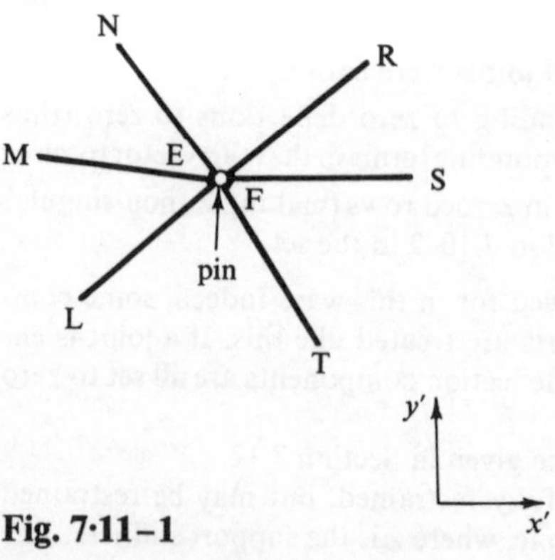

Fig. 7·11–1

7.11 Internal pins

A structure which is primarily rigid jointed, may frequently contain a number of joints which must be regarded as pinned, as the design will not allow them to transmit moments. Such a joint is shown in Fig. 7.11–1. Note that the two sides of the joint have been lettered separately E and F.

The two groups of members EN, EM, EL and FR, FS, FT are rigid-jointed within themselves, but are pinned together at EF. Such a joint has a total of four degrees of freedom, and therefore can take a total of four load components. It is here assumed that the external direct loads are applied to joint E, although this choice is quite arbitrary and has no effect on the final solution. The pin transmits a direct force having components q_x and q_y between E and F (Fig. 7.11–2). The stiffness matrix is assembled in the usual way,

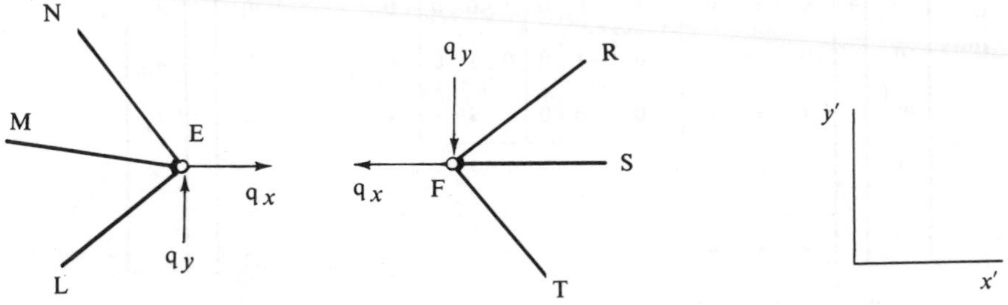

Fig. 7·11–2

but the load vector must take account of q_x and q_y. The relevant equilibrium equations are in Eqn 7.11–1. Asterisks represent terms associated with other joints.

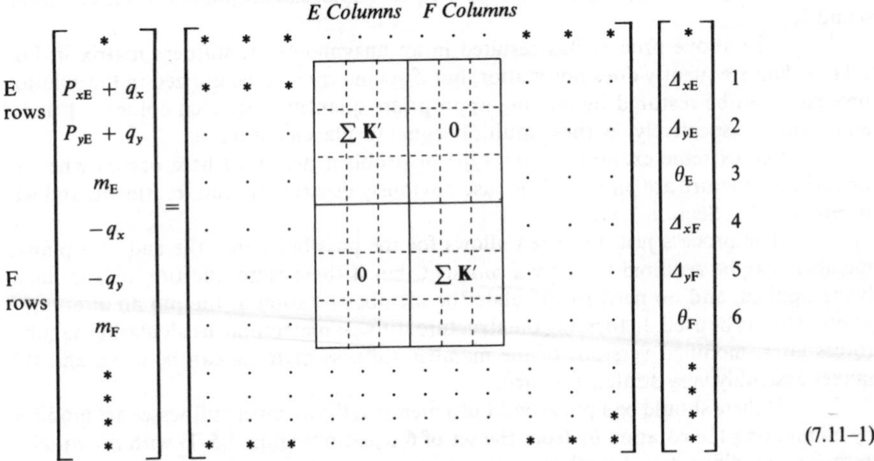

(7.11–1)

Blocks EE and FF are in general full, but EF and FE are zero.
Equation 4 is added to 1.
Equation 5 is added to 2.

This removes the q_x and q_y terms from 1 and 2. Equations 4 and 5 are now removed, and replaced by the condition that the linear deflections of E and F are equal.

$$0 = -\Delta_{xE} + \Delta_{xF}$$

$$0 = -\Delta_{yE} + \Delta_{yF} \tag{7.11–2}$$

This amounts to placing zeroes in the load vector at positions 4 and 5, +1 in positions 4,4 and 5,5 of the stiffness matrix and −1 in positions 4,1 and 5,2. The form of the final set of equations is:

$$
\begin{array}{c}
\\ \text{E} \\ \text{rows} \\ \\ \text{F} \\ \text{rows} \\ \\ \\ \\ \\ \\
\end{array}
\begin{bmatrix} * \\ P_{xE} \\ P_{yE} \\ m_E \\ 0 \\ 0 \\ m_F \\ * \\ \cdot \\ \cdot \\ \cdot \end{bmatrix}
=
\overset{\begin{array}{cccccc} & E\ Columns & & & F\ Columns & \\ 1 & 2 & 3 & 4 & 5 & 6 \end{array}}{
\begin{bmatrix}
* & * & * & \cdot & \cdot & \cdot & \cdot & \cdot & \cdot & \cdot & \cdot & \cdot & \cdot \\
\multicolumn{3}{c}{\sum 1 \text{ and } 4} & & \text{original} & & \multicolumn{7}{c}{\sum \text{original Eqns 1 and 4}} \\
\multicolumn{3}{c}{\sum 2 \text{ and } 5} & & \sum \mathbf{K}' & & \multicolumn{7}{c}{\sum \text{original Eqns 2 and 5}} \\
* & * & * & & & & 0 & 0 & 0 & * & * & * & * \\
0 & \cdot & 0 & -1 & 0 & 0 & 1 & 0 & 0 & 0 & \cdot & \cdot & 0 \\
0 & \cdot & 0 & 0 & -1 & 0 & 0 & 1 & 0 & 0 & \cdot & \cdot & 0 \\
* & * & * & 0 & 0 & 0 & * & * & * & * & * & * & * \\
\cdot & \cdot & \cdot & \cdot & \cdot & \cdot & \cdot & \cdot & \cdot & \cdot & \cdot & \cdot & \cdot \\
\cdot & \cdot & \cdot & \cdot & \cdot & \cdot & \cdot & \cdot & \cdot & \cdot & \cdot & \cdot & \cdot \\
\cdot & \cdot & \cdot & \cdot & \cdot & \cdot & \cdot & \cdot & \cdot & \cdot & \cdot & \cdot & \cdot \\
* & * & * & \cdot & \cdot & \cdot & \cdot & \cdot & \cdot & \cdot & \cdot & \cdot & *
\end{bmatrix}}
\begin{bmatrix} * \\ \Delta_{xE} \\ \Delta_{yE} \\ \theta_E \\ \Delta_{xF} \\ \Delta_{yF} \\ \theta_F \\ * \\ \cdot \\ \cdot \\ \cdot \end{bmatrix}
\begin{array}{c} \\ 1 \\ 2 \\ 3 \\ 4 \\ 5 \\ 6 \\ \\ \\ \\ \\ \end{array}
\qquad (7.11\text{–}3)
$$

This set of equations may now be solved in the normal way. It is a slightly larger set than required, as there are really four unknowns only associated with E and F, but as suggested previously it is more economical on a computer to handle the slightly larger set than to reduce the stiffness matrix. If hand solution is involved, however, any reduction in equation number is useful, and this can be achieved easily by adding columns 4 to 1 and 5 to 2 in Eqn 7.11–3 (they both multiply equal displacements) and omitting equations 4 and 5.

The above process has resulted in an unsymmetrical stiffness matrix in Eqn 7.11–3. This frequently does not matter, but if symmetry is to be utilized in the solution process it can be restored by moving appropriate quantities between columns 1 and 4 and 2 and 5 respectively, as these multiply equal displacement terms.

An extreme example of the type of problem discussed here occurs when all joints in a structure are pinned. This case obviously deserves special treatment, and will be discussed in Section 7.13.

The process just discussed allows for the possibility that the end of a pinned member may be required to carry a moment, but if there is no question of a moment being applied, and no need to calculate the member rotation at the pin an alternative approach may be used. In forming the structure stiffness matrix and in calculating member forces later, modified versions of the member stiffness matrices can be used, and this makes assembly very straightforward.

If there should be a pin at end 1 of a member, the member stiffnesses are modified by eliminating the rotation θ_1 from the set of 6 equations (Eqn 7.5–1) with m_1 equal to zero. The modified 3 × 3 matrices are:

$$
\mathbf{K}_{11} = \begin{bmatrix} \dfrac{EA}{L} & 0 & 0 \\ 0 & \dfrac{3EI}{L^3} & 0 \\ 0 & 0 & 0 \end{bmatrix}
\qquad
\mathbf{K}_{22} = \begin{bmatrix} \dfrac{EA}{L} & 0 & 0 \\ 0 & \dfrac{3EI}{L^3} & -\dfrac{3EI}{L^2} \\ 0 & -\dfrac{3EI}{L^2} & \dfrac{3EI}{L} \end{bmatrix}
$$

$$\mathbf{K}_{12} = \mathbf{K}_{21}^T = \begin{bmatrix} -\frac{EA}{L} & 0 & 0 \\ 0 & -\frac{3EI}{L^3} & \frac{3EI}{L^2} \\ 0 & 0 & 0 \end{bmatrix} \qquad (7.11\text{–}4)$$

(Although all coefficients of θ_1 are zero in the modified matrices the order must be kept at 3 × 3 to fit the normal assembly process for $\mathbf{K}_s$.)

If the pin occurs at end 2 of a member, the member matrices are:

$$\mathbf{K}_{11} = \begin{bmatrix} \frac{EA}{L} & 0 & 0 \\ 0 & \frac{3EI}{L^3} & \frac{3EI}{L^2} \\ 0 & \frac{3EI}{L^2} & \frac{3EI}{L} \end{bmatrix} \qquad \mathbf{K}_{22} = \begin{bmatrix} \frac{EA}{L} & 0 & 0 \\ 0 & \frac{3EI}{L^3} & 0 \\ 0 & 0 & 0 \end{bmatrix}$$

$$\mathbf{K}_{12} = \mathbf{K}_{21}^T = \begin{bmatrix} -\frac{EA}{L} & 0 & 0 \\ 0 & -\frac{3EI}{L^3} & 0 \\ 0 & -\frac{3EI}{L^2} & 0 \end{bmatrix} \qquad (7.11\text{–}5)$$

A pin at both ends and no end moments implies also zero shear, and the only terms remaining in the member stiffness matrices are then those in (EA/L).

The techniques discussed here for dealing with pinned joints within a rigid-jointed structure could be adopted to deal with other examples of lack of continuity (a sliding joint, perhaps), but a pin is the kind most frequently met.

7.12 Worked example on a rigid frame

In the structure of Fig. 7.12–1 the forces in the members AB and OB are required.

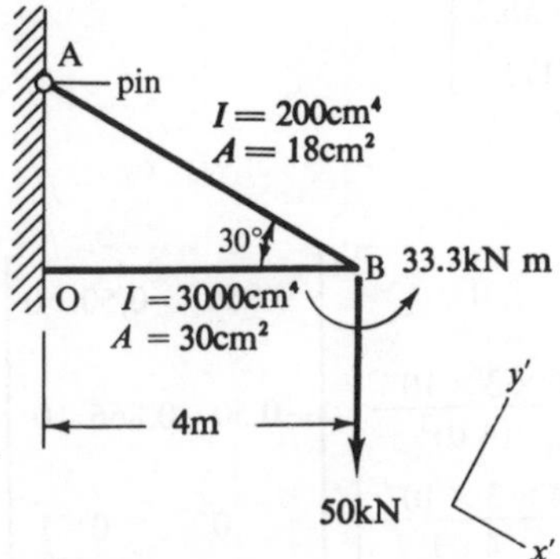

Fig. 7·12–1

The structure stiffness matrix is first set up in general terms, on the assumption that A is free. Hence

$$\begin{bmatrix} \mathbf{P}_A \\ = \\ \mathbf{P}_B \end{bmatrix} = \begin{bmatrix} (\mathbf{K}'_{11})_{AB} & (\mathbf{K}'_{12})_{AB} \\ (\mathbf{K}'_{21})_{AB} & (\mathbf{K}'_{22})_{AB} + (\mathbf{K}'_{22})_{OB} \end{bmatrix} \begin{bmatrix} \Delta_A \\ \Delta_B \end{bmatrix}$$

The structure axes are chosen as indicated. If a computer solution is involved, the choice depends mainly on the ease with which the input data can be set up, but for a hand solution the choice depends more on the number of $\mathbf{K}'$ matrices that have to be separately calculated. The choice shown keeps the axes transformation problem to a minimum.

Member AB

$$L = \frac{4.00}{\cos 30^\circ} = 4.619 \text{ m}$$

$$\mathbf{K}'_{11} = \mathbf{K}_{11} = E \begin{bmatrix} \dfrac{18 \times 10^{-4}}{4.619} & 0 & 0 \\ 0 & \dfrac{12 \times 200 \times 10^{-8}}{(4.619)^3} & \dfrac{6 \times 200 \times 10^{-8}}{(4.619)^2} \\ 0 & \dfrac{6 \times 200 \times 10^{-8}}{(4.619)^2} & \dfrac{4 \times 200 \times 10^{-8}}{4.619} \end{bmatrix}$$

$$= 10^{-8} E \begin{bmatrix} 39000 & 0 & 0 \\ 0 & 24.4 & 56.2 \\ 0 & 56.2 & 173 \end{bmatrix}$$

The remaining matrices can be filled up by inspection:

$$\mathbf{K}'_{12} = \mathbf{K}'^{T}_{21} = \mathbf{K}_{12} = 10^{-8} E \begin{bmatrix} -39000 & 0 & 0 \\ 0 & -24.4 & 56.2 \\ 0 & -56.2 & 86.6 \end{bmatrix}$$

and

$$\mathbf{K}'_{22} = \mathbf{K}_{22} = 10^{-8} E \begin{bmatrix} 39000 & 0 & 0 \\ 0 & 24.4 & -56.2 \\ 0 & -56.2 & 173 \end{bmatrix}$$

Member OB

$$\mathbf{K}'_{22} = \mathbf{T}^T \mathbf{K}_{22} \mathbf{T}$$

$$\mathbf{K}_{22}\mathbf{T} = E \begin{bmatrix} \dfrac{30 \times 10^{-4}}{4.0} & 0 & 0 \\ 0 & \dfrac{12 \times 3 \times 10^{-5}}{(4.0)^3} & -\dfrac{6 \times 3 \times 10^{-5}}{(4.0)^2} \\ 0 & -\dfrac{6 \times 3 \times 10^{-5}}{(4.0)^2} & \dfrac{4 \times 3 \times 10^{-5}}{4.00} \end{bmatrix} \begin{bmatrix} 0.866 & 0.50 & 0 \\ -0.50 & 0.866 & 0 \\ 0 & 0 & 1 \end{bmatrix}$$

$$\mathbf{K}'_{22} = 10^{-5}\,E\begin{bmatrix} 0.866 & -0.50 & 0 \\ 0.50 & 0.866 & 0 \\ 0 & 0 & 1 \end{bmatrix}\begin{bmatrix} 65.0 & 37.5 & 0 \\ -0.281 & 0.487 & -1.125 \\ 0.562 & -0.974 & 3.00 \end{bmatrix}$$

$$= 10^{-5}\,E\begin{bmatrix} 56.4 & 32.2 & 0.563 \\ 32.2 & 19.2 & -0.974 \\ 0.563 & -0.974 & 3.00 \end{bmatrix}$$

Hence

$$\mathbf{K}_s = 10^{-6}E\begin{bmatrix} 390 & 0 & 0 & -390 & 0 & 0 \\ 0 & 0.244 & 0.562 & 0 & -0.244 & 0.562 \\ 0 & 0.562 & 1.73 & 0 & -0.562 & 0.866 \\ -390 & 0 & 0 & 954 & 322 & 5.63 \\ 0 & -0.244 & -0.562 & 322 & 192 & -10.3 \\ 0 & 0.562 & 0.866 & 5.63 & -10.3 & 31.7 \end{bmatrix}$$

To cater for the pin at A, the first two rows and columns are now set to zero, with 1 inserted on the leading diagonal (see Section 7.10). The full set of equations, using the loads of Fig. 7.12–1 resolved into the coordinate directions, is:

$$\begin{bmatrix} 0 \\ 0 \\ 0 \\ 25.0 \\ -43.3 \\ 33.3 \end{bmatrix} = 10^{-6}E\begin{bmatrix} 1 & 0 & 0 & 0 & 0 & 0 \\ 0 & 1 & 0 & 0 & 0 & 0 \\ 0 & 0 & 1.73 & 0 & -0.562 & 0.866 \\ 0 & 0 & 0 & 954 & 322 & 5.63 \\ 0 & 0 & -0.562 & 322 & 192 & -10.3 \\ 0 & 0 & 0.866 & 5.63 & -10.3 & 31.7 \end{bmatrix}\begin{bmatrix} \Delta_{xA} \\ \Delta_{yA} \\ \theta_A \\ \Delta_{xB} \\ \Delta_{yB} \\ \theta_B \end{bmatrix}$$

The solution to this set of equations is:

$$\begin{bmatrix} \Delta_{xA} \\ \Delta_{yA} \\ \theta_A \\ \Delta_{xB} \\ \Delta_{yB} \\ \theta_B \end{bmatrix} = \frac{10^6}{E}\begin{bmatrix} 0 \\ 0 \\ -0.597 \\ 0.190 \\ -0.500 \\ 0.870 \end{bmatrix}\begin{matrix} \\ \\ \text{rad} \\ \text{m} \\ \text{m} \\ \text{rad} \end{matrix}$$

The member forces are now obtained by selecting the appropriate set from this displacement vector, and fitting it into the member stiffness equations.

Member AB

$$\mathbf{p}'_1 = \mathbf{p}_1 = 10^{-2}\begin{bmatrix} 39000 & 0 & 0 \\ 0 & 24.4 & 56.2 \\ 0 & 56.2 & 173 \end{bmatrix}\begin{bmatrix} 0 \\ 0 \\ -0.597 \end{bmatrix}$$

$$+ 10^{-2}\begin{bmatrix} -39000 & 0 & 0 \\ 0 & -24.4 & 56.2 \\ 0 & -56.2 & 86.6 \end{bmatrix}\begin{bmatrix} 0.190 \\ -0.500 \\ 0.870 \end{bmatrix}$$

$$= 10^{-2}\begin{bmatrix} 0 \\ -33.6 \\ -103 \end{bmatrix} + 10^{-2}\begin{bmatrix} -7400 \\ 61.1 \\ 103 \end{bmatrix} = \begin{bmatrix} -74.0 \\ 0.275 \\ 0 \end{bmatrix}\begin{matrix} \text{kN} \\ \text{kN} \\ \\ \end{matrix}$$

Similarly

$$\mathbf{p}'_2 = \mathbf{p}_2 = 10^{-2}\begin{bmatrix} 74.0 \\ -0.275 \\ 1.27 \end{bmatrix}\begin{matrix} \text{kN} \\ \text{kN} \\ \text{kN m} \end{matrix}$$

Member OB

$$\mathbf{p}'_2 = \mathbf{K}'_{22}\,\Delta_B = 10\begin{bmatrix} 56.4 & 32.2 & 0.563 \\ 32.2 & 19.2 & -0.974 \\ 0.563 & -0.974 & 3.00 \end{bmatrix}\begin{bmatrix} 0.190 \\ -0.500 \\ 0.870 \end{bmatrix}$$

$$= \begin{bmatrix} -49.0 \\ -43.0 \\ 32.0 \end{bmatrix}\begin{matrix} \text{kN} \\ \text{kN} \\ \text{kN m} \end{matrix}$$

$$\mathbf{p}_2 = \mathbf{T}\mathbf{p}'_2 \text{ (by Eqn 7.6–1)} = \begin{bmatrix} 0.866 & +0.50 & 0 \\ -0.50 & 0.866 & 0 \\ 0 & 0 & 1 \end{bmatrix}\begin{bmatrix} -49.0 \\ -43.0 \\ 32.0 \end{bmatrix}$$

$$= \begin{bmatrix} -64.0 \\ -12.8 \\ 32.0 \end{bmatrix}\begin{matrix} \text{kN} \\ \text{kN} \\ \text{kN m} \end{matrix}$$

To calculate the forces at O ($\mathbf{p}_1$) a knowledge of $\mathbf{K}'_{12}$ for OB is required, but this can be set up easily from $\mathbf{K}'_{22}$, hence

$$\mathbf{p}'_1 = \mathbf{K}'_{12}\Delta_B = 10\begin{bmatrix} -56.4 & -32.2 & -0.563 \\ -32.2 & -19.2 & 0.974 \\ 0.563 & -0.974 & 1.50 \end{bmatrix}\begin{bmatrix} 0.190 \\ -0.500 \\ 0.870 \end{bmatrix} = \begin{bmatrix} 49.0 \\ 43.0 \\ 19.0 \end{bmatrix}\begin{matrix} \text{kN} \\ \text{kN} \\ \text{kN m} \end{matrix}$$

and

$$\mathbf{P}_1 = \mathbf{T}\mathbf{p}_1' = \begin{bmatrix} 64.0 \\ 12.8 \\ 19.0 \end{bmatrix} \begin{matrix} \text{kN} \\ \text{kN} \\ \text{kN m} \end{matrix}$$

7.13 Pin-jointed structures

The analysis so far described is suitable for two-dimensional structures which are predominantly rigid-jointed. A method for dealing with pins has been discussed, but its use would not be economical for structures which have pin-joints only. The analysis of such structures will now be described.

Pin-jointed members can carry one force only—an axial force—and suffer one distortion—a change of length. The member stiffness relations (Eqns 7.5–1) can therefore be reduced to:

$$\begin{bmatrix} p_{x1} \\ p_{x2} \end{bmatrix} = \begin{bmatrix} \dfrac{EA}{L} & -\dfrac{EA}{L} \\ -\dfrac{EA}{L} & \dfrac{EA}{L} \end{bmatrix} \begin{bmatrix} d_{x1} \\ d_{x2} \end{bmatrix}$$

or

$$\begin{bmatrix} p_{x1} \\ p_{x2} \end{bmatrix} = \begin{bmatrix} K & -K \\ -K & K \end{bmatrix} \begin{bmatrix} d_{x1} \\ d_{x2} \end{bmatrix} \tag{7.13–1}$$

where $K = EA/L$ (i.e. $\mathbf{K}_{11} = \mathbf{K}_{22} = K$ and $\mathbf{K}_{12} = \mathbf{K}_{21} = -K$).

A positive value of p_{x2} indicates tension in the member, and therefore it is p_{x2} which indicates the member force in the normal (positive for tension) sign convention.

Similarly, the coordinate transformation matrix (Eqn 7.6–2) is now reduced to:

$$\mathbf{T} = [L_x/L \quad L_y/L] = [\cos\beta \quad \sin\beta] \tag{7.13–2}$$

β is the angle between the structure and member axes (Fig. 7.6–1).

$\mathbf{T}$ is no longer square, and cannot therefore be described as orthogonal, but Eqn 7.6–1 states that

$$\mathbf{p} = \mathbf{T}\mathbf{p}'$$

and it is still true that

$$\mathbf{p}' = \mathbf{T}^T\mathbf{p}$$

Also,

$$\mathbf{d} = \mathbf{T}\mathbf{d}' \quad \text{and} \quad \mathbf{d}' = \mathbf{T}^T\mathbf{d}$$

As the member stiffness matrices are now all scalar, the product $\mathbf{T}^T\mathbf{K}\mathbf{T}$ is simply:

$$\frac{EA}{L}\begin{bmatrix} \left(\dfrac{L_x}{L}\right)^2 & \dfrac{L_xL_y}{L^2} \\ \dfrac{L_xL_y}{L^2} & \left(\dfrac{L_y}{L}\right)^2 \end{bmatrix} = \frac{EA}{L}\begin{bmatrix} \cos^2\beta & \cos\beta\sin\beta \\ \cos\beta\sin\beta & \sin^2\beta \end{bmatrix} \tag{7.13–3}$$

7.14 Example on a pin-jointed truss

The plane truss in Fig. 7.14–1 is pinned at all joints. All members have the same stiffness EA. The joint A is pinned to a support, and B can move in a vertical direction only. Structure axes have been chosen as shown.

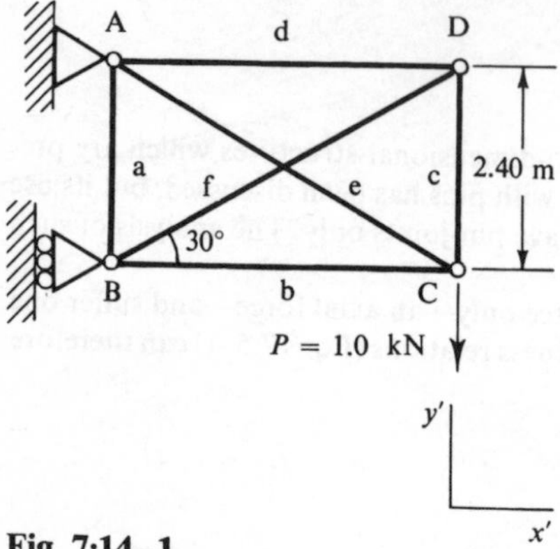

Fig. 7·14–1

The structure stiffness matrix is first set up in general terms, on the assumption that B is free:

$$\begin{bmatrix} \mathbf{P}_B \\ \mathbf{P}_C \\ \mathbf{P}_D \end{bmatrix} = \begin{bmatrix} \mathbf{K}'_a + \mathbf{K}'_b + \mathbf{K}'_f & -\mathbf{K}'_b & -\mathbf{K}'_f \\ -\mathbf{K}'_b & \mathbf{K}'_b + \mathbf{K}'_c + \mathbf{K}'_e & -\mathbf{K}'_c \\ -\mathbf{K}'_f & -\mathbf{K}'_c & \mathbf{K}'_c + \mathbf{K}'_d + \mathbf{K}'_f \end{bmatrix} \begin{bmatrix} \boldsymbol{\Delta}_B \\ \boldsymbol{\Delta}_C \\ \boldsymbol{\Delta}_D \end{bmatrix}$$

(columns B, C, D)

The members carry lower-case letters to simplify the notation. The ends 1 and 2 of each member are indicated in the figure.

The matrix is symmetrical as before, and in general will be banded if numbered carefully. There is no banding here on this small structure as all joints are interconnected.

Member a $\quad \mathbf{K}' = \dfrac{EA}{2.40}\begin{bmatrix} 0 & 0 \\ 0 & 1 \end{bmatrix}$

Member b $\quad \mathbf{K}' = \dfrac{EA}{2.40\sqrt{3}}\begin{bmatrix} 1 & 0 \\ 0 & 0 \end{bmatrix}$

Member c $\quad \mathbf{K}' = \dfrac{EA}{2.40}\begin{bmatrix} 0 & 0 \\ 0 & 1 \end{bmatrix}$

Member d $\quad \mathbf{K}' = \dfrac{EA}{2.40\sqrt{3}}\begin{bmatrix} 1 & 0 \\ 0 & 0 \end{bmatrix}$

Member e $\quad \mathbf{K}' = \dfrac{EA}{4.80}\begin{bmatrix} \dfrac{3}{4} & \dfrac{-\sqrt{3}}{4} \\[2ex] \dfrac{-\sqrt{3}}{4} & \dfrac{1}{4} \end{bmatrix}$

Member f $\quad \mathbf{K}' = \dfrac{EA}{4.80}\begin{bmatrix} \dfrac{3}{4} & \dfrac{\sqrt{3}}{4} \\[2ex] \dfrac{\sqrt{3}}{4} & \dfrac{1}{4} \end{bmatrix}$

$$
\begin{array}{c} \\ \mathrm{B} \\ \\ \mathrm{C} \\ \\ \mathrm{D} \\ \\ \end{array}
\begin{bmatrix} P_{x\mathrm{B}} \\ 0 \\ 0 \\ -1.0 \\ 0 \\ 0 \end{bmatrix}
= \frac{EA}{2.40}
\overset{\qquad\mathrm{B}\qquad\qquad\qquad\qquad\mathrm{C}\qquad\qquad\qquad\qquad\mathrm{D}}{
\begin{bmatrix}
\frac{1}{\sqrt{3}}+\frac{3}{8} & \frac{\sqrt{3}}{8} & -\frac{1}{\sqrt{3}} & 0 & -\frac{3}{8} & -\frac{\sqrt{3}}{8} \\
\frac{\sqrt{3}}{8} & 1+\frac{1}{8} & 0 & 0 & \frac{-\sqrt{3}}{8} & -\frac{1}{8} \\
-\frac{1}{\sqrt{3}} & 0 & \frac{1}{\sqrt{3}}+\frac{3}{8} & \frac{-\sqrt{3}}{8} & 0 & 0 \\
0 & 0 & \frac{-\sqrt{3}}{8} & 1+\frac{1}{8} & 0 & -1 \\
-\frac{3}{8} & \frac{-\sqrt{3}}{8} & 0 & 0 & \frac{1}{\sqrt{3}}+\frac{3}{8} & \frac{\sqrt{3}}{8} \\
\frac{-\sqrt{3}}{8} & -\frac{1}{8} & 0 & -1 & \frac{\sqrt{3}}{8} & 1+\frac{1}{8}
\end{bmatrix}}
\begin{bmatrix} \Delta_{x\mathrm{B}} \\ \Delta_{y\mathrm{B}} \\ \Delta_{x\mathrm{C}} \\ \Delta_{y\mathrm{C}} \\ \Delta_{x\mathrm{D}} \\ \Delta_{y\mathrm{D}} \end{bmatrix}
$$

The first row and column (corresponding to the zero horizontal deflection at B) are now eliminated, and the equations become:

$$
\begin{bmatrix} 0 \\ 0 \\ 0 \\ -1.0 \\ 0 \\ 0 \end{bmatrix}
= \frac{EA}{2.40}
\begin{bmatrix}
1 & 0 & 0 & 0 & 0 & 0 \\
0 & 1.125 & 0 & 0 & -0.217 & -0.125 \\
0 & 0 & 0.954 & -0.217 & 0 & 0 \\
0 & 0 & -0.217 & 1.125 & 0 & -1 \\
0 & -0.217 & 0 & 0 & 0.954 & 0.217 \\
0 & -0.125 & 0 & -1 & 0.217 & 1.125
\end{bmatrix}
\begin{bmatrix} \Delta_{x\mathrm{B}} \\ \Delta_{y\mathrm{B}} \\ \Delta_{x\mathrm{C}} \\ \Delta_{y\mathrm{C}} \\ \Delta_{x\mathrm{D}} \\ \Delta_{y\mathrm{D}} \end{bmatrix}
$$

The solution to this set of equations is:

$$
\begin{bmatrix} \Delta_{x\mathrm{B}} \\ \Delta_{y\mathrm{B}} \\ \Delta_{x\mathrm{C}} \\ \Delta_{y\mathrm{C}} \\ \Delta_{x\mathrm{D}} \\ \Delta_{y\mathrm{D}} \end{bmatrix}
= \frac{1}{EA}
\begin{bmatrix} 0 \\ -1.12 \\ -3.86 \\ -17.0 \\ 3.35 \\ -15.9 \end{bmatrix} \mathrm{m}
$$

The member forces are obtained by returning to the member stiffness relations (in structure axes):

Member a

$$\mathbf{p}_1' = -\mathbf{K}_a'\boldsymbol{\Delta}_B = -\begin{bmatrix} 0 & 0 \\ 0 & 1 \end{bmatrix}\frac{EA}{2.40}\cdot\frac{1}{EA}\begin{bmatrix} 0 \\ -1.12 \end{bmatrix} = \begin{bmatrix} 0 \\ 0.465 \end{bmatrix}\text{kN}$$

$$\mathbf{p}_2' = \mathbf{K}_a'\boldsymbol{\Delta}_B = \begin{bmatrix} 0 & 0 \\ 0 & 1 \end{bmatrix}\frac{EA}{2.40}\cdot\frac{1}{EA}\begin{bmatrix} 0 \\ -1.12 \end{bmatrix} = \begin{bmatrix} 0 \\ -0.465 \end{bmatrix}\text{kN}$$

Then

$$\mathbf{p}_2 = \mathbf{T}\mathbf{p}_2' = [0 \quad -1]\begin{bmatrix} 0 \\ -0.465 \end{bmatrix} = 0.465\text{ kN} \quad (= -p_1)$$

Member b $p_2 = -0.928$ kN

Member c $p_2 = 0.465$ kN

Member d $p_2 = 0.806$ kN

Member e $p_2 = 1.071$ kN

Member f $p_2 = -0.931$ kN

This analysis will not give the reaction forces at A and B directly. They can be determined by summing the member forces in structure co-ordinates (either $\mathbf{p}_1'$ or $\mathbf{p}_2'$ as appropriate) at A and B respectively, and the results are:

	Force in x′-direction	*Force in y′-direction*
At A	−1.73 kN	1.00 kN
At B	1.73 kN	0

7.15 Loads between joints

So far only structures loaded at joints have been considered, but in rigid-jointed structures this is generally not the case. In order to deal with this problem, the whole solution process must be reviewed. First, the concept of **kinematic indeterminacy** must be introduced. Static indeterminacy has been discussed in earlier chapters, and kinematic indeterminacy is a similar idea associated with joint displacements, rather than forces. The 'degree' of kinematic indeterminacy of a structure is equal to the number of degrees of freedom. For instance, a propped cantilever is of static indeterminacy 1, but of kinematic indeterminacy 2—lengthwise and rotational displacements are possible. If no joint displacements are possible the structure is **kinematically determinate**.

The first step in a full stiffness analysis is to render the structure kinematically determinate by imagining all the joints of the structure clamped against displacement. A 'particular' solution of this fully restrained structure is then carried out. This is straightforward, as no displacements occur at joints and all members behave as separate fixed-ended beams or columns.

The second step is to carry out a 'complementary' analysis of the structure, using at the joints a set of loads equal and opposite to the fixed-end moments and forces exerted by the clamps in the particular solution, Step 1.

The third step is to superimpose the results of the first two analyses. The result is the set of member end forces required. The process will be illustrated in an example shown in Fig. 7.15–1.

The structure is the same one as used in the earlier example on the stiffness method (Section 7.12).

(1) *The particular solution.* A and B are clamped before application of W and there are therefore no forces in AB. OB is a fixed-ended beam as shown in Fig. 7.15–2.

(2) *The complementary solution.* The loads to be used are shown in Fig. 7.15–3. The loads at O have no effect on the structure, and can be ignored at this stage. The loads

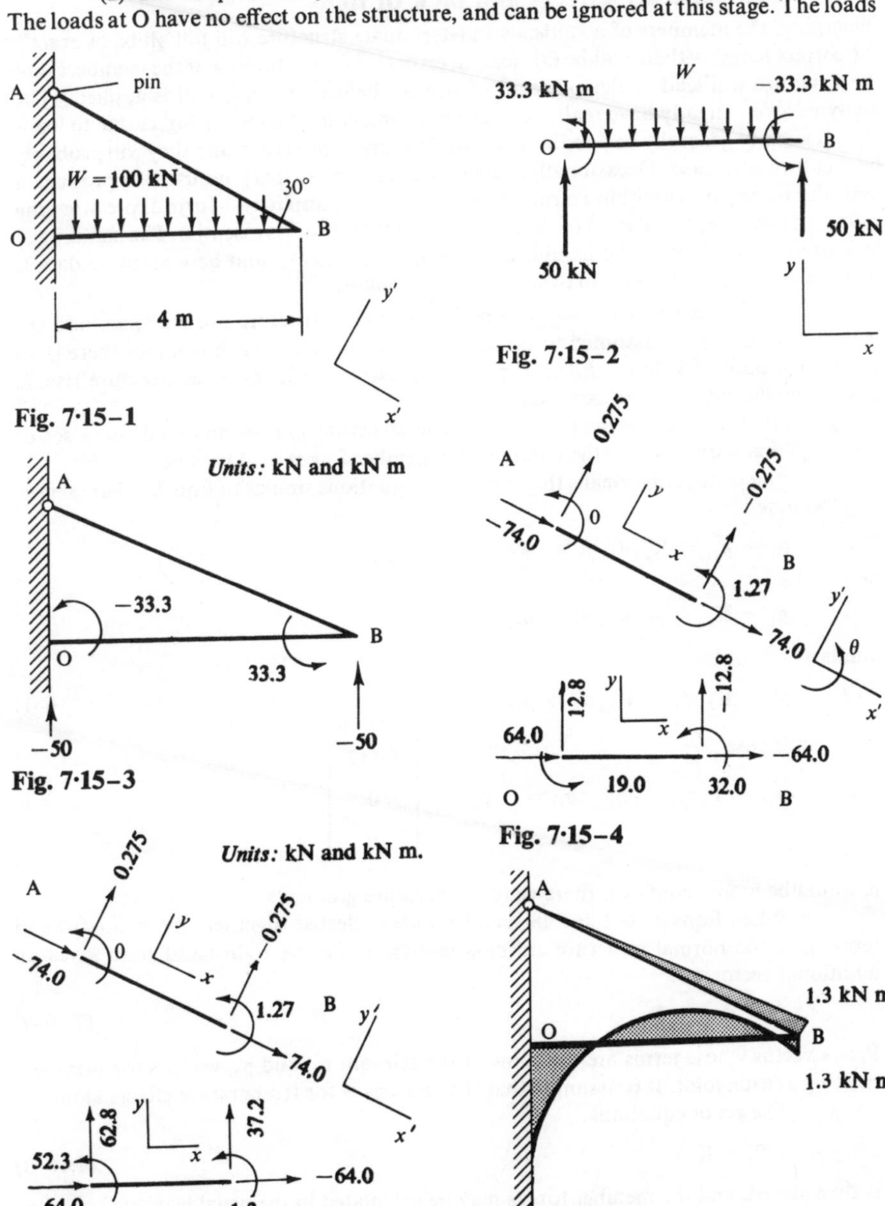

Fig. 7·15–1

Fig. 7·15–2

Fig. 7·15–3

Fig. 7·15–4

Fig. 7·15–5

Fig. 7·15–6

at B are those already used in the earlier example, and the member forces resulting are as shown in Fig. 7.15–4.

(3) *The complete solution.* The results of the first two stages are summed as shown in Fig. 7.15–5.

The bending moment diagram for the complete structure is shown in Fig. 7.15–6.

7.16 Temperature effects and lack of fit

Inevitably, the members of a statically indeterminate structure will not all be of exactly the correct length—there will be fabrication errors—and the forcing of the members into the structure will lead to the existence of stresses before external load is applied. (The analyses given already have really only given the *increase* of member forces due to loading.) It will not generally be possible to easily determine the errors, and they will probably be small in any case. Occasionally, intentional lack of fit may be used to produce a desirable force pattern within a structure before loads are applied—a form of pre-stressing—and, in this case, calculation of the resulting member forces is essential. The handling of lack of fit is identical to the handling of temperature effects, and here accurate data is more readily available. The analysis steps are as follows:

(1) *The particular solution.* The joints of the structure are clamped, and the temperature change is assumed to occur. The result is that in each member there is an axial force only of value $-EA\alpha T$ (i.e. a compressive load) for a temperature rise T. α is the coefficient of linear expansion.

(2) *The complementary solution.* The structure is now analysed for a set of loads equal and opposite to the compressive member forces in (1) above.

Practically, this means that if member equations similar to Eqn 7.6–5 are set up, they become:

$$\mathbf{p}'_1 - \mathbf{p}'_{1t} = \mathbf{K}'_{12}\mathbf{d}'_1 + \mathbf{K}'_{11}\mathbf{d}'_2$$

or

$$\mathbf{p}'_1 = \mathbf{K}'_{11}\mathbf{d}'_1 + \mathbf{K}'_{12}\mathbf{d}'_2 + \mathbf{p}'_{1t}$$

Similarly,

$$\mathbf{p}'_2 = \mathbf{K}'_{21}\mathbf{d}'_1 + \mathbf{K}'_{22}\mathbf{d}'_2 + \mathbf{p}'_{2t} \qquad (7.16\text{–}1)$$

In a simple case

$$\mathbf{p}'_{1t} = \mathbf{T}^T\mathbf{p}_{1t}, \quad \text{and} \quad \mathbf{p}_{1t} = -\mathbf{p}_{2t} = \begin{bmatrix} EA\alpha T \\ 0 \\ 0 \end{bmatrix}$$

$\mathbf{p}_t$ would be more complex if there were temperature gradients.

When Eqns 7.16–1 for all members are collected together, the two left-hand terms give the normal structure stiffness matrix, while the right-hand term gives an additional vector, i.e.

$$0 = \mathbf{K}_s\mathbf{\Delta} + \mathbf{P}_t \qquad (7.16\text{–}2)$$

$\mathbf{P}_t$ is a vector whose terms are the sums of the relevant $\mathbf{p}'_{1t}$ and $\mathbf{p}'_{2t}$ vectors for members meeting at each joint. It is assumed that the analysis is for temperature effects alone.

The set of equations:

$$-\mathbf{P}_t = \mathbf{K}_s\mathbf{\Delta} \qquad (7.16\text{–}3)$$

is then solved, and the member forces may be calculated in the usual way.

Lack of fit will be handled in an identical manner, but the vectors corresponding to $\mathbf{p}_{1t}$ and $\mathbf{p}_{2t}$ may be more complex.

7.17 Continuous beams

Like pin-jointed structures, continuous beams carrying only normal loads form a class for which a modified solution process can be used to great advantage. A normal rigid-joint analysis could be used, but it would be an unnecessarily weighty solution process.

With a system of continuous beams, if joints are at supports, the only deflections occurring are rotations. The member stiffness relation (Eqn 7.5–1) may be modified to:

$$\begin{bmatrix} m_1 \\ m_2 \end{bmatrix} = \begin{bmatrix} 4EI/L & 2EI/L \\ 2EI/L & 4EI/L \end{bmatrix} \begin{bmatrix} \theta_1 \\ \theta_2 \end{bmatrix}$$

$$= \frac{EI}{L} \begin{bmatrix} 4 & 2 \\ 2 & 4 \end{bmatrix} \begin{bmatrix} \theta_1 \\ \theta_2 \end{bmatrix} \qquad (7.17\text{–}1)$$

or

$$K_{11} = K_{22} = \frac{4EI}{L} \quad \text{and} \quad K_{12} = K_{21} = \frac{2EI}{L}$$

No coordinate transformation is required, with a suitable choice of axes. The method will be illustrated by the following example:

Example 7.17–1. The continuous beam of Fig. 7.17–1 carries a uniformly distributed load w between B and C, and is of constant stiffness EI throughout its length.

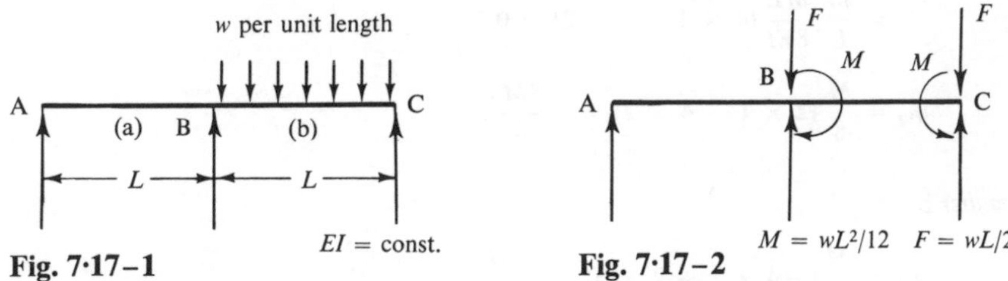

Fig. 7·17–1 **Fig. 7·17–2**

The structure axes x'–y' are chosen to coincide with the member axes. As a result of a particular solution, the fixed-end moments $wL^2/12$ act on member BC. A general solution with the loading of Fig. 7.17–2 is now carried out. The structure stiffness matrix can be set up according to the rules of Section 7.9.

$$\begin{bmatrix} m_A \\ m_B \\ m_C \end{bmatrix} = \begin{bmatrix} (K'_{11})_a & (K'_{12})_a & 0 \\ (K'_{21})_a & (K'_{22})_a + (K'_{11})_b & (K'_{12})_b \\ 0 & (K'_{21})_b & (K'_{22})_b \end{bmatrix} \begin{bmatrix} \theta_A \\ \theta_B \\ \theta_C \end{bmatrix} \qquad (7.17\text{–}2)$$

For both members:

$$K'_{11} = K'_{22} = K_{11} = K_{22} = \frac{4EI}{L}$$

$$K'_{12} = K'_{21} = K_{12} = K_{21} = \frac{2EI}{L}$$

Hence

$$\begin{bmatrix} 0 \\ -M \\ M \end{bmatrix} = \frac{EI}{L} \begin{bmatrix} 4 & 2 & 0 \\ 2 & 8 & 2 \\ 0 & 2 & 4 \end{bmatrix} \begin{bmatrix} \theta_A \\ \theta_B \\ \theta_C \end{bmatrix} \tag{7.17–3}$$

Inversion of the stiffness matrix gives the result:

$$\begin{bmatrix} \theta_A \\ \theta_B \\ \theta_C \end{bmatrix} = \frac{L}{24EI} \begin{bmatrix} 7 & -2 & 1 \\ -2 & 4 & -2 \\ 1 & -2 & 7 \end{bmatrix} \begin{bmatrix} 0 \\ -M \\ M \end{bmatrix}$$

$$= \frac{ML}{8EI} \begin{bmatrix} 1 \\ -2 \\ 3 \end{bmatrix} \tag{7.17–4}$$

Member a

$$m_1 = K_{11}\theta_A + K_{12}\theta_B$$

$$= \frac{EI}{L} \cdot \frac{ML}{8EI} (4 \times 1 - 2 \times 2) = 0$$

$$m_2 = \frac{M}{8} (2 \times 1 - 4 \times 2) = -\frac{3M}{4}$$

Member b

$$m_1 = \frac{M}{8} (-4 \times 2 + 2 \times 3) = -\frac{M}{4}$$

$$m_2 = \frac{M}{8} (-2 \times 2 + 4 \times 3) = M$$

Addition of the moments of the particular solution leads to the final set of bending moments of the complete solution shown in Fig. 7.17–3. Also shown are the end reactions, which may be determined by calculation of the shears on each member.

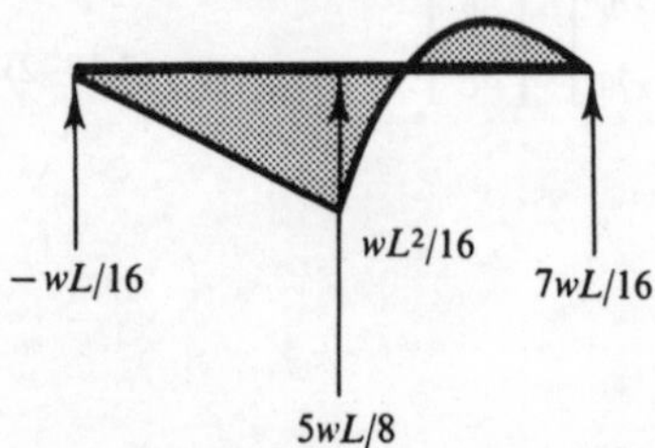

Fig. 7·17–3

Member a

From Eqn 7.5–1:

$$p_{y1} = \frac{6EI}{L^2}\theta_A + \frac{6EI}{L^2}\theta_B$$

$$= \frac{6EI}{L^2}\frac{ML}{8EI}(1 - 2) = -\frac{3M}{4L}$$

$$p_{y2} = -p_{y1} = \frac{3M}{4L}$$

Member b

$$p_{y1} = \frac{3M}{4L}(-2 + 3) = \frac{3M}{4L}$$

hence

$$p_{y2} = -p_{y1} = -\frac{3M}{4L}$$

The final shears of the complete solution are obtained by adding the shears of the particular solution, and the reactions are the sums of the shears at each support.

7.18 Three-dimensional structures

The process of analysis already presented for two-dimensional rigid frames applies equally to space frames. The consideration of a third dimension involves alterations to the member stiffness matrices and the transformation matrix only.

In a rigid-jointed space frame each end of each member has six degrees of freedom as shown in Fig. 7.18–1—three linear displacements and three rotations.

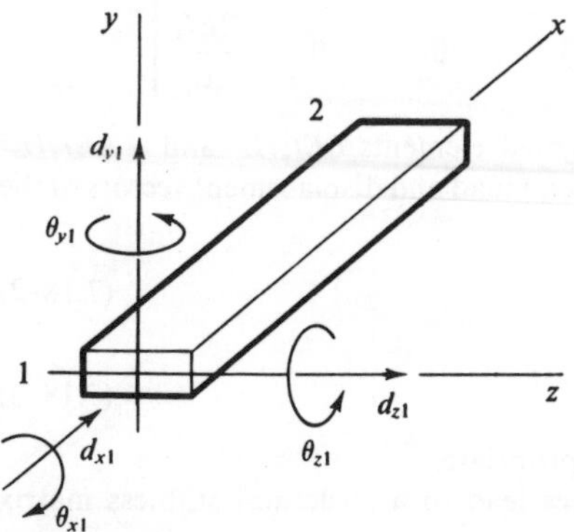

Fig. 7·18–1

The x-axis is along the length of the member (as with two dimensions), and the y- and z-axes coincide with the principal bending axes for the member. The rotation θ_x is a twist, and the corresponding force m_x is therefore a torque. (The occurrence of torsion raises some problems and these will be discussed in Section 11.11.) The member stiffness matrices taken from Eqn 5.2–1 become:

$$\mathbf{K}_{11} = \begin{bmatrix} \frac{EA}{L} & 0 & 0 & 0 & 0 & 0 \\ 0 & \frac{12EI_z}{L^3} & 0 & 0 & 0 & \frac{6EI_z}{L^2} \\ 0 & 0 & \frac{12EI_y}{L^3} & 0 & \frac{-6EI_y}{L^2} & 0 \\ 0 & 0 & 0 & \frac{GJ}{L} & 0 & 0 \\ 0 & 0 & \frac{-6EI_y}{L^2} & 0 & \frac{4EI_y}{L} & 0 \\ 0 & \frac{6EI_z}{L^2} & 0 & 0 & 0 & \frac{4EI_z}{L} \end{bmatrix}$$

and

$$\mathbf{K}_{12} = \mathbf{K}_{21}^T = \begin{bmatrix} -\frac{EA}{L} & 0 & 0 & 0 & 0 & 0 \\ 0 & \frac{-12EI_z}{L^3} & 0 & 0 & 0 & \frac{6EI_z}{L^2} \\ 0 & 0 & \frac{-12EI_y}{L^3} & 0 & \frac{-6EI_y}{L^2} & 0 \\ 0 & 0 & 0 & \frac{-GJ}{L} & 0 & 0 \\ 0 & 0 & \frac{6EI_y}{L^2} & 0 & \frac{2EI_y}{L} & 0 \\ 0 & \frac{-6EI_z}{L^2} & 0 & 0 & 0 & \frac{2EI_z}{L} \end{bmatrix} \tag{7.18–1}$$

and $\mathbf{K}_{22} = \mathbf{K}_{11}$, with the signs of the off-diagonal elements $6EI_z/L^2$ and $-6EI_y/L^2$ reversed. These matrices correspond to member end load and displacement vectors of the form:

$$\mathbf{p} = [p_x \; p_y \; p_z \; m_x \; m_y \; m_z]^T \tag{7.18–2}$$

and

$$\mathbf{d} = [d_x \; d_y \; d_z \; \theta_x \; \theta_y \; \theta_z]^T \tag{7.18–3}$$

the additional suffices 1 and 2 being used as appropriate.

These larger member stiffness matrices lead to a structural stiffness matrix $6j \times 6j$, for a structure of j joints.

Like the member stiffness matrices, the transformation matrix **T** is now 6×6 and this question has already been discussed in Section 5.3. The form is:

$$\mathbf{T} = \begin{bmatrix} \mathbf{R}_0 & \\ & \mathbf{R}_0 \end{bmatrix} \tag{7.18–4}$$

where the rotation matrix $\mathbf{R}_0$ is defined in Eqn 5.3–16. The special case noted in Example 5.3–2 should be remembered because it can occur very frequently.

The member stiffness matrices Eqn 7.18–1 can be adjusted to cope with pins, as can be done in two dimensions (there are, of course, three different types of pin to consider), but if a structure is effectively pin-jointed throughout, the complete analysis can be reduced in scale. The member stiffness equations (as in two dimensions) become:

$$\begin{bmatrix} p_{x1} \\ p_{x2} \end{bmatrix} = \begin{bmatrix} K & -K \\ -K & K \end{bmatrix} \begin{bmatrix} d_{x1} \\ d_{x2} \end{bmatrix} \tag{7.18–5}$$

where $K = EA/L$. The transformation matrix $\mathbf{T}$ is:

$$\begin{bmatrix} \dfrac{L_x}{L} & \dfrac{L_y}{L} & \dfrac{L_z}{L} \end{bmatrix} \tag{7.18–6}$$

(the first row of $\mathbf{R}_0$ in Eqn 5.3–16).

7.19 Grillages

A common class of structure is one in which the members all lie in one plane, but the loads are applied normal to this plane—e.g. floor systems in buildings, or bridge decks (Fig. 7.19–1). A space frame analysis could be used to determine forces and displacements, but would involve a great deal of unnecessary calculation. If it can be assumed that 'in plane' direct forces and moments are negligible (which in practice is usually fully justified) a modified analysis is possible.

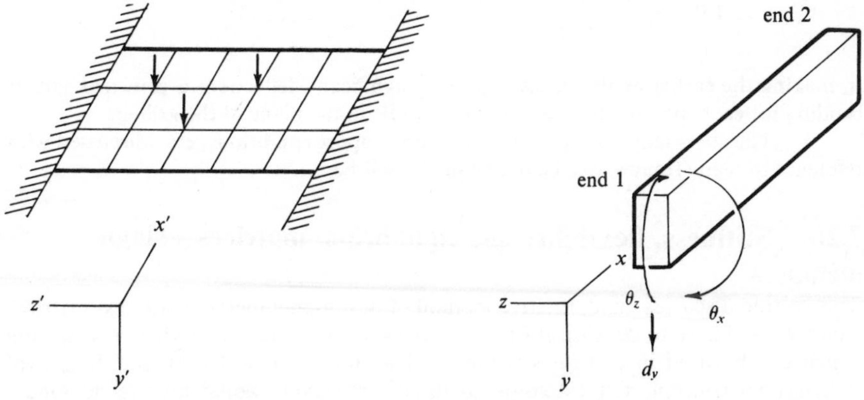

Fig. 7·19–1 **Fig. 7·19–2**

The significant member forces and deflections are:

$$\begin{bmatrix} p_y \\ m_x \\ m_z \end{bmatrix} \text{ and } \begin{bmatrix} d_y \\ \theta_x \\ \theta_z \end{bmatrix} \quad \text{(see Fig. 7.19–2)}$$

The member stiffness matrices are:

$$\mathbf{K}_{11} = \begin{bmatrix} \dfrac{12EI}{L^3} & 0 & \dfrac{6EI}{L^2} \\ 0 & \dfrac{GJ}{L} & 0 \\ \dfrac{6EI}{L^2} & 0 & \dfrac{4EI}{L} \end{bmatrix} \qquad \mathbf{K}_{22} = \begin{bmatrix} \dfrac{12EI}{L^3} & 0 & \dfrac{-6EI}{L^2} \\ 0 & \dfrac{GJ}{L} & 0 \\ \dfrac{-6EI}{L^2} & 0 & \dfrac{4EI}{L} \end{bmatrix}$$

$$\mathbf{K}_{12} = \mathbf{K}_{21}^T = \begin{bmatrix} \dfrac{-12EI}{L^3} & 0 & \dfrac{6EI}{L^2} \\ 0 & \dfrac{-GJ}{L} & 0 \\ \dfrac{-6EI}{L^2} & 0 & \dfrac{2EI}{L} \end{bmatrix} \tag{7.19–1}$$

(The relevant terms may be extracted from Eqn 7.18–1.)

The transformation matrix is 3 × 3 and the terms may be chosen from the full 6 × 6 version for the space frame (Eqns 7.18–4 and 5.3–16).

$$\mathbf{T} = \begin{bmatrix} 1 & 0 & 0 \\ 0 & \dfrac{L_x}{L} & \dfrac{L_z}{L} \\ 0 & \dfrac{-L_z}{L} & \dfrac{L_x}{L} \end{bmatrix} \tag{7.19–2}$$

In making the earlier assumptions regarding significant deflections a principal axis of bending for each member has been assumed to lie in the plane of the grillage.

The steps taken to deal with differing support conditions, etc., discussed with reference to plane frames, can be used equally well here.

7.20 Stiffness, flexibility and equilibrium matrices—single members

This section discusses an alternative method of developing member stiffness matrices which is used here to develop stiffness matrices for members of varying cross-section which can be used in a stiffness analysis. The discussion will be based on a two-dimensional concept, but extension to three dimensions would involve no fundamentally new ideas.

The singularity of the member stiffness matrix **K**, where

$$\mathbf{K} = \begin{bmatrix} \mathbf{K}_{11} & \mathbf{K}_{12} \\ \mathbf{K}_{21} & \mathbf{K}_{22} \end{bmatrix}$$

has been discussed in Section 5.2. The concept of member distortion was introduced, and the distortion vector **e** was defined as the displacement at end 2 of a member relative to end 1. The member flexibility matrix **F** was defined in Eqn 5.2–5:

$$\mathbf{e} = \mathbf{F}\mathbf{p}_2$$

In two dimensions **F** has the form (taken from Eqn 5.2–6):

$$\mathbf{F} = \begin{bmatrix} \dfrac{L}{EA} & 0 & 0 \\ 0 & \dfrac{L^3}{3EI} & \dfrac{L^2}{2EI} \\ 0 & \dfrac{L^2}{2EI} & \dfrac{L}{EI} \end{bmatrix} \tag{7.20–1}$$

e and **F** may be expressed in terms of structure axes

$$\mathbf{e}' = \mathbf{T}^T\mathbf{e} \quad \text{and} \quad \mathbf{F}' = \mathbf{T}^T\mathbf{F}\mathbf{T} \tag{7.20–2}$$

The equilibrium relations existing between the member end forces $\mathbf{p}_1$ and $\mathbf{p}_2$ have been discussed in Section 5.1. The equations are:

$$p_{x1} + p_{x2} = 0$$

$$p_{y1} + p_{y2} = 0$$

and by taking moments about end 1

$$m_1 + m_2 + p_{y2}L = 0 \tag{7.20–3}$$

(L is the member length), or, in matrix terms,

$$\begin{bmatrix} p_{x1} \\ p_{y1} \\ m_1 \end{bmatrix} + \begin{bmatrix} 1 & 0 & 0 \\ 0 & 1 & 0 \\ 0 & L & 1 \end{bmatrix} \begin{bmatrix} p_{x2} \\ p_{y2} \\ m_2 \end{bmatrix} = 0 \tag{7.20–4}$$

or

$$\mathbf{p}_1 + \mathbf{H}\mathbf{p}_2 = 0 \tag{7.20–5}$$

The 3×3 matrix **H** is termed the **equilibrium matrix** for the member. Its association with $\mathbf{p}_2$ rather than $\mathbf{p}_1$ is to a large extent arbitrary, but the vector $\mathbf{p}_2$ is more conveniently a vector of 'member force' than $\mathbf{p}_1$ as a positive value for p_{x2} corresponds to tension in the member. Tensile forces are conventionally regarded as of positive sign in structural analysis.

H is not singular, so that either $\mathbf{p}_1$ or $\mathbf{p}_2$ can be calculated if the other is known. It should be noted that **H** depends only on the length of the member, and is independent of its other properties.

Eqn 7.20–5 is stated in terms of a member set of coordinates, but can be converted to a structure system. Pre-multiplication of Eqn 7.20–5 by $\mathbf{T}^T$ leads to:

$$\mathbf{T}^T\mathbf{p}_1 + \mathbf{T}^T\mathbf{H}\mathbf{p}_2 = 0$$

or

$$\mathbf{p}'_1 + (\mathbf{T}^T\mathbf{H}\mathbf{T})\mathbf{p}'_2 = 0$$

or

$$\mathbf{p}'_1 + \mathbf{H}'\mathbf{p}'_2 = 0 \tag{7.20–6}$$

When the matrix **H**′ is multiplied out it has the form:

$$\mathbf{H}' = \begin{bmatrix} 1 & 0 & 0 \\ 0 & 1 & 0 \\ -L_y & L_x & 1 \end{bmatrix} \tag{7.20–7}$$

where L_x and L_y are the projections of L on the x'- and y'-axes respectively.

The matrix **H** serves a useful purpose in relating the end displacements of a member during a rigid body movement. These rigid body displacements are denoted by $\mathbf{d}_1^*$ and $\mathbf{d}_2^*$. If while constant end forces $\mathbf{p}_1$ and $\mathbf{p}_2$ act on a member, its ends undergo virtual, rigid-body movements $\bar{\mathbf{d}}_1^*$ and $\bar{\mathbf{d}}_2^*$, the total work done must be zero. Hence

$$\mathbf{p}_1^T\bar{\mathbf{d}}_1^* + \mathbf{p}_2^T\bar{\mathbf{d}}_2^* = 0$$

or

$$-(\mathbf{H}\mathbf{p}_2)^T\bar{\mathbf{d}}_1^* + \mathbf{p}_2^T\bar{\mathbf{d}}_2^* = 0$$

or

$$-\mathbf{p}_2^T\mathbf{H}^T\bar{\mathbf{d}}_1^* + \mathbf{p}_2^T\bar{\mathbf{d}}_2^* = 0$$

If this is to be true for any arbitrary value of $\mathbf{p}_2$, and if the actual displacements are small, then

$$\mathbf{H}^T\mathbf{d}_1^* = \mathbf{d}_2^* \tag{7.20–8}$$

The distortion of a member is related to the end displacements. If a member has both rigid body displacements ($\mathbf{d}_1^*$ and $\mathbf{d}_2^*$) and carries forces which produce a distortion **e**, then the end displacements can be written:

$$\mathbf{d}_1 = \mathbf{d}_1^*$$

and

$$\mathbf{d}_2 = \mathbf{d}_2^* + \mathbf{e}$$

or

$$\mathbf{d}_2 = \mathbf{H}^T\mathbf{d}_1^* + \mathbf{e} \tag{7.20–9}$$

Elimination of $\mathbf{d}_1^*$ leads to

$$\mathbf{e} = \mathbf{d}_2 - \mathbf{H}^T\mathbf{d}_1 \tag{7.20–10}$$

(see Fig. 5.2–2).

If the earlier relations are placed in Eqn 7.20–10, member stiffness equations follow.

$$\mathbf{d}_2 = \mathbf{H}^T\mathbf{d}_1 + \mathbf{e} = \mathbf{H}^T\mathbf{d}_1 + \mathbf{F}\mathbf{p}_2$$

then

$$\mathbf{p}_2 = -(\mathbf{K}\mathbf{H}^T)\mathbf{d}_1 + \mathbf{K}\mathbf{d}_2 \tag{7.20–11}$$

and it can be seen that

$$\mathbf{K} = \mathbf{K}_{22}$$

and

$$\mathbf{K}_{21} = -\mathbf{K}_{22}\mathbf{H}^T \tag{7.20–12}$$

Also,

$$\mathbf{p}_1 = -\mathbf{H}\mathbf{p}_2$$

then

$$\mathbf{p}_1 = (\mathbf{H}\mathbf{K}\mathbf{H}^T)\mathbf{d}_1 - \mathbf{H}\mathbf{K}\mathbf{d}_2 \tag{7.20–13}$$

and

$$\mathbf{K}_{11} = (\mathbf{H}\mathbf{K}_{22}\mathbf{H}^T) \quad \text{and} \quad \mathbf{K}_{12} = -\mathbf{H}\mathbf{K}_{22} \tag{7.20–14}$$

The matrices introduced in this section may be used to set up the member stiffness matrix **K** for a member of non-uniform section. This might be much simpler in a structural analysis than breaking the member down into a large number of short members assumed constant in cross-section, although there might be difficulties in determining effective end loads.

In Fig. 7.20–1 PZ is a member of varying properties within a structure. The distortion of PZ is the displacement of Z relative to P, and if the equilibrium matrix $\mathbf{H}'$ and stiffness matrix $\mathbf{K}'_{22}$ for PZ are set up, $\mathbf{K}'_{11}$, $\mathbf{K}'_{12}$, and $\mathbf{K}'_{21}$ can be determined. The analysis *must be carried out in terms of structure axes*, but the primes will be omitted for clarity. Forces are considered first, and end P is regarded as fixed.

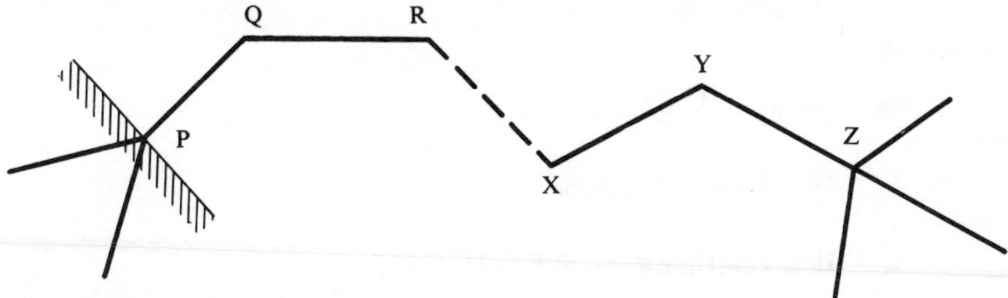

Fig. 7·20–1

The loads applied to point Z by other members and external loads is $\mathbf{p}_{2(YZ)}$. Then

$$\mathbf{p}_{1(YZ)} = -\mathbf{H}_{(YZ)}\mathbf{p}_{2(YZ)}$$

For equilibrium at Y,

$$\mathbf{p}_{2(XY)} + \mathbf{p}_{1(YZ)} = 0$$

hence

$$\mathbf{p}_{2(XY)} = \mathbf{H}_{(YZ)}\mathbf{p}_{2(YZ)}$$

and

$$\mathbf{p}_{1(XY)} = -\mathbf{H}_{(XY)}\mathbf{H}_{(YZ)}\mathbf{p}_{2(YZ)}$$

Or for the composite member X–Z,

$$\mathbf{p}_{1(XZ)} + \mathbf{H}_{(XZ)}\mathbf{p}_{2(XZ)} = 0 \tag{7.20–15}$$

The equilibrium matrix for the composite member is merely the product of the equilibrium matrices of the constituent parts, or of the form of Eqn 7.20–7 where L_x and L_y are the projections of the composite member. That is to say:

$$\mathbf{H}_{(XZ)} = \begin{bmatrix} 1 & 0 & 0 \\ 0 & 1 & 0 \\ -L_{y(XZ)} & L_{x(XZ)} & 1 \end{bmatrix}$$

$$= \begin{bmatrix} 1 & 0 & 0 \\ 0 & 1 & 0 \\ -(L_{y(XY)} + L_{y(YZ)}) & (L_{x(XY)} + L_{x(YZ)}) & 1 \end{bmatrix}$$

$$= \begin{bmatrix} 1 & 0 & 0 \\ 0 & 1 & 0 \\ -L_{y(XY)} & L_{x(XY)} & 1 \end{bmatrix} \begin{bmatrix} 1 & 0 & 0 \\ 0 & 1 & 0 \\ -L_{y(YZ)} & L_{x(YZ)} & 1 \end{bmatrix} \tag{7.20–16}$$

hence (Fig. 7.20–1)

$$\mathbf{p}_{1(PQ)} + \mathbf{H}_{(PQ)} \cdots \mathbf{H}_{(XY)} \cdot \mathbf{H}_{(YZ)}\mathbf{p}_{2(YZ)} = 0 \tag{7.20–17}$$

Distortions are considered secondly.

Member PQ

$$\mathbf{e}_{PQ} = \mathbf{d}_{2(PQ)} - \mathbf{H}^T_{(PQ)}\mathbf{d}_{1(PQ)} \qquad (\mathbf{d}_{1(PQ)} = 0)$$

then

$$\begin{aligned}\Delta_Q &= \mathbf{F}_{(PQ)}\mathbf{p}_{2(PQ)} \\ &= (\mathbf{F}_{(PQ)}\mathbf{H}_{(QZ)})\mathbf{p}_{2(YZ)}\end{aligned}$$

Member QR

$$\mathbf{e}_{QR} = \mathbf{d}_{2(QR)} - \mathbf{H}^T_{(QR)}\mathbf{d}_{1(QR)}$$

then

$$\Delta_R = \mathbf{H}^T_{(QR)}\mathbf{d}_{1(QR)} + \mathbf{F}_{(QR)}\mathbf{p}_{2(QR)}$$

since

$$\begin{aligned}\Delta_R &= \mathbf{H}^T_{(QR)}\mathbf{F}_{(PQ)}\mathbf{H}_{(QZ)}\mathbf{p}_{2(YZ)} + \mathbf{F}_{(QR)}\mathbf{H}_{(RZ)}\mathbf{p}_{2(YZ)} \\ &= (\mathbf{H}^T_{(QR)}\mathbf{F}_{(PQ)}\mathbf{H}_{(QZ)} + \mathbf{F}_{(QR)}\mathbf{H}_{(RZ)})\mathbf{p}_{2(YZ)}\end{aligned}$$

Repetition of this process along the length of the member leads to:

$$\begin{aligned}\mathbf{d}_{2(YZ)} = (&\mathbf{H}^T_{(QZ)}\mathbf{F}_{(PQ)}\mathbf{H}_{(QZ)} + \mathbf{H}^T_{(RZ)}\mathbf{F}_{(QR)}\mathbf{H}_{(RZ)} + \cdots \\ &+ \mathbf{H}^T_{(YZ)}\mathbf{F}_{(XY)}\mathbf{H}_{(YZ)} + \mathbf{F}_{(YZ)})\mathbf{p}_{2(YZ)}\end{aligned}$$

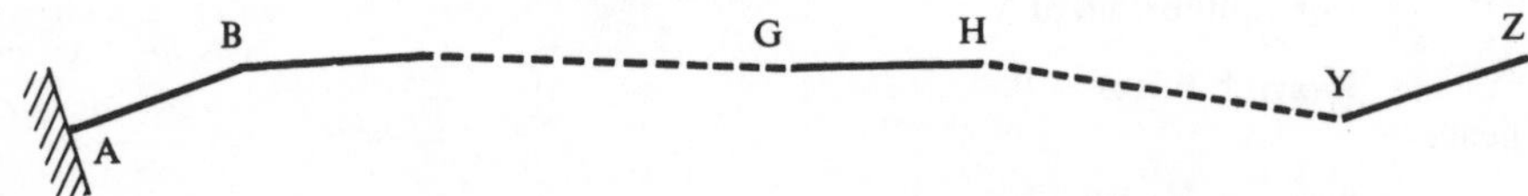

Fig. 7·20–2

Figure 7.20–2 may be used to generalize this result.

$$\Delta_Z = (\sum \mathbf{H}^T_{(HZ)}\mathbf{F}_{(GH)}\mathbf{H}_{(HZ)})\mathbf{p}_{2(YZ)}$$

or the flexibility matrix $\mathbf{F}_{(AZ)}$ of A–Z is

$$\mathbf{F}_{(AZ)} = \sum \mathbf{H}^T_{(HZ)}\mathbf{F}_{(GH)}\mathbf{H}_{(HZ)}$$

The summation is carried out for all sections like GH between A and Z. This matrix **F** is of course $\mathbf{K}_{22}^{-1}$ for the member A–Z. The $\mathbf{K}_{12}$, $\mathbf{K}_{21}$, and $\mathbf{K}_{11}$ matrices can be determined using Eqns 7.20–12 and 7.20–14.

(Care must be exercised in using this process. Obviously the fixed moments for such a member are not those for a prismatic member. If the composite member carries a series of point loads, additional 'joints' can be inserted at the loads. A distributed load is likely to be difficult to handle.)

7.21 Concluding remarks

This chapter has, in Section 7.3, a list of assumptions made in developing the matrix stiffness method of structural analysis. At first sight, this list is very restrictive, but later sections introduce methods of generalizing the process, and most of the restrictions are removed—internal pins are catered for, and the analysis is extended to three dimensions, for example. Some restrictions still remain, and their treatment is felt to be outside the scope of this book, but the one major question of the influence of large axial forces on bending stiffness will be discussed, in Chapter 9.

The handling of the stiffness method on the computer will be discussed in Chapter 13.

General Reference

Livesley, R. K. *Matrix Methods of Structural Analysis.* Pergamon Press, Oxford, 1975.

Problems

7.1. Set up the structure stiffness matrix for the plane frame in the figure in terms of the individual member stiffnesses.

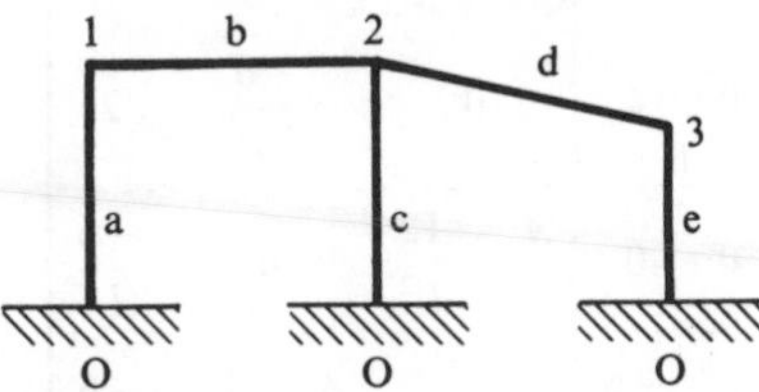

Problem 7·1

$$Ans.\quad \mathbf{K}_s = \left[\begin{array}{c:c:c} (\mathbf{K}'_{22})_a + (\mathbf{K}'_{11})_b & (\mathbf{K}'_{12})_b & 0 \\ \hdashline (\mathbf{K}'_{21})_b & (\mathbf{K}'_{22})_b + (\mathbf{K}'_{22})_c + (\mathbf{K}'_{11})_d & (\mathbf{K}'_{12})_d \\ \hdashline 0 & (\mathbf{K}'_{21})_d & (\mathbf{K}'_{22})_d + (\mathbf{K}'_{22})_e \end{array}\right]$$

7.2. The plane frame in the figure is encastré at A and C and rigidly jointed at B. The section is constant throughout, with the properties shown. Show that the displacements at B are:

$$\Delta_x = -448 \times 10^{-3} \text{ mm}$$
$$\Delta_y = -10.2 \times 10^{-3} \text{ mm}$$
$$\theta = -0.822 \times 10^{-3} \text{ rad}$$

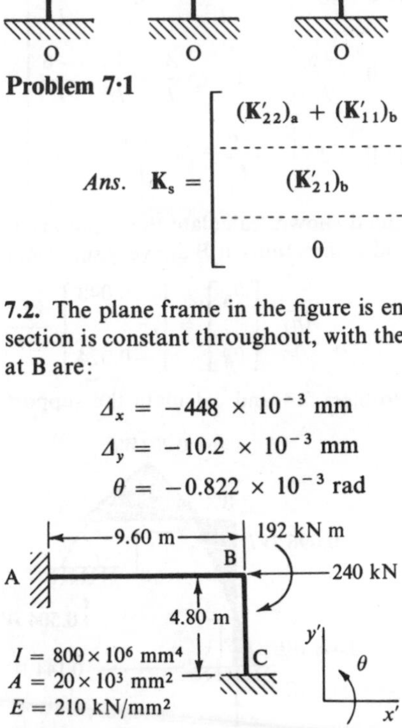

9.60 m
192 kN m
A
B
240 kN
4.80 m
C
y'
θ
x'

$I = 800 \times 10^6$ mm^4
$A = 20 \times 10^3$ mm^2
$E = 210$ kN/mm^2

Problem 7·2

7.3. The plane frame in the figure is encastré at C and pinned at A. Choose suitable structure axes and set up the structure stiffness matrix assuming the section properties E, I, and A constant.

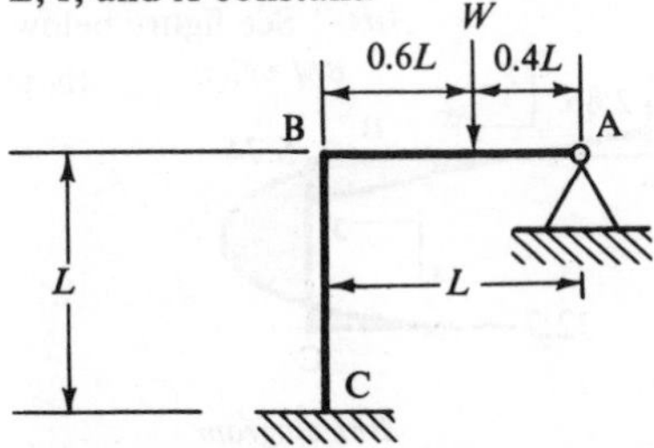

Problem 7·3

Ans. The computation required is least if positive x' is chosen acting from right to left, and positive y' vertically downwards.

$$\mathbf{K}_s = \frac{EI}{L}\begin{array}{cc} \begin{array}{cc} \qquad\quad A \qquad\qquad & \qquad\qquad B \qquad\quad \end{array} & \\ \left[\begin{array}{ccc|ccc} 1 & 0 & 0 & 0 & 0 & 0 \\ 0 & 1 & 0 & 0 & 0 & 0 \\ 0 & 0 & 4 & 0 & \frac{-6}{L} & 2 \\ \hline 0 & 0 & 0 & \frac{A}{I}+\frac{12}{L^2} & 0 & \frac{-6}{L} \\ 0 & 0 & \frac{-6}{L} & 0 & \frac{A}{I}+\frac{12}{L^2} & \frac{-6}{L} \\ 0 & 0 & 2 & \frac{-6}{L} & \frac{-6}{L} & 8 \end{array}\right] & \begin{array}{c} A \\ \\ \\ B \end{array} \end{array}$$

7.4. If the structure of Problem 7.3 carries the load shown, calculate the rotations at A and B. Assume that the displacements in x'- and y'-directions at B are very small and can be neglected.

$$Ans. \quad \begin{bmatrix} \theta_A \\ \theta_B \end{bmatrix} = \begin{bmatrix} 0.048 \\ -0.024 \end{bmatrix} \frac{WL^2}{EI}$$

7.5. Draw the bending moment diagram for Problem 7.4 and calculate the support reactions.

Ans. See figure below

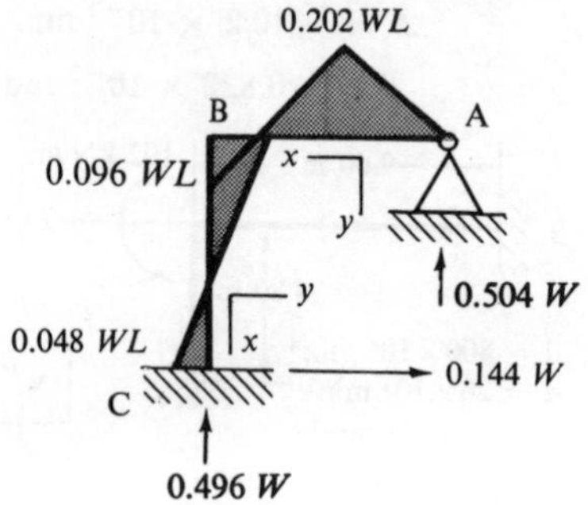

Answer to Problem 7·5

7.6. Draw the bending moment diagram for the structure of Problem 7.2, using the uniformly distributed load of the figure. (Use the results of Problem 7.2.)

Ans. See figure below

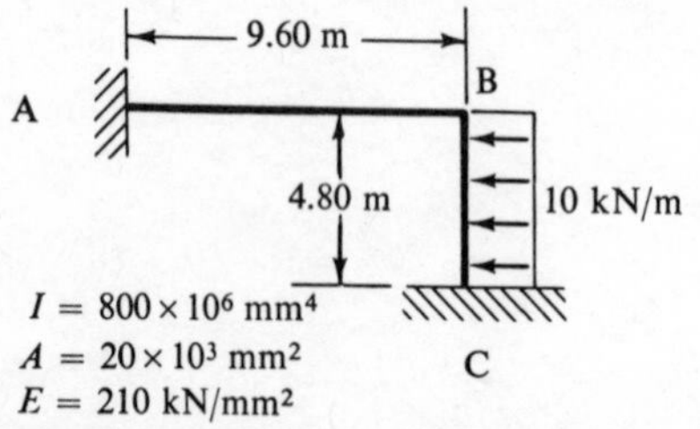

$I = 800 \times 10^6$ mm^4
$A = 20 \times 10^3$ mm^2
$E = 210$ kN/mm^2

Problem 7·6

BM units: kN m

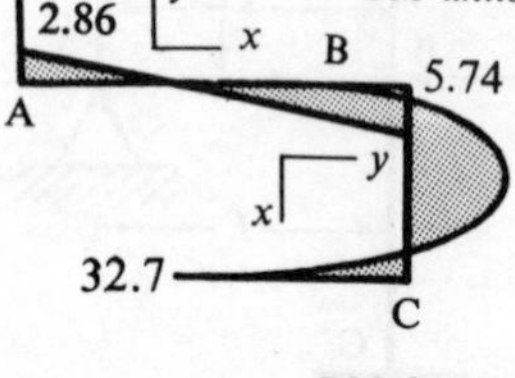

BM diagram

Answer to Problem 7·6

7.7. The symmetrical pitched portal frame in the figure is of constant section throughout and is symmetrically loaded. Assemble the structure stiffness matrix in terms of the E, I, and A of the members, where these terms have their usual meanings. Assume the $A/I \gg 200/s^2$ and that $h = s\sqrt{(5)}/8$.

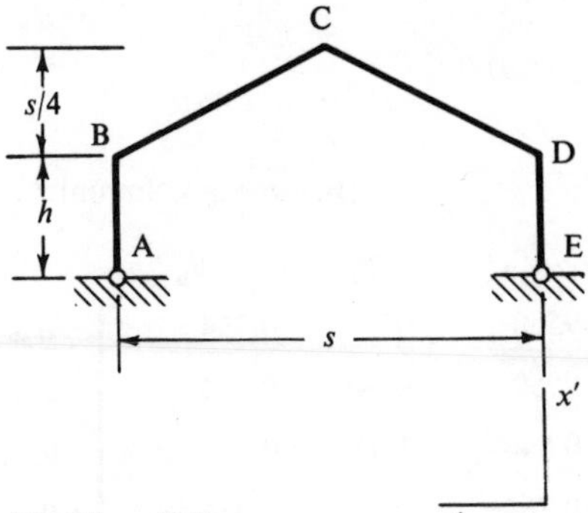

Problem 7·7 y'

$$
\textit{Ans. } \mathbf{K}_s = \frac{EI}{s}
\begin{array}{c}
\begin{array}{ccc|ccc|ccc}
 & \text{A} & & & \text{B} & & & \text{C} & \\
\end{array}\\
\left[\begin{array}{ccc|ccc|ccc}
1 & 0 & 0 & 0 & 0 & 0 & 0 & 0 & 0 \\
 & 1 & 0 & 0 & 0 & 0 & 0 & 0 & 0 \\
 & & \dfrac{32}{\sqrt{5}} & 0 & -\dfrac{384}{5s} & \dfrac{16}{\sqrt{5}} & 0 & 0 & 0 \\
\hline
 & & & \dfrac{8.8}{\sqrt{5}}\left(\dfrac{A}{I}\right) & -\dfrac{16}{\sqrt{5}}\left(\dfrac{A}{I}\right) & \dfrac{192}{5s\sqrt{5}} & -\dfrac{4}{5\sqrt{5}}\left(\dfrac{A}{I}\right) & 0 & 0 \\
 & & & & \dfrac{32}{\sqrt{5}}\left(\dfrac{A}{I}\right) & \dfrac{96}{5s}\left(4 + \dfrac{1}{\sqrt{5}}\right) & \dfrac{16}{\sqrt{5}}\left(\dfrac{A}{I}\right) & 0 & 0 \\
 & \text{Symmetric} & & & & \dfrac{48}{\sqrt{5}} & -\dfrac{192}{5s\sqrt{5}} & 0 & 0 \\
\hline
 & & & & & & \dfrac{4}{5\sqrt{5}}\left(\dfrac{A}{I}\right) & 0 & 0 \\
 & & & & & & & 1 & 0 \\
 & & & & & & & & 1
\end{array}\right]
\begin{array}{c}
\\ \text{A} \\ \\ \\ \text{B} \\ \\ \\ \text{C} \\ \\
\end{array}
\end{array}
$$

7.8. The plane structure of the figure is rigidly supported at A and C and has a pin at B. The section properties are constant throughout, and are:

$I = 0.72 \times 10^6\ \text{mm}^4$

$A = 27.0 \times 10^3\ \text{mm}^2$

$E = 200\ \text{kN/mm}^2$

Assemble a 4×4 stiffness matrix for the structure.

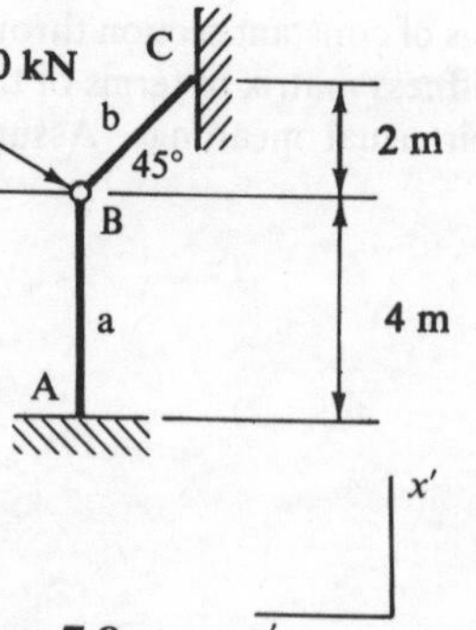

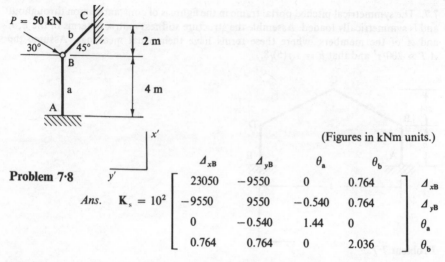

Problem 7·8

(Figures in kNm units.)

$$\textit{Ans.}\quad \mathbf{K}_s = 10^2 \begin{array}{c} \begin{array}{cccc} \Delta_{xB} & \Delta_{yB} & \theta_a & \theta_b \end{array} \\ \left[\begin{array}{cccc} 23050 & -9550 & 0 & 0.764 \\ -9550 & 9550 & -0.540 & 0.764 \\ 0 & -0.540 & 1.44 & 0 \\ 0.764 & 0.764 & 0 & 2.036 \end{array}\right] \end{array} \begin{array}{c} \Delta_{xB} \\ \Delta_{yB} \\ \theta_a \\ \theta_b \end{array}$$

7.9. On the assumption that no moments are applied at joint B of the structure of Problem 7.8, and that only translational displacements of B are of interest, assemble a 2×2 stiffness matrix for the structure, and show that the displacements under the load P shown are

$$\Delta_{xB} = -50.6 \times 10^{-6}\ \text{m}$$
$$\Delta_{yB} = -95.9 \times 10^{-6}\ \text{m}$$

7.10. Assemble the stiffness matrix for the pin-jointed structure shown in the figure, using the joint numbering and axes shown. All members are of the same cross-sectional area—10^3 mm²—and $E = 200$ kN/mm².

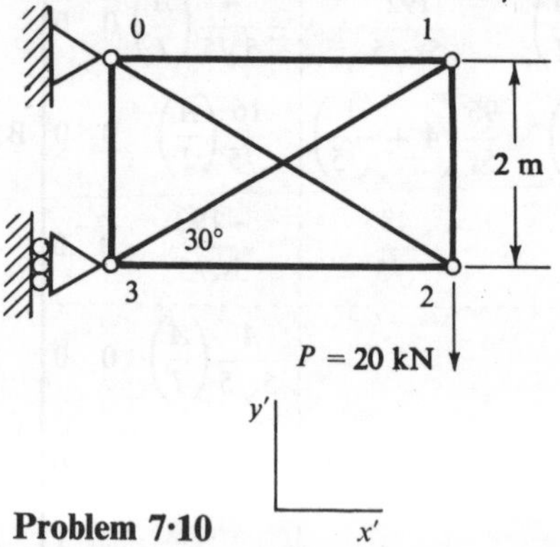

Problem 7·10

(Figures in kNm units.)

$$\textit{Ans.}\quad \mathbf{K}_s = 10^4 \begin{array}{c} \begin{array}{ccc} 1 & 2 & 3 \end{array} \\ \left[\begin{array}{cc:cc:cc} 9.53 & 2.17 & 0 & 0 & 0 & -2.17 \\ 2.17 & 11.25 & 0 & -10.0 & 0 & -1.25 \\ \hdashline 0 & 0 & 9.52 & -2.17 & 0 & 0 \\ 0 & -10.0 & -2.17 & 11.25 & 0 & 0 \\ \hdashline 0 & 0 & 0 & 0 & 1 & 0 \\ -2.17 & -1.25 & 0 & 0 & 0 & 11.25 \end{array}\right] \end{array} \begin{array}{c} 1 \\ \\ 2 \\ \\ 3 \end{array}$$

7.11. Calculate the member forces of the structure of Problem 7.10 if it carries a single vertical load $P = 20$ kN as shown. The flexibility matrix is:

$$K_5^{-1} = \begin{bmatrix} 1.42 & -1.50 & -0.32 & -1.39 & 0 & 0.11 \\ -1.50 & 7.10 & 1.50 & 6.60 & 0 & 0.50 \\ -0.32 & 1.50 & 1.42 & 1.61 & 0 & 0.11 \\ -1.39 & 6.60 & 1.61 & 7.06 & 0 & 0.46 \\ 0 & 0 & 0 & 0 & 1 & 0 \\ 0.11 & 0.50 & 0.11 & 0.46 & 0 & 0.96 \end{bmatrix} \times 10^{-5}$$

Ans.

Member	*Force* (kN)
0–1	16.1
2–3	−18.5
1–2	9.3
1–3	−18.6
0–2	21.4
0–3	9.3

7.12. Calculate, using the stiffness method, the member forces of the symmetrical pin-jointed structure in the figure.

For members AB, BD, AC, and CD, $EA/L = K$.
For member BC, $EA/L = 2K$.

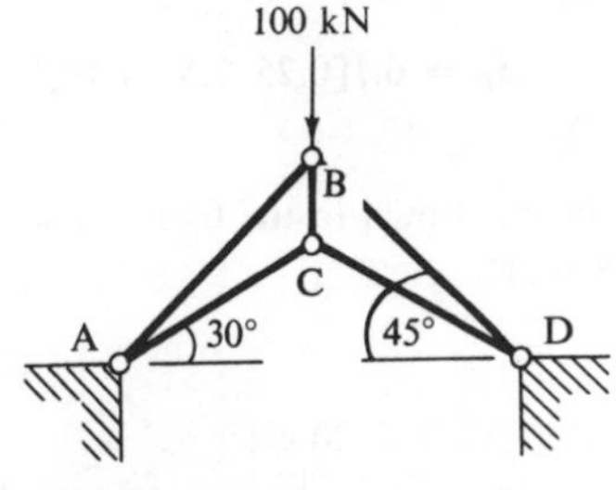

Problem 7·12

Ans.

Member	*Force* (kN)
AB, BD	$-71.4/\sqrt{2}$
AC, CD	−28.6
BC	−28.6

7.13. Calculate the member forces in the pin-jointed structure in the figure. The product AE is constant for all members.

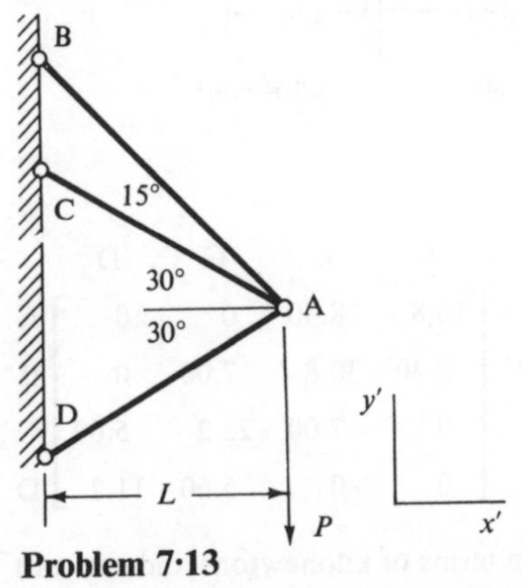

Problem 7·13

Ans.

Member	*Force*
AB	$0.55P$
AC	$0.38P$
AD	$-0.83P$

7.14. Calculate the displacement of the joint D of the space frame in the figure. The supports A, B, and C all lie in the $x'y'$ plane. The cross-sectional areas of the members are: AD and BD, 200 mm^2; CD, 600 mm^2. $E = 210$ kN/mm^2.

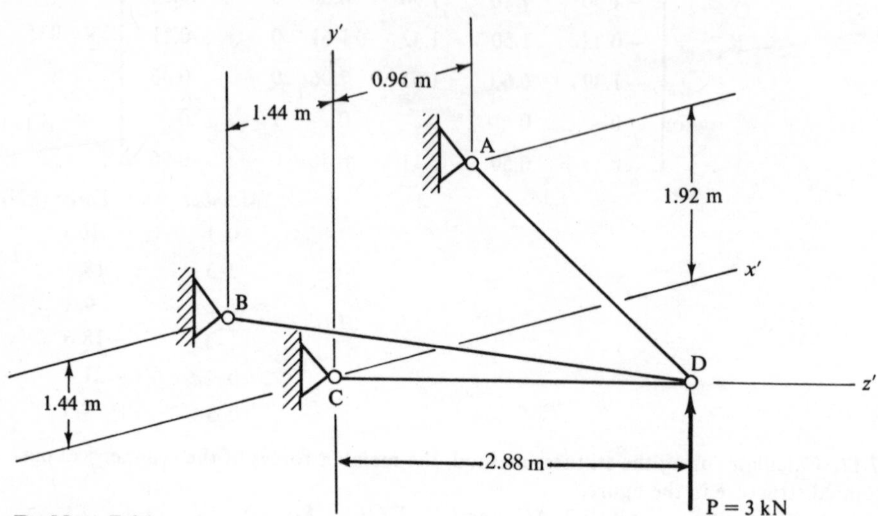

Problem 7·14

Ans. $\Delta_D = 0.1[0.25 \ 7.57 \ 1.14]^T$.

7.15. Assemble the stiffness matrix for the continuous beam shown in the figure. The second moments of area are as shown, and $E = 210$ kN/mm^2.

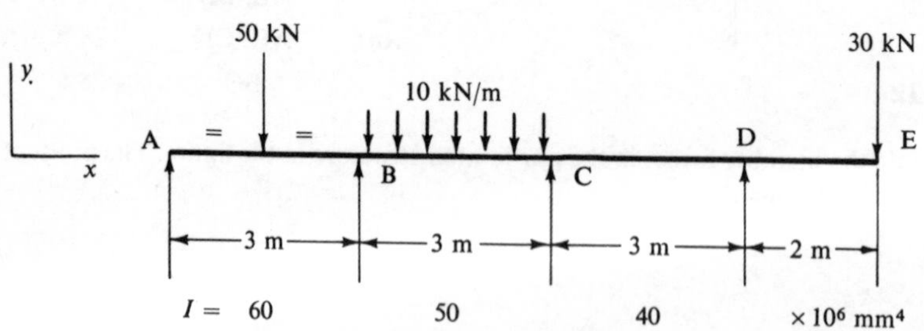

Problem 7·15

Ans.
$$\mathbf{K}_s = 10^3 \begin{bmatrix} 16.8 & 8.40 & 0 & 0 \\ 8.40 & 30.8 & 7.00 & 0 \\ 0 & 7.00 & 25.2 & 5.60 \\ 0 & 0 & 5.60 & 11.2 \end{bmatrix} \begin{matrix} A \\ B \\ C \\ D \end{matrix}$$

(columns A B C D)

(Written in terms of kilonewtons and metres.)

7.16. Draw the bending moment diagram for the structure in Problem 7.15 with the loading shown. The flexibility matrix is:

$$\mathbf{K}_s^{-1} = 10^{-5}\begin{bmatrix} 6.98 & -2.05 & 0.640 & -0.320 \\ -2.05 & 4.10 & -1.28 & 0.640 \\ 0.640 & -1.28 & 4.86 & -2.43 \\ -0.320 & 0.640 & -2.43 & 10.14 \end{bmatrix}$$

(Written in kilonewtons and metre units.)

Ans. See figure below.

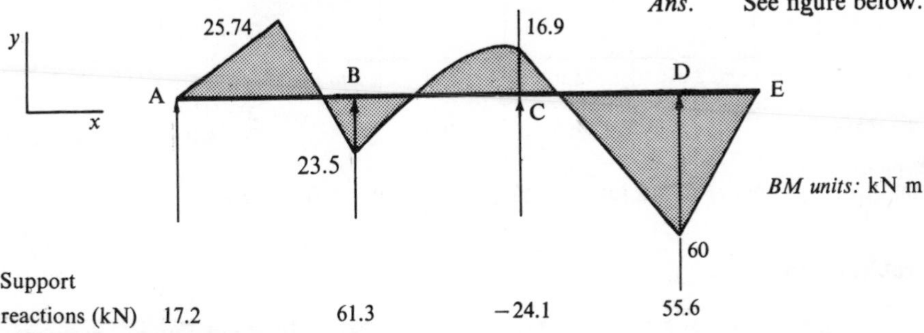

Solution to Problem 7·16

7.17. The continuous beam ABCD shown in the figure is simply supported at B and C and encastré at A and D. The spans, section properties, and loading are as indicated. Analyse the structure, and draw the bending moment diagram, ignoring second-order effects.

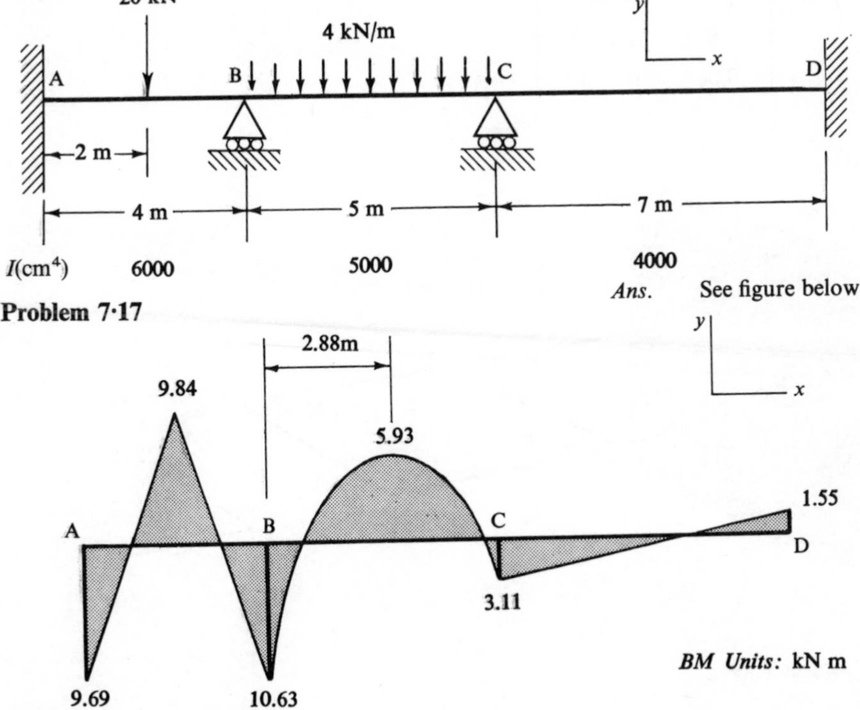

Problem 7·17

Ans. See figure below.

Solution to Problem 7·17

7.18. The continuous beam in the figure is supported at A, B, C, and D by bearings which are rigid against vertical displacement, but are able to exert restraining moments which are proportional to the rotations. $E = 200$ kN/mm^2. Determine the reaction moments at the supports.

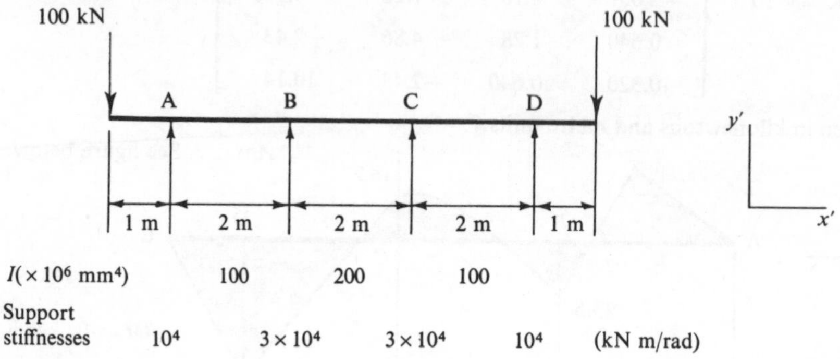

Problem 7·18

Ans. $\theta_A = 2.16 \times 10^{-3}$ rad; $\theta_B = -0.392 \times 10^{-3}$ rad. Support moments $M_A = -M_D = -21.6$ kN m; $M_B = -M_C = +11.8$ kN m

7.19. Assemble the complete structure stiffness matrix for the grillage in the figure. All members have the same properties, and the $\mathbf{K}_{11}$ matrix for each is

$$\begin{bmatrix} 1 \times 10^{-3} & 0 & 0.087 \\ 0 & 3 & 0 \\ 0.087 & 0 & 10 \end{bmatrix}$$

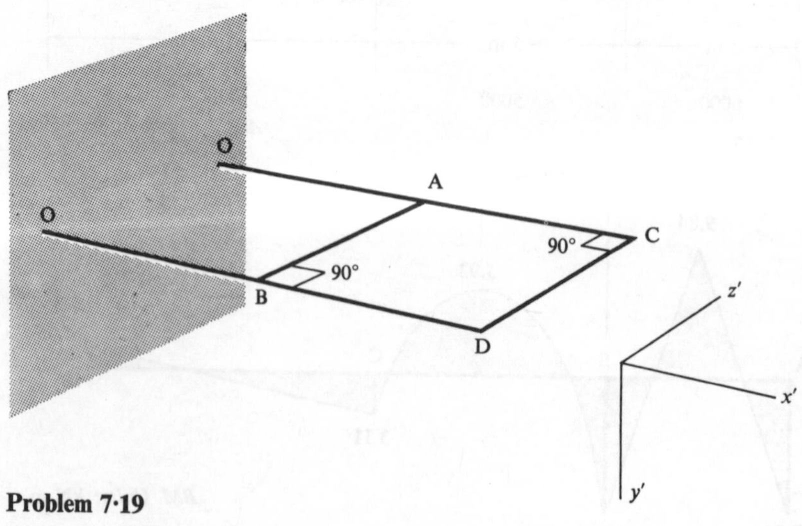

Problem 7·19

Ans. See p. 291.

7.19. *Ans.*

$$
\mathbf{K}_s = \begin{array}{c}
\begin{array}{ccc:ccc:ccc:ccc}
& \text{A} & & & \text{B} & & & \text{C} & & & \text{D} &
\end{array} \\
\left[\begin{array}{ccc:ccc:ccc:ccc}
3 \times 10^{-3} & 0.087 & 0 & -10^{-3} & 0.087 & 0 & -10^{-3} & 0 & 0.087 & & & \\
 & 16 & 0 & -0.087 & 5 & 0 & 0 & -3 & 0 & & 0 & \\
 & & 23 & 0 & 0 & -3 & -0.087 & 0 & 5 & & & \\
\hdashline
 & & & 3 \times 10^{-3} & 0 & -0.087 & & & & -10^{-3} & 0 & 0.087 \\
 & & & & 16 & 0 & & 0 & & 0 & -3 & 0 \\
 & & & & & 23 & & & & -0.087 & 0 & 5 \\
\hdashline
 & & & & & & 2 \times 10^{-3} & 0.087 & -0.087 & -10^{-3} & 0.087 & 0 \\
 & & & & \textit{Symmetric} & & & 13 & 0 & -0.087 & 5 & 0 \\
 & & & & & & & & 13 & 0 & 0 & -3 \\
\hdashline
 & & & & & & & & & 2 \times 10^{-3} & -0.087 & -0.087 \\
 & & & & & & & & & & 13 & 0 \\
 & & & & & & & & & & & 13
\end{array}\right]
\begin{array}{c} \\ \text{A} \\ \\ \\ \text{B} \\ \\ \\ \text{C} \\ \\ \\ \text{D} \\ \\ \end{array}
\end{array}
$$

7.20. Assemble the three matrices $\mathbf{K}_{11}$, $\mathbf{K}_{12}$, and $\mathbf{K}_{22}$ for the member AB of variable section shown in the figure. The depth is 200 mm throughout the length of 4 m, but the width varies linearly from 240 mm at one end to 720 mm at the other. (Assume AB may be divided into four equal parts of constant section, the properties at the centre being used for the whole part.). $E = 210$ kN/mm^2.

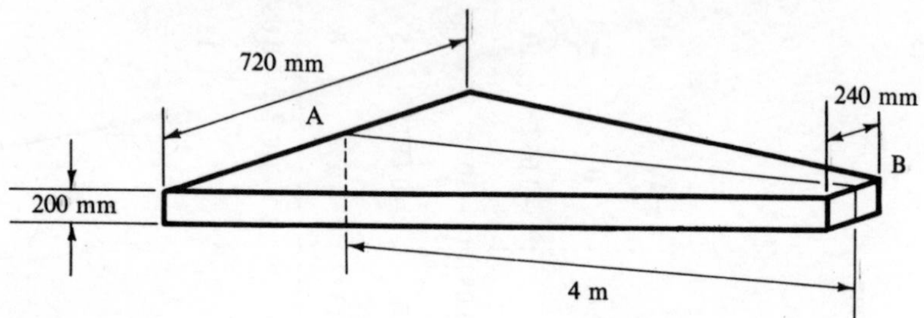

Problem 7·20

Ans.

$$\mathbf{K}_{11} = \begin{bmatrix} 4640 & 0 & 0 \\ 0 & 12.0 & -20.1 \times 10^3 \\ 0 & -20.1 \times 10^3 & 49.0 \times 10^6 \end{bmatrix}$$

$$\mathbf{K}_{12} = \begin{bmatrix} -4640 & 0 & 0 \\ 0 & -12.0 & 20.1 \times 10^3 \\ 0 & -27.9 \times 10^3 & 31.4 \times 10^6 \end{bmatrix}$$

$$\mathbf{K}_{22} = \begin{bmatrix} 4640 & 0 & 0 \\ 0 & 12.0 & 27.9 \times 10^3 \\ 0 & 27.9 \times 10^3 & 80.2 \times 10^6 \end{bmatrix}$$

(Matrices in terms of kN and mm)

Chapter 8
Matrix flexibility method

8.1 Basic form

The **flexibility method** has already been used in Section 4.6 to solve simple statically indeterminate structures. Here, the formal steps implied in that section will be discussed more fully so that the method can be extended to structures having many redundancies. As in the discussion of the stiffness method in Chapter 7 the description of the flexibility method will be based on plane structures having only prismatic members of material obeying Hooke's law and it will be assumed that all loads are applied at joints. If the loads were not applied at joints the solution would be carried out in accordance with the approach of Section 7.15. The principle described in that section is a general one, and is independent of the analysis method used.

In the flexibility method there are four main steps. The statically indeterminate plane frame of Fig. 8.1–1 will be used as an example.

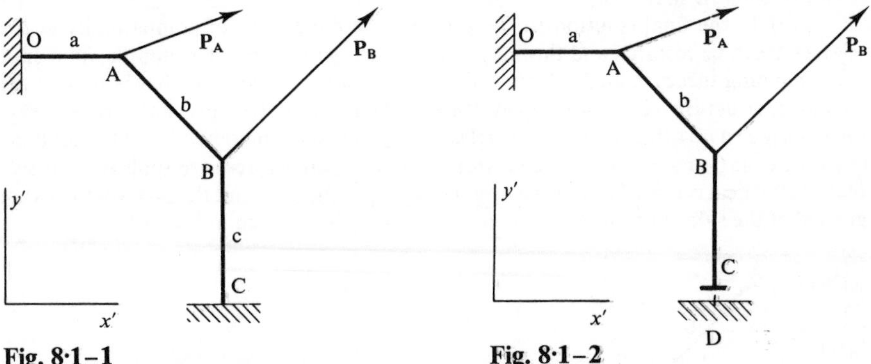

Fig. 8·1–1

Fig. 8·1–2

(1) The structure is rendered determinate by the insertion of suitable releases, and is now called the primary structure. In the present example the degree of redundancy is three, and the insertion of three releases at C gives the primary structure of Fig. 8.1–2. It is convenient to retain the letter C for the foot of the column—now free—and to introduce an additional letter, say D, for the encastré support itself.

In order to leave a statically determinate primary structure it is only necessary to introduce releases of the same number as the degree of redundancy. These can be chosen to occur individually at several distributed points, but it is usual to group as large a number as possible at a few convenient points. In this case the structure—initially redundant to the third degree—has been made determinate by the introduction of three releases at C, but the releases could have been made equally well at O, A, or B or (less conveniently) at some intermediate point. One consequence of the insertion of releases at points such as A or B of Fig. 8.1–1 would be the breaking of the structure into two parts, as shown in the alternative release pattern of Fig. 8.1–3. A new joint letter must be introduced (D), and the convention is that the original load vector $\mathbf{P}_A$ is taken to act on the joint of the same letter on the primary structure.

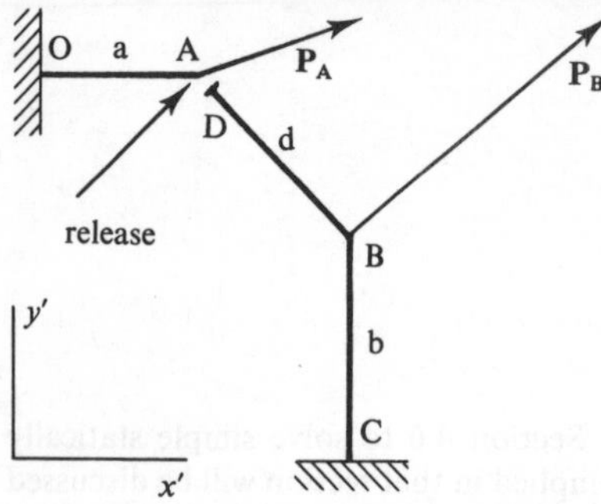

Fig. 8·1–3

(2) By inserting a release, a condition of compatibility at that point is abandoned. In the primary structure of Fig. 8.1–2, three displacement components due to the applied loads can therefore occur at C.

A particular solution of the primary structure is carried out, and as the structure is statically determinate the member forces can be calculated by applying equilibrium conditions only. The displacements at the releases (Δ_{rp}) due to the applied loads (**P**) are calculated by any suitable method—the unit load method of Section 4.16(a) and (b) would be suitable here.

(3) In the final solution to the problem the conditions of compatibility at the releases are to be restored and this may be achieved by insertion of a suitable number of corresponding forces, called **release forces** (**r**). Figure 8.1–4 shows what is termed the secondary structure—it is statically identical to the primary structure, but carries release forces only and no other loading. The release forces must be thought of as self-equilibrating force pairs (+**r** and −**r**). The convention is that negative forces are applied to the side of the release carrying a letter of the original structure (here C) and the positive forces are applied at the side carrying the new letter (here D). This is seen in Fig. 8.1–4.

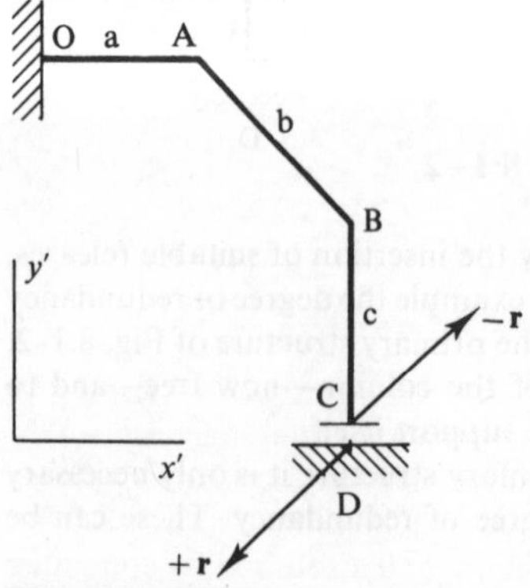

Fig. 8·1–4

A complementary solution of the secondary structure is now carried out. In this the displacements at the releases due to the release forces only ($\mathbf{\Delta}_{rr}$) are calculated. The values of the release forces required to ensure compatibility at the releases are of course unknown, and the displacements can only be obtained as functions of these forces. In matrix form:

$$\mathbf{\Delta}_{rr} = \mathbf{F}_{rr}\mathbf{r} \tag{8.1–1}$$

Here $\mathbf{F}_{rr}$ is a matrix of flexibility-influence coefficients (see Section 4.4) or flexibility matrix whose terms are the displacements at the releases for unit values of each of the release forces taken in turn. It is a square, non-singular matrix of an order equal to the number of degrees of indeterminacy in the original structure and, because of the reciprocality of flexibility-influence coefficients (see Section 4.11(a)), will be symmetrical about the leading diagonal. As with the particular solution the member forces of the complementary solution can be obtained by the application of equilibrium conditions only. The unit load method of Section 4.16(a) and (b) could again be used to calculate the terms of $\mathbf{F}_{rr}$.

(4) The results of Steps 2 and 3 can now be combined. The total displacement at the releases due to both applied loads and release forces ($\mathbf{\Delta}_r$) is given by:

$$\begin{aligned}\mathbf{\Delta}_r &= \mathbf{\Delta}_{rp} + \mathbf{\Delta}_{rr} \\ &= \mathbf{\Delta}_{rp} + \mathbf{F}_{rr}\mathbf{r}\end{aligned} \tag{8.1–2}$$

The displacement vector $\mathbf{\Delta}_r$ is a set of *relative* displacements at the releases. This makes no difference where a release is at a support (as in Fig. 8.1–2), but is an important point to remember when releases are at internal joints (see Example 8.1–1).

In the original structure of Fig. 8.1–1 no relative displacements can take place at the positions chosen for the releases (at C) and the vector $\mathbf{\Delta}_r$ can therefore be taken as zero. Hence

$$\mathbf{\Delta}_r = \mathbf{\Delta}_{rp} + \mathbf{F}_{rr}\mathbf{r} = 0 \tag{8.1–3}$$

and since

$$\mathbf{F}_{rr}\mathbf{r} = -\mathbf{\Delta}_{rp} \tag{8.1–4}$$

then

$$\mathbf{r} = -\mathbf{F}_{rr}^{-1}\mathbf{\Delta}_{rp} \tag{8.1–5}$$

(Or the equations (8.1–4) may be solved for $\mathbf{r}$.)

The only situation in which $\mathbf{\Delta}_r$ would not be equal to zero, would be in the case of some kind of elastic joint, or support settlement. In the former case the non-zero terms in $\mathbf{\Delta}_r$ would be proportional to terms in $\mathbf{r}$ and in the latter case the non-zero terms of $\mathbf{\Delta}_r$ would be known. In either case an equation similar to Eqn 8.1–4 could be obtained.

Once $\mathbf{r}$ is known the member forces in the original structure may be obtained by the superposition of effects from the particular and complementary solutions. As the solution has yielded values for member forces rather than joint displacements the method is sometimes given the alternative title of the **force method**.

Had the structure of Fig. 8.1–1 been analysed using the stiffness method a stiffness matrix of order 6 × 6 would have had to be inverted, or six equations solved. By using the flexibility method the order of inversion required is reduced to 3 × 3 because the actual structure has six degrees of freedom but only three degrees of redundancy. Whenever a structure to be analysed has more degrees of freedom than indeterminacies a flexibility analysis leads to the solution of a smaller number of equations than a stiffness analysis requires. For this reason it is frequently preferred in hand calculations, but if the structure is too large to be analysed economically by hand, and a solution by computer is necessary, then the general answer is to use the stiffness method because it can be so easily programmed and because the data is required in a very simple form. However, if the number of degrees of freedom is so large that the available computer storage proves inadequate then the answer may lie in the use of the flexibility method—if the number of degrees of indeterminacy is small.

The aim is to present an approach to the flexibility method which takes advantage of the small degree of indeterminacy a structure may have, and avoids where possible the assembly of large matrices. For instance, one method of calculating the structural displacements in Steps 2 and 3 would be to use the stiffness method. However, this would involve setting up a stiffness matrix even larger than that required to analyse the actual structure using the stiffness method throughout (9 × 9 as against 6 × 6 in this example). A more economical method of analysing the determinate primary and secondary structures is given by Livesley.

Example 8.1–1. The structure of Fig. 8.1–5 will be analysed using the flexibility method, the steps being numbered to correspond with Section 8.1. It will be assumed that the cross-sectional areas of the members are sufficiently large to prevent axial shortening or lengthening from having a significant influence on joint displacements, and that joint displacements are small enough for changes of geometry to be ignored.

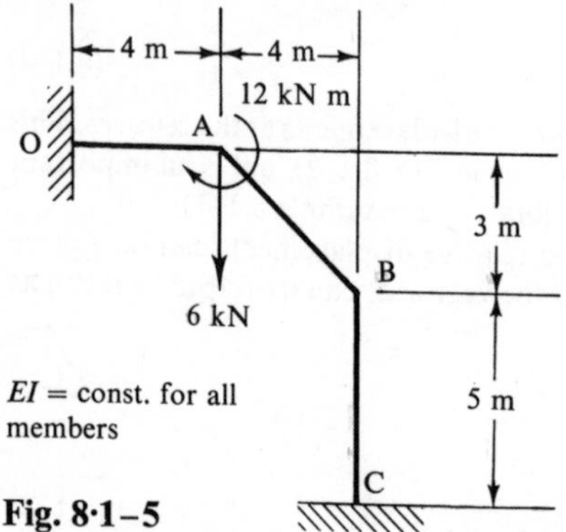

Fig. 8·1–5

(1) A release is inserted at A, giving the primary and secondary structures of Figs. 8.1–6 and 8.1–7 respectively. There is the possibility of two direct load components at A in the coordinate directions indicated—p_x and p_y—and a moment $-M$, i.e.

$$\mathbf{P}_A = \begin{bmatrix} P_x \\ P_y \\ M \end{bmatrix} = \begin{bmatrix} 0 \\ -6 \\ -12 \end{bmatrix} \begin{matrix} \\ \text{kN} \\ \text{kN m} \end{matrix} \qquad (8.1\text{–}6)$$

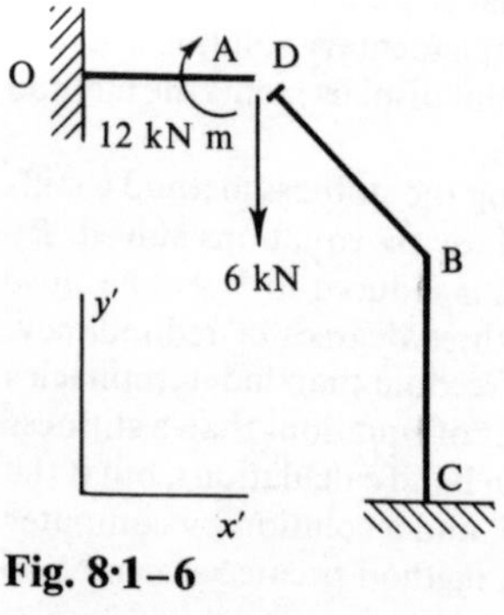

Fig. 8·1–6

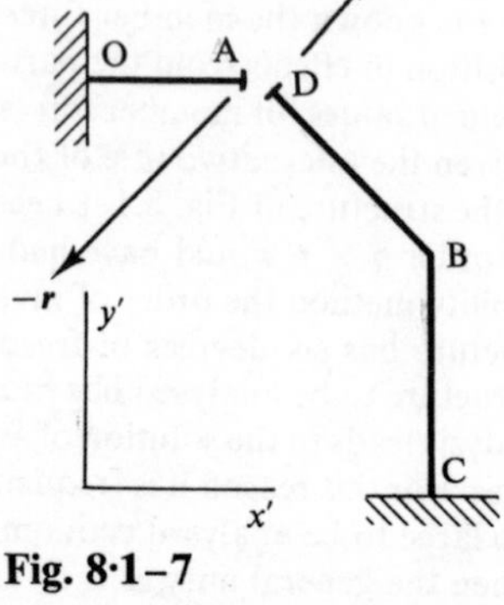

Fig. 8·1–7

Similarly, the release force has three components, i.e.

$$\mathbf{r} = [r_x \; r_y \; m]^T \tag{8.1–7}$$

(2) The unit load method of Section 4.16 will now be used to calculate the relative displacements due to applied load at A and D (Δ_{rp}). Figure 8.1–8 shows the three load cases to be analysed, together with the three separate unit loads required in each case. The results are

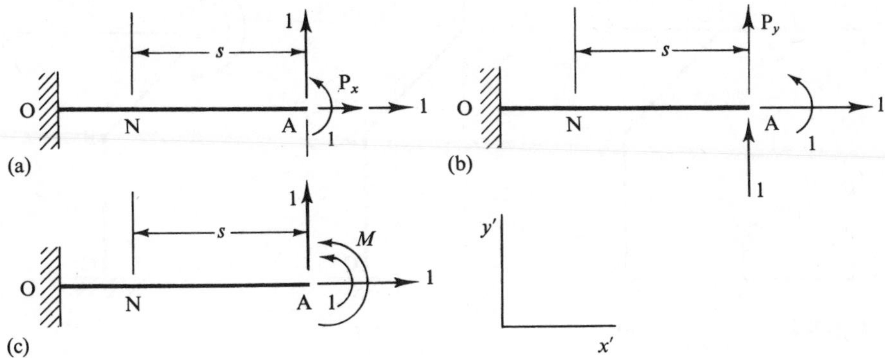

Fig. 8·1–8

summarized in Table 8.1, where $M_{(N)}$ and $M_{U(N)}$ represent the bending moment at a point N due to applied load and unit load respectively.

Table 8.1

	$\int \frac{M_{(N)} M_{U(N)}\, ds}{EI}$					
Loading	$P_x = 1$	$P_y = 1$	$M = 1$	Δ_{xA}	Δ_{yA}	θ_A
P_x	0	0	0	0	0	0
P_y	0	$\int_0^4 \frac{s^2 P_y\, ds}{EI}$	$\int_0^4 \frac{s P_y\, ds}{EI}$	0	$\frac{64P_y}{3EI}$	$\frac{8P_y}{EI}$
M	0	$\int_0^4 \frac{sM\, ds}{EI}$	$\int_0^4 \frac{M\, ds}{EI}$	0	$\frac{8M}{EI}$	$\frac{4M}{EI}$

Δ_{xA} is zero because of the assumption regarding cross-sectional areas. No displacements due to applied loading occur at D. Hence

$$\Delta_{rp} = \text{displacement at D } (= 0) - \text{displacement at A}$$

$$\Delta_{rp} = -\frac{1}{EI}\begin{bmatrix} 0 & 0 & 0 \\ 0 & \frac{64}{3} & 8 \\ 0 & 8 & 4 \end{bmatrix}\begin{bmatrix} 0 \\ -6 \\ -12 \end{bmatrix} = \frac{1}{EI}\begin{bmatrix} 0 \\ 224 \\ 96 \end{bmatrix} \tag{8.1–8}$$

or

$$\Delta_{rp} = \mathbf{F}_{rp}\mathbf{P}_A \tag{8.1–9}$$

$\mathbf{F}_{rp}$ is a flexibility matrix for displacements at releases due to applied loads, and Δ_{rp} is a vector of *relative* displacements.

(3) In the complementary solution of the secondary structure the displacements due to $-\mathbf{r}$ acting on member OA ($\mathbf{\Delta}_{Ar}$) will be determined first. As this is merely a repetition of Step 2, the result can be taken from Eqn 8.1–9. Then

$$\mathbf{\Delta}_{Ar} = \mathbf{F}_{rp}(-\mathbf{r}) = -\mathbf{F}_{rp}\mathbf{r} \tag{8.1–10}$$

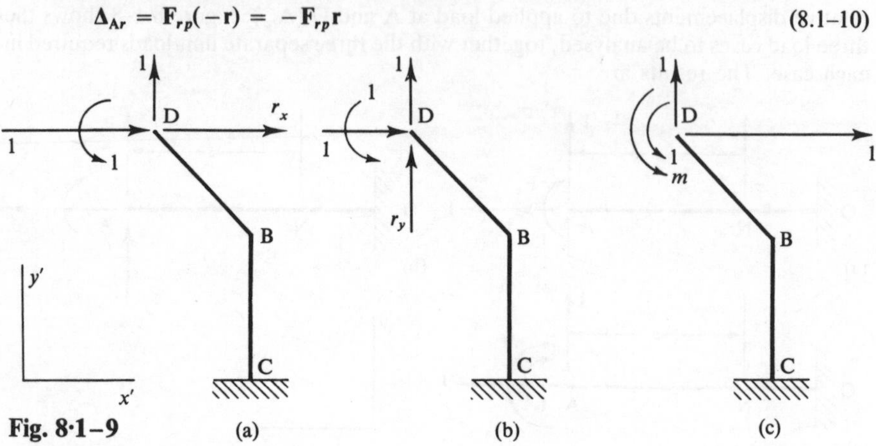

Fig. 8·1–9 (a) (b) (c)

Figure 8.1–9 shows the three load cases to be considered for the analysis of structure D–B–C, and the results are summarized in Table 8.2. If the displacements at D due to $\mathbf{r}$ are $\mathbf{\Delta}_{Dr}$, then:

$$\mathbf{\Delta}_{Dr} = \frac{1}{EI}\begin{bmatrix} \frac{530}{3} & 130 & -35 \\ 130 & \frac{320}{3} & -30 \\ -35 & -30 & 10 \end{bmatrix}\begin{bmatrix} r_x \\ r_y \\ m \end{bmatrix} \tag{8.1–11}$$

The relative displacements at the releases ($\mathbf{\Delta}_{rr}$) can now be calculated from Eqns 8.1–10 and 8.1–11.

Hence

$$\mathbf{\Delta}_{rr} = \mathbf{\Delta}_{Dr} - \mathbf{\Delta}_{Ar}$$

$$= \frac{1}{EI}\begin{bmatrix} \frac{530}{3} & 130 & -35 \\ 130 & \frac{320}{3} & -30 \\ -35 & -30 & 10 \end{bmatrix}\mathbf{r} + \frac{1}{EI}\begin{bmatrix} 0 & 0 & 0 \\ 0 & \frac{64}{3} & 8 \\ 0 & 8 & 4 \end{bmatrix}\mathbf{r}$$

$$= \frac{1}{EI}\begin{bmatrix} \frac{530}{3} & 130 & -35 \\ 130 & 128 & -22 \\ -35 & -22 & 14 \end{bmatrix}\mathbf{r}$$

$$= \mathbf{F}_{rr}\mathbf{r} \tag{8.1–12}$$

Table 8.2

	$\int \frac{M_{(N)} M_{U(N)}\, ds}{EI}$					
Loading	$r_x = 1$	$r_y = 1$	$m = 1$	Δ_{xD}	Δ_{yD}	θ_D
r_x	$\int_0^5 \frac{r_x s^2 \cos^2\theta\, ds}{EI} + \int_3^8 \frac{r_x s^2\, ds}{EI}$	$\int_0^5 \frac{r_x s^2 \cos\theta \sin\theta\, ds}{EI} + \int_3^8 \frac{r_x s.4\, ds}{EI}$	$-\int_0^5 \frac{r_x s \cos\theta\, ds}{EI} - \int_3^8 \frac{r_x s\, ds}{EI}$	$\frac{530 r_x}{3EI}$	$\frac{130 r_x}{EI}$	$\frac{-35 r_x}{EI}$
r_y	$\int_0^5 \frac{r_y s^2 \sin\theta \cos\theta\, ds}{EI} + \int_3^8 \frac{r_y s.4\, ds}{EI}$	$\int_0^5 \frac{r_y s^2 \sin^2\theta\, ds}{EI} + \int_3^8 \frac{r_y.4.4\, ds}{EI}$	$-\int_0^5 \frac{r_y s \sin\theta\, ds}{EI} - \int_3^8 \frac{r_y.4\, ds}{EI}$	$\frac{130 r_y}{EI}$	$\frac{320 r_y}{3EI}$	$\frac{-30 r_y}{EI}$
m	$-\int_0^5 \frac{ms \cos\theta\, ds}{EI} - \int_3^8 \frac{ms\, ds}{EI}$	$-\int_0^5 \frac{ms \sin\theta\, ds}{EI} - \int_3^8 \frac{m.4\, ds}{EI}$	$\int_0^5 \frac{m\, ds}{EI} + \int_3^8 \frac{m\, ds}{EI}$	$\frac{-35m}{EI}$	$\frac{-30m}{EI}$	$\frac{10m}{EI}$

(4) The total relative displacement at the releases is now calculated from Eqns 8.1–9 and 8.1–12. Then

$$\mathbf{u} = \mathbf{F}_{rp}\mathbf{P}_A + \mathbf{F}_{rr}\mathbf{r}$$

or

$$\mathbf{u} = \frac{1}{EI}\begin{bmatrix} 0 \\ 224 \\ 96 \end{bmatrix} + \frac{1}{EI}\begin{bmatrix} \frac{530}{3} & 130 & -35 \\ 130 & 128 & -22 \\ -35 & -22 & 14 \end{bmatrix}\begin{bmatrix} r_x \\ r_y \\ m \end{bmatrix} = 0 \tag{8.1–13}$$

Setting $\mathbf{u}$ equal to zero to ensure compatibility at the release provides a set of three linear equations whose solution yields:

$$\begin{bmatrix} r_x \\ r_y \\ m \end{bmatrix} = \begin{bmatrix} 2.10 \\ -5.70 \\ -10.56 \end{bmatrix} \begin{matrix} \text{kN} \\ \text{kN} \\ \text{kN m} \end{matrix} \tag{8.1–14}$$

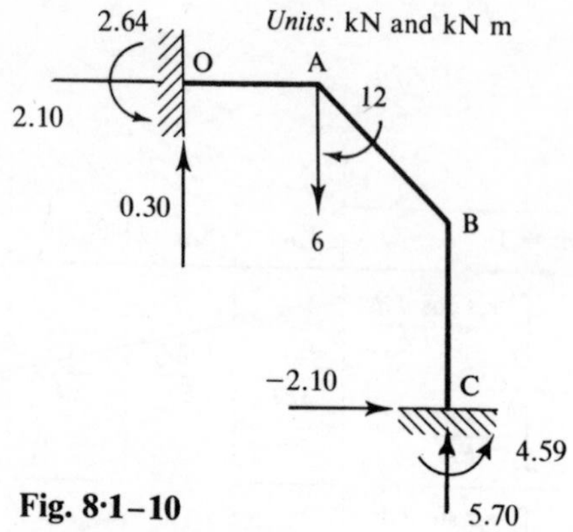

Fig. 8·1–10

The support forces are shown in Fig. 8.1–10 and the displacements at A can be calculated from Eqn 8.1–11.

$$\Delta_A = \Delta_{Dr} = \frac{1}{EI}\begin{bmatrix} \frac{530}{3} & 130 & -35 \\ 130 & \frac{320}{3} & -30 \\ -35 & -30 & 10 \end{bmatrix}\begin{bmatrix} 2.10 \\ -5.70 \\ -10.56 \end{bmatrix} = \frac{1}{EI}\begin{bmatrix} 0 \\ -17.9 \\ -8.2 \end{bmatrix} \quad (8.1\text{–}15)$$

8.2 Relative merits of flexibility and stiffness methods

It was shown in Chapter 7 that one basic form of the stiffness method could be applied to a wide range of structures, with only minor adjustments to cope with each variant. The advantages of the method can be summarized as:

(a) A general-purpose program is easy to write.

(b) It requires a minimum of input data.

(c) It can be made entirely automatic. Its use requires no understanding of structural mechanics.

The method has a major disadvantage in that no account is taken of the degree of indeterminacy, and therefore there is little opportunity to benefit from the structural expertise of the operator. Equally, this will be seen as an essential concomitant of the advantage listed in (c) above. The time required to perform an analysis and the amount of computer storage necessary depends almost entirely on the number of degrees of freedom involved.

The flexibility method has the potential advantage that it recognizes the state of indeterminacy of a structure, and regards the number of degrees of static indeterminacy as the unknown. It allows the engineer to exercise his judgement over the details of the solution—such as the choice of releases. Structures having many degrees of freedom but few degrees of static indeterminacy should be much more economically analysed by the flexibility rather than the stiffness method. However, the flexibility method has the severe disadvantage that a general program is difficult to write. The full advantage of the method for skeletal analysis is only achieved if the matrix $\mathbf{F}_{rr}$ can be set up directly, and although this can be done for plane frames, it is by no means as straightforward for other types of structure.

General Reference

Livesley, R. K. *Matrix Methods of Structural Analysis*. Pergamon Press, Oxford, 1975.

Problems

8.1. Determine the displacements for the joints of the pin-jointed structure in the figure. The product EA is the same for all members.

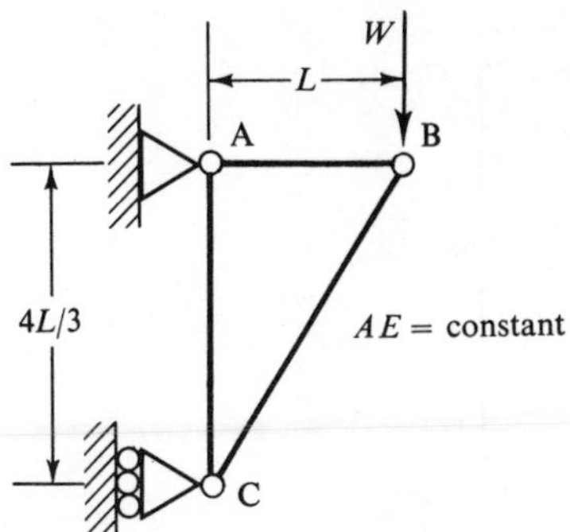

Problem 8·1

Ans.

	A	B	C
Vertical displacement	—	$\frac{9WL}{2EA}$	$\frac{4WL}{3EA}$
Horizontal displacement	—	$\frac{3WL}{4EA}$	—

8.2. Determine the member forces in the pin-jointed structure in the figure.

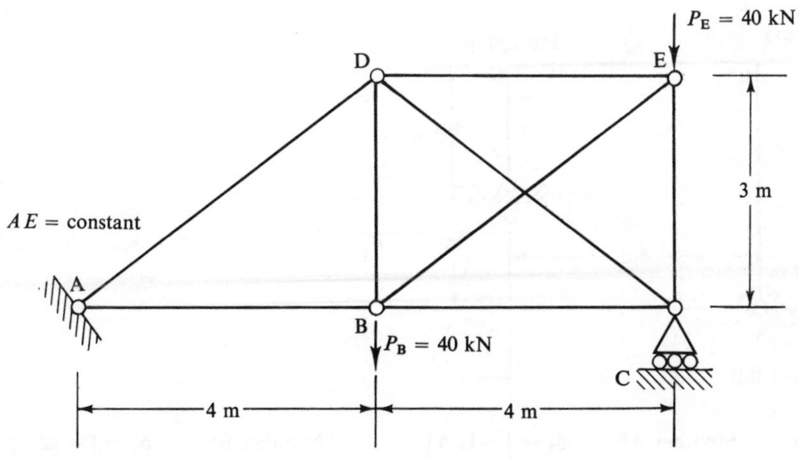

Problem 8·2

Ans.

Member	AB	AD	BD	DC	BE	DE	BC	EC
Forces (kN)	26.6	−33.2	31.2	−18.7	14.4	−11.6	15.0	−48.8

8.3. Determine the member forces in the pin-jointed structure in the figure.

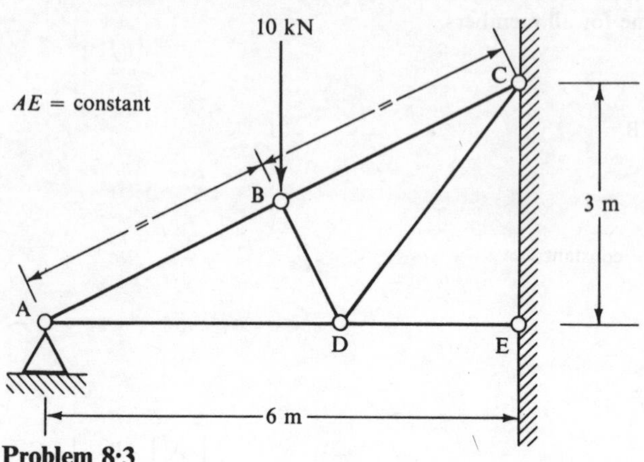

Problem 8·3

Ans.

Member	AB	AD	BD	BC	DC	DE
Forces (kN)	−2.24	3.75	−8.94	2.23	10.0	−6.25

8.4. Determine the member forces of the structure of Problem 7.12.

8.5. Calculate the member forces in the structure in the figure.

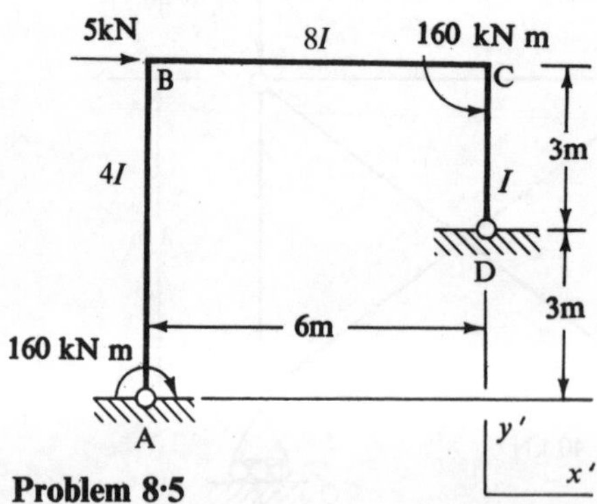

Problem 8·5

Ans. Member AB $\mathbf{p}_2 = \begin{bmatrix} -11.4 \\ 27.8 \\ -6.87 \end{bmatrix}$ Member BC $\mathbf{p}_2 = \begin{bmatrix} -32.8 \\ -11.4 \\ 61.6 \end{bmatrix}$

Member CD $\mathbf{p}_2 = \begin{bmatrix} 11.4 \\ -32.8 \\ 0 \end{bmatrix}$

8.6. Calculate the reaction moments for the structure in the figure.

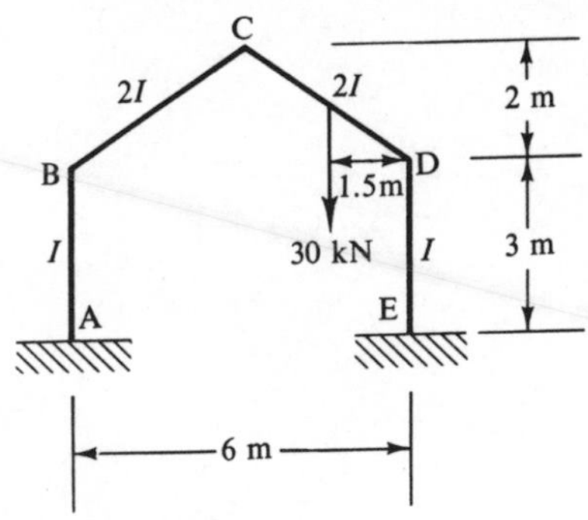

Problem 8·6

Ans. $M_{AB} = 8.45$ kN m; $M_{ED} = 5.64$ kN m.

Chapter 9
Instability of struts and frameworks

9.1 Pin-ended strut and Euler buckling load

Situations in which superposition can, or cannot be applied have already been discussed in Chapter 4. This chapter is concerned with the latter, and Fig. 9.1–1 shows a **beam-column**—a member carrying both axial and lateral loads. A 'column' or 'strut' carries axial loads only.

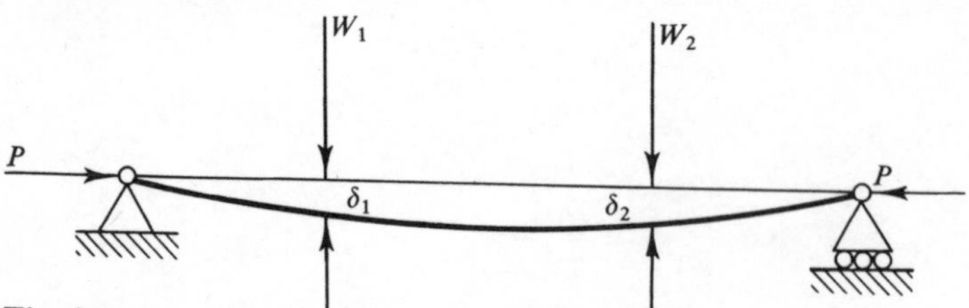

Fig. 9·1–1

If the member is perfectly straight, axial load P acting on its own would produce no lateral deflections. Lateral loads W_1 and W_2 acting on their own would produce lateral deflections, and these would be increased if the member were to carry both axial compressive and lateral loads at the same time, or decreased for an axial tensile load. The deflections of the one loading case cannot therefore be superimposed on those of the other. Even without the presence of forces W, lateral deflections would be caused by P if it were eccentric to the column centroid (the column then would carry an axial load, plus an end moment) or if the column had an initial curvature. In all these cases the lateral deflections would be related to the value of the axial load in a nonlinear manner, even if the material of the column were linearly elastic, as was shown in Section 4.5.

Columns are very common in structural engineering, and an understanding of their behaviour is vital to the designer. He may try to load columns centrally, but fabrication tolerances prevent this from being achieved exactly and no column is perfectly straight. In general, he recognizes that columns are imperfect, and may be required to carry moments, and designs accordingly.

The real problem is therefore the study of imperfect, eccentrically loaded columns, but to gain a full understanding it is instructive to look first at the behaviour of columns which are ideal. It will be assumed in the following analysis that:

(a) The ends of the strut are mounted in perfectly smooth pinned supports.

(b) The strut is perfectly straight, and uniform in cross-section throughout its length.

(c) It carries only a perfectly axial compressive load.

(d) The material of the strut is linearly elastic, and the magnitude of the load is such that no question of yielding arises.

(e) The lateral deflections of the strut remain small in relation to its length.

The strut 1–2 is shown in Fig. 9.1–2 carrying an axial compressive load P. The purpose of the following analysis is to investigate whether the strut can remain in equi-

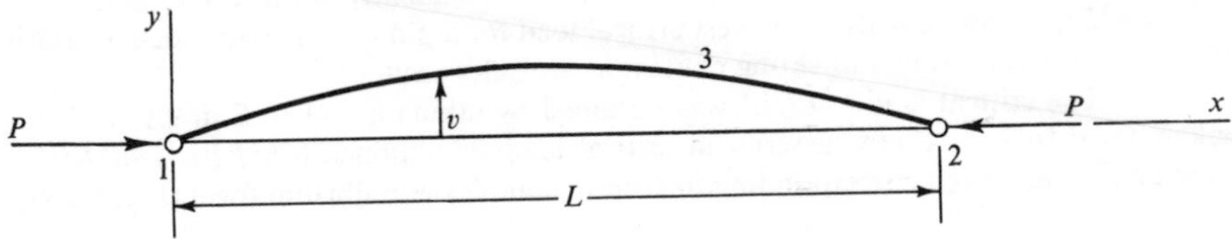

Fig. 9·1–2

librium in any non-straight configuration, such as that shown by curve 1–3–2, under a suitable value of axial force P. Let it be assumed that this is possible. The bending moment M at any distance x from the left-hand end is:

$$EI\frac{d^2v}{dx^2} = M = -Pv \qquad (9.1\text{–}1)$$

hence

$$\frac{d^2v}{dx^2} + \left(\frac{P}{EI}\right)v = 0 \qquad (9.1\text{–}2)$$

The solution of this differential equation is:

$$v = A\sin\sqrt{\left(\frac{P}{EI}\right)}x + B\cos\sqrt{\left(\frac{P}{EI}\right)}x \qquad (9.1\text{–}3)$$

where A and B are integration constants.

The boundary conditions are:

$$\text{(a) } v = 0 \text{ at } x = 0 \qquad \text{(b) } v = 0 \text{ at } x = L$$

From (a)

$$B = 0 \quad \text{and} \quad v = A\sin\sqrt{\left(\frac{P}{EI}\right)}x \qquad (9.1\text{–}4)$$

and from (b)

$$0 = A\sin\sqrt{\left(\frac{P}{EI}\right)}L$$

Then either

$$A = 0 \quad \text{or} \quad \sqrt{\left(\frac{P}{EI}\right)}L = n\pi \quad (n = 0, 1, 2, \text{etc.}) \qquad (9.1\text{–}5)$$

Equation 9.1–4 shows that the deflection curve is sinusoidal, with end 1 as origin. If $A = 0$ or the integer $n = 0$ (Eqn 9.1–5) then the strut remains straight. If n is taken as 1 then:

$$P = \pi^2\frac{EI}{L^2} \qquad (9.1\text{–}6)$$

and

$$v = A \sin \frac{\pi x}{L} \tag{9.1–7}$$

The deflection form is a half sine curve of undefined magnitude. The value of P given by Eqn. 9.1–6 is the smallest value for which this non-straight equilibrium form can exist. If P is smaller than this value, then the strut can be in equilibrium in the straight form only ($A = 0$). The value of P given by Eqn 9.1–6 is therefore a very special value in that it marks the boundary between the straight equilibrium state and the non-straight given in Eqn 9.1–4, and it is termed the **critical load** or **buckling load**. This case was first solved by the Swiss mathematician Euler, and is therefore frequently termed the **Euler load** ($P_E = \pi^2 EI/L^2$) and equals the lowest critical load for a pin-ended strut. Such a strut is frequently referred to in a buckling context as an **Euler strut**.

The critical load $\pi^2 EI/L^2$ was obtained by taking $n = 1$ in Eqn 9.1–5. If n is taken equal to 2, 3, 4, etc., a series of critical loads is obtained ($4\pi^2 EI/L^2$, $9\pi^2 EI/L^2$, $16\pi^2 EI/L^2$, etc.), each corresponding to a more complex equilibrium form (Fig. 9.1–3):

$$v = A \sin \frac{2\pi x}{L}, \qquad v = A \sin \frac{3\pi x}{L}, \qquad v = A \sin \frac{4\pi x}{L} \tag{9.1–8}$$

Each of these curves is termed an **instability** or **buckling mode**.

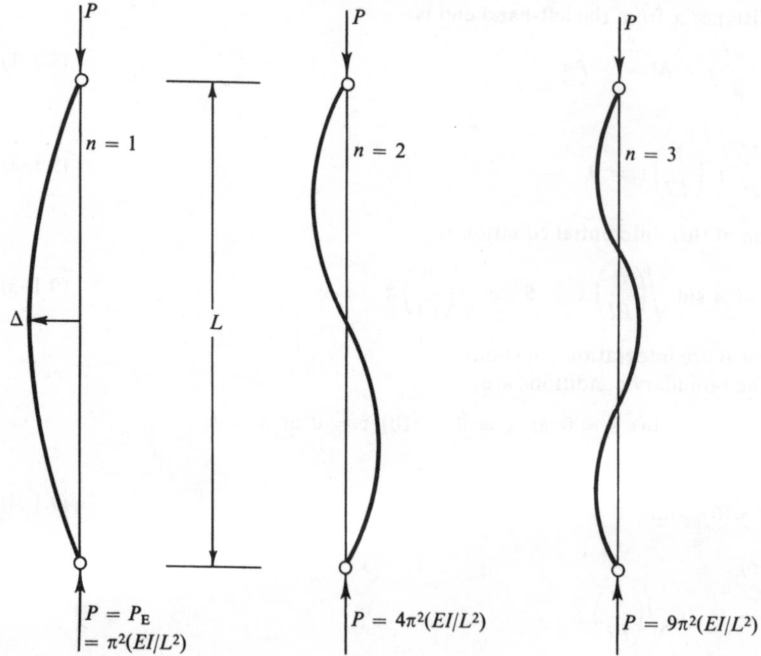

Fig. 9·1–3

The behaviour of the strut can be compared to that of the sphere shown in Fig. 9.1–4(a), (b), (c).

In Fig. 9.1–4(a) a sphere is shown resting on a concave surface—it is in a state of stable equilibrium. If displaced from its initial 'at rest' position in the centre of the surface, it will return to that position (it may oscillate initially, but friction will ensure that it eventually comes to rest). This is similar to the column carrying an axial load less than the Euler load—if slightly displaced laterally, it returns to its initial straight configuration.

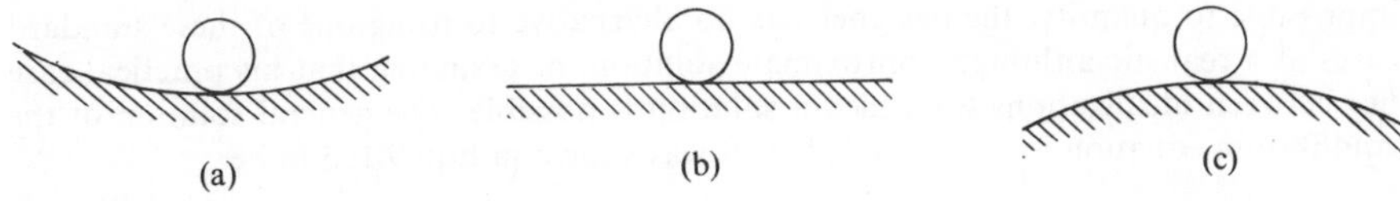

Fig. 9·1–4

Figure 9.1–4(c) shows the sphere resting on a convex surface. A sphere balanced in this way, displaced by the smallest disturbance, would roll further and further from the initial position—it is in a state of unstable equilibrium. (Although one may theoretically conceive of a sphere balanced on a convex surface in this way, one's experience in fact suggests that if the two surfaces were of such a perfect form that the balance could be upset by the minutest disturbance, a balance would not, in practice, be achieved.) The position of Fig. 9.1–4(c) is similar to that of the column loaded beyond the Euler load. While in theory it may remain straight ($A = 0$ in Eqn 9.1–5) the minutest disturbance only is required to cause lateral deflections ($A \neq 0$). In practice such disturbances are always present, so that it is not possible for a strut to be loaded beyond the Euler load unless supports constrain the strut to follow one of the 'higher' modes of Fig. 9.1–3.

Figure 9.1–4(b) shows a state intermediate between the stable and unstable states of (a) and (c). The sphere rests on a flat surface, and if disturbed to a new position merely remains there—it is in neutral equilibrium. This corresponds to the strut loaded to exactly the Euler load—it can be in equilibrium in an infinite number of positions, as constant A in Eqn 9.1–4 is not defined in magnitude. Figure 9.1–5 shows a graph of axial load P against central lateral deflection Δ.

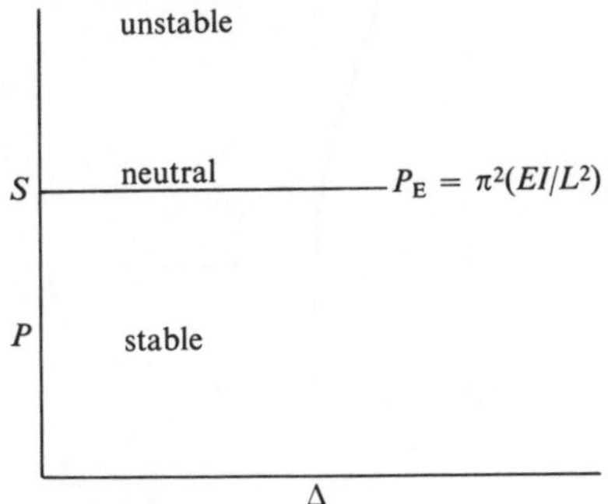

Fig. 9·1–5

Although the curve in Fig. 9.1–5 is for an ideal strut and the curve followed in practice by a real strut is modified, its form is a significant help in the understanding of instability problems.

9.2 Other simple strut problems

A strut having both ends pinned has already been analysed in Section 9.1, but other end conditions are met in structural engineering. Three further cases are illustrated in Fig. 9.2–1. They are:

(a) Both ends fixed (against lateral deflection and rotation).

(b) One end fixed, and the other end completely free.

(c) One end fixed and the other end pinned.

Real struts, in fact, rarely fit any of these categories—their ends are usually supported in some flexible manner. But as the true value of the flexibility is generally impossible to quantify, the designer has no alternative to using one of these standard cases as a realistic although approximate solution, or assuming that his practical case lies between idealizations for which a solution is possible. The general solution of the equilibrium equation for a pin-ended strut was shown in Eqn 9.1–3 to be

$$v = A \sin \sqrt{\left(\frac{P}{EI}\right)} x + B \cos \sqrt{\left(\frac{P}{EI}\right)} x$$

This could be rewritten in the form:

$$v = C \sin \left[\sqrt{\left(\frac{P}{EI}\right)} x + D \right] \tag{9.2–1}$$

where C and D are integration constants.

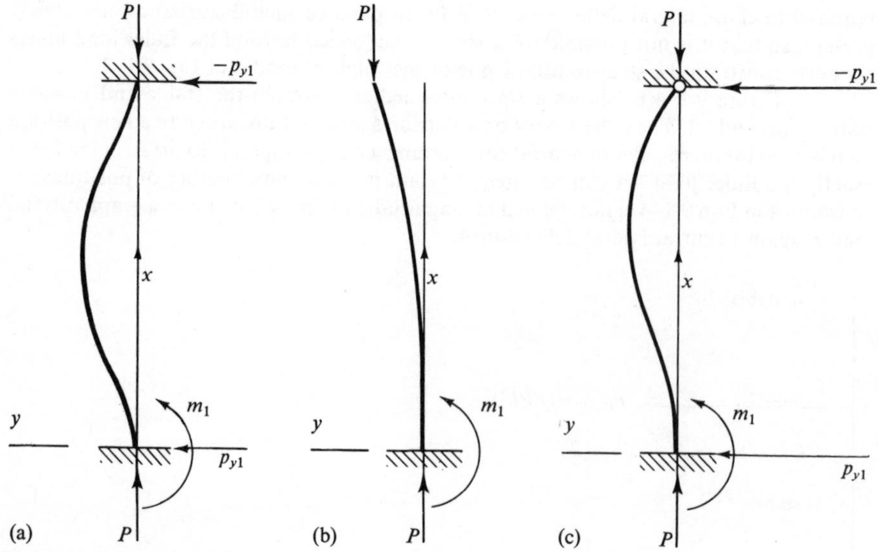

Fig. 9·2–1

For the more complex end conditions considered in this section, the equilibrium equation is still of the form of Eqn 9.1–2, but may have additional terms, as there can be both a moment and a shear at the end support. For instance, the equation for a fixed-pinned column (Fig. 9.2–1(c)) is

$$\frac{d^2v}{dx^2} + \left(\frac{P}{EI}\right) v - \left(\frac{p_{y1} x - m_1}{EI}\right) = 0 \tag{9.2–2}$$

(using the generalized force notation of Chapter 5).

The general form, for any end condition is

$$\frac{d^2v}{dx^2} + \left(\frac{P}{EI}\right) v - \frac{(Mx + N)}{EI} = 0 \tag{9.2–3}$$

where $(Mx + N)$ represents the bending moment, at a distance x from the origin O, due to end reactions such as p_{y1} and m_1.

The general solution is therefore similar to Eqn 9.2–1 but with the addition of a particular integral, i.e.

$$v = C \sin\left[\sqrt{\left(\frac{P}{EI}\right)}\, x + D\right] + \frac{Mx + N}{P} \tag{9.2–4}$$

i.e. v = a sine wave + a linear function in x.

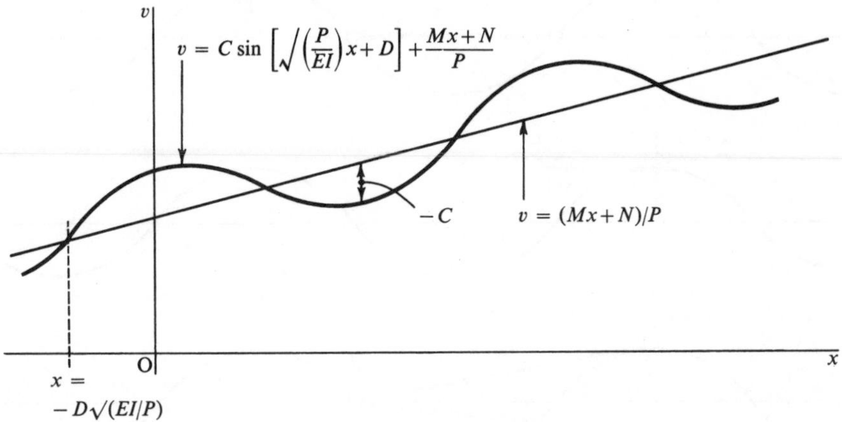

Fig. 9·2–2

The value of D determines the relative position of the sine curve in the x-direction. The function $(Mx + N)$ is the base line for the sine curve, and the value of C is the maximum departure of the curve from it. The value of $\sqrt{(P/EI)}$ determines the *scale* of the sine curve in the x-direction. This diagram (Fig. 9.2–2) may be used to obtain the solutions to the special cases listed at the head of the following sections.

(a) **Both ends fixed** (Fig. 9.2–1(a)). From symmetry it can be argued that $p_{y1} = 0$. $(Mx + N)$ is therefore equal to a constant, and the base line for the sine curve is parallel to the x-axis. It is the heavy line marked (ii) in Fig. 9.2–3(a). The base is tangent to the curve where both slope and deflection are zero. (Curve (i) is the base line for the Euler strut.) Therefore

$$\sqrt{\left(\frac{P_{cr}}{EI}\right)} L = 2\pi \quad \text{then} \quad P_{cr} = 4\pi^2 \frac{EI}{L^2} = 4P_E \tag{9.2–5}$$

where P_{cr} = critical load, P_E = Euler load.

(b) **One end fixed, other end free** (Fig. 9.2–1(b)). Here again $p_{y1} = 0$ and $(Mx + N)$ = const. The solution is indicated by curve (iii) in Fig. 9.2–3(b), for which the deflection and slope are zero at $x = 0$ and the curvature is zero at $x = L$, where the deflection is non-zero. Hence

$$\sqrt{\left(\frac{P_{cr}}{EI}\right)} L = \frac{\pi}{2} \quad \text{since} \quad P_{cr} = \frac{\pi^2 EI}{4L^2} = \frac{P_E}{4} \tag{9.2–6}$$

(c) **One end fixed, other end pinned** (Fig. 9.2–1(c)). In this case $(Mx + N)$ is of the form $(p_{y1}x - m_1)$, and the boundary conditions require that deflection and slope are zero at

$x = 0$ and deflection and curvature are zero at $x = L$. The solution is indicated by solution (iv) in Fig. 9.2–3(c). Hence

$$\sqrt{\left(\frac{P_{cr}}{EI}\right)} L \approx \frac{3\pi}{2} \quad \text{since} \quad P_{cr} \approx 2.25 \frac{\pi^2 EI}{L^2}$$

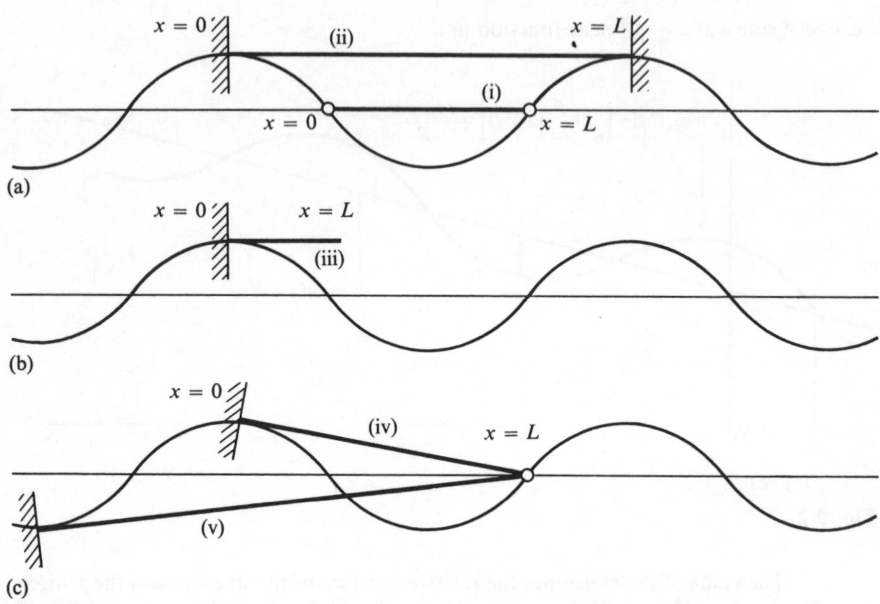

Fig. 9·2–3

This over-estimates the length slightly, and therefore over-estimates the actual buckling load. A correct analysis gives the solution:

$$P_{cr} = 2.04 \frac{\pi^2 EI}{L^2} = 2.04 P_E \qquad (9.2\text{–}7)$$

The solutions (i), (ii), (iii), and (iv) chosen in Fig. 9.2–3 are the simplest modes giving the lowest critical loads. Solution (v) in Fig. 9.2–3(c) is a higher order mode, corresponding to a higher critical load for case (c), and a series of critical loads could be similarly obtained for all the cases considered. It should also be noted that in each case the instability mode is defined in form, but not in magnitude—constant C is never determined.

As has already been pointed out, the cases of column instability considered here by no means cover all the possibilities, but do cover the ideal cases to which the structural designer most frequently approximates. More complex problems will be considered in a later section.

The preceding analyses have given values of critical load for various types of column, but the designer usually works with stresses. In the fundamental case of the pin-ended column, the critical stress is given by $\sigma_{cr} = P_{cr}/A$ (A = cross-sectional area) and

$$\frac{I}{A} = r^2 \quad (r = \text{radius of gyration})$$

then

$$\sigma_{cr} = \frac{\pi^2 EIr^2}{L^2 I} = \frac{\pi^2 E}{(L/r)^2} \qquad (9.2\text{–}8)$$

The critical stress in a column is therefore dependent only on the Young's modulus of the column material, and the **slenderness ratio** (L/r) of the column, and the form of the relationship is shown in Fig. 9.2–4 in which the non-dimensional ratio σ_{cr}/E is plotted against (L/r). The curve is asymptotic to both axes, but as the study so far applies to elastic stresses only, there is a horizontal limit corresponding to the ratio of (yield point/Young's modulus) for the material considered. This will be discussed further in Section 9.3.

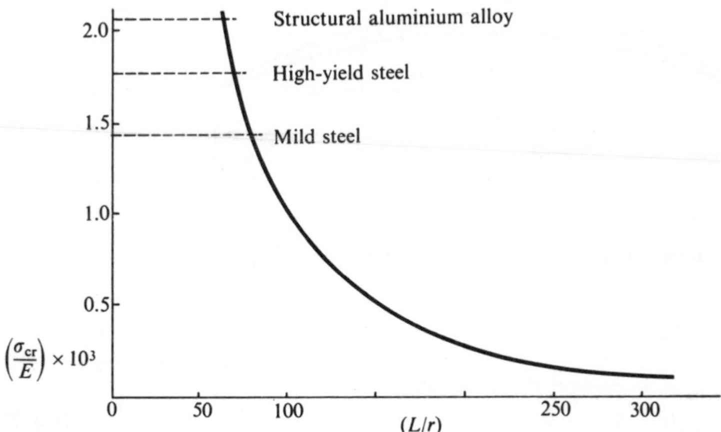

Fig. 9·2–4

Similar expressions to Eqn 9.2–8 can be set up for other end conditions. For the fixed-end column,

$$\sigma_{cr} = \frac{4\pi^2 E}{(L/r)^2} = \frac{\pi^2 E}{(L/2r)^2} \qquad (9.2\text{–}9)$$

and for the fixed-free column,

$$\sigma_{cr} = \frac{\pi^2 E}{4(L/r)^2} = \frac{\pi^2 E}{(2L/r)^2} \qquad (9.2\text{–}10)$$

The curve of Fig. 9.2–4 can therefore be used for a strut having any end conditions, so long as a modified value for the length is used. This modified value is termed the 'effective length' (l). Equation 9.2–8 is then written:

$$\sigma_{cr} = \frac{\pi^2 E}{(l/r)^2} \qquad (9.2\text{–}11)$$

and for the pin-ended column $l = L$; for the fixed-fixed column $l = L/2$; for the fixed-free column $l = 2L$; and for the fixed-pinned column $l = 0.699L$. These values of effective length are clearly indicated by the curves (i) to (v) in Fig. 9.2–3. Suitable values of effective length for design purposes for columns having other end conditions are listed in the design manuals and codes of practice.

9.3 Real behaviour of struts

The preceding sections have been concerned with the behaviour of ideal struts, but as was pointed out in Section 9.1, the real problem involves struts which are initially curved, are eccentrically loaded, and which may be stressed beyond the proportional limit.

Initial curvature. This is perhaps the most likely departure from the ideal in a real strut, and its effect will be examined first.

Figure 9.3–1 shows a pin-ended strut similar to that of Fig. 9.1–2. The only difference is that it has some departure from the straight (v_0) in the unloaded state. The problem, as before, is to study the state of equilibrium of the deflected form 1–3–2—can this be a neutral equilibrium state? $\bar{v}$ is the additional deflection due to P.

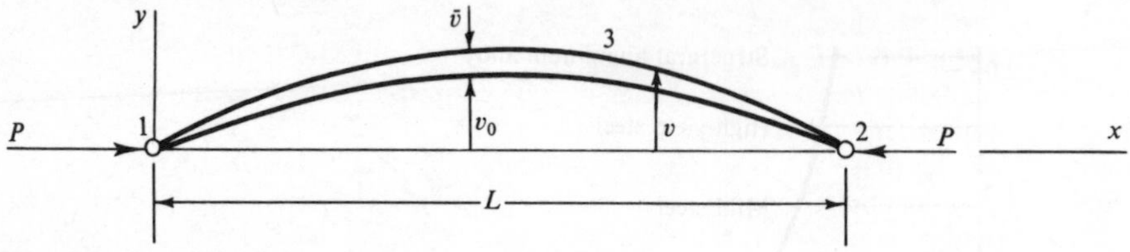

Fig. 9·3–1

The equation governing the equilibrium of the column is:

$$\frac{d^2\bar{v}}{dx^2} = -\frac{Pv}{EI} \tag{9.3–1}$$

or

$$\frac{d^2\bar{v}}{dx^2} + \frac{P}{EI}(\bar{v} + v_0) = 0 \tag{9.3–2}$$

or

$$\frac{d^2\bar{v}}{dx^2} + \frac{P\bar{v}}{EI} = -\frac{Pv_0}{EI} \tag{9.3–3}$$

The solution to this equation is therefore dependent on the initial deflection (v_0) of the column. Although, strictly speaking, it could only be determined by measurement, in general it could be expressed as an infinite trigonometric series. It is assumed here that it can be adequately represented by the first term:

$$v_0 = V \sin \frac{\pi x}{L} \tag{9.3–4}$$

where V is the central initial deflection. (This expression satisfies the boundary conditions, and is a not unreasonable approximation to the type of curve that might be expected in practice.) The solution to Eqn 9.3–3 is now:

$$\bar{v} = A \sin \sqrt{\left(\frac{P}{EI}\right)} x + B \cos \sqrt{\left(\frac{P}{EI}\right)} x + \frac{(P/EI)}{(\pi/L)^2 - (P/EI)} V \sin \frac{\pi x}{L} \tag{9.3–5}$$

Application of the boundary conditions gives:

$$B = 0 \quad (\bar{v} = 0 \text{ at } x = 0) \qquad A = 0 \quad (\bar{v} = 0 \text{ at } x = L)$$

$$\bar{v} = \frac{(P/EI)}{(\pi/L)^2 - (P/EI)} V \sin \frac{\pi x}{L} \tag{9.3–6}$$

If the simplification $\rho = P/P_E$ is made ($P_E = \pi^2 EI/L^2$) then Eqn 9.3–6 becomes:

$$\bar{v} = \frac{\rho}{1 - \rho} V \sin \frac{\pi x}{L} \tag{9.3–7}$$

If the total deflection of the strut is calculated it is

$$v = v_0 + \bar{v} = V \sin \frac{\pi x}{L} + \frac{\rho}{1 - \rho} V \sin \frac{\pi x}{L}$$

or

$$v = \frac{V}{1 - \rho} \sin \frac{\pi x}{L} \tag{9.3–8}$$

The initial deflection at the bar centre is thus multiplied by the ratio $1/(1 - \rho)$ as a result of axial compressive load, and tends to infinity as P approaches P_E. A typical series of plots of central lateral deflection against axial load is shown in Fig. 9.3–2. It can be seen from this that there is no discontinuity (point S in Fig. 9.1–5) in this case, and for all values of P less than P_E there is a unique value of lateral deflection: thus there is no neutral equilibrium situation. Moreover, as P approaches P_E the deflection will cease to be small compared to the length and yield of the strut material would begin. The influence of these effects will be discussed more fully later in this section.

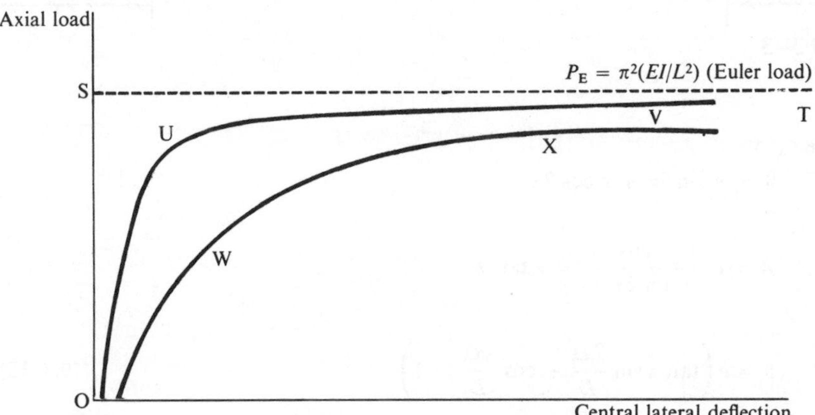

Fig. 9·3–2

In Fig. 9.3–2 the actual curve followed would depend on the magnitude of the initial deflection parameter V. The curve OWX would represent quite large initial distortions, while curve OUV would represent a much straighter strut. The curve OST can be seen as the plot for the ideal case of the perfectly straight strut.

Figure 9.3–2 shows the value of a knowledge of the Euler load for a strut, despite the fact that it does not represent real behaviour. It is an upper bound—real struts always have a lower ultimate load-carrying capacity, and the same fact will be seen later in other examples. In the solution of any instability problem the determination of the instability load corresponding to a perfectly elastic structure, of perfect geometrical form, is a valuable first step.

An eccentrically loaded strut. (Fig. 9.3–3) The equation of equilibrium is (cf. Eqn 9.1–1):

$$\frac{d^2v}{dx^2} = -\frac{P(v + e)}{EI} = -\frac{Pv}{EI} - \frac{Pe}{EI} \tag{9.3–9}$$

The solution is therefore (cf. Eqn 9.1–3):

$$v = A \sin \sqrt{\left(\frac{P}{EI}\right)} x + B \cos \sqrt{\left(\frac{P}{EI}\right)} x - e$$

This form may be greatly simplified by using the substitution:

$$\alpha = \frac{\pi}{2}\sqrt{\rho} = \frac{\pi}{2}\sqrt{\left(\frac{P}{P_{\mathrm{E}}}\right)} = \frac{\pi}{2}\sqrt{\left(\frac{P}{\pi^2 EI}\right)}L = \frac{1}{2}\sqrt{\left(\frac{P}{EI}\right)}L \tag{9.3–10}$$

then

$$v = A \sin\frac{2\alpha x}{L} + B\cos\frac{2\alpha x}{L} - e \tag{9.3–11}$$

The boundary conditions are: (1) $x = 0$, $v = 0$, hence $B = e$; (2) $x = L$, $v = 0$.

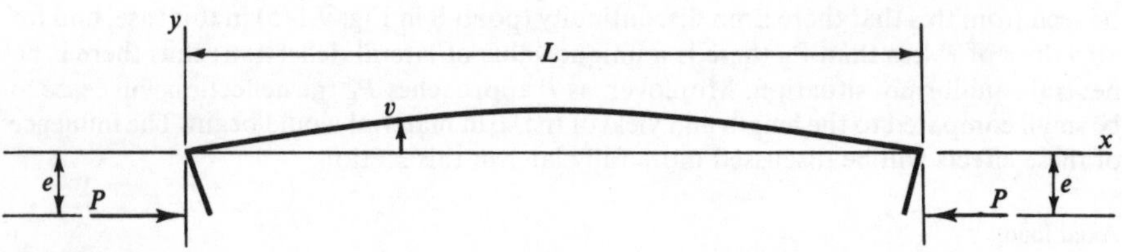

Fig. 9·3–3

Hence

$$0 = A\sin 2\alpha + e\cos 2\alpha - e$$

then

$$A = e\frac{(1 - \cos 2\alpha)}{\sin 2\alpha} = e\tan\alpha$$

and

$$v = e\left(\tan\alpha\sin\frac{2\alpha x}{L} + \cos\frac{2\alpha x}{L} - 1\right) \tag{9.3–12}$$

If, in addition to an eccentric end load the column has an initial curvature, it can be shown (see Example 9.3–1) that the total deflection is a combination of the results of Eqns 9.3–8 and 9.3–12.

$$v = \frac{V}{1-\rho}\sin\frac{\pi x}{L} + e\left(\tan\alpha\sin\frac{2\alpha x}{L} + \cos\frac{2\alpha x}{L} - 1\right) \tag{9.3–13}$$

For every value of axial load there is a unique value for lateral deflection, and there is thus no neutral equilibrium state. However, as P approaches P_{E}, α approaches $\pi/2$, ρ approaches unity, and the lateral deflection tends to infinity.

The maximum moment which can occur in the strut is at mid-span and is:

$$M_{\max} = P(e + v_{\max})$$

$$= P\left(\frac{V}{1-\rho} + \frac{e}{\cos\alpha}\right) \tag{9.3–14}$$

Example 9.3–1. Calculate the total lateral deflection of a pin-ended strut having an initial curvature v_0, where

$$v_0 = V\sin\frac{\pi x}{L}$$

(V is the initial central lateral deflection), and loaded by an eccentric longitudinal end load P (eccentricity e)

Initial lateral deflection $= v_0$.

Increase in lateral deflection due to $P = \bar{v}$.

Total lateral deflection $= \bar{v} + v_0 = v$ (see Fig. 9.3–1). Then

$$EI\frac{d^2\bar{v}}{dx^2} = \text{total bending moment}$$

or

$$\frac{d^2\bar{v}}{dx^2} = \frac{-P(v + e)}{EI} = \frac{-P(\bar{v} + v_0 + e)}{EI}$$

$$\frac{d^2\bar{v}}{dx^2} = \frac{-P\bar{v}}{EI} - \frac{PV}{EI}\sin\frac{\pi x}{L} - \frac{Pe}{EI}$$

The solution to this equation is:

$$\bar{v} = A\sin\sqrt{\left(\frac{P}{EI}\right)}x + B\cos\sqrt{\left(\frac{P}{EI}\right)}x + \frac{(P/EI)}{(\pi/L)^2 - (P/EI)}V\sin\frac{\pi x}{L} - e$$

where A and B are integration constants.

Making the substitution $\rho = P/P_E$ and $\alpha = \pi\sqrt{\rho}/2$, $\bar{v}$ becomes

$$\bar{v} = A\sin\frac{2\alpha x}{L} + B\cos\frac{2\alpha x}{L} + \frac{\rho}{1-\rho}V\sin\frac{\pi x}{L} - e$$

The boundary conditions are: (1) $x = 0$, $\bar{v} = 0$, hence $B = e$; (2) $x = L$, $\bar{v} = 0$. Hence

$$0 = A\sin 2\alpha + e\cos 2\alpha - e$$

thus

$$A = \frac{e(1 - \cos 2\alpha)}{\sin 2\alpha} = e\tan\alpha$$

then

$$\bar{v} = e\left[\tan\alpha\sin\frac{2\alpha x}{L} + \cos\frac{2\alpha x}{L} - 1\right] + \frac{\rho}{1-\rho}V\sin\frac{\pi x}{L}$$

Total lateral deflection $v = v_0 + \bar{v}$, then

$$v = e\left[\tan\alpha\sin\frac{2\alpha x}{L} + \cos\frac{2\alpha x}{L} - 1\right] + \frac{V}{(1-\rho)}\sin\frac{\pi x}{L}$$

i.e.

total deflection = (deflection of an initially straight strut loaded eccentrically) + (deflection of an initially curved strut loaded centrally)

Inelastic behaviour. So far in this chapter it has been assumed that the strut material remains linearly elastic, but the onset of nonlinear material behaviour is an obvious result of the growth of lateral deflections in a real strut. Figure 9.3–4 shows a simplified stress–strain curve for mild steel in tension and compression.

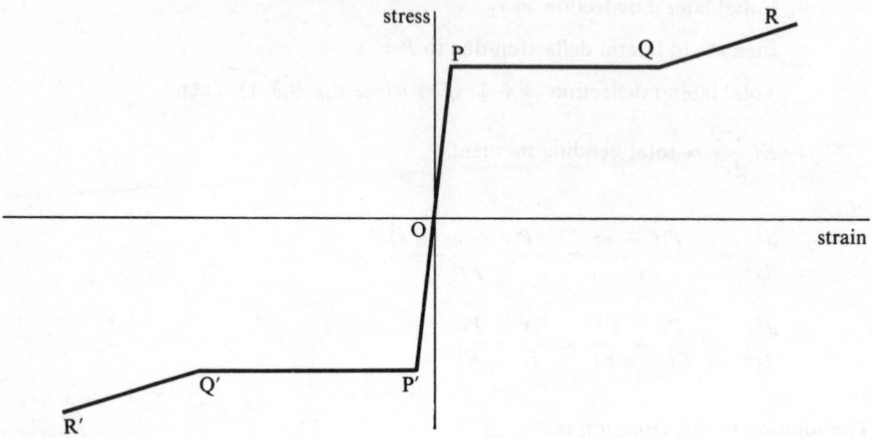

Fig. 9·3–4

In the range OP Hooke's law is obeyed, and the slope is Young's modulus (E). P is the yield point and PQ represents the large increase of strain on yielding with no increase of stress. Beyond P the material is plastic until Q, where the strain is more than ten times that at P, after which strain hardening occurs and there is again a rise of stress for increased strain, although at a lower rate than that of OP. The deflections that result when the material of a structure enters the range PQ are often so large that the structure becomes unserviceable, and there is therefore no need to consider the influence of strain hardening. For many purposes, steel can be regarded, therefore, as **elastic-plastic** as in Fig. 9.3–5.

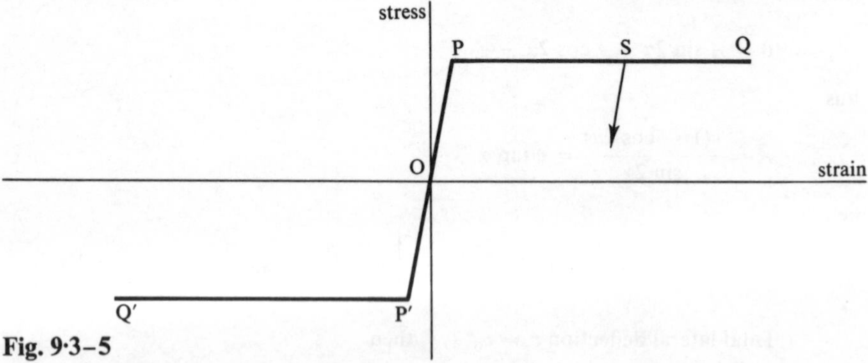

Fig. 9·3–5

The behaviour of mild steel in tension and compression is virtually the same, and a curve P′Q′, similar to PQ, can be drawn.

Most materials such as aluminium alloys and even some steels, do not have a sharply defined yield, as does mild steel. They follow stress–strain curves more like that of Fig. 9.3–6.

P and P′ define the end of a linearly elastic range—the **limits of proportionality**. Thereafter strain hardening begins, and the material is described as **elastic-strain hardening.** The slope of the curve OP (or OP′) is Young's modulus, and that beyond P (or P′) is termed the **tangent modulus** (E_T), and this will vary from point to point.

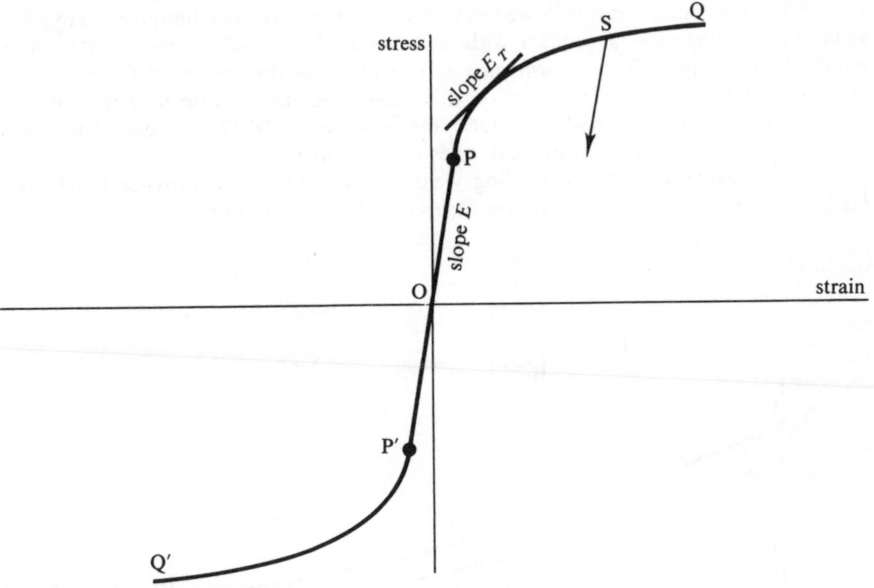

Fig. 9·3–6

If, for either elastic-plastic, or elastic-strain-hardening material, the stress is reduced after the point P has been reached (point S), the return curve is a straight line parallel to OP (of slope equal to E).

The post-elastic behaviour of a structure will obviously depend to some extent on the behaviour of its material, and subsequent remarks will be confined to the elastic-plastic case.

Figure 9.1–5 shows a graph of end load (P) against central deflection (Δ) for a perfectly straight pin-ended strut. There is no lateral deflection until the end load reaches the Euler value, but at this point any infinitesimal disturbance produces a deflection. However, as soon as a deflection occurs, the strut is subject to both axial and bending stress, and the highest stress will occur at mid-span on the concave side. If the stress here reaches yield, a curve such as ABCD results (Fig. 9.3–7).

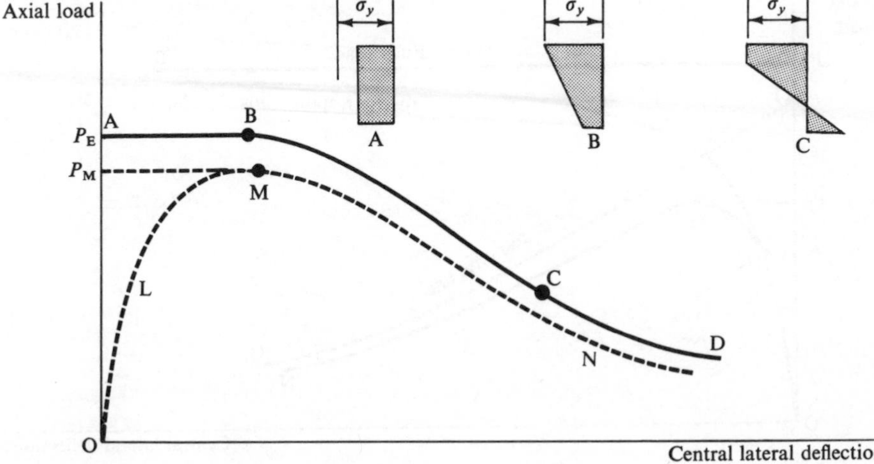

Fig. 9·3–7

A horizontal line is followed as in Fig. 9.1–5 until the maximum stress reaches yield. As straining now takes place without increase of stress, the stress distribution is modified from that at B to a form such as that at C. As the stresses at C represent a smaller total force on the cross-section than that at B, the increase of deflection can occur with a reduction in end load—hence the falling curve BCD. At higher deflections than point C, yield may occur on both sides of the strut.

If a strut is such that on loading, yield occurs over the whole cross-section before the Euler load is reached, it will follow a curve such as that in Fig. 9.3–8.

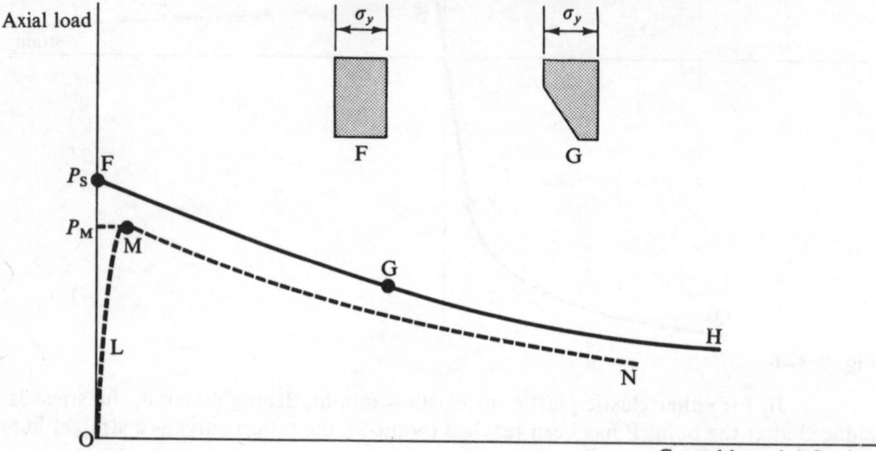

Fig. 9·3–8

Yield is reached at point F, and the load here is referred to as the **squash load**. If the slenderness ratio is extremely small (as in a compression strength test specimen) the load may increase further due to strain hardening, but if l/r is not so small, and if yield is reached sooner on one side of the member than the other, then lateral deflections can be expected to ensue. The resulting combination of bending and axial load produce the stress distributions and falling curve FGH in Fig. 9.3–8.

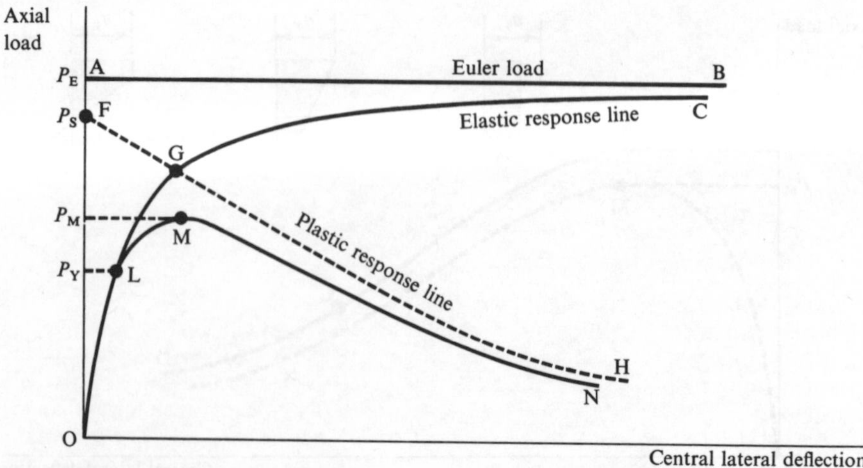

Fig. 9·3–9

The behaviour illustrated in Figs. 9.3–7 and 9.3–8 refers to perfectly straight struts, but the presence of initial imperfections leads to the modifications in these figures shown in dotted line (LMN). The maximum load the strut can carry (P_M) is reduced in both cases.

The foregoing results are combined in Fig. 9.3–9. The perfectly straight, elastic strut would follow the response curve OAB, while a strut initially non-straight would follow the elastic response curve OLGC. A straight strut which yielded below P_E would follow the plastic response line FGH, while such a strut with an initial curvature would follow OLGC as far as L only. At L yield stress is reached, and the curve falls away, reaching a maximum at M and then following a path similar to MN of Fig. 9.3–8 to N. OLMN represents the behaviour of the type of strut normally met in structural engineering.

Large deflections. It has been assumed so far that the lateral deflections of the strut at all times remain 'small', but the meaning of 'small' in this context has not been defined. The word is used to imply that the definition of curvature in Eqn 9.1–1 is given sufficiently accurately by d^2v/dx^2. When the exact equation for curvature is used, i.e.

$$\frac{d^2v/dx^2}{[1 + (dv/dx)^2]^{3/2}}$$

the shape of the elastic curve, known as the **elastica** and investigated by Euler, is as in Fig. 9.3–10. It can be seen that the load does rise above the Euler value, but only at deflections which would normally be quite large enough to produce yielding. To increase P only 1% above P_E would require the maximum deflection of a pin-ended strut to be about one-twelfth of the span.

In practice the assumptions of 'small-deflection' theory give an adequate description of actual behaviour.

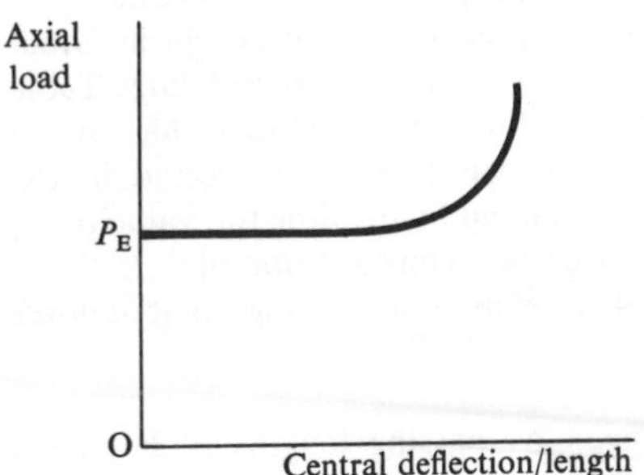

Fig. 9·3–10

9.4 The design of steel struts

Although this text is primarily concerned with analysis, the design of members carrying compressive loads is so dependent upon the ideas already discussed here, that it would seem most inappropriate not to mention it. As an example steel struts are considered, but many of the ideas can be applied to other materials. Early tests indicated that pin-ended struts, especially those having small l/r ratios, fail at loads much lower than the Euler load, and in a manner rather different from that indicated in Fig. 9.2–4. Many attempts were made to explain this and to obtain a suitable analytical expression, and one by Ayrton and Perry[1] is of particular importance. Their explanation was that shorter struts fail when the maximum stress at some point along the length (owing to axial load plus bending) reaches the yield stress (σ_y), i.e.

$$\sigma_y = \frac{P_{cr}}{A} + \frac{Mc}{I} \qquad (9.4\text{–}1)$$

where I is the least second moment of area, A the cross-sectional area, M the bending moment, c is the distance from the neutral axis to the extreme fibre, and P_{cr} is the failure load.

If the strut is assumed to be initially curved, but eccentricity of loading is ignored, then Eqn 9.4–1 becomes:

$$\sigma_y = \sigma_{cr}\left[1 + \frac{cV}{r^2(1 - \sigma_{cr}/\sigma_E)}\right] \qquad (9.4\text{–}2)$$

or

$$\sigma_{cr}^2 - \sigma_{cr}[\sigma_y + \sigma_E(1 + \eta)] + \sigma_E\sigma_y = 0 \qquad (9.4\text{–}3)$$

σ_{cr} and σ_E are the stresses corresponding to P_{cr} and P_E respectively; $\eta = (Vc/r^2)$ is termed the initial curvature parameter. Although eccentricity of loading is ignored, Perry assumed that the influence of a small, unavoidable eccentricity was equivalent to an increased initial curvature. The smaller root of Eqn 9.4–3 is:

$$\sigma_{cr} = \frac{\sigma_y + (1 + \eta)\sigma_E}{2} - \sqrt{\left\{\left[\frac{\sigma_y + (1 + \eta)\sigma_E}{2}\right]^2 - \sigma_y\sigma_E\right\}} \qquad (9.4\text{–}4)$$

This stress corresponds to the point L of Fig. 9.3–9.

Robertson,[2] as a result of an extensive series of very carefully conducted tests, suggested that η should be taken as proportional to the slenderness ratio, thus recognizing that slender struts are likely to be less straight than stiff ones and found a minimum value of 0.003 (l/r). Although η has been defined as Vc/r^2 this value of 0.003 (l/r) seems to give an extremely good approximation to actual test results in the range of practical sizes. It is important to recognize that Eqn 9.4–4 has nothing to do with instability—it only works because of the experimental determination of η. Equation 9.4–4 has come to be known as the Perry-Robertson formula and has been used for strut design in Great Britain for many years (BS449:1969—*The Use of Structural Steel in Building*, Table 17)[3]. Although η was initially taken as 0.003 (l/r), later work by Duthiel has shown that a better value is 0.3 $(l/100r)^2$, and this value was later incorporated in the standard. The actual allowable stress in BS449 is obtained by dividing the value of σ_{cr} in Eqn 9.4–4 by a 'load factor', and the value used is a constant one of 1.7.

Although the background to Eqn 9.4–4 seems entirely logical it ignores several additional and important factors.

a) Residual stresses. Steel columns are produced by passing lengths of hot steel through sets of profiled rollers. These gradually change the cross-sectional form from a solid square or rectangle of steel to the final shape required—frequently an **I** shape (to ensure a large value of r—Eqn 9.2–8). The steel is then allowed to cool to ambient temperature. The cooling rate is not uniform, however, with the toes of the flanges cooling more rapidly than more solid parts at the flange/web junction. This differential cooling results in stresses being 'locked' into the finished section. These are termed *residual stresses*. A typical residual stress pattern for a Universal Column is shown in Fig. 9.4–1(a) in which **T** denotes tension and **C** compression. A compressive stress of 100 N/mm^2 at the flange toes is not untypical. It must be remembered that these stresses are in addition to those resulting from the application of external loads.

The residual stresses in a member depend not only on the type of section but also on the manufacturing process. The residual stress pattern for a rectangular hollow section (RHS) is shown in Fig. 9.4–1(b). Stemenkovic[4] reported stresses of 75 N/mm^2 for such a section. Again, columns fabricated by the welding of individual plate

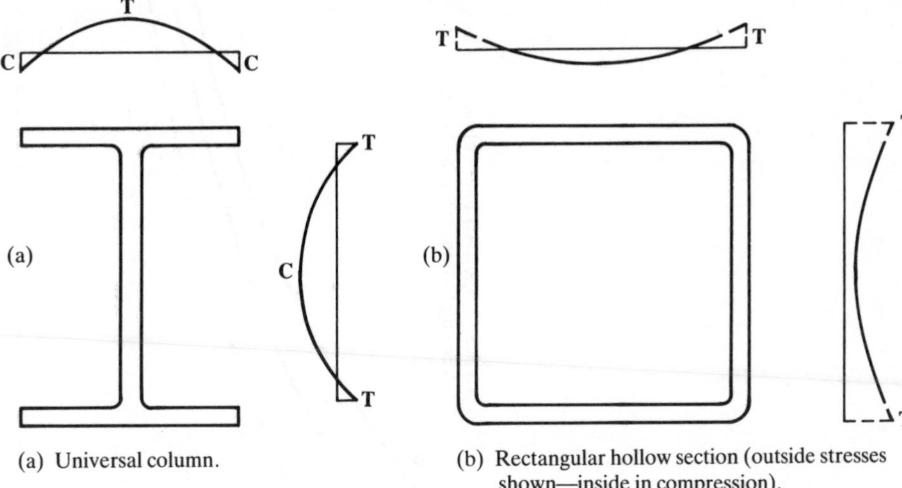

(a) Universal column.

(b) Rectangular hollow section (outside stresses shown—inside in compression).

Fig. 9·4–1

elements will have another type of residual stress pattern, and stress levels close to yield are possible.

b) Strain hardening. This effect is ignored in the Perry-Robertson formula. The result is that the curve is falling (if slowly) for the smallest values of slenderness ratio $\lambda(l/r)$, and the plateau shown for small λ in Fig. 9.2–4 is not represented.

c) Cross-section shape. Although an **I** section is convenient and widely used many other shapes are used—box section, angles, tees. Robertson proposed a value of $\eta = Vc/r^2 = 0.003\ (l/r)$. This implies $V/l = 0.003(r/c)$, suggesting that the initial lack of straightness for a column would depend on the shape of the cross-section. For typical sections the range of values of the ratio (r/c) can vary quite widely (a 1–2.5 variation).

As a result of international cooperation a set of European strut curves has been proposed, to provide a more rational basis for design than earlier rules. To cover the shortcomings of the Perry-Robertson equation, and other similar equations used in other countries, three different curves were provided, covering the range of sections and steel grades commonly used, and the axis about which buckling could occur. The curves were based on a complex theoretical analysis, using assumed residual stress patterns, backed by a very extensive testing programme.

Although the European curves were comprehensive, answering the deficiencies of the earlier design curves, they could not be represented by a simple mathematical expression. Designers must be grateful to Dwight[5] at Cambridge, who proposed a modified form of the Perry-Robertson equation that matched the European curves very closely, and this has been incorporated into the replacement to BS449–BS5950:1985—*Structural use of steelwork in building*[6]. This new standard provides four tables of compressive strengths for columns in a range of steels (σ_y = 225 N/mm^2 to 450 N/mm^2, corresponding to steels of grades 43, 50 and 55), based on Eqn 9.4–4. The factor η, or 'Perry Factor', is obtained from

$$\eta = 0.001a(\lambda - \lambda_0) \quad (\eta \nless 0) \qquad \text{and } \lambda_0 = 0.2\sqrt{\pi E/\sigma_y}$$

The term '*a*' is the 'Robertson constant', and taken the value of 2.0, 3.5, 5.5 and 8.0 respectively in the four tables. (It will be recalled that Robertson's own value was 3.0.)

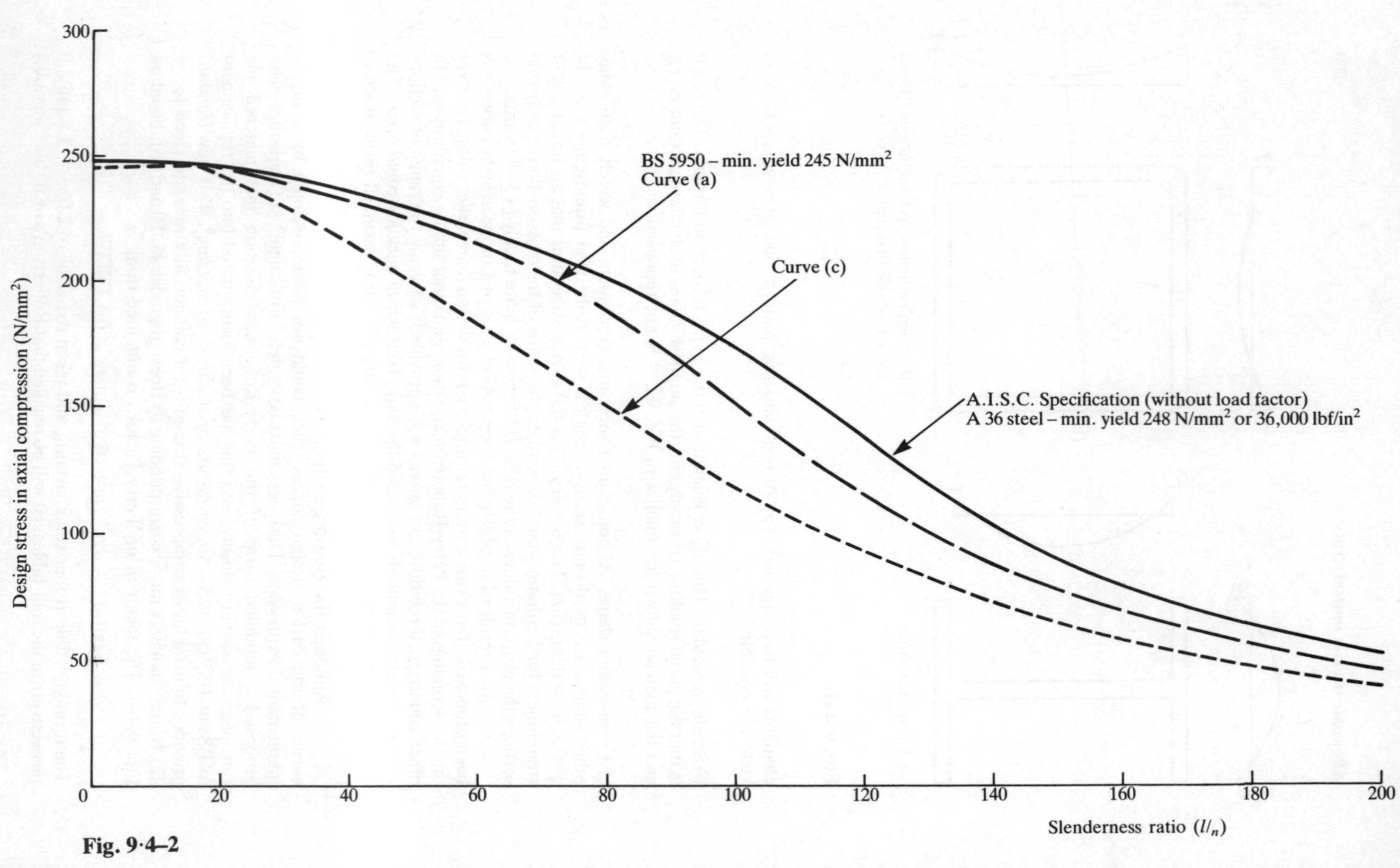
BS 5950 – min. yield 245 N/mm²
Curve (a)
Curve (c)
A.I.S.C. Specification (without load factor)
A 36 steel – min. yield 248 N/mm² or 36,000 lbf/in²
Design stress in axial compression (N/mm²)
300
250
200
150
100
50
0
20
40
60
80
100
120
140
160
180
200
Slenderness ratio (l/n)

Fig. 9·4–2

The term λ_0 is introduced to provide the required plateau on the curve for small λ. The fourth curve has been introduced to allow for the very high residual stresses present in particularly heavy column sections. Two curves are illustrated in Fig. 9.4–2. No 'load factor' is required with the BS5950 equation as this is a 'limit state' standard—the standard recognizes that structural elements may become unfit to fulfil their designed purpose in a variety of ways. Eqn 9.4–4 would define a 'limit-state' of yield in a column.

In the United States a different approach has been used. It has been recognized that the failure of very slender struts is due primarily to instability, and the form for the Euler curve is used for the higher l/r values. Shorter struts, however, fail as a result of yielding in the column material, and United States practice has always recognized that large residual stresses, due to both manufacture and fabrication, may exist in column sections. In addition, nonlinear strut behaviour begins when the limit of proportionality is reached, and this may be significantly lower than the yield stress. To include residual stress, limit of proportionality, and yield stress in one formula would, of course, be too complex, and the formula used by the American Institute for Steel Construction in its *Specification for the Design, Fabrication and Erection of Structural Steel* is of a parabolic form, which is an approximation suggested by Bleich.[7] It is;

$$\sigma_{cr} = \left[1 - \frac{(l/r)^2}{2C^2}\right]\sigma_y \tag{9.4–5}$$

where

$$C^2 = \frac{2\pi^2 E}{\sigma_y}$$

C is the value of (l/r) for which $\sigma_{cr} = \sigma_y/2$, and this is the assumed value of the limit of proportionality. The parabolic and Euler curves are tangential at this point.

In choosing a load factor, the AISC has recognized that the onset of yield in very stiff columns does not lead to immediate collapse, as strain hardening may provide a substantial strength reserve. No such reserve exists where failure is primarily due to instability. A cubic expression in (l/r) is given which leads to a variation from 1.67 at $(l/r) = 0$ to 1.92 at $(l/r) = C$. For all higher (l/r) values 1.92 is used.

The British and American design curves are compared in Fig. 9.4–2. American A36 steel has been used, and the nearest tabulated yield stress of BS5950, corresponding to Grade 43, although this figure is at the low end of the range for this steel. As the British standard is 'limit state' and the American 'allowable stress', the normal load factor has been omitted from the latter to allow a comparison to be made.

The formulae discussed here are all based on struts with pinned ends, but they can be applied to struts having other end conditions by use of the appropriate effective length. A useful survey of column design formulae has been carried out by Godfrey[8] and much work has been done in the United States by the Structural Stability Research Council.[9]

9.5 Complex struts

The method used so far for determining elastic buckling loads is quite suitable for struts with simple end conditions. For struts with complex supports a modified approach is preferable. Only perfect struts will be considered, as it has been pointed out that the elastic critical load is an upper limit to the actual collapse load, and its determination is a valuable starting-point in the study of any particular problem. Out-of-straightness, and eccentricity can be introduced as in Section 9.3. The solution of a single differential equation as in Section 9.2 is not practicable in complex problems, and a computer-based process is more advantageous.

As an example, the strut 1–3–2 in Fig. 9.5–1 will be considered. The member is initially straight with a pin at 2, but supported at 1 by an elastic foundation (stiffness k_1) which provides a restoring moment proportional to the rotation θ_1. Similarly at 3 there is a support which provides a lateral reaction force proportional to the displacement v_3 (stiffness k_3). For simplicity in this particular case the generally used notation for forces has been modified, as in Fig. 9.5–1. The problem is to determine the lowest value of axial

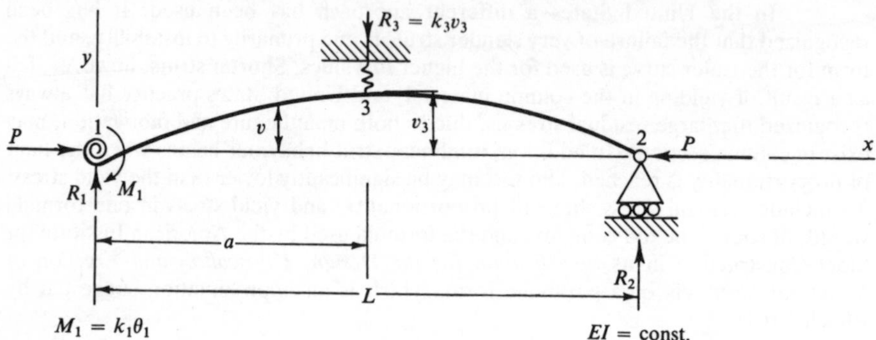

Fig. 9·5–1

force P which will cause the strut to become elastically unstable. The process is similar to that used already in Section 9.2—the equilibrium of a possible non-straight configuration is investigated. The bending moment equation for 1–3–2 changes at 3, and the two parts, 1–3 and 3–2 must be handled separately.

Member 1–3

$$\frac{d^2v}{dx^2} = -\frac{Pv}{EI} + \frac{R_1x}{EI} + \frac{M_1}{EI}$$

or

$$\frac{d^2v}{dx^2} + \frac{Pv}{EI} - \frac{R_1x}{EI} - \frac{M_1}{EI} = 0 \qquad (9.5\text{–}1)$$

then

$$v = A_1 \sin\frac{2\alpha x}{L} + B_1 \cos\frac{2\alpha x}{L} + \frac{R_1x}{P} + \frac{M_1}{P} \qquad (9.5\text{–}2)$$

A_1 and B_1 are integration constants, and α was defined in Eqn 9.3–10 as $[(\pi/2).\sqrt{(P/P_E)}]$. There are three boundary conditions:

$$x = 0,\quad v = 0 \qquad x = 0,\quad dv/dx = \theta_1 \qquad x = a,\quad v = v_3$$

and their insertion leads to three equations in A_1, B_1, R_1, M_1, and v_3 (see Eqn 9.5–4).

Member 3–2. The equilibrium equation is the same as that for 1–3, but a term must be added for R_3. The solution is:

$$v = A_2 \sin\frac{2\alpha x}{L} + B_2 \cos\frac{2\alpha x}{L} + \frac{R_1x}{P} + \frac{M_1}{P} - \frac{R_3(x-a)}{P} \qquad (9.5\text{–}3)$$

The two boundary conditions:

$$x = a,\quad v = v_3 \quad \text{and} \quad x = L,\quad v = 0$$

provide a further two equations. There are a total of seven unknown quantities (except for P) and the two further equations necessary are obtained by recognizing the continuity of slope at 3, and the existence of a zero bending moment at 2. The seven equations are summarized in Eqn 9.5–4.

$$\begin{bmatrix} 0 & 1 & 0 & 0 & 0 & \frac{k_1}{P} & 0 \\ \frac{2\alpha}{L} & 0 & 0 & 0 & \frac{1}{P} & -1 & 0 \\ \sin\frac{2\alpha a}{L} & \cos\frac{2\alpha a}{L} & 0 & 0 & \frac{a}{P} & \frac{k_1}{P} & -1 \\ 0 & 0 & \sin\frac{2\alpha a}{L} & \cos\frac{2\alpha a}{L} & \frac{a}{P} & \frac{k_1}{P} & -1 \\ 0 & 0 & \sin 2\alpha & \cos 2\alpha & \frac{L}{P} & \frac{k_1}{P} & \frac{-k_3(L-a)}{P} \\ \cos\frac{2\alpha a}{L} & -\sin\frac{2\alpha a}{L} & -\cos\frac{2\alpha a}{L} & \sin\frac{2\alpha a}{L} & 0 & 0 & \frac{Lk_3}{2\alpha P} \\ 0 & 0 & 0 & 0 & L & k_1 & -k_3(L-a) \end{bmatrix} \begin{bmatrix} A_1 \\ B_1 \\ A_2 \\ B_2 \\ R_1 \\ \theta_1 \\ v_3 \end{bmatrix} = 0$$

or

$$\mathbf{CX} = \mathbf{0} \qquad (9.5\text{–}4)$$

This is a homogeneous set of equations (right-hand sides all zero) and if non-zero values are to exist for the unknowns **X** then the determinant of the coefficient matrix **C** must equal zero. If this can be so, then a non-straight equilibrium form can exist. In any actual problem the values of L, a, k_1, k_3, E, and I would be known, and the problem therefore reduces to that of finding a value of P (or α) which would make the determinant of **C** vanish. This value of P is P_{cr} for the strut.

It would be difficult to determine a general solution to Eqn 9.5–4 and a suitable approach using the computer is to calculate the determinant of **C**, varying the axial load in a step-by-step process until a zero value is reached. The value of the determinant is being used here to check the singularity of matrix **C** and alternative approaches are discussed in Section 13.5.

A suggested flow chart is shown in Fig. 9.5–2. It requires the values of k_1, k_3, L, a, E, and I and an initial value of P and a step size as data, and the form of **C** must be set up in the program.

Fig. 9.5–3 shows a series of curves illustrating the variation of P_{cr} with varying k_1 and k_3 in the problem of Fig. 9.5–1. The properties are given on the figure.

A number of interesting conclusions may be drawn from the results of this example.

(1) If k_1 is zero, the strut is pin-ended, with a central lateral support of stiffness k_3. With k_3 also zero the critical load (P_{cr}) is the Euler load (246.7 N)—point A in Fig. 9.5–3. As k_3 is increased P_{cr} rises, the relationship being practically linear. At k_3 = 4 N/mm it reaches a peak value of 987 N at point B, and rises no further for increased support stiffness (987 N is $4P_E$, the second critical load for a pin-ended strut). It corresponds to buckling in a full sine wave and there is therefore no displacement at support 3. This is the maximum value that can be achieved. A comparatively small value of k_3 is thus enough to raise the buckling load by the maximum possible of four times.

(2) Figure 9.5–3 also shows a series of curves for steadily increasing values of

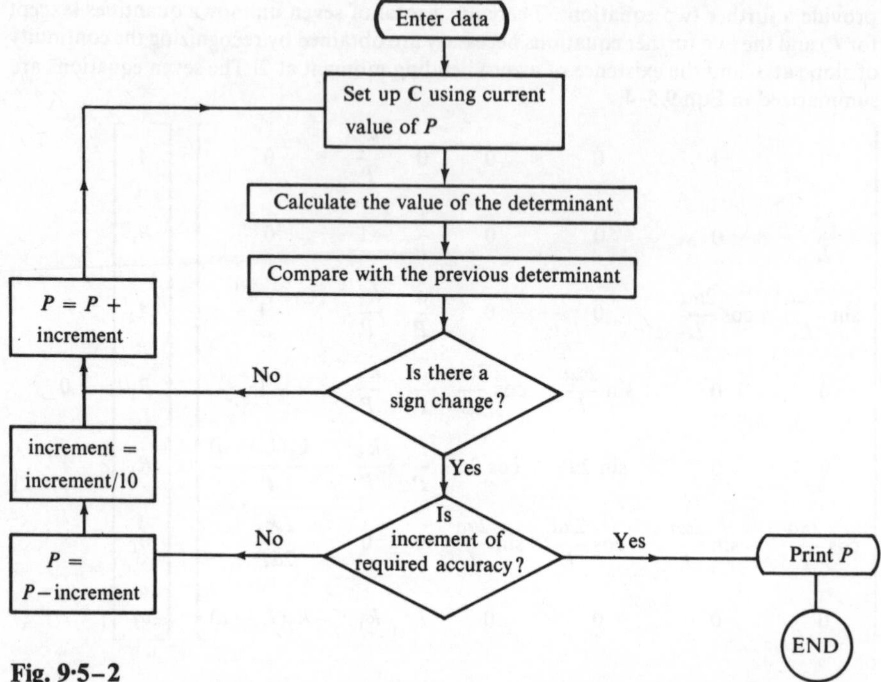

Fig. 9·5–2

support stiffness k_1. They all pass through the same point (B) but continue to rise for larger k_3, although fairly slowly. The significant point here is that the values used for k_1 are extremely large, and the influence on P_{cr} (at least for the range of k_3 shown) is very small. If k_3 is zero, P_{cr} for $k_1 = \infty$ is 505 N (Eqn 9.2–7) (point C in Fig. 9.5–3), but a value of $k_1 = 10^5$ N mm/rad gives a P_{cr} of 366 N only (point D). An end which is nominally built in must thus be very stiff indeed to give anything approaching a fully encastré effect. The value of P_{cr} corresponding to full rigidity at 1 and 3 ($k_1 = k_3 = \infty$) is indicated on the right of the figure. It can be seen that only with very high values of k_1 and k_3 indeed could this limit be approached.

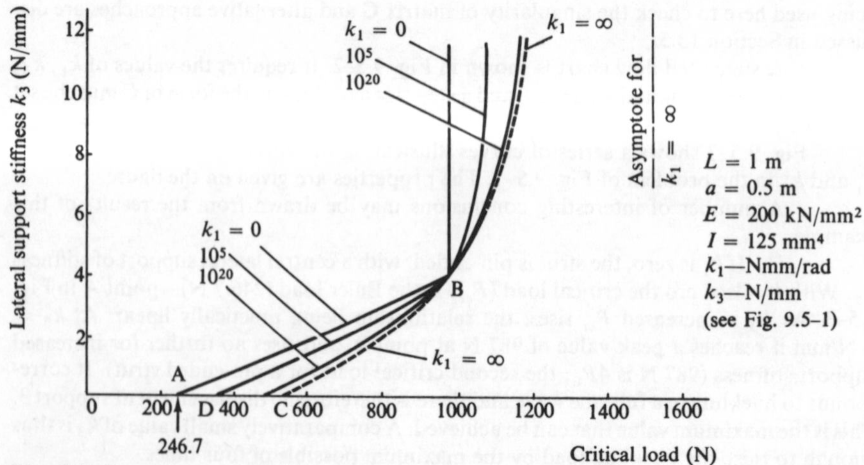

Fig. 9·5–3

9.6 Lateral torsional buckling of beams

Beams are members designed to carry loads by bending. In Fig. 9.6–1 an **I** beam simply supported at the ends is shown carrying a central point load. There is no support to the beam laterally. The top flange is in compression and can therefore be likened to the struts discussed in the earlier parts of the Chapter. Like a strut it can fail by buckling. Such a flange, however, cannot buckle upwards because of the presence of the web—it must move sideways—and in doing so the web and lower flange must rotate. The lower flange cannot, of course, move sideways freely as it is restrained by the tensile force it carries, like a taut string. The section therefore twists, being restrained in some ways at the support, leading to a form of instability known as **lateral torsional buckling**. Fig. 9.6–2 shows an **I** beam cantilever that has failed in this way. It could be regarded as representing the behaviour of half of the beam of Fig. 9.6–1 (reversed).

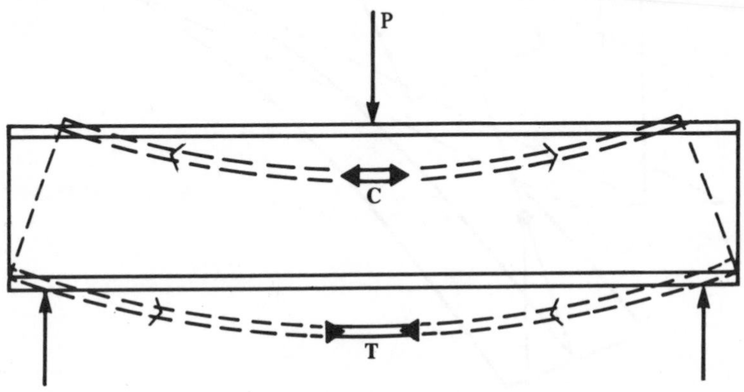

Fig. 9·6–1

To determine the critical buckling moment for a beam consideration will be given to a simply supported **I** beam symmetrical about *xx* and *yy* axes, carrying a pure moment (Fig. 9.6–3(a)). If it is assumed that there is some non-straight equilibrium configuration (adopting the approach of Section 9.1), then a plan view of the deflected shape is shown in Fig. 9.6–3(b). A typical cross-section is shown in fig. 9.6–3(c) where a new set of axes ($\xi\eta\zeta$) based on the delected position with origin at the centroid and coincident shear centre, is defined. Using the relations of Section 4.18, the following equations are obtained

$$EI_\xi \frac{d^2v}{dz^2} = -M_\xi \quad \text{and} \quad EI_\eta \frac{d^2u}{dz^2} = M_\eta \tag{9.6–1}$$

From Fig. 9.6–3(c) it is seen that if the end moment M_0 acts in the *yz* plane then there is a component acting in the $\xi\eta$ plane producting a torsional moment about the ζ axis, M_T. M_T can be considered as having two components, M_{T1} and M_{T2}. M_{T1} is the torque associated with normal St Venant torsion (Section 11.11) and, from Eqn 11.11–19,

$$M_{T1} = GJ\frac{d\phi}{dz} \tag{9.6–2}$$

where J is the torsion constant for the cross-section.

The second part of the torque is found by consideration of the bending of the flanges in their own plane. Fig. 9.6–4(a) shows an elevation of an **I** section cantilever fixed at end B and carrying a torque T at end A. Fig. 9.6–4(b) shows a plan view of the top flange, indicating that the end A must rotate about the *y* axis. Fig. 9.6–4(c) shows a

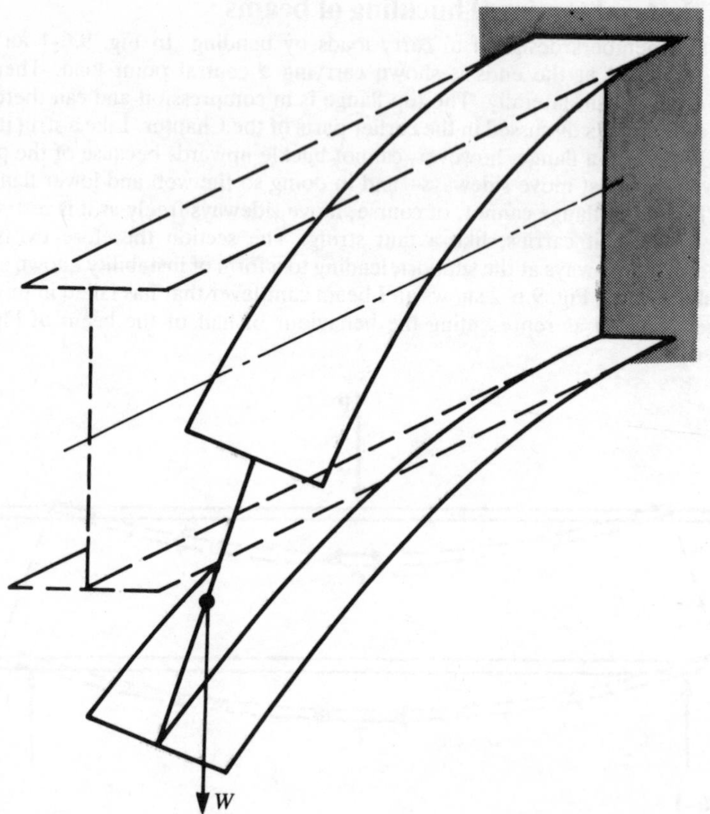

Fig. 9·6–2

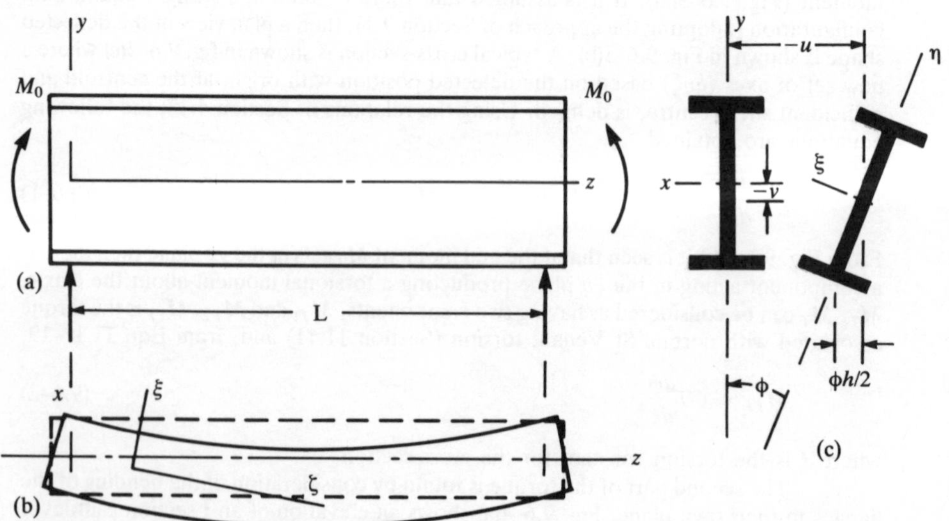

Fig. 9·6–3

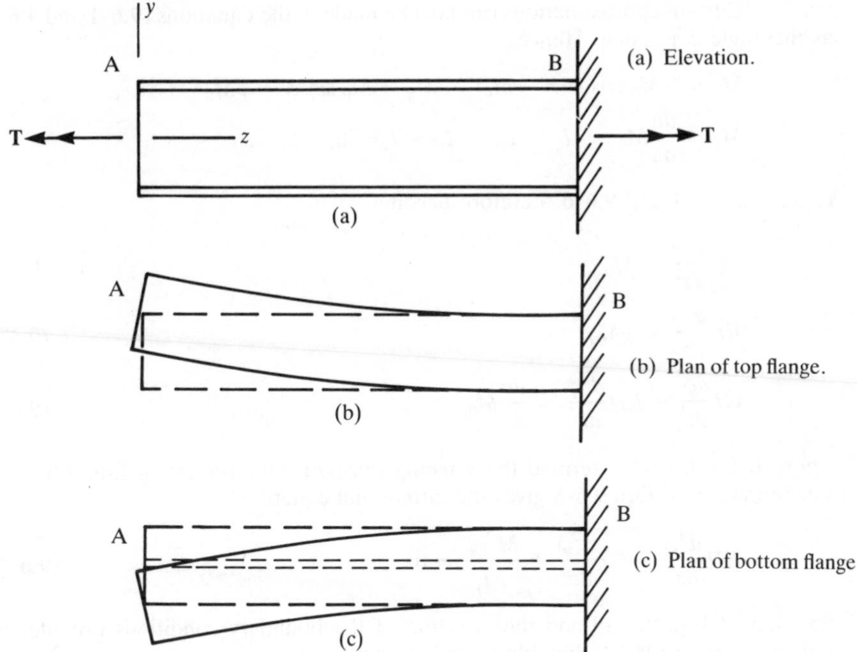

Fig. 9·6–4

plan view of the bottom flange. Again the end A must rotate about the y axis, but this time in the opposite sense. The cross-section is clearly deformed and the initially plane cross-section at A is now no longer plane. This phenomenon, associated with the twisting of all non-circular sections, is known as **warping**. In the case of the cantilever of Fig. 9.6–4, however, the warping is not free to occur at all cross-sections—none can occur at end B, for instance, where the beam is rigidly held. The warping has been suppressed and this results in the creation of additional longitudinal stresses, or warping stresses, in the flanges. If the depth of the section is h, then the lateral deflection of the lower flange from the shear centre is $\phi h/2$ (Fig. 9.6–3(c)). If M_f is the moment in the flange, and I_f is the second moment of area of the flange about the y axis, then

$$M_f = EI_f \frac{h}{2} \frac{d^2\phi}{dz^2} \tag{9.6–3}$$

The associated shear force in the flange V_f is then

$$V_f = -\frac{dM_f}{dz} = -EI_f \frac{h}{2} \frac{d^3\phi}{dz^3} \tag{9.6–4}$$

In the top flange there is an indentical shear force, but of opposite sign, and together they form the second part of the torsional moment

$$M_{T2} = -EI_f \frac{h^2}{2} \frac{d^3\phi}{dz^3} \tag{9.6–5}$$

The total torsional moment is therefore

$$M_T = GJ\frac{d\phi}{dz} - EI_f \frac{h^2}{2} \frac{d^3\phi}{dz^3} \tag{9.6–6}$$

Certain approximations can now be made in the equations (9.6–1 and 9.6–6) as the angle ϕ is small. Hence

$$M_\xi = -M_0 \cos\phi = -M_0 \qquad M_\eta = M_0 \sin\phi = \phi M_0$$

$$M_\zeta = \frac{du}{dz}M_0 \qquad I_\xi = I_x \qquad I_\eta = I_y = 2I_f$$

Equations 9.6–1 and 9.6–6 therefore become

$$EI_x\frac{d^2v}{dz^2} = M_0 \tag{9.6–7}$$

$$EI_y\frac{d^2u}{dz^2} = \phi M_0 \tag{9.6–8}$$

$$GJ\frac{d\phi}{dz} - EH\frac{d^3\phi}{dz^3} = \frac{du}{dz}M_0 \tag{9.6–9}$$

where H $(= I_y h^2/4)$ is termed the **warping constant**. Differentiating Eqn 9.6–9 and substituting from Eqn 9.6–8 gives the differential equation

$$EH\frac{d^4\phi}{dz^4} - GJ\frac{d^2\phi}{dz^2} + \frac{M^2{}_0\phi}{EI_y} = 0 \tag{9.6–10}$$

Solution of Eqn 9.6–10 and the insertion of the boundary conditions provides the critical moment $(M_0)_{cr}$. Suitable boundary conditions are

$$\phi = \frac{d^2\phi}{dz^2} = 0 \text{ at } z = 0 \text{ and } L \tag{9.6–11}$$

These imply that the ends of the beam cannot rotate about the z axis, but are free to warp. The solution gives

$$(M_0)_{cr} = \frac{\pi}{L}\sqrt{EI_yGJ}\sqrt{1 + \frac{\pi^2EH}{L^2GJ}} \tag{9.6–12}$$

If I_y is taken as $Ar_y{}^2$ (r_y is the radius of gyration of the section about the y–y axis) then

$$(M_0)_{cr} = \frac{\pi}{(L/r_y)}\sqrt{EAGJ}\sqrt{1 + \frac{\pi^2EH}{L^2GJ}} \tag{9.6–13}$$

and it can be seen that, just as for a strut, the capacity of a slender beam is inversely proportional to the slenderness ratio, here (L/r_y) for bending about the x–x axis. The same arguments concerning slenderness ratio that are applied to struts can be applied to beams.

Eqn 9.6–13 gives the critical moment for elastic buckling of a symmetrical **I** section under pure moment. It corresponds, for beams, to the Euler buckling equation (Eqn 9.1–6) for struts. Clearly, however, modifications are required to cover the wide range of variables that can occur in practice, and are not included in the analysis here (for instance, load distribution, support conditions—affecting the boundary conditions of Eqn. 9.6–11 or unequal flanges) before it can be applied to design. With certain approximations the equation was the basis for beam design in BS449. In the limit state BS5950 the beam design formula is based, like the strut formula, on a modified 'Perry' approach. This covers the whole range of sections from the slender ones that can fail by lateral torsional buckling to the stocky ones that fail by yielding, as discussed in Section 14.3. For a full treatment of the background to the BS5950 equation readers are referred to more specialist texts.[10,11]

9.7 Solution of stability problems—virtual work approach

The instability problems examined so far have been solved by means of a second-order differential equation. An examination of its solution has led to a value for critical load. The use of virtual work allows solutions to be obtained for much more involved problems than have been tackled so far—e.g. plates, arches, shells. In these cases any differential equation of equilibrium which can be set up (similar to Eqn 9.5–1) may be too involved to solve explicitly.

The pin-ended prismatic strut 1–2 of Fig. 9.7–1 is assumed to be loaded to the critical level $P = P_{cr}$, and to be in a state of neutral equilibrium. If it is given a virtual lateral displacement, the ends approach by an amount δ, and the consequent bending of 1–2 results in a change in the amount of strain energy U. If the strut is assumed to have a deflection curve $v = f(x)$, as yet unknown, and deflections are assumed small, then:

$$U = \int_0^L \frac{M^2}{2EI}\,dx = \int_0^L \frac{(EIv'')^2\,dx}{2EI} = \frac{EI}{2}\int_0^L (v'')^2\,dx \qquad (9.7\text{–}1)$$

where $v'' = d^2v/dx^2$. (If EI is retained within the integral sign, then the expression applies to non-uniform members.)

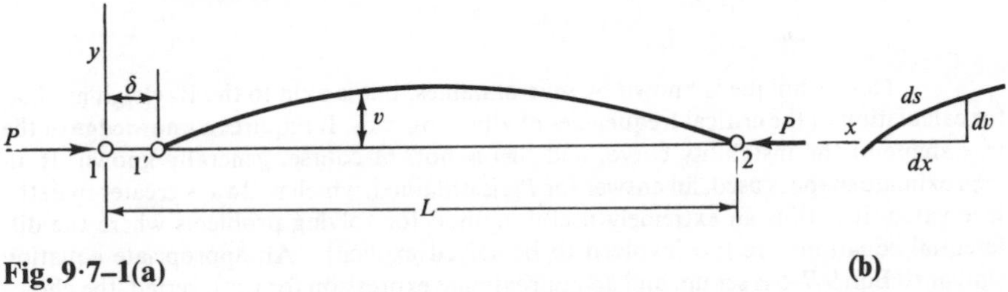

Fig. 9·7–1(a) **(b)**

The energy of compression ($P^2L/2AE$) should really be added to the bending energy, but as it remains constant so long as P is constant, it need not be considered.

When the end 1 of the strut moves to 1′, through the distance δ, the curve 1′–2 must remain the same length as in the straight position 1–2, as the average compressive stress in each (P/A) is the same. Therefore, for the element ds of Fig. 9.7–1(b),

$$ds = \sqrt{(dx^2 + dv^2)} = dx\sqrt{\left[1 + \left(\frac{dv}{dx}\right)^2\right]} = \sqrt{[1 + (v')^2]}\,dx$$

then

$$L = \int_0^L ds = \int_\delta^L dx[1 + (v')^2]^{1/2}$$

$$= \int_0^L [1 + (v')^2]^{1/2}\,dx - \int_0^\delta [1 + (v')^2]^{1/2}\,dx$$

$$\approx \int_0^L \left[1 + \frac{(v')^2}{2}\right] dx - \int_0^\delta \left[1 + \frac{(v')^2}{2}\right] dx$$

if $v' \ll 1$ and powers of v' higher than the second are ignored. Hence

$$L = L + \frac{1}{2}\int_0^L (v')^2\,dx - \delta - \frac{1}{2}\int_0^\delta (v')^2\,dx$$

The last term may be ignored as v' is small, and the integration is carried out over the interval 1 to 1′ only. Then

$$\delta = \frac{1}{2}\int_0^L (v')^2\,dx \tag{9.7–2}$$

The work done by the external loads is thus:

$$\text{work} = P\delta = \frac{P}{2}\int_0^L (v')^2\,dx \tag{9.7–3}$$

(This can be looked upon as a loss of 'potential energy' on the part of the loads—V.)

By the principle of virtual work, this work is equal to the internal strain energy gained ($V = U$), thus

$$\frac{P}{2}\int_0^L (v')^2\,dx = \frac{EI}{2}\int_0^L (v'')^2\,dx \tag{9.7–4}$$

and

$$P_{cr} = \frac{\displaystyle\int_0^L EI(v'')^2\,dx}{\displaystyle\int_0^L (v')^2\,dx} \tag{9.7–5}$$

This technique is known by several names, but is akin to the Rayleigh method for calculation of the critical frequencies of vibrating rods. It requires a knowledge of the true shape of the instability curve, and this is not, of course, generally known. If an approximate shape is used, an answer for P_{cr} is obtained, which is always greater than the true value. It is thus an extremely useful method for solving problems where the differential equations are too involved to be solved explicitly. An appropriate equation similar to Eqn 9.7–5 is set up, and an approximate expression for v is inserted, the choice being based on experimental observation and the known boundary conditions. The process is illustrated in Example 9.7–1.

Proof of the statement that the approximate value is an upper bound to the true value (provided that the approximate solution satisfies the geometrical boundary conditions of the problem) is too involved to be given here, and readers are referred to more advanced texts on instability (see Bleich,[4] p. 69).

Example 9.7–1. The critical compressive load for the strut already shown in Fig. 9.5–1 will be calculated using the virtual work approach. The end 1 is supported on an elastic foundation (stiffness k_1) which provides a restraining moment M_1 proportional to the rotation θ_1, and a lateral support (stiffness k_3) taken at mid-span in this case provides a restraining force proportional to the displacement v_3.

It will be assumed that the buckled form of the strut can be represented by a fourth-order polynomial:

$$v = Ax + Bx^2 + Cx^3 + Dx^4 \tag{9.7–6}$$

and the first stage of the calculation is to determine ratios between the constants A, B, C, and D which will allow this expression to satisfy the geometrical boundary conditions of zero displacement at either end, zero curvature at end 2, and a curvature proportional to the restoring moment at end 1.

$x = 0,\quad v = 0$ (satisfied by choice of v)

$$x = L,\quad v = 0 \quad \text{hence} \quad 0 = AL + BL^2 + CL^3 + DL^4 \tag{9.7–7}$$

$x = L,\quad v'' = 0$

(end 2 is pinned, and can thus carry no bending moment). Then

$$0 = 2B + 6CL + 12DL^2 \tag{9.7–8}$$

$$x = 0, \quad v' = \theta_1 \quad \text{hence} \quad \theta_1 = A \quad \text{(see Fig. 9.5–1)} \tag{9.7–9}$$

$$x = 0, \quad EIv'' = \text{bending moment at end 1} = -M_1 = -k_1\theta_1$$

then

$$EI.2B = -k_1 A$$

and

$$A = -\frac{2EI}{k_1} B \tag{9.7–10}$$

A is now substituted in Eqn 9.7–7, and D is eliminated between Eqns 9.7–7 and 9.7–8, providing

$$B = -\frac{3L^2 k_1 C}{5Lk_1 - 12EI} \tag{9.7–11}$$

From Eqn 9.7–10

$$A = \frac{6EIL^2 C}{5Lk_1 - 12EI} \tag{9.7–12}$$

And from Eqn 9.7–7

$$D = -\frac{2(Lk_1 - 3EI)C}{L(5Lk_1 - 12EI)} \tag{9.7–13}$$

The assumed displacement can now be rewritten:

$$v = \frac{C}{L(5Lk_1 - 12EI)} [6EIL^3 x - 3L^3 k_1 x^2 + L(5Lk_1 - 12EI)x^3 - 2(Lk_1 - 3EI)x^4] \tag{9.7–14}$$

As shown in Eqn 9.7–3 the work done by the end load during buckling is

$$\frac{P}{2}\int_0^L (v')^2\, dx$$

and the energy stored in the strut is

$$\frac{EI}{2}\int_0^L (v'')^2\, dx \quad \text{(Eqn 9.7–1)}$$

There are additional terms in this case, however, as energy is also stored in the elastic supports. That is, energy stored in support 1 is

$$\tfrac{1}{2}M_1\theta_1 = \frac{k_1\theta_1^2}{2}$$

and energy stored in support 3 is

$$\tfrac{1}{2}R_3 v_3 = \frac{k_3 v_3^2}{2}$$

The equating of the work done by the load P and the stored energy, gives the relation

$$\frac{P_{cr}}{2}\int_0^L (v')^2\,dx = \frac{EI}{2}\int_0^L (v'')^2\,dx + \frac{k_1}{2}\theta_1^2 + \frac{k_3}{2}v_3^2 \tag{9.7–15}$$

where P is assumed to be at its critical level for this to be true. The quantities,

$$\int_0^L (v')^2\,dx, \quad \int_0^L (v'')^2\,dx, \quad \theta_1, \quad \text{and} \quad v_3$$

are thus required, and these are calculated from Eqn 9.7–14:

$$\int_0^L (v')^2\,dx = \frac{L^5C^2}{(5Lk_1 - 12EI)^2}\,[17.5(EI)^2 - 4.5EILk_1 + 0.34(Lk_1)^2]$$

$$\int_0^L (v'')^2\,dx = \frac{L^3C^2}{(5Lk_1 - 12EI)^2}\,[173(EI)^2 - 43.2EILk_1 + 7.2(Lk_1)^2] \tag{9.7–16}$$

$$\theta_1 = \frac{6EIL^2C}{5Lk_1 - 12EI} \quad \text{(Eqns 9.7–9 and 9.7–12)} \tag{9.7–17}$$

$$v_3 = \frac{(15EI - 2Lk_1)L^3C}{8(5Lk_1 - 12EI)} \tag{9.7–18}$$

Substitution of these quantities leads to the final equation in P_{cr}:

$$P_{cr} = \frac{EI}{L^2}\,\frac{[173(EI)^2 - 7.2EILk_1 + 7.2(Lk_1)^2 + (k_3L^3/64EI)(15EI - 2Lk_1)^2]}{[17.5(EI)^2 - 4.5EILk_1 + 0.34(Lk_1)^2]} \tag{9.7–19}$$

The simplest case occurs when $k_1 = k_3 = 0$, and the member is then a pin-ended strut. The result is

$$P_{cr} = 9.89\frac{EI}{L^2}$$

which differs from the Euler load by about 0.1%, but is on the high side. Another simple case occurs when $k_1 = \infty$ and $k_3 = 0$ (case (c) of Section 9.2). Here, Eqn 9.7–19 gives

$$P_{cr} = 21.2\frac{EI}{L^2}$$

or about 5% too large. A range of other values for P_{cr} for various values of the stiffnesses k_1 and k_3 using the data of Fig. 9.5–3 are given in the following table.

Table 9.7–1

	k_1			
k_3	0	100	10^{20}	∞
0	247	247	530	530
2	655	655	897	897
4	1053	1053	1267	1267
10	2260	2260	2380	2380
∞	∞			

Values of P_{cr} in newtons. Comparison between these figures and the curves of Fig. 9.5–3 shows that the assumed displacement polynomial (Eqn 9.7–7) predicts the buckling load fairly well in the range A–B of k_3, the approximation being best for small values of k_1. Beyond the point B, however, it can be seen that the prediction is much poorer. It is important to note that all the values of Table 9.7–1 are greater than the correct values of Fig. 9.5–3.

9.8 Post-buckling behaviour

The coverage of this Chapter has been primarily concerned so far with the determination of elastic critical buckling loads for struts. Fig. 9.1–5 shows a graph of the central lateral deflection of a pin-ended strut under axial loading. When the buckling load is reached (P_E in this case) the lateral deflection is shown as increasing without change in load. The structure remains in neutral equilibrium, certainly within the limits of small deflection theory. Only when material yielding is considered (for example Fig. 9.3–8) is there the possibility of a fall in capacity. There are cases, however, where the equilibrium after elastic buckling may not be neutral, but may be either stable or unstable. An example given by Thompson[12] is shown in Fig. 9.8–1. It can be likened to a shallow arch, pinned at A, B and C, and having springs of equal stiffness k in AB and BC. A vertical load is applied at B, as shown.

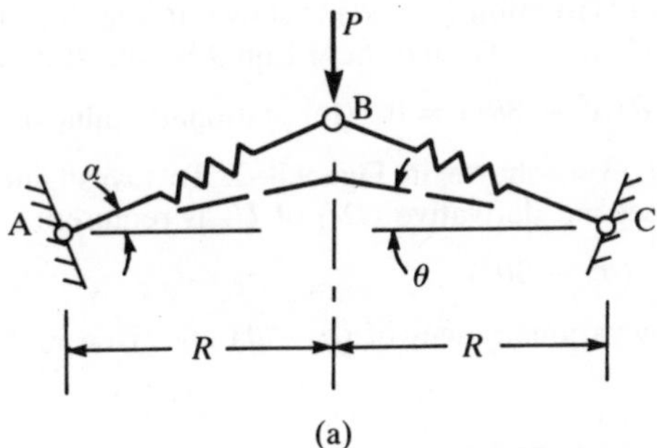

Fig. 9·8–1

If the point B is deflected to the position defined by angle θ the strain energy in the springs (U) is

$$U = 2 \times 1/2 \times k \times (R/\cos\alpha - R/\cos\theta)^2$$

$$1/\cos\alpha = 1 + \alpha^2/2 \qquad 1/\cos\theta = 1 + \theta^2/2$$

Hence $U = kR^2(\alpha^2 - \theta^2)/4$

The deflection (δ) of the load P is

$$\delta = R(\tan\alpha - \tan\theta) = R(\alpha - \theta)$$

The total potential energy (U_T) of the system is

$$U_T = kR^2(\alpha^2 - \theta^2)/4 - PR(\alpha - \theta) \tag{9.8–1}$$

If the structure is in equilibrium in this deflected position, the potential energy must have a stationary value (cf. Eqn 9.7–4, where change of strain energy is equated to change of potential energy), and, in particular, if this equilibrium is stable, the value must be a minimum. The first differential with respect to $\theta(D_1)$ is

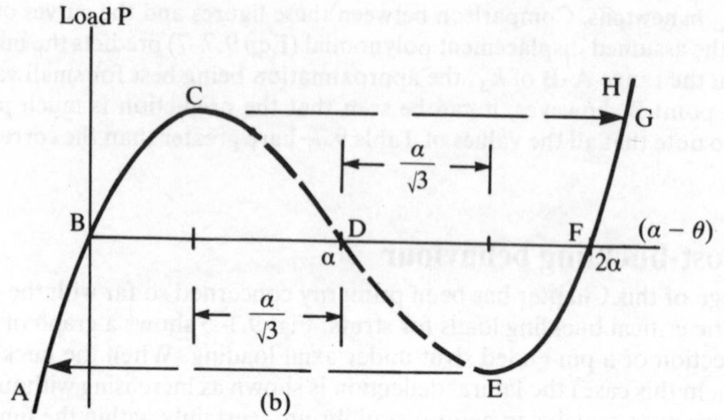

Fig. 9·8–2

$$D_1 = -kR^2(\alpha^2 - \theta^2)\theta + PR = 0 \text{ (for a minimum)}$$

Hence $P = kR\theta(\alpha^2 - \theta^2)$ (9.8–2)

The variation of P with rotation $(\alpha - \theta)$ is shown in Fig. 9.8–2 and this in turn has stationary values at C and E. To find these Eqn 9.8–2 must be differentiated

$$\partial P/\partial\theta = kR(\alpha^2 - 3\theta^2) = 0 \quad \text{(for stationary values)}$$

This gives $\theta = \pm\alpha/\sqrt{3}$ as shown in Fig. 9.8–2. To investigate the stability of the deflection path the second derivative (D_2) of U_T is required.

$$D_2 = -kR^2(\alpha^2 - 3\theta^2)$$

The values of D_2 for various ranges of $(\alpha - \theta)$ are given in Table 9.8–1.

Table 9.8–1

$(\alpha - \theta)$		D_2	Equilibrium Type
$(\alpha - \theta) < \alpha(1 - 1/\sqrt{3})$	(A to C)	positive	stable
$\alpha(1 - 1/\sqrt{3})$	(at C)	zero	neutral
$\alpha(1 - 1/\sqrt{3}) < (\alpha - \theta) < \alpha(1 + 1/\sqrt{3})$	(C to E)	negative	unstable
$\alpha(1 + 1/\sqrt{3})$	(at E)	zero	neutral
$(\alpha - \theta) > \alpha(1 + 1/\sqrt{3})$	(E to H)	positive	stable

The practical consequence of the results in Table 9.8–1 is that in a loading test the rotation $(\alpha - \theta)$ would increase with increasing P, following the path A to C of Fig. 9.8–2 in a stable manner. To continue following the falling curve C to D to E the load would have to be reduced under perfect control. In practice this would never be possible on account of the inevitable presence of small accidental disturbances, and the rotation would 'snap' over to the alternative equilibrium position G on EH. The load position would drop sharply, with loss of potential energy, and the system would oscillate about the position G. Natural damping would eventually bring the load to rest in position G, if the structure had not been damaged by the sudden release of energy. If the load was then to be removed the falling curve GF would be followed. This phenomenon, whereby the structure can suddenly jump from one position (as at C) to an alternative, lower energy position, as at G, is termed **snap through** and is common in more complex structures such as plates and shells. Instances of its occurrence in shallow dome structures have been recorded.

Two further examples quoted by Thompson are shown in Fig. 9.8–3 and Fig. 9.8–5. In the former case the post buckling curve is unstable (Fig. 9.8–4) while in the latter case it is stable (Fig. 9.8–6). The point of departure (C) of the deflection curve from the load axis at the critical load level is known as the **bifurcation point**.

The question of post buckling behaviour is an important one for the designer. If the post-buckling equilibrium is stable, the designer can allow the structural forces to approach the buckling level in the knowledge that sudden increases of deflection

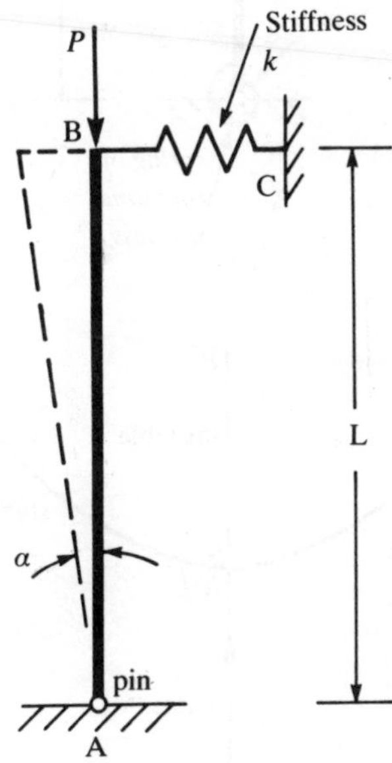

Fig. 9·8–3

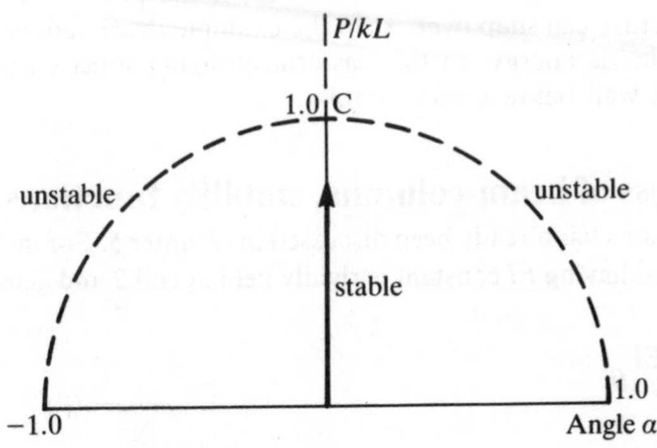

Fig. 9·8–4

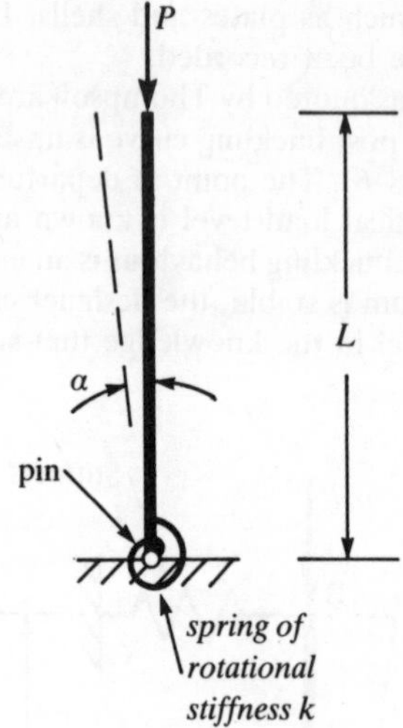

Fig. 9·8–5

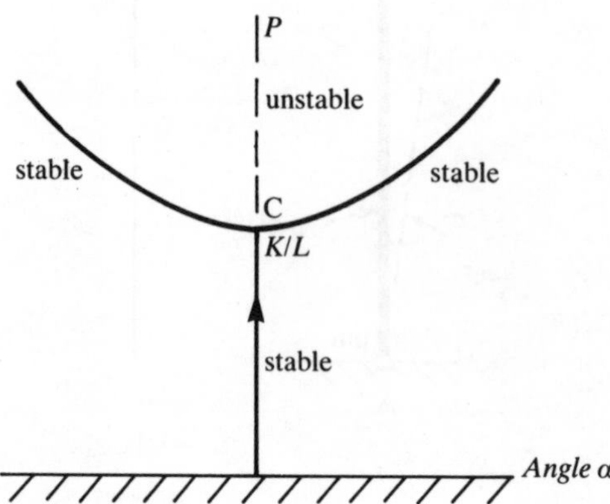

Fig. 9·8–6

will not occur on buckling. On the other hand, if the post buckling equilibrium is unstable the structure can snap over, as in the example discussed, with the conversion of potential to kinetic energy. In this case the designer must ensure that structural forces are always well below critical levels.

9.9 Stiffness of beam-columns, stability functions

The stiffness of beams has already been discussed in Chapter 5. For instance, if a member 1–2 of length L and having EI constant is rigidly held at end 2 and acted on by a moment at end 1, then

$$m_1 = \frac{4EI}{L}\theta_1$$

and

$$m_2 = \frac{2EI}{L}\theta_1 \tag{9.9–1}$$

The development of these relations was based on the assumption that the member 1–2 carried bending moments and shear forces only, but if the influence of an axial force is considered, modified expressions are obtained. These two relations can be rewritten:

$$m_1 = sk\theta_1$$

and

$$\frac{m_2}{m_1} = c$$

then

$$m_2 = sck\theta_1 \qquad (9.9\text{–}2)$$

where $k = EI/L$. s is thus a factor modifying stiffness k, while c is a carry-over factor (see Section 6.1). Symbols s and c are termed **stability functions**, and depend on the axial load. They obviously have the values 4 and $\frac{1}{2}$ respectively when the axial load is zero. The word 'stability' is used because positive or negative values for s define states of stable or unstable equilibrium respectively in the member.

The use of the stability functions allows a form of the member stiffness equation (Eqn 7.5–1) to be developed which includes the influence of axial load on bending stiffness. Each of the stiffness influence coefficients will be determined separately. End rotation is considered first.

The member 1–2 of Fig. 9.9–1 is of constant EI, and of length L. Its longitudinal axis coincides with the x-axis. The end 1 is acted on by a moment m_1, and rotates through an angle θ_1, while the end 2 is rigidly held in position and direction. The restraining

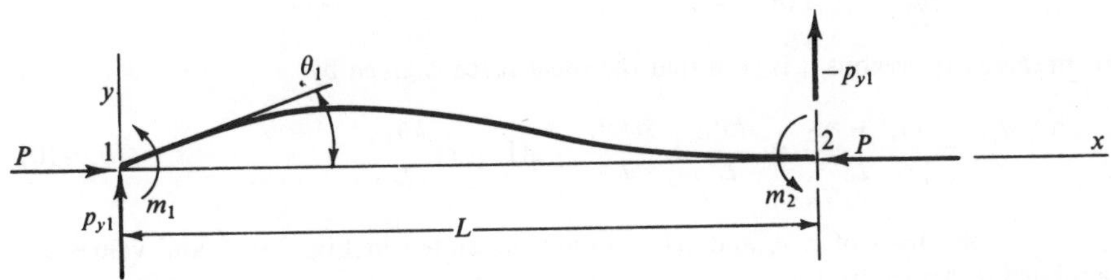

Fig. 9·9–1

moment at 2 is m_2. Member 1–2 carries an axial compressive load P. There is a uniform shear force p_{y1}, which is obtained by taking moments about end 2:

$$p_{y1} = \left(\frac{m_1 + m_2}{L}\right) \qquad (9.9\text{–}3)$$

The equation of flexure (cf. Eqn 9.2–2) is

$$\frac{d^2 v}{dx^2} + \left(\frac{P}{EI}\right) v - \left(\frac{p_{y1} x - m_1}{EI}\right) = 0$$

If ρ is defined as P/P_E and α is substituted for $\pi\sqrt{\rho}/2$ as in Section 9.3, this becomes:

$$\frac{d^2 v}{dx^2} + \left(\frac{4\alpha^2}{L^2}\right) v - \left(\frac{p_{y1} x - m_1}{EI}\right) = 0 \qquad (9.9\text{–}4)$$

The solution is:

$$v = A \sin \frac{2\alpha x}{L} + B \cos \frac{2\alpha x}{L} + \frac{L}{4\alpha^2 k}\left[\frac{(m_1 + m_2)x}{L} - m_1\right] \tag{9.9–5}$$

The integration constants A and B may be obtained from the boundary conditions, $v = 0$ at $x = L$, and $v = 0$ at $x = 0$.

$$A = -\frac{L}{4\alpha^2 k}[m_1 \cot 2\alpha + m_2 \operatorname{cosec} 2\alpha]$$

$$B = \frac{L}{4\alpha^2 k} m_1 \tag{9.9–6}$$

Differentiation of Eqn 9.9–5 gives the slope:

$$\frac{dv}{dx} = \frac{2\alpha A}{L} \cos \frac{2\alpha x}{L} - \frac{2\alpha}{L} B \sin \frac{2\alpha x}{L} + \frac{m_1 + m_2}{4\alpha^2 k} \tag{9.9–7}$$

The boundary condition, $dv/dx = 0$ at $x = L$, provides the ratio between m_1 and m_2 required for the stability function c (Eqn 9.9–2).

$$c = \frac{m_2}{m_1} = \frac{2\alpha - \sin 2\alpha}{\sin 2\alpha - 2\alpha \cos 2\alpha} \qquad \text{where } \alpha = \frac{\pi}{2}\sqrt{\rho} \tag{9.9–8}$$

Finally, the boundary condition, $dv/dx = \theta_1$ at $x = 0$, provides the stability function s.

$$s = \frac{m_1}{k\theta_1} = \frac{\alpha(1 - 2\alpha \cot 2\alpha)}{\tan \alpha - \alpha} \tag{9.9–9}$$

From these expressions it is seen that the shear force is given by

$$p_{y1} = \frac{m_1 + m_2}{L} = \frac{sk\theta_1}{L} + \frac{sck\theta_1}{L} = s(1 + c)\frac{k\theta_1}{L} \tag{9.9–10}$$

The forms of s, c, and $s(1 + c)$ are illustrated in Fig. 9.9–3, and values are tabulated in Appendix 1.

Expressions equivalent to Eqn 9.9–8, 9.9–9, and 9.9–10 can be set up for the moments and shears arising when end 1 is held, and end 2 rotates. The forces resulting from a simultaneous rotation at both ends are obtained by superposition, as there is a linear relation between force and deflection for a given value of axial load.

If the axial force P is not compressive as has been assumed, but tensile, the term ι in Eqn 9.9–4 becomes negative. This leads to a solution in terms of hyperbolic functions, and the resulting forms for s and c can be shown to be:

$$s = \gamma \frac{1 - 2\gamma \coth 2\gamma}{\tanh \gamma - \gamma}$$

$$c = \frac{2\gamma - \sinh 2\gamma}{\sinh 2\gamma - 2\gamma \cosh 2\gamma}$$

where $\gamma = (\pi/2)\sqrt{-\rho}$.

A relative lateral deflection between ends 1 and 2$(d_{y2} - d_{y1})$ is considered next.

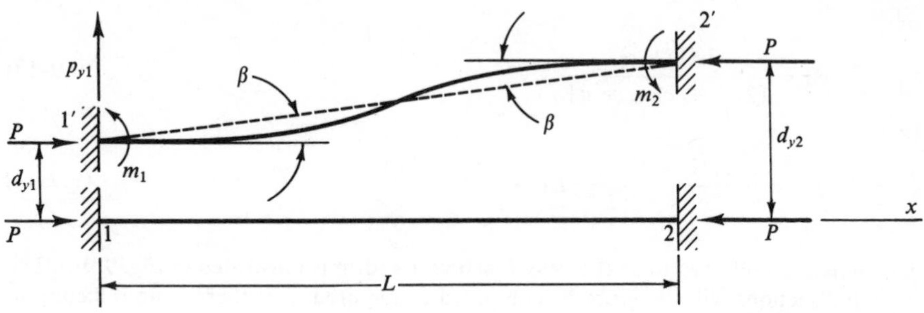

Fig. 9·9–2

Figure 9.9–2 shows the member 1–2 initially carrying end loads P, deflected to the position 1′–2′ without end rotations. The deflections can conveniently be thought of as having taken place in two stages:

(a) A rigid body movement of 1–2 to the position 1′–2′ indicated by the dotted line. No end moments result from this movement.

(b) Equal rotations through angles $-\beta$ at each end to bring 1–2 to its final configuration.

Provided that P is constant during the stages 1 and 2, the final result can be obtained by superposition.

	m_1	m_2
(a) Rigid body movement	0	0
(b) Rotation $-\beta$ at end 1	$-sk\beta$	$-sck\beta$
Rotation $-\beta$ at end 2	$-sck\beta$	$-sk\beta$
	$-s(1+c)k\beta$	$-s(1+c)k\beta$

then

$$m_1 = m_2 = -s(1+c)k\beta = -\frac{s(1+c)k}{L}(d_{y2} - d_{y1}) \tag{9.9–11}$$

The shear force p_{y1} may be obtained by taking moments about end 2′:

$$p_{y1} = \frac{m_1 + m_2}{L} + \frac{P}{L}(d_{y2} - d_{y1})$$

$$= \left(-\frac{2s(1+c)k}{L} + P\right)\frac{(d_{y2} - d_{y1})}{L} \tag{9.9–12}$$

When $P = 0$, then $m_1 = m_2 = (L/2)p_{y1}$.

It is convenient to define a new stability function m such that:

$$m_1 = m_2 = \frac{mL}{2}p_{y1}$$

hence

$$m = \frac{2m_1}{Lp_{y1}} = \frac{2s(1+c)k}{[2s(1+c)k - PL]}$$

or

$$m = \frac{2s(1 + c)}{[2s(1 + c) - \pi^2 \rho]} \tag{9.9–13}$$

then

$$p_{y1} = -\frac{2s(1 + c)}{m} \cdot \frac{k}{L^2} (d_{y2} - d_{y1}) \tag{9.9–14}$$

m is sometimes referred to as the **sway function**. Its form is illustrated in Fig. 9.9–3. The stability functions will therefore be computed as required, and there is no difficulty in

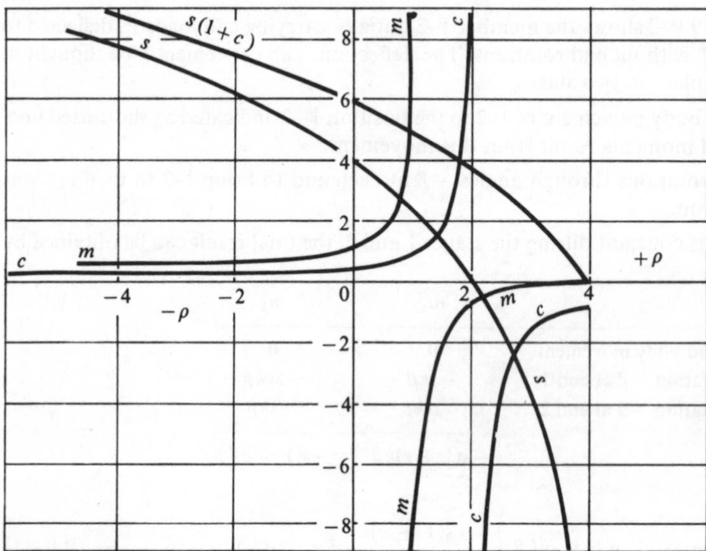

Fig. 9·9–3

this so long as the appropriate form is used for positive and negative axial forces respectively. However, if the axial force is zero, the expressions for s and c become indeterminate. Although they can be expressed in a series form which removes this difficulty, it is usually satisfactory to assign the values 4 and $\frac{1}{2}$ respectively when ρ is less than a suitably small limit.

In order to complete the member stiffness equations, the axial force must be expressed in terms of the end displacements:

$$P = p_{x1} = \frac{EA}{L} (d_{x1} - d_{x2})$$

(This neglects the influence of curvature on end shortening.)

The member stiffness equation, 7.5–1, may now be restated. The various ϕ terms are convenient groupings of s, c, and m.

$$\begin{bmatrix} p_{x1} \\ p_{y1} \\ m_1 \\ p_{x2} \\ p_{y2} \\ m_2 \end{bmatrix} = \left[\begin{array}{ccc|ccc} \frac{EA}{L} & 0 & 0 & -\frac{EA}{L} & 0 & 0 \\ 0 & \frac{EI\phi_1}{L^3} & \frac{EI\phi_2}{L^2} & 0 & -\frac{EI\phi_1}{L^3} & \frac{EI\phi_2}{L^2} \\ 0 & \frac{EI\phi_2}{L^2} & \frac{EI\phi_4}{L} & 0 & -\frac{EI\phi_2}{L^2} & \frac{EI\phi_3}{L} \\ \hline -\frac{EA}{L} & 0 & 0 & \frac{EA}{L} & 0 & 0 \\ 0 & -\frac{EI\phi_1}{L^3} & -\frac{EI\phi_2}{L^2} & 0 & \frac{EI\phi_1}{L^3} & -\frac{EI\phi_2}{L^2} \\ 0 & \frac{EI\phi_2}{L^2} & \frac{EI\phi_3}{L} & 0 & -\frac{EI\phi_2}{L^2} & \frac{EI\phi_4}{L} \end{array}\right] \begin{bmatrix} d_{x1} \\ d_{y1} \\ \theta_1 \\ d_{x2} \\ d_{y2} \\ \theta_2 \end{bmatrix} \tag{9.9–15}$$

$$\phi_4 = s \qquad \phi_3 = sc$$

$$\phi_2 = s(1 + c) = \phi_4 + \phi_3 \qquad \phi_1 = 2s(1 + c) - \pi^2\rho = 2\phi_2 - \pi^2\rho$$

$$= \frac{2\phi_2}{m}$$

The use of m is not essential but may sometimes be convenient. The stiffness matrix may be partitioned into $\mathbf{K}_{11}$, $\mathbf{K}_{12}$, $\mathbf{K}_{21}$, and $\mathbf{K}_{22}$ as in Eqn 7.5–2.

The stiffness matrices can be modified to allow for the presence of a pin at either end 1 or 2, as in Section 7.11. If there is a pin at end 1, then:

$$\mathbf{K}_{11} = \begin{bmatrix} \frac{EA}{L} & 0 & 0 \\ 0 & \frac{EI\phi_6}{L^3} & 0 \\ 0 & 0 & 0 \end{bmatrix} \qquad \mathbf{K}_{22} = \begin{bmatrix} \frac{EA}{L} & 0 & 0 \\ 0 & \frac{EI\phi_6}{L^3} & -\frac{EI\phi_5}{L^2} \\ 0 & -\frac{EI\phi_5}{L^2} & \frac{EI\phi_5}{L} \end{bmatrix}$$

$$\mathbf{K}_{12} = \mathbf{K}_{21}^T \begin{bmatrix} -\frac{EA}{L} & 0 & 0 \\ 0 & -\frac{EI\phi_6}{L^3} & \frac{EI\phi_5}{L^2} \\ 0 & 0 & 0 \end{bmatrix} \tag{9.9–16}$$

$$\phi_5 = s(1 - c^2) \qquad \phi_6 = \phi_5 - \pi^2\rho$$

If there is a pin at end 2, then:

$$\mathbf{K}_{11} = \begin{bmatrix} \dfrac{EA}{L} & 0 & 0 \\ 0 & \dfrac{EI\phi_6}{L^3} & \dfrac{EI\phi_5}{L^2} \\ 0 & \dfrac{EI\phi_5}{L^2} & \dfrac{EI\phi_5}{L} \end{bmatrix} \qquad \mathbf{K}_{22} = \begin{bmatrix} \dfrac{EA}{L} & 0 & 0 \\ 0 & \dfrac{EI\phi_6}{L^3} & 0 \\ 0 & 0 & 0 \end{bmatrix}$$

$$\mathbf{K}_{12} = \mathbf{K}_{21}^T = \begin{bmatrix} -\dfrac{EA}{L} & 0 & 0 \\ 0 & -\dfrac{EI\phi_6}{L^3} & 0 \\ 0 & -\dfrac{EI\phi_5}{L^2} & 0 \end{bmatrix} \tag{9.9–17}$$

If $\rho = 0$, then $\phi_6 = \phi_5 = 3$, and these stiffness matrices become identical to those of Eqns 7.11–4 and 7.11–5.

If there are pins at both ends 1 and 2, only the terms in EA/L remain, and there are no stability functions present in the stiffness matrices.

This revised form of the member stiffness matrix can be used in the analysis of complete structures, and will be discussed in Section 9.11.

Although no particular type of cross-section has been mentioned in this section, it has been assumed that the member could buckle in one plane only (the x–y plane) and that it could not fail by lateral torsional buckling (Section 9.6), either on account of its cross-sectional shape or because of lateral support. The member has been assumed prismatic, and to have both centroid and shear centre in the x–y plane.

9.10 Influence of axial loads on end moments of fixed-ended beams

In the analysis of skeletal structures by the stiffness method described in Chapter 7, it was observed that the loading vector might contain fixed-end forces due to loads applied between joints (Section 7.15). It is found that the presence of an axial load in a member affects the values of the fixed-end forces, and this will be studied in this section.

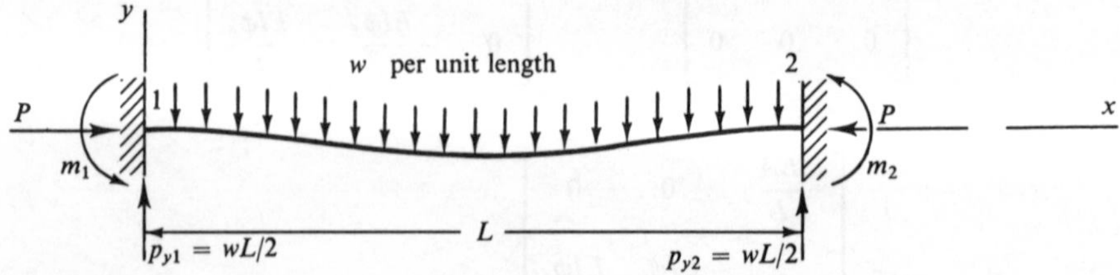

Fig. 9·10–1

Uniformly distributed load. The structure is symmetrical: $p_{y1} = p_{y2}$ and $m_1 = -m_2$. The equation of flexure is

$$\frac{d^2v}{dx^2} = -\left(\frac{P}{EI}\right)v + \frac{m_1}{EI} - p_{y1}\frac{x}{EI} + \frac{wx^2}{2EI}$$

hence

$$\frac{d^2v}{dx^2} + \left(\frac{P}{EI}\right)v - \frac{1}{EI}\left(m_1 - \frac{wLx}{2} + \frac{wx^2}{2}\right) = 0 \tag{9.10–1}$$

The solution to this equation is:

$$v = A \sin\frac{2\alpha x}{L} + B \cos\frac{2\alpha x}{L} + \frac{L}{4\alpha^2 k}\left(m_1 - \frac{wx}{2}(L - x) - \frac{wL^2}{4\alpha^2}\right) \tag{9.10–2}$$

Substitution of the two boundary conditions $v = 0$ and $dv/dx = 0$ at $x = 0$ gives the two integration constants A and B. Then

$$A = \frac{wL^3}{16\alpha^3 k}$$

$$B = \frac{wL^3}{16\alpha^4 k}\left(1 - \frac{4\alpha^2 m_1}{wL^2}\right) \tag{9.10–3}$$

Introduction of the third condition ($v = 0$ at $x = L$) gives an answer for the remaining unknown, m_1:

$$m_1 = \left[\frac{3}{\alpha^2}(1 - \alpha \cot \alpha)\right]\frac{wL^2}{12} \tag{9.10–4}$$

or

$$m_1 = f \cdot \frac{wL^2}{12}$$

This solution is valid for all compressive axial loads, except $P = 0$. In this case $f = 1$. The variation of f with p is illustrated in Fig. 9.10–2, and is tabulated in Appendix 2.

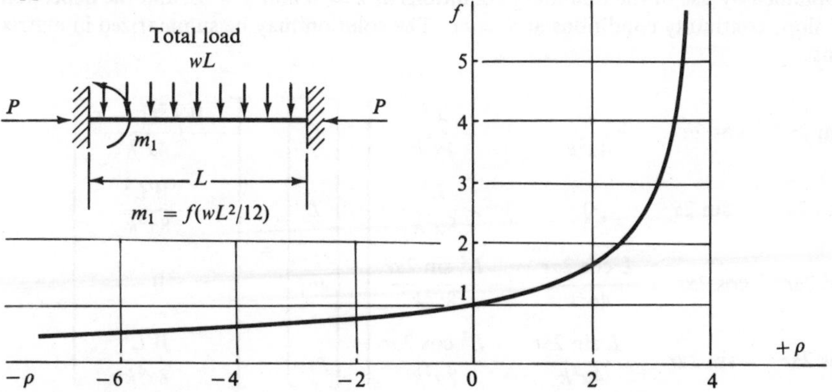

Fig. 9·10–2

Single-point load. (Fig. 9.10–3) The solution must be performed in two parts. $0 \leqslant x \leqslant rL$

The equation of flexure is:

$$\frac{d^2v}{dx^2} + \left(\frac{P}{EI}\right)v - \frac{1}{EI}(m_1 - P_{y1}x) = 0 \tag{9.10–5}$$

The solution is:

$$v = A \sin \frac{2\alpha x}{L} + B \cos \frac{2\alpha x}{L} + \frac{L}{4x^2 k}(m_1 - p_{y1} x) \tag{9.10–6}$$

The two boundary conditions at end 1 give values for the integration constants A and B, and the solution becomes:

$$v = \frac{L}{4\alpha^2 k}\left[\frac{p_{y1} L}{2\alpha} \sin \frac{2\alpha x}{L} - m_1 \cos \frac{2\alpha x}{L} + (m_1 - p_{y1} x)\right] \tag{9.10–7}$$

$rL \leqslant x \leqslant L$

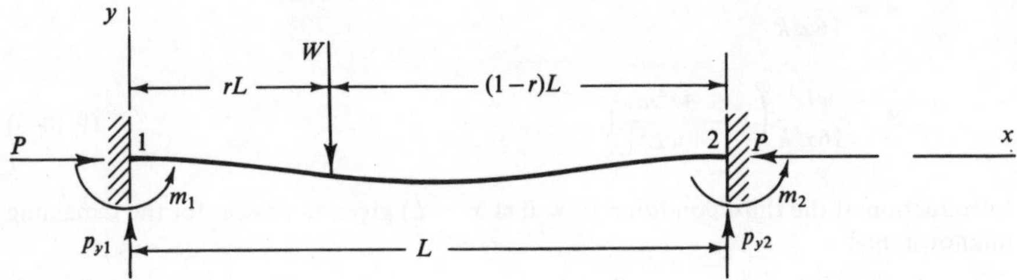

Fig. 9·10–3

The solution here is similar to Eqn 9.10–6, with an additional term:

$$v = C \sin \frac{2\alpha x}{L} + D \cos \frac{2\alpha x}{L} + \frac{L}{4\alpha^2 k}[m_1 - p_{y1} x + W(x - rL)] \tag{9.10–8}$$

There are four unknowns—C, D, m_1, and p_{y1} in these equations, and they may be obtained by use of the boundary conditions at $x = 0$ and $x = L$, and the deflection and slope continuity conditions at $x = rL$. The solution may be summarized in matrix terms:

$$\begin{bmatrix} \sin 2\alpha & \cos 2\alpha & \dfrac{L}{4\alpha^2 k} & -\dfrac{L^2}{4\alpha^2 k} \\ \cos 2\alpha & -\sin 2\alpha & 0 & -\dfrac{L^2}{8\alpha^3 k} \\ \sin 2\alpha r & \cos 2\alpha r & \dfrac{L \cos 2\alpha r}{4\alpha^2 k} & -\dfrac{L^2 \sin 2\alpha r}{8\alpha^3 k} \\ \cos 2\alpha r & -\sin 2\alpha r & -\dfrac{L \sin 2\alpha r}{4\alpha^2 k} & -\dfrac{L^2 \cos 2\alpha r}{8\alpha^3 k} \end{bmatrix} \begin{bmatrix} C \\ D \\ m_1 \\ p_{y1} \end{bmatrix} = \begin{bmatrix} -\dfrac{WL^2(1-r)}{4\alpha^2 k} \\ -\dfrac{WL^2}{8\alpha^3 k} \\ 0 \\ -\dfrac{WL^2}{8\alpha^3 k} \end{bmatrix} \tag{9.10–9}$$

An explicit solution for m_1 is possible, but it is of such an involved nature that it is not sensible to calculate values manually. The results of a computer solution are illustrated in Fig. 9.10–4 and Appendix 2. The moment m_2 can be obtained by using the distance $(1 - r)$ in place of r, and shears can be obtained by taking moments about either end.

End moments for series of point loads may be obtained by superposition, as the moment–load relation is linear for a constant P.

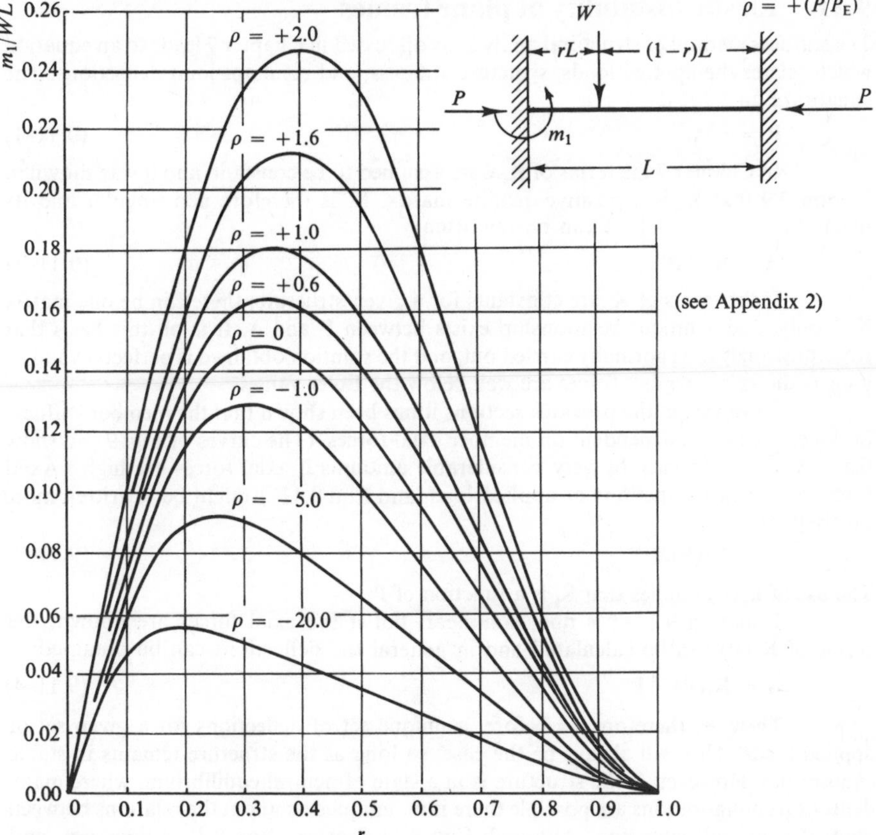

Fig. 9·10–4

Mention has been made in this section of axial compressive forces only. If the axial force is tensile, the solutions are similar to Eqns 9.10–4 and 9.10–9 respectively, but are in terms of the equivalent hyperbolic functions with some sign changes, i.e.

$$f = -3(1 - \gamma \coth \gamma)/\gamma^2 \tag{9.10–10}$$

where $\gamma = (\pi/2)\sqrt{-\rho}$ as in Section 9.9, and

$$\begin{bmatrix} \sinh 2\gamma & \cosh 2\gamma & -\dfrac{L}{4\gamma^2 k} & \dfrac{L^2}{4\gamma^2 k} \\ \cosh 2\gamma & \sinh 2\gamma & 0 & \dfrac{L^2}{8\gamma^3 k} \\ \sinh 2\gamma r & \cosh 2\gamma r & -\dfrac{L \cosh 2\gamma r}{4\gamma^2 k} & \dfrac{L^2 \sinh 2\gamma r}{8\gamma^3 k} \\ \cosh 2\gamma r & \sinh 2\gamma r & -\dfrac{L \sinh 2\gamma r}{4\gamma^2 k} & \dfrac{L^2 \cosh 2\gamma r}{8\gamma^3 k} \end{bmatrix} \begin{bmatrix} C \\ D \\ m_1 \\ p_{y1} \end{bmatrix} = \begin{bmatrix} \dfrac{WL^2(1-r)}{4\gamma^2 k} \\ \dfrac{WL^2}{8\gamma^3 k} \\ 0 \\ \dfrac{WL^2}{8\gamma^3 k} \end{bmatrix} \tag{9.10–11}$$

9.11 Elastic instability of plane frames

The stiffness method of structural analysis as discussed in Chapter 7 leads to an equation which relates the applied loads, structure stiffness, and resultant joint deflections. The equation Eqn 7.9–2 is:

$$\mathbf{P} = \mathbf{K}_s\mathbf{\Delta} \tag{9.11–1}$$

In Chapter 7 the terms of $\mathbf{K}_s$ were assumed to be constant, and it was shown in Section 7.9 that $\mathbf{K}_s$ is a positive-definite matrix. $\mathbf{K}_s$ is therefore non-singular and its inverse exists. Eqn 9.11–1 can be rewritten:

$$\mathbf{\Delta} = \mathbf{K}_s^{-1}\mathbf{P} \tag{9.11–2}$$

If the terms of $\mathbf{K}_s$ are constants for a given structure, there can be one matrix $\mathbf{K}_s^{-1}$ only, and a unique relationship exists between $\mathbf{P}$ and $\mathbf{\Delta}$. It is on this basis that structural analysis is normally carried out, and the solution obtained is perfectly valid so long as the axial member forces are well below the Euler values.

However, in the previous sections it has been shown that the member stiffness matrices are in fact dependent on member axial forces. (The curves of Fig. 9.9–3 show that the terms can vary by very considerable amounts if axial forces are high.) Axial forces are in turn functions of applied load, and Eqn 9.11–1 might be rewritten more precisely as:

$$\mathbf{P} = \mathbf{K}_s(\mathbf{P})\mathbf{\Delta} \tag{9.11–3}$$

The use of $\mathbf{K}_s(\mathbf{P})$ implies that $\mathbf{K}_s$ is a function of $\mathbf{P}$.

Equation 9.11–3 is now nonlinear, but if the axial forces are known, the terms of $\mathbf{K}_s(\mathbf{P})$ can be calculated and in general the deflections can be obtained:

$$\mathbf{\Delta} = \mathbf{K}_s(\mathbf{P})^{-1}\mathbf{P} \tag{9.11–4}$$

There is, therefore, as before, a unique set of deflections for a given set of applied loads. This will always be the case, so long as the structure remains in stable equilibrium. However, if the structure is in a state of neutral equilibrium, where many deflection configurations are possible there is no uniqueness about the relations between applied load and deflection. Although Eqn 9.11–3 exists, Eqn 9.11–4 does not, and this is only possible if matrix $\mathbf{K}_s(\mathbf{P})$ is singular. The singularity of $\mathbf{K}_s(\mathbf{P})$ is equivalent to the structure stiffness being zero.

A test of the singularity of the matrix $\mathbf{K}_s(\mathbf{P})$ can therefore be used as a check on stability—if it is non-singular and positive-definite the structure is stable—if it is singular the structure is on the verge of instability. This point will be illustrated in an example.

Example 9.11–1. The plane structure of Fig. 9.11–1 is rigidly jointed at C and encastré at A and B. The structure and its loading are symmetrical, and therefore both

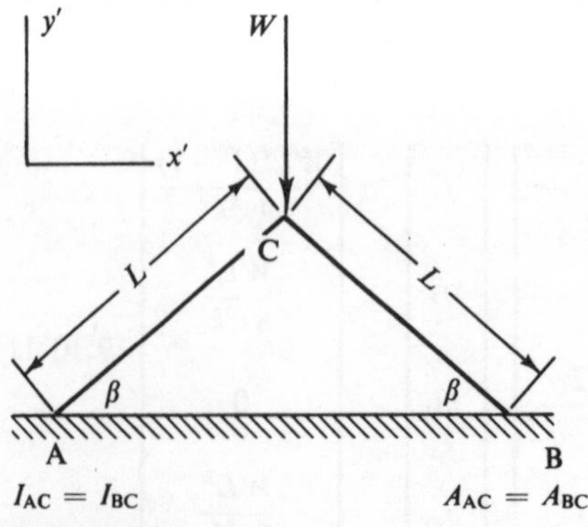

Fig. 9·11–1

members carry the same axial forces. The value of W required to produce elastic instability in the frame is required.

The form of the 3 × 3 stiffness matrix is:

$$\mathbf{K}_s(\mathbf{P}) = [(\mathbf{K}'_{22})_{AC} + (\mathbf{K}'_{22})_{BC}]$$

If the structure axes shown in the figure are used then the stiffness matrix becomes:

$$\mathbf{K}_s(\mathbf{P}) = \frac{2EI}{L}\begin{bmatrix} \left(\frac{A}{I}\right)\cos^2\beta + \frac{\phi_1}{L^2}\sin^2\beta & 0 & \frac{\phi_2}{L}\sin\beta \\ 0 & \left(\frac{A}{I}\right)\sin^2\beta + \frac{\phi_1}{L^2}\cos^2\beta & 0 \\ \frac{\phi_2}{L}\sin\beta & 0 & \phi_4 \end{bmatrix} \qquad (9.11\text{–}5)$$

The ϕ terms are given in Eqn 9.9–15 and apply to both members.

If $K_s(P)$ is singular its determinant is zero. Hence

$$\frac{2EI}{L}\left[\frac{A}{I}\sin^2\beta + \frac{\phi_1}{L^2}\cos^2\beta\right]\left[\left(\frac{A}{I}\cos^2\beta + \frac{\phi_1}{L^2}\sin^2\beta\right)\phi_4 - \frac{\phi_2^2}{L^2}\sin^2\beta\right] = 0 \qquad (9.11\text{–}6)$$

Either square bracket term may be zero. Taking the first gives:

$$-\frac{AL^2}{I}\tan^2\beta = \phi_1 = \frac{2s(1+c)}{m} = 2s(1+c) - \pi^2\rho$$

or

$$\left(\frac{L}{r}\right)^2\tan^2\beta = \pi^2\rho - 2s(1+c) \qquad (9.11\text{–}7)$$

where r = radius of gyration = $\sqrt{(I/A)}$.

The left-hand side of Eqn 9.11–7 is necessarily positive and is likely to be large for angles β of about the order shown in Fig. 9.11–1 (approximately 45°) as (L/r) must be large if buckling is to be elastic. Examination of the stability function tables in Appendix 1 shows that the right-hand side of Eqn 9.11–7 can only be positive for values of p between 1 and 4. A solution is therefore possible if the left-hand side is no larger than $4\pi^2$ or 39.5. Otherwise, this equation provides no solution.

If the second square bracket term of Eqn 9.11–6 is zero, then

$$\frac{sA}{I}\cos^2\beta = \left[\frac{s^2(1+c)^2}{L^2} - \frac{2s^2(1+c)}{mL^2}\right]\sin^2\beta$$

Hence either $s = 0$ and $\rho = 2.05$, or

$$\left(\frac{L}{r}\right)\cot^2\beta = [s(c^2 - 1) + \pi^2\rho] \qquad (9.11\text{–}8)$$

The left-hand term is similar to that of Eqn 9.11–7, and again it can be argued that it must be large for buckling to be elastic, and for angles β near 45°. Examination of the curves of Fig. 9.9–3 shows that the right-hand side can only be large and positive when c is large, or $\rho \approx 2$, but less than 2.05 (cf. $s = 0$ above). This result is likely to be smaller than that of Eqn 9.11–7.

For actual values of (L/r) the critical value of ρ can be determined by trial and error from Eqn 9.11–8. For instance, if (L/r) is 100 and β is 45°, then Eqn 9.11–8 becomes:

$$10^4 = s(c^2 - 1) + \pi^2 \rho$$

and $\quad \rho_{cr} > 2.0 \quad$ and $\quad < 2.05$

or

$$W_{cr} \approx 4.10 P_E \sin \beta \tag{9.11–9}$$

where P_E is the Euler load for member AC.

This result corresponds to the buckling of a strut encastré at one end and pinned at the other. It is a result one might expect and is clearly associated with a rotation at joint C. While W is increased from zero to its final value the resulting axial forces in AC and BC lead to a reduction in their bending stiffness until at instability this stiffness is zero.

9.12 Calculation of critical loads of plane frames

The solution of the problem in the previous section was perfectly straightforward because of the symmetry of the structure, and its loading. The axial forces were known. In this section a computer-based method for determining instability loads for a general plane frame will be considered.

The structure of Fig. 9.12–1 will be considered as an example. It carries four loads as shown. These have the initial values $\mathbf{P}_A$, $\mathbf{P}_B$, $\mathbf{P}_D$, and $\mathbf{P}_E$, and the problem is to find the value of the load factor λ by which they must be increased to produce elastic instability. It is assumed that the frame is adequately restrained against buckling out of its plane.

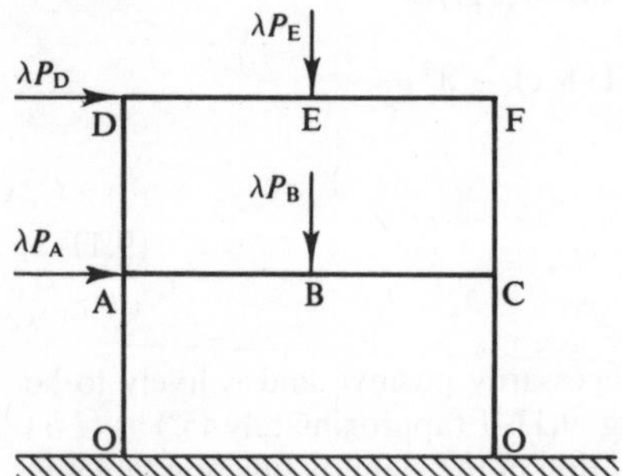

Fig. 9·12–1

The elastic behaviour of the structure is governed by the equation:

$$\mathbf{P} = \mathbf{K}_s \boldsymbol{\Delta}$$

or more precisely:

$$\lambda \mathbf{P} = \mathbf{K}_s(\lambda \mathbf{P}) \boldsymbol{\Delta} \tag{9.12–1}$$

The use of $\mathbf{K}_s(\lambda \mathbf{P})$ implies that $\mathbf{K}_s$ is a function of the applied load $\lambda \mathbf{P}$. This equation is nonlinear.

To determine the value of λ_{cr} —the critical load factor—the problem is linearized by carrying out a doubly iterative process. The value of λ is increased in a step-by-step manner, and at each load level the singularity of $\mathbf{K}_s(\lambda \mathbf{P})$ is checked. At each load level, also, an inner iteration is performed before the singularity check to find the correct values of the member axial forces—Eqn 9.12–1, is solved repeatedly until a consistent set of deflections is obtained. The number of iterations required here depends on how near the structure is to instability, and how good a guess of axial force can be made initially. It is

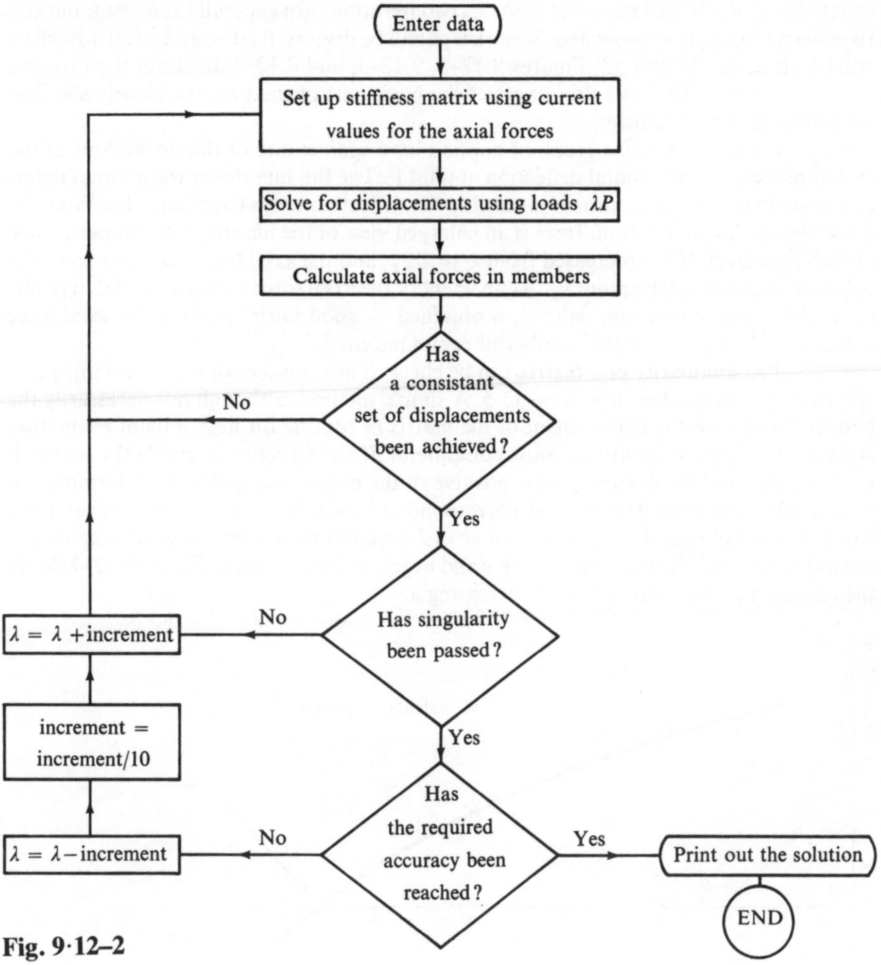

Enter data

Set up stiffness matrix using current values for the axial forces

Solve for displacements using loads λP

Calculate axial forces in members

Has a consistant set of displacements been achieved? (No → back to set up stiffness matrix; Yes ↓)

Has singularity been passed? (No → $\lambda = \lambda$ + increment; Yes ↓)

Has the required accuracy been reached? (No → $\lambda = \lambda -$ increment → increment = increment/10 → $\lambda = \lambda$ + increment; Yes → Print out the solution → END)

Fig. 9·12–2

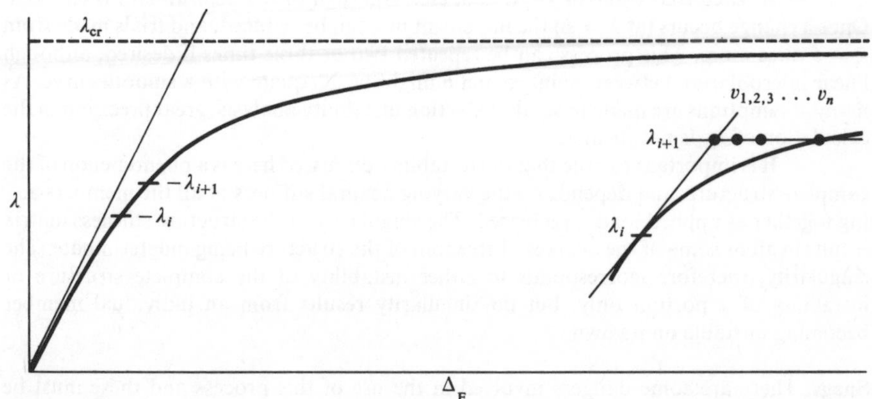

Fig. 9·12–3

found that at loads well removed from λ_{cr} two iterations are generally sufficient, but this rises sharply as λ_{cr} is approached. Some snags will be discussed later, and a full flow chart will be given in Chapter 13, Figures 9.12–2, 9.12–3, and 9.12–4 illustrate the process.

Figure 9.12–2 is a flow chart of the basic part of the process, clearly showing its doubly iterative nature.

Figure 9.12–3 is a graph of applied load against one of the deflections of the structure—say the horizontal deflection at joint F. The full line shows the gradual reduction in structure stiffness, until at instability the deflection tends to infinity. The faint line illustrates the linear solution. Inset is an enlarged view of the iterative process performed at each load level. If λ is increased from λ_i to λ_{i+1}, and the axial forces at λ_i are used, the solution obtained will be point v_1. Refinement of the axial force values will lead to points v_2, v_3, etc., until a constant value v_n is obtained. A good initial guess at the axial force values will clearly reduce the number of cycles required.

The singularity of a matrix can be checked in a number of ways, and this point will be discussed further in Section 13.5. A simple method, although not necessarily the best, is to examine the determinant of the matrix (a routine for determinant evaluation is a standard library facility on most computers). If the structure is stable the matrix is non-singular, and the determinant is positive; if the matrix is singular the determinant is zero, while a negative determinant corresponds to unstable equilibrium. In practice it will not be possible to determine the value of λ required to produce an exact singularity, but only values producing small positive and negative determinants. Figure 9.12–4 shows the variation of determinants with increasing λ.

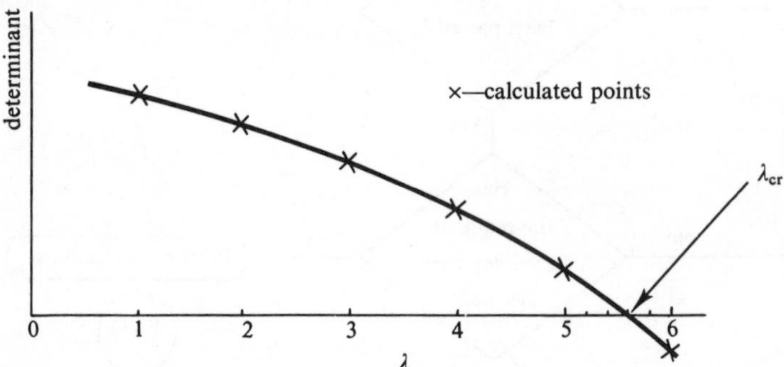

Fig. 9·12–4

At each trial value of λ—1, 2, 3, etc.—the sign of the determinant is checked. Once a change occurs (at $\lambda = 6$) the increment in λ can be reduced, and trials made from $\lambda = 5$ once more. This process can be repeated two or three times if desired, although linear interpolation between points 5 and 6 might be adequate with a smooth curve. As many assumptions are made in small-deflection instability analysis, great precision in the calculation of λ_{cr} is not justified.

It is important to note that the instability discussed here is a phenomenon of the complete structure, and depends on the varying flexural stiffness of all the members acting together as applied load is increased. The singularity of the structure stiffness matrix results in all or some of the degrees of freedom of the structure being indeterminate. The singularity, therefore, corresponds to either instability of the complete structure or instability of a portion only, but no singularity results from an individual member becoming unstable on its own.

Snags. There are some dangers involved in the use of this process and these must be clearly recognized.

(1) There is a series of possible buckling modes, each corresponding to a value of buckling load as with individual struts. This point will be further discussed later. If the determinant of the stiffness matrix $\mathbf{K}_s(\lambda\mathbf{P})$ of order $n \times n$ is expanded, a polynomial of the nth order in $\lambda\mathbf{P}$ is obtained. There are, therefore, n possible values of $\lambda\mathbf{P}$ for which $\mathbf{K}_s(\lambda\mathbf{P})$ is singular. Although the engineer will in general be interested in the smallest value of λ_{cr} only, he might be interested to know if other values are close as in Fig. 9.12–5.

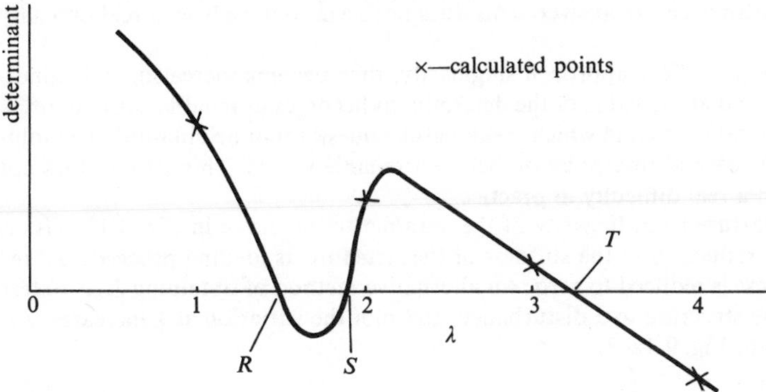

Fig. 9·12–5

If the initial value of λ used is too near the first critical value, or the increment used is too large, there is a danger that the lowest λ_{cr} will be missed altogether. In Fig. 9.12–5 it can be seen that the third smallest value of λ_{cr} might well be picked up as the smallest (point T). There seems no way of predicting in advance the form the curve will have. The only safe solution is to keep the initial value of λ low, keep the increment small, and accept the fact that a fairly large number of trials must be performed. An alternative check on singularity is through the use of eigenvalues. The stiffness matrix would have one negative eigenvalue after point R, two after point S, and three after point T. This will be discussed further in Section 13.5.

(2) The determinant of a matrix is particularly sensitive to the magnitude of the terms of the matrix. For instance, an alteration in the magnitude of the terms by a factor of 100 (by a change in the units used for instance) would alter the determinant by a factor of 100^n for a $n \times n$ matrix. This could well take the determinant outside the range of

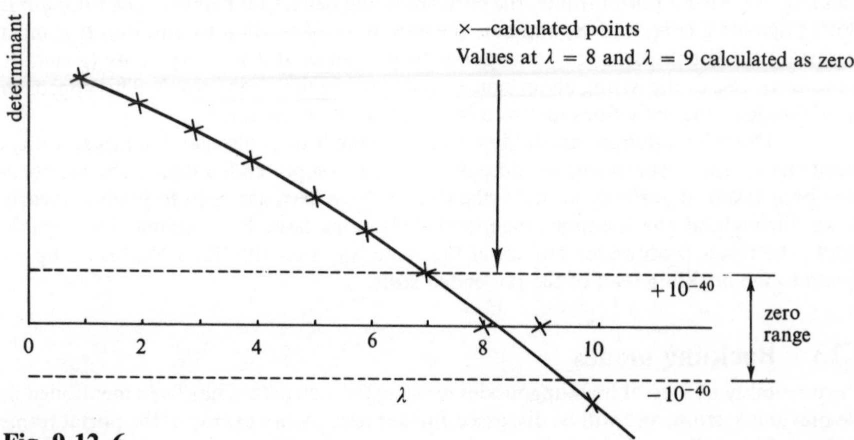

Fig. 9·12–6

positive numbers which a computer can hold (typically 10^{40} to 10^{-40}). Scaling of the matrix to remove this difficulty will be discussed in Section 13.3.

(3) Even if the value of λ giving exact singularity of matrix $\mathbf{K}_s(\lambda\mathbf{P})$ were chosen, the analyst would not know this. In the computer there is a range of small numbers which are all called zero. This point is illustrated in Fig. 9.12–6.

If there was a sharp change of slope between the last positive and first negative determinants, and the slope in itself was small, great care might have to be taken in interpolation to ensure an accurate answer. This does not seem to have been a real difficulty in practice.

(4) As Eqns 9.12–1 approach singularity, they become increasingly ill-conditioned. The result is that the value of the determinant becomes unreliable. As in point 3, there is a zone around the axis in which meaningful values cannot be obtained. The solution is, once more, careful interpolation between reliable values. This, again, does not seem to have been a real difficulty in practice.

The departure from linearity of the load/deflection curve in Fig. 9.12–3 is an illustration of the reduction in the stiffness of the structure as loading proceeds. At instability the stiffness is reduced to zero. An alternative method of obtaining λ_{cr} is to test the response of the structure to a disturbance, and plot the variation as λ increases. An example is shown in Fig. 9.12–7.

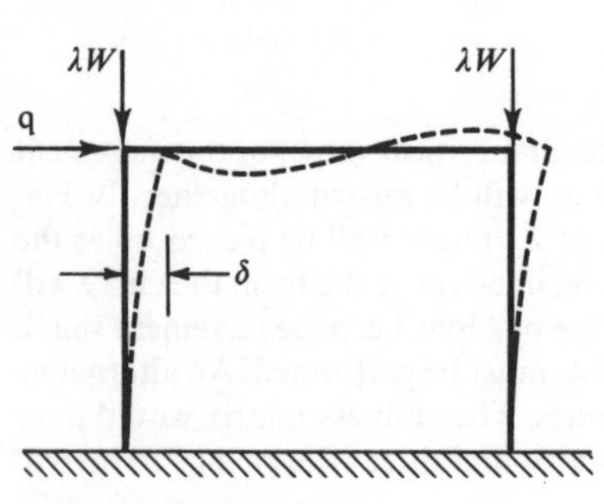

Fig. 9·12–7

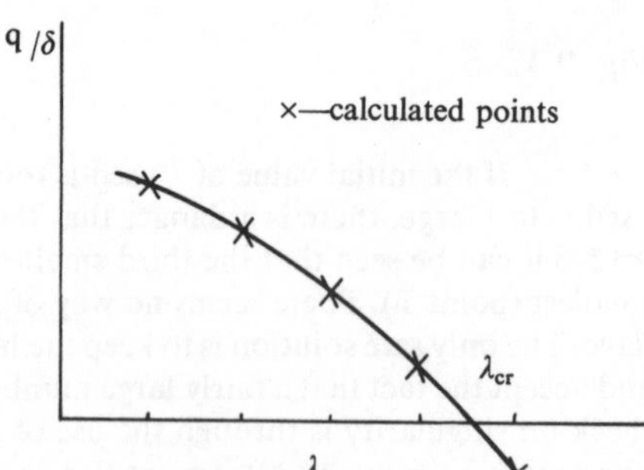

Fig. 9·12–8

The portal frame shown carries equal loads λW at the head of the columns; q is a disturbing force chosen to deflect the structure into a 'sway' mode, which is the lowest buckling mode for a portal frame, the corresponding deflection being δ. The ratio q/δ is plotted against λ (Fig. 9.12–8) and cuts the axis at λ_{cr} (δ tending to infinity). It is most important the the disturbance corresponds to the required instability mode (generally the lowest). Use of the wrong disturbance would lead to the wrong result. This approach avoids some of the difficulties involved in use of the determinant.

The calculation of instability loads for plane frames discussed in this section is based on the same assumptions as those of Section 9.1 on pin-ended struts. The members have been assumed perfectly straight, the stresses have been assumed to remain linearly elastic throughout the loading range, and deflections have been assumed to remain small. The result is an upper bound, in the same way that the Euler load is an upper bound to the buckling load of the pin-ended strut.

9.13 Buckling modes

The possibility of several buckling modes existing for a structure has been mentioned in the previous section, and will be discussed further here. As an example, the portal frame of Fig. 9.13–1 will be considered.

The span is equal to the height, and the section of the members is constant throughout. Up to instability the equilibrium of the structure is governed by the equation:

$$\lambda \mathbf{P} = \mathbf{K}_s(\lambda \mathbf{P})\boldsymbol{\Delta} \tag{9.13–1}$$

At the verge of instability, the structure is in a state of neutral equilibrium. This is a property of the members of the structure, and the axial forces they carry. So long as the members do have the necessary axial forces, Eqn 9.13–1 could be written:

$$\mathbf{K}_s(\lambda \mathbf{P})\boldsymbol{\Delta} = 0 \tag{9.13–2}$$

The equations represented by Eqn 9.13–2 are not linearly independent, and cannot therefore be solved for $\boldsymbol{\Delta}$. However, a displacement vector can be obtained, and this represents the buckling mode for the appropriate value of critical load.

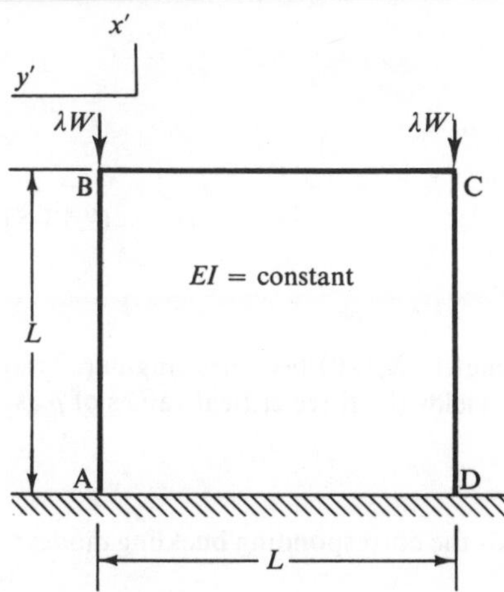

Fig. 9·13–1

In the portal frame of Fig. 9.13–1 the columns carry a compressive axial load λW, while the axial load in the beam is small. Although there is a total of six degrees of freedom, there are three significant ones only—the rotation at the heads of the columns (θ), and the side-sway of the complete frame (δ). If equations of the type (9.13–2) are set up, they become:

$$\begin{bmatrix} \dfrac{4s(1+c)}{m} & -s(1+c) & -s(1+c) \\ -s(1+c) & 4+s & 2 \\ -s(1+c) & 2 & 4+s \end{bmatrix} \begin{bmatrix} \dfrac{\delta}{L} \\ \theta_B \\ \theta_C \end{bmatrix} = 0 \tag{9.13–3}$$

s, c, and m are stability functions for the columns.

If the final equation is omitted, and the remaining two are rearranged Eqn 9.13–3 becomes:

$$\begin{bmatrix} \dfrac{4s(1+c)}{m} & -s(1+c) \\ -s(1+c) & 4+s \end{bmatrix} \begin{bmatrix} \dfrac{\delta}{L} \\ \theta_B \end{bmatrix} = \begin{bmatrix} s(1+c) \\ -2 \end{bmatrix} \theta_C \tag{9.13–4}$$

This set can be solved for the ratios $\delta/L\theta_C$ and θ_B/θ_C. Using Cramer's rule,

$$\frac{\delta}{L\theta_C} = \frac{\begin{vmatrix} s(1+c) & -s(1+c) \\ -2 & 4+s \end{vmatrix}}{\begin{vmatrix} \dfrac{4s(1+c)}{m} & -s(1+c) \\ -s(1+c) & 4+s \end{vmatrix}}$$

and

$$\frac{\theta_B}{\theta_C} = \frac{\begin{vmatrix} 4s(1+c) & s(1+c) \\ -s(1+c) & -2 \end{vmatrix}}{\begin{vmatrix} \dfrac{4s(1+c)}{m} & -s(1+c) \\ -s(1+c) & 4+s \end{vmatrix}} \qquad (9.13\text{–}5)$$

The structure becomes unstable when matrix $\mathbf{K}_s(\lambda\mathbf{P})$ becomes singular. Trial and error, as previously explained in Section 9.9, yields the three critical values of ρ as:

$$\rho_{cr} = 0.75,\ 2.55,\ \text{and}\ 3.11$$

Substitution of these values into Eqns 9.13–5 yields the corresponding buckling modes:

$\rho = 0.75$

$$\frac{\delta}{L\theta_C} = 1.14 \qquad \frac{\theta_B}{\theta_C} = 1.00$$

$$(\delta/L:\theta_B:\theta_C) = (1.75:1:1)$$

$\rho = 2.55$

$$(\delta/L:\theta_B:\theta_C) = (0:-1:1)$$

$\rho = 3.11$

$$(\delta/L:\theta_B:\theta_C) = (-0.0025:1:1)$$

The three results are illustrated in Fig. 9.13–2.

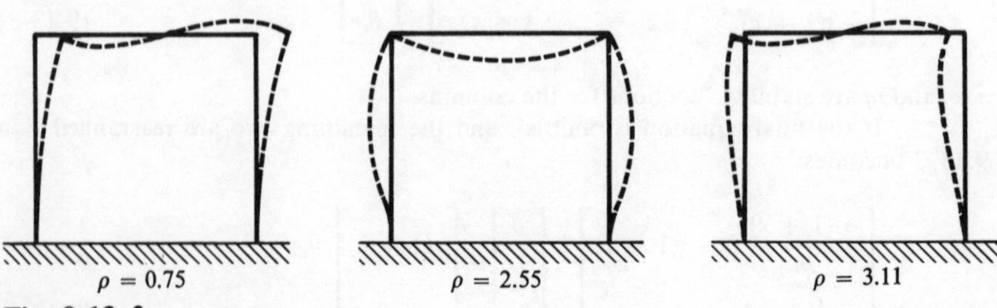

Fig. 9·13–2

9.14 Ultimate load analysis of structures

A method has been described in Section 9.12 of calculating the elastic critical load for a structure. It has been pointed out that this is an upper bound to the actual collapse load, and various factors may contribute towards collapse at a lower load. One of the most important of these factors is the onset of yield in the material of the structure.

Elementary studies of plastic behaviour in frames assume that collapse occurs when sufficient plastic hinges have formed to turn the structure into a mechanism. In practice, collapse will generally be due to a combination of plastic effects and instability —the concept of a deterioration of stability (see Wood[13]). The elastic analysis of section 9.12 can be modified to cater for plastic effects as follows.

Repeated analyses are performed at a series of gradually increasing load factors. The first load factor is chosen to give only elastic forces, but at each increment the bending moments obtained are compared with the fully plastic moments for the respective members. When a bending moment is found to have reached the plastic moment level, a plastic hinge has formed. The member stiffness matrices involved are modified to maintain the moment at this point at a constant level, and to allow continuous rotation, and the loading is then increased again. In this way the development of hinges is traced with increasing load. At each stage the positive definiteness of the stiffness matrix is checked, until eventually singularity is reached.

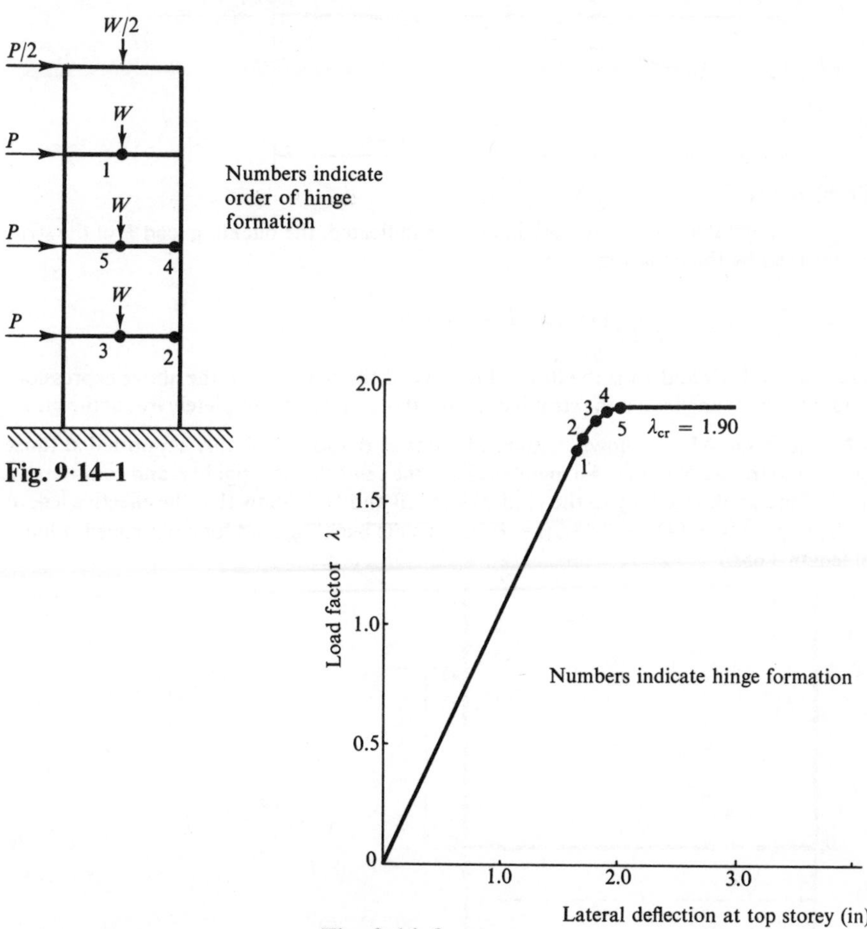

Fig. 9·14–1

Fig. 9·14–2

The results of such an analysis on the plane rigid-jointed frame of Fig. 9.14–1 described by Wood[13] are given in Fig. 9.14–2. The order and position of hinge formation are shown. It can be seen that the top storey lateral displacement becomes very large at a load factor of 1.90, indicating that collapse is imminent although the analysis showed that the stiffness matrix was still just positive-definite after the formation of the fifth hinge. This figure of 1.90 compares with a collapse load factor of 2.15 when instability is ignored.

A more detailed discussion of this topic will be found in Horne.[14]

Problems

In all problems it is assumed that the structure described can become unstable in one plane only, and that it is adequately restrained against displacement in the normal plane.

9.1. The initially straight strut AB shown in the figure is encastré at B and restrained transversely at A by an elastic support of stiffness *k* (load per unit displacement).

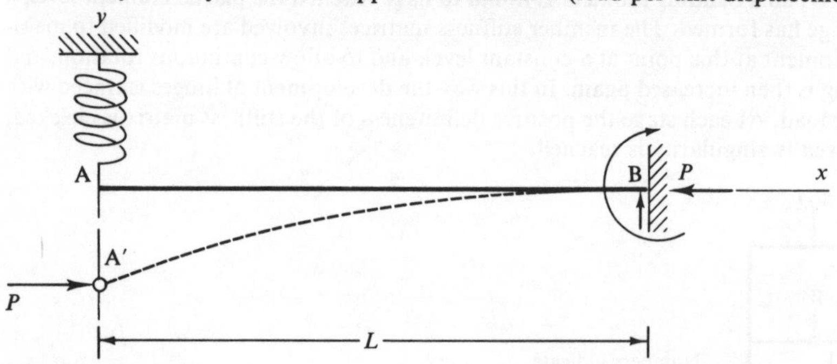

Problem 9·1

Show that, taking the origin at A′ as indicated, the buckling load P of the strut is governed by the equation

$$\alpha \cot \alpha L - \frac{k}{\alpha^2 EI}(\alpha L \cot \alpha L - 1) = 0$$

where $\alpha^2 = P/EI$ and EI is the flexural rigidity of the strut. From the above expression, obtain the buckling load if the strut is encastré at one end and completely free at the other.

9.2. The frame ABCD shown, with rigid joints at B and C is simply supported at those points and free at A and D. All members have the same flexural rigidity, and compressive loads P are applied axially to the vertical legs AB and DC. Show that the effective length of the struts AB and DC is $3.65L$ (i.e. P_{cr} equals the buckling load for a pin-ended column of length $3.65L$).

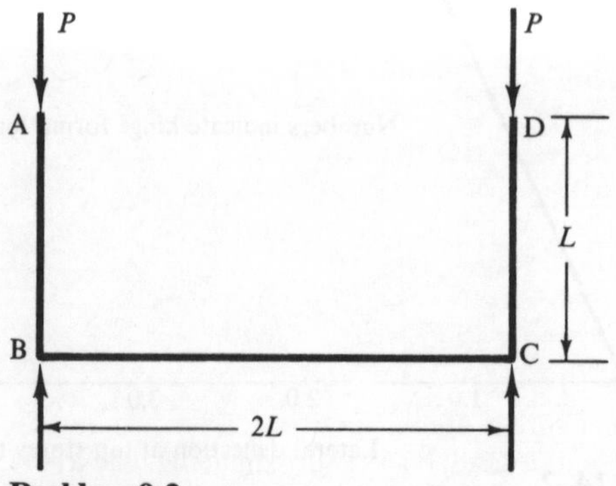

Problem 9·2

9.3. A strut of length L and uniform flexural rigidity EI carries an axial load of P and a central lateral load of W.

Show that buckling will take place when $P = \pi^2 EI/L^2$, and that if P is one-quarter of the buckling load the maximum bending moment in the strut is WL/π.

9.4. A strut AB of length L and of uniform flexural rigidity EI is pinned in position at both A and B. Rotation can occur freely at A, but is resisted at B by a restoring moment of kEI/L per radian. A compressive end load, P, is applied at A and B in the direction AB. Show that the corresponding critical load, P_1, is given by the roots of the equation

$$\tan \beta = \frac{k\beta,}{k + \beta^2}$$

where $\beta^2 = P_1 L^2/EI$.

9.5. A pin-ended strut of length L and uniform flexural rigidity EI is subjected to an axial force of P and a lateral point force of W at the span centre. Show that the maximum deflection is

$$\frac{WL^3}{16EI\alpha^2}\left(\frac{\tan \alpha}{\alpha} - 1\right) \quad \text{where} \quad \alpha = \frac{L}{2}\sqrt{\frac{P}{EI}} < \frac{\pi}{2}$$

9.6. A pin-ended strut of length L is initially curved and may be assumed to have a sinusoidal shape about one end as origin, with a maximum amplitude of a. Show that, when subjected to a thrust of 0.9 times the Euler buckling load, the maximum departure of the strut centre line from the line of action of the thrust is $10a$.

9.7. A straight, pin-ended strut of length 4 m and second moment of area 6×10^6 mm^4 is supported laterally at its mid-point by a spring of stiffness 2.0 kN/mm. Show that the minimum value of axial compressive load which causes elastic instability is approximately 2330 kN ($E = 210$ kN/mm^2).

9.8. The plane frame ABC shown in the figure is of constant second moment of area (2.8×10^6 mm^4) and cross-sectional area (3230 mm^2) throughout. Show that it could not become elastically unstable at a load P less than 720 kN ($E = 210$ kN/mm^2).

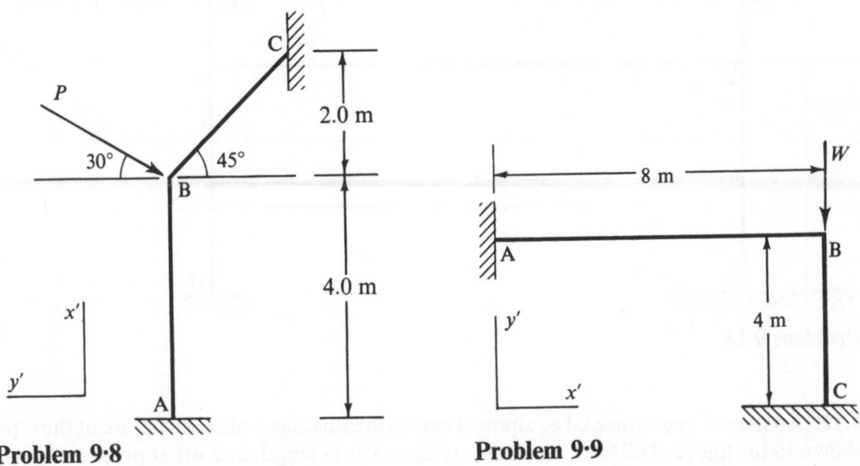

Problem 9·8 **Problem 9·9**

9.9. The plane frame ABC shown in the figure is of constant section throughout. Show that it would become elastically unstable at a load W of approximately 2000 kN ($I = 6 \times 10^6$ mm^4; $E = 210$ kN/mm^2).

9.10. The plane frame ABC shown in the figure is pinned at joint A, and carries a load λW at B. Show that the frame would become elastically unstable at a value of the load factor λ of between 9.00 and 9.50. (Assume that the effects of rotation only at the joints need be considered in the elastic analysis.)

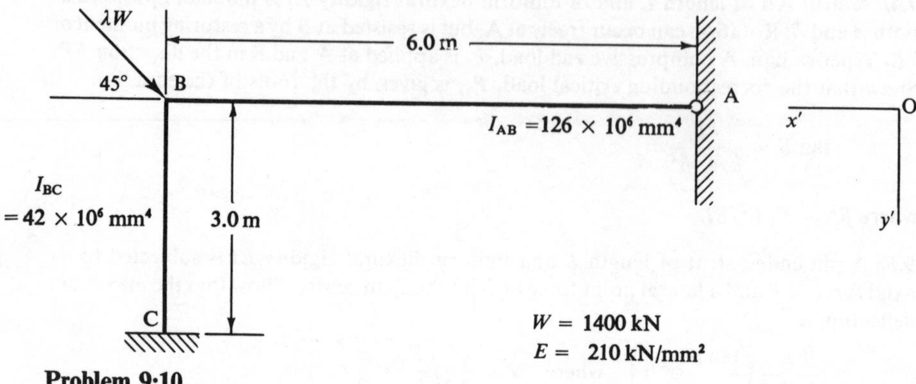

Problem 9·10

9.11. The portal frame ABCD in the figure carries symmetrical vertical loads at B and C. Show that the frame becomes unstable when the ratio of the axial load in a column to its Euler load is 0.61. Show that at this load the instability is of a sway mode, with displacement ratios

$$\frac{\Delta}{h} : \theta_B : \theta_C = 1.14:1:1$$

where Δ is the lateral displacement of B or C. (It may be assumed that only rotation, and the sway of the frame are significant in determining the frame stiffness.)

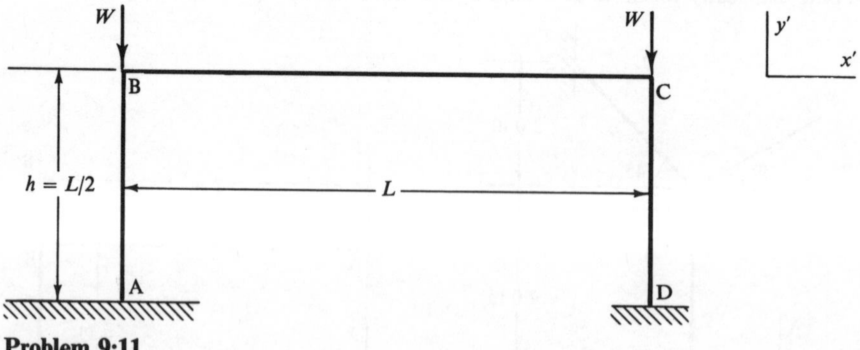

Problem 9·11

9.12. An item of experimental equipment can be idealized as a plane structure of the type shown in the figure. The four members are identical in length and other properties, and are identically loaded with a uniformly distributed load w. Members 1 and 3 are perpendicular to 2 and 4, all are rigidly connected together at the centre, and each is rigidly held at its other end. If the temperature of the assembly is raised, show that the structure will become unstable at 62°C above ambient.

The properties of each member are: $A = 50\,\text{mm}^2$; $I = 220\,\text{mm}^4$; $L = 250$ mm; $E = 70\ \text{kN/mm}^2$.

The coefficient of linear expansion $= 23 \times 10^{-6}$/degC

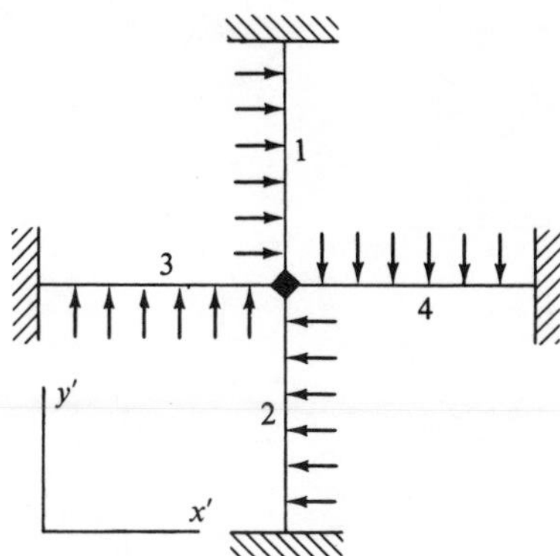

Problem 9·12

References

1 Ayrton, W.E. and J. Perry. On struts. *Engineer*, 10 and 24 December, 1886.

2 Robertson, A. The strength of struts. *Selected Engineering Paper No. 28.* Institution of Civil Engineers, 1925.

3 BS449 *The Use of Structural Steel in Building*, Part 2. British Standards Institution.

4 Stamenkovic, A. and Gardener, M.J. Effect of residual stresses on the column behaviour of hot-finished steel structural hollow sections. *Proc. Inst. Civ. Engrs*, Part 2, 1983, **75**, Dec, 599–616.

5 Dwight, J.B. Adaptation of Perry formula to represent the new European steel column-curves. *Steel Construction*, AISC, 9(1), 1975.

6 BS5950: Part 1: 1985. *Structural Use of Steelwork in Building*. British Standards Institution.

7 Bleich, F. *Buckling Strength of Metal Structures*, McGraw-Hill, 1952.

8 Godfrey, G.B. The allowable stresses in axially loaded steel struts. *Structural Engineer*, March, 1962, 97–112.

9 Johnston, Bruce, G. *Guide to Stability Design Criteria for Metal Structures*, Wiley, 1976.

10 Timoshenko, S. and Gere, J.M. *Theory of Elastic Stability*, McGraw-Hill, 1961.

11 Kirby, P.A. and Nethercot, D.A. *Design for Structural Stability*, Granada, 1979.

12 Thompson, J.M.T. and Hunt, G.W. *A General Theory of Elastic Stability*, Wiley, 1973.

13 Wood, R.H. The stability of tall buildings. *Proc. Inst. Civ. Engrs.*, 1958, **11**, 69.

14 Horne, M.R. *Plastic Theory of Structures*, Nelson, 1971.

General Reference

Horne, M. R. and W. Merchant. *The Stability of Frames*. Pergamon Press, Oxford, 1965.

Chapter 10
Structural dynamics

10.1 Aim

The aim of this chapter is to analyse the dynamical behaviour of structures with many degrees of freedom. The term 'degrees of freedom' has already been referred to in previous chapters. Briefly, it is equal to the number of independent coordinates or measurements required to define completely the configuration of the structure at any instant of time; in other words, it is equal to the number of independent types of motion possible in the structure.

Structural engineers often have to study the dynamical behaviour of structures with many degrees of freedom. For example, in the design of a tall building, the engineer would wish to know its natural frequencies of vibration in order to estimate the likelihood of resonance due to gusty winds, earthquakes, or other dynamic disturbances. He may also wish to determine the maximum acceleration or displacement which the building may experience when it is subjected to dynamic disturbances that are likely to occur during the lifetime of the structure.

In this chapter, it is assumed that the characteristics of the dynamic disturbances are known, and the emphasis is on the presentation of a matrix method of analysis for structures with any finite number of degrees of freedom, the method being intended for computer application.

It is true that structural engineers are generally concerned with structures that possess infinite numbers of degrees of freedom. However, in most cases approximate but useful information can be obtained by idealizing such structures into structures with finite numbers of degrees of freedom, as discussed in Section 10.7.

10.2 Structures with one degree of freedom

By definition, a structure with one degree of freedom is one whose configuration at any instant of time can be completely defined by one coordinate, i.e. by one measurement such as a displacement or an angle. Practical structures usually have many degrees of freedom. However, with some simplifying assumptions, many such structures can be regarded approximately as one-degree-of-freedom structures. Examples are shown in Fig. 10.2–1.

The mass M in Fig. 10.2–1(a) is constrained to move vertically. If the magnitude of M is large compared with the mass of the spring, then the latter can be assumed to be massless, in which case the structure becomes a one-degree-of-freedom structure and the position of the mass M is defined completely by the coordinate x measured from the equilibrium position. Similarly, in the portal frame in Fig. 10.2–1(b), if the vertical members are assumed to have negligible mass compared with the mass M of the horizontal member, which is assumed to be rigid, then the swaying motion of the horizontal member is defined by the displacement x, and the frame can be regarded as having only one degree of freedom. Similarly, in Fig. 10.2–1(c), if the distributed moment of inertia of the shaft is negligible compared with that of the disc, the structure becomes a one-degree-of-freedom structure and the torsional vibration is completely defined by the coordinate θ which is the angular displacement measured from the position of rest.

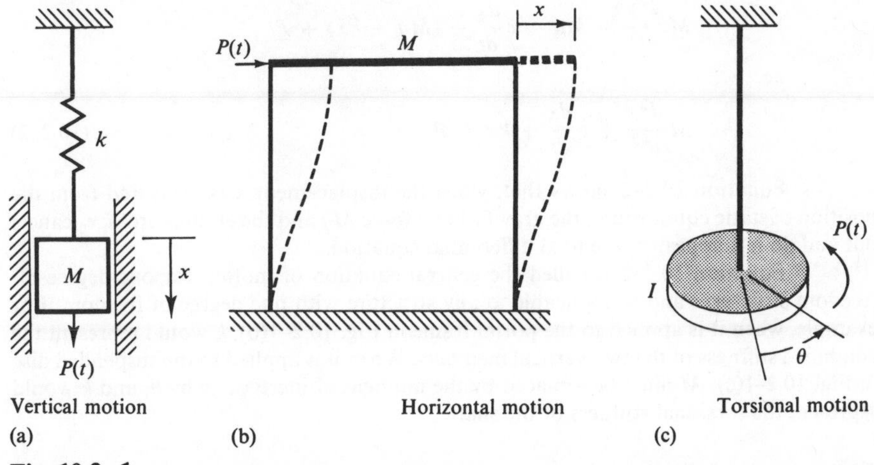

Fig. 10·2–1

In each of the structures in Fig. 10.2–1, the mass is assumed to be concentrated at one place. Such structures are sometimes called **lumped-mass structures.**

We shall begin the study of the dynamical behaviour of one-degree-of-freedom structures by consideration of the system in Fig. 10.2–1(a), the analysis being based simply on Newton's second law of motion.

$$\text{Mass} \times \text{acceleration} = \text{force} \tag{10.2–1}$$

The 'force' in the above equation consists of:

(a) The gravitation force Mg acting on the mass, where g is the acceleration due to gravity.

(b) The retarding force. For moderate velocities it is usual to assume that the retarding force is proportional to the velocity; such a force is called **viscous damping.**

$$\text{Viscous damping} = -c\frac{dx}{dt}$$

where t represents time, c is the **damping coefficient** which is usually assumed to be a constant, and the negative sign indicates that the damping force always acts in the opposite direction to that of the velocity dx/dt.

(c) The elastic force in the spring. If the stiffness of the spring is k, i.e. k absolute units of force are required to produce a unit extension, then

$$\text{elastic force} = -k(x_0 + x)$$

where x_0 is the extension of the spring when the system is in static equilibrium, and x is measured from the position of static equilibrium, so that the actual extension is $x_0 + x$. Since $x_0 = Mg/k$ at static equilibrium, the above equation becomes

$$\text{elastic force} = -k\left(\frac{Mg}{k} + x\right)$$
$$= -(Mg + kx)$$

(d) Any applied force P that may be acting on the mass. P is, in general, a function of the time t and is often referred to as the **forcing function**. Hence Eqn 10.2–1 becomes

$$M\frac{d^2x}{dt^2} = Mg - c\frac{dx}{dt} - (Mg + kx) + P$$

i.e.

$$M\frac{d^2x}{dt^2} + c\frac{dx}{dt} + kx = P \tag{10.2–2}$$

Equation 10.2–2 shows that, when the displacement x is measured from the position of static equilibrium, the gravitational force Mg and the elastic force kx_0 cancel out and do not appear in the final differential equation.

Equation 10.2–2 is called the general equation of motion for one-degree-of-freedom structures and is applicable to any structure with one degree of freedom. For example, when it is applied to the portal frame in Fig. 10.2–1(b), k would represent the combined stiffness of the two vertical members. When it is applied to the suspended disc in Fig. 10.2–1(c), M must be replaced by the moment of inertia I, x by θ, and k would represent the torsional stiffness of the shaft.

10.3 Free vibration with one degree of freedom

If in the above general equation of motion (Eqn 10.2–2) the term P representing the applied force is deleted, the equation becomes

$$M\frac{d^2x}{dt^2} + c\frac{dx}{dt} + kx = 0 \tag{10.3–1}$$

Equation 10.3–1 describes the behaviour of the system in the absence of applied disturbance; such intrinsic, or natural, behaviour of the system is called the free motion or **free vibration**. Free vibration occurs when, for example, the mass is displaced and then suddenly released.

The solution of Eqn 10.3–1 is

$$x = A_1\,e^{m_1t} + A_2\,e^{m_2t} \tag{10.3–2}$$

where m_1, m_2, and the constants of integration A_1 and A_2, may be real or complex, depending on the relative values of c, k, and M, as explained in Cases 1 to 4 below. The constants of integration A_1 and A_2 can be evaluated from the initial conditions, e.g. the known displacements at two different times, or the known velocities at two different times, or the known velocity at one time and the known displacement at another time. By substituting Eqn 10.3–2 into 10.3–1, it will be found that m_1 and m_2 are given by the roots of the following equation, called the auxiliary equation:

$$Mm^2 + cm + k = 0$$

or

$$m_1 = -\frac{c}{2M} + \frac{1}{2M}\sqrt{(c^2 - 4kM)} \tag{10.3–3(a)}$$

$$m_2 = -\frac{c}{2M} - \frac{1}{2M}\sqrt{(c^2 - 4kM)} \tag{10.3–3(b))}$$

There are four cases to consider:

$$c^2 - 4kM \begin{cases} > 0 & \text{(Case 1)} \\ = 0 & \text{(Case 2)} \\ < 0 & \text{(Case 3)} \end{cases} \qquad \begin{matrix} (10.3\text{–}4(a)) \\ (10.3\text{–}4(b)) \\ (10.3\text{–}4(c)) \end{matrix}$$

and

$$c = 0 \quad \text{(Case 4)} \qquad (10.3\text{–}4(d))$$

Case 1: *$c^2 - 4kM > 0$.* In this case, there is a comparatively large amount of damping, and, naturally enough, the system is said to be **overdamped**. Since $c^2 - 4kM > 0$, both m_1 and m_2 in Eqn 10.3–3 are real (so are A_1 and A_2) and negative, so that the equation of motion (Eqn 10.3–2)

$$x = A_1 \exp(m_1 t) + A_2 \exp(m_2 t) \qquad (10.3\text{–}5)$$

represents a gradual creeping back of the mass towards the position of static equilibrium (Fig. 10.3–1). For overdamping to occur, the damping has in general to be much greater than that occurring in practical civil engineering structures, although dash-pot devices producing substantial damping are occasionally included in machine systems to prevent vibration. The behaviour of overdamped systems is of comparatively little practical interest to the structural engineer and will not be investigated any further.

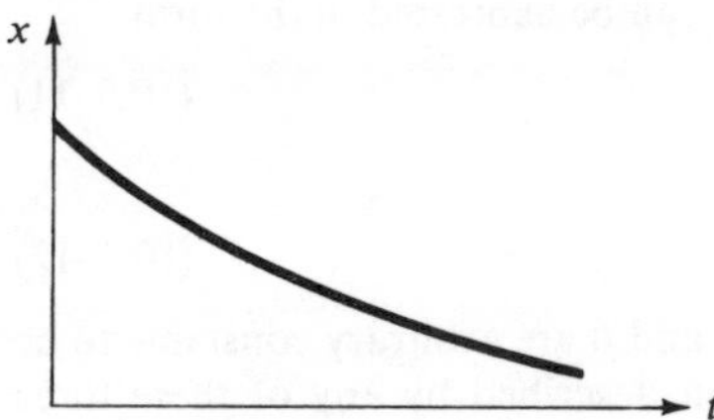

Typical displacement-time curve for an overdamped system

Fig. 10·3–1

Case 2: *$c^2 - 4kM = 0$.* In this case the motion is said to be **critically damped**, and the particular value of the damping coefficient,

$$c_c = 2\sqrt{(kM)} \qquad (10.3\text{–}6)$$

is called the **critical damping coefficient**.

When $c = c_c$, the values of m_1, m_2, A_1, and A_2 are all real, m_1 and m_2 (Eqn 10.3–3) being each equal to $-c/(2M)$. In such a case (see Sokolnikoff and Redheffer[1]), the solution of Eqn 10.3–1 takes the form

$$\begin{aligned} x &= A_1 \exp[-(c/2M)t] + A_2 t \exp[-(c/2M)t] \\ &= \exp[-(c/2M)t](A_1 + A_2 t) \end{aligned} \qquad (10.3\text{–}7)$$

Again, Eqn 10.3–7 shows that there is no vibratory motion; critical damping is just a particular case of overdamping, so that the displacement–time curve would be similar to that in Fig. 10.3–1. The case of critical damping is of little practical importance in itself, but the magnitude of the critical damping coefficient is a useful measure of the damping capacity of a structure. It is often convenient to express the damping coefficient of a structure as a percentage of the critical damping coefficient. For example, a structure with 3% critical damping will have a damping coefficient of

$$c = 0.03c_c = 0.03[2\sqrt{(kM)}] = 0.06\sqrt{(kM)}$$

Case 3: $c^2 - 4kM < 0$. In this case, the motion is said to be **underdamped**, or simply **damped**. The values of m_1 and m_2 in Eqn 10.3–3 are the conjugate complex numbers

$$\left.\begin{matrix} m_1 \\ m_2 \end{matrix}\right\} = -\frac{c}{2M} \pm i\frac{1}{2M}\sqrt{(4kM - c^2)} \quad \text{where } i = \sqrt{-1}$$

$$= -\frac{c}{2M} \pm i\omega_d \tag{10.3–8}$$

where ω_d is sometimes called the **circular natural frequency with damping** and is usually expressed in radians per second. The equation of motion (Eqn 10.3–2) now becomes

$$x = \exp\left[-(c/2M)t\right]\left[A_1 \exp(i\omega_d t) + A_2 \exp(-i\omega_d t)\right] \tag{10.3–9}$$

Expressing $\exp(i\omega_d t)$ and $\exp(-i\omega_d t)$ in terms of trigonometric functions, we have

$$x = \exp(-ct/2M)\left[i(A_1 - A_2)\sin\omega_d t + (A_1 + A_2)\cos\omega_d t\right]$$

Thus it is seen that A_1 and A_2 are complex conjugates, so that $A_1 - A_2$ is imaginary while $A_1 + A_2$ is real. Replacing the quantity $i(A_1 - A_2)$ by A, and $(A_1 + A_2)$ by B we have

$$x = \exp\left[-(c/2M)t\right](A\sin\omega_d t + B\cos\omega_d t) \tag{10.3–10}$$

The reader should verify that Eqn 10.3–10 can be expressed in the form

$$x = C\exp(-ct/2M)\sin(\omega_d t + \alpha) \tag{10.3–11}$$

or in the form

$$x = D\exp(-ct/2M)\cos(\omega_d t + \beta) \tag{10.3–12}$$

In Eqns 10.3–10 to 10.3–12, A, B, C, D, α, and β are arbitrary constants to be determined from the initial conditions. The motion described by any of these three equations is known as a **damped free vibration** and a typical displacement–time plot is shown in Fig. 10.3–2, using Eqn 10.3–11.

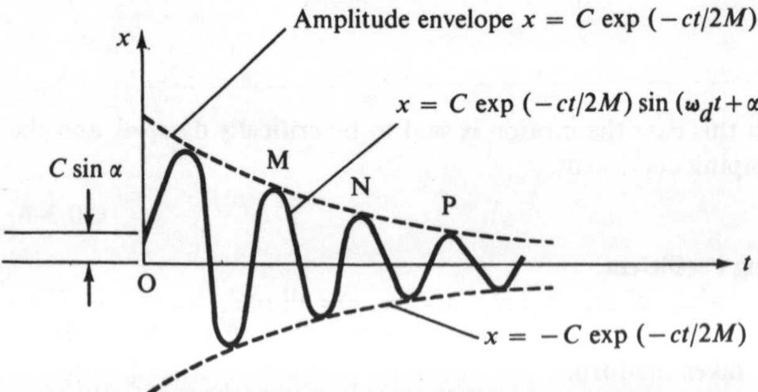

Fig. 10·3–2 Damped free vibration

Note that the motion is not truly periodic, because the multiplying factor $\exp(-ct/2M)$ is continuously decreasing. Figure 10.3–2 shows that the displacement x is zero at intervals, i.e. the mass M passes through the position of static equilibrium at intervals. From Eqn 10.3–11, x is zero whenever $\sin(\omega_d t + \alpha)$ is zero, i.e. whenever

$$\omega_d t + \alpha = \pi + n\pi$$

or

$$t = \frac{1}{\omega_d}(\pi - \alpha) + n\frac{\pi}{\omega_d} \qquad (n = 0, 1, 2 \ldots) \tag{10.3–13}$$

That is, x is zero at *regular* intervals of π/ω_d. The period T_d of the damped free vibration is twice this interval, i.e.

$$T_d = \frac{2\pi}{\omega_d} \tag{10.3–14}$$

The frequency f_d of the vibration is the reciprocal of T_d:

$$f_d = \frac{1}{T_d} = \frac{\omega_d}{2\pi} \tag{10.3–15}$$

Referring to Fig. 10.3–2, the amplitudes at points M, N, P... occur at successive intervals of T_d. From Eqn 10.3–11 or 10.3–12 the ratios of the amplitudes are constant:

$$\frac{x_M}{x_N} = \frac{x_N}{x_P} = \cdots = \frac{\exp(-ct/2M)}{\exp[-c(t + T_d)/2M]} = \exp(cT_d/2M)$$

The natural logarithm of the ratio is called the **logarithmic decrement**, i.e.

$$\text{logarithmic decrement } \delta = \frac{cT_d}{2M} \tag{10.3–16}$$

The logarithmic decrement is often used as an indication of the damping capacity of a structure.

Case 4: $c = 0$. If $c = 0$, the system is undamped. From Eqn 10.3–3,

$$\begin{matrix} m_1 \\ m_2 \end{matrix} = \pm i\sqrt{\frac{k}{M}} = \pm i\omega \tag{10.3–17}$$

so that the equation of motion (Eqn 10.3–2) becomes

$$x = A_1 e^{i\omega t} + A_2 e^{-i\omega t} \tag{10.3–18}$$

or, in terms of trigonometric functions,

$$x = A \sin \omega t + B \cos \omega t \tag{10.3–19}$$

The quantity ω in Eqns 10.3–17 to 10.3–19 is called the **natural circular frequency**; the natural period of vibration is

$$T = \frac{2\pi}{\omega} \tag{10.3–20}$$

and the natural frequency of the vibration is

$$f = \frac{1}{T} = \frac{\omega}{2\pi} \tag{10.3–21}$$

It can be seen from Eqns 10.3–17 and 10.3–8 that ω is always greater than ω_d; hence the frequency f is always higher than the frequency f_d.

Eqn 10.3–19 can be expressed as

$$x = C \sin(\omega t + \alpha) \tag{10.3–22}$$

where $C = (A^2 + B^2)^{1/2}$ and the phase angle $\alpha = \tan^{-1}(-B/A)$, or as

$$x = D\cos(\omega t + \beta) \tag{10.3–23}$$

where $D = (A^2 + B^2)^{1/2}$ and the phase angle $\beta = \tan^{-1}(-A/B)$.

The motion described by Eqns 10.3–19, 10.3–22, or 10.3–23 is known as **undamped free vibration** and is strictly periodic; its general appearance is shown in Fig. 10.3–3.

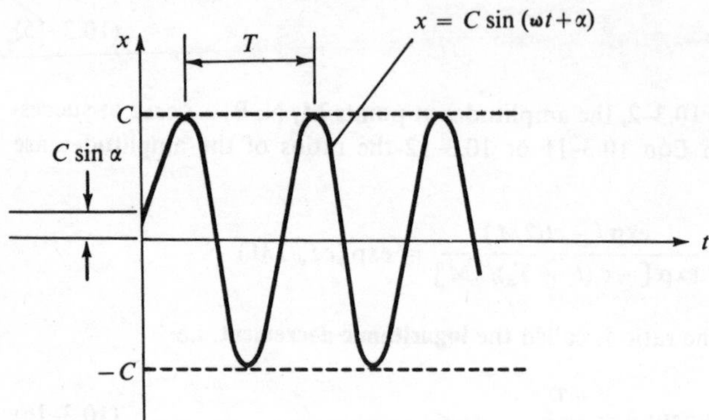

Fig. 10·3–3 Undamped free vibration

The arbitrary constants A and B in Eqn 10.3–19, or C and α in Eqn 10.3–22 or D and β in Eqn 10.3–23, are determined from the initial conditions. For example, if it is known that at time $t = 0$, the displacement and velocity of the mass M are x_0 and $\dot{x}_0$ respectively, then substituting $x = x_0$ and $t = 0$ into Eqn 10.3–19,

$$x_0 = 0 + B, \quad \text{i.e.} \quad B = x_0$$

Differentiating Eqn 10.3–19 with respect to t and substituting $dx/dt = \dot{x}_0$ at $t = 0$,

$$x_0 = A\omega + 0, \quad \text{i.e.} \quad A = \frac{\dot{x}_0}{\omega}$$

Hence, the equation of motion is

$$x = \frac{\dot{x}_0}{\omega}\sin\omega t + x_0\cos\omega t \tag{10.3–24}$$

(for $x = x_0$ and $dx/dt = \dot{x}_0$ at $t = 0$). Similarly, if at $t = \tau$, $x = x_\tau$ and $dx/dt = \dot{x}_\tau$, it is easy to determine A and B and show that

$$x = \frac{\dot{x}_\tau}{\omega}\sin\omega(t - \tau) + x_\tau\cos\omega(t - \tau) \tag{10.3–25}$$

(for $x = x_\tau$ and $dx/dt = \dot{x}_\tau$ at $t = \tau$).

Example 10.3–1. (a) Determine the units of the logarithmic decrement

(b) If the damping coefficient c of a structure is r times the critical damping coefficient c_c, determine the logarithmic decrement.

SOLUTION (a)

It is clear from Eqn 10.2–2 that the unit of the quantity $c(dx/dt)$ is N, hence

$$\text{(units of } c) \times (\text{ms}^{-1}) = \text{N}$$
$$= \text{kg m s}^{-2}$$

then

$$\text{units of } c = (\text{kg m s}^{-2})\,\text{m}^{-1}\,\text{s}$$
$$= \text{kg s}^{-1} \quad \text{(i.e. kilograms per second)}$$

From Eqn 10.3–8,

$$\text{units of } \omega_d = \frac{\sqrt{[(\text{N m}^{-1})\text{ kg} - (\text{kg s}^{-1})^2]}}{\text{kg}}$$
$$= \sqrt{(\text{N m}^{-1}\text{ kg}^{-1} - \text{s}^{-2})}$$
$$= \sqrt{[(\text{kg m s}^{-2})\text{m}^{-1}\text{ kg}^{-1} - \text{s}^{-2}]} = \sqrt{\text{s}^{-2}}$$
$$= \text{s}^{-1} \quad \text{(radians per second)}$$

then

$$\text{units of period } T_d = 1/\text{s}^{-1} = \text{s} \quad \text{(seconds)}$$

From Eqn 10.3–16

$$\text{units of } \delta = \frac{(\text{units of } c)(\text{units of } T_d)}{(\text{units of } M)}$$
$$= \frac{(\text{kg s}^{-1})\text{ (s)}}{\text{kg}}$$
$$= \underline{\text{dimensionless}}$$

i.e. as expected, the logarithmic decrement δ is a number, without units.

(b) From Eqn 10.3–16,

$$\delta = \frac{cT_d}{2M} \quad \text{where} \quad T_d = \frac{2\pi}{\omega_d} \quad \text{(see Eqn 10.3–14)}$$

and

$$\omega_d = \frac{(4kM - c^2)^{1/2}}{2M} \quad \text{(see Eqn 10.3–8)}$$

then

$$\delta = \frac{2\pi c}{(4kM - c^2)^{1/2}} \quad \text{and} \quad c = rc_c \quad \text{(given)}$$
$$= r \times 2\sqrt{(kM)} \quad \text{(from Eqn 10.3–6)}$$

Hence

$$\delta = \frac{2\pi r[2\sqrt{(kM)}]}{(4kM - r^2 4kM)^{1/2}} = \underline{2\pi r(1 - r^2)^{-1/2}}$$

For practical structures, r often lies between a few per cent to 20%, so that $\delta \approx 2\pi r$.

Example 10.3–2. Figure 10.3–4 shows a portal frame ABCD; it may be assumed that the horizontal member is infinitely stiff and that the vertical members have negligible mass compared with that of the horizontal member. If there is no damping, determine

(a) the natural frequency f and the natural period T;
(b) the displacement at time t seconds if the member BC is displaced a distance l millimetres and then suddenly released at $t = 1$ s.

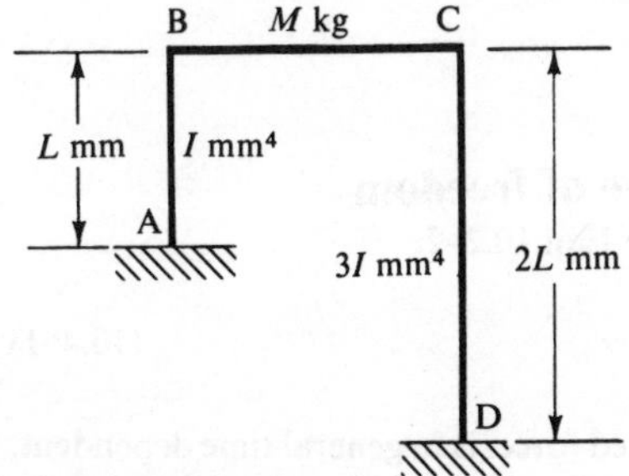

Fig. 10·3–4

SOLUTION (1) From Section 5.2, Eqn 5.2–1, the stiffnesses of the members AB and CD are:

$$k_{AB} = \frac{12EI}{L^3}\ \text{N/mm}; \qquad k_{CD} = \frac{12E(3I)}{(2L)^3} = 4.5\frac{EI}{L^3}\ \text{N/mm}$$

Hence

$$k = k_{AB} + k_{CD} = 16.5\frac{EI}{L^3}\ \text{N/mm}$$

$$= 16\,500\frac{EI}{L^3}\ \text{N/m}$$

From Eqn 10.3–17,

$$\omega = \sqrt{\frac{k}{M}}$$

(where k is in newtons per metre and M in kilograms), then

$$\omega = \sqrt{\left(\frac{16\,500EI}{ML^3}\right)} = 128\sqrt{\left(\frac{EI}{ML^3}\right)} \quad \text{radians per second}$$

From Eqn 10.3–20,

$$T = \frac{2\pi}{\omega} = \frac{2\pi}{128}\sqrt{\left(\frac{ML^3}{EI}\right)} \quad \text{seconds}$$

$$f = \frac{1}{T} = \frac{128}{2\pi}\sqrt{\left(\frac{EI}{ML^3}\right)} \quad \text{cycles per second}$$

(2) From Eqn 10.3–25, if the initial conditions are $x = x_\tau$ m and $\dot{x} = \dot{x}_\tau$ m/s at $t = \tau$ seconds, then

$$x = \frac{\dot{x}_\tau}{\omega}\sin\omega(t-\tau) + x_\tau\cos\omega(t-\tau) \quad \text{m} \qquad (10.3\text{–}26)$$

In this example, $x_\tau = l$ mm $= 0.001\,l$ m, $\dot{x}_\tau = 0$, $\tau = 1$ s, and $\omega = 128\sqrt{[(EI)/(ML^3)]}$, hence

$$x = 0.001\,l\cos\left[128\sqrt{\left(\frac{EI}{ML^3}\right)}(t-1)\right] \text{m}$$

The reader should pay particular attention to units in structural dynamics problems. The stiffness k should be in absolute units, e.g. N/m; the mass M should be in mass units, e.g. kg. Similarly, in using Eqn 10.3–26, x should be expressed in metres, $\dot{x}$ in m/s, and t and τ in seconds.

10.4 Forced vibration with one degree of freedom

The general equation of forced motion is given by Eqn 10.2–2

$$M\frac{d^2x}{dt^2} + c\frac{dx}{dt} + kx = P \qquad (10.4\text{–}1)$$

where the **forcing function** P, representing the applied force, is in general time dependent. Equation 10.4–1 is known as a homogeneous differential equation of the second order and its general solution consists of two parts: (a) a 'complementary function' x_c, which is the solution of Eqn 10.4–1 when P is zero, and which contains two arbitrary constants;

and (b) a 'particular integral' x_p, which is a solution of Eqn 10.4–1 in its entirety, but which does not contain any arbitrary constants. Thus the solution of Eqn 10.4–1 is

$$x = x_c + x_p \tag{10.4–2}$$

Structural engineers are concerned mainly with underdamped vibrations, in which case x_c is given by Eqn 10.3–10 as

$$x_c = \exp(-ct/2M)(A \sin \omega_d t + B \cos \omega_d t) \tag{10.4–3}$$

where ω_d is defined by Eqn 10.3–8.

That is, the general solution of Eqn 10.4–1 for the case of underdamped vibrations is

$$x = \exp(-ct/2M)(A \sin \omega_d t + B \cos \omega_d t) + x_p \tag{10.4–4}$$

x_p would depend on P, but regardless of the nature of x_p it is possible to conclude that, as time t increases, the motion represented by x_c will die away exponentially (see Fig. 10.3–2). Hence x_c is often called the **transient response** and x_p the **steady-state response**.

We shall now study the motion of the structure under various forcing functions.

Case 1: The harmonic force $P = P_0 \cos \omega_l t$. The equation of motion for this classical problem of forced vibration under a pulsating load of circular frequency ω_l is

$$M\frac{d^2x}{dt^2} + c\frac{dx}{dt} + kx = P_0 \cos \omega_l t \tag{10.4–5}$$

Using the differential operator D, the particular integral is given by

$$x_p = \frac{1}{M\,D^2 + c\,D + k} P_0 \cos \omega_l t$$

where $D \equiv d\ /dt$. After some manipulation, this reduces to

$$x_p = \frac{P_0 \cos(\omega_l t - \phi)}{[(k - M\omega_l^2)^2 + c^2\omega_l^2]^{1/2}}$$

where

$$\phi = \tan^{-1}\frac{c\omega_l}{k - M\omega_l^2} \tag{10.4–6}$$

ϕ is called the **phase angle** or the **angle of lag** of the response. With the complementary function x_c as given by Eqn 10.4–3, the complete solution is

$$x = \exp(-ct/2M)(A \sin \omega_d t + B \cos \omega_d t) + \frac{P_0 \cos(\omega_l t - \phi)}{[(k - M\omega_l^2)^2 + c^2\omega_l^2]^{1/2}} \tag{10.4–7}$$

As explained above (see also Section 10.3, Case 3), the transient response x_c dies away exponentially; after some time, only the steady-state response x_p will remain so that the motion is described by Eqn 10.4–6. The circular frequency of the remaining motion is the same as that of the excitation load P, and the amplitude of the displacement is

$$X = \frac{P_0}{[(k - M\omega_l^2)^2 + c^2\omega_l^2]^{1/2}}$$

The displacement of the mass under a static force P_0 is $\Delta = P_0/k$. The ratio X/Δ is called the **dynamic magnification factor** or simply the magnification factor:

$$\frac{X}{\Delta} = \frac{P_0}{[(k - M\omega_l^2)^2 + c^2\omega_l^2]^{1/2}} \bigg/ \left(\frac{P_0}{k}\right)$$

This can be reduced to

$$\frac{X}{\Delta} = \frac{1}{\{[1 - (\omega_l/\omega)^2]^2 + 4(c/c_c)^2(\omega_l/\omega)^2\}^{1/2}} \tag{10.4–8}$$

where the natural circular frequency $\omega = \sqrt{(k/M)}$ (from Eqn 10.3–17), and the critical damping coefficient $c_c = 2\sqrt{(kM)}$ (from Eqn 10.3–6).

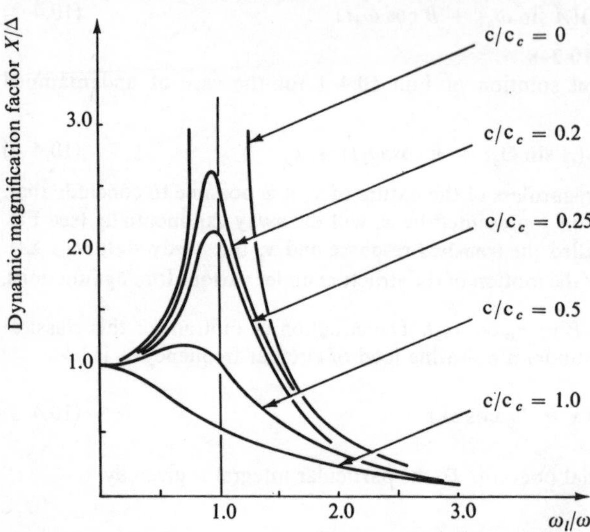

Fig. 10·4–1

Figure 10.4–1 shows curves of the dynamic magnification factor X/Δ plotted against the frequency ratio ω_l/ω for different damping ratios c/c_c. The following broad conclusions can be drawn from these curves:

(1) The magnification factor approaches unity if ω_l/ω is small, regardless of the amount of damping. This indicates that the definition of a load as static or dynamic is of relative rather than fundamental significance. If the frequency f_l $(= \omega_l/2\pi)$ of the load is small *compared* with the natural frequency f $(= \omega/2\pi)$ of the structure, the load may be considered static.

(2) Regardless of the amount of damping, the magnification factor approaches zero when the frequency of the load is large compared with the natural frequency of the structure. That is, if ω_l/ω is large, the load produces virtually no displacement.

(3) For damping ratios c/c_c less than about 0.5, the magnification factor is largest when ω_l/ω is just under 1.

(4) When the amount of damping is small, the magnification factor becomes very large for ω_l/ω near to unity. Equation 10.4–8 would easily give the false impression that for $c = 0$ and $\omega_l = \omega$, the magnification factor becomes infinity. Such a condition, where $X/\Delta \to \infty$, is known as **mathematical resonance**. Mathematical resonance does not occur in practice, unless the pulsating force acts for an infinitely long time. If we now re-examine Eqn 10.4–7, it would become clear that when $c = 0$, the term $\exp(-ct/2M)$ becomes unity, and the complementary function will no longer die away with time. The motion represented by this complementary function interferes with that represented by the particular integral so that the displacement x remains finite as long as t remains finite (see Example 10.4–1). However, the important point to note is that, in practical structures, X/Δ can become objectionably large if the frequency of the pulsating force is equal to, or very nearly equal to, the natural frequency of the structure.

Case 2: Suddenly applied constant force $P = P_0$ $(c = 0)$. In this case, and in subsequent cases, the damping coefficient c will be assumed to be zero, with the result that the analysis is simplified considerably. Another justification for the assumption is that, in practice, the effect of damping on dynamical behaviour is often unimportant. This is because dynamic disturbances often act on a structure for a relatively short time, and the amount of damping in many practical structures is sufficiently small (for steel frames c/c_c is usually about 1 or 2%) to have negligible effect on dynamical behaviour during the critical time interval of the first few cycles of vibration. In any case, where it is suspected that damping effects may be significant, the damping coefficient c can be included in the analysis, which will be similar to an analysis without damping but may involve much more algebra in some cases.

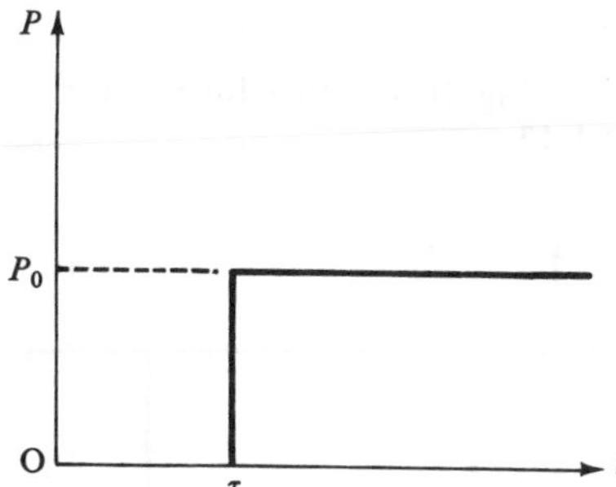

Fig. 10·4–2

Let us now return to the case of a structure subjected to a constant force P_0 suddenly applied at time $t = \tau$ (Fig. 10.4–2). The governing differential equation (Eqn 10.4–1) reduces to

$$M\frac{d^2x}{dt^2} + kx = P_0 \tag{10.4–9}$$

From Eqn 10.3–19 the complementary function is

$$x_c = A \sin \omega t + B \cos \omega t \tag{10.4–10}$$

The particular integral is, by inspection,

$$x_p = \frac{P_0}{k} \tag{10.4–11}$$

Hence the complete solution is $x = x_c + x_p$, i.e.

$$x = A \sin \omega t + B \cos \omega t + \frac{P_0}{k} \tag{10.4–12}$$

Suppose the initial conditions are:

$$\text{at} \quad t = \tau \quad \begin{cases} x = x_\tau \\ \dfrac{dx}{dt} = \dot{x}_\tau \end{cases}$$

The reader should use these two conditions to determine the arbitrary constants A and B and verify that

$$x = \frac{\dot{x}_\tau}{\omega} \sin \omega(t - \tau) + \left(x_\tau - \frac{P_0}{k}\right) \cos \omega(t - \tau) + \frac{P_0}{k} \tag{10.4–13}$$

10.4–21 and it gives $x = dx/dt = 0$ at $t = \tau'$; Eqn 10.4–13 is the solution of Eqn 10.4–20 and, at $t = \tau'$, it gives the values of x and dx/dt as required by Eqn 10.4–19. Hence the sum of Eqn 10.4–13 and the modified Eqn 10.4–15 is the solution of Eqn 10.4–19. Hence the superposition Fig. 10.4–5(a) = Fig. 10.4–5(b) + (c) is valid.

Case 3: Impulse force ($c = 0$). Suppose at time $t = \tau$ a structure is at rest, i.e. $x(\tau) = 0$, $dx/dt(\tau) = 0$. If at this instant, an impulse I is applied to the mass M, then from the law of conservation of momentum, the velocity of the mass immediately after the application of the momentum is

$$\frac{dx}{dt} = \frac{I}{M}$$

while the displacement remains instantaneously zero. After the application of I, the motion will be undamped free vibration (Eqn 10.3–19):

$$x = A \sin \omega t + B \cos \omega t \tag{10.4–22}$$

where the initial conditions are, at $t = \tau$ just after application of I, $x(\tau) = 0$, $dx/dt(\tau) = I/M$. The reader should determine A and B from these conditions and verify that

$$x = \frac{I}{M\omega} \sin \omega(t - \tau) \tag{10.4–23}$$

Next consider a more general case. Suppose a structure is vibrating freely such that at time $t = \tau$, the displacement and velocity are $x(\tau) = x_\tau$, $dx/dt(\tau) = \dot{x}_\tau$. If at this instant $t = \tau$, an impulse I is applied, then the subsequent motion will be undamped free vibration (Eqn 10.4–22) with the initial conditions that at $t = \tau$ just after application of the impulse,

$$x(\tau) = x_\tau; \qquad \frac{dx}{dt}(\tau) = \dot{x}_\tau + \frac{I}{M}$$

From these conditions, A and B in Eqn 10.4–22 can be determined, and the motion shown to be

$$\begin{aligned} x &= \left[\frac{\dot{x}_\tau}{\omega} \sin \omega(t - \tau) + x_\tau \cos \omega(t - \tau)\right] + \left[\frac{I \sin \omega(t - \tau)}{M\omega}\right] \\ &= (\text{Eqn } 10.3\text{–}25) + (\text{Eqn } 10.4\text{–}23) \end{aligned} \tag{10.4–24}$$

Suppose at a later time $t = \tau'$, another impulse I' is applied, then it can be shown that (see Problem 10.3) the subsequent motion is

$$\begin{aligned} x &= \left[\frac{\dot{x}_\tau}{\omega} \sin \omega(t - \tau) + x_\tau \cos \omega(t - \tau)\right] + \left[\frac{I \sin \omega(t - \tau)}{M\omega}\right] \\ &\quad + \left[\frac{I' \sin \omega(t - \tau')}{M\omega}\right] \\ &= (\text{Eqn } 10.3\text{–}25) + (\text{Eqn } 10.4\text{–}23) \\ &\quad + (\text{Eqn } 10.4\text{–}23 \text{ with } \tau' \text{ replacing } \tau \text{ and } I' \text{ replacing } I) \end{aligned} \tag{10.4–25}$$

The superposition in Eqn 10.4–25 is based on the fact that the governing differential equations are always linear (see discussion at end of Case 2). Such superposition can be extended indefinitely, so that we can conclude as follows: If a structure is freely vibrating such that at time $t = \tau$ its displacement and velocity are $x(\tau) = x_\tau$.

$dx/dt(\tau) = \dot{x}_\tau$, and if at later times $\tau_1, \tau_2, \tau_3 \ldots \tau_n$, impulses $I_1, I_2, I_3 \ldots I_n$ are applied. then for $t > \tau_n$ the motion is obtained by superposition, namely

$$x = \frac{\dot{x}_\tau}{\omega} \sin \omega(t - \tau) + x_\tau \cos \omega(t - \tau) + \sum_{i=1}^{i=n} \frac{I_i \sin \omega(t - \tau_i)}{M\omega} \quad (10.4\text{–}26)$$

At an intermediate time, say, $t_4 > t > t_3$ we would have

$$x = \frac{\dot{x}_\tau}{\omega} \sin \omega(t - \tau) + x_\tau \cos \omega(t - \tau) + \sum_{i=1}^{i=3} \frac{I_i \sin \omega(t - \tau_i)}{M\omega} \quad (10.4\text{–}27)$$

Case 4: The arbitrary force ($c = 0$). Any arbitrary-force diagram can be regarded as the superposition of a series of impulses. The shaded area in Fig. 10.4–6, $I = P\,d\tau$, can be regarded as a typical impulse occurring at time $t = \tau$. If the initial conditions at $t = \tau_1$, are: $x = x_{\tau_1}$, $dx/dt = x_{\tau_1}$, then, from Eqn 10.4–26, the motion for $t \geqslant \tau_2$ is

$$x = \frac{\dot{x}_{\tau_1}}{\omega} \sin \omega(t - \tau_1) + x_{\tau_1} \cos \omega(t - \tau_1) + \int_{\tau_1}^{\tau_2} \frac{P \sin \omega(t - \tau)\, d\tau}{M\omega} \quad (10.4\text{–}28)$$

Similarly, for $t = \tau'$ $(\tau_1 < \tau' < \tau_2)$:

$$x = \frac{\dot{x}_{\tau_1}}{\omega} \sin \omega(t - \tau_1) + x_{\tau_1} \cos \omega(t - \tau_1) + \int_{\tau_1}^{\tau'} \frac{P \sin \omega(t - \tau)\, d\tau}{M\omega} \quad (10.4\text{–}29)$$

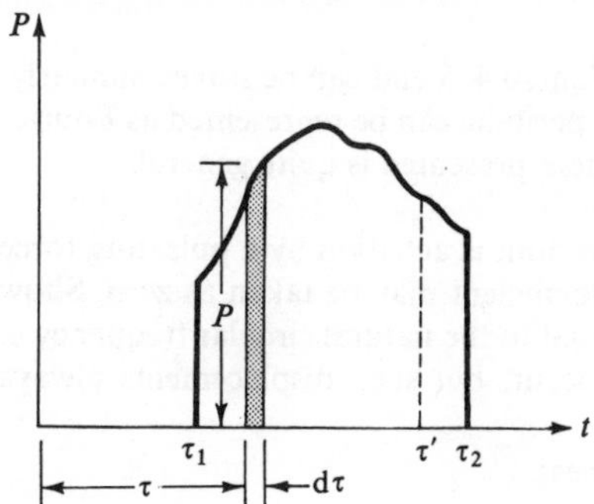

Fig. 10·4–6

Case 5: Support movements ($c = 0$). Referring to the frame in Fig. 10.2–1(b), suppose there is a **periodic foundation displacement** defined by

$$x_f = s \cos \omega_f t \quad (10.4\text{–}30)$$

i.e. the amplitude of the foundation movement is s and its frequency is $2\pi/\omega_f$.

Under the action of this foundation disturbance alone, the motion of the mass M is

$$M \frac{d^2x}{dt^2} + k(x - x_f) = 0 \quad (10.4\text{–}31)$$

Note that the inertia force, $M\,d^2x/dt^2$, depends on the absolute acceleration of the mass M, while the elastic force, $k(x - x_f)$, depends on the displacement of the mass M relative to that of the foundation.

Using Eqn 10.4–30, Eqn 10.4–31 can be written as

$$M \frac{d^2x}{dt^2} + kx = ks \cos \omega_f t \quad (10.4\text{–}32)$$

This equation is of the same form as Eqn 10.4–5 and can be handled similarly.

If, instead of a periodic foundation displacement, there is a **periodic foundation acceleration,** defined by

$$\frac{d^2x_f}{dt^2} = a \cos \omega_f t \tag{10.4–33}$$

then we introduce the relative displacement x' such that $x' = x - x_f$. Equation 10.4–31 then becomes

$$M\frac{d^2x}{dt^2} + kx' = 0 \tag{10.4–34}$$

which is now the governing differential equation. Writing $x = x' + x_f$, Eqn 10.4–34 can be written

$$M\frac{d^2}{dt^2}(x' + x_f) + kx' = 0$$

i.e.

$$M\frac{d^2x'}{dt^2} + kx' = -M\frac{d^2x_f}{dt^2}$$

$$M\frac{d^2x'}{dt^2} + kx' = -Ma \cos \omega_f t \tag{10.4–35}$$

Equation 10.4–35 is again of the same form as Eqn 10.4–5 and can be solved similarly.

Foundation disturbances which are not periodic can be represented as Fourier series of periodic functions, so that the method here presented is quite general.

Example 10.4–1. A structure of one degree of freedom is acted on by a pulsating force $P = P_0 \cos \omega_l t$. It is known that the damping coefficient may be taken as zero. Show that, if the circular frequency ω_l of the force is equal to the natural circular frequency ω of the structure, very large displacements can occur, but such displacements always remain finite as long as the time t remains finite.

SOLUTION For $c = 0$, Eqn 10.4–7 becomes:

$$x = A \sin \omega t + B \cos \omega t + \frac{P_0 \cos \omega_l t}{k - M\omega_l^2}$$

$$= A \sin \omega t + B \cos \omega t + \frac{P_0 \cos \omega_l t}{\omega^2 M[(k/\omega^2 M) - (\omega_l^2/\omega^2)]}$$

$$= A \sin \omega t + B \cos \omega t + \frac{P_0 \cos \omega_l t}{\omega^2 M[1 - (\omega_l^2/\omega^2)]} \tag{10.4–36}$$

(since $k/M = \omega^2$ from Eqn 10.3–17).

If the initial conditions are:

$$\text{at} \quad t = 0 \quad \begin{cases} x = x_0 \\ \dfrac{dx}{dt} = \dot{x}_0 \end{cases}$$

then the reader should verify that

$$A = \frac{\dot{x}_0}{\omega}; \qquad B = x_0 - \frac{P_0}{\omega^2 M[1 - (\omega_l^2/\omega^2)]}$$

i.e. Eqn 10.4–36 becomes:

$$x = \frac{\dot{x}_0}{\omega}\sin\omega t + x_0\cos\omega t + \frac{P_0(\cos\omega_l t - \cos\omega t)}{\omega^2 M[1 - (\omega_l^2/\omega^2)]}$$

$$\lim_{\omega_l\to\omega} x = \frac{\dot{x}_0}{\omega}\sin\omega t + x_0\cos\omega t + \frac{0}{0}$$

Using L'Hospital's rule, we take derivatives of the numerator and denominator of the last term with respect to ω_l and then let $\omega_l \to \omega$:

$$\lim_{\omega_l\to\omega} x = \frac{\dot{x}_0}{\omega}\sin\omega t + x_0\cos\omega t + \frac{P_0(-t\sin\omega t - 0)}{\omega^2 M[0 - (2\omega/\omega^2)]}$$

$$= \frac{\dot{x}_0}{\omega}\sin\omega t + x_0\cos\omega t + \frac{P_0 t\sin\omega t}{2\omega M} \qquad (10.4\text{–}37)$$

Equation 10.4–37 shows that, as t increases, x becomes progressively larger, but that it remains finite as long as t remains finite; that is, infinitely large displacements do not occur unless the pulsating load acts for an infinitely long time. (The reader should note, however, that if the load acts for a sufficiently long time for the displacement to become so large that the structure is no longer linearly elastic, then Eqn 10.4–7 would no longer apply.)

10.5 Free vibration with many degrees of freedom

A structure is said to have many degrees of freedom when two or more independent coordinates are required to define its configuration. For example, the mass–spring system in Fig. 10.5–1 requires two independent coordinates, x_1 and x_2, to define completely the positions of the two masses (and hence to define completely the configuration of the system); therefore it is said to have two degrees of freedom. Another example of a structure with many degrees of freedom is the n-storey frame in Fig. 10.5–2. The horizontal members are assumed to be infinitely rigid compared with vertical members,

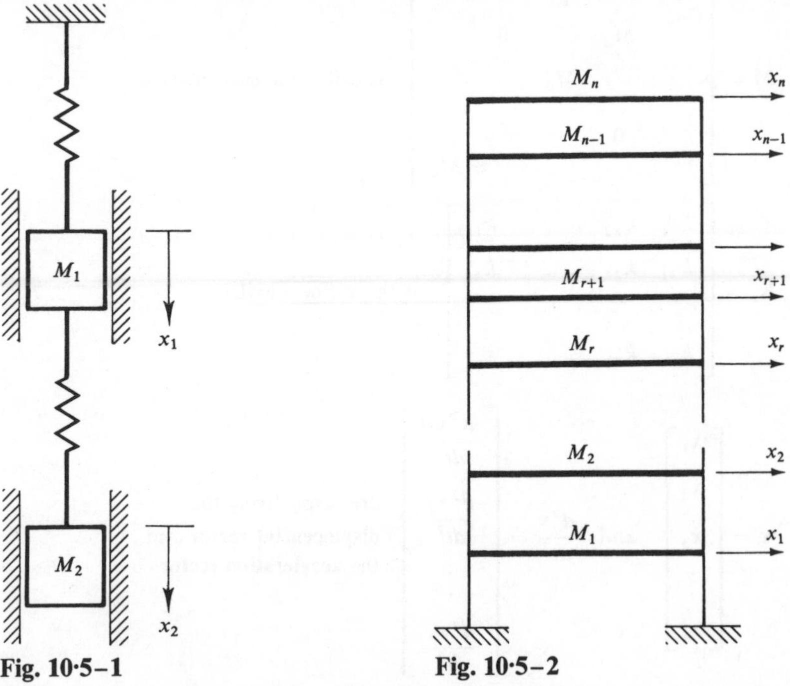

Fig. 10·5–1

Fig. 10·5–2

which are assumed to be massless and inextensible. Such a frame is referred to in structural dynamics as a **shear building** or a shear frame. In a shear building, rotation of joints is assumed not to occur and the structure is assumed to sway only in its plane. To define completely the configuration of the n-storey shear building, n independent coordinates (x_1 to x_n inclusively) are required; hence it has n degrees of freedom. In both the mass–spring system in Fig. 10.5–1 and the shear building in Fig. 10.5–2, the masses are assumed to be concentrated or lumped at specific positions; hence they are both lumped-mass structures.

The general theory of the free undamped vibration of structures with many degrees of freedom will now be developed, and the shear building in Fig. 10.5–2 will be used as a typical example of such a structure. Using the concept of stiffness coefficient introduced in the matrix stiffness method in Chapter 7, let k_{ij} represent the *restoring* force at the ith storey due to unit displacement at the jth storey. If at a certain instant of time, the displacements at the various storeys are $x_1, x_2, \ldots, x_n$ respectively, then the equations of motion for the n masses are:

$$M_1 \frac{d^2x_1}{dt^2} + k_{11}x_1 + k_{12}x_2 + \cdots + k_{1n}x_n = 0$$

$$M_2 \frac{d^2x_2}{dt^2} + k_{21}x_1 + k_{22}x_2 + \cdots + k_{2n}x_n = 0$$

$$\vdots$$

$$M_n \frac{d^2x_n}{dt^2} + k_{n1}x_1 + k_{n2}x_2 + \cdots + k_{nn}x_n = 0 \qquad (10.5\text{–}1)$$

i.e.

$$\mathbf{M}\frac{d^2\mathbf{x}}{dt^2} + \mathbf{K}_s\mathbf{x} = 0$$

where

$$\mathbf{M} = \begin{bmatrix} M_1 & & & & \\ & M_2 & & 0 & \\ & & M_3 & & \\ & 0 & & \ddots & \\ & & & & M_n \end{bmatrix}$$ is called the **mass matrix**

$$\mathbf{K}_s = \begin{bmatrix} k_{11} & k_{12} & \cdots & k_{1n} \\ k_{21} & k_{22} & \cdots & k_{2n} \\ \vdots & & & \\ k_{n1} & k_{n2} & \cdots & k_{nn} \end{bmatrix}$$ is the **stiffness matrix**

and

$$\mathbf{x} = \begin{bmatrix} x_1 \\ x_2 \\ x_3 \\ \vdots \\ x_n \end{bmatrix} \quad \text{and} \quad \frac{d^2\mathbf{x}}{dt^2} = \begin{bmatrix} \frac{d^2x_1}{dt^2} \\ \frac{d^2x_2}{dt^2} \\ \vdots \\ \frac{d^2x_n}{dt^2} \end{bmatrix}$$ are respectively the **displacement vector** and the **acceleration vector**

A set of displacements, $x_1, x_2, \ldots, x_n$, satisfying Eqns 10.5–1 or 10.5–2 would be a possible motion for the structure. However, the set of equations (10.5–1) are *coupled*, i.e. each equation contains more than one unknown, so that the set of equations cannot be easily solved directly. In what follows we shall develop a method to *decouple* these equations; that is, we shall transform this set of n coupled equations into a set of n *uncoupled* equations such that each transformed equation would contain only one unknown.

Let us investigate a solution of Eqn 10.5–1 of the form

$$x_r = e_r \sin(\omega t + \alpha) \quad (r = 1, 2, 3, \ldots, n; \alpha \text{ is a constant}) \tag{10.5–3}$$

where e_r is the amplitude of the vibration of the mass M_r (the e here must not be confused with that in Eqn 10.3–2, where e was the base of the natural logarithm). Note that an assumed solution of the form of Eqn 10.5–3 would imply that the frequency of vibration of every mass is $\omega/2\pi$ (cycles per second, if ω is in radians per second). The reader will realize later (see Example 10.5–2) that it is only for individual 'normal modes' of vibration (which term will be explained later in this section) that the frequency is the same for every mass.

Using Eqn 10.5–3, the set of equations (10.5–1) can be written as:

$$\begin{aligned}
-\omega^2 M_1 e_1 + k_{11} e_1 + k_{12} e_2 + \cdots + k_{1n} e_n &= 0 \\
-\omega^2 M_2 e_2 + k_{21} e_1 + k_{22} e_2 + \cdots + k_{2n} e_n &= 0 \\
\vdots \\
-\omega^2 M_n e_n + k_{n1} e_1 + k_{n2} e_2 + \cdots + k_{nn} e_n &= 0
\end{aligned} \tag{10.5–4}$$

That is,

$$-\omega^2 \mathbf{M}\mathbf{e} + \mathbf{K}_s \mathbf{e} = 0$$

i.e.

$$\{\mathbf{K}_s - \omega^2 \mathbf{M}\}\mathbf{e} = 0 \tag{10.5–5}$$

where $\mathbf{e} = [e_1 \; e_2 \; e_3 \; \ldots \; e_n]^T$ is the vector of the amplitudes of the vibrations of the n masses.

Equation 10.5–5 represents a set of n simultaneous homogeneous equations; non-trivial solutions for $\mathbf{e}$ exist only if the determinant

$$|\mathbf{K}_s - \omega^2 \mathbf{M}| = 0$$

i.e.

$$|\mathbf{M}\{\mathbf{M}^{-1} \mathbf{K}_s - \omega^2 \mathbf{I}\}| = 0 \tag{10.5–6}$$

where $\mathbf{I}$ is an $n \times n$ unit matrix.

Since the determinant of the product of two matrices is equal to the product of the determinants of the two matrices, Eqn 10.5–6 can be written:

$$|\mathbf{M}||\mathbf{M}^{-1} \mathbf{K}_s - \omega^2 \mathbf{I}| = 0 \tag{10.5–7}$$

Since $\mathbf{M}$ is a diagonal matrix whose elements are all non-zero, $\mathbf{M}$ is non-singular; hence for Eqn 10.5–7 to hold, the $n \times n$ matrix $\{\mathbf{M}^{-1} \mathbf{K}_s - \omega^2 \mathbf{I}\}$ must be singular, i.e.

$$|\mathbf{M}^{-1} \mathbf{K}_s - \omega^2 \mathbf{I}| = 0 \tag{10.5–8}$$

This equation is satisfied if ω^2 is an eigenvalue of the matrix $\{\mathbf{M}^{-1} \mathbf{K}_s\}$, known as the **dynamic matrix**. In other words, if non-trivial solutions to Eqn 10.5–5 exist, then the values of ω^2 are the eigenvalues of the dynamic matrix $\mathbf{M}^{-1} \mathbf{K}_s$, and the corresponding amplitude vectors $\mathbf{e}$ satisfying Eqn 10.5–5 are the eigenvectors of the dynamic matrix.

For a structure of n degrees of freedom, the dynamic matrix is of order $n \times n$, and hence it will have n eigenvalues and n eigenvectors. Suppose the n eigenvalues are

$$\omega_1^2, \omega_2^2, \omega_3^2 \ldots \omega_i^2 \ldots \omega_n^2$$

where

$$\omega_1^2 < \omega_2^2 < \omega_3^2 \ldots < \omega_n^2$$

and the corresponding eigenvectors are

$$_1\mathbf{e},\ _2\mathbf{e}, \ldots,\ _i\mathbf{e}, \ldots,\ _n\mathbf{e}$$

where

$$_1\mathbf{e} = \begin{bmatrix} _1e_1 \\ _1e_2 \\ _1e_3 \\ \vdots \\ _1e_n \end{bmatrix} \quad \text{and} \quad _i\mathbf{e} = \begin{bmatrix} _ie_1 \\ _ie_2 \\ _ie_3 \\ \vdots \\ _ie_n \end{bmatrix} \quad \text{etc.}$$

Each of the n values of ω corresponds to a **natural mode frequency** ($\omega/2\pi$). The smallest natural mode frequency, $\omega_1/2\pi$, is called the first natural mode frequency, or simply the first natural frequency; similarly $\omega_2/2\pi$ is the second natural frequency and so on. Each of the amplitude vectors $\mathbf{e}$ is called a **normal mode shape**; $_1\mathbf{e}$ is the first normal mode shape, $_2\mathbf{e}$ is the second normal mode shape, and so on.

Using ω_i^2 and $_i\mathbf{e}$ in Eqn 10.5–5,

$$\{\mathbf{K}_s - \omega_i^2 \mathbf{M}\}_i\mathbf{e} = 0$$

i.e.

$$\mathbf{K}_{s\,i}\mathbf{e} = \omega_i^2 \mathbf{M}_i\mathbf{e} \qquad (10.5\text{–}9)$$

Similarly, using ω_j^2 and $_j\mathbf{e}$ in Eqn 10.5–5 would give

$$\mathbf{K}_{s\,j}\mathbf{e} = \omega_j^2 \mathbf{M}_j\mathbf{e} \qquad (10.5\text{–}10)$$

Transposing each side of Eqn 10.5–9,

$$_i\mathbf{e}^T \mathbf{K}_s^T = \omega_i^2\ _i\mathbf{e}^T \mathbf{M}^T$$

As explained in Chapters 4 and 7, the stiffness matrix $\mathbf{K}_s$ is symmetrical, i.e. $\mathbf{K}_s^T = \mathbf{K}_s$; similarly, the mass matrix $\mathbf{M}$, being diagonal, is also symmetrical, i.e. $\mathbf{M}^T = \mathbf{M}$. Hence the above equation can be written

$$_i\mathbf{e}^T \mathbf{K}_s = \omega_i^2\ _i\mathbf{e}^T \mathbf{M}$$

Post-multiplying each side by $_j\mathbf{e}$,

$$_i\mathbf{e}^T \mathbf{K}_{s\,j}\mathbf{e} = \omega_i^2\ _i\mathbf{e}^T \mathbf{M}_j\mathbf{e} \qquad (10.5\text{–}11)$$

Pre-multiplying each side of Eqn 10.5–10 by $_i\mathbf{e}^T$,

$$_i\mathbf{e}^T \mathbf{K}_{s\,j}\mathbf{e} = \omega_j^2\ _i\mathbf{e}^T \mathbf{M}_j\mathbf{e} \qquad (10.5\text{–}12)$$

Subtracting Eqn 10.5–12 from Eqn 10.5–11,

$$(\omega_i^2 - \omega_j^2)\ _i\mathbf{e}^T \mathbf{M}_j\mathbf{e} = 0 \qquad (10.5\text{–}13)$$

In the case of two different normal modes, $\omega_i^2 \neq \omega_j^2$. Therefore, from Eqn 10.5–13,

$$_i\mathbf{e}^T \mathbf{M}_j\mathbf{e} = 0 \qquad (10.5\text{–}14)$$

which means that the *normal modes are orthogonal to one another with respect to the mass matrix*. Substituting Eqn 10.5–14 into Eqn 10.5–11 we have

$$_i\mathbf{e}^T \mathbf{K}_{s\,j}\mathbf{e} = 0 \qquad (10.5\text{–}15)$$

which means that the normal modes are also orthogonal with respect to the stiffness matrix.

In Eqn 10.5–13, for $\omega_i^2 = \omega_j^2$, and ${}_i\mathbf{e} = {}_j\mathbf{e}$, then

$${}_i\mathbf{e}^T\mathbf{M}\,{}_i\mathbf{e}$$

is not necessarily zero. In fact, the quantity ${}_i\mathbf{e}^T\mathbf{M}\,{}_i\mathbf{e}$, known as a quadratic form in matrix algebra, is equal to

$${}_ie_1^2 M_1 + {}_ie_2^2 M_2 + {}_ie_3^2 M_3 + \cdots + {}_ie_n^2 M_n$$

and is always positive, since all the masses are positive. Let

$${}_i\mathbf{e}^T\mathbf{M}\,{}_i\mathbf{e} = L_i^2 \tag{10.5–16}$$

and write

$${}_i\mathbf{z} = \frac{{}_i\mathbf{e}}{L_i} \tag{10.5–17}$$

Then, we have

$${}_i\mathbf{z}^T\mathbf{M}\,{}_i\mathbf{z} = \frac{{}_i\mathbf{e}^T\mathbf{M}\,{}_i\mathbf{e}}{L_i^2} = 1 \tag{10.5–18}$$

and

$$\begin{aligned}{}_i\mathbf{z}^T\mathbf{M}\,{}_j\mathbf{z} &= \frac{{}_i\mathbf{e}^T\mathbf{M}\,{}_j\mathbf{e}}{L_iL_j}\\ &= \frac{0}{L_iL_j} = 0\end{aligned} \tag{10.5–19}$$

The matrix $\mathbf{Z}$, the columns of which are the normalized vectors ${}_1\mathbf{z}, {}_2\mathbf{z}, \ldots, {}_n\mathbf{z}$, is called the **modal matrix** of the dynamic matrix $\mathbf{M}^{-1}\mathbf{K}_s$:

$$\mathbf{Z} = \begin{bmatrix} {}_1z_1 & {}_2z_1 & {}_3z_1 & {}_nz_1 \\ {}_1z_2 & {}_2z_2 & \vdots & {}_nz_2 \\ \vdots & \vdots & & \vdots \\ {}_1z_n & {}_2z_n & & {}_nz_n \end{bmatrix} \tag{10.5–20}$$

For a given ω, say ω_1, Eqn 10.5–5 can be written as

$$\mathbf{K}_s \begin{bmatrix} {}_1z_1 \\ {}_1z_2 \\ \vdots \\ {}_1z_n \end{bmatrix} = \mathbf{M} \begin{bmatrix} {}_1z_1 \\ {}_1z_2 \\ \vdots \\ {}_1z_n \end{bmatrix} \omega_1^2 \quad (n \text{ equations})$$

since ${}_1\mathbf{z}$ is equal to ${}_1\mathbf{e}$ divided by a constant. Similarly, for $\omega = \omega_2$,

$$\mathbf{K}_s \begin{bmatrix} {}_2z_1 \\ {}_2z_2 \\ \vdots \\ {}_2z_n \end{bmatrix} = \mathbf{M} \begin{bmatrix} {}_2z_1 \\ {}_2z_2 \\ \vdots \\ {}_2z_n \end{bmatrix} \omega_2^2 \quad (n \text{ equations})$$

In all, there are $n \times n$ equations, corresponding to the n values of ω. These $n \times n$ equations can be written in an all-inclusive form as

$$\mathbf{K}_s\mathbf{Z} = \mathbf{M}\mathbf{Z}\boldsymbol{\omega}^2 \tag{10.5–21}$$

where

$$\boldsymbol{\omega}^2 = \begin{bmatrix} \omega_1^2 & & & 0 \\ & \omega_2^2 & & \\ & & \ddots & \\ 0 & & & \omega_n^2 \end{bmatrix} \tag{10.5–22}$$

Pre-multiplying Eqn 10.5–21 by $\mathbf{Z}^T$,

$$\mathbf{Z}^T\mathbf{K}_s\mathbf{Z} = \mathbf{Z}^T\mathbf{M}\mathbf{Z}\boldsymbol{\omega}^2 \tag{10.5–23}$$

The reader should use Eqns 10.5–18 and 10.5–19 to verify that

$$\mathbf{Z}^T\mathbf{M}\mathbf{Z} = \underset{\text{unit matrix}}{\mathbf{I}} \tag{10.5–24}$$

Substituting this into Eqn 10.5–23 gives

$$\begin{aligned} \mathbf{Z}^T\mathbf{K}_s\mathbf{Z} &= \mathbf{I}\boldsymbol{\omega}^2 \\ &= \boldsymbol{\omega}^2 \end{aligned} \tag{10.5–25}$$

We are now in a position to decouple the original set of n simultaneous differential equations (Eqns 10.5–1 and 10.5–2):

$$\mathbf{M}\frac{d^2\mathbf{x}}{dt^2} + \mathbf{K}_s\mathbf{x} = 0 \tag{10.5–26}$$

If we introduce a change of coordinates defined by

$$\mathbf{x} = \mathbf{Z}\mathbf{q} \tag{10.5–27}$$

where $\mathbf{q} = [q_1 q_2 \cdots q_n]^T$, then Eqn 10.5–26 becomes

$$\mathbf{M}\mathbf{Z}\frac{d^2\mathbf{q}}{dt^2} + \mathbf{K}_s\mathbf{Z}\mathbf{q} = 0$$

Pre-multiplying by $\mathbf{Z}^T$,

$$\mathbf{Z}^T\mathbf{M}\mathbf{Z}\frac{d^2\mathbf{q}}{dt^2} + \mathbf{Z}^T\mathbf{K}_s\mathbf{Z}\mathbf{q} = 0$$

Since

$$\mathbf{Z}^T\mathbf{M}\mathbf{Z} = \mathbf{I} \quad \text{(Eqn 10.5–24)}$$

and

$$\mathbf{Z}^T\mathbf{K}_s\mathbf{Z} = \boldsymbol{\omega}^2 \quad \text{(Eqn 10.5–25)}$$

we have

$$\frac{d^2\mathbf{q}}{dt^2} + \boldsymbol{\omega}^2\mathbf{q} = 0 \tag{10.5–28}$$

Eqn 10.5–28 represents a set of n uncoupled equations:

$$\begin{aligned} &\frac{d^2q_1}{dt^2} + \omega_1^2 q_1 = 0 \\ &\frac{d^2q_2}{dt^2} + \omega_2^2 q_2 = 0 \\ &\vdots \\ &\frac{d^2q_n}{dt^2} + \omega_n^2 q_n = 0 \end{aligned} \tag{10.5–29}$$

We have thus reduced the set of n coupled equations (Eqn 10.5–1) to a set of n uncoupled equations which can readily be solved. In fact, since each equation is of the same form as Eqn 10.3–1 with $c = 0$, the solution must be of the same form as Eqns 10.3–19 or 10.3–22 or 10.3–23:

$$
\begin{aligned}
q_1 &= A_1 \sin \omega_1 t + B_1 \cos \omega_1 t, \text{ or } C_1 \sin (\omega_1 t + \alpha_1), \text{ or } D_1 \cos (\omega_1 t + \beta_1) \\
q_2 &= A_2 \sin \omega_2 t + B_2 \cos \omega_2 t, \text{ or } C_2 \sin (\omega_2 t + \alpha_2), \text{ or } D_2 \cos (\omega_2 t + \beta_2) \\
&\vdots \\
q_n &= A_n \sin \omega_n t + B_n \cos \omega_n t, \text{ or } C_n \sin (\omega_n t + \alpha_n), \text{ or } D_n \cos (\omega_n t + \beta_n)
\end{aligned}
\tag{10.5–30}
$$

The set of equations (10.5–30) give each of $q_1, q_2, \ldots, q_n$ in terms of two arbitrary constants. The next step is to determine the initial conditions on $\mathbf{q}$ and $d\mathbf{q}/dt$ from the initial conditions prescribed on $\mathbf{x}$ and $d\mathbf{x}/dt$, using the modified forms of Eqn 10.5–27:†

$$\mathbf{q} = \mathbf{Z}^T \mathbf{M} \mathbf{x} \tag{10.5–31(a)}$$

$$\frac{d\mathbf{q}}{dt} = \mathbf{Z}^T \mathbf{M} \frac{d\mathbf{x}}{dt} \tag{10.5–31(b)}$$

After the two arbitrary constants in each of the n equations in Eqn 10.5–30 have been found, $\mathbf{x}$ is determined immediately from Eqn 10.5–27, which defines the co-ordinate transformation:

$$\mathbf{x} = \mathbf{Z} \mathbf{q} \tag{10.5–32}$$

Equation 10.5–32 shows that the *actual* displacement vector $\mathbf{x}$ for the n masses is a combination of the various normal mode shapes. It was stated also, immediately after Eqn 10.5–3, that the frequency of vibration ω was the same for every mass. (Note carefully that this is true only for individual normal modes of vibration and not necessarily true for the actual vibration of the masses (Example 10.5–2).)

The reader should now re-examine Eqns 10.5–5 and 10.5–6 and note that the vibration problem of a lumped-mass system is essentially an eigenvalue problem, in that the natural mode frequencies are given by the eigenvalues of the dynamic matrix and the normal mode shapes by the corresponding eigenvectors. Equation 10.5–25 further shows that the modal matrix $\mathbf{Z}$, whose columns are the normalized eigenvectors of the dynamic matrix, transforms the stiffness matrix into a diagonal matrix, the elements of which are the squares of the natural circular frequencies. When the reader has studied Section 11.5 on principal stresses, he will realize that the principal stress problem is also essentially an eigenvalue problem, in that the principal stresses are the eigenvalues of the stress tensor, and the principal directions are the corresponding eigenvectors. The equation $\mathbf{T}\boldsymbol{\sigma}\mathbf{T}^T = \boldsymbol{\sigma}'$ in Example 11.5–1 is markedly similar to Eqn 10.5–25: the direction cosine matrix $\mathbf{T}$ of unit eigenvectors of the stress tensor transforms the stress tensor into a diagonal matrix, the elements of which are the principal stresses.

Standard computer procedures for determining eigenvalues and eigenvectors are available at most computing centres, so that the matrix method presented in this section readily lends itself to computer application.

Example 10.5–1. In the three-storey shear building in Fig. 10.5–3, the masses are $M_1 = 4000$ kg, $M_2 = 4000$ kg, $M_3 = 2000$ kg, and the flexural stiffnesses of the vertical members are:

$$k_{(AC)} = k_{(BD)} = 1.5 \times 10^6 \text{ N/m}, \qquad k_{(CE)} = k_{(DF)} = 1.0 \times 10^6 \text{ N/m}$$

and

$$k_{(EG)} = k_{(FH)} = 0.75 \times 10^6 \text{ N/m}$$

† Since $\mathbf{x} = \mathbf{Z}\mathbf{q}$, $\mathbf{M}\mathbf{x} = \mathbf{M}\mathbf{Z}\mathbf{q}$, and $\mathbf{Z}^T\mathbf{M}\mathbf{x} = \mathbf{Z}^T\mathbf{M}\mathbf{Z}\mathbf{q} = \mathbf{I}\mathbf{q} = \mathbf{q}$ from Eqn 10.5–24.

(i) Determine the natural mode frequencies, the normal mode shapes, and the modal matrix.

(ii) Hence express the equations of motion of the three masses in uncoupled form.

(iii) If the initial conditions are (at $t = \tau$ seconds):

$$\begin{bmatrix} x_1 \\ x_2 \\ x_3 \end{bmatrix} = \begin{bmatrix} x_{1(\tau)} \\ x_{2(\tau)} \\ x_{3(\tau)} \end{bmatrix} \text{ m, and } \begin{bmatrix} \dot{x}_1 \\ \dot{x}_2 \\ \dot{x}_3 \end{bmatrix} = \begin{bmatrix} \dot{x}_{1(\tau)} \\ \dot{x}_{2(\tau)} \\ \dot{x}_{3(\tau)} \end{bmatrix} \text{ m/s}$$

determine the initial conditions on the displacements and velocities in the generalized coordinates in the uncoupled equations of motion. *Show all units clearly.*

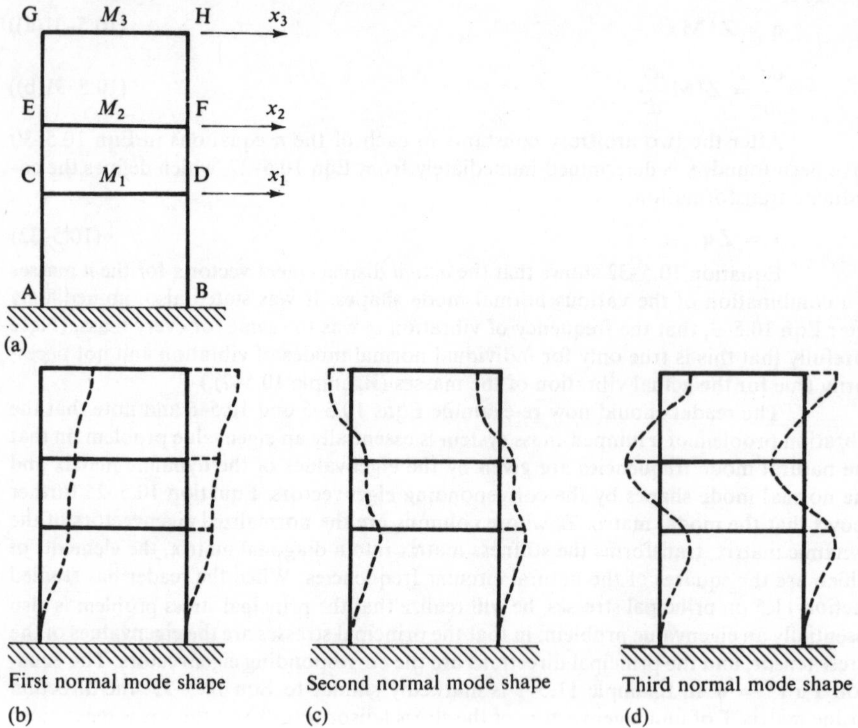

Fig. 10·5–3

SOLUTION (i) Using k_{ij} to represent the restoring force acting at the ith floor due to unit displacement at the jth floor,

$$k_{11} = (2 \times 1.5 \times 10^6 + 2 \times 1 \times 10^6) \text{ N/m} = 5 \times 10^6 \text{ N/m}$$

$$k_{12} = -2 \times 1 \times 10^6 \text{ N/m} \quad \text{and so on}$$

Whence the stiffness matrix can be shown to be

$$\mathbf{K}_s = \begin{bmatrix} k_{11} & k_{12} & k_{13} \\ k_{21} & k_{22} & k_{23} \\ k_{31} & k_{32} & k_{33} \end{bmatrix} = \begin{bmatrix} 5 & -2 & 0 \\ -2 & 3.5 & -1.5 \\ 0 & -1.5 & 1.5 \end{bmatrix} \times 10^6 \text{ N/m} \qquad (10.5\text{–}33)$$

The mass matrix is

$$\mathbf{M} = \begin{bmatrix} M_1 & 0 & 0 \\ 0 & M_2 & 0 \\ 0 & 0 & M_3 \end{bmatrix} = \begin{bmatrix} 4 & 0 & 0 \\ 0 & 4 & 0 \\ 0 & 0 & 2 \end{bmatrix} \times 10^3 \text{ kg} \tag{10.5–34}$$

Whence the dynamic matrix $[\mathbf{M}^{-1}\mathbf{K}]$ is

$$\begin{bmatrix} \frac{1}{4} & 0 & 0 \\ 0 & \frac{1}{4} & 0 \\ 0 & 0 & \frac{1}{2} \end{bmatrix} \times 10^{-3} \text{ kg}^{-1} \times \begin{bmatrix} 5 & -2 & 0 \\ -2 & 3.5 & -1.5 \\ 0 & -1.5 & 1.5 \end{bmatrix} \times 10^6 \text{ N/m}$$

$$= \begin{bmatrix} 1250 & -500 & 0 \\ -500 & 875 & -375 \\ 0 & -750 & 750 \end{bmatrix} \text{N m}^{-1}\text{ kg}^{-1} \quad \text{(i.e. second}^{-2}\text{)} \tag{10.5–35}$$

(Note: Since the unit N = kg m s^{-2}, the unit N m^{-1} kg^{-1} = (kg m s^{-2}) m^{-1} kg^{-1} = s^{-2}.) The eigenvalues ω_1^2, ω_2^2, and ω_3^2 and the eigenvectors ${}_1\mathbf{e}$, ${}_2\mathbf{e}$, and ${}_3\mathbf{e}$ of the dynamic matrix are determined readily by standard computer routines available in practically all computing centres:

$$\boldsymbol{\omega}^2 = \begin{bmatrix} \omega_1^2 & 0 & 0 \\ 0 & \omega_2^2 & 0 \\ 0 & 0 & \omega_3^2 \end{bmatrix} = \begin{bmatrix} 167 & 0 & 0 \\ 0 & 1000 & 0 \\ 0 & 0 & 1710 \end{bmatrix} \text{s}^{-2} \tag{10.5–36}$$

and

$${}_1\mathbf{e} = \begin{bmatrix} 1.00C_1 \\ 2.17C_1 \\ 2.78C_1 \end{bmatrix}; \quad {}_2\mathbf{e} = \begin{bmatrix} 1.00C_2 \\ 0.50C_2 \\ -1.50C_2 \end{bmatrix}; \quad {}_3\mathbf{e} = \begin{bmatrix} 1.00C_3 \\ -0.92C_3 \\ 0.72C_3 \end{bmatrix} \tag{10.5–37}$$

where C_1, C_2, and C_3 are arbitrary non-zero constants. From Eqn 10.5–36 the natural mode frequencies are:

$$f_1 = \frac{\omega_1}{2\pi} = \frac{\sqrt{167}}{2\pi} \text{ s}^{-1} = \underline{2.03 \text{ Hz}}$$

$$f_2 = \frac{\omega_2}{2\pi} = \frac{\sqrt{1000}}{2\pi} \text{ s}^{-1} = \underline{5.04 \text{ Hz}}$$

$$f_3 = \frac{\omega_3}{2\pi} = \frac{\sqrt{1710}}{2\pi} \text{ s}^{-1} = \underline{6.59 \text{ Hz}}$$

The three normal mode shapes, as given in Eqns 10.5–37 are plotted in Figs 10.5–3(b) to (d).

To determine the modal matrix **Z**, we note from Eqn 10.5–16 that

$$L_1^2 = {}_1\mathbf{e}^T \mathbf{M}\, {}_1\mathbf{e}$$

$$= [1.00C_1 \quad 2.17C_1 \quad 2.78C_1] \begin{bmatrix} 4000 & 0 & 0 \\ 0 & 4000 & 0 \\ 0 & 0 & 2000 \end{bmatrix} \begin{bmatrix} 1.00C_1 \\ 2.17C_1 \\ 2.78C_1 \end{bmatrix}$$

$$= 38\,320C_1^2 \text{ kg m}^2 \quad \text{then} \quad L_1 = 196C_1 \text{ kg}^{1/2} \text{ m}$$

$$L_2^2 = [1.00C_2 \quad 0.50C_2 \quad -1.50C_2] \begin{bmatrix} 4000 & 0 & 0 \\ 0 & 4000 & 0 \\ 0 & 0 & 2000 \end{bmatrix} \begin{bmatrix} 1.00C_2 \\ 0.50C_2 \\ -1.50C_2 \end{bmatrix}$$

$$= 9500C_2^2 \text{ kg m}^2 \quad \text{then} \quad L_2 = 97.5C_2 \text{ kg}^{1/2} \text{ m}$$

$$L_3^2 = [1.00C_3 \quad -0.92C_3 \quad 0.72C_3] \begin{bmatrix} 4000 & 0 & 0 \\ 0 & 4000 & 0 \\ 0 & 0 & 2000 \end{bmatrix} \begin{bmatrix} 1.00C_3 \\ -0.92C_3 \\ 0.72C_3 \end{bmatrix}$$

$$= 8410C_3^2 \text{ kg m}^2 \quad \text{then} \quad L_3 = 91.9C_3 \text{ kg}^{1/2} \text{ m}$$

Using Eqn 10.5–17, the modal matrix is

$$\mathbf{Z} = \begin{bmatrix} \frac{{}_1\mathbf{e}_1}{L_1} & \frac{{}_2\mathbf{e}_1}{L_2} & \frac{{}_3\mathbf{e}_1}{L_3} \\ \frac{{}_1\mathbf{e}_2}{L_1} & \frac{{}_2\mathbf{e}_2}{L_2} & \frac{{}_3\mathbf{e}_2}{L_3} \\ \frac{{}_1\mathbf{e}_3}{L_1} & \frac{{}_2\mathbf{e}_3}{L_2} & \frac{{}_3\mathbf{e}_3}{L_3} \end{bmatrix} = \begin{bmatrix} 5.1 & 10.3 & 10.9 \\ 11.1 & 5.1 & -10.0 \\ 14.2 & -15.4 & 7.9 \end{bmatrix} \times 10^{-3} \text{ kg}^{-1/2} \tag{10.5–38}$$

(ii) In terms of the coordinates x_1, x_2, x_3 the equations of motion are coupled:

$$\mathbf{M}\ddot{\mathbf{x}} + \mathbf{K}_s\mathbf{x} = 0$$

To decouple them, introduce a change of coordinates defined by Eqn 10.5–27,

$$\mathbf{x} = \mathbf{Z}\mathbf{q}$$

resulting in the three uncoupled equations:

$$\ddot{\mathbf{q}} + \boldsymbol{\omega}^2\mathbf{q} = 0 \quad \text{(see Eqn 10.5–28)}$$

i.e., with $\boldsymbol{\omega}^2$ as given by Eqn 10.5–36, we have

$$\frac{d^2q_1}{dt^2} + 167q_1 = 0$$

$$\frac{d^2q_2}{dt^2} + 1000q_2 = 0$$

$$\frac{d^2q_3}{dt^2} + 1710q_3 = 0 \tag{10.5–39}$$

(iii) Using Eqns 10.5–31,

$$\mathbf{q} = \mathbf{Z}^T\mathbf{M}\mathbf{x}$$

$$\dot{\mathbf{q}} = \mathbf{Z}^T\mathbf{M}\dot{\mathbf{x}}$$

that is,

$$\begin{bmatrix} q_{1(\tau)} \\ q_{2(\tau)} \\ q_{3(\tau)} \end{bmatrix} = \begin{bmatrix} 5.1 & 10.3 & 10.9 \\ 11.1 & 5.1 & -10.0 \\ 14.2 & -15.4 & 7.9 \end{bmatrix}^T \begin{bmatrix} 4 & 0 & 0 \\ 0 & 4 & 0 \\ 0 & 0 & 2 \end{bmatrix} \begin{bmatrix} x_{1(\tau)} \\ x_{2(\tau)} \\ x_{3(\tau)} \end{bmatrix} \text{kg}^{1/2} \text{ m} \tag{10.5–40}$$

and similarly for $[\dot{q}_{1(\tau)} \; \dot{q}_{2(\tau)} \; \dot{q}_{3(\tau)}]^T$.

Example 10.5–2. A lumped mass structure of n degrees of freedom is undergoing undamped free vibration. Determine the *actual* period of vibration of the ith mass.

Are the actual periods of vibration of the n masses equal to one another?

SOLUTION The solution to the uncoupled equations of motion is (Eqns 10.5–30):

$$\begin{bmatrix} q_1 \\ q_2 \\ \vdots \\ q_i \\ \vdots \\ q_n \end{bmatrix} = \begin{bmatrix} C_1 \sin(\omega_1 t + \alpha_1) \\ C_2 \sin(\omega_2 t + \alpha_2) \\ \vdots \\ C_i \sin(\omega_i t + \alpha_i) \\ \vdots \\ C_n \sin(\omega_n t + \alpha_n) \end{bmatrix} \tag{10.5–41}$$

Using the coordinate transformation in Eqn 10.5–32, the actual displacements are:

$$\begin{bmatrix} x_1 \\ x_2 \\ \vdots \\ x_i \\ \vdots \\ x_n \end{bmatrix} = \begin{bmatrix} {}_1z_1 & {}_2z_1 & \cdots & {}_nz_1 \\ {}_1z_2 & {}_2z_2 & & \\ \vdots & & & \\ {}_1z_i & {}_2z_i & & \\ \vdots & & & \\ {}_1z_n & & & {}_nz_n \end{bmatrix} \begin{bmatrix} q_1 \\ q_2 \\ \vdots \\ q_i \\ \vdots \\ q_n \end{bmatrix}$$

That is,

$$\begin{aligned} x_i = {} & {}_1z_iC_1 \sin(\omega_1 t + \alpha_1) + {}_2z_iC_2 \sin(\omega_2 t + \alpha_2) + \cdots \\ & + {}_iz_iC_i \sin(\omega_i t + \alpha_i) + \cdots + {}_nz_iC_n \sin(\omega_n t + \alpha_n) \end{aligned} \tag{10.5–42}$$

where ω_1 is the natural circular frequency corresponding to the first normal mode shape, ω_2 that corresponding to the second normal mode shape, and so on.

In Eqn 10.5–42 the period of the first term on the right-hand side is $2\pi/\omega_1$, that of the second term is $2\pi/\omega_2$, and so on. Hence the actual period of the displacement x_i must be the least common multiple of all the periods on the right-hand side. That is,

$$T_i\,(\text{actual}) = \text{LCM of } \frac{2\pi}{\omega_1}, \frac{2\pi}{\omega_2}, \frac{2\pi}{\omega_3}, \ldots, \frac{2\pi}{\omega_n} \tag{10.5–43}$$

provided none of the coefficients ${}_1z_iC_1, {}_2z_iC_2, \ldots, {}_nz_iC_n$ are zero. If any of these coefficients is zero, say ${}_rz_iC_r = 0$, then the term $2\pi/\omega_r$ will be omitted in determining the LCM. Thus if none of the $n \times n$ quantities,

$$\begin{matrix} {}_1z_1C_1 & {}_2z_1C_2 & {}_3z_1C_3 & \cdots & {}_nz_1C_n \\ {}_2z_1C_1 & {}_2z_2C_2 & {}_3z_2C_3 & \cdots & {}_nz_2C_n \\ \vdots & & & & \\ {}_nz_1C_1 & {}_nz_2C_2 & {}_nz_3C_3 & \cdots & {}_nz_nC_n \end{matrix}$$

are zero, then the actual periods of vibration of all the masses must be the same, i.e.

$$T_1\,(\text{actual}) = T_2\,(\text{actual}) = \cdots = T_n\,(\text{actual})$$

each being given by Eqn 10.5–43.

Since some of these $n \times n$ quantities may be zero, the actual periods of vibration need not be the same for all the n masses (even though, for each individual normal mode shape, the period of vibration is the same for all the masses).

10.6 Forced vibration with many degrees of freedom

For forced vibration, the equations of motion will be similar to those in Eqn 10.5–1 except that the right-hand-side terms are no longer zero. In matrix form, the equations are represented as (see Eqn 10.5–2)

$$\mathbf{M}\frac{d^2\mathbf{x}}{dt^2} + \mathbf{K}_s\mathbf{x} = \mathbf{P} \tag{10.6–1}$$

where $\mathbf{P} = [P_1\ P_2\ P_3\ \ldots\ P_n]^T$ is the dynamic force vector, the components of which are, in general, functions of the time t.

Introducing the change of coordinates defined by Eqn 10.5–27, we have

$$\mathbf{MZ}\frac{d^2\mathbf{q}}{dt^2} + \mathbf{K}_s\mathbf{Zq} = \mathbf{P}$$

Pre-multiplying each side by $\mathbf{Z}^T$ and simplifying as done previously for Eqn 10.5–28, we have

$$\frac{d^2\mathbf{q}}{dt^2} + \boldsymbol{\omega}^2\mathbf{q} = \mathbf{Z}^T\mathbf{P} \tag{10.6–2}$$

Equation 10.6–2 represents a set of n uncoupled equations:

$$\frac{d^2q_1}{dt^2} + \omega_1^2 q_1 = {}_1z_1P_1 + {}_1z_2P_2 + \cdots + {}_1z_nP_n$$

$$\frac{d^2q_2}{dt^2} + \omega_2^2 q_2 = {}_2z_1P_1 + {}_2z_2P_2 + \cdots + {}_2z_nP_n$$

$$\vdots$$

$$\frac{d^2q_n}{dt^2} + \omega_n^2 q_n = {}_nz_1P_1 + {}_nz_2P_2 + \cdots + {}_nz_nP_n \tag{10.6–3}$$

Each of these n equations contains only one unknown, and its solution consists of a complementary function (see Eqn 10.5–30) and a particular integral, which is the algebraic sum of the n particular integrals corresponding to ${}_iz_1P_1$, ${}_iz_2P_2$, ${}_iz_3P_3 \ldots$ and ${}_iz_nP_n$, respectively. In other words, each of the equations in Eqn 10.6–3 is similar to the equation of motion for the forced vibration of a one-degree-of-freedom structure and can be handled as explained in Section 10.4.

After the vector $\mathbf{q}$ has been determined, the displacement vector $\mathbf{x}$ is determined, using the same method as explained for free vibrations in Section 10.5 (see Example 10.6–1).

Example 10.6–1. Suppose the shear building of Example 10.5–1 is acted on by the force vector

$$\mathbf{P} = \begin{bmatrix} P_1 \\ P_2 \\ P_3 \end{bmatrix} = \begin{bmatrix} 0 \\ 100(1-t) \\ 0 \end{bmatrix} \text{ kN}$$

during the time interval $0 \leqslant t \leqslant 1$ s.

Determine the motion of each lumped mass within the interval, if it is known that at $t = 0$, the structure is at rest.

SOLUTION The uncoupled equations are as given by Eqns 10.6–3, in which the force components should be expressed in newtons and not kilonewtons (in imperial units, they should be expressed in the pound-force unit lbf). Using the value of ω^2 and $\mathbf{Z}$ in Example 10.5–1, Eqns 10.5–36 and 10.5–38, we have

$$\ddot{q}_1 + 167q_1 = 1110(1 - t) \tag{10.6–4(a)}$$

$$\ddot{q}_2 + 1000q_2 = 510(1 - t) \tag{10.6–4(b)}$$

$$\ddot{q}_3 + 1710q_3 = -1000(1 - t) \tag{10.6–4(c)}$$

The solution of each of these three equations consists of a complementary function (of the standard form in Eqns 10.5–30) and a particular integral, namely

$$\begin{aligned} q_1 &= A_1 \sin \sqrt{(167)}t + B_1 \cos \sqrt{(167)}t + \frac{1110(1 - t)}{167} \\ q_2 &= A_2 \sin \sqrt{(1000)}t + B_2 \cos \sqrt{(1000)}t + \frac{510(1 - t)}{1000} \\ q_3 &= A_3 \sin \sqrt{(1710)}t + B_3 \cos \sqrt{(1710)}t - \frac{1000(1 - t)}{1710} \end{aligned} \tag{10.6–5}$$

Differentiating with respect to t,

$$\begin{aligned} \dot{q}_1 &= A_1 \sqrt{(167)} \cos \sqrt{(167)}t - B_1 \sqrt{(167)} \sin \sqrt{(167)}t - 6.65 \\ \dot{q}_2 &= A_2 \sqrt{(1000)} \cos \sqrt{(1000)}t - B_2 \sqrt{(1000)} \sin \sqrt{(1000)}t - 0.51 \\ \dot{q}_3 &= A_3 \sqrt{(1710)} \cos \sqrt{(1710)}t - B_3 \sqrt{(1710)} \sin \sqrt{(1710)}t + 0.59 \end{aligned} \tag{10.6–6}$$

The initial conditions on $\mathbf{q}$ and $\dot{\mathbf{q}}$ are determined from those prescribed on $\mathbf{x}$ and $\dot{\mathbf{x}}$, using Eqns 10.5–40 of Example 10.5–1. In this particular case, $\mathbf{x}$ and $\dot{\mathbf{x}}$ are both zero at $t = 0$; hence by inspection, at $t = 0$:

$$\begin{bmatrix} q_1 \\ q_2 \\ q_3 \end{bmatrix} = 0; \qquad \begin{bmatrix} \dot{q}_1 \\ \dot{q}_2 \\ \dot{q}_3 \end{bmatrix} = 0 \tag{10.6–7}$$

Substituting $q_1 = 0$ and $t = 0$ into the first equation in Eqns 10.6–5,

$$B_1 = -6.65$$

Substituting $\dot{q}_1 = 0$, and $t = 0$ into the first equation in Eqns 10.6–6,

$$A_1 = 6.65/\sqrt{(167)} = 0.515$$

The other four constants of integrations can be determined in a similar way, resulting in:

$$\begin{bmatrix} q_1 \\ q_2 \\ q_3 \end{bmatrix} = \begin{bmatrix} 0.515 \sin \sqrt{(167)}t - 6.65 \cos \sqrt{(167)}t + 6.65(1 - t) \\ 0.016 \sin \sqrt{(1000)}t - 0.51 \cos \sqrt{(1000)}t + 0.51(1 - t) \\ -0.014 \sin \sqrt{(1710)}t + 0.59 \cos \sqrt{(1710)}t - 0.59(1 - t) \end{bmatrix} \text{kg}^{1/2}\,\text{m} \tag{10.6–8}$$

whence, from Eqns 10.5–32 and 10.5–38

$$\begin{bmatrix} x_1 \\ x_2 \\ x_3 \end{bmatrix} = \begin{bmatrix} 5.1 & 10.3 & 10.9 \\ 11.1 & 5.1 & -10.0 \\ 14.2 & -15.4 & 7.9 \end{bmatrix} \times 10^{-3} \begin{bmatrix} q_1 \\ q_2 \\ q_3 \end{bmatrix} \text{m} \tag{10.6–9}$$

where $\mathbf{q}$ is as given in Eqn 10.6–8.

10.7 Vibrations with infinite number of degrees of freedom

So far the dynamic response of lumped-mass structures only has been dealt with. In most practical engineering structures the mass is distributed and not lumped at specific places; in other words such structures have an infinite number of degrees of freedom. A formal discussion of the dynamical behaviour of such structures is beyond the scope of this book. However, in many cases, approximate but useful information regarding the dynamical behaviour of these structures can be obtained by analysing them as idealized lumped-mass structures. For example, the stepped beam in Fig. 10.7–1(a) can be idealized as a system of, say, seven lumped masses supported by a massless beam (Fig. 10.7–1(b)).

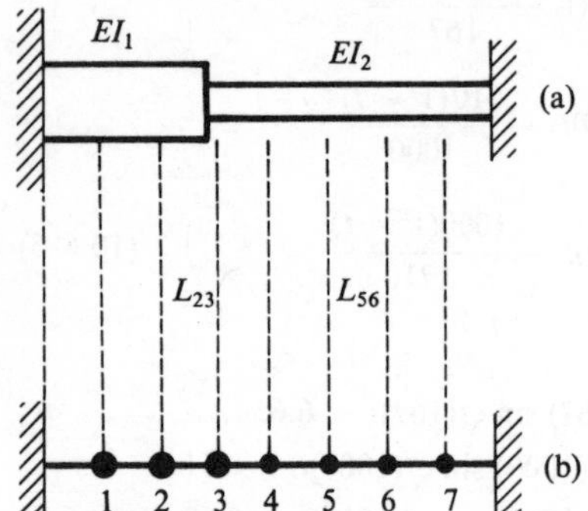

Fig. 10·7–1

The stiffness coefficients for the idealized lumped-mass system would be calculated from the properties of the original beam. If the flexural stiffness of the beam is as shown in Fig. 10.7–1(a), then for the idealized system,

$$k_{23} = -\frac{12EI_1}{(L_{23})^3}; \qquad k_{56} = -\frac{12EI_2}{(L_{56})^3}, \quad \text{etc.}$$

Using a larger number of masses would lead to better results, but would at the same time increase the amount of work required to analyse the idealized system. Warburton[2] has suggested that the number of masses used should be at least four times the number of normal modes to be studied. In practice structural engineers are most interested in the first two or three normal modes because the frequencies of the higher normal modes are usually too high to be of consequence (see Fig. 10.4–1 on dynamic magnification factor). Hence the use of 8 to 12 masses would seem suitable.

Many other structures may be similarly idealized as lumped-mass systems. For example, a multistorey building frame may, as a first approximation, be idealized as a shear building and analysed by the method explained in Sections 10.5 and 10.6. The mass of the frame, together with the mass of any dead or live load carried, will be lumped at the floor levels.

For detailed treatments of distributed-mass structures, the student should consult texts exclusively devoted to structural dynamics, such as those by Warburton,[2] Hurty and Rubinstein,[3] and Rogers.[4]

Problems

10.1. At time $t = \tau$ seconds a force P_0 newton is suddenly applied to a one-degree-of-freedom structure of mass M kilograms and stiffness k newtons per metre, and the force is then sustained. If damping is neglected, and if the initial conditions are, at $t = \tau$, $x = x_\tau$, and $\dot{x} = \dot{x}_\tau$, show that the motion is defined by

$$x = \frac{\dot{x}_\tau}{\omega} \sin \omega(t - \tau) + \left(x_\tau - \frac{P_0}{k}\right) \cos \omega(t - \tau) + \frac{P_0}{k}$$

$$\frac{\dot{x}}{\omega} = -\left(x_\tau - \frac{P_0}{k}\right) \sin \omega(t - \tau) + \frac{\dot{x}_\tau}{\omega} \cos \omega(t - \tau)$$

where $\omega = \sqrt{(k/M)}$ radians per second is the natural circular frequency.

10.2. A portal frame ABCD, in which the mass of the rigid horizontal member BC is 3000 kg, and the combined flexural stiffness of the light vertical members AB and CD is 12 000 π^2 N/m. The frame is acted upon by a two-step force P as shown, and it is given that the displacement and velocity are both zero at time $t = 0.14$ s.

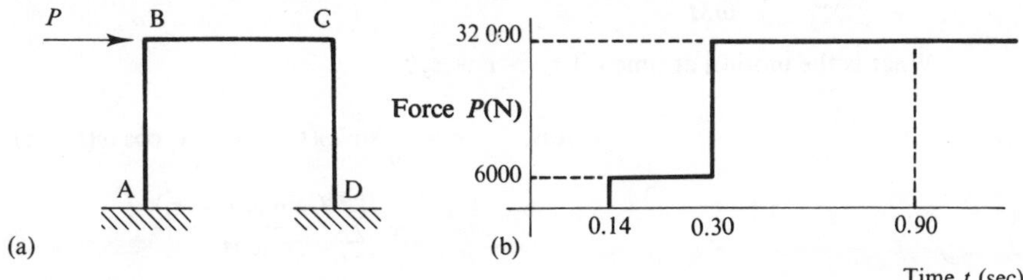

Problem 10·2

Determine the displacement and velocity at time $t = 0.9$ s. Work from first principles. Neglect damping. You are NOT allowed to use the method of superposition.

Ans. $x = 0.44$ m; $\dot{x} = -1.13$ m/s.

10.3. (a) An undamped structure of one degree of freedom is vibrating freely with a natural circular frequency of ω radians per second, and the initial conditions are $x = x_\tau$ m and $\dot{x} = \dot{x}_\tau$ metres per second at time $t = \tau$ seconds. Show that the motion is defined by:

$$x = \frac{\dot{x}_\tau}{\omega} \sin \omega(t - \tau) + x_\tau \cos \omega(t - \tau) \quad \text{metres}$$

(b) If the above structure is *at rest* and an impulse I N s is applied at time $t = \tau$ seconds, show that the subsequent motion is

$$x = \frac{I \sin \omega(t - \tau)}{\omega M} \text{ metres}$$

where M kilograms is the mass of the structure.

(c) If the structure is vibrating with motion defined in (a), and if at an instant $t = \tau$ seconds, an impulse I newton seconds is applied, show that the subsequent motion is

$$x = \frac{\dot{x}_\tau}{\omega} \sin \omega(t - \tau) + x_\tau \cos \omega(t - \tau) + \frac{I \sin \omega(t - \tau)}{\omega M}$$

i.e.

$$(c) = (a) + (b)$$

(d) If, subsequent to (c), at time $t = \tau'$ seconds, another impulse I' Newton seconds is applied, show that the motion for $t > \tau'$ seconds is

$$x = \frac{\dot{x}_\tau}{\omega} \sin \omega(t - \tau) + x_\tau \cos \omega(t - \tau) + \frac{I \sin \omega(t - \tau)}{\omega M} + \frac{I' \sin \omega(t - \tau')}{\omega M}$$

i.e.

$$(\text{d}) = (\text{a}) + (\text{b}) + \frac{I' \sin \omega(t - \tau')}{\omega M}$$

(e) Hence satisfy yourself with the useful concept that, if an undamped one-degree-of-freedom structure of mass M and natural circular frequency ω is freely vibrating such that $x = x_\tau$ and $\dot{x} = \dot{x}_\tau$ at $t = \tau$, and if at later times $\tau_1, \tau_2, \tau_3, \ldots, \tau_i, \ldots, \tau_n$, impulses $I_1, I_2, I_3, \ldots, I_i, \ldots, I_n$ respectively are applied, then, at $t > \tau_n$, the motion is

$$x = \frac{\dot{x}_\tau}{\omega} \sin \omega(t - \tau) + x_\tau \cos \omega(t - \tau) + \sum_{i=1}^{i=n} \frac{I_i \sin \omega(t - \tau_i)}{\omega M}$$

What is the motion at time t if $\tau_4 > t > \tau_3$?

Ans.
$$x = \frac{\dot{x}_\tau}{\omega} \sin \omega(t - \tau) + x_\tau \cos \omega(t - \tau) + \sum_{i=1}^{i=3} \frac{I_i \sin \omega(t - \tau_i)}{\omega M}$$

10.4. Solve Problem 10.2 by considering the force diagram as the superposition of (a) a force of 6000 N suddenly applied to the structure at $t = 0.14$ s and sustained; and (b) a force of 32 000 − 6000 = 26 000 N suddenly applied to the structure *at rest* at $t =$ 0.30 s and sustained.

Ans. As for Problem 10.2.

10.5. An undamped structure of one degree of freedom is subjected to the dynamic disturbance shown in the diagram. If the structure is at rest at $t = 0$, show that the displacement at time $t = \tau_1$ will be

$$x = \frac{1}{\omega M} \int_0^{\tau_1} P \sin \omega(t - \tau)\, d\tau$$

where ω is the natural circular frequency and M the mass.

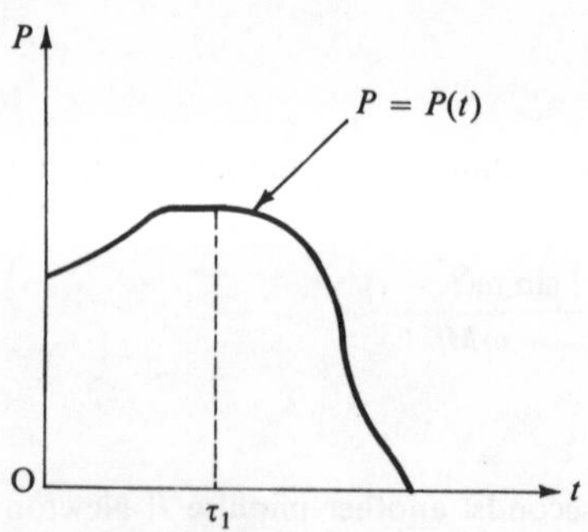

Problem 10·5

10.6. For a structure of n degrees of freedom, there are n equations of motion. Neglecting damping, explain:

(a) How the n natural frequencies are obtained.

(b) How the element ${}_iz_j$ in the modal matrix $\mathbf{Z}$ is related to the ith normal mode shape ${}_i\mathbf{e}$ and the mass matrix $\mathbf{M}$.

(c) How the n coupled equations

$$\mathbf{M}\ddot{\mathbf{x}} + \mathbf{K}_s\mathbf{x} = 0$$

where $\mathbf{K}_S$ is the stiffness matrix, are reduced to the n uncoupled equations

$$\ddot{\mathbf{q}} + \boldsymbol{\omega}^2\mathbf{q} = 0$$

where $\boldsymbol{\omega}^2$ is an $n \times n$ diagonal matrix of the squares of the n natural circular frequencies.

(d) If the structurè is acted on by a dynamic force vector $\mathbf{P}$, how the set of n coupled equations

$$\mathbf{M}\ddot{\mathbf{x}} + \mathbf{K}_s\mathbf{x} = \mathbf{P}$$

can be reduced to the set of n uncoupled equations:

$$\ddot{\mathbf{q}} + \boldsymbol{\omega}^2\mathbf{q} = \mathbf{Z}^T\mathbf{P}$$

(e) How the $2n$ arbitrary constants in the solution of the uncoupled equations in (d) can be determined from the $2n$ initial conditions prescribed on the actual displacements $\mathbf{x}$ and on the actual velocities $\dot{\mathbf{x}}$.

10.7. The equation of motion of a structure with n degrees of freedom is:

$$\mathbf{M}\ddot{\mathbf{x}} + \mathbf{K}_s\mathbf{x} = \mathbf{P}$$

where $\mathbf{M}$ is the mass matrix, $\mathbf{K}_s$ the stiffness matrix, and $\mathbf{P}$ the dynamic force vector.

(a) Determine a matrix $\mathbf{T}$ such that

$$\mathbf{TMT}^T = \text{unit matrix}$$
$$\mathbf{TK}_s\mathbf{T}^T = \text{diagonal matrix}$$

(b) If

$$\mathbf{M} = \begin{bmatrix} 5 & 0 \\ 0 & 5 \end{bmatrix} \text{kg}, \qquad \mathbf{K}_s = \begin{bmatrix} 10\,000 & -5000 \\ -5000 & 5000 \end{bmatrix} \text{Nm}^{-1}$$

determine $\mathbf{T}$.

Ans. (a) $\mathbf{T} = \mathbf{Z}^T$; (b) $\begin{bmatrix} 0.235 & 0.381 \\ 0.381 & -0.235 \end{bmatrix} \text{kg}^{-1/2}$

10.8. Two masses, each of mass M equal to 5 kg, are connected by springs having equal stiffness k of 5000 N/m each. The masses are free to slide along a straight line AB on a frictionless table. Neglecting damping, determine: (a) the natural frequencies; (b) the normal mode shapes: (c) the modal matrix that transforms he coordinates x_1, x_2 into generalized coordinates which will yield uncoupled equations of motion.

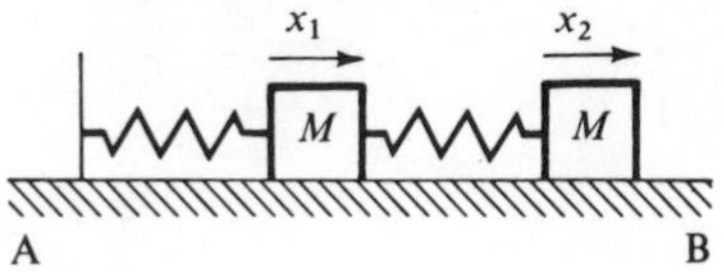

Problem 10·8

Ans. $f_1 = 3.11$ Hz, $f_2 = 8.14$ Hz,

${}_1\mathbf{e} = [1C_1 \; 1.62C_1]^T$ m, ${}_2\mathbf{e} = [1C_2 - 0.62C_2]^T$ m,

$$\mathbf{Z} = \begin{bmatrix} 0.235 & 0.381 \\ 0.381 & -0.235 \end{bmatrix} \text{kg}^{-1/2}$$

10.9. If in Problem 10.8, the second mass is now connected to B by a spring of stiffness k, and if the initial conditions are $x_1 = 0$ and $x_2 = x_\tau$ at $t = 0$, show that the motion for $t > 0$ is defined by

$$\begin{bmatrix} x_1 \\ x_2 \end{bmatrix} = \tfrac{1}{2}x_\tau \begin{bmatrix} \cos\sqrt{\left(\frac{k}{M}\right)}t - \cos\sqrt{\left(\frac{3k}{M}\right)}t \\ \cos\sqrt{\left(\frac{k}{M}\right)}t + \cos\sqrt{\left(\frac{3k}{M}\right)}t \end{bmatrix}$$

10.10 (to be attempted after reading Section 11.5 on principal stresses). In principal stress analysis, a transformation matrix **T** of unit eigenvectors of the stress tensor $\boldsymbol{\sigma}$ transforms the stress tensor into a diagonal matrix of eigenvalues, which are the principal stresses:

$$\mathbf{T}\boldsymbol{\sigma}\mathbf{T}^T = \boldsymbol{\sigma}_{\text{principal}}$$

That is, the principal stress problem is essentially an eigenvalue problem.

Show that, in structural dynamics, the natural frequency problem is also in essence an eigenvalue problem. That is, a transformation matrix of normalized eigenvectors of the dynamic matrix $[\mathbf{M}^{-1}\mathbf{K}_s]$ transforms the stiffness matrix $\mathbf{K}_s$ into a diagonal matrix of eigenvalues, which are the squares of the natural circular frequencies:

$$\mathbf{T}\mathbf{K}_s\mathbf{T}^T = \boldsymbol{\omega}^2$$

Do the normalized eigenvectors form the rows or columns of **T**?

Ans. The rows, because they form the columns of the modal matrix, which is the transpose of the transformation matrix **T**.

References

1 Sokolnikoff, I.S. and R.M. Redheffer. *Mathematics of Physics and Modern Engineering*, McGraw-Hill, 2nd edition, 1966, p. 173.

2 Warburton, G.B. The *Dynamical Behaviour of Structures*. Pergamon Press, Oxford, 2nd edition, 1976.

3 Hurty, W.C. and M.F. Rubinstein. *Dynamics of Structures*. Prentice-Hall, Englewood Cliffs, New Jersey, 1964.

4 Rogers, G.L. *Dynamics of Framed Structures*. John Wiley, New York, 1959.

Chapter 11
Elasticity problems and the finite difference method

The twofold aim of this chapter is (a) to present the fundamentals of the theory of elasticity, leading to the development of the Laplace and biharmonic equations, which have important applications in structural engineering, and (b) to present the finite difference method as a powerful tool for the numerical solution of these equations, the exact solution of which is often very difficult and sometimes impossible.

A grasp of the fundamentals of the theory of elasticity is rapidly becoming essential to the structural engineer. There is a need for a concise account of the essentials of the subject and the authors are attempting to fulfil this need by presenting, in simple terms and in a coherent manner, those topics (only) of the theory of elasticity, a knowledge of which is a real asset to the structural engineer.

11.1 Differential equations of equilibrium

Consider an arbitrary body (Fig. 11.1–1) in equilibrium under a system of **surface forces** and **body forces**. 'Surface forces' is a term for stresses acting on the surface or boundary of a body; examples of surface forces include hydrostatic pressure between a solid and

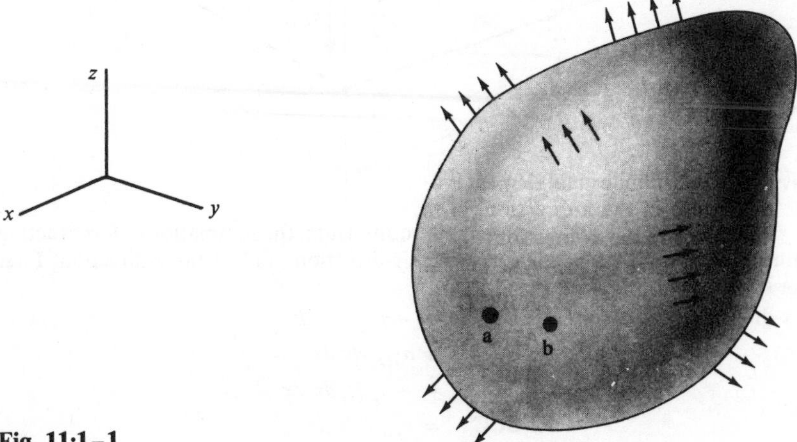

Fig. 11·1–1

a fluid and contact pressures and frictional forces between solid bodies. 'Body forces' are distributed over the volume or mass of the body; examples of body forces include gravitational forces and magnetic forces.

Suppose at an arbitrary point 'a' (x, y, z) the stresses are given by the elements of the following matrix:

$$\begin{bmatrix} \sigma_x & \tau_{xy} & \tau_{xz} \\ \tau_{yx} & \sigma_y & \tau_{yz} \\ \tau_{zx} & \tau_{zy} & \sigma_z \end{bmatrix}$$

This matrix is called a **stress tensor**, for a reason to be explained in Section 11.2, Eqn 11.2–9. The elements in the first row of the stress tensor are the stresses acting on a plane normal to the x-axis, those in the second row act on a plane normal to the y-axis, and those in the third row act on a plane normal to the z-axis (see Fig. 1.5–4).

Figure 11.1–2 shows an elementary parallelepiped with its centroid at point 'a' of the body in Figure 11.1–1, and with each face normal to a coordinate axis. The stresses acting at the centroids of the faces ABCD, ADD′A′ and A′B′BA are $\sigma_{x(1)}$, $\tau_{xy(1)}$, $\tau_{xz(1)}$, $\sigma_{y(1)}$, . . ., etc.; these are given a suffix (1) and their directions are indicated by the dotted-line arrows. The stresses acting on the faces A′B′C′D′, B′BCC′, and D′C′CD are given a suffix (2) and their directions are indicated by the full-line arrows. Note that all stresses are shown in their positive directions, in accordance with the stress notation and sign convention in Chapter 1, Section 1.5.

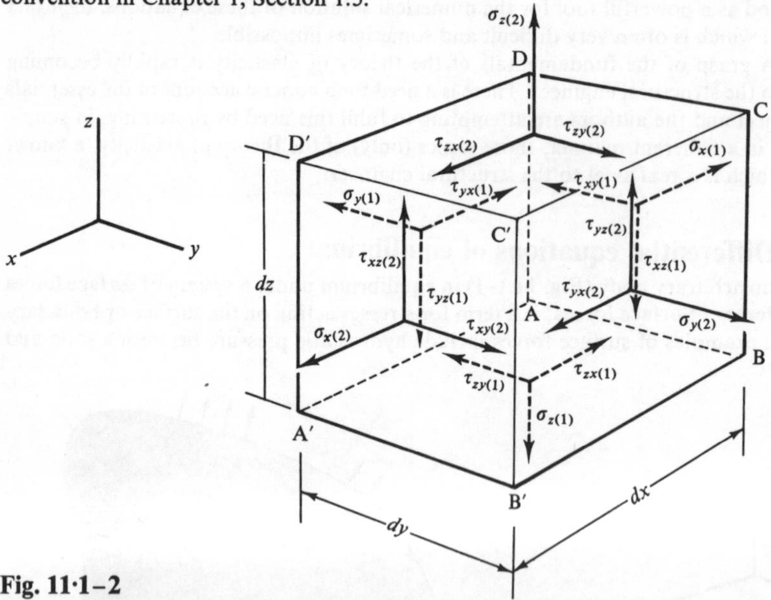

Fig. 11·1–2

Since this infinitesimal element is in equilibrium, the summation of forces acting on it must be zero in the x-direction, in the y-direction, and in the z-direction. First consider the forces in the x-direction:

$$\begin{aligned} x\text{-direction force on face ABCD} &= -\sigma_{x(1)}\,dy\,dz \\ \text{A}'\text{B}'\text{C}'\text{D}' &= \sigma_{x(2)}\,dy\,dz \\ \text{ADD}'\text{A}' &= -\tau_{yx(1)}\,dz\,dx \\ \text{B}'\text{BCC}' &= \tau_{yx(2)}\,dz\,dx \\ \text{A}'\text{B}'\text{BA} &= -\tau_{zx(1)}\,dx\,dy \\ \text{D}'\text{C}'\text{CD} &= \tau_{zx(2)}\,dx\,dy \end{aligned}$$

The above forces are all surface forces, i.e. they act on the surface or boundary of the element. Summation of these forces on the six faces gives:

Total surface force in x-direction is

$$\left[\frac{\sigma_{x(2)} - \sigma_{x(1)}}{dx} + \frac{\tau_{yx(2)} - \tau_{yx(1)}}{dy} + \frac{\tau_{zx(2)} - \tau_{zx(1)}}{dz}\right] dx\, dy\, dz$$

Now

$$\frac{\sigma_{x(2)} - \sigma_{x(1)}}{dx} = \frac{\partial \sigma_x}{\partial x}$$

$$\frac{\tau_{yx(2)} - \tau_{yx(1)}}{dy} = \frac{\partial \tau_{yx}}{\partial y}$$

$$\frac{\tau_{zx(2)} - \tau_{zx(1)}}{dz} = \frac{\partial \tau_{zx}}{\partial z}$$

Hence, total surface force in x-direction is

$$\left[\frac{\partial \sigma_x}{\partial x} + \frac{\partial \tau_{yx}}{\partial y} + \frac{\partial \tau_{zx}}{\partial z}\right] dx\, dy\, dz \tag{11.1–1}$$

In addition to surface forces, the element may be subjected to body forces, of magnitude, say, X, Y, Z per unit volume in the respective coordinate directions. Then,

$$\text{total body force in } x\text{-direction} = X\, dx\, dy\, dz \tag{11.1–2}$$

Since the element is in equilibrium, we must have

Total surface force in x-direction (Eqn 11.1–1) + total body force in x-direction (Eqn 11.1–2) = 0

i.e.

$$\left[\frac{\partial \sigma_x}{\partial x} + \frac{\partial \tau_{yx}}{\partial y} + \frac{\partial \tau_{zx}}{\partial z} + X\right] dx\, dy\, dz = 0$$

or

$$\frac{\partial \sigma_x}{\partial x} + \frac{\partial \tau_{yx}}{\partial y} + \frac{\partial \tau_{zx}}{\partial z} + X = 0 \tag{11.1–3(a)}$$

Similarly, for equilibrium in the y- and z-directions, we have

$$\frac{\partial \tau_{xy}}{\partial x} + \frac{\partial \sigma_y}{\partial y} + \frac{\partial \tau_{zy}}{\partial z} + Y = 0 \tag{11.1–3(b)}$$

$$\frac{\partial \tau_{xz}}{\partial x} + \frac{\partial \tau_{yz}}{\partial y} + \frac{\partial \sigma_z}{\partial z} + Z = 0 \tag{11.1–3(c)}$$

Equations 11.1–3(a), (b), and (c) are called the **differential equations of equilibrium** and refer to a body at rest or moving under uniform velocity. The reader should verify that, if the body is not in equilibrium, then Eqns 11.1–3(a), (b), and (c) should be modified to

$$\frac{\partial \sigma_x}{\partial x} + \frac{\partial \tau_{yx}}{\partial y} + \frac{\partial \tau_{zx}}{\partial z} + X = \rho \frac{d^2 u}{dt^2} \tag{11.1–4(a)}$$

$$\frac{\partial \tau_{xy}}{\partial x} + \frac{\partial \sigma_y}{\partial y} + \frac{\partial \tau_{zy}}{\partial z} + Y = \rho \frac{d^2 v}{dt^2} \tag{11.1–4(b)}$$

$$\frac{\partial \tau_{xz}}{\partial x} + \frac{\partial \tau_{yz}}{\partial y} + \frac{\partial \sigma_z}{\partial z} + Z = \rho \frac{d^2 w}{dt^2} \tag{11.1–4(c)}$$

where ρ is the density of the material of the body;
u is the displacement component in the x-direction;
v is that in the y-direction;
w is that in the z-direction, and t denotes time.

Equation 11.1–4(a), (b), and (c) are called the **differential equations of motion.**

Note that the validity of Eqns 11.1–3 and 11.1–4 is general and *in no sense limited by the specific mechanical properties of the material of the body.*

11.2 Specification of stress at a point

Particular case : two dimensions. Figure 11.2–1 shows a wedge OAB, of uniform thickness, acted on by normal and shear stresses (which are all shown in their positive directions as defined in Chapter 1, Section 1.5). The side AB is normal to an arbitrary direction X.

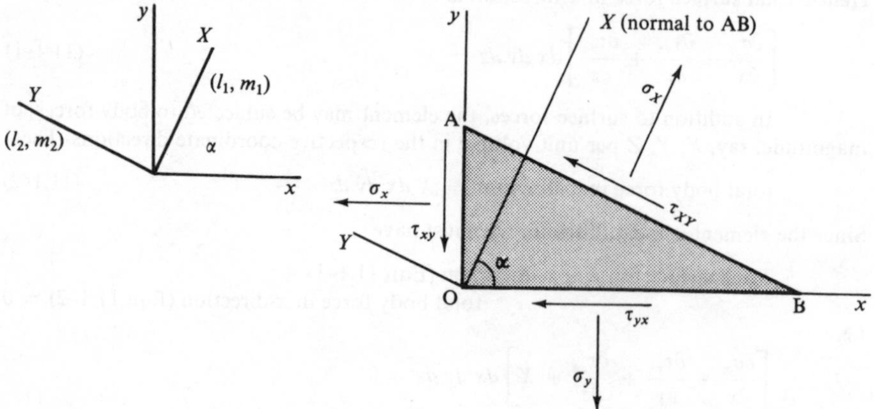

Fig. 11·2–1

If the stresses σ_x, σ_y, τ_{xy} $(= \tau_{yx})$ are given, then the normal stress σ_X and the shear stress τ_{XY} can be readily shown, by considering the equilibrium of the wedge in the x- and y-directions, to be

$$\sigma_X = l_1^2 \sigma_x + m_1^2 \sigma_y + 2 l_1 m_1 \tau_{xy} \tag{11.2–1(a)}$$

$$\tau_{XY} = l_1 l_2 \sigma_x + m_1 m_2 \sigma_y + (l_1 m_2 + l_2 m_1)\tau_{xy} \tag{11.2–1(b)}$$

where l_1 and m_1 are the direction cosines of OX with respect to xOy, and l_2, m_2 are those of OY. That is

$$\begin{bmatrix} l_1 & m_1 \\ l_2 & m_2 \end{bmatrix} = \begin{bmatrix} \cos\alpha & \sin\alpha \\ -\sin\alpha & \cos\alpha \end{bmatrix} \tag{11.2–2}$$

Equation 11.2–1(a) and (b) still hold when the wedge OAB is of infinitesimal size. Thus, by considering OAB to be infinitely small, it is seen that if the stresses at a point O, referred to axes xOy, are σ_x, σ_y, τ_{xy} and τ_{yx} then the stresses at that point referred to new cartesian axes XOY are as given by Eqns 11.2–1. Actually, the reader will recognize that Eqns 11.2–1 are in fact the results of the well-known **Mohr-circle analysis.**

General case: three dimensions. Suppose at a certain point O in a body the stresses, referred to rectangular axes $Oxyz$, are given by the elements of the following stress tensor:

$$\begin{bmatrix} \sigma_x & \tau_{xy} & \tau_{xz} \\ \tau_{yx} & \sigma_y & \tau_{yz} \\ \tau_{zx} & \tau_{zy} & \sigma_z \end{bmatrix}$$

As explained in the previous section, the elements in the first row of the stress tensor are the stresses acting on a plane normal to the x-axis, those in the second row act on a plane normal to the y-axis, and those in the third row act on a plane normal to the z-axis. The *stress tensor is symmetrical*, because, as explained in Section 1.5, $\tau_{xy} = \tau_{yx}$, $\tau_{yz} = \tau_{zy}$, and $\tau_{zx} = \tau_{xz}$.

Consider an infinitesimal tetrahedron OABC (Fig. 11.2–2) where the three mutually perpendicular faces are each normal to a coordinate axis, and the normal **p** to

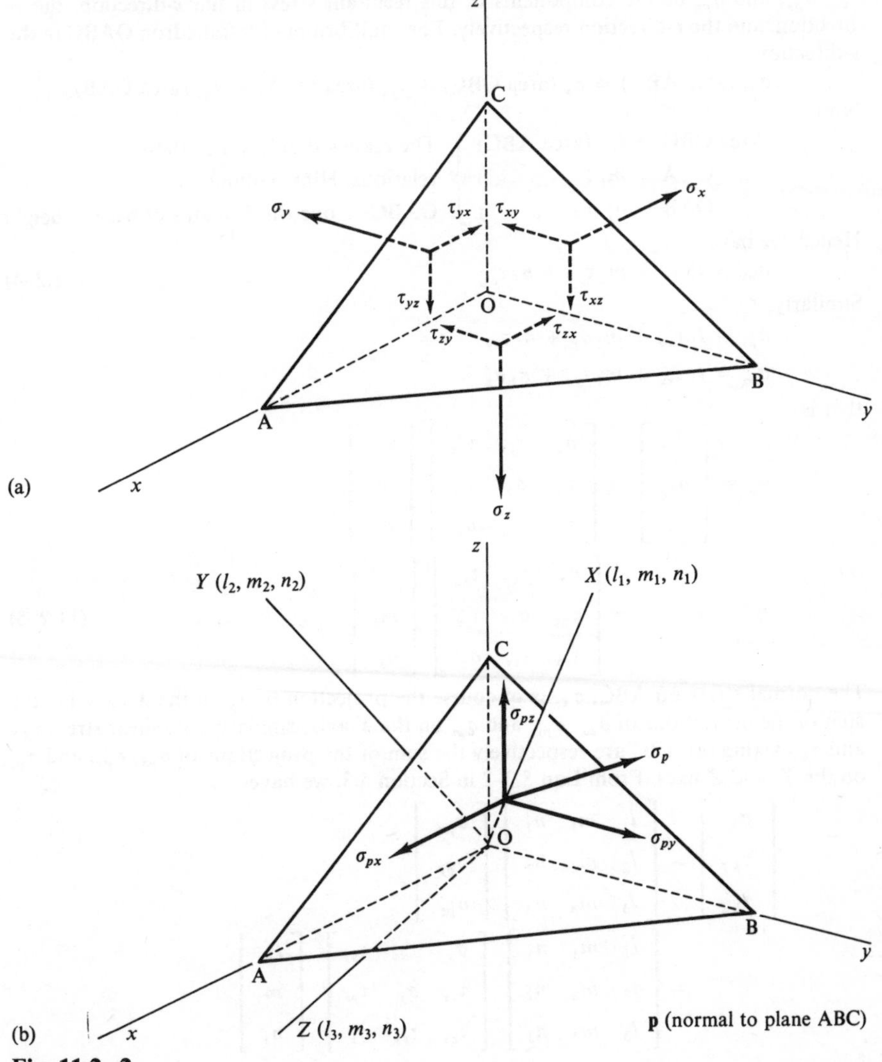

Fig. 11·2–2

the inclined face ABC is in the direction of axis OX of another arbitrary rectangular system $OXYZ$, the orientation of which is defined by the following direction-cosine matrix:

$$\begin{bmatrix} l_1 & m_1 & n_1 \\ l_2 & m_2 & n_2 \\ l_3 & m_3 & n_3 \end{bmatrix} = \begin{bmatrix} \cos X\hat{O}x & \cos X\hat{O}y & \cos X\hat{O}z \\ \cos Y\hat{O}x & \cos Y\hat{O}y & \cos Y\hat{O}z \\ \cos Z\hat{O}x & \cos Z\hat{O}y & \cos Z\hat{O}z \end{bmatrix} \tag{11.2–3}$$

For clarity, the stresses acting on the faces OBC, OCA, and OAB are shown in Fig. 11.2–2(a) and those on the inclined face ABC are shown in Fig. 11.2–2(b).

The stresses acting on face OBC are σ_x, τ_{xy}, τ_{xz} respectively.
" " " " " OCA are τ_{yx}, σ_y, τ_{yz} "
" " " " " OAB are τ_{zx}, τ_{zy}, σ_z "

Also, the resultant stress $\boldsymbol{\sigma}_p$ on face ABC is in general oblique to the face; let σ_{px}, σ_{py}, and σ_{pz} be the components of this resultant stress in the x-direction, the y-direction, and the z-direction respectively. For equilibrium of tetrahedron OABC in the x-direction,

$$\sigma_{px}\,(\text{area ABC}) = \sigma_x\,(\text{area OBC}) + \tau_{yx}\,(\text{area OCA}) + \tau_{zx}\,(\text{area OAB})$$

Now,

$$\left.\begin{aligned} \text{Area OBC} &= l_1\ (\text{area ABC}) \\ \text{,, OCA} &= m_1\ (\ \ ,,\ \ ) \\ \text{,, OAB} &= n_1\ (\ \ ,,\ \ ) \end{aligned}\right\}$$ The reader should verify these relations. Hint: volume of OABC = one-third of area of base × height

Hence, we have

$$\sigma_{px} = l_1\sigma_x + m_1\tau_{yx} + n_1\tau_{zx} \tag{11.2–4}$$

Similarly,

$$\sigma_{py} = l_1\tau_{xy} + m_1\sigma_y + n_1\tau_{zy}$$

$$\sigma_{pz} = l_1\tau_{xz} + m_1\tau_{yz} + n_1\sigma_z$$

that is

$$\boldsymbol{\sigma}_p = \begin{bmatrix} \sigma_{px} \\ \sigma_{py} \\ \sigma_{pz} \end{bmatrix} = \begin{bmatrix} \sigma_x & \tau_{yx} & \tau_{zx} \\ \tau_{xy} & \sigma_y & \tau_{zy} \\ \tau_{xz} & \tau_{yz} & \sigma_z \end{bmatrix} \begin{bmatrix} l_1 \\ m_1 \\ n_1 \end{bmatrix}$$

$$= \begin{bmatrix} \sigma_x & \tau_{xy} & \tau_{xz} \\ \tau_{yx} & \sigma_y & \tau_{yz} \\ \tau_{zx} & \tau_{zy} & \sigma_z \end{bmatrix}^T \begin{bmatrix} l_1 \\ m_1 \\ n_1 \end{bmatrix} \tag{11.2–5}$$

The normal stress on ABC, σ_X, is of course the projection of $\boldsymbol{\sigma}_p$ on the X-axis, i.e. the sum of the projections of σ_{px}, σ_{py}, and σ_{pz} on the X-axis; similarly, the shear stress τ_{XY} and τ_{XZ} acting on ABC are respectively the sum of the projections of σ_{px}, σ_{py}, and σ_{pz} on the Y- and Z-axes. From Eqn 5.3–2 in Section 5.3, we have

$$\begin{bmatrix} \sigma_X \\ \tau_{XY} \\ \tau_{XZ} \end{bmatrix} = \begin{bmatrix} l_1 & m_1 & n_1 \\ l_2 & m_2 & n_2 \\ l_3 & m_3 & n_3 \end{bmatrix} \begin{bmatrix} \sigma_{px} \\ \sigma_{py} \\ \sigma_{pz} \end{bmatrix}$$

$$= \begin{bmatrix} l_1 & m_1 & n_1 \\ l_2 & m_2 & n_2 \\ l_3 & m_3 & n_3 \end{bmatrix} \begin{bmatrix} \sigma_x & \tau_{xy} & \tau_{xz} \\ \tau_{yx} & \sigma_y & \tau_{yz} \\ \tau_{zx} & \tau_{zy} & \sigma_z \end{bmatrix}^T \begin{bmatrix} l_1 \\ m_1 \\ n_1 \end{bmatrix}$$

(from Eqn 11.2–5)

i.e.

$$\begin{bmatrix} \sigma_X \\ \tau_{XY} \\ \tau_{XZ} \end{bmatrix} = \mathbf{T}\boldsymbol{\sigma}^T \begin{bmatrix} l_1 \\ m_1 \\ n_1 \end{bmatrix}$$

or

$$[\sigma_X \tau_{XY} \tau_{XZ}] = [l_1 m_1 n_1] \underset{(Oxyz)}{\boldsymbol{\sigma}} \mathbf{T}^T \tag{11.2–6}$$

Thus, at an arbitrary point O, if the stress tensor $\boldsymbol{\sigma}$ referred to any given set of rectangular axes $Oxyz$ is known, then the normal and shear stresses acting on any plane through O normal to axis OX of an arbitrary set of rectangular axes $OXYZ$ are given by Eqn 11.2–6. Equation 11.2–6 refers to a plane normal to OX; if we consider a plane normal to OY then a reasoning similar to the above would yield the stresses σ_Y, τ_{YX}, and τ_{YZ}:

$$[\tau_{YX} \sigma_Y \tau_{YZ}] = [l_2 m_2 n_2] \underset{(Oxyz)}{\boldsymbol{\sigma}} \mathbf{T}^T \tag{11.2–7}$$

Similarly,

$$[\tau_{ZX} \tau_{ZY} \sigma_Z] = [l_3 m_3 n_3] \underset{(Oxyz)}{\boldsymbol{\sigma}} \mathbf{T}^T \tag{11.2–8}$$

Equations 11.2–6, 11.2–7, and 11.2–8 can be written in a general form:

$$\begin{bmatrix} \sigma_X & \tau_{XY} & \tau_{XZ} \\ \tau_{YX} & \sigma_Y & \tau_{YZ} \\ \tau_{ZX} & \tau_{ZY} & \sigma_Z \end{bmatrix} = \begin{bmatrix} l_1 & m_1 & n_1 \\ l_2 & m_2 & n_2 \\ l_3 & m_3 & n_3 \end{bmatrix} \begin{bmatrix} \sigma_x & \tau_{xy} & \tau_{xz} \\ \tau_{yx} & \sigma_y & \tau_{yz} \\ \tau_{zx} & \tau_{zy} & \sigma_z \end{bmatrix} \begin{bmatrix} l_1 & m_1 & n_1 \\ l_2 & m_2 & n_2 \\ l_3 & m_3 & n_3 \end{bmatrix}^T$$

i.e.

$$\underset{(OXYZ)}{\boldsymbol{\sigma}} = \mathbf{T} \underset{(Oxyz)}{\boldsymbol{\sigma}} \mathbf{T}^T \tag{11.2–9}$$

That is, at any point O, the stress tensor referred to coordinate axes $OXYZ$ is related to the stress tensor referred to axes $Oxyz$ by Eqn 11.2–9. The transformation represented by Eqn 11.2–9 is known as the **tensor transformation law**. If a matrix can be transformed according to this law it is a tensor. The stress matrix can always be so transformed and can therefore be called a stress tensor.

Given $\boldsymbol{\sigma}(Oxyz)$, $\boldsymbol{\sigma}(OXYZ)$ can be obtained readily from Eqn 11.2–9, using standard computer routines. Of course, explicit forms of σ_X, τ_{XY}, etc., can easily be obtained by expanding Eqn 11.2–9 or Eqns 11.2–6, 11.2–7, and 11.2–8. The reader should verify that

$$\sigma_X = l_1^2\sigma_x + m_1^2\sigma_y + n_1^2\sigma_z + 2l_1m_1\tau_{xy} + 2m_1n_1\tau_{yz} + 2n_1l_1\tau_{zx} \tag{11.2–10(a)}$$

$$\begin{aligned} \tau_{XY} = {} & l_1l_2\sigma_x + m_1m_2\sigma_y + n_1n_2\sigma_z + (l_1m_2 + l_2m_1)\tau_{xy} \\ & + (m_1n_2 + m_2n_1)\tau_{yz} + (n_1l_2 + n_2l_1)\tau_{zx} \end{aligned} \tag{11.2–10(b)}$$

$$\begin{aligned} \tau_{XZ} = {} & l_1l_3\sigma_x + m_1m_3\sigma_y + n_1n_3\sigma_z + (l_1m_3 + l_3m_1)\tau_{xy} \\ & + (m_1n_3 + m_3n_1)\tau_{yz} + (n_1l_3 + n_3l_1)\tau_{zx} \end{aligned} \tag{11.2–10(c)}$$

plus six similar equations for $\tau_{YX}, \sigma_Y, \tau_{YZ}, \tau_{ZX}, \tau_{ZY}, \sigma_Z$.

In the case of plane stress, $\sigma_z = \tau_{yz} = \tau_{zx} = 0$ (Section 1.7); in the case of plane strain, $\varepsilon_z = \gamma_{yz} = \gamma_{zx} = 0$ resulting in $\sigma_z = \nu(\sigma_x + \sigma_y)$ and $\tau_{yz} = \tau_{zx} = 0$ (Section 1.7). Hence in both cases, the problem reduces to a two-dimensional one, provided the axes Oz and OZ coincide. The reader should verify that, in such cases, $n_1 = n_2 = l_3 = m_3 = 0$ and $n_3 = 1$ and that Eqns 11.2–10 reduce to Eqns 11.2–1. Note also that Eqn 11.2–9, and indeed all equations in this section, are *valid irrespective of the mechanical properties of the material.*

11.3 Boundary conditions

We have shown that if a body is in equilibrium then the stresses at any point of the body must satisfy the differential equations of equilibrium, Eqns 11.1–3. In general, these stresses vary from point to point over the volume of the body, and when we arrive at the boundary surface, these stresses must be in equilibrium with the external forces acting on the boundary. To obtain the equations of equilibrium at the point on the boundary, we recall that Eqns 11.2–4 give the three components σ_{px}, σ_{py}, and σ_{pz} of the stress $\boldsymbol{\sigma}_p$ on an oblique plane through a point in the body. Therefore, if this oblique plane is tangential to the boundary surface, then $\boldsymbol{\sigma}_p$ can be considered as the external stresses on the boundary. Hence for equilibrium of an element on the boundary, such as the infinitesimal tetrahedron in Fig. 11.2–2 with plane ABC representing part of the boundary surface, the **boundary conditions** are

$$\begin{aligned} \sigma_{px} &= l\sigma_x + m\tau_{yx} + n\tau_{zx} \\ \sigma_{py} &= l\tau_{xy} + m\sigma_y + n\tau_{zy} \\ \sigma_{pz} &= l\tau_{xz} + m\tau_{yz} + n\sigma_z \end{aligned} \tag{11.3–1(a)}$$

where σ_{px}, σ_{py}, and σ_{pz} are now the components of the external stress acting on the boundary point concerned, and l, m, n are the direction cosines of the outward normal to the boundary surface at that point.

It is customary to denote the components of the surface forces in the x-, y-, and z-directions respectively as $\bar{X}$, $\bar{Y}$, $\bar{Z}$ and to write Eqn 11.3–1(a) as

$$\begin{aligned} l\sigma_x + m\tau_{yx} + n\tau_{zx} &= \bar{X} \\ l\tau_{xy} + m\sigma_y + n\tau_{zy} &= \bar{Y} \\ l\tau_{xz} + m\tau_{yz} + n\sigma_z &= \bar{Z} \end{aligned} \tag{11.3–1(b)}$$

Note that the surface components $\bar{X}$, $\bar{Y}$, $\bar{Z}$ are expressed as forces per unit area of the boundary surface, i.e. they are stresses.

11.4 Large-displacement definitions of normal and shear strains; specification of strain at a point

In Chapter 1, expressions for the normal strains ε_x, ε_y, ε_z and the shear strains γ_{xy}, γ_{yz}, and γ_{zx} were given in Eqns 1.6–5 and 1.6–7, and it was pointed out that these equations applied only when displacements were small. In this section, the **large-displacement definitions** of normal and shear strains will be introduced. Note that though these definitions are called large-displacement (or large-deflection) definitions, they are valid irrespective of whether displacements are large or small. It will be shown that when displacements are small, the large-displacement definitions would yield the expressions in Eqns 1.6–5 and 1.6–7.

It is true that in this book we are concerned mainly with small displacements so that Eqns 1.6–5 and 1.6–7 apply, unless otherwise stated. However, the large-displacement definitions are very useful in that they enable us to determine the strains for an arbitrary set of axes from the strains for a given set of axes.

Consider an infinitesimal line **ds** in an arbitrary direction $X\,(l_1\, m_1\, n_1)$ in a body (Fig. 11.4–1). Let the ends of the line element **ds** be A (x, y, z) and B $(x + dx, y + dy, z + dz)$. After deformation A moves to A′ and B to B′, i.e. **ds** becomes **ds**′. The components of the displacement AA′ are (u, v, w) in the x-, y-, and z-directions respectively; the components of BB′ are (u', v', w') where

$$u' = u + \frac{\partial u}{\partial x}dx + \frac{\partial u}{\partial y}dy + \frac{\partial u}{\partial z}dz \tag{11.4–1(a)}$$

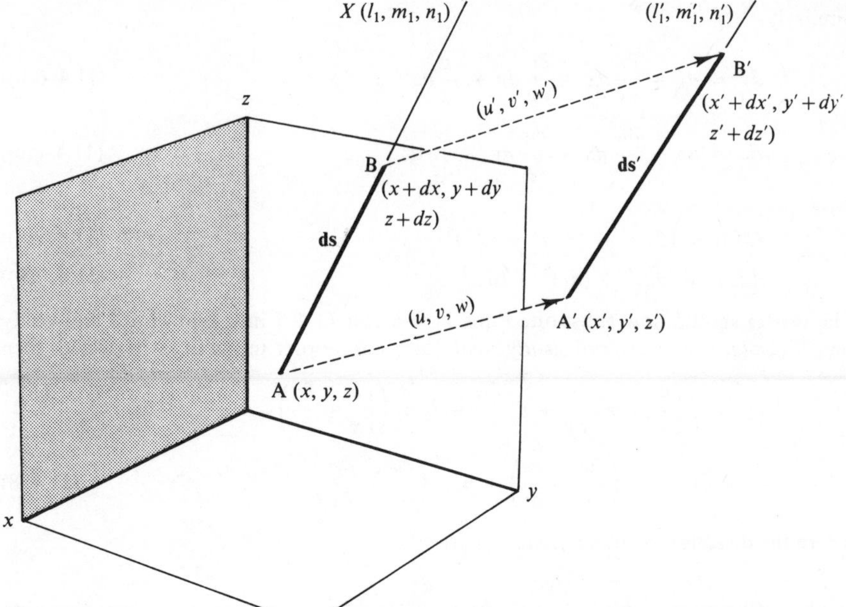

Fig. 11·4–1

$$v' = v + \frac{\partial v}{\partial x}dx + \frac{\partial v}{\partial y}dy + \frac{\partial v}{\partial z}dz \qquad (11.4\text{–}1(b))$$

$$w' = w + \frac{\partial w}{\partial x}dx + \frac{\partial w}{\partial y}dy + \frac{\partial w}{\partial z}dz \qquad (11.4\text{–}1(c))$$

The large-displacement definition of the normal strain of the line element **ds** *is*

$$\varepsilon_X = \tfrac{1}{2}\left[\frac{\mathbf{ds}'.\mathbf{ds}' - \mathbf{ds}.\mathbf{ds}}{(ds)^2}\right] \qquad (11.4\text{–}2)$$

or

$$\varepsilon_X = \tfrac{1}{2}\left[\frac{(ds')^2 - (ds)^2}{(ds)^2}\right] \qquad (11.4\text{–}3)$$

To evaluate ε_X, we note from Fig. 11.4–1 that:

$$\begin{aligned} x\text{-coordinate of A}' &= (x\text{-coordinate of A}) + u \\ &= x + u \qquad (11.4\text{–}4) \end{aligned}$$

$$\begin{aligned} x\text{-coordinate of B}' &= (x\text{-coordinate of B}) + u' \\ &= (x + dx) + u' \\ &= x + dx + u + \frac{\partial u}{\partial x}dx + \frac{\partial u}{\partial y}dy + \frac{\partial u}{\partial z}dz \\ &\qquad \text{(from Eqn 11.4–1(a))} \quad (11.4\text{–}5) \end{aligned}$$

$$\begin{aligned} dx' &= (x\text{-coordinate of B}') - (x\text{-coordinate of A}') \\ &= dx + \frac{\partial u}{\partial x}dx + \frac{\partial u}{\partial y}dy + \frac{\partial u}{\partial z}dz \quad \text{(from Eqns 11.4–4 and 11.4–5)} \\ &\qquad (11.4\text{–}6(a)) \end{aligned}$$

Similarly,

$$dy' = dy + \frac{\partial v}{\partial x} dx + \frac{\partial v}{\partial y} dy + \frac{\partial v}{\partial z} dz \qquad (11.4\text{–}6(b))$$

$$dz' = dz + \frac{\partial w}{\partial x} dx + \frac{\partial w}{\partial y} dy + \frac{\partial w}{\partial z} dz \qquad (11.4\text{–}6(c))$$

Now,

$$(ds')^2 = (dx')^2 + (dy')^2 + (dz')^2 \qquad (11.4\text{–}7(a))$$

$$(ds)^2 = (dx)^2 + (dy)^2 + (dz)^2 \qquad (11.4\text{–}7(b))$$

The reader should now substitute Eqns 11.4–6 and 11.4–7 into Eqn 11.4–3 and verify that *if displacements are sufficiently small* for second-order terms to be neglected, then

$$\varepsilon_X = l_1^2 \frac{\partial u}{\partial x} + m_1^2 \frac{\partial v}{\partial y} + n_1^2 \frac{\partial w}{\partial z} + l_1 m_1 \left(\frac{\partial v}{\partial x} + \frac{\partial u}{\partial y}\right) + m_1 n_1 \left(\frac{\partial w}{\partial y} + \frac{\partial v}{\partial z}\right) + n_1 l_1 \left(\frac{\partial u}{\partial z} + \frac{\partial w}{\partial x}\right) \qquad (11.4\text{–}8)$$

where the direction cosines (l_1, m_1, n_1) are

$$l_1 = \frac{dx}{ds}; \qquad m_1 = \frac{dy}{ds}; \qquad n_1 = \frac{dz}{ds}$$

Equation 11.4–8 expresses the normal strain ε_X in an arbitrary direction X (l_1, m_1, n_1) in terms of the displacements u, v, w provided such displacements are small.

Consider the particular case when the direction of X coincides with the x-axis, i.e. $l_1 = 1$, $m_1 = 0$, $n_1 = 0$; then Eqn 11.4–8 reduces to

$$\varepsilon_x = \frac{\partial u}{\partial x} \qquad (11.4\text{–}9(a))$$

Similarly,

$$\varepsilon_y = \frac{\partial v}{\partial y} \qquad (11.4\text{–}9(b))$$

$$\varepsilon_z = \frac{\partial w}{\partial z} \qquad (11.4\text{–}9(c))$$

Equation 11.4–9(a) gives the normal strain of a line element initially in the x-direction, and Eqns 11.4–9(b) and (c) give the normal strains of line elements initially in the y- and z-directions respectively. These equations agree with Eqns 1.6–5 in Chapter 1.

We shall next evaluate the direction cosines (l_1', m_1', n_1') of $\mathbf{ds}'$ in Fig. 11.4–1. By definition,

$$l_1' = \frac{dx'}{ds'}; \qquad m_1' = \frac{dy'}{ds'}; \qquad n_1' = \frac{dz'}{ds'} \qquad (11.4\text{–}10)$$

From Eqn 11.4–6(a),

$$dx' = dx + \frac{\partial u}{\partial x} dx + \frac{\partial u}{\partial y} dy + \frac{\partial u}{\partial z} dz$$

From Eqn 11.4–3,

$$ds' = ds\,\sqrt{(1 + 2\varepsilon_X)}$$

Hence

$$l_1' = \frac{dx'}{ds'} = \frac{1}{\sqrt{(1 + 2\varepsilon_X)}}\left\{\frac{dx}{ds}\left(1 + \frac{\partial u}{\partial x}\right) + \frac{dy}{ds}\frac{\partial u}{\partial y} + \frac{dz}{ds}\frac{\partial u}{\partial z}\right\}$$

$$= \frac{l_1[1 + (\partial u/\partial x)] + m_1(\partial u/\partial y) + n_1(\partial u/\partial z)}{\sqrt{(1 + 2\varepsilon_X)}} \qquad (11.4\text{–}11(a))$$

Similarly,

$$m_1' = \frac{l_1(\partial v/\partial x) + m_1[1 + (\partial v/\partial y)] + n_1(\partial v/\partial z)}{\sqrt{(1 + 2\varepsilon_X)}} \qquad (11.4\text{–}11(b))$$

$$n_1' = \frac{l_1(\partial w/\partial x) + m_1(\partial w/\partial y) + n_1[1 + (\partial w/\partial z)]}{\sqrt{(1 + 2\varepsilon_X)}} \qquad (11.4\text{–}11(c))$$

Equations 11.4–11 express the final direction cosines (l_1', m_1', n_1') of a line element in terms of the initial direction cosines (l_1, m_1, n_1), the displacements u, v, w and the normal strain ε_X (Eqns 11.4–3 and 11.4–8).

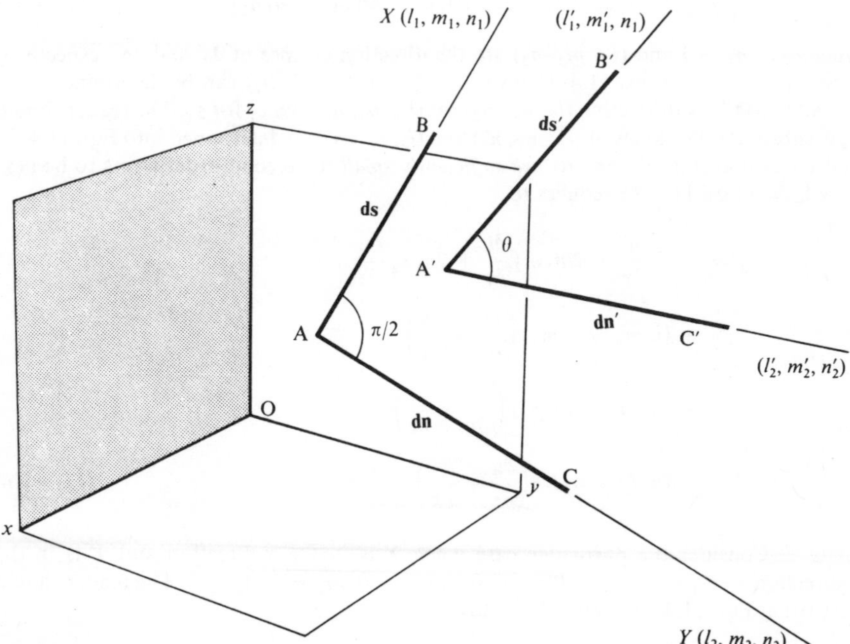

Fig. 11·4–2

Figure 11.4–2 shows two line elements **ds** and **dn** which are initially perpendicular to each other; **ds** is in the $X(l_1, m_1, n_1)$-direction and **dn** is in the $Y(l_2, m_2, n_2)$-direction. After deformation, **ds** becomes **ds**′ and **dn** becomes **dn**′. *The large-displacement definition of the shear strain γ_{XY} between two initially perpendicular directions X and Y* is

$$\gamma_{XY} = \frac{\mathbf{ds}'.\mathbf{dn}' - \mathbf{ds}.\mathbf{dn}}{(ds)(dn)} \qquad (11.4\text{–}12)$$

(If X and Y are not perpendicular, then γ_{XY} is not defined.) Equation 11.4–12 is equivalent to

$$\gamma_{XY} = \frac{(ds')(dn')}{(ds)(dn)} \cos\theta \quad \text{since} \quad \mathbf{ds}.\mathbf{dn} = 0 \tag{11.4–13}$$

where θ is the $B'\hat{A}'C'$ in Fig. 11.4–2. From Eqn 11.4–3

$$ds' = ds\sqrt{(1 + 2\varepsilon_X)}$$

By similar reasoning

$$dn' = dn\sqrt{(1 + 2\varepsilon_Y)}$$

Substituting these two expressions into Eqn 11.4–13,

$$\gamma_{XY} = \sqrt{[(1 + 2\varepsilon_X)(1 + 2\varepsilon_Y)]}\cos\theta \tag{11.4–14}$$

Since we know from coordinate geometry that

$$\cos\theta = l_1' l_2' + m_1' m_2' + n_1' n_2'$$

we have

$$\gamma_{XY} = \sqrt{[(1 + 2\varepsilon_X)(1 + 2\varepsilon_Y)]}\{l_1' l_2' + m_1' m_2' + n_1' n_2'\} \tag{11.4–15}$$

where (l_1', m_1', n_1') and (l_2', m_2', n_2') are the direction cosines of $\mathbf{ds}'$ and $\mathbf{dn}'$ respectively (Fig. 11.4–2). Equations 11.4–11 give (l_1', m_1', n_1'); (l_2', m_2', n_2') can be determined from Eqns 11.4–11 by substituting (l_2, m_2, n_2) for (l_1, m_1, n_1) and ε_Y for ε_X. The reader should now substitute the values of (l_1', m_1', n_1') and (l_2', m_2', n_2') so determined into Eqn 11.4–15 and verify that *if displacements are sufficiently small* for second-order terms to be neglected, then Eqn 11.4–15 reduces to

$$\begin{aligned}\gamma_{XY} = {} & 2l_1 l_2 \frac{\partial u}{\partial x} + 2m_1 m_2 \frac{\partial v}{\partial y} + 2n_1 n_2 \frac{\partial w}{\partial z} \\ & + (l_1 m_2 + l_2 m_1)\left(\frac{\partial v}{\partial x} + \frac{\partial u}{\partial y}\right) \\ & + (m_1 n_2 + m_2 n_1)\left(\frac{\partial w}{\partial y} + \frac{\partial v}{\partial z}\right) \\ & + (n_1 l_2 + n_2 l_1)\left(\frac{\partial u}{\partial z} + \frac{\partial w}{\partial x}\right)\end{aligned} \tag{11.4–16}$$

Consider the particular case when X is in the x-direction and Y is in the y-direction, i.e. $l_1 = 1, m_1 = 0, n_1 = 0$ and $l_2 = 0, m_2 = 1, n_2 = 0$. The reader should verify that Eqn 11.4–16 then reduces to

$$\gamma_{xy} = \frac{\partial v}{\partial x} + \frac{\partial u}{\partial y} \tag{11.4–17(a)}$$

Similarly, show that

$$\gamma_{yz} = \frac{\partial w}{\partial y} + \frac{\partial v}{\partial z} \tag{11.4–17(b)}$$

$$\gamma_{zx} = \frac{\partial u}{\partial z} + \frac{\partial w}{\partial x} \tag{11.4–17(c)}$$

Equation 11.4–17(a) gives the shear strain for line elements initially parallel to the x- and y-axes; similarly Eqns 11.4–17(b) and (c) give the shear strains for line elements initially parallel to the y- and z-axes and to the z- and x-axes respectively. Of course, these equations are only applicable when displacements are small; they agree with Eqns 1.6–7 in Chapter 1.

We are now in a position to simplify Eqns 11.4–8 and 11.4–16. Using Eqns 11.4–9 and 11.4–17, Eqns 11.4–8 and 11.4–16 can be simplified to the following forms:

$$\varepsilon_X = l_1^2\varepsilon_x + m_1^2\varepsilon_y + n_1^2\varepsilon_z + 2l_1m_1(\tfrac{1}{2}\gamma_{xy}) + 2m_1n_1(\tfrac{1}{2}\gamma_{yz}) + 2n_1l_1(\tfrac{1}{2}\gamma_{zx}) \qquad (11.4\text{–}18(a))$$

$$\tfrac{1}{2}\gamma_{XY} = l_1l_2\varepsilon_x + m_1m_2\varepsilon_y + n_1n_2\varepsilon_z + (l_1m_2 + l_2m_1)(\tfrac{1}{2}\gamma_{xy}) + (m_1n_2 + m_2n_1)(\tfrac{1}{2}\gamma_{yz}) + (n_1l_2 + n_2l_1)(\tfrac{1}{2}\gamma_{zx}) \qquad (11.4\text{–}18(b))$$

Thus, given the strains ε_x, ε_y, ε_z, γ_{xy}, γ_{yz}, and γ_{zx}, we can determine the normal strain ε_X in an arbitrary direction $X(l_1, m_1, n_1)$ and the shear strain γ_{XY}, where $Y(l_2, m_2, n_2)$ is another arbitrary direction perpendicular to X. Also, a comparison of Eqns 11.4–18 with Eqns 11.2–10 shows that they are of the same form. Thus, Eqn 11.4–18(a) can be obtained from Eqn 11.2–10(a) by writing ε's for σ's and $\frac{1}{2}\gamma$'s for τ's. The reader should use this fact to verify that the following relation, which is similar to Eqn 11.2–9, holds for strains:

$$\begin{bmatrix} \varepsilon_X & \frac{1}{2}\gamma_{XY} & \frac{1}{2}\gamma_{XZ} \\ \frac{1}{2}\gamma_{YX} & \varepsilon_Y & \frac{1}{2}\gamma_{YZ} \\ \frac{1}{2}\gamma_{ZX} & \frac{1}{2}\gamma_{ZY} & \varepsilon_Z \end{bmatrix} = \begin{bmatrix} l_1 & m_1 & n_1 \\ l_2 & m_2 & n_2 \\ l_3 & m_3 & n_3 \end{bmatrix} \begin{bmatrix} \varepsilon_x & \frac{1}{2}\gamma_{xy} & \frac{1}{2}\gamma_{xz} \\ \frac{1}{2}\gamma_{yx} & \varepsilon_y & \frac{1}{2}\gamma_{yz} \\ \frac{1}{2}\gamma_{zx} & \frac{1}{2}\gamma_{zy} & \varepsilon_z \end{bmatrix} \begin{bmatrix} l_1 & m_1 & n_1 \\ l_2 & m_2 & n_2 \\ l_3 & m_3 & n_3 \end{bmatrix}^T$$

i.e.

$$\underset{(OXYZ)}{\boldsymbol{\varepsilon}} = \mathbf{T} \underset{(Oxyz)}{\boldsymbol{\varepsilon}} \mathbf{T}^T \qquad (11.4\text{–}19)$$

That is, if at any point O the **strain tensor** ε for rectangular axes $Oxyz$ is given, then the strain tensor at that point for an arbitrary set of rectangular axes $OXYZ$ is given by Eqn 11.4–19. We are entitled to use the term 'strain tensor' because Eqn 11.4–19 represents the tensor transformation law; any 3 × 3 matrix which transforms in accordance with this law can be called a tensor. Note in particular that the elements γ_{xy}, γ_{yz}, γ_{zx}, etc., in Eqn 11.4–19 must have coefficients of $\frac{1}{2}$; otherwise ε will not be a tensor. The reader should refer to Eqns 11.4–16 and 11.4–17 and satisfy himself that $\gamma_{xy} = \gamma_{yx}$, $\gamma_{yz} = \gamma_{zy}$, and $\gamma_{zx} = \gamma_{xz}$, i.e. that the strain tensor is symmetrical, just as the stress tensor is.

We shall conclude this section with a comparison of the small-displacement definition of strains as given in Chapter 1 with the large-displacement definition of strain as given in Eqn 11.4–3, and Eqns 11.4–12 and 11.4–14.

Consider the line element AB in Fig. 11.4–1.

Let $\varepsilon_{X(1)}$ be the normal strain as defined in Chapter 1, Section 1.6.

$$\varepsilon_{X(1)} = \frac{\text{length A'B'} - \text{length AB}}{\text{length AB}} = \frac{ds' - ds}{ds} = \frac{ds'}{ds} - 1$$

The reader should verify that

$$\varepsilon_{X(1)} + \tfrac{1}{2}\varepsilon_{X(1)}^2 = \tfrac{1}{2}\left[\frac{(ds')^2 - (ds)^2}{(ds)^2}\right] = \varepsilon_X \text{ of Eqn 11.4–3} \qquad (11.4\text{–}20)$$

Therefore, as long as displacements are sufficiently small for $\frac{1}{2}\varepsilon_{X(1)}^2$ to be negligible compared with $\varepsilon_{X(1)}$, the small-displacement definition of normal strain in Section 1.6 will give the same result as the large-displacement definition of Eqn 11.4–3.

The same is true of shear strains. Equation 11.4–14 shows that the shear strain according to the large-displacement definition is

$$\gamma_{XY} = \sqrt{[(1 + 2\varepsilon_X)(1 + 2\varepsilon_Y)]} \cos\theta$$

When displacements are small, ε_X and ε_Y are small compared with unity; hence γ_{XY} reduces to $\gamma_{XY} = \cos\theta$, where θ is the angle $\widehat{B'A'C'}$ in Fig. 11.4–2, i.e.

$$\gamma_{XY} = \cos\theta = \sin\left(\frac{\pi}{2} - \theta\right) = \left(\frac{\pi}{2} - \theta\right) \qquad (11.4\text{–}21)$$

when $\pi/2 - \theta$ is small. That is, γ_{XY} becomes equal to the change in angle of the initially right-angle BAC in Fig. 11.4–2; this agrees with the definition of shear strain in Section 1.6 of Chapter 1.

11.5 Principal stresses

It will be shown in this section that for any given state of stress, through any point O in a body, there exists either three mutually perpendicular planes or else an infinite number of planes, on which the shear stresses vanish. Each such plane is called a **principal plane** and the direction of the normal to a principal plane is called a **principal direction**. The normal stress on a principal plane is called a **principal stress.**

Referring to Fig. 11.2–2(b), suppose $X(l_1, m_1, n_1)$ is a principal direction; since OABC is an infinitesimal tetrahedron, it is sufficiently accurate to regard the normal stress $\boldsymbol{\sigma}_p$ on plane ABC as a principal stress at point O. We have to determine the direction of $X(l_1, m_1, n_1)$ and the magnitude of the normal† stress $\boldsymbol{\sigma}_p$ acting on ABC. Since, by definition, no shear stresses act on a principal plane, σ_{px} in Fig. 11.2–2(b) is just the projection of $\boldsymbol{\sigma}_p$ on the x-direction, i.e.

$$\sigma_{px} = l_1\sigma_p; \qquad \text{similarly} \quad \sigma_{py} = m_1\sigma_p, \quad \sigma_{pz} = n_1\sigma_p$$

Substituting into Eqn 11.2–5,

$$\sigma_p \begin{bmatrix} l_1 \\ m_1 \\ n_1 \end{bmatrix} = \boldsymbol{\sigma}^T \begin{bmatrix} l_1 \\ m_1 \\ n_1 \end{bmatrix}$$

where $\boldsymbol{\sigma}$ is the stress tensor.

Since the stress tensor $\boldsymbol{\sigma}$ is symmetrical, $\boldsymbol{\sigma}^T = \boldsymbol{\sigma}$, and we are entitled to write:

$$\sigma_p \begin{bmatrix} l_1 \\ m_1 \\ n_1 \end{bmatrix} = \boldsymbol{\sigma} \begin{bmatrix} l_1 \\ m_1 \\ n_1 \end{bmatrix}$$

or

$$\{\boldsymbol{\sigma} - \sigma_p\mathbf{I}\} \begin{bmatrix} l_1 \\ m_1 \\ n_1 \end{bmatrix} = 0 \qquad (11.5\text{–}1)$$

where **I** is a unit matrix.

The trivial solution $l_1 = m_1 = n_1 = 0$ violates the condition (see Jeffrey[1]) $l_1^2 + m_1^2 + n_1^2 = 1$ and is hence not acceptable as a solution to Eqn 11.5–1.

† Since ABC is now assumed to be a principal plane, the stress $\boldsymbol{\sigma}_p$ will be a normal stress.

From Eqn 11.5–1, it is seen that the values of σ_p that would yield non-trivial solutions for (l_1, m_1, n_1) are the eigenvalues of the stress tensor $\boldsymbol{\sigma}$, and the non-trivial solutions for the direction cosines, (l_1, m_1, n_1), are the eigenvectors† of $\boldsymbol{\sigma}$. It is known from matrix algebra that (a) for any 3 × 3 symmetrical matrix (such as the stress tensor $\boldsymbol{\sigma}$) there are three eigenvalues, which are all real; (b) the eigenvectors corresponding to distinct eigenvalues are orthogonal. Hence it can be concluded that:

(1) At any point in a body, the principal stresses are the eigenvalues of the stress tensor at that point, and the principal directions are the eigenvectors of the stress tensor.

(2) There are three principal stresses, which are all real, but which may or may not be different in magnitude.

(3) If the three principal stresses say, $\sigma_1, \sigma_2, \sigma_3$, are all different in magnitude, then the three principal directions (and hence the three principal planes) are mutually perpendicular.

(4) If any two of the principal stresses are equal in magnitude but different from the third one, for example $\sigma_1 = \sigma_2 \neq \sigma_3$, then the direction of σ_3 is perpendicular to that of σ_2 and also to that of σ_1. That is, σ_1 and σ_2 lie on a plane perpendicular to the direction of σ_3; on this plane any direction is a principal direction and the principal stresses on all such principal directions are equal to σ_1 (and $= \sigma_2$).

(5) If all three principal stresses are equal then any direction is a principal direction, i.e. the stress on any plane through the point is a purely normal stress and its magnitude is independent of the orientation of the plane.

In two-dimensional analysis, we know from elementary mechanics of solids that if the two principal stresses are σ_1 and σ_2, then the maximum shear stress is

$$\tau_{\max} = \frac{\sigma_1 - \sigma_2}{2} \quad \text{(assuming } \sigma_1 > \sigma_2\text{)} \tag{11.5–2}$$

and $\tau_{\max}$ acts on a plane whose normal bisects the right angle between the two principal directions. In a three-dimensional stress system, it can be shown that if the principal stresses at a point O are σ_1, σ_2, and σ_3 (assume $\sigma_1 > \sigma_2 > \sigma_3$) then the absolute maximum shear stress is

$$\tau_{\max} = \frac{\sigma_1 - \sigma_3}{2} \tag{11.5–3}$$

and it acts on a plane whose normal bisects the right angle between the directions of σ_1 and σ_3. We shall further state without proof that for isotropic materials, shear strains also vanish on principal planes, but on no other planes. The normal strains in the principal directions are called **principal strains**. In other words, the terms 'principal planes' and 'principal directions' apply equally to both stresses and strains; that is, principal planes for stresses are also principal planes for strains and vice versa.

Example 11.5–1. In Chapter 10, Structural Dynamics, it was shown that the matrix $\mathbf{Z}$ of eigenvectors of $\mathbf{M}^{-1}\mathbf{K}_s$, normalized with respect to the mass matrix $\mathbf{M}$, transforms the stiffness matrix $\mathbf{K}_s$ into a diagonal matrix of eigenvalues, which are the squares of the natural circular frequencies of vibration of the structure; that is

$$\mathbf{Z}^T\mathbf{K}_s\mathbf{Z} = \boldsymbol{\omega}^2 \quad \text{(see Eqn 10.5–25)}$$

or

$$\mathbf{T}\mathbf{K}_s\mathbf{T}^T = \boldsymbol{\omega}^2$$

if we write $\mathbf{T}$ for $\mathbf{Z}^T$.

† Strictly speaking, they should be called *unit* eigenvectors.

Show that in principal-stress analysis, a matrix of eigenvectors of the stress tensor $\boldsymbol{\sigma}$ would similarly transform the stress tensor into a diagonal matrix of eigenvalues, which are the principal stresses.

SOLUTION Let σ_1, σ_2, and σ_3 be the principal stresses as determined from Eqn 11.5–1, and (l_1, m_1, n_1), (l_2, m_2, n_2), and (l_3, m_3, n_3) the corresponding principal directions. Of course, as previously explained, the principal stresses are the eigenvalues of $\boldsymbol{\sigma}$, and the principal directions are the eigenvectors.

Equation 11.5–1 states that

$$\boldsymbol{\sigma}\begin{bmatrix} l_1 \\ m_1 \\ n_1 \end{bmatrix} = \sigma_1 \begin{bmatrix} l_1 \\ m_1 \\ n_1 \end{bmatrix}; \quad \boldsymbol{\sigma}\begin{bmatrix} l_2 \\ m_2 \\ n_2 \end{bmatrix} = \sigma_2 \begin{bmatrix} l_2 \\ m_2 \\ n_2 \end{bmatrix}; \quad \boldsymbol{\sigma}\begin{bmatrix} l_3 \\ m_3 \\ n_3 \end{bmatrix} = \sigma_3 \begin{bmatrix} l_3 \\ m_3 \\ n_3 \end{bmatrix}$$

These three equations can be summed up as

$$\boldsymbol{\sigma}\begin{bmatrix} l_1 & l_2 & l_3 \\ m_1 & m_2 & m_3 \\ n_1 & n_2 & n_3 \end{bmatrix} = \begin{bmatrix} l_1 & l_2 & l_3 \\ m_1 & m_2 & m_3 \\ n_1 & n_2 & n_3 \end{bmatrix}\begin{bmatrix} \sigma_1 & 0 & 0 \\ 0 & \sigma_2 & 0 \\ 0 & 0 & \sigma_3 \end{bmatrix}$$

or

$$\boldsymbol{\sigma}\mathbf{T}^T = \mathbf{T}^T\boldsymbol{\sigma}'$$

where

$$\mathbf{T} = \begin{bmatrix} l_1 & m_1 & n_1 \\ l_2 & m_2 & n_2 \\ l_3 & m_3 & n_3 \end{bmatrix} \quad \text{and} \quad \boldsymbol{\sigma}' = \begin{bmatrix} \sigma_1 & 0 & 0 \\ 0 & \sigma_2 & 0 \\ 0 & 0 & \sigma_3 \end{bmatrix}$$

Pre-multiplying by $\mathbf{T}$, we have

$$\mathbf{T}\boldsymbol{\sigma}\mathbf{T}^T = \mathbf{T}\mathbf{T}^T\boldsymbol{\sigma}'$$

Since the eigenvectors (or principal directions) are orthogonal,

$$\mathbf{T}\mathbf{T}^T = \mathbf{I}$$

(see proof in Section 5.3). Hence

$$\mathbf{T}\boldsymbol{\sigma}\mathbf{T}^T = \boldsymbol{\sigma}' \tag{11.5–4}$$

It is interesting to recall that with a change of axes the stress tensor transforms in accordance with Eqn 11.2–9, where $\mathbf{T}$ is of course the matrix of direction cosines of the new axes; i.e. the matrix of the unit vectors in the directions of the new axes. Equation 11.5–4 in this worked example shows that when $\mathbf{T}$ is a matrix of unit eigenvectors of $\boldsymbol{\sigma}$, then $\boldsymbol{\sigma}$ transforms into diagonal matrix $\boldsymbol{\sigma}'$ whose elements are the principal stresses. Hence the principal-stress problem is essentially an eigenvalue problem.

Example 11.5–2. Given that the stresses at a point are

$$\boldsymbol{\sigma} = \begin{bmatrix} \sigma_x & \tau_{xy} & \tau_{xz} \\ \tau_{yx} & \sigma_y & \tau_{yz} \\ \tau_{zx} & \tau_{zy} & \sigma_z \end{bmatrix} = \begin{bmatrix} 127 & -60 & 0 \\ -60 & 113 & -53 \\ 0 & -53 & 53 \end{bmatrix} \text{N/mm}^2$$

Determine the principal stresses, the principal directions, and the maximum shear stress.

SOLUTION The principal stresses are the eigenvalues of the stress tensor $\boldsymbol{\sigma}$ and the principal directions are the eigenvectors of $\boldsymbol{\sigma}$. The direction cosines of the principal directions are simply the unit eigenvectors. They are

$$\sigma_1 = 191 \text{ N/mm}^2; \qquad \sigma_2 = 90 \text{ N/mm}^2; \qquad \sigma_3 = 12 \text{ N/mm}^2$$

and

$$\begin{bmatrix} l_1 \\ m_1 \\ n_1 \end{bmatrix} = \begin{bmatrix} 0.66 \\ -0.70 \\ 0.27 \end{bmatrix}; \quad \begin{bmatrix} l_2 \\ m_2 \\ n_2 \end{bmatrix} = \begin{bmatrix} 0.69 \\ 0.42 \\ -0.60 \end{bmatrix} \quad \begin{bmatrix} l_3 \\ m_3 \\ n_3 \end{bmatrix} = \begin{bmatrix} 0.30 \\ 0.58 \\ 0.76 \end{bmatrix}$$

Of course, σ_1, σ_2, and σ_3 are the principal stresses and $[l_1, m_1, n_1]^T$, etc., are the direction cosines of the principal directions. The maximum shear stress is

$$\tau_{\max} = \frac{\sigma_{\max} - \sigma_{\min}}{2}$$

$$= \frac{191 - 12}{2} \text{ N/mm}^2 = \underline{89.5 \text{ N/mm}^2}$$

In the above solution it has been assumed that the eigenvalues and unit eigenvectors are determined by standard computer procedures (which are available in almost all computing centres). Since the stress tensor is only a 3 × 3 matrix, it is also feasible to determine the eigenvalues and eigenvectors by hand. The eigenvalues of $\boldsymbol{\sigma}$ are given by the determinantal equation

$$\begin{vmatrix} 127 - \sigma_i & -60 & 0 \\ -60 & 113 - \sigma_i & -53 \\ 0 & -53 & 53 - \sigma_i \end{vmatrix} = 0$$

i.e.

$$\sigma_i^3 - 293\sigma_i^2 + 20\,700\sigma_i - 213\,300 = 0$$

This cubic equation can be conveniently solved graphically, or else by systematic trial and error (which takes much less time than many people think, and which is preferable, for practical purposes, to the formal algebraic solution) giving the three values of σ_i as before. To determine the direction corresponding to σ_1, use Eqn 11.5–1:

$$\begin{bmatrix} (127 - 191) & -60 & 0 \\ -60 & (113 - 191) & -53 \\ 0 & -53 & (53 - 191) \end{bmatrix} \begin{bmatrix} l_1 \\ m_1 \\ n_1 \end{bmatrix} = 0$$

which gives

$$\frac{l_1}{n_1} = 2.44; \qquad \frac{m_1}{n_1} = -2.59$$

From the condition $l_1^2 + m_1^2 + n_1^2 = 1$, we have

$$(2.44n_1)^2 + (-2.59n_1)^2 + n_1^2 = 1$$

giving $n_1 = 0.27$; whence

$$l_1 = 0.66, \qquad m_1 = -0.70 \qquad \text{as before}$$

The other two sets of direction cosines are similarly determined from Eqn 11.5–1 using σ_2 and σ_3 respectively.

Example 11.9–1. Figure 11.9–1 shows a uniform circular shaft of radius r, polar second moment of area J, modulus of rigidity G; it is subjected to a torque T which produces an angle of twist θ per unit length of shaft. The stresses at any point in the shaft are given by the following well-known relations which are based on the assumption that no warping of cross-sections occurs, i.e. plane sections remain plane after twisting:

$$\sigma_x = \sigma_y = \sigma_z = \tau_{xy} = 0 \tag{11.9–1}$$

and

$$\frac{\tau}{r} = \frac{T}{J} = G\theta \tag{11.9–2}$$

where τ $(= \sqrt{(\tau_{zx}^2 + \tau_{zy}^2)})$ is the shear stress at radius r.

Explain whether the stresses in Eqns 11.9–1 and 11.9–2 are a correct solution to the problem.

SOLUTION From Fig. 11.9–1,

$$\tau_{zx} = -\tau \sin\alpha = -\frac{\tau}{r}y \tag{11.9–3}$$

$$\tau_{zy} = \tau\cos\alpha = \frac{\tau}{r}x$$

Hence, from Eqn 11.9–2,

$$\tau_{zx} = -\left(\frac{T}{J}\right)y; \qquad \tau_{zy} = \left(\frac{T}{J}\right)x \tag{11.9–4}$$

where the quantity (T/J) is of course constant for a given shaft subjected to a given torque.

Hence, from Eqns 11.9–1 and 11.9–4 the stresses at a point (x, y, z) at a cross-section z from the origin O are independent of z, and are

$$\sigma_x = \sigma_y = \sigma_z = \tau_{xy} = 0$$

$$\tau_{zx} = -\left(\frac{T}{J}\right)y, \qquad \tau_{zy} = \left(\frac{T}{J}\right)x \tag{11.9–5}$$

From Eqns 1.7–8, the corresponding strains are

$$\varepsilon_x = \varepsilon_y = \varepsilon_z = 0, \qquad \gamma_{xy} = 0$$

$$\gamma_{zx} = -\frac{T}{GJ}y, \qquad \gamma_{zy} = \frac{T}{GJ}x \tag{11.9–6}$$

As an exercise the reader should verify that the stresses in Eqns 11.9–5 satisfy the equilibrium conditions in Eqns 11.1–3, and that the strains in Eqns 11.9–6 satisfy the compatibility conditions in Eqns 11.6–1.

The boundary conditions (Eqns 11.3–1) are:

On lateral boundary (Fig. 11.9–2)

$$l\sigma_x + m\tau_{yx} + n\tau_{zx} = 0$$

$$l\tau_{xy} + m\sigma_y + n\tau_{zy} = 0$$

$$l\tau_{xz} + m\tau_{yz} + n\sigma_z = 0 \tag{11.9–7}$$

On the ends (Fig. 11.9–3)

$$\int (x\tau_{zy} - y\tau_{zx})\,dA = \text{applied torque } T \tag{11.9–8}$$

We note that the first two equations of Eqns 11.9–7 are identically satisfied since $\sigma_x = \sigma_y = \tau_{xy} = \tau_{yx} = 0$ and $n = 0$; the third one reduces to

$$l\tau_{xz} + m\tau_{yz} = 0 \tag{11.9–9}$$

since both n and σ_z are zero.

From Fig. 11.9–2, $l = \cos\alpha = x/a$, $m = \sin\alpha = y/a$; from Eqn 11.9–4, $\tau_{xz} = -(T/J)y$, $\tau_{yz} = (T/J)x$; hence Eqn 11.9–9 becomes

$$\frac{x}{a}\left(-\frac{T}{J}y\right) + \frac{y}{a}\left(\frac{T}{J}x\right) = 0 \tag{11.9–10}$$

which is identically satisfied.

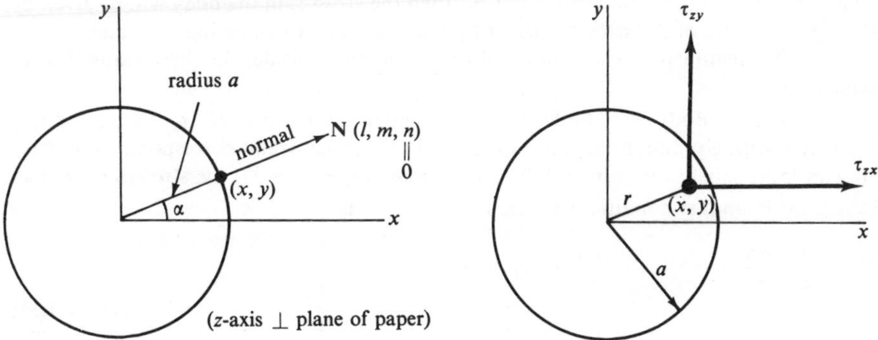

Fig. 11·9–2 **Fig. 11·9–3**

Also, substituting Eqns 11.9–5 into 11.9–8, we have:

$$\begin{aligned}
\text{left-hand side} &= \int (x\tau_{zy} - y\tau_{zx})\,dA \\
&= \int \left(x\frac{T}{J}x + y\frac{T}{J}y\right) dA \\
&= \frac{T}{J}\int (x^2 + y^2)\,dA \\
&= \frac{T}{J}J \\
&= T = \text{right-hand side}
\end{aligned} \tag{11.9–11}$$

Equations 11.9–10 and 11.9–11 show that the boundary conditions are satisfied. Since the equilibrium and compatibility conditions are also satisfied, the stresses in Eqns 11.9–1 and 11.9–2 form a solution to the problem, and from the uniqueness theorem, it is the only correct solution.

Note carefully, however, that at the ends the torque T must be applied through shear stresses τ distributed over the end planes in accordance with Eqn 11.9–2. If, for example, the torque T is applied as two equal and opposite forces P at a lever arm d apart such that $Pd = T$, then the stresses in Eqns 11.9–1 and 11.9–2 will not be the correct solution. However, if the shaft is long compared with the diameter, then by Saint-Venant's principle (Section 11.8) such stresses still apply reasonably accurately to points further away from each end than a distance equal to the diameter of the shaft.

11.10 Principle of superposition

The **principle of superposition**, which has wide applications in structural analysis, has previously been discussed in Section 4.4; a more rigorous proof is given in this section.

Consider an elastic body of arbitrary size and shape. Let σ_{x1}, σ_{y1}, σ_{z1}, τ_{xy1}, τ_{yz1}, and τ_{zx1} be the stresses produced by the application of force system F_1 consisting of body forces X_1, Y_1, Z_1 per unit volume of the body and of surface forces $\overline{X}_1$, $\overline{Y}_1$, $\overline{Z}_1$ per unit surface area of the body. Of course, both the body forces and the surface forces may vary from point to point; similarly the stresses $\sigma_{x1}, \sigma_{y1} \ldots \tau_{zx1}$ usually also vary from point to point.

Now let σ_{x2}, σ_{y2}, σ_{z2}, τ_{xy2}, τ_{yz2}, τ_{zx2} be the stresses in the same body produced by the application of force system F_2 consisting of body forces X_2, Y_2, Z_2, and surface forces $\overline{X}_2$, $\overline{Y}_2$, $\overline{Z}_2$. The principle of superposition states that if this elastic body is subjected simultaneously to the action of F_1 and F_2, then the stresses in the body will be given by the algebraic sum of the stresses due to F_1 and those due to F_2 acting separately.

To demonstrate the validity of this principle, consider the three loading cases separately:

CASE 1: Body acted on by F_1 only. The stresses σ_{x1}, σ_{y1}, σ_{z1}, τ_{xy1}, τ_{yz1}, and τ_{zx1} must satisfy the equilibrium conditions (Eqns 11.1–3); the strains corresponding to these stresses must satisfy the compatibility conditions (Eqns 11.6–1); the stresses must also satisfy the boundary conditions (Eqns 11.3–1). That is

$$\left.\begin{array}{l}\dfrac{\partial \sigma_{x1}}{\partial x} + \dfrac{\partial \tau_{yx1}}{\partial y} + \dfrac{\partial \tau_{zx1}}{\partial z} + X_1 = 0 \\ \ldots \\ \text{three equations of equilibrium (Eqns 11.1–3)}\end{array}\right\} \quad (11.10\text{–}1(a))$$

$$\left.\begin{array}{l}\dfrac{\partial^2 \varepsilon_{x1}}{\partial y^2} + \dfrac{\partial^2 \varepsilon_{y1}}{\partial x^2} = \dfrac{\partial^2 \gamma_{xy1}}{\partial x\, \partial y} \\ \ldots \\ \text{six equations of compatibility (Eqns 11.6–1)}\end{array}\right\} \quad (11.10\text{–}1(b))$$

$$\left.\begin{array}{l}l\sigma_{x1} + m\tau_{yx1} + n\tau_{zx1} = \overline{X}_1 \\ \ldots \\ \text{three equations of boundary equilibrium (Eqns 11.3–1)}\end{array}\right\} \quad (11.10\text{–}1(c))$$

The strains $\varepsilon_{x1}, \varepsilon_{y1}, \ldots, \gamma_{zx1}$ are of course related to the stresses $\sigma_{x1} \ldots \tau_{zx1}$ by Eqn 1.7–8.

CASE 2: Body acted on by F_2 alone. In this case, the equations are, similarly:

$$\left.\begin{array}{l}\dfrac{\partial \sigma_{x2}}{\partial x} + \dfrac{\partial \tau_{yx2}}{\partial y} + \dfrac{\partial \tau_{zx2}}{\partial z} + X_2 = 0 \\ \ldots \\ \dfrac{\partial^2 \varepsilon_{x2}}{\partial y^2} + \dfrac{\partial^2 \varepsilon_{y2}}{\partial x^2} = \dfrac{\partial^2 \gamma_{xy2}}{\partial x\, \partial y} \\ \ldots \\ l\sigma_{x2} + m\tau_{yx2} + n\tau_{zx2} = \overline{X}_2 \\ \ldots\end{array}\right\} \quad (11.10\text{–}2)$$

CASE 3: Body acted on by $F = F_1 + F_2$. If the stresses and strains are σ_x, $\sigma_y, \ldots, \tau_{zx}$ and $\varepsilon_x, \varepsilon_y, \ldots, \gamma_{zx}$ respectively, the set of equations will be

$$\left.\begin{aligned}
&\frac{\partial \sigma_x}{\partial x} + \frac{\partial \tau_{yx}}{\partial y} + \frac{\partial \tau_{zx}}{\partial z} + X_1 + X_2 = 0 \\
&\ldots \\
&\frac{\partial^2 \varepsilon_x}{\partial y^2} + \frac{\partial^2 \varepsilon_y}{\partial x^2} = \frac{\partial^2 \gamma_{xy}}{\partial x\, \partial y} \\
&\ldots \\
&l\sigma_x + m\tau_{yx} + n\tau_{zx} = \bar{X}_1 + \bar{X}_2 \\
&\ldots
\end{aligned}\right\} \tag{11.10–3}$$

From Eqns 11.10–1 and 11.10–2, it can be seen that a solution to Eqns 11.10–3 is

$$\begin{aligned}
&\sigma_x = \sigma_{x1} + \sigma_{x2}, \qquad \sigma_y = \sigma_{y1} + \sigma_{y2}, \ldots, \qquad \tau_{zx} = \tau_{zx1} + \tau_{zx2} \\
&\varepsilon_x = \varepsilon_{x1} + \varepsilon_{x2}, \qquad \varepsilon_y = \varepsilon_{y1} + \varepsilon_{y2}, \ldots, \qquad \gamma_{zx} = \gamma_{zx1} + \gamma_{zx2}
\end{aligned} \tag{11.10–4}$$

And from the uniqueness theorem (Section 11.8) we know that this is the only solution. Hence the principle of superposition is correct. Note, however, that (1) the compatibility conditions in Eqns 11.10–1 to 11.10–3 apply only when *displacements are small*; (2) in deriving the differential equations of equilibrium and the boundary conditions it was tacitly assumed that the *position and shape of the element, and hence of the body, remained sensibly the same after loading*; and (3) the stresses $\sigma_x, \sigma_y, \ldots, \tau_{zx}$ and the strains ε_x, $\varepsilon_y, \ldots, \gamma_{zx}$ in Eqns 11.10–3 are related by the stress–strain relation of the material of the body. If the material is linearly elastic, then the relation is that of Eqn 1.7–8, which is compatible with the stresses and strains in Eqns 11.10–4. If the material is not linearly elastic, then, even though the Eqns 11.10–3 still hold (as long as conditions (1) and (2) above are satisfied), the stresses $\sigma_x \ldots \tau_{zx}$ and the strains $\varepsilon_x \ldots \gamma_{zx}$ are now related by the nonlinear stress–strain relation of the material, and in general a nonlinear stress–strain relation cannot be compatible with the values of stresses and strains in Eqns 11.10–4, i.e. if $\sigma_x = \sigma_{x1} + \sigma_{x2}, \ldots$, etc., then $\varepsilon_x \neq \varepsilon_{x1} + \varepsilon_{x2}, \ldots$, etc. *unless the stress–strain relation is linear*. Hence we conclude that the principle of superposition holds only when the material is linearly elastic and when the displacements are small.

11.11 Torsion of non-circular shafts

It will be shown in this section that the solution of the problem of torsion of a prismatic shaft of arbitrary (but uniform) cross-section can be reduced to the solution of the **Laplace equation** (sometimes called the **harmonic equation**):

$$\nabla^2 \phi = -2G\theta \tag{11.11–1}$$

where ϕ is a particular stress function, called **Prandtl's torsion function**, which satisfies certain boundary conditions. The operator ∇^2 is called the **Laplacian operator** or the **harmonic operator**, and is defined by

$$\begin{aligned}
\nabla^2 &= \nabla . \nabla \\
&= \left(\mathbf{i}\frac{\partial}{\partial x} + \mathbf{j}\frac{\partial}{\partial y}\right)\cdot\left(\mathbf{i}\frac{\partial}{\partial x} + \mathbf{j}\frac{\partial}{\partial y}\right)† \\
&= \frac{\partial^2}{\partial x^2} + \frac{\partial^2}{\partial y^2}
\end{aligned}$$

† Where **i** and **j** are unit vectors in the x and y directions, respectively.

i.e. the harmonic operator ∇^2 is the scalar product of the operator 'del', ∇, with itself. (Note that ∇ is sometimes called 'nabla'.) In Eqn 11.11–1, G is the modulus of rigidity and θ the angle of twist per unit length of shaft.

We shall begin by examining why the usual formula for circular shafts is not applicable to non-circular shafts. Example 11.9–1 in Section 11.9 shows that the stresses in Eqns 11.9–1 and 11.9–2 of that worked example were correct for circular shafts. Let us now examine the boundary conditions in Eqn 11.9–7 for a non-circular section. Figure 11.9–2 of that example shows that for a circular section, the direction cosines of the normal to the lateral boundary are $l = \cos\alpha = x/a$, $m = \sin\alpha = y/a$, and $n = 0$; these values of l, m, and n make Eqn 11.9–9 identically satisfied. However, if the cross-section is non-circular, then, even though n is still zero, l and m would not in general have these values, and, consequently, Eqn 11.9–9 cannot be satisfied for all points on the lateral boundary and hence the solution

$$\frac{\tau}{r} = \frac{T}{J} = G\theta$$

is not valid for non-circular shafts.

Consider a uniform shaft (Fig. 11.11–1(a)) having a cross-section of arbitrary shape (Fig. 11.11–1(b)). Choose coordinate axes as shown; the origin O being so placed that the z-axis is the 'axis of twist', i.e. the axis about which the shaft rotates when twisted. For a shaft of arbitrary cross-section it is admittedly difficult to see which is the axis of twist, but fortunately, it can be proved that as long as the z-axis is parallel to the

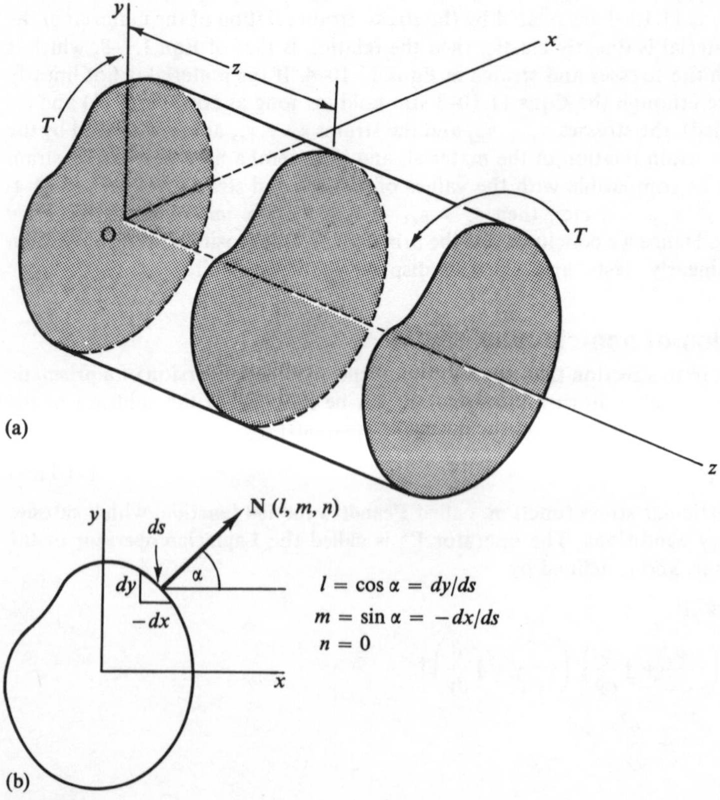

Fig. 11·11–1

axis of twist then the position of the origin O is immaterial in the sense that stresses and strains are not affected by it.

The shaft is subjected to torques T at the ends and it is required to solve for (a) the stresses at any point in the shaft, and (b) the angle of twist per unit length. The direct solution of this three-dimensional problem is very difficult, and Saint-Venant has used an inverse approach as discussed in Section 11.9 by assuming that the displacements (u, v, w) of a general point (x, y, z) are

$$u = -\theta zy, \qquad v = \theta zx, \qquad w = \theta\psi(x, y) \qquad (11.11\text{–}2)$$

where θ is the angle of twist per unit length and ψ is a function of (x, y) only and is called 'Saint-Venant's warping function'. Note that w, the displacement in the z-direction, is now assumed to be a function of x, y; hence plane sections are not assumed to remain plane after twisting, i.e. warping is assumed to take place. From Eqn 11.4–9,

$$\varepsilon_x = \varepsilon_y = \varepsilon_z = 0 \qquad (11.11\text{–}3(a))$$

From Eqn 11.4–17,

$$\gamma_{xy} = \gamma_{yx} = \frac{\partial v}{\partial x} + \frac{\partial u}{\partial y} = 0 \qquad (11.11\text{–}3(b))$$

$$\gamma_{yz} = \gamma_{zy} = \frac{\partial w}{\partial y} + \frac{\partial v}{\partial z} = \theta\left(\frac{\partial \psi}{\partial y} + x\right) \qquad (11.11\text{–}3(c))$$

$$\gamma_{zx} = \gamma_{xz} = \frac{\partial u}{\partial z} + \frac{\partial w}{\partial x} = \theta\left(\frac{\partial \psi}{\partial x} - y\right) \qquad (11.11\text{–}3(d))$$

Since these strains are derived from the displacements (u, v, w) which have been tacitly assumed to be single-valued and continuous, they would automatically satisfy the compatibility conditions (Eqns 11.6–1). The reader, of course, can check that this is so by substituting the strains in Eqns 11.11–3 into Eqns 11.6–1.

From Eqn 1.7–9, the stresses are

$$\sigma_x = \sigma_y = \sigma_z = 0 \qquad (11.11\text{–}4(a))$$

$$\tau_{xy} = \tau_{yx} = 0 \qquad (11.11\text{–}4(b))$$

$$\tau_{zy} = \tau_{yz} = G\theta\left(\frac{\partial \psi}{\partial y} + x\right) \qquad (11.11\text{–}4(c))$$

$$\tau_{zx} = \tau_{xz} = G\theta\left(\frac{\partial \psi}{\partial x} - y\right) \qquad (11.11\text{–}4(d))$$

These stresses identically satisfy the first two differential equations of equilibrium (Eqns 11.1–3), where body forces X, Y, Z are zero.

Equation 11.1–3(c) reduces to

$$\frac{\partial \tau_{xz}}{\partial x} + \frac{\partial \tau_{yz}}{\partial y} = 0 \qquad (11.11\text{–}5)$$

This equation can be identically satisfied by a function $\phi(x, y)$, Prandtl's torsion function, (Eqn 11.11–1), which is defined by

$$\tau_{zx} = \frac{\partial \phi}{\partial y}; \qquad \tau_{zy} = -\frac{\partial \phi}{\partial x} \qquad (11.11\text{–}6)$$

Equations 11.11–4(c) and (d) may now be written as

$$-\frac{\partial \phi}{\partial x} = G\theta\left(\frac{\partial \psi}{\partial y} + x\right) \qquad (11.11\text{–}7(a))$$

$$\frac{\partial \phi}{\partial y} = G\theta\left(\frac{\partial \psi}{\partial x} - y\right) \qquad (11.11\text{–}7(b))$$

Differentiating Eqn 11.11–7(a) with respect to x and Eqn 11.11–7(b) with respect to y and then subtracting the one from the other, we have

$$\frac{\partial^2 \phi}{\partial x^2} + \frac{\partial^2 \phi}{\partial y^2} = -2G\theta \qquad (11.11\text{–}8)$$

That is, any torsion function $\phi(x, y)$ satisfying the above Laplace's equation will automatically give stresses which satisfy the equilibrium conditions; the strains corresponding to such stresses will automatically satisfy the compatibility conditions.

Consider the boundary conditions on the lateral boundary (Fig. 11.11–1(b)). Since no surface forces are applied to the lateral boundary, Eqns 11.3–1 may be written:

$$l\sigma_x + m\tau_{yx} + n\tau_{zx} = 0 \qquad (11.11\text{–}9(a))$$

$$l\tau_{xy} + m\sigma_y + n\tau_{zy} = 0 \qquad (11.11\text{–}9(b))$$

$$l\tau_{xz} + m\tau_{yz} + n\sigma_z = 0 \qquad (11.11\text{–}9(c))$$

Equations 11.11–9(a) and (b) are identically satisfied by the stresses in Eqns 11.11–4 and the values of (l, m, n) shown in Fig. 11.11–1(b); using Eqns 11.11–6 and noting that $l = dy/ds$, $m = -(dx/ds)$, $n = 0$, Eqn 11.11–9(c) is reduced to

$$\left(\frac{dy}{ds}\right)\frac{\partial \phi}{\partial y} + \left(\frac{dx}{ds}\right)\frac{\partial \phi}{\partial x} = 0$$

i.e.

$$\frac{\partial \phi}{\partial x}\frac{dx}{ds} + \frac{\partial \phi}{\partial y}\frac{dy}{ds} = 0$$

i.e.

$$\frac{\partial \phi}{\partial s} = 0$$

i.e.

$$\begin{aligned}\phi &= \text{constant along lateral boundary}\\ &= \phi_c, \text{ say}\end{aligned} \qquad (11.11\text{–}10)$$

The boundary conditions at the ends of the shaft are (Fig. 11.11–2):

$$\int_A \tau_{zx}\, dA = 0 \text{ since no end force is applied in } x\text{-direction} \qquad (11.11\text{–}11(a))$$

$$\int_A \tau_{zy}\, dA = 0 \text{ similarly, in } y\text{-direction} \qquad (11.11\text{–}11(b))$$

$$\int_A \sigma_z\, dA = 0 \text{ similarly, in } z\text{-direction} \qquad (11.11\text{–}11(c))$$

$$\int_A x\tau_{zy}\, dA - \int_A y\tau_{zx}\, dA = \text{applied torque } T \qquad (11.11\text{–}11(d))$$

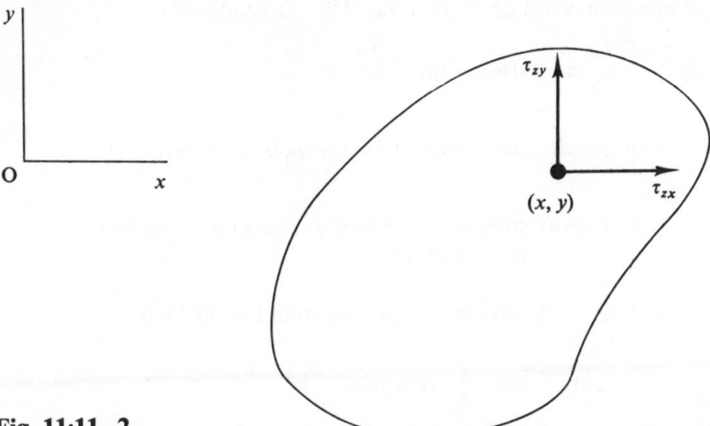

Fig. 11·11–2

Before we show that the left-hand side is equal to the right-hand side in each of the above equations, we first briefly refer to the following relations which are known as **Green's lemma** or **Green's theorem** the proof of which can be found in mathematics texts (see Wylie[2]):

$$\int_A \frac{\partial Q}{\partial x}\,dA = \oint_C Q\,dy;\dagger \qquad \int_A \frac{\partial Q}{\partial y}\,dA = -\oint_C Q\,dx$$

where Q is any function of (x, y) and A is the plane area enclosed by the curve C in the xy plane (Fig. 11.11–3). Since $dy = l\,ds$ and $-dx = m\,ds$, where l and m are the direction cosines of the outward normal to the curve C, Green's theorem can be expressed as

$$\int_A \frac{\partial Q}{\partial x}\,dA = \oint_C lQ\,ds \tag{11.11–12(a)}$$

$$\int_A \frac{\partial Q}{\partial y}\,dA = \oint_C mQ\,ds \tag{11.11–12(b)}$$

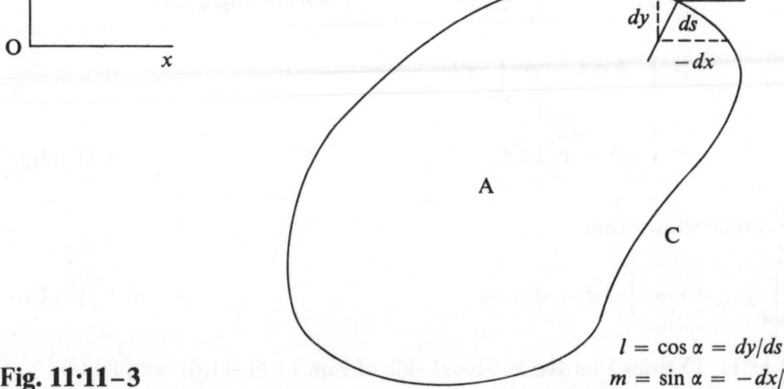

Fig. 11·11–3

† Line integrals around closed curves are denoted by the symbol $\oint$, with the convention that the path of integration is taken anti-clockwise.

We shall now return to Eqn 11.11–11(a). The left-hand side is

$$\int_A \tau_{zx}\,dA = \int_A \frac{\partial\phi}{\partial y}\,dA \quad \text{(from Eqn 11.11–6)}$$

$$= \oint_C m\phi\,ds \quad \text{(from Eqn 11.11–12(b)), by considering } Q \text{ as } \phi$$

$$= \phi_c \oint_C m\,ds \quad \text{since } \phi = \text{constant } \phi_c \text{ along boundary from Eqn 11.11–10}$$

$$= \phi_c \oint_C -\,dx \quad \text{since } m = -(dx/ds) \text{ from Fig. 11.11–3}$$

$$= \phi_c \times 0 \qquad \text{since } \oint_C dx \text{ is zero}$$

$$= 0$$

Hence Eqn 11.11–11(a) is identically satisfied; similarly, it can be seen that Eqn 11.11–11(b) is identically satisfied. Eqn 11.11–11(c) is also identically satisfied because $\sigma_z = 0$. Consider Eqn 11.11–11(d);

$$\int_A x\tau_{zy}\,dA = -\int x\frac{\partial\phi}{\partial x}\,dA \qquad \text{(from Eqn 11.11–6)}$$

$$= -\left\{\int_A \frac{\partial(\phi x)}{\partial x}\,dA - \int_A \phi\,dA\right\} \quad \text{integrating by parts}$$

$$= \int_A \phi\,dA - \int_A \frac{\partial(\phi x)}{\partial x}\,dA$$

$$= \int_A \phi\,dA - \oint_C l\phi x\,ds \qquad \text{(From Eqn 11.11–12(a)) by considering } Q \text{ as } (\phi x)$$

$$= \int_A \phi\,dA - \phi_c\oint_C lx\,ds \qquad \text{since } \phi = \text{constant} = \phi_c \text{ along boundary (Eqn 11.11–10)}$$

$$= \int_A \phi\,dA - \phi_c\int_A \frac{\partial x}{\partial x}\,dA \qquad \text{(from Eqn 11.11–12(a)) by considering } Q \text{ as } x$$

$$= \int_A \phi\,dA - \phi_c\int_A dA$$

$$= \int_A (\phi - \phi_c)\,dA \qquad (11.11\text{–}13(a))$$

Similarly, it can be shown that

$$-\int_A y\tau_{zx}\,dA = \int_A (\phi - \phi_c)\,dA \qquad (11.11\text{–}13(b))$$

Hence Eqns 11.11–13 show that the left-hand side of Eqn 11.11–11(d) is equal to

$$2\int_A (\phi - \phi_c)\,dA$$

Hence for the boundary condition in Eqn 11.11–11(d) to be identically satisfied we must have

$$2\int_A (\phi - \phi_c)\, dA = T \qquad (11.11\text{–}14)$$

Regarding the value of ϕ_c, we note that Eqn 11.11–10 states that ϕ must be constant along the boundary, but the value of the constant ϕ_c is not specified. Note also that Eqn 11.11–6 states that the shear stresses τ_{zx} and τ_{zy} depend only on the rate of change of ϕ and are independent of its absolute value; similarly the torque

$$T = \int_A x\tau_{zy}\, dA - \int_A y\tau_{zx}\, dA$$

is also independent of the absolute value of ϕ. In other words we are entitled to assign an arbitrary value to ϕ_c; for convenience we shall henceforth set $\phi_c = 0$, and rewrite Eqns 11.11–8, 11.11–10, and 11.11–14 as follows:

$$\nabla^2\phi = -2G\theta \qquad (11.11\text{–}15(a))$$

$$\phi = 0 \quad \text{on lateral boundary} \qquad (11.11\text{–}15(b))$$

$$2\int \phi\, dA = T \qquad (11.11\text{–}15(c))$$

Suppose we now let

$$\phi_1 = \frac{\phi}{G\theta} \qquad (11.11\text{–}16)$$

so that ϕ_1 is the torsion function corresponding to unit $G\theta$; let T_1 be the torque which produces this unit value of $G\theta$. Then Eqns 11.11–15 become:

$$\nabla^2\phi_1 = -2 \qquad (11.11\text{–}17(a))$$

$$\phi_1 = 0 \quad \text{on lateral boundary} \qquad (11.11\text{–}17(b))$$

$$2\int \phi_1\, dA = T_1 \qquad (11.11\text{–}17(c))$$

Since T denotes the actual torque applied to the section, the actual value of $G\theta$ produced by T is in general not unity, but given by

$$G\theta = \frac{T}{T_1} = \frac{T}{2\int \phi_1\, dA} \qquad (11.11\text{–}18(a))$$

Also,

$$\tau_{zx} = \frac{\partial\phi}{\partial y} = G\theta\frac{\partial\phi_1}{\partial y} \qquad (11.11\text{–}18(b))$$

$$\tau_{zy} = -\frac{\partial\phi}{\partial x} = -G\theta\frac{\partial\phi_1}{\partial x} \qquad (11.11\text{–}18(c))$$

Equations 11.11–18 may be summed up as

$$\frac{\tau_{zx}}{\partial\phi_1/\partial y} = \frac{\tau_{zy}}{-(\partial\phi_1/\partial x)} = \frac{T}{2\int \phi_1\, dA} = G\theta \qquad (11.11\text{–}19)$$

Equation 11.11–19 corresponds to the familiar equation $\tau/r = T/J = G\theta$ for a circular section. The quantity $2\int \phi_1\, dA$ is called the **torsion constant** of the section.

Hence the solution of the problem of torsion of an arbitrary section reduces to the solution of the Laplace equation 11.11–17(a) for ϕ_1 which satisfies the condition that ϕ_1 is zero on the lateral boundary. As soon as ϕ_1 is determined, Eqn 11.11–19 can be applied. (Note also that Eqns 11.11–17 and 11.11–19 are equally applicable to circular and non-circular sections.)

11.12 Plane stress and plane strain problems

It will be shown that, in the absence of body forces, the solution of **plane stress** and **plane strain** problems can be reduced to the solution of the equation

$$\nabla^4 \phi = 0 \tag{11.12–1}$$

where $\phi(x, y)$ is a stress function—the **Airy stress function**, which satisfies certain boundary conditions and which is defined by

$$\sigma_x = \frac{\partial^2 \phi}{\partial y^2}; \qquad \sigma_y = \frac{\partial^2 \phi}{\partial x^2}; \qquad \tau_{xy} = -\frac{\partial^2 \phi}{\partial x\, \partial y} \tag{11.12–2}$$

The operator ∇^4 is defined by

$$\nabla^4 = \nabla^2\, \nabla^2 = \left(\frac{\partial^2}{\partial x^2} + \frac{\partial^2}{\partial y^2}\right)\left(\frac{\partial^2}{\partial x^2} + \frac{\partial^2}{\partial y^2}\right)$$

$$= \frac{\partial^4}{\partial x^4} + 2\frac{\partial^4}{\partial x^2\, \partial y^2} + \frac{\partial^4}{\partial y^4}$$

i.e. the operator ∇^4 is equal to the harmonic operator ∇^2 operating twice (see Eqn 11.11–1); hence it is called the **biharmonic operator** and Eqn 11.12–1 is called the **biharmonic equation.**

Plane stress problems. As explained in Section 1.7, a state of plane stress is one in which $\sigma_z = \tau_{zx} = \tau_{zy} = 0$, and in which σ_x, σ_y, and τ_{xy} are functions of x and y only. A thin uniform sheet subjected only to in-plane forces can be said to be approximately in a state of plane stress.

For a state of plane stress with body forces equal to zero, the equilibrium conditions (Eqns 11.1–3) reduce to

$$\left.\begin{aligned} \frac{\partial \sigma_x}{\partial x} + \frac{\partial \tau_{yx}}{\partial y} = 0 \\ \frac{\partial \tau_{xy}}{\partial x} + \frac{\partial \sigma_y}{\partial y} = 0 \end{aligned}\right\} \tag{11.12–3}$$

which can be seen to be identically satisfied by stresses corresponding to any Airy stress function (Eqns 11.12–2). Equation 1.7–8 also shows that $\gamma_{zx} = \gamma_{zy} = 0$ for $\tau_{zx} = \tau_{zy} = 0$, and that the other strains are functions of (x, y) only. Hence the compatibility conditions (Eqns 11.6–1) reduce to:

$$\frac{\partial^2 \varepsilon_x}{\partial y^2} + \frac{\partial^2 \varepsilon_y}{\partial x^2} = \frac{\partial^2 \gamma_{xy}}{\partial x\, \partial y} \tag{11.12–4(a)}$$

$$\frac{\partial^2 \varepsilon_z}{\partial y^2} = 0 \tag{11.12–4(b)}$$

$$\frac{\partial^2 \varepsilon_z}{\partial x^2} = 0 \tag{11.12–4(c)}$$

$$\frac{\partial^2 \varepsilon_z}{\partial x\, \partial y} = 0 \tag{11.12–4(d)}$$

In practical problems about thin plates subjected to in-plane forces, the compatibility equations (11.12–4(b) to (d)) can be neglected without introducing significant error. Hence, as an approximation, Eqn 11.12–4(a) is our compatibility condition.

From Eqns 1.7–10 and 11.12–2:

$$\varepsilon_x = \frac{1}{E}[\sigma_x - \nu\sigma_y] = \frac{1}{E}\left[\frac{\partial^2\phi}{\partial y^2} - \nu\frac{\partial^2\phi}{\partial x^2}\right]$$

$$\varepsilon_y = \frac{1}{E}[-\nu\sigma_x + \sigma_y] = \frac{1}{E}\left[-\nu\frac{\partial^2\phi}{\partial y^2} + \frac{\partial^2\phi}{\partial x^2}\right]$$

$$\gamma_{xy} = \frac{\tau_{xy}}{G} = \frac{2(1+\nu)}{E}\tau_{xy} = -\frac{2(1+\nu)}{E}\frac{\partial^2\phi}{\partial x\,\partial y}$$

Substituting into Eqn 11.12–4(a), we have

$$\frac{\partial^4\phi}{\partial x^4} + 2\frac{\partial^4\phi}{\partial x^2\,\partial y^2} + \frac{\partial^4\phi}{\partial y^4} = 0$$

i.e.

$$\nabla^4\phi = 0 \tag{11.12–5(a)}$$

Plane strain problems. As explained in Section 1.7, a state of plane strain is one in which $\varepsilon_z = \gamma_{zx} = \gamma_{zy} = 0$ and in which ε_x, ε_y, γ_{xy} are functions of x and y only. If a long rod of uniform cross-section is subjected to a lateral loading uniformly distributed along the axis of the rod, and if the ends of the rod are rigidly held, then a state of plane strain can be said to exist in sections far away from the ends.

Equation 1.7–9 shows that $\tau_{zx} = \tau_{zy} = 0$ when $\gamma_{zx} = \gamma_{zy} = 0$, and from Eqn 1.7–15

$$\sigma_z = \nu(\sigma_x + \sigma_y), \quad \text{i.e. } \sigma_z \text{ is a function of } (x, y) \text{ only}$$

Hence, in the absence of body forces, the equilibrium conditions (Eqns 11.1–3) again reduce to Eqns 11.12–3, which are identically satisfied by stresses corresponding to any Airy stress function. Also, the compatibility conditions (Eqns 11.6–1) reduce to Eqn 11.12–4(a) only, while Eqns 11.12–4(b) to (d) are identically satisfied.

From Eqns 1.7–14 and 11.12–2,

$$\varepsilon_x = \frac{1+\nu}{E}[(1-\nu)\sigma_x - \nu\sigma_y]$$

$$= \frac{1+\nu}{E}\left[(1-\nu)\frac{\partial^2\phi}{\partial y^2} - \nu\frac{\partial^2\phi}{\partial x^2}\right]$$

Similarly,

$$\varepsilon_y = \frac{1+\nu}{E}\left[-\nu\frac{\partial^2\phi}{\partial y^2} + (1-\nu)\frac{\partial^2\phi}{\partial x^2}\right]$$

Also

$$\gamma_{xy} = \frac{\tau_{xy}}{G} = \frac{2(1+\nu)}{E}\tau_{xy} = -\frac{2(1+\nu)}{E}\frac{\partial^2\phi}{\partial x\,\partial y}$$

Substituting into Eqn 11.12–4(a), we again have

$$\nabla^4\phi = 0 \tag{11.12–5(b)}$$

Hence, in the absence of body forces, any Airy stress function satisfying the biharmonic equation (11.12–5) will automatically give stresses which satisfy the equilibrium conditions, and the strains corresponding to such stresses will automatically satisfy the compatibility conditions—approximately in the case of plane stress. Hence the solution of plane stress or plane strain problems can be said to reduce to the solution of the biharmonic equation for $\phi(x, y)$ which will also satisfy the boundary conditions.

11.13 Bending of thin plates

Consider a thin plate of uniform thickness t, which is small compared with the other dimensions. Choose a coordinate system $Oxyz$ as shown in Fig. 11.13–1(a), with origin O in the middle plane of the plate. The plate is initially flat, so that a section 1–1 before loading is as shown in Fig. 11.13–1(b). The only applied loadings considered are transverse forces in the z-direction and moments about the x- or y-axis. Figure 11.13–1(c) shows section 1–1 of the plate after loading. The point p displaces to p′ such that pp′ is w. It has been found that when the deflection w of any point on the middle plane is small compared with the plate thickness t, then the following assumptions are valid:

(a) Normals to the middle plane before bending remain normal to the middle plane after bending.

(b) The middle plane remains a neutral plane during bending, i.e. it is not strained.

(c) The stress σ_z is negligible compared with other stresses.

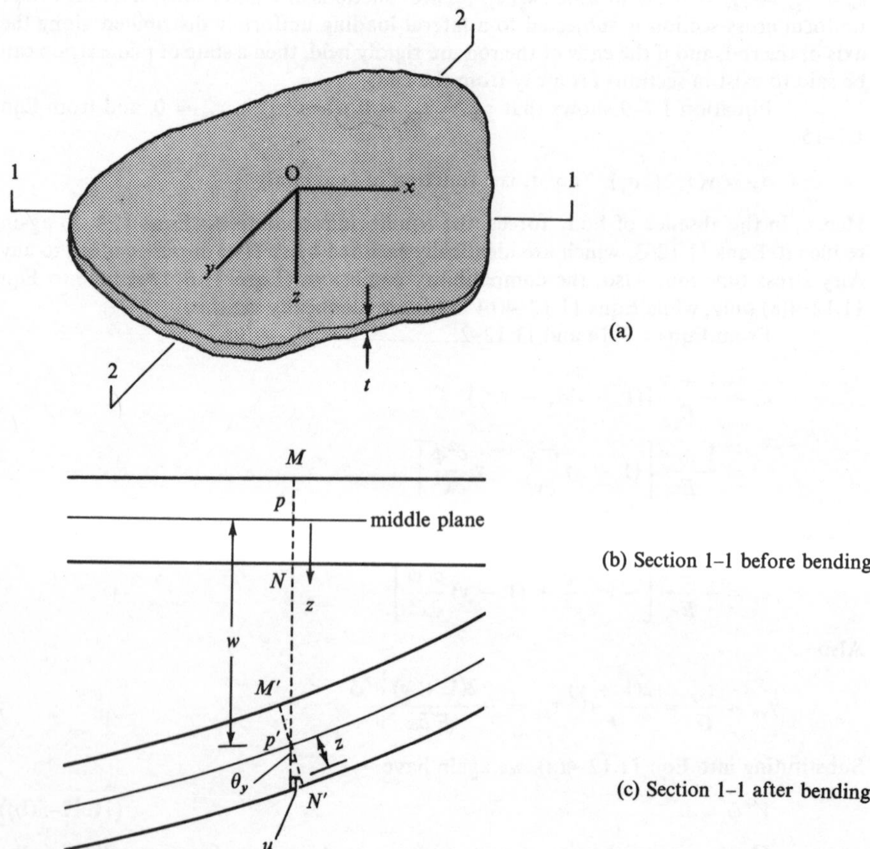

Fig. 11·13–1

From assumption (a), the normal MpN deforms to M′p′N′ which is perpendicular to the middle plane at p′. Note also that points such as p(x, y, 0) which is on the middle plane is displaced in the z-direction only, i.e. using the usual notation,

$$\text{deflection of p} = w$$

For a point at distance z from p, the displacement in the x-direction is very nearly given by

$$u = z\theta_y \tag{11.13–1}$$

where θ_y is the rotation of the middle plane at p about the y-axis. We shall adhere strictly to the right-hand-screw rule for vectors, so that positive rotations about the x- and y-axes are as shown in Fig. 11.13–2.

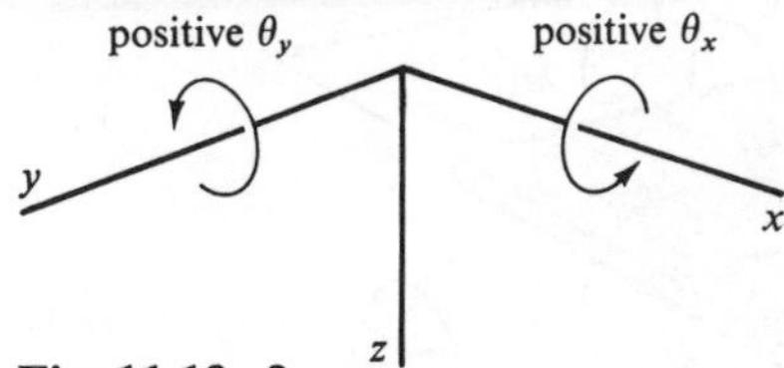

Fig. 11·13–2

By similar consideration of the deformation of a section 2–2 parallel to the y-axis, it can be shown that the displacement in the y-direction is given very nearly by

$$v = -z\theta_x \tag{11.13–2}$$

Figure 11.13–1(c) shows that

$$\theta_y = -\frac{\partial w}{\partial x}; \qquad \text{similarly} \quad \theta_x = \frac{\partial w}{\partial y} \tag{11.13–3}$$

hence

$$u = -z\frac{\partial w}{\partial x}; \qquad v = -z\frac{\partial w}{\partial y} \tag{11.13–4}$$

and the strains ε_x, ε_y, and γ_{xy} are then

$$\varepsilon_x = \frac{\partial u}{\partial x} = -z\frac{\partial^2 w}{\partial x^2} \tag{11.13–5(a)}$$

$$\varepsilon_y = \frac{\partial v}{\partial y} = -z\frac{\partial^2 w}{\partial y^2} \tag{11.13–5(b)}$$

$$\gamma_{xy} = \frac{\partial v}{\partial x} + \frac{\partial u}{\partial y} = -2z\frac{\partial^2 w}{\partial x\,\partial y} \tag{11.13–5(c)}$$

From assumption (c), σ_z is taken as zero, and the stress–strain relations in Eqn 1.7–8 become

$$\varepsilon_x = \frac{1}{E}[\sigma_x - \nu\sigma_y]; \qquad \varepsilon_y = \frac{1}{E}[-\nu\sigma_x + \sigma_y]; \qquad \gamma_{xy} = \frac{\tau_{xy}}{G} \tag{11.13–6}$$

from which,

$$\sigma_x = \frac{E}{1-\nu^2}[\varepsilon_x + \nu\varepsilon_y] = -\frac{Ez}{1-\nu^2}\left[\frac{\partial^2 w}{\partial x^2} + \nu\frac{\partial^2 w}{\partial y^2}\right] \tag{11.13–7(a)}$$

$$\sigma_y = \frac{E}{1-\nu^2}[\varepsilon_y + \nu\varepsilon_x] = -\frac{Ez}{1-\nu^2}\left[\frac{\partial^2 w}{\partial y^2} + \nu\frac{\partial^2 w}{\partial x^2}\right] \tag{11.13–7(b)}$$

$$\tau_{xy} = -2Gz\frac{\partial^2 w}{\partial x\,\partial y} = -\frac{Ez}{1+\nu}\frac{\partial^2 w}{\partial x\,\partial y} \tag{11.13–7(c)}$$

We shall now consider the bending and twisting moments acting on an infinitesimal element of the plate. Consider an element $(dx) \times (dy)$ with sides parallel to the x- and y-axes (Fig. 11.13–3).

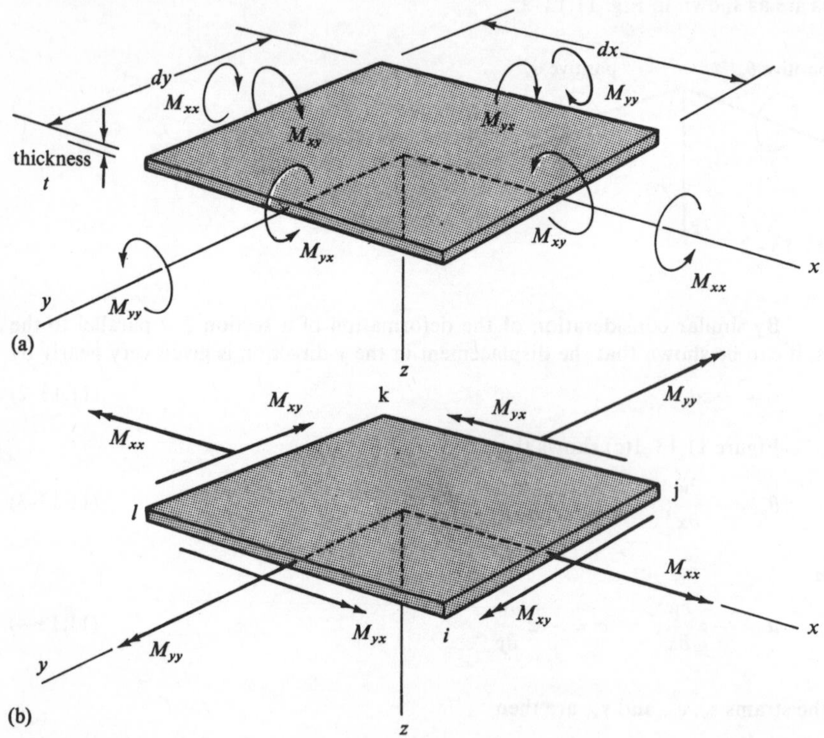

Fig. 11·13–3

In the figure, the bending moments M_{yx}, M_{xy}, and the twisting moments M_{xx}, M_{yy} are moments per unit width of the plate element. The meanings of the suffixes are similar to those for stresses (see Section 1.5). That is, the first suffix denotes the direction of the normal to the plane on which the moment acts, and the second suffix denotes the axis about which the moment acts. Thus,

M_{yx} represents a bending moment acting on a face normal to the y-axis, and it acts about the x-axis;

M_{xy} represents a bending moment acting on a face normal to the x-axis, and it acts about the y-axis;

M_{xx} represents a twisting moment acting on a face normal to the x-axis, and it acts about the x-axis; and

M_{yy} represents a twisting moment acting on a face normal to the y-axis, and it acts about the y-axis.

These moments are represented vectorially in Fig. 11.13–3(b) using the same notation as in Fig. 2.8–1 in Chapter 2.

The sign convention is similar to that for stresses as explained in Section 1.5. namely if the normal to a face is in the positive direction of the coordinate axis then the positive directions of the moment vectors are also in the positive directions of the coordinate axis. For example, on face ij the normal is in the positive x-direction; hence the vector for a positive M_{xy} is in the positive direction of the y-axis. On face jk the normal is in the negative y-direction; hence the vector for positive M_{yy} is in the negative y-direction, and so on.

In Figs. 11.13–3(a) and (b) all moments are shown acting in the positive directions, as defined above. It is appropriate to point out that many well-known writers use M_{xy} to represent a twisting moment, and M_{xx} to represent a bending moment about the y-axis, acting on a face normal to the x-axis; they use M_{yy} to represent a bending moment about the x-axis on a face normal to the y-axis; moreover they often use a sign convention which is inconsistent with the right-hand-screw rule. The present authors have chosen to give the suffixes xy, xx, yy, etc., meanings which are consistent with their meanings when they are used with stress components; also, they have chosen to adhere strictly to a sign convention (Fig. 11.13–3(b)) which is consistent with that for stresses. It is believed that such consistency will reduce the chance of making mistakes, especially in computer work.

With the notation in Fig. 11.13–3 we must have

$$\begin{aligned} M_{yx} &= -\int_{-t/2}^{t/2} \sigma_y z \, dz \\ &= \int_{-t/2}^{t/2} \frac{Ez^2}{1-\nu^2}\left[\frac{\partial^2 w}{\partial y^2} + \nu\frac{\partial^2 w}{\partial x^2}\right] dz \quad \text{(from Eqn 11.13–7(b))} \\ &= +\frac{Et^3}{12(1-\nu^2)}\left[\frac{\partial^2 w}{\partial y^2} + \nu\frac{\partial^2 w}{\partial x^2}\right] \end{aligned} \qquad (11.13\text{–}8(a))$$

Similarly,

$$M_{xy} = +\int_{-t/2}^{t/2} \sigma_x z \, dz = -\frac{Et^3}{12(1-\nu^2)}\left[\nu\frac{\partial^2 w}{\partial y^2} + \frac{\partial^2 w}{\partial x^2}\right] \qquad (11.13\text{–}8(b))$$

$$M_{xx} = -\int_{-t/2}^{t/2} \tau_{xy} z \, dz = +\frac{Et^3}{12(1-\nu^2)}(1-\nu)\frac{\partial^2 w}{\partial x \, \partial y} \qquad (11.13\text{–}8(c))$$

$$M_{yy} = +\int_{-t/2}^{t/2} \tau_{yx} z \, dz = +\int_{-t/2}^{t/2} \tau_{xy} z \, dz = -M_{xx} \qquad (11.13\text{–}8(d))$$

We shall now study the equilibrium of the element $dx\,dy$ (Fig. 11.13–4). The notation and sign convention in Fig. 11.13–3 will be used, but this time we shall take into consideration:

(a) The applied load *per unit area*, $q(x, y)$, which is in the z-direction and which may vary from point to point.

(b) The shear forces *per unit width* of plate, Q_{xz} and Q_{yz}. Again the suffixes xz and yz have similar meanings as when they are used with stress components. Thus, Q_{xz} is a shear force acting on a face normal to the x-axis and the shear force itself acts in the z-direction. On a face with the normal in the positive x-direction, the positive Q_{xz} will be in the positive z-direction, and so on.

(c) The fact that both moments and shear forces are functions of (x, y). They would in general vary with the distances dx and dy. (Thus the condition in Fig. 11.13–4 is similar to that in Fig. 11.1–2, in which the stresses σ_x, τ_{xy}, etc., vary with dx. etc.)

Consider equilibrium about the x-axis. The summation of moments about the x-axis must be zero. Therefore,

$$\left(\frac{\partial M_{xx}}{\partial x}\,dx\right)dy + \left(\frac{\partial M_{yx}}{\partial y}\,dy\right)dx + (Q_{yz}\,dx)\,dy = 0$$

Note that the contribution of the shear force increment, $(\partial Q_{yz}/\partial y)\,dy$, to moment about the x-axis is an infinitesimal quantity of a higher order than that of the terms in the above equation and hence can be neglected.

Cancelling the quantities $dx\,dy$ in the above equation, we have

$$Q_{yz} = -\frac{\partial M_{xx}}{\partial x} - \frac{\partial M_{yx}}{\partial y} \tag{11.13–9}$$

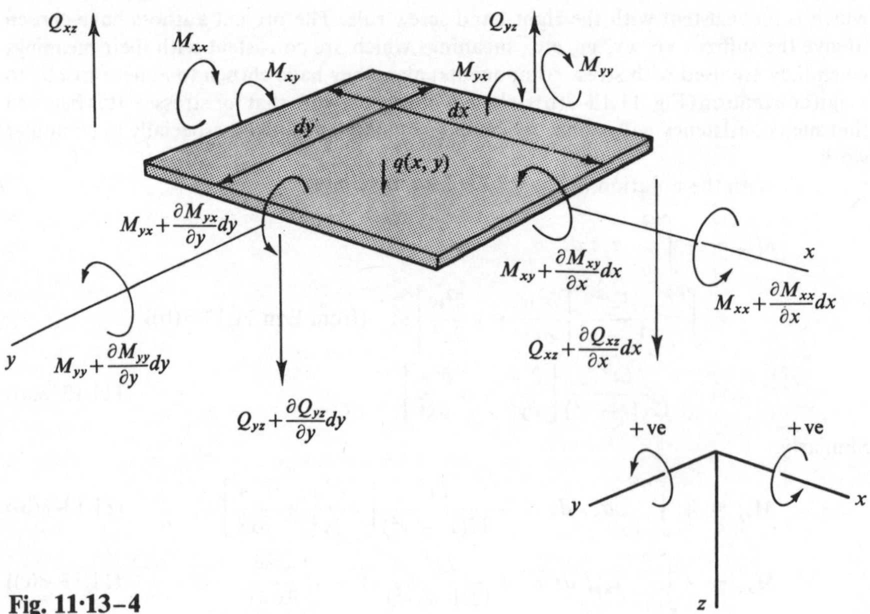

Fig. 11·13–4

Similarly, for equilibrium about the y-axis,

$$\left(\frac{\partial M_{yy}}{\partial y}\,dy\right)dx + \left(\frac{\partial M_{xy}}{\partial x}\,dx\right)dy - (Q_{xz}\,dy)\,dx = 0$$

i.e.

$$Q_{xz} = \frac{\partial M_{yy}}{\partial y} + \frac{\partial M_{xy}}{\partial x} \tag{11.13–10}$$

For equilibrium in the z-direction,

$$\left(\frac{\partial Q_{xz}}{\partial x}\,dx\right)dy + \left(\frac{\partial Q_{yz}}{\partial y}\,dy\right)dx + q\,dx\,dy = 0$$

i.e.

$$\frac{\partial Q_{xz}}{\partial x} + \frac{\partial Q_{yz}}{\partial y} + q = 0 \tag{11.13–11}$$

Substituting Eqns 11.13–9 and 11.13–10 into Eqn 11.13–11,

$$\frac{\partial^2 M_{yy}}{\partial x\,\partial y} + \frac{\partial^2 M_{xy}}{\partial x^2} - \frac{\partial^2 M_{xx}}{\partial y\,\partial x} - \frac{\partial^2 M_{yx}}{\partial y^2} + q = 0$$

Using Eqns 11.13–8 to express the moments M_{yy}, etc., in terms of w, this equation reduces to

$$\frac{\partial^4 w}{\partial x^4} + 2\frac{\partial^4 w}{\partial x^2\,\partial y^2} + \frac{\partial^4 w}{\partial y^4} = \frac{q}{\mathscr{D}}$$

i.e.

$$\nabla^4 w = \frac{q}{\mathscr{D}} \tag{11.13–12}$$

where $\mathscr{D}$ represents the quantity $Et^3/[12(1 - \nu^2)]$ and is called the **flexural rigidity of the plate** (see Example 11.14–2 about units of $\mathscr{D}$).

In Eqn 11.13–12, the value of $\mathscr{D}$ is uniquely determined from the plate thickness and from the mechanical properties of the material. For a specified loading $q(x, y)$ and specified boundary conditions, the solution of the plate problem is reduced to finding a deflection function $w(x, y)$ which satisfies the biharmonic equation $\nabla^4 w = q/\mathscr{D}$ and which also satisfies the boundary conditions.

11.14 Finite difference method

In sections 11.11 to 11.13, we derive the Laplace equation

$$\nabla^2 \phi = -2G\theta$$

for torsion problems, the biharmonic equation

$$\nabla^4 \phi = 0$$

for plane stress and plane strain problems, and the biharmonic equation

$$\nabla^4 \phi = \frac{q}{\mathscr{D}}$$

for plate bending.

The analytical solutions of these equations for functions ϕ which satisfy the prescribed boundary conditions are very often impossible. In this section the finite difference method will be presented as a powerful tool for solving these (and similar) differential equations numerically. Before converting the above differential equations into finite difference equations, some basic concepts about finite differences will be mentioned.

Figure 11.14–1 shows the values of a function of x at finite intervals $h/2$. The differences

$$f(x + h) - f(x), \quad f(x) - f(x - h), \quad \text{and} \quad f(x + h/2) - f(x - h/2)$$

are the differences between the values of the function at a finite interval h, and are called **finite differences**. The finite difference $f(x + h) - f(x)$ is called the **forward difference** at point i, the finite difference $f(x) - f(x - h)$ is called the **backward difference** at point i, and $f(x + h/2) - f(x - h/2)$ is called the **central difference** at point i. The use of central differences generally results in smaller truncation errors (see Section 11.15) than when forward or backward differences are used. Hence for the rest of this chapter, only central differences will be used.

The first central difference of the function at point i (Fig. 11.14–1) is defined as

$$\delta f(x) = f(x + h/2) - f(x - h/2) \tag{11.14–1}$$

That is $\delta f(x)$ represents the difference between the value of the function at $(x + h/2)$ and that at $(x - h/2)$, where the symbol δ is called the **central difference operator**. Like the differential operator $D(\equiv d/dx)$, *the operator δ satisfies the formal laws of algebra and can be used as though it were a number or a variable*, in the same way as the operator D can be used.

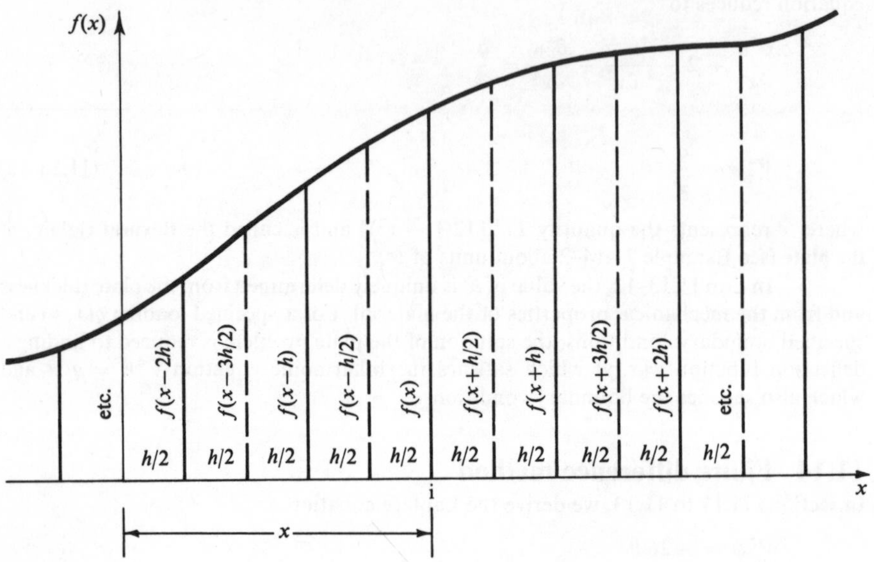

Fig. 11·14–1

The second central difference of the function at point i is the difference of the first central difference:

$$\delta^2 f(x) = \delta[\delta f(x)]$$

$$= \delta\left[f\left(x + \frac{h}{2}\right) - f\left(x - \frac{h}{2}\right)\right] \quad \text{(from Eqn 11.14–1)}$$

$$= \delta f\left(x + \frac{h}{2}\right) - \delta f\left(x - \frac{h}{2}\right)$$

$$= \left[f\left(x + \frac{h}{2} + \frac{h}{2}\right) - f\left(x + \frac{h}{2} - \frac{h}{2}\right)\right]$$

$$- \left[f\left(x - \frac{h}{2} + \frac{h}{2}\right) - f\left(x - \frac{h}{2} - \frac{h}{2}\right)\right] \quad \text{(from Eqn 11.14–1)}$$

$$= f(x + h) - 2f(x) + f(x - h) \tag{11.14–2}$$

Similarly, the reader can verify that

$$\delta^3 f(x) = \delta[\delta^2 f(x)]$$

$$= \delta[f(x + h) - 2f(x) + f(x - h)] \quad \text{(from Eqn 11.14–2)}$$

$$= f\left(x + \frac{3h}{2}\right) - 3f\left(x + \frac{h}{2}\right) + 3f\left(x - \frac{h}{2}\right) - f\left(x - \frac{3h}{2}\right) \tag{11.14–3}$$

Similarly,

$$\delta^4 f(x) = \delta[\delta^3 f(x)]$$

$$= \delta\left[f\left(x + \frac{3h}{2}\right) - 3f\left(x + \frac{h}{2}\right) + 3f\left(x - \frac{h}{2}\right) - f\left(x - \frac{3h}{2}\right)\right]$$

(from Eqn 12.14–3)

$$= f(x + 2h) - 4f(x + h) + 6f(x) - 4f(x - h) + f(x - 2h) \tag{11.14–4}$$

Of course, we could have expanded $\delta^4 f(x)$ by $\delta^4 f(x) = \delta^2[\delta^2 f(x)]$ and using Eqn 11.14–2 twice, and obtained the same result as that in Eqn 11.14–4.

In general,

$$\delta^n f(x) = \delta[\delta^{n-1} f(x)] = \delta^r[\delta^{n-r} f(x)] \tag{11.14–5}$$

The reader should verify by induction that the coefficients in the expansion of $\delta^n f(x)$ are the coefficients in the binomial expansion of $(1 - a)^n$, namely,

$$\delta^n f(x) = f\left(x + \frac{n}{2}h\right) - {}^nC_1 f\left(x + \frac{n}{2}h - h\right) + {}^nC_2 f\left(x + \frac{n}{2}h - 2h\right) + \cdots$$

$$+ (-1)^r\,{}^nC_r f\left(x + \frac{n}{2}h - rh\right) + \cdots + (-1)^n f\left(x + \frac{n}{2}h - nh\right) \tag{11.14–6}$$

where, of course,

$${}^nC_r = \frac{n!}{r!(n - r)!}$$

In Eqn 11.14–1, $\delta f(x)$, the first central difference of $f(x)$, is expressed in terms of the values of the function at $x + h/2$ and $x - h/2$. Usually, it is more convenient to work with values of the function at full intervals h rather than at half intervals $h/2$. Thus, instead of using the first central difference, it is generally preferable to use the **averaged first central difference**, $\mu\, \delta f(x)$, defined by

$$\mu\, \delta f(x) = \frac{1}{2}\left[\delta f\left(x + \frac{h}{2}\right) + \delta f\left(x - \frac{h}{2}\right)\right]$$

$$= \frac{1}{2}\left\{\left[f\left(x + \frac{h}{2} + \frac{h}{2}\right) - f\left(x + \frac{h}{2} - \frac{h}{2}\right)\right]\right.$$

$$\left. + \left[f\left(x - \frac{h}{2} + \frac{h}{2}\right) - f\left(x - \frac{h}{2} - \frac{h}{2}\right)\right]\right\} \quad \text{(from Eqn 11.14–1)}$$

i.e.

$$\mu\, \delta f(x) = \tfrac{1}{2}[f(x + h) - f(x - h)] \tag{11.14–7}$$

Values of differential coefficients by finite differences. Let us now investigate the relations between

$$\mu\, \delta f(x) \quad \text{and} \quad h\, Df(x)$$

$$\delta f(x) \quad \text{and} \quad h\, Df(x)$$

$$\delta^2 f(x) \quad \text{and} \quad h^2\, D^2 f(x)$$

$$\delta^4 f(x) \quad \text{and} \quad h^4\, D^4 f(x)$$

where the differential operator $D \equiv d/dx$. We have chosen to investigate the first, the second, and the fourth differences because of their important application in solving the Laplace and biharmonic equations, as will be shown later in this section.

Taylor's expansion of $f(x + h)$ is

$$f(x+h) = f(x) + h\,Df(x) + \frac{h^2 D^2}{2!} f(x) + \frac{h^3 D^3}{3!} f(x) + \cdots + \frac{h^n D^n}{n!} f(x) + \cdots$$

$$= \left[1 + hD + \frac{h^2 D^2}{2!} + \frac{h^3 D^3}{3!} + \cdots + \frac{h^n D^n}{n!} + \cdots\right] f(x)$$

$$= e^{hD} f(x) \tag{11.14-8}$$

i.e. Taylor's expansion of $f(x + h)$ is given by the exponential series e^{hD} operating on $f(x)$. From Eqn 11.14–8,

$$f(x-h) = e^{-hD} f(x)$$

hence

$$\mu\,\delta f(x) = \tfrac{1}{2}[f(x+h) - f(x-h)]$$
$$= \tfrac{1}{2}[e^{hD} - e^{-hD}] f(x) = \sinh(h\,D) f(x)$$

Expanding sinh $(h\,D)$,

$$\mu\,\delta f(x) = \left[h\,D + \frac{h^3 D^3}{6} + \frac{h^5 D^5}{120} + \cdots\right] f(x)$$

That is, operationally, $\mu\,\delta$ is equivalent to the quantity in the brackets on the right-hand side, i.e.

$$\mu\,\delta = \tfrac{1}{2}[e^{hD} - e^{-hD}]$$
$$= \sinh(h\,D)$$
$$= h\,D + \frac{h^3 D^3}{6} + \frac{h^5 D^5}{120} + \ldots$$
$$= h\,D + \text{error } \mu E_1 \tag{11.14-9}$$

From Eqn (11.14–8)

$$f\left(x + \frac{h}{2}\right) = e^{(h\,D)/2} f(x)$$

$$f\left(x - \frac{h}{2}\right) = e^{(-h\,D)/2} f(x)$$

hence

$$\delta f(x) = f\left(x + \frac{h}{2}\right) - f\left(x - \frac{h}{2}\right)$$
$$= [e^{h\,D/2} - e^{(-h\,D)/2}] f(x)$$

i.e. operationally,

$$\delta = e^{hD/2} - e^{(-hD)/2}$$

$$= 2 \sinh \frac{hD}{2}$$

$$= hD + \frac{h^3 D^3}{24} + \frac{h^5 D^5}{1920} + \ldots$$

$$= hD + \text{error } E_1 \tag{11.14–10}$$

Thus,

$$\delta^2 = \delta\,\delta$$

$$= (hD + \text{error } E_1)(hD + \text{error } E_1)$$

$$= h^2 D^2 + 2h D(\text{error } E_1) + (\text{error } E_1)^2$$

$$= h^2 D^2 + \frac{h^4 D^4}{12} + \frac{h^6 D^6}{360} + \ldots$$

$$= h^2 D^2 + \text{error } E_2 \tag{11.14–11}$$

Similarly, δ^4 can be obtained by raising the right-hand side of Eqn 11.14–10 to the fourth power, but it is easier to use Eqn 11.14–11

$$\delta^4 = \delta^2\,\delta^2$$

$$= (h^2 D^2 + \text{error } E_2)^2 \text{ (from Eqn 11.14–11)}$$

$$= h^4 D^4 + \frac{h^6 D^6}{6} + \frac{h^8 D^8}{80} + \ldots$$

$$= h^4 D^4 + \text{error } E_4 \tag{11.14–12}$$

Up to now functions of x only have been considered. For a function $f(x, y)$ of x and y, it is necessary only to write

$$D_x \text{ for } \frac{\partial}{\partial x} \quad \text{and} \quad D_y \text{ for } \frac{\partial}{\partial y}$$

Then Eqn 11.14–9 to 11.14–12 becomes:

$$\left.\begin{aligned}
\mu\,\delta_x &= h D_x + \frac{h^3 D_x^3}{6} + \frac{h^5 D_x^5}{120} + \ldots \\
&= h D_x + \text{error } \mu E_{x1} \\
\mu\,\delta_y &= h D_y + \text{error } \mu E_{y1} \\
&\text{Similarly,} \\
\delta_x &= h D_x + E_{x1} \\
\delta_y &= h D_y + E_{y1} \\
\delta_x^2 &= h^2 D_x^2 + E_{x2} \\
\delta_y^2 &= h^2 D_y^2 + E_{y2} \\
\delta_x^4 &= h^4 D_x^4 + E_{x4} \\
\delta_y^4 &= h^4 D_y^4 + E_{y4}
\end{aligned}\right\} \tag{11.14–13}$$

where the errors $\mu E_{x1}, \mu E_{y1}, \cdots E_{x4}, E_{y4}$ are those given by Eqns 11.14–9 to 11.14–12 with the relevant subscript x or y.

From Eqns 11.14–13, it is seen that

$$\delta_x^2\,\delta_y^2 = (h^2\,D_x^2 + E_{x2})(h^2\,D_y^2 + E_{y2})$$

$$= \left[h^2\,D_x^2 + \frac{h^4\,D_x^4}{12} + \frac{h^6\,D_x^6}{360} + \ldots\right]\left[h^2\,D_y^2 + \frac{h^4\,D_y^4}{12} + \frac{h^6\,D_y^6}{360} + \ldots\right]$$

(from Eqn 11.14–11)

$$= h^4\,D_x^2\,D_y^2 + \frac{h^6\,D_x^4\,D_y^2}{12} + \frac{h^6\,D_x^2\,D_y^4}{12} + \ldots$$

$$= h^4\,D_x^2\,D_y^2 + \text{error } E_{xy2} \qquad (11.14\text{–}14)$$

Equations 11.14–13 show that if

$$\frac{\mu\,\delta_x f(x, y)}{h} \quad \text{is used to represent} \quad \frac{\partial f(x, y)}{\partial x}$$

then the error is

$$\frac{h^2\,D_x^3}{6} + \frac{h^4\,D_x^5}{120} + \ldots$$

i.e. the error is of the order (h^2). Similarly, if

$$\frac{\delta_x^2 f(x, y)}{h^2} \quad \text{is used to represent} \quad \frac{\partial^2 f(x, y)}{\partial x^2}$$

or

$$\frac{\delta_x^4 f(x, y)}{h^4} \quad \text{is used to represent} \quad \frac{\partial^4 f(x, y)}{\partial x^4}$$

or

$$\frac{\delta_x^2\,\delta_y^2 f(x, y)}{h^4} \quad \text{is used to represent} \quad \frac{\partial^4 f(x, y)}{\partial x^2\,\partial y^2}$$

the error is in each case of the order (h^2). The derivatives $D_x f(x, y)$, $D_y^2 f(x, y)$, etc., are based on values of the function $f(x, y)$ at infinitesimal distances from the point (x, y); the central differences $\delta_x f(x, y)$, $\delta_x^2 f(x, y)$, $\delta_y^4 f(x, y)$, etc. (Eqn 11.14–6) are based on values of the function at finite distances from the point (x, y). However, if the finite distance interval h is kept small, the central difference expressions will be nearly equal to the corresponding derivatives.

For the rest of this section we shall disregard the errors $E_{x1}, E_{y1}, \cdots, E_{x4}, E_{y4}$, etc., but Section 11.15 explains how these errors, called **truncation errors**, can be reduced.

Laplace operator and biharmonic operator. The central finite difference expressions in Eqns 11.14–1 to 11.14–6 refer to Fig. 11.14–1 where $f(x)$ is a function of x only. For a function, $f(x, y)$, of both x and y, the central difference expressions will be given in terms of the values of the function at finite distances from point i in both the x- and y-directions. In Fig. 11.14–2, if point i is (x, y) then point 1 is $(x + h, y)$, point 5 is $(x + h, y + h)$, and so on.

Thus, from Eqn 11.14–7,

$$\mu\,\delta_x f(\mathrm{i}) = \tfrac{1}{2}[f(1) - f(3)] \qquad (11.14\text{–}15)$$

$$\mu\,\delta_y f(\mathrm{i}) = \tfrac{1}{2}[f(2) - f(4)] \qquad (11.14\text{–}16)$$

Similarly, from Eqn 11.14–2,

$$\delta_x^2 f(\text{i}) = f(1) - 2f(\text{i}) + f(3) \tag{11.14–17}$$

$$\delta_y^2 f(\text{i}) = f(2) - 2f(\text{i}) + f(4) \tag{11.14–18}$$

Hence $\delta_x^2\, \delta_y^2 f(\text{i})$ is

$$\begin{aligned}\delta_x^2\, \delta_y^2 f(\text{i}) &= \delta_x^2[f(2) - 2f(\text{i}) + f(4)]\\ &= \delta_x^2 f(2) - 2\delta_x^2 f(\text{i}) + \delta_x^2 f(4) \quad \text{(from Eqn 11.14–2)}\\ &= [f(5) - 2f(2) + f(6)]\\ &\quad -2[f(1) - 2f(\text{i}) + f(3)]\\ &\quad + [f(8) - 2f(4) + f(7)] \quad \text{(from Eqn 11.14–2)}\end{aligned} \tag{11.14–19}$$

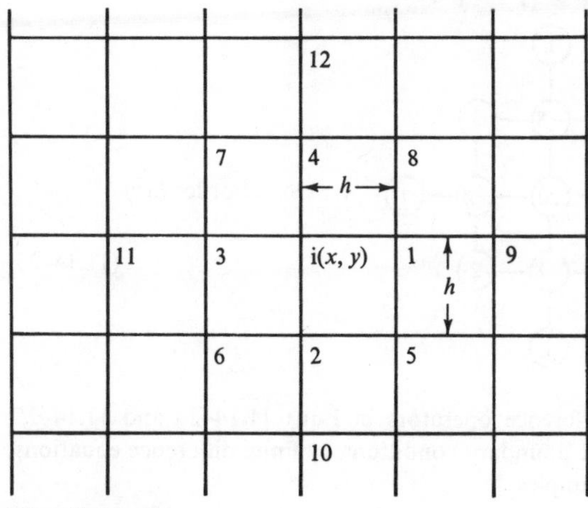

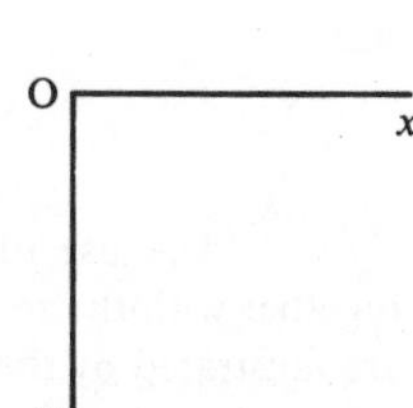

Fig. 11·14–2

From Eqn 11.14–4

$$\delta_x^4 f(\text{i}) = f(9) - 4f(1) + 6f(\text{i}) - 4f(3) + f(11) \tag{11.14–20}$$

$$\delta_y^4 f(\text{i}) = f(10) - 4f(2) + 6f(\text{i}) - 4f(4) + f(12) \tag{11.14–21}$$

From Eqns 11.14–17 and 11.14–18, the Laplace operator is

$$\begin{aligned}\nabla^2 &= \frac{\partial^2}{\partial x^2} + \frac{\partial^2}{\partial y^2}\\ &\approx \frac{1}{h^2}[\delta_x^2 + \delta_y^2]\\ &= \frac{1}{h^2}[f(1) + f(3) + f(2) + f(4) - 4f(\text{i})]\end{aligned} \tag{11.14–22}$$

Eqn 11.14–22 can be represented by the pattern:

$$\nabla^2 = \frac{1}{h^2} \begin{matrix} & (1) & \\ (1) & (-4) & (1) \\ & (1) & \end{matrix} + \text{error of order } (h^2) \tag{11.14–23}$$

Similarly, using Eqns 11.14–19, 11.14–20, and 11.14–21, the biharmonic operator is

$$\nabla^4 = \frac{\partial^4}{\partial x^4} + 2\frac{\partial^4}{\partial x^2\,\partial y^2} + \frac{\partial^4}{\partial y^4}$$

$$= \frac{1}{h^4}[\delta_x^4 + 2\delta_x^2\,\delta_y^2 + \delta_y^4]$$

$$= \frac{1}{h^4}\{20f(\mathrm{i}) - 8[f(1) + f(2) + f(3) + f(4)] + 2[f(5) + f(6) + f(7) + f(8)] + [f(9) + f(10) + f(11) + f(12)]\} \quad (11.14\text{–}24)$$

Equation 11.14–24 can be represented by the pattern:

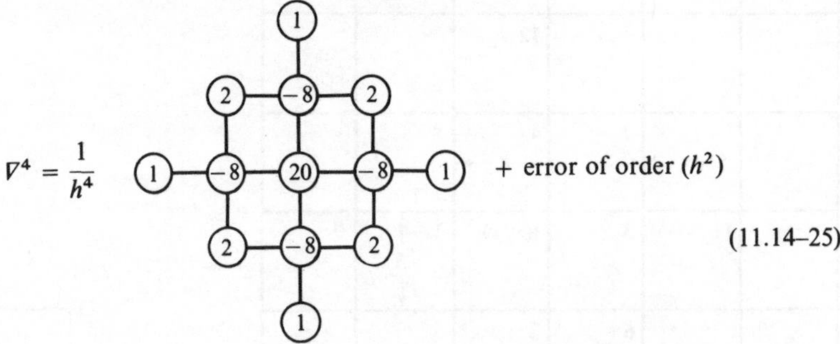

(11.14–25)

The use of the finite difference operators in Eqns 11.14–23 and 11.14–25, together with the representation of boundary conditions by finite difference equations, are illustrated by the following examples:†

Example 11.14–1. A shaft of 100 mm × 100 mm square cross-section (Fig. 11.14–3) is acted on by a torque of 100 000 N mm. Using the method of finite difference:

(a) Express the Laplace equation $\nabla^2\,\phi_1 = -2$ as a system of simultaneous equations and determine the values of the torsion function ϕ_1 at the nodal points.

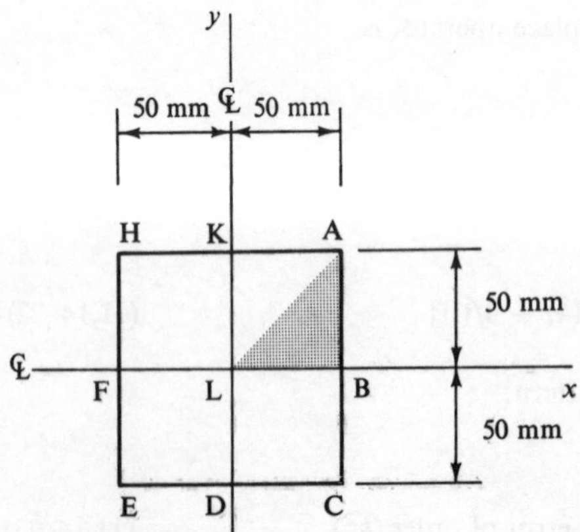

Fig. 11·14–3

† Thanks are due to Mr D. Hayon for providing the solution to Example 11.14–1.

(b) Determine the torsion constant $J = 2\int \phi_1 \, dA$ for the section.

(c) Determine the angle of twist produced by the torque, if $G = 80\,000$ N/mm^2.

(d) Determine shear stresses at points A, B, K, and L.

SOLUTION Because of symmetry, it is necessary only to consider one-eighth pf the cross-section, i.e. the shaded portion in Fig. 11.14–3. For the purpose of explaining the method, let us choose a mesh interval of 10 mm generally, and 5 mm in the region near the middle of an edge. Let the nodal points be numbered as shown in Fig. 11.14–4.

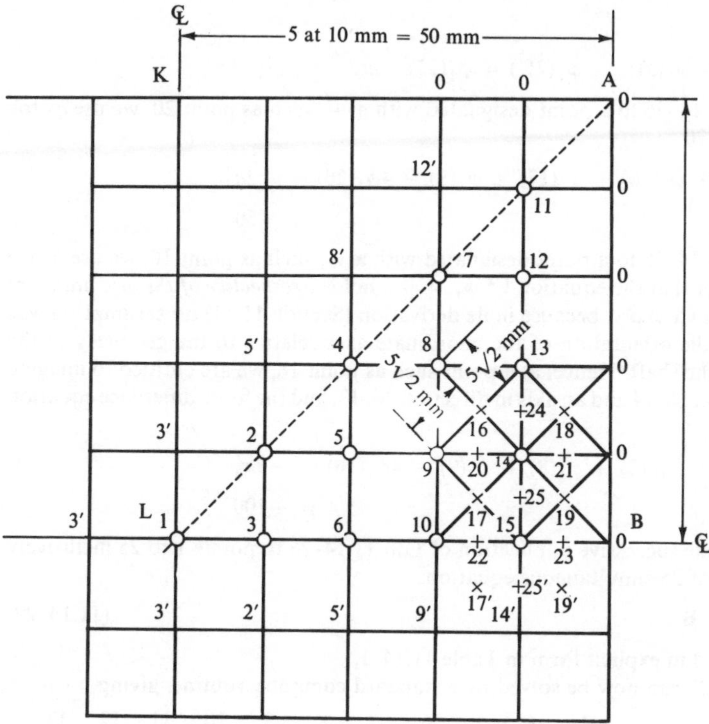

Fig. 11·14–4

Note that the mesh interval h_1 between points 5 and 9 or points 12 and 13, etc., is 10 mm; the interval h_2 between points 9 and 20 or points 14 and 24 is 5 mm; the interval h_3 between points 8 and 16 or points 16 and 13 is $5\sqrt{2}$ mm. The significance of h_3 will be explained later.

(a) The finite difference form of the Laplace equation is given by Eqn 11.14–23:

$$\nabla^2 \phi_1 = \frac{1}{h^2} \left\{ \begin{matrix} & 1 & \\ 1 & -4 & 1 \\ & 1 & \end{matrix} \right\} \phi_1 = -2 \qquad (11.14\text{–}26)$$

where the mesh interval h is 10 mm (h_1) or 5 mm (h_2), or $5\sqrt{2}$ mm (h_3) as the case may be.

Applying Eqn 11.14–26 to a point designated with a dot (.), such as point 1, we use h_1 for h; thus for point 1

$$\phi_1(3) + \phi_1(3') + \phi_1(3') + \phi_1(3') - 4\phi_1(1) = -2h_1^2 = -200$$

i.e.

$$4\phi_1(3) - 4\phi_1(1) = -200$$

Note that, by symmetry,

$$\phi_1(3') = \phi_1(3)$$

Similarly,

$$\phi_1(5') = \phi_1(5); \qquad \phi_1(25') = \phi_1(25), \text{ etc.}$$

Applying Eqn 11.14–26 to a point designated with a +, such as point 20, we use h_2 for h; thus for point 20,

$$\phi_1(14) + \phi_1(16) + \phi_1(17) + \phi_1(9) - 4\phi_1(20) = -2h_2^2 = -50$$

To apply Eqn 11.14–26 to a point designated with a ×, such as point 16, we use h_3 for h. We note further that the equation $\nabla^2 \phi_1 = -2$ holds *irrespective of the orientation of the coordinate axes x and y*, because in its derivation (Section 11.11) no assumption was made regarding the orientation of the coordinate axes relative to the geometry of the cross-section of the shaft. Hence, at a point such as point 16, we are entitled to imagine an axis in direction 16–14 and an axis in direction 16–13, and the finite difference equation is then

$$\phi_1(14) + \phi_1(13) + \phi_1(8) + \phi_1(9) - 4\phi_1(16) = -2h_3^2 = -100$$

In other words, the successive application of Eqn 11.14–26 to points 1 to 25 inclusively would give a set of 25 simultaneous equations:

$$\mathbf{A}\,\boldsymbol{\phi}_1 = \mathbf{B} \tag{11.14–27}$$

which is expressed in explicit form in Table 11.14–1.
Equation 11.14–27 can now be solved by a standard computer routine, giving

$$\boldsymbol{\phi}_1 = \begin{matrix} 1 & 2 & 3 & 4 & 5 & 6 & 7 & 8 & 9 & 10 & 11 & 12 & 13 \\ [14.6 & 13.6 & 14.1 & 10.9 & 12.2 & 12.6 & 6.86 & 8.59 & 9.56 & 9.86 & 2.56 & 4.13 & 5.08 \\ 5.60 & 5.77 & 7.46 & 7.95 & 2.92 & 3.09 & 7.76 & 3.03 & 8.01 & 3.11 & 5.39 & 5.73]^T \times 10^2 \\ 14 & 15 & 16 & 17 & 18 & 19 & 20 & 21 & 22 & 23 & 24 & 25 \end{matrix}$$

(b) The value of $\int \phi_1\, dA$ over the whole section is eight times that over region ABL. To compute $\int \phi_1\, dA$ over area ABL, let the area be divided into triangles (Fig. 11.14–5).

The volume $\int \phi_1\, dA$ over any triangle, whose nodes are, say, i, j, k is

$$\text{vol. } (i, j, k) = (\text{area of } \triangle ijk) \times \left\{\frac{\phi_1(i) + \phi_1(j) + \phi_1(k)}{3}\right\}$$

i.e.

$$\text{vol.} = \text{area of base triangle} \times \text{average height} \tag{11.14–28}$$

(Note that this formula is exact, and is correct irrespective of the shape of $\triangle$ijk, but the ordinates $\phi_1(i)$, $\phi_1(j)$, and $\phi_1(k)$ must be normal to it. As an exercise the reader should use the well-known prismoidal formula in earthwork computation to derive a proof of Eqn 11.14–28.)

Table 11.14–1

	1	2	3	4	5	6	7	8	9	10	11	12	13	14	15	16	17	18	19	20	21	22	23	24	25			
1	−4		4																							$\phi_1(1)$		−200
2		−4	2		2																					$\phi_1(2)$		−200
3	1	2	−4			1																				$\phi_1(3)$		−200
4				−4	2			2																		$\phi_1(4)$		−200
5		1		1	−4	1			1																	$\phi_1(5)$		−200
6			1		2	−4				1																$\phi_1(6)$		−200
7							−4	2				2														$\phi_1(7)$		−200
8				1			1	−4	1				1													$\phi_1(8)$		−200
9					1			1	−4	1				1												$\phi_1(9)$		−200
10						1			2	−4					1											$\phi_1(10)$		−200
11											−4	2														$\phi_1(11)$		−200
12						1					1	−4	1													$\phi_1(12)$		−200
13								1				1	−4	1												$\phi_1(13)$	=	−200
14									1				1	−4	1											$\phi_1(14)$		−200
15										1				2	−4											$\phi_1(15)$		−200
16								1	1				1	1		−4										$\phi_1(16)$		−100
17									1	1				1	1		−4									$\phi_1(17)$		−100
18													1	1				−4								$\phi_1(18)$		−100
19														1	1				−4							$\phi_1(19)$		−100
20									1					1		1	1			−4						$\phi_1(20)$		−50
21														1				1	1		−4					$\phi_1(21)$		−50
22										1					1		2					−4				$\phi_1(22)$		−50
23															1				2				−4			$\phi_1(23)$		−50
24													1	1		1		1						−4		$\phi_1(24)$		−50
25														1	1		1		1						−4	$\phi_1(25)$		−50

All other elements are zero

In this example, the area of each base triangle is $10^2/2$, so that we have (Fig. 11.14–5)

$$\text{vol. }(1, 2, 3) = \frac{100}{6}\{\phi_1(1) + \phi_1(2) + \phi_1(3)\}$$

$$\text{vol. }(2, 4, 5) = \frac{100}{6}\{\phi_1(2) + \phi_1(4) + \phi_1(5)\}$$

$$\text{vol. }(2, 3, 5) = \frac{100}{6}\{\phi_1(2) + \phi_1(3) + \phi_1(5)\}, \quad \text{etc.}$$

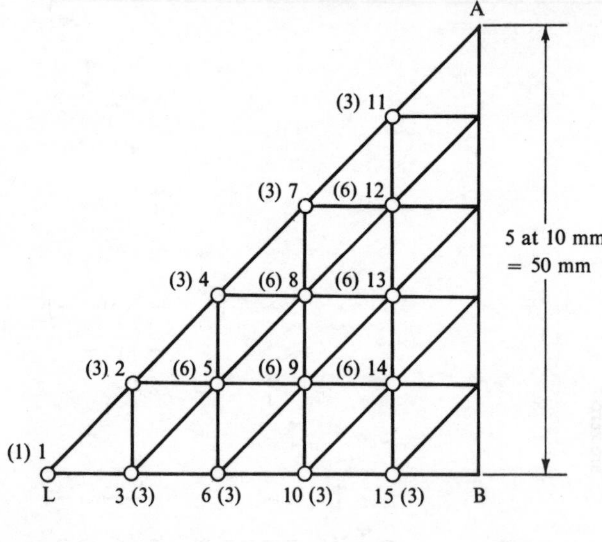

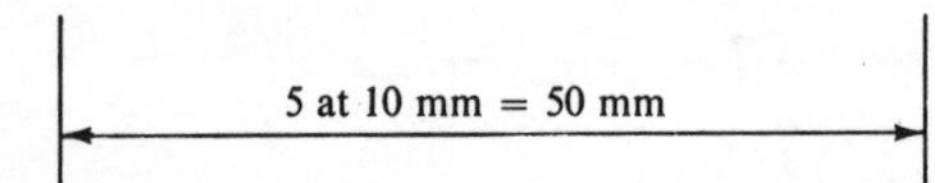

Fig. 11·14–5

In other words

$$\text{vol. over ABL} = \frac{100}{6}\{\phi_1(1) + 3\phi_1(2) + 3\phi_1(3) + 3\phi_1(4) + 6\phi_1(5) + \ldots + 6\phi_1(14) + 3\phi_1(15)\}$$

where the coefficients of the ϕ_1's are as shown within brackets in Fig. 11.14–5. Hence

$$J = 2\int \phi_1\, dA = 16 \times \text{vol over ABL}$$

Using the values of ϕ_1 determined previously

$$J = 1.36 \times 10^7\ \text{mm}^4 \quad \text{(see Problem 11.7 about units)}$$

(c) From Eqn 11.11–19,

$$\theta = \frac{T}{GJ} \quad \text{where} \quad J = 2\int \phi_1\, dA = 1.36 \times 10^7\ \text{mm}^4$$

$$= \frac{100\,000\ \text{Nmm}}{80\,000\ \text{N/mm}^2 \times 1.36 \times 10^7\ \text{mm}^4} = \underline{9.16 \times 10^{-8}\ \text{rad/mm}}$$

(d) From Eqn 11.11–19

$$\tau_{zx} = \frac{T\partial\phi_1}{J\partial y}$$

$$\tau_{zy} = -\frac{T\partial\phi_1}{J\partial x}$$

Then

$$\tau_{zx}(\mathrm{B}) = 0 \quad \text{since} \quad \frac{\partial\phi_1}{\partial y} = 0 \quad \text{at} \quad \mathrm{B}$$

$$\tau_{zy}(\mathrm{B}) \approx -\frac{T}{J}\{\phi_1(\mathrm{B}) - \phi_1(23)\}\frac{1}{h_2}$$

$$= +\frac{100\,000 \text{ Nmm}}{1.36 \times 10^7 \text{ mm}^4 \times 5 \text{ mm}} \times 3.11 \times 10^2 \text{ mm}^2 = \underline{+0.46 \text{ N/mm}^2}$$

(since $\phi_1(\mathrm{B}) = 0$ and $h_2 = 5$ mm).

The positive sign indicates that $\tau_{zy}(\mathrm{B})$ acts in the positive y-direction, that is, if the torsional moment T is positive as assumed—i.e., anticlockwise, so that the moment vector is in the positive z-direction.

$$\tau_{zx}(\mathrm{A}) = 0 \quad \text{since} \quad \frac{\partial\phi_1}{\partial y} = 0 \quad \text{at} \quad \mathrm{A}$$

$$\tau_{zy}(\mathrm{A}) = 0 \quad \text{since} \quad \frac{\partial\phi_1}{\partial x} = 0 \quad \text{at} \quad \mathrm{A}$$

By symmetry

$$\tau_{zx}(\mathrm{K}) = -\tau_{zy}(\mathrm{B}) = \underline{-0.46 \text{ N/mm}^2}$$

$$\tau_{zy}(\mathrm{K}) = \tau_{zx}(\mathrm{B}) = 0$$

$$\tau_{zx}(\mathrm{L}) = \frac{T\partial\phi_1}{J\partial y} = \frac{T\mu\delta_x\phi_1}{Jh_1}$$

$$= \frac{T}{J} \times \frac{1}{2h_1} \times \{\phi_1(3') - \phi_1(3)\} = \underline{0}$$

Similarly, $\tau_{xy}(\mathrm{L})$ is also $\underline{0}$.

Example 11.14–2. Figure 11.14–6 shows a slab with fixed edges along AB, BC, simply supported along AF, FE, and ED, and partially fixed along DC. The partial fixity along DC offers complete restraint against vertical deflection and a resistance of k newton-metres per metre width for each radian of rotation about the edge. The column at G provides vertical support only, with no other restraints on the slab. The flexural rigidity of the slab is $\mathscr{D}$ newton-metres and the total vertical force (in z direction) acting on the slab is q newtons per metre^2. Using a 5 m grid network, explain how the finite difference method can be used to convert the biharmonic equation into a system of simultaneous equations which can be solved for the grid-point deflections caused by q. Express the bending moments M_{xy}, M_{yx}, M_{xx}, and M_{yy} at the point H in terms of these deflections.

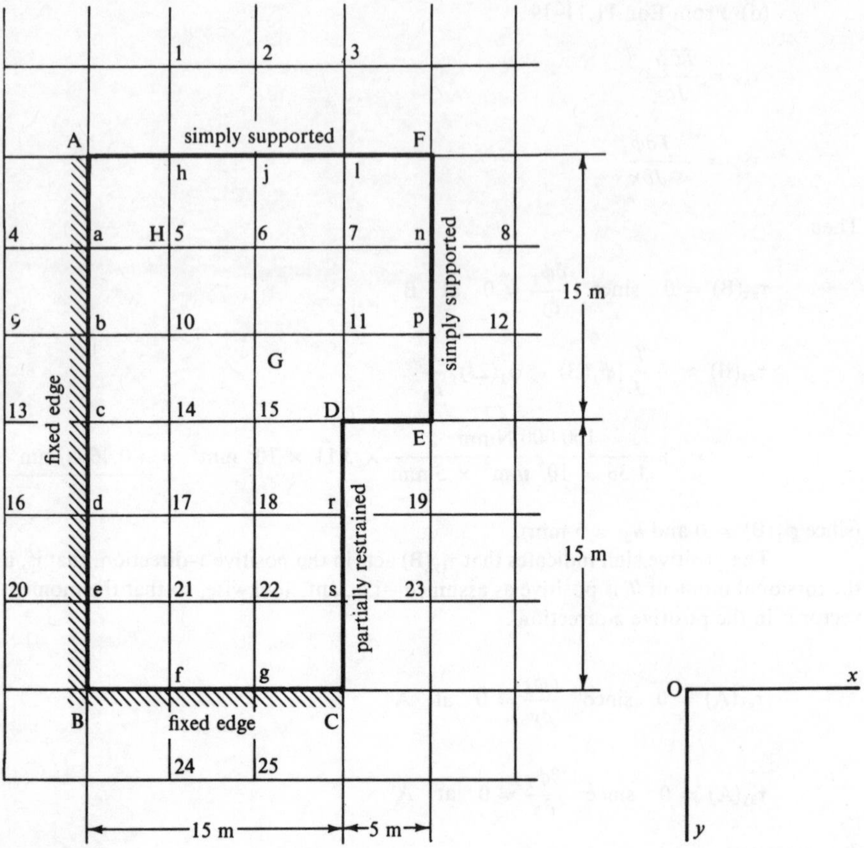

Fig. 11·14–6

SOLUTION The biharmonic equation is

$$\nabla^4 w = \frac{q}{\mathscr{D}}$$

where w = deflection at point considered (m);
q = force intensity at point considered (N/m^2);
$\mathscr{D}$ = flexural rigidity (Nm).†

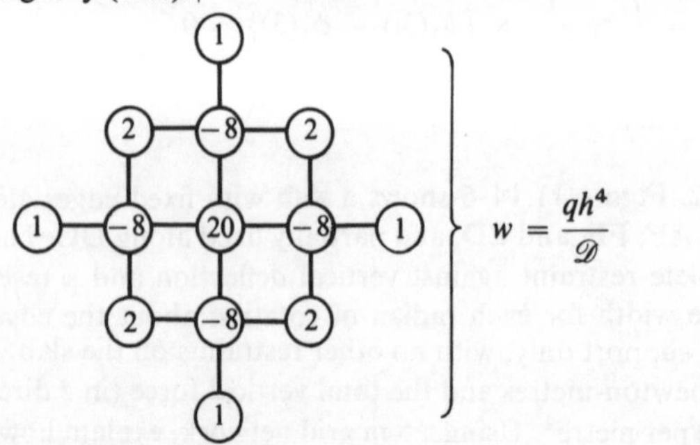

$$w = \frac{qh^4}{\mathscr{D}} \tag{11.14–29}$$

where the grid spacing h is 5 m in this case.

† The flexural rigidity of 1m width of the slab will be 1m × $\mathscr{D}$ Nm = $\mathscr{D}$ Nm^2.

Suppose the grid points, or nodes, are numbered as shown in the figure. Since the equation $\nabla^4 w = q/\mathscr{D}$ holds at every point in the slab, it means that the finite difference equation above must hold at every nodal point in the slab (but not necessarily at the *fictitious points* outside the slab, such as points 1, 2, 3, or 4, 12, etc.). Thus applying Eqn 11.14–29 to point 5 would result in the equation

$$20w_5 - 8(w_6 + w_{10} + w_a + w_h) + 2(w_j + w_G + w_b + w_A) + (w_7 + w_{14} + w_4 + w_1) = qh^4/\mathscr{D}$$

Now the edges are all completely restrained against vertical deflection, then

$$w_A = w_a = w_b = w_h = w_j, \text{ etc.} = 0$$

Also $w_G = 0$. Hence the above equation becomes

$$20w_5 - 8(w_6 + w_{10}) + (w_7 + w_{14} + w_4 + w_1) = \frac{qh^4}{\mathscr{D}} \tag{11.14–30}$$

In a similar way, Eqn 11.14–29 is applied successively to the other ten points in the slab, viz. points 6, 7, 10, 11, 14, 15, 17, 18, 21, and 22, resulting in a total of 11 simultaneous equations (including Eqn 11.14–30 above). These 11 equations contain also the unknown deflections w_1, w_2, w_3, w_4, $w_9, \cdots$, etc., of the 14 fictitious points outside the slab. In other words there are a total of 25 unknown w's and so far we have only 11 equations. The other 14 equations are obtained from consideration of boundary conditions:

Fully fixed edge. At the fully fixed edges AB and BC, both deflections and slopes are zero, i.e.

$$w = 0; \qquad \frac{\partial w}{\partial x} = 0; \qquad \frac{\partial w}{\partial y} = 0$$

The condition $w = 0$ along all the edges has already been used earlier to eliminate w_a, w_b, etc.

The condition $\partial w/\partial x = 0$ can be expressed in finite difference form, using Eqn 11.14–7. Thus, at point a, we have

$$\left(\frac{\partial w}{\partial x}\right)_{\text{point a}} = \frac{1}{2h}\{w_5 - w_4\} = 0$$

i.e.

$$w_5 - w_4 = 0 \tag{11.14–31}$$

That is, the condition of zero slope gives rise to seven similar equations at points a, b, $\cdots$, g along the fully fixed edges AB and BC.

Simply supported edge. At the simply supported edges AF, FE, and ED, besides the condition $w = 0$ (which has already been utilized) there is the condition of zero bending moment. Along FE, for example,

$$M_{xy} = 0$$

i.e.

$$-\mathscr{D}\left\{\nu\frac{\partial^2 w}{\partial y^2} + \frac{\partial^2 w}{\partial x^2}\right\} = 0 \quad \text{(from Eqn 11.13–8(b))}$$

The quantity $\partial^2 w/\partial y^2$ is the curvature of the slab along direction FE and is zero because no deflection occurs along the edge (i.e. $w = 0$, $\partial w/\partial y = 0$ and $(\partial/\partial y)(\partial w/\partial y) = 0$). Hence the condition $M_{xy} = 0$ reduces to

$$\frac{\partial^2 w}{\partial x^2} = 0$$

which can be expressed in finite difference form, using Eqn 11.14–2. Thus, at point n, we have

$$\frac{1}{h^2}(w_8 - 2w_n + w_7) = 0$$

i.e.

$$w_8 + w_7 = 0 \quad \text{since} \quad w_n = 0 \tag{11.14–32}$$

Similarly, at a point such as h, we have

$$w_5 + w_1 = 0$$

Thus the condition of zero bending moment along simply supported edges gives rise to a total of five equations at points h, j, l, n, and p.

Partially fixed edge. The condition $w = 0$ has already been utilized. The other condition relates the rotation of the slab to the stiffness of the support. Consider the slab at point r (Fig. 11.14–7). For an anticlockwise (i.e. positive) rotation θ_y at r, the anticlockwise (positive) moment acting on the supporting column is

$$M = k\theta_y \text{ Nm per metre width}$$

$$= k\left(-\frac{\partial w}{\partial x}\right) \quad \text{(from Eqn 11.13–3)}$$

$$= -\frac{k}{2h}(w_{19} - w_{18}) \quad \text{(from Eqn 11.14–7)}$$

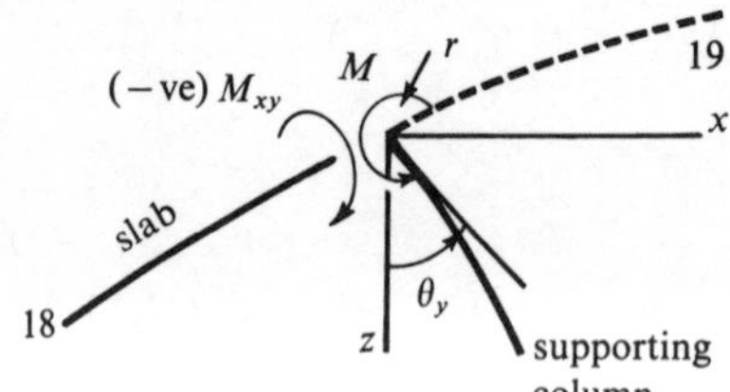

Fig. 11·14–7

From Eqn 11.13–8(b), the positive M_{xy} is

$$M_{xy} = -\mathscr{D}\left\{\nu\frac{\partial^2 w}{\partial y^2} + \frac{\partial^2 w}{\partial x^2}\right\}$$

$$= -\mathscr{D}\frac{\partial^2 w}{\partial x^2} \quad \text{since} \quad \frac{\partial^2 w}{\partial y^2} = 0$$

for zero curvature as explained previously.

Now the outward normal to the face of the slab at r is in the positive x-direction; hence the positive M_{xy} moment vector is in the positive y-direction, i.e. $+M_{xy}$ is anti-clockwise. Hence the M_{xy} acting on the slab in Fig. 11.14–7 is negative;

$$-M_{xy} = +\mathscr{D}\frac{\partial^2 w}{\partial x^2}$$

$$= \mathscr{D}\frac{w_{19} - 2w_r + w_{18}}{h^2} \quad \text{(from Eqn 11.14–2)}$$

Figure 11.14–7 shows that

$$M \text{ on column} = -M_{xy} \text{ on slab}$$

i.e.

$$-\frac{k}{2h}(w_{19} - w_{18}) = \mathscr{D}\frac{w_{19} + w_{18}}{h^2} \quad \text{since} \quad w_r = 0$$

i.e.

$$w_{19} = w_{18}\left\{\frac{1 - 2\mathscr{D}/kh}{1 + 2\mathscr{D}/kh}\right\} \tag{11.14–33}$$

The rotation restraint condition along the partially fixed edge DC gives rise to two equations for points r and s (see also Problem 11.11).

Therefore we have, in all, 25 equations:

(i) eleven equations obtained from applying the pattern in Eqn 12.14–29 to the 11 internal points;

(ii) seven equations from the boundary condition along the fully fixed edges;

(iii) five equations from the boundary condition along the simply supported edges; and

(iv) two equations from the boundary condition along the partially fixed edges.

These 25 equations can be solved by standard computer routine for the 25 unknown w's. Assuming that the 25 w's are now known, then the moments at point H are determined as follows. From Eqn 11.13–8,

$$M_{xy} = -\mathscr{D}\left\{\nu\frac{\partial^2 w}{\partial y^2} + \frac{\partial^2 w}{\partial x^2}\right\}_H \quad \text{Nm per m width}$$

$$M_{yx} = \mathscr{D}\left\{\frac{\partial^2 w}{\partial y^2} + \nu\frac{\partial^2 w}{\partial x^2}\right\}_H \quad \text{Nm/m}$$

$$M_{xx} = \mathscr{D}(1 - \nu)\left\{\frac{\partial^2 w}{\partial x\,\partial y}\right\}_H = -M_{yy} \quad \text{Nm/m}$$

Using Eqns 11.14–17 and 11.14–18,

$$\left\{\frac{\partial^2 w}{\partial x^2}\right\}_H \approx \frac{1}{h^2}\{\delta_x^2 w\}_5 = \frac{1}{h^2}\{w_6 - 2w_5 + w_a\}$$

$$= \frac{1}{h^2}\{w_6 - 2w_5\} \quad \text{since} \quad w_a = 0$$

$$\left\{\frac{\partial^2 w}{\partial y^2}\right\}_H \approx \frac{1}{h^2}\{\delta_y^2 w\}_5 = \frac{1}{h^2}\{w_{10} - 2w_5\}$$

Using Eqns 11.14–15 and 11.14–16,

$$\left\{\frac{\partial w}{\partial y}\right\}_H \approx \{\mu\,\delta_y w\}_5 = \frac{1}{2h}\{w_{10} - w_h\}$$

$$\left\{\frac{\partial^2 w}{\partial x\,\partial y}\right\}_H \approx \{\mu\,\delta_x(\mu\,\delta_y w)\}_5 = \frac{1}{4h^2}\{(w_G - w_b) - (w_j - w_A)\} = 0$$

Hence

$$M_{xy} = -\frac{\mathscr{D}}{h^2}\{\nu(w_{10} - 2w_5) + (w_6 - 2w_5)\}\ \text{Nm/m}$$

$$M_{yx} = \frac{\mathscr{D}}{h^2}\{(w_{10} - 2w_5) + \nu(w_6 - 2w_5)\}\quad \text{Nm/m}$$

$$M_{xx} = -M_{yy} = 0$$

11.15 Truncation errors and finite difference operators for uneven mesh intervals

From Eqns 11.14–13,

$$\mu\,\delta_x = h\,D_x + \frac{h^3\,D_x^3}{6} + \frac{h^5\,D_x^5}{120} + \ldots$$

Therefore, writing $\mu\,\delta_x = h\,D_x$ amounts to truncating the infinite series after the term $h\,D_x$. Hence the errors $\mu\,E_{x1}, \mu\,E_{y1}, \cdots, E_{y4}$, etc., in Eqns 11.14–13 are called **truncation errors**. The use of the finite difference expressions in Eqns 11.14–15 to 11.14–21 introduces truncation errors as given in Eqns 11.14–9 to 11.14–12, and methods of reducing such truncation errors will now be studied. From Eqn 11.14–9

$$\mu\,\delta = \sinh(h\,D)$$

i.e.

$$h\,D = \sinh^{-1}(\mu\,\delta)$$

$$= (\mu\,\delta) - \frac{(\mu\,\delta)^3}{6} + \frac{3(\mu\,\delta)^5}{40} - \frac{5(\mu\,\delta)^7}{112} + \ldots \tag{11.15–1}$$

That is, if $h\,D$ is represented by $\mu\,\delta$ the error is

$$-\frac{(\mu\,\delta)^3}{6} + \frac{3(\mu\,\delta)^5}{40} - \ldots$$

and the error will be reduced to the order of $(\mu\,\delta)^5$ if $h\,D$ is represented by

$$\mu\,\delta - \frac{(\mu\,\delta)^3}{6}$$

Eqn 11.14–7 gives $\mu\,\delta f(x)$ as

$$\mu\,\delta f(x) = \tfrac{1}{2}[f(x+h) - f(x-h)]$$

Of course $(\mu\,\delta)^3 f(x)$ can be evaluated by

$$\begin{aligned}(\mu\,\delta)^3 f(x) &= (\mu\,\delta)^2 \mu\,\delta f(x)\\ &= (\mu\,\delta)^2\{\tfrac{1}{2}[f(x+h) - f(x-h)]\}\\ &= \tfrac{1}{2}\mu\,\delta[\mu\,\delta f(x+h) - \mu\,\delta f(x-h)]\\ &= \tfrac{1}{8}[f(x+3h) - 3f(x+h) + 3f(x-h) - f(x-3h)]\end{aligned}$$

by repeated applications of Eqn 11.14–7.

This expression for $(\mu\,\delta)^3 f(x)$ involves values of the function at points $x + 3h$ and $x - 3h$, and the resulting expression for $h\,D = \mu\,\delta - (\mu\,\delta)^3/6$ is also rather cumbersome. Therefore, it is customary to take $(\mu\,\delta)^3$ approximately as $\mu\,\delta(\delta^2)$ instead of $\mu\,\delta(\mu\,\delta)^2$, i.e.

$$\begin{aligned}(\mu\,\delta)^3 f(x) &\approx \mu\,\delta[\delta^2 f(x)]\\ &= \mu\,\delta[f(x+h) - 2f(x) + f(x-h)] \qquad \text{(from Eqn 11.14–2)}\\ &= \mu\,\delta f(x+h) - 2\mu\,\delta f(x) + \mu\,\delta f(x-h)\\ &= \tfrac{1}{2}[f(x+2h) - 2f(x+h) + 2f(x-h) - f(x-2h)]\\ &\qquad \text{(from Eqn 11.14–7)}\end{aligned}$$

Hence the improved finite difference expression for $h\,D$ is

$$\begin{aligned}h\,Df(x) &= \left[\mu\,\delta - \frac{(\mu\,\delta)^3}{6}\right]f(x) = \mu\,\delta f(x) - \frac{(\mu\,\delta)^3}{6}f(x)\\ &\approx \tfrac{1}{2}[f(x+h) - f(x-h)]\\ &\quad - \tfrac{1}{12}[f(x+2h) - 2f(x+h) + 2f(x-h) - f(x-2h)]\\ &= \tfrac{1}{12}[-f(x+2h) + 8f(x+h) - 8f(x-h) + f(x-2h)]\end{aligned} \tag{11.15–2}$$

From Eqn 11.14–10

$$\delta = 2\sinh\frac{h\,D}{2}$$

hence

$$\begin{aligned}\frac{h\,D}{2} &= \sinh^{-1}\left(\frac{\delta}{2}\right)\\ &= \left(\frac{\delta}{2}\right) - \frac{1}{6}\left(\frac{\delta}{2}\right)^3 + \frac{3}{40}\left(\frac{\delta}{2}\right)^5 - \frac{5}{112}\left(\frac{\delta}{2}\right)^7 + \dots\end{aligned}$$

i.e.

$$h\,D = \delta - \frac{1}{3}\left(\frac{\delta}{2}\right)^3 + \frac{3}{20}\left(\frac{\delta}{2}\right)^5 - \frac{5}{56}\left(\frac{\delta}{2}\right)^7 + \dots$$

squaring

$$h^2\,D^2 = \delta^2 - \frac{\delta^4}{12} + \frac{\delta^6}{90} - \dots \tag{11.15–3}$$

squaring

$$h^4\,D^4 = \delta^4 - \frac{\delta^6}{6} + \frac{7\delta^8}{240} - \dots \tag{11.15–4}$$

Hence an improved finite difference expression for $h^2\,D^2 f(x)$

$$\begin{aligned}h^2\,D^2 f(x) &= \left[\delta^2 - \frac{\delta^4}{12}\right]f(x)\\ &= \tfrac{1}{12}[-f(x+2h) + 16f(x+h) - 30f(x) + 16f(x-h)\\ &\quad - f(x-2h)]\end{aligned} \tag{11.15–5}$$

as the reader can verify from Eqns 11.14–2 and 11.14–4.

Similarly, an improved finite difference expression for $h^4\,D^4f(x)$ can be shown to be

$$h^4\,D^4f(x) = \left[\delta^4 - \frac{\delta^6}{6}\right]f(x)$$

$$= \tfrac{1}{6}[-f(x+3h) + 12f(x+2h) - 39f(x+h) + 56f(x) - 39f(x-h) + 12f(x-2h) - f(x-3h)] \qquad (11.15\text{–}6)$$

Equations 11.14–7, 11.14–2, 11.14–4 are summarized in the following patterns:

$$\begin{array}{lcccccc} & & (x-2h) & (x-h) & \overset{i}{(x)} & (x+h) & (x+2h) \\ D = & \dfrac{1}{2h}\{ & & -1 & 0 & 1 & \} \\ D^2 = & \dfrac{1}{h^2}\{ & & 1 & -2 & 1 & \} \\ D^4 = & \dfrac{1}{h^4}\{ & 1 & -4 & 6 & -4 & 1\ \} \end{array} \quad \text{errors of order } (h^2)$$

$$(11.15\text{–}7)$$

The improved Eqns 11.15–2, 11.15–5, and 11.15–6 are summarized as

$$\begin{array}{lcccccccc} & & (x-3h) & (x-2h) & (x-h) & \overset{i}{(x)} & (x+h) & (x+2h) & (x+3h) \\ D = & \dfrac{1}{12h}\{ & & 1 & -8 & 0 & 8 & -1 & \} \\ D^2 = & \dfrac{1}{12h^2}\{ & & -1 & 16 & -30 & 16 & -1 & \} \\ D^4 = & \dfrac{1}{6h^4}\{ & -1 & 12 & -39 & 56 & -39 & 12 & -1\ \} \end{array} \quad \text{errors of order } (h^4)$$

$$(11.15\text{–}8)$$

Uneven mesh intervals. Equations 11.15–7 and 11.15–8 refer to a regular mesh of interval h. If the mesh intervals are not regular (Fig. 11.15–1), then by Taylor's expansion

$$f(x+\alpha h) = f(x) + \alpha h\,Df(x) + (\alpha^2)\frac{h^2\,D^2}{2!}f(x) + (\alpha^3)\frac{h^3\,D^3}{3!}f(x) + \ldots \qquad (11.15\text{–}9)$$

Similarly,

$$f(x-h) = f(x) - h\,Df(x) + \frac{h^2\,D^2}{2!}f(x) - \frac{h^3\,D^3}{3!}f(x) + \ldots \qquad (11.15\text{–}10)$$

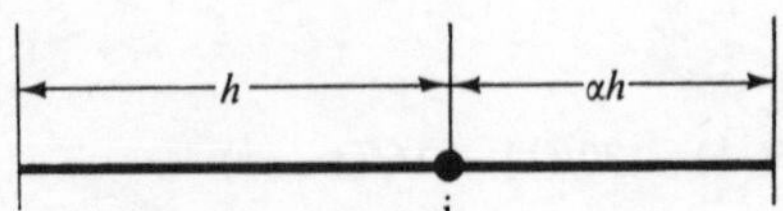

Fig. 11·15–1

Eliminating $h^2\ D^2f(x)$ from Eqns 11.15–9 and 11.15–10, we have

$$Df(x) = \frac{1}{\alpha(1+\alpha)h}\left[f(x+\alpha h) - (1-\alpha^2)f(x) - \alpha^2 f(x-h)\right]$$
$$+ \text{ error of order } h^2 \qquad (11.15\text{–}11)$$

Similarly, by eliminating $h\ Df(x)$ from Eqns 11.15–9 and 11.15–10,

$$D^2f(x) = \frac{2}{\alpha(1+\alpha)h^2}\left[f(x+\alpha h) - (1+\alpha)f(x) + \alpha f(x-h)\right]$$
$$+ \text{ error of order } h \text{ for } \alpha \neq 1 \text{ (and } h^2 \text{ for } \alpha = 1) \qquad (11.15\text{–}12)$$

Using Eqn 11.15–12, the reader should verify that for irregular mesh intervals in two perpendicular directions the pattern for the Laplace operator is

$$\nabla^2 = \frac{2}{\alpha\beta(1+\alpha)(1+\beta)h^2} \quad \begin{array}{ccccc} & & \boxed{\alpha(1+\alpha)} & & \\ & & \beta h & & \\ \boxed{\alpha\beta(1+\beta)} & h & \boxed{-(1+\alpha)(1+\beta)(\alpha+\beta)} & \alpha h & \boxed{\beta(1+\beta)} \\ & & h & & \\ & & \boxed{\alpha\beta(1+\alpha)} & & \end{array}$$

with error of order (h)

(11.15–13)

The above finite difference pattern is sometimes referred to as an **irregular star**.

For a rectangular mesh of intervals h in the x-direction and intervals αh in the y-direction (Fig. 11.15–2), Eqns 11.14–20 and 11.14–21 are rewritten as

$$\frac{1}{h^4}\,\delta_x^4 f(\mathrm{i}) = \frac{1}{h^4}\left[f(9) - 4f(1) + 6f(\mathrm{i}) - 4f(3) + f(11)\right] \qquad (11.15\text{–}14)$$

$$\frac{1}{\alpha^4h^4}\,\delta_y^4 f(\mathrm{i}) = \frac{1}{\alpha^4h^4}\left[f(10) - 4f(2) + 6f(\mathrm{i}) - 4f(4) + f(12)\right] \qquad (11.15\text{–}15)$$

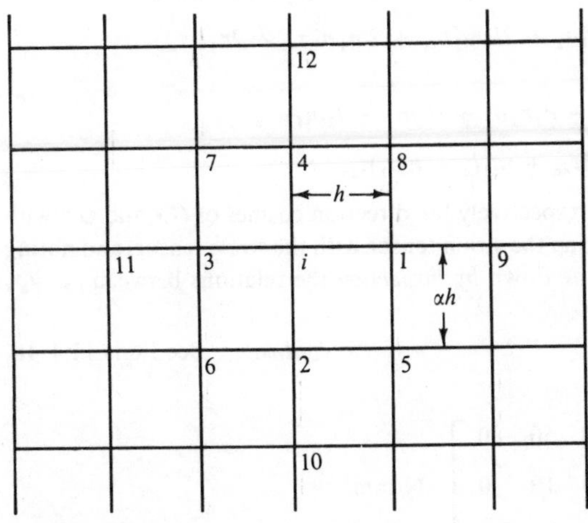

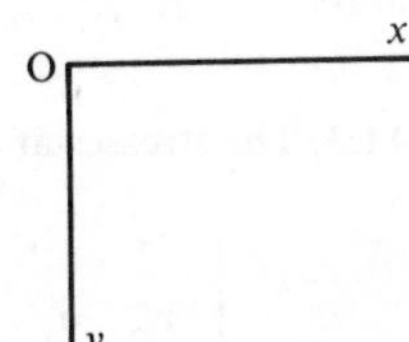

Fig. 11·15–2

Equation 11.14–19 becomes

$$\left(\frac{1}{h^2}\delta_x^2\right)\left(\frac{1}{\alpha^2h^2}\delta_y^2\right) = \frac{1}{h^2}\delta_x^2\frac{1}{\alpha^2h^2}[f(2) - 2f(\mathrm{i}) + f(4)]$$

$$= \frac{1}{\alpha^2h^4}[\delta_x^2 f(2) - 2\delta_x^2 f(\mathrm{i}) + \delta_x^2 f(4)]$$

$$= \frac{1}{\alpha^2h^4}\{[f(5) - 2f(2) + f(6)] - 2[f(1) - 2f(\mathrm{i}) + f(3)]$$

$$+ [f(8) - 2f(4) + f(7)]\} \qquad (11.15\text{–}16)$$

Using Eqns 11.15–14 to 11.15–16, the biharmonic operator for the mesh in Fig. 11.15–2 becomes:

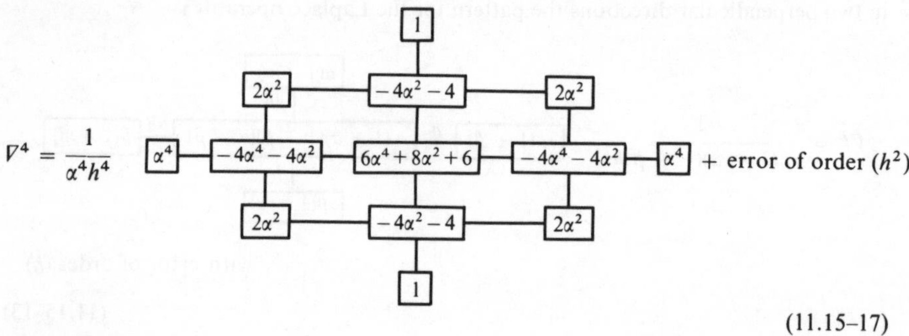

(11.15–17)

Problems

11.1. (a) Express each element of the strain tensor in terms of the elements of the stress tensor and vice versa.

(b) Does ε_x depend on τ_{xy}, or τ_{yz}, or τ_{zx}?

(c) Does γ_{xy} depend on σ_x, or σ_y, or σ_z?

(d) Does ε_x depend on σ_y?

(e) Does γ_{xy} depend on τ_{yz}?

Ans. (a) see Eqns 1.7–8 and 1.7–9; (b) No; (c) No; (d) Yes; (e) No.

11.2. With reference to cartesian systems *Oxyz* and *OXYZ*, it is given that

$$\sigma_X = l_1^2\sigma_x + m_1^2\sigma_y + n_1^2\sigma_z + 2l_1m_1\tau_{xy} + 2m_1n_1\tau_{yz} + 2n_1l_1\tau_{zx}$$

and

$$\tau_{XY} = l_1l_2\sigma_x + m_1m_2\sigma_y + n_1n_2\sigma_z + (l_1m_2 + l_2m_1)\tau_{xy}$$
$$+ (m_1n_2 + m_2n_1)\tau_{yz} + (n_1l_2 + n_2l_1)\tau_{zx}$$

where l_1, m_1, n_1 and l_2, m_2, n_2 are respectively the direction cosines of *OX* and *OY* with respect to system *Oxyz*. By comparing the strain tensor with the stress tensor, and noting the tensor transformation law, write down *by inspection* the relations between ε_X, γ_{XY} and $\varepsilon_x, \varepsilon_y, \varepsilon_z, \gamma_{xy}, \gamma_{yz}, \gamma_{zx}$.

Ans. See Eqns 11.4–18.

11.3. The stresses at a point are:

$$\begin{bmatrix} \sigma_x & \tau_{xy} & \tau_{xz} \\ \tau_{yx} & \sigma_y & \tau_{yz} \\ \tau_{zx} & \tau_{zy} & \sigma_z \end{bmatrix} = \begin{bmatrix} 30 & 50 & 80 \\ 50 & 10 & 0 \\ 80 & 0 & 20 \end{bmatrix} \text{ N/mm}^2$$

Determine (a) the principal stresses, (b) the direction cosines of the maximum principal stress.

Ans. $\sigma_1 = 118.4$ N/mm^2; $\sigma_2 = 12.8$ N/mm^2; $\sigma_3 = -71.1$ N/mm^2; the direction cosines are $l = 0.731$, $m = 0.338$, $n = 0.595$.

11.4. If the normal and shear stresses at a point O, referred to cartesian system *Oxyz*, are $\sigma_x, \sigma_y, \sigma_z, \tau_{xy}, \tau_{yz}$, and τ_{zx}, show that, if the particle at O is in equilibrium and there are no body forces acting, then

$$\begin{bmatrix} \dfrac{\partial}{\partial x} & \dfrac{\partial}{\partial y} & \dfrac{\partial}{\partial z} \end{bmatrix} \begin{bmatrix} \sigma_x & \tau_{xy} & \tau_{xz} \\ \tau_{yx} & \sigma_y & \tau_{yz} \\ \tau_{zx} & \tau_{zy} & \sigma_z \end{bmatrix} = 0$$

11.5. A plane element through the point O in Problem 11.4 is defined by the direction cosines l, m, n, of its normal. Show that

$$[\sigma_{px} \quad \sigma_{py} \quad \sigma_{pz}] = [l \quad m \quad n] \begin{bmatrix} \sigma_x & \tau_{xy} & \tau_{xz} \\ \tau_{yx} & \sigma_y & \tau_{yz} \\ \tau_{zx} & \tau_{zy} & \sigma_z \end{bmatrix}$$

where σ_{px}, σ_{py}, and σ_{pz} are the components of the stress acting on the plane element in the directions of the respective coordinate axes.

11.6. Given the following strain tensor, determine the stress tensor, if $E = 207$ kN/mm^2 and $\nu = 0.29$:

$$\varepsilon = \begin{bmatrix} 800 & 80 & 320 \\ 80 & 400 & 400 \\ 320 & 400 & 560 \end{bmatrix} \times 10^{-6}$$

Ans. $$\sigma = \begin{bmatrix} 323.4 & 12.8 & 51.2 \\ 12.8 & 258.6 & 64.0 \\ 51.2 & 64.0 & 285.7 \end{bmatrix} \text{N/mm}^2$$

Hint: The element in the first row and second column of ε is $\frac{1}{2}\gamma_{xy}$ and not γ_{xy}.

11.7. (a) Explain why the formula

$$\frac{\tau}{r} = \frac{T}{J} = G\theta$$

is not applicable to the torsion of non-circular sections.

(b) The torsion formula for a general section (circular or non-circular) is

$$\frac{\tau_{zx}}{\partial\phi_1/\partial y} = -\frac{\tau_{zy}}{\partial\phi_1/\partial x} = \frac{T}{2\int \phi_1 \, dA} = G\theta$$

Explain how the torsion function ϕ_1 can be determined for a given section. If forces are to be expressed in newtons, linear dimensions in millimetres, and rotations in radians, what would be the units for ϕ_1? What then would be the units for ϕ in Eqns 11.11–15?

Ans. (b) ϕ_1 is in square millimetres (mm^2) units; ϕ is in newtons per millimetre units (N/mm).

11.8. The bending and twisting moments acting at a point (x, y) in a thin uniform plate subjected to small deflections may be expressed in the form:

$$M_{yx} = \mathscr{D}\left\{\frac{\partial^2 w}{\partial y^2} + \nu\frac{\partial^2 w}{\partial x^2}\right\}$$

$$M_{xy} = -\mathscr{D}\left\{\nu\frac{\partial^2 w}{\partial y^2} + \frac{\partial^2 w}{\partial x^2}\right\}$$

$$M_{xx} = -M_{yy} = \mathscr{D}(1 - \nu)\frac{\partial^2 w}{\partial x\, \partial y}$$

(a) Draw a sketch of an element $dx \times dy$ and show clearly how these moments would act on the faces of the element. Show all moments in the positive sense and explain your sign convention.

(b) By considering the equilibrium of an element, derive the biharmonic equation which governs the deflection of the plate. Explain why the validity of the biharmonic equation does not depend on the support conditions and why it is applicable to plates of any shape and not to rectangular plates only.

11.9. Express the Laplace equation $\nabla^2\phi_1 = -2$ in finite difference form, using central differences and a regular mesh interval h.

Determine the error in the representation of

$$\frac{\partial^2\phi_1}{\partial x^2} \text{ by } \frac{\delta_x^2\phi_1}{h^2}$$

Ans. $\frac{1}{12}h^2\frac{\partial^4\phi_1}{\partial x^4} + \frac{1}{360}h^4\frac{\partial^6\phi_1}{\partial x^6} + \ldots$

11.10. (a) Express the biharmonic equation $\nabla^4 w = q/\mathscr{D}$ in finite difference form using a regular mesh interval h.

(b) Determine the errors in the representation of

$$\frac{\partial^4 w}{\partial x^4} \text{ by } \frac{\delta_x^4 w}{h^4}$$

and

$$\frac{\partial^4 w}{\partial x^2\, \partial y^2} \text{ by } \frac{\delta_x^2\,\delta_y^2\, w}{h^4}$$

(c) Determine the finite difference equation for the boundary condition at (i) a simply supported edge, (ii) a fully fixed edge of a thin plate in bending.

Ans. (a) See Eqn 11.14–25.

(b) $\frac{1}{6}h^2\frac{\partial^6 w}{\partial x^6} + \frac{1}{80}h^6\frac{\partial^8 w}{\partial x^8} + \ldots$

and $\frac{1}{12}h^2\frac{\partial^6 w}{\partial x^4\,\partial y^2} + \frac{1}{12}h^2\frac{\partial^6 w}{\partial x^2\,\partial y^4} + \ldots$

(c) (i) $w_{i+h} + w_{i-h} = 0; \quad w_i = 0$

(ii) $w_{i+h} - w_{i-h} = 0; \quad w_i = 0$

where i is a nodal point on the edge.

11.11. Equation 11.14–33 of Example 11.14–2 shows that, on an edge with full restraint against z-displacement and partial restraint against y-rotation, such as edge PQ in the following figure,

$$\frac{w_{i+h}}{w_{i-h}} = \frac{1 - 2\mathscr{D}/kh}{1 + 2\mathscr{D}/kh}$$

This relation applies only to a 'right-hand-side' edge, where the fictitious point $(i + h)$ is at a positive distance $+h$ from point i on the edge. Show that for a 'left-hand-side' edge such as P′Q′, where the fictitious point $(i' - h)$ is at a negative distance $-h$ from the point i′, the relation would be

$$\frac{w_{i'+h}}{w_{i'-h}} = \frac{1 + 2\mathscr{D}/kh}{1 - 2\mathscr{D}/kh}$$

Or, in general, for both 'left-hand-side' and 'right-hand-side' edges,

$$\frac{w \text{ of fictitious point}}{w \text{ of real point}} = \frac{1 - 2\mathscr{D}/kh}{1 + 2\mathscr{D}/kh}$$

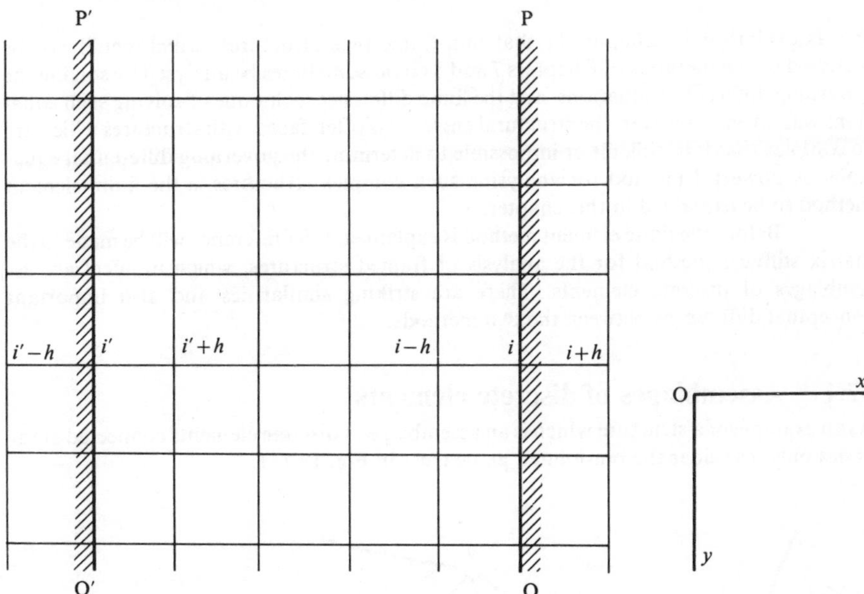

Problem 11·11

References

1 Jeffrey, A. *Mathematics for Engineers and Scientists*, Van Nostrand Reinhold, 3rd edition, 1985, p. 128.

2 Wylie, C.R. and Barrett, L.C. *Advanced Engineering Mathematics*, McGraw-Hill, New York, 5th edition, 1982, pp. 795 and 804.

General references

Fenner, R.T. *Engineering elasticity—application of numerical and analytical technique*, Halstead Press, New York, 1986.

Wang, C.T. *Applied Elasticity*. McGraw-Hill, New York, 1953.

Wah, T. and Calcote, L.R. *Structural Analysis by Finite Difference Calculus*. Van Nostrand Reinhold Co., New York, 1970.

Chapter 12
The finite element method

It was explained in Chapter 11 that non-frame-type structures, which could not be analysed by the methods of Chapters 7 and 8 could sometimes be analysed by solving the governing differential equations, and the finite difference technique of solving such equations was given. However, the structural engineer is often faced with structures which are so complex that it is difficult or impossible to determine the governing differential equations. A powerful method for analysing such complex structures is the finite element method to be explained in this chapter.

Before the finite element method is explained, brief reference will be made to the matrix stiffness method for the analysis of framed structures, which in effect are assemblages of discrete elements. There are striking similarities and also important conceptual differences between the two methods.

12.1 Assemblages of discrete elements

As an example of a structure which is an assemblage of discrete elements connected at the nodes only, consider the pin-jointed plane truss in Fig. 12.1–1.

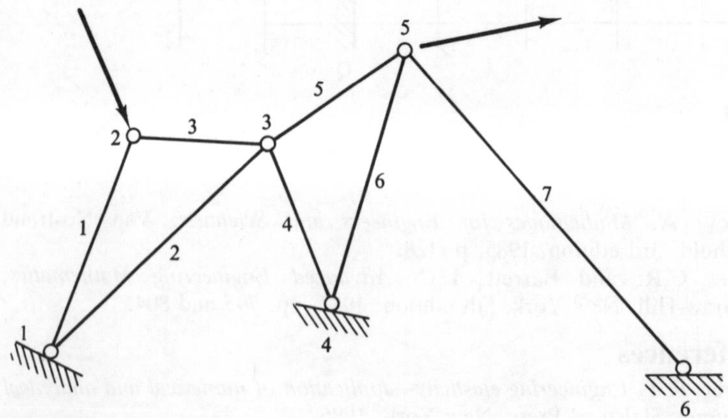

Fig. 12·1–1

This truss consists of seven elements, each element being a line member having two nodes; because nodes 1, 2, 3, 4, and 5 are common to two or more elements, the structure has 6 nodes only. Nodes 2, 3, and 5 are free to displace under load, while nodes 1, 4, and 6 are restrained against displacements.

Using the matrix stiffness method explained in Chapter 7, we first determine the stiffness matrices of the seven elements or members, then build up the overall structure stiffness matrix for the whole truss. Next we determine the nodal displacements (that is the displacements of the joints) by solving an equation of the type

$$\mathbf{P} = \mathbf{K}_s \mathbf{\Delta} \tag{12.1–1}$$

where **P** is the nodal load vector, **Δ** the nodal-displacement vector, and $\mathbf{K}_s$ the overall structure stiffness matrix modified to allow for the restraints at nodes 1, 4, and 6. Having found the nodal displacements, we proceed to determine the stresses in each element or member and then we determine the support reactions. All these were explained in Chapter 7.

Next consider the structure in Fig. 12.1–2 which is also an assemblage of seven discrete elements connected at the nodes only, except that this time the elements are not line members but are rectangular and triangular plates.

However, provided the stiffness matrix of each individual element can be determined, we can build up the overall structure stiffness matrix and analyse the structure in the same way as we would analyse the truss in Fig. 12.1–1; that is, the general approach explained in Chapter 7 is equally applicable to structures in which the elements are not line members but are plates of triangular or other shapes. For two-dimensional elements such as those in Fig. 12.1–2, each element stiffness matrix relates the nodal forces acting on that element to the nodal displacements. Consider, for example, one of the elements in Fig. 12.1–2, having nodes, say, *i, j, k* (Fig. 12.1–3).

We are assuming pin connections at the nodes; therefore at each node there are two displacement components, one in the *x*-direction, the other in the *y*-direction. Similarly, there are two nodal force components at each node. The nodal forces and nodal displacements are related by the equation

$$\mathbf{p} = \mathbf{K}\mathbf{\Delta} \tag{12.1–2}$$

where

$$\mathbf{p} = \begin{bmatrix} p_{xi} \\ p_{yi} \\ p_{xj} \\ p_{yj} \\ p_{xk} \\ p_{yk} \end{bmatrix} \qquad \mathbf{\Delta} = \begin{bmatrix} \Delta_{xi} \\ \Delta_{yi} \\ \Delta_{xj} \\ \Delta_{yj} \\ \Delta_{xk} \\ \Delta_{yk} \end{bmatrix} \tag{12.1–3}$$

and the matrix **K**, called the **element stiffness matrix**, is

$$\mathbf{K} = \begin{bmatrix} K_{xi,\,xi} & K_{xi,\,yi} & \dots & K_{xi,\,yk} \\ K_{yi,\,xi} & K_{yi,\,yi} & \dots & K_{yi,\,yk} \\ & & & \\ K_{yk,\,xi} & K_{yk,\,yi} & \dots & K_{yk,\,yk} \end{bmatrix} \tag{12.1–4}$$

We shall further consider the stiffness matrices of such two-dimensional elements in Sections 12.2 and 12.3.

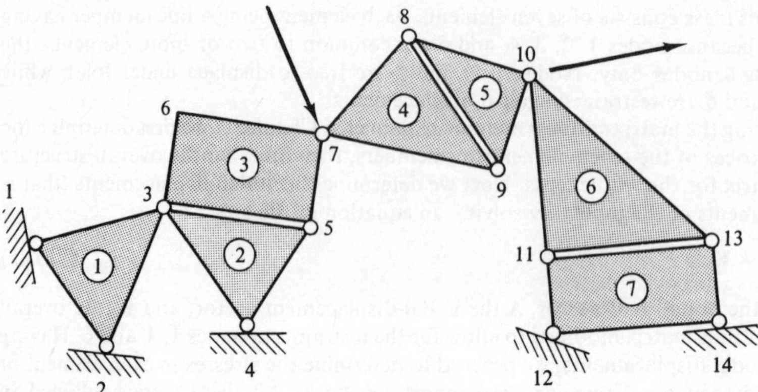

Fig. 12·1–2

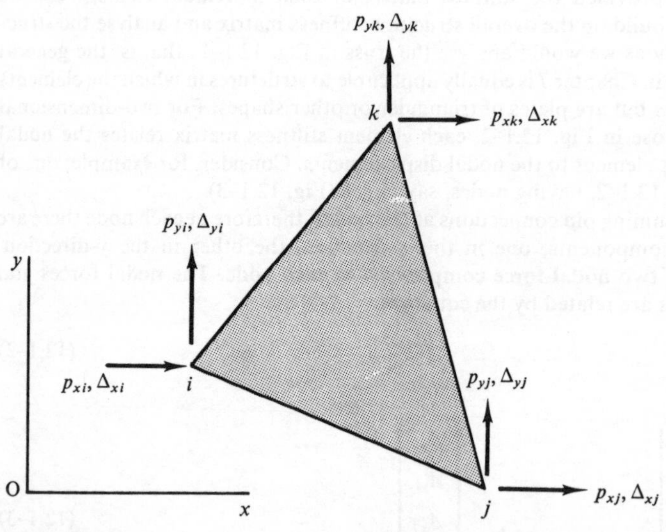

Fig. 12·1–3

12.2 Elastic continua

In the previous section, we studied structures which are assemblages of discrete elements. In Fig. 12.1–1, the structure is an assemblage of discrete line members; in Fig. 12.1–2, it is an assemblage of discrete plates of triangular or rectangular shape. We shall now study a different type of structure, which is not an assemblage of discrete elements, but is a continuous medium, or 'continuum'. Consider, for example, a steel plate (Fig. 12.2–1) which is arbitrarily supported and subjected to *in-plane* forces.

As the structure in this case is a continuum and not an assemblage of discrete elements connected at the nodes, it would appear impossible at first sight to apply the same techniques which were applied to the structures in Fig. 12.1–1 and Fig. 12.1–2. However, Turner,[1] Argyris,[2], Clough,[3] and others have developed the **finite element**

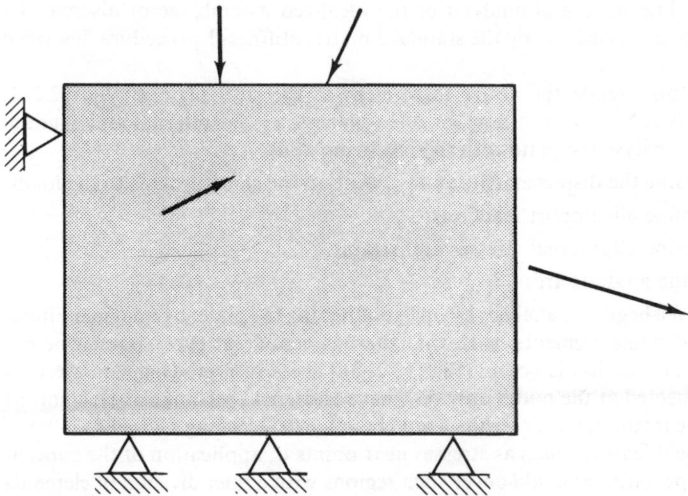

Fig. 12·2–1

method, which enables these standard techniques to be applied to continua. Basically, the finite element analysis of a continuum consists of three steps:

STEP 1: Structural idealization, whereby the continuum is idealized as an assemblage of a number, often a large number, of discrete elements connected at the nodes only.

STEP 2: Specifying the relation between the internal displacements of each element and its nodal displacements. This is done by using a displacement function to specify the pattern in which the element is to deform. (Displacement functions will be discussed in greater detail in subsequent sections of this chapter.) On the basis of this displacement function, we derive the element stiffness matrix, which relates the element nodal forces to the element nodal displacements.

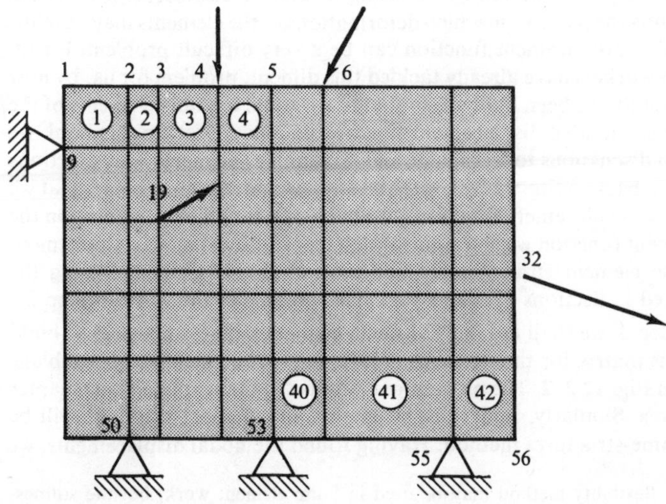

Fig. 12·2–2

STEP 3: The structural analysis of the idealized assemblage of discrete elements. This analysis is carried out by the standard matrix stiffness† procedure described in Chapter 7.

We shall now relate the above three steps to the steel plate in Fig. 12.2–1. Suppose the plate is subjected to a number of in-plane loads, at internal and external points. We wish to analyse the plate with the following aims:

(a) To determine the displacements at any point on the structure due to the loads.

(b) To determine all support reactions.

(c) To determine all internal stresses and strains.

The three steps in the analysis are:

STEP 1: We begin the analysis by first subdividing the plate, by imaginary lines, into a number of discrete elements, such that there is a node at every point where a concentrated load or reaction is acting (Fig. 12.2–2). These discrete elements are now assumed to be connected at the nodes only. We have chosen 42 rectangular elements as an illustration of the method. Better results would be achieved by using a larger number of elements, and, if local features such as stresses near points of application of the concentrated loads are important, we could cover such regions with a finer division of elements than elsewhere in the structure. Also, we have chosen rectangular elements, but if preferred, we could have chosen elements of triangular or other shapes, and indeed could have chosen a combination of elements of different shapes or sizes. In practice, of course, we tend to choose elements which are of convenient shapes for representing the boundaries of the structure and which are known to be satisfactory. Note that we are not just replacing the steel plate by an assemblage of 42 smaller plates connected at the nodes only. Indeed, if we do so, we would find that the behaviour of the original plate is quite different from that of the assemblage of elements. In effect, what we do is to idealize the steel plate in Fig. 12.2–1 into an assemblage of discrete elements (Fig. 12.2–2) which are no longer made of steel, but made of such an imaginary material that the elements have stiffness characteristics as specified in Step 2 (see Example 12.3–1).

STEP 2: In Step 2, we want to ensure that the load behaviour of the assemblage of discrete elements will reasonably represent that of the steel plate. That is, each discrete element is required to deform reasonably similarly to the deformation developed in the corresponding region of the steel plate. Also, we want the finite element solution to converge to the true solution for the steel plate as the number of elements is successively increased. These aims may be achieved by choosing a suitable displacement function, i.e. by suitably specifying the patterns in which deformations of the elements may develop. The choice of a suitable displacement function can be a very difficult problem. Fortunately, many research workers have already tackled this difficult problem for us; by now quite a catalogue of results has been built up, so that if we would use only elements of the more usual shapes, then suitable displacement functions already exist. In this book we shall restrict our main discussions to triangular and rectangular elements only (Sections 12.3 and 12.4). Coming back to Fig. 12.2–2, we shall suppose, for the time being, that we have already chosen the displacement function suitable for rectangular elements. On the basis of this displacement function we determine the element stiffness matrix, the element stress matrix, and the element strain matrix, for each of the 42 elements, using the method to be explained in Sections 12.3 and 12.4. This completes the work in Step 2.

STEP 3: In Step 3, we shall use the 42 known element stiffness matrices to build up the overall stiffness matrix for the whole structure, i.e. for the idealized assemblage of discrete elements in Fig. 12.2–2. The procedure is similar to that explained in Chapter 7 for framed structures. Similarly, nodal displacements and support reactions will be determined by the frame-structures method. Having found the nodal displacements, we

† The matrix flexibility method may be used in finite element work, but the stiffness method is usually preferred and is used here.

can use the element stress matrix and element strain matrix to determine the stresses and strains at any point in each element; this point will be made clearer in subsequent sections of this chapter.

12.3 Triangular elements for plane stress

Consider a triangular element *ijk*, with nodal coordinates (x_i, y_i), (x_j, y_j) and (x_k, y_k) respectively (Fig. 12.3–1). The displacement at any point (x, y) has two components u and v in the x- and y-directions, respectively. These displacements, u and v, uniquely define the internal displacements of the element.

It is known that for triangular elements, a suitable displacement function is

$$\left.\begin{aligned} u &= \alpha_1 + \alpha_2 x + \alpha_3 y \\ v &= \alpha_4 + \alpha_5 x + \alpha_6 y \end{aligned}\right\} \tag{12.3–1}$$

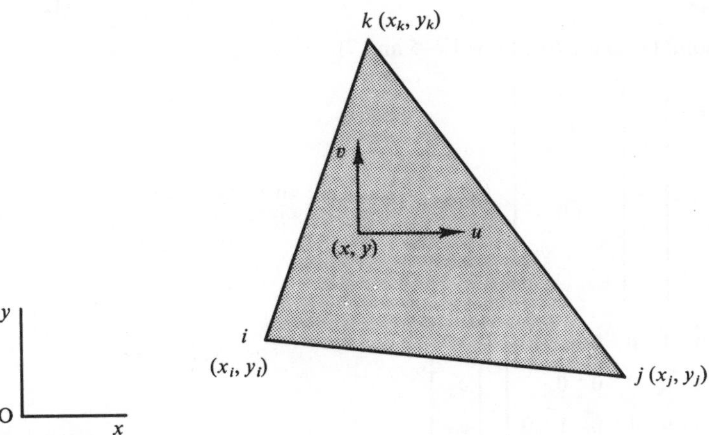

Fig. 12·3–1

where α_1 to α_6 are constants whose values depend on the nodal displacement (see Eqn 12.3–3). Previous research has also shown that a good displacement function should satisfy the following conditions:

(1) The displacement function, and its first derivatives, should be continuous within the element.

(2) The displacement function should allow nodal displacements caused by rigid-body translations and rotations to occur without straining the element, i.e. without changing the strain energy in the element.

(3) The displacement function should allow for all states of uniform strain within the element.

(4) The displacement function should satisfy internal compatibility within the element, and should also maintain compatibility of displacements between adjacent elements at the nodes and along the boundaries. If all such compatibility conditions are satisfied, the displacement function is said to be a **conforming function**.

The displacement function in Eqn 12.3–1 obviously satisfies condition 1. To test whether it satisfies condition 2, let us first consider the nodal displacements and internal strains.

From Eqn 12.3–1, the displacements at each node are found by substituting in the nodal coordinates. Thus, the x-component of the displacement at node i is:

$$\Delta_{xi} = \alpha_1 + \alpha_2 x_i + \alpha_3 y_i$$

Proceeding in a similar manner, we have, for all the nodes,

$$\mathbf{\Delta} = \begin{bmatrix} \Delta_{xi} \\ \Delta_{yi} \\ \Delta_{xj} \\ \Delta_{yj} \\ \Delta_{xk} \\ \Delta_{yk} \end{bmatrix} = \begin{bmatrix} 1 & x_i & y_i & 0 & 0 & 0 \\ 0 & 0 & 0 & 1 & x_i & y_i \\ 1 & x_j & y_j & 0 & 0 & 0 \\ 0 & 0 & 0 & 1 & x_j & y_j \\ 1 & x_k & y_k & 0 & 0 & 0 \\ 0 & 0 & 0 & 1 & x_k & y_k \end{bmatrix} \begin{bmatrix} \alpha_1 \\ \alpha_2 \\ \alpha_3 \\ \alpha_4 \\ \alpha_5 \\ \alpha_6 \end{bmatrix} \tag{12.3–2}$$

that is,

$$\mathbf{\Delta} = \mathbf{A}\boldsymbol{\alpha} \tag{12.3–3}$$

The strains at point (x, y) are (see Eqns 1.6–5 and 7):

$$\boldsymbol{\varepsilon} = \begin{bmatrix} \varepsilon_x \\ \varepsilon_y \\ \gamma_{xy} \end{bmatrix} = \begin{bmatrix} \dfrac{\partial u}{\partial x} \\ \dfrac{\partial v}{\partial y} \\ \dfrac{\partial v}{\partial x} + \dfrac{\partial u}{\partial y} \end{bmatrix} \quad \text{(where } u \text{ and } v \text{ are given by Eqn 12.3–1)}$$

$$= \begin{bmatrix} 0 & 1 & 0 & 0 & 0 & 0 \\ 0 & 0 & 0 & 0 & 0 & 1 \\ 0 & 0 & 1 & 0 & 1 & 0 \end{bmatrix} \begin{bmatrix} \alpha_1 \\ \alpha_2 \\ \alpha_3 \\ \alpha_4 \\ \alpha_5 \\ \alpha_6 \end{bmatrix} \tag{12.3–4}$$

That is,

$$\boldsymbol{\varepsilon} = \mathbf{B}\boldsymbol{\alpha} \tag{12.3–5}$$

$$= \mathbf{BA}^{-1}\mathbf{\Delta} \quad \text{(from Eqn 12.3–3)}$$

where the matrix $\{\mathbf{BA}^{-1}\}$ is sometimes referred to as the **element strain matrix**. In this particular case, the **B** matrix, and hence the element strain matrix, are independent of the position within the element; therefore the triangular element is under uniform strain.

Now consider the particular case when $\alpha_2 = \alpha_6 = 0$ and $\alpha_5 = -\alpha_3$. From Eqn 12.3–2, we see that the nodal displacements are:

$$\mathbf{\Delta} = \begin{bmatrix} \Delta_{xi} \\ \Delta_{yi} \\ \Delta_{xj} \\ \Delta_{yj} \\ \Delta_{xk} \\ \Delta_{yk} \end{bmatrix} = \begin{bmatrix} \alpha_1 + \alpha_3 y_i \\ \alpha_4 - \alpha_3 x_i \\ \alpha_1 + \alpha_3 y_j \\ \alpha_4 - \alpha_3 x_j \\ \alpha_1 + \alpha_3 y_k \\ \alpha_4 - \alpha_3 x_k \end{bmatrix} \tag{12.3–6}$$

These nodal displacements correspond to a rigid-body translation of α_1 in the x-direction, α_4 in the y-direction, and a rigid-body rotation of α_3 radians clockwise.† If α_3 is zero, the motion is pure rigid-body translation; if α_1 and α_4 are zero, it is pure rigid-body rotation. Also, from Eqn 12.3–4, the internal strains are

$$\boldsymbol{\varepsilon} = \begin{bmatrix} \varepsilon_x \\ \varepsilon_y \\ \gamma_{xy} \end{bmatrix} = \begin{bmatrix} \alpha_2 \\ \alpha_6 \\ \alpha_3 + \alpha_5 \end{bmatrix} = 0 \qquad (12.3\text{–}7)$$

Hence the chosen displacement function does not cause strains in the element, when the nodal displacements are due solely to rigid-body motion. Hence condition 2 is satisfied.

Referring again to Eqn 12.3–4, we see that the internal strains ε are in fact independent of the coordinates (x, y), that is they are uniform throughout the element. Hence condition 3 is automatically satisfied.

Regarding condition 4 about compatibility, it is obvious from Eqn 12.3–4 that

$$\frac{\partial^2 \varepsilon_x}{\partial y^2} + \frac{\partial^2 \varepsilon_y}{\partial x^2} = \frac{\partial^2 \gamma_{xy}}{\partial x\, \partial y} \quad \text{(see Eqn 11.6–1(a))}$$

for internal compatibility is satisfied. Also, since the elements are connected at the nodes, which remain connected during deformation of the structure, nodal compatibility is automatically guaranteed. From Eqn 12.3–1 we see that the chosen displacement function specifies that displacements at all points vary linearly in two perpendicular directions —x and y. Then any straight line in the undeformed element will remain straight during deformation, and, consequently, the element boundaries will remain in contact as the elements deform. Hence condition 4 is satisfied, and the displacement function in Eqn 12.3–1 is therefore a conforming function; and, when such a conforming function is used, the element is a **conforming element**.

Having thus satisfied ourselves of the suitability of the displacement function in Eqn 12.3–1, we shall proceed to determine the element stiffness matrix.

Suppose the triangular element in Fig. 12.3–1 is subjected to arbitrary nodal displacements $\boldsymbol{\Delta}$, where

$$\boldsymbol{\Delta} = \begin{bmatrix} \Delta_{xi} \\ \Delta_{yi} \\ \Delta_{xj} \\ \Delta_{yj} \\ \Delta_{xk} \\ \Delta_{yk} \end{bmatrix} \qquad (12.3\text{–}8)$$

The nodal forces required to produce these nodal displacements are $\mathbf{p}$, where

$$\mathbf{p} = \begin{bmatrix} p_{xi} \\ p_{yi} \\ p_{xj} \\ p_{yj} \\ p_{xk} \\ p_{yk} \end{bmatrix} \qquad (12.3\text{–}9)$$

† That is, the rotation vector is in the negative z-direction (Fig. 12.3–1).

Of course, the nodal force vector $\mathbf{p}$ is related to nodal displacement vector $\boldsymbol{\Delta}$ by the equation

$$\mathbf{p} = \mathbf{K}\boldsymbol{\Delta} \tag{12.3–10}$$

where $\mathbf{K}$ is the element stiffness matrix yet to be determined.

Also, from Eqn 12.3–5, the nodal displacements $\boldsymbol{\Delta}$ would cause internal strains $\boldsymbol{\varepsilon}$ in the element, where

$$\boldsymbol{\varepsilon} = \mathbf{B}\mathbf{A}^{-1}\boldsymbol{\Delta} \tag{12.3–11}$$

For plane stress in an isotropic material of Young's modulus E and Poisson's ratio ν, stresses and strains are related by Eqn 1.7–11 in Chapter 1, namely,

$$\begin{bmatrix} \sigma_x \\ \sigma_y \\ \tau_{xy} \end{bmatrix} = \frac{E}{1-\nu^2}\begin{bmatrix} 1 & \nu & 0 \\ \nu & 1 & 0 \\ 0 & 0 & \dfrac{1-\nu}{2} \end{bmatrix}\begin{bmatrix} \varepsilon_x \\ \varepsilon_y \\ \gamma_{xy} \end{bmatrix}$$

that is

$$\boldsymbol{\sigma} = \mathbf{D}\boldsymbol{\varepsilon} \tag{12.3–12}$$

$$= \mathbf{D}\mathbf{B}\mathbf{A}^{-1}\boldsymbol{\Delta} \quad \text{(from Eqn 12.3–11)}$$

These internal stresses $\boldsymbol{\sigma}$ are caused by the nodal displacement $\boldsymbol{\Delta}$.

Let us now impose virtual nodal displacements $\overline{\boldsymbol{\Delta}}$, where $\overline{\boldsymbol{\Delta}}$ of course is a six-component vector. These virtual nodal displacements will produce virtual strains $\overline{\boldsymbol{\varepsilon}}$ in the element given by Eqn 12.3–11, namely,

$$\overline{\boldsymbol{\varepsilon}} = \mathbf{B}\mathbf{A}^{-1}\overline{\boldsymbol{\Delta}} \tag{12.3–13}$$

Hence the virtual strain energy in the element is

$$\int_V \overline{\boldsymbol{\varepsilon}}^T \boldsymbol{\sigma}\, dV$$

where V is the volume of the triangular element.

Substituting in the expression for $\overline{\boldsymbol{\varepsilon}}$ from Eqn 12.3–13 and the expression for $\boldsymbol{\sigma}$ from Eqn 12.3–12, we have

$$\text{virtual strain energy} = \int_V \{\mathbf{B}\mathbf{A}^{-1}\overline{\boldsymbol{\Delta}}\}^T \mathbf{D}\mathbf{B}\mathbf{A}^{-1}\boldsymbol{\Delta}\, dV$$

$$= \int_V \overline{\boldsymbol{\Delta}}^T \mathbf{A}^{-T}\mathbf{B}^T\mathbf{D}\mathbf{B}\mathbf{A}^{-1}\boldsymbol{\Delta}\, dV \tag{12.3–14}†$$

In this equation, $\overline{\boldsymbol{\Delta}}$ and $\boldsymbol{\Delta}$ are independent of x and y and can therefore be taken out of the integration sign. Also Eqns 12.3–2 and 12.3–3 show that matrix $\mathbf{A}$ contains nodal coordinates and constant terms only and is hence independent of x and y. Similarly Eqns 12.3–4 and 12.3–5 show that, for a triangular element, $\mathbf{B}$ also contains constant terms only. Hence we can write Eqn 12.3–14 as

$$\text{virtual strain energy} = \overline{\boldsymbol{\Delta}}^T \mathbf{A}^{-T}\left\{\int_V \mathbf{B}^T\mathbf{D}\mathbf{B}\, dV\right\}\mathbf{A}^{-1}\boldsymbol{\Delta}$$

$$= \overline{\boldsymbol{\Delta}}^T\{\mathbf{A}^{-T}\mathbf{B}^T\mathbf{D}\mathbf{B}\mathbf{A}^{-1}V\}\boldsymbol{\Delta} \tag{12.3–15}$$

where V represents the volume of the triangular element.

† The symbol $\mathbf{A}^{-T}$ is used to represent $(\mathbf{A}^{-1})^T$. The reader is reminded that $(\mathbf{A}^{-1})^T = (\mathbf{A}^T)^{-1}$.

Now the external virtual work done by the real nodal forces $\mathbf{p}$ in undergoing virtual nodal displacements $\overline{\mathbf{\Delta}}$ is

$$\overline{\mathbf{\Delta}}^T\mathbf{p}, \quad \text{i.e.} \quad \overline{\mathbf{\Delta}}^T\mathbf{K\Delta} \quad \text{(from Eqn 12.3–10)}$$

Equating external virtual work and virtual strain energy, we have

$$\overline{\mathbf{\Delta}}^T\mathbf{K\Delta} = \overline{\mathbf{\Delta}}^T\{\mathbf{A}^{-T}\mathbf{B}^T\mathbf{DBA}^{-1}V\}\mathbf{\Delta} \tag{12.3–16}$$

Since the above equality holds for any virtual nodal displacements $\overline{\mathbf{\Delta}}$ and for any real nodal displacement $\mathbf{\Delta}$, we know from the theory of matrix algebra that the following equality must exist:

$$\mathbf{K} = \mathbf{A}^{-T}\mathbf{B}^T\mathbf{DBA}^{-1}V \tag{12.3–17}$$

It can be shown by coordinate geometry that if the nodal coordinates of the triangular element are (x_i, y_i), (x_j, y_j) and (x_k, y_k) and the thickness is t, then its volume is given by the following determinantal equation:

$$V = \frac{1}{2}\begin{vmatrix} 1 & x_i & y_i \\ 1 & x_j & y_j \\ 1 & x_k & y_k \end{vmatrix} t$$

where

$$\frac{1}{2}\begin{vmatrix} 1 & x_i & y_i \\ 1 & x_j & y_j \\ 1 & x_k & y_k \end{vmatrix} = \frac{1}{2}\{(x_iy_j - x_jy_i) + (x_jy_k - x_ky_j) + (x_ky_i - x_iy_k)\}$$

is the area of the triangle ijk.

Hence, for a triangular element with the displacement function in Eqn 12.3–1, the element stiffness matrix is

$$\mathbf{K} = \frac{1}{2}\mathbf{A}^{-T}\mathbf{B}^T\mathbf{DBA}^{-1}\{(x_iy_j - x_jy_i) + (x_jy_k - x_ky_j) + (x_ky_i - x_iy_k)\}\, t \tag{12.3–18}$$

From Eqn 12.3–18, it is of course possible to carry out the matrix multiplications manually and express the element stiffness matrix in explicit form. However, in the solution of practical problems, it is not necessary to know the explicit form of $\mathbf{K}$; the computer can easily generate the various matrices and carry out the matrix operations in the right-hand side of Eqn 12.3–18.

Distributed load on element. We shall determine the nodal forces produced by an in-plane distributed load, when the nodes are restrained against displacement. Consider the element ijk (Fig. 12.3–2) subjected to a distributed load

$$\mathbf{q} = \begin{bmatrix} q_x \\ q_y \end{bmatrix} \quad \text{per unit volume} \tag{12.3–19}$$

where the x and y components of $\mathbf{q}$ are functions of the coordinates (x, y); that is $\mathbf{q}$ may vary from point to point.

Suppose $\mathbf{q}$ produces nodal forces $\mathbf{p}_0$ when the nodes are not allowed to displace.

$$\mathbf{p}_0 = \begin{bmatrix} (p_0)_{xi} \\ (p_0)_{yi} \\ (p_0)_{xj} \\ (p_0)_{yj} \\ (p_0)_{xk} \\ (p_0)_{yk} \end{bmatrix} \tag{12.3–20}$$

This means that under the simultaneous action of $\mathbf{q}$ and $\mathbf{p}_0$, no nodal displacement occurs.

Suppose we now superimpose nodal forces $\mathbf{p}^*$; the superposition of $\mathbf{p}^*$ will upset the balance under $\mathbf{p}_0$ and $\mathbf{q}$ and will produce nodal displacements $\mathbf{\Delta}^*$ and internal displacements $\boldsymbol{\delta}^*$, where

$$\mathbf{p}^* = \begin{bmatrix} p^*_{xi} \\ p^*_{yi} \\ p^*_{xj} \\ p^*_{yj} \\ p^*_{xk} \\ p^*_{yk} \end{bmatrix} \qquad \mathbf{\Delta}^* = \begin{bmatrix} \Delta^*_{xi} \\ \Delta^*_{yi} \\ \Delta^*_{xj} \\ \Delta^*_{yj} \\ \Delta^*_{xk} \\ \Delta^*_{yk} \end{bmatrix} \qquad \boldsymbol{\delta}^* = \begin{bmatrix} u^* \\ v^* \end{bmatrix} \tag{12.3–21}$$

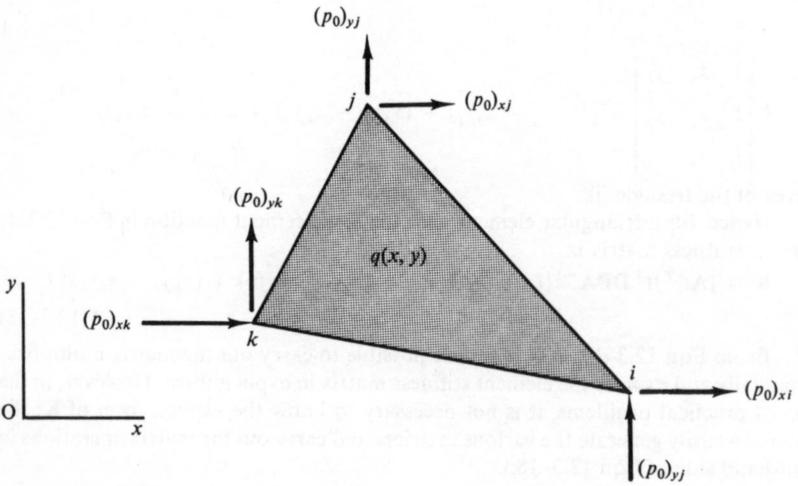

Fig. 12·3–2

By Betti's theorem (see Chapter 4, Section 4.12),

System $\mathbf{p}_0 \sim \mathbf{q}$

$$\mathbf{p}^{*T}\mathbf{0} = \mathbf{p}_0{}^T\mathbf{\Delta}^* + \int_V \mathbf{q}^T\boldsymbol{\delta}^*\, dV$$

System $\mathbf{p}^*$

that is,

$$\mathbf{p}_0{}^T\mathbf{\Delta}^* + \int_V \mathbf{q}^T\boldsymbol{\delta}^*\, dV = 0 \tag{12.3–22}$$

Now from Eqn 12.3–1,

$$\boldsymbol{\delta} = \begin{bmatrix} u \\ v \end{bmatrix} = \begin{bmatrix} 1 & x & y & 0 & 0 & 0 \\ 0 & 0 & 0 & 1 & x & y \end{bmatrix} \begin{bmatrix} \alpha_1 \\ \alpha_2 \\ \alpha_3 \\ \alpha_4 \\ \alpha_5 \\ \alpha_6 \end{bmatrix}$$

That is,

$$\boldsymbol{\delta} = \mathbf{N}_q \boldsymbol{\alpha} \qquad (12.3\text{–}23)$$

$$= \mathbf{N}_q \mathbf{A}^{-1} \boldsymbol{\Delta} \quad \text{(from Eqn 12.3–3)}$$

Hence

$$\boldsymbol{\delta}^* = \mathbf{N}_q \mathbf{A}^{-1} \boldsymbol{\Delta}^* \qquad (12.3\text{–}24)$$

Substituting Eqn 12.3–24 into 12.3–22,

$$\mathbf{p}_0^T \boldsymbol{\Delta}^* + \int_V \mathbf{q}^T \mathbf{N}_q \mathbf{A}^{-1} \boldsymbol{\Delta}^* \, dV = 0$$

Taking $\boldsymbol{\Delta}^*$, which is independent of (x, y), out of the integration sign,

$$\{\mathbf{p}_0^T + \int_V \mathbf{q}^T \mathbf{N}_q \mathbf{A}^{-1} \, dV\} \boldsymbol{\Delta}^* = 0$$

Since this equation holds for all possible nodal displacements $\boldsymbol{\Delta}^*$, we must have

$$\mathbf{p}_0^T + \int_V \mathbf{q}^T \mathbf{N}_q \mathbf{A}^{-1} \, dV = 0$$

Transposing

$$\mathbf{p}_0 + \int_V \mathbf{A}^{-T} \mathbf{N}_q^T \mathbf{q} \, dV = 0$$

i.e.

$$\mathbf{p}_0 = -\mathbf{A}^{-T} \int_V \mathbf{N}_q^T \mathbf{q} \, dV \qquad (12.3\text{–}25)$$

Equation 12.3–25 gives the nodal forces due to the distributed in-plane load, when the nodes are restrained against displacement.

Point loads on element. Suppose the element ijk in Fig. 12.3–2 is acted on by internal point loads $\mathbf{Q}$:

$$\mathbf{Q} = \begin{bmatrix} Q_{x1} \\ Q_{y1} \\ Q_{x2} \\ Q_{y2} \\ \vdots \\ Q_{xn} \\ Q_{yn} \end{bmatrix}$$

at points 1, 2, 3, . . ., *n*. In this case, the internal displacements corresponding to **Q** will be (see Eqns 12.3–21 and 12.3–23)

$$\boldsymbol{\delta}^* = \begin{bmatrix} u_1^* \\ v_1^* \\ u_2^* \\ v_2^* \\ \vdots \\ u_n^* \\ v_n^* \end{bmatrix} = \begin{bmatrix} 1 & x_1 & y_1 & 0 & 0 & 0 \\ 0 & 0 & 0 & 1 & x_1 & y_1 \\ 1 & x_2 & y_2 & 0 & 0 & 0 \\ 0 & 0 & 0 & 1 & x_2 & y_2 \\ \vdots & & & & & \\ 1 & x_n & y_n & 0 & 0 & 0 \\ 0 & 0 & 0 & 1 & x_n & y_n \end{bmatrix} \begin{bmatrix} \alpha_1 \\ \alpha_2 \\ \alpha_3 \\ \alpha_4 \\ \alpha_5 \\ \alpha_6 \end{bmatrix}$$

That is,

$$\boldsymbol{\delta}^* = \mathbf{N}_Q \boldsymbol{\alpha} \tag{12.3–26}$$

$$= \mathbf{N}_Q \mathbf{A}^{-1} \boldsymbol{\Delta}^* \quad \text{(from Eqn 12.3–3)}$$

Hence, for point loads, Eqn 12.3–25 becomes

$$\mathbf{p}_0 = -\mathbf{A}^{-T} \mathbf{N}_Q^T \mathbf{Q} \tag{12.3–27}$$

Equation 12.3–27 gives the nodal forces $\mathbf{p}_0$ due to internal point loads **Q**, when the nodes are restrained against displacements.

Example 12.3–1. A uniform steel plate subjected to in plane forces is to be analysed by the finite element method, using triangular elements; the element stiffness matrices **K** are to be derived from the displacement function in Eqn 12.3–1. If, instead of using the element stiffness matrices so derived, precise tests are carried out on steel triangular plates to determine accurately the element stiffness matrices, explain whether the use of such experimental stiffness matrices would improve the accuracy of the finite element analysis.

SOLUTION The answer is that it definitely would not, and the reason is as follows.

It is true that for each triangular steel plate, a unique relationship exists between nodal forces and nodal displacements, and this relationship can be determined, at least experimentally. However, a stiffness matrix based on such a true relationship is the true stiffness matrix of the steel element, and should not be used in finite element work. Using such true stiffness matrices would mean replacing the steel plate by an assemblage of discrete steel elements connected at the nodes only, and such replacement can result in serious errors. What we do in finite element analysis is not only to replace the actual steel plate by an assemblage of discrete elements, but also to replace the true stiffness matrix of each element by a *fictitious* element stiffness matrix, which is so chosen (by suitable selection of a displacement function) as to give acceptable final results for the actual steel plate. That is, the discrete elements are not made of steel, but of a fictitious material that behaves in a fictitious way under load.

In order to appreciate further the fictitious properties of the elements used in the analysis, consider the element ijk in Fig. 12.3–3, which is acted on by distributed in-plane forces **q**. Suppose the nodes are restrained against displacement. Note that Eqn 12.3–5 states that the internal strains $\boldsymbol{\varepsilon}$ are zero when nodal displacements $\boldsymbol{\Delta}$ are zero. Therefore, since the nodes are restrained, **q** produces no strain anywhere in the element and consequently no strain energy. That is, as long as the nodes are held in position, no amount of load acting on the element can cause any deformation of the element or cause any stresses $\boldsymbol{\sigma}$ in the element (see Eqn 12.3–29 in Example 12.3–2).

On the other hand, if any node is displaced (e.g. node j in Fig. 12.3–3), then stresses $\boldsymbol{\sigma}$ will exist at all points of the element, as given by Eqn 12.3–12, including points such as 'a' on the boundary, even when the boundaries are in contact with the atmosphere. The existence of such stresses (which is impossible if the element is made of any real material such as steel) means that in general we cannot expect equilibrium of stresses along the boundaries of the elements.

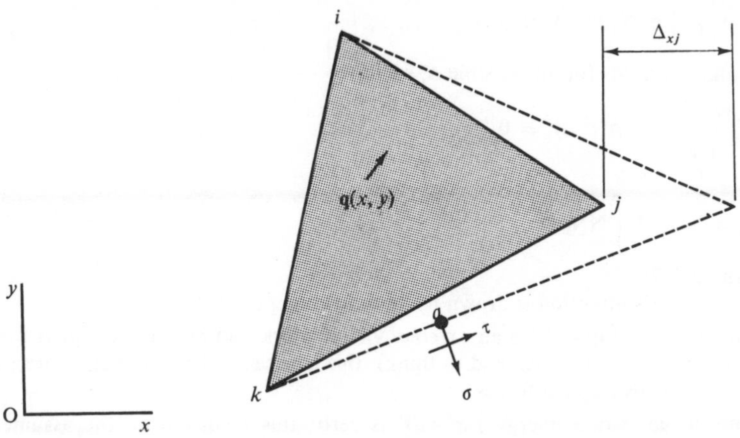

Fig. 12·3–3

It should be noted, however, that the nodal forces acting on any individual element are always in equilibrium (see Problem 12.9).

Example 12.3–2. Use the principle of virtual work to derive Eqn 12.3–25:

$$\mathbf{p}_0 = -\mathbf{A}^{-T} \int \mathbf{N}_q^T \mathbf{q} \, dV$$

SOLUTION With reference to the element ijk in Fig. 12.3–2, if virtual displacements $\overline{\boldsymbol{\Delta}}$ occur, then:

external virtual work done by the nodal forces $\mathbf{p}_0 = \mathbf{p}_0^T \overline{\boldsymbol{\Delta}}$

external virtual work done by the distributed forces $\mathbf{q}$

$$= \int \mathbf{q}^T \overline{\boldsymbol{\delta}} \, dV$$

$$= \int \mathbf{q}^T [\mathbf{N}_q \mathbf{A}^{-1} \overline{\boldsymbol{\Delta}}] \, dV \quad \text{(from Eqn 12.3–24)}$$

virtual strain energy due to the stresses $\boldsymbol{\sigma}$ caused by $\mathbf{q}$ is

$$\int \boldsymbol{\sigma}^T \overline{\boldsymbol{\varepsilon}} \, dV$$

where $\overline{\boldsymbol{\varepsilon}}$ are the virtual strains compatible with the virtual nodal displacements $\overline{\boldsymbol{\Delta}}$.

The principle of virtual work (Eqn 4.14–4) states that

$$\mathbf{p}_0^T \overline{\boldsymbol{\Delta}} + \int \mathbf{q}^T \mathbf{N}_q \mathbf{A}^{-1} \overline{\boldsymbol{\Delta}} \, dV = \int \boldsymbol{\sigma}^T \overline{\boldsymbol{\varepsilon}} \, dV \qquad (12.3\text{–}28)$$

As pointed out in Example 12.3–1, under fixed-node condition, the applied forces $\mathbf{q}$ cannot cause any internal stress $\boldsymbol{\sigma}$. Hence

$$\int \boldsymbol{\sigma}^T \bar{\boldsymbol{\varepsilon}}\, dV = \int 0\, \bar{\boldsymbol{\varepsilon}}\, dV = 0 \tag{12.3–29}$$

Hence Eqn 12.3–28 becomes

$$\mathbf{p}_0^T \overline{\boldsymbol{\Delta}} + \int \mathbf{q}^T \mathbf{N}_q \mathbf{A}^{-1} \overline{\boldsymbol{\Delta}}\, dV = 0$$

Since this equation holds for all possible $\overline{\boldsymbol{\Delta}}$, we have

$$\mathbf{p}_0^T + \int \mathbf{q}^T \mathbf{N}_q \mathbf{A}^{-1}\, dV = 0$$

or

$$\mathbf{p}_0 = -\mathbf{A}^{-T} \int \mathbf{N}_q^T \mathbf{q}\, dV$$

which is Eqn 12.3–25.

The reader's attention is drawn to the following points:

(a) The quantity $\int \mathbf{q}^T \bar{\boldsymbol{\delta}}\, dV$ is an *external* virtual work and not an *internal* virtual work (as many students tend to think); this is because the distributed forces $\mathbf{q}$ are externally applied forces.

(b) The virtual strain energy $\int \boldsymbol{\sigma}^T \bar{\boldsymbol{\varepsilon}}\, dV$ is zero; this results from the assumed properties of the fictitious element, as explained in Example 12.3–1. An element made of a real material cannot, obviously, behave this way.

12.4 Rectangular elements for plane stress

Figure 12.4–1 shows a rectangular element *ijkl*, of size $2a \times 2b$ and uniform thickness t. Choose local coordinate system as shown, with axes parallel to the sides of the rectangle and with the origin at the centroid of the element; the coordinates of the nodes are then as shown in the figure. As will be seen later, placing the origin at the centroid simplifies subsequent computation.

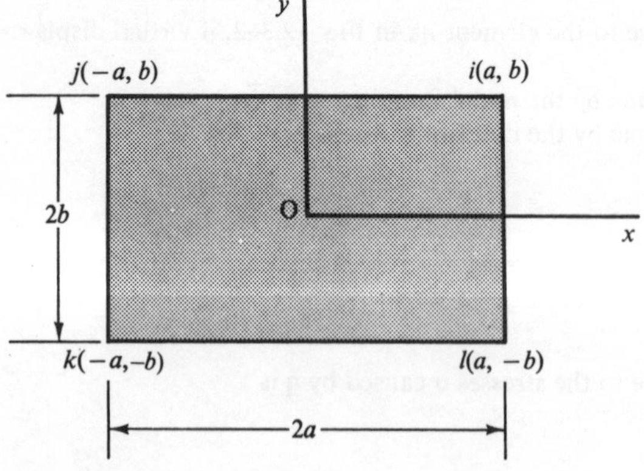

Fig. 12·4–1

For rectangular elements with sides parallel to the coordinate axes, it is known that a suitable displacement function is

$$\left.\begin{aligned} u &= \alpha_1 + \alpha_2 x + \alpha_3 y + \alpha_4 xy \\ v &= \alpha_5 + \alpha_6 x + \alpha_7 y + \alpha_8 xy \end{aligned}\right\} \tag{12.4–1}$$

As an exercise, the reader should show that this displacement function satisfies the four conditions in Section 12.3. Note, however, that if the sides of the element were not parallel to the coordinate axes, the condition of boundary compatibility (see condition 4) would not be satisfied.

From Eqn 12.4–1, we can express the nodal displacements in terms of the α's; by substituting in the local nodal coordinates, we obtain the following equation, which is similar to Eqn 12.3–3:

$$\boldsymbol{\Delta} = \begin{bmatrix} \Delta_{xi} \\ \Delta_{yi} \\ \Delta_{xj} \\ \Delta_{yj} \\ \Delta_{xk} \\ \Delta_{yk} \\ \Delta_{xl} \\ \Delta_{yl} \end{bmatrix} = \begin{bmatrix} 1 & a & b & ab & 0 & 0 & 0 & 0 \\ 0 & 0 & 0 & 0 & 1 & a & b & ab \\ 1 & -a & b & -ab & 0 & 0 & 0 & 0 \\ 0 & 0 & 0 & 0 & 1 & -a & b & -ab \\ 1 & -a & -b & ab & 0 & 0 & 0 & 0 \\ 0 & 0 & 0 & 0 & 1 & -a & -b & ab \\ 1 & a & -b & -ab & 0 & 0 & 0 & 0 \\ 0 & 0 & 0 & 0 & 1 & a & -b & -ab \end{bmatrix} \begin{bmatrix} \alpha_1 \\ \alpha_2 \\ \alpha_3 \\ \alpha_4 \\ \alpha_5 \\ \alpha_6 \\ \alpha_7 \\ \alpha_8 \end{bmatrix}$$

That is,

$$\boldsymbol{\Delta} = \mathbf{A}\boldsymbol{\alpha} \tag{12.4–2}$$

The strains at any point (x, y) in the element are:

$$\boldsymbol{\varepsilon} = \begin{bmatrix} \varepsilon_x \\ \varepsilon_y \\ \gamma_{xy} \end{bmatrix} = \begin{bmatrix} \dfrac{\partial u}{\partial x} \\ \dfrac{\partial v}{\partial y} \\ \dfrac{\partial v}{\partial x} + \dfrac{\partial u}{\partial y} \end{bmatrix} \quad \text{where } u \text{ and } v \text{ are given by Eqn 12.4–1}$$

$$= \begin{bmatrix} 0 & 1 & 0 & y & 0 & 0 & 0 & 0 \\ 0 & 0 & 0 & 0 & 0 & 0 & 1 & x \\ 0 & 0 & 1 & x & 0 & 1 & 0 & y \end{bmatrix} \begin{bmatrix} \alpha_1 \\ \alpha_2 \\ \alpha_3 \\ \alpha_4 \\ \alpha_5 \\ \alpha_6 \\ \alpha_7 \\ \alpha_8 \end{bmatrix}$$

That is,

$$\boldsymbol{\varepsilon} = \mathbf{B}\boldsymbol{\alpha}$$

$$= \mathbf{B}\mathbf{A}^{-1}\boldsymbol{\Delta} \quad \text{(from Eqn 12.4–2)} \tag{12.4–3}$$

Since the **B** matrix is a linear function of x and y, it follows that internal strains are also linear functions of x and y, and are not uniformly distributed, as they are in triangular elements (see Eqn 12.3–4).

The internal stresses are given by

$$\begin{aligned}\boldsymbol{\sigma} &= \mathbf{D}\boldsymbol{\varepsilon}\\ &= \mathbf{DBA}^{-1}\boldsymbol{\Delta} \quad \text{(from Eqn 12.4–3)}\end{aligned} \qquad (12.4\text{–}4)$$

where the matrix **D** is as defined in Eqn 12.3–12.

We can now derive an expression for the element stiffness matrix **K** by imposing virtual nodal displacements $\bar{\boldsymbol{\Delta}}$ and equating the virtual external work with the virtual strain energy, as we did for the triangular element in Section 12.3. Using the same arguments, we would end up with the following expression for the virtual strain energy (see Eqn 12.3–14):

$$\text{virtual strain energy} = \bar{\boldsymbol{\Delta}}^T\mathbf{A}^{-T}\left\{\int_V \mathbf{B}^T\mathbf{DB}\,dV\right\}\mathbf{A}^{-1}\boldsymbol{\Delta} \qquad (12.4\text{–}5)$$

Note that this time the matrices $\mathbf{B}^T$ and **B** are no longer independent of x and y—as were those for the triangular element in Eqn 12.3–14—and hence they cannot be taken out of the integral sign.

Equating the above expression for virtual strain energy with expression $\bar{\boldsymbol{\Delta}}^T\mathbf{K}\boldsymbol{\Delta}$ for virtual external work (see Eqn 12.3–16), and using the argument in the paragraph following Eqn 12.3–16, we have

$$\text{element stiffness matrix } \mathbf{K} = \mathbf{A}^{-T}\left\{\int_V \mathbf{B}^T\mathbf{DB}\,dV\right\}\mathbf{A}^{-1} \qquad (12.4\text{–}6)$$

where **A** is defined by Eqn 12.4–2; **B** is defined by Eqn 12.4–3; **D** is defined by Eqn 12.3–12; and

$$\int_V \mathbf{B}^T\mathbf{DB}\,dV$$

is to be integrated over the entire volume V of the element *ijkl*.

In the solution of practical problems, the element stiffness matrix **K** is generated by the computer. However, unless a sophisticated computer program is available, it is necessary first to determine the explicit form of $\int_V \mathbf{B}^T\mathbf{DB}\,dV$ by hand. Once the explicit form is known, the matrix operations in Eqn 12.4–6 can be carried out by standard computer procedure. As an exercise the reader should verify that

$$\int_V \mathbf{B}^T\mathbf{DB}\,dV = t\int_{-b}^{b}\int_{-a}^{a} \mathbf{B}^T\mathbf{DB}\,dx\,dy$$

$$= \frac{4abtE}{1-\nu^2}\begin{array}{c}\begin{array}{cccccccc} 1 & 2 & 3 & 4 & 5 & 6 & 7 & 8 \end{array}\\ \left[\begin{array}{cccccccc} 0 & 0 & 0 & 0 & 0 & 0 & 0 & 0 \\ & 1 & 0 & 0 & 0 & 0 & \nu & 0 \\ & & \dfrac{1-\nu}{2} & 0 & 0 & \dfrac{1-\nu}{2} & 0 & 0 \\ & & & \dfrac{(1-\nu)a^2+2b^2}{6} & 0 & 0 & 0 & 0 \\ & & & & 0 & 0 & 0 & 0 \\ & \textit{symmetrical} & & & & \dfrac{1-\nu}{2} & 0 & 0 \\ & & & & & & 1 & 0 \\ & & & & & & & \dfrac{2a^2+(1-\nu)b^2}{6} \end{array}\right]\end{array}\begin{array}{c} 1\\2\\3\\4\\5\\6\\7\\8\end{array} \qquad (12.4\text{–}7)$$

The many zero elements in the above matrix simplify input of computer data: they result from the choice of the element centroid as the origin of coordinates. If we had chosen some other point as origin, say a corner of the rectangle, we would not have obtained so many zero elements.

Distributed load on element. For a rectangular element with nodes restrained against displacement, the nodal forces $\mathbf{p}_0$ produced by an internal distributed load $\mathbf{q}$ are evidently given by an equation of the same form as Eqn 12.3–25:

$$\mathbf{p}_0 = -\mathbf{A}^{-T}\int_V \mathbf{N}_q^T\mathbf{q}\,dV \tag{12.4–8}$$

where $\mathbf{p}_0$ now has eight components; $\mathbf{A}$ is given by Eqn 12.4–2; and $\mathbf{N}_q$ is now (see Eqn 12.4–1) defined by

$$\mathbf{N}_q = \begin{bmatrix} 1 & x & y & xy & 0 & 0 & 0 & 0 \\ 0 & 0 & 0 & 0 & 1 & x & y & xy \end{bmatrix} \tag{12.4–9}$$

Point loads on element. For a rectangular element with nodes restrained against displacement, the nodal forces $\mathbf{p}_0$ produced by n point loads $\mathbf{Q}$ are evidently given by an equation of the same form as Eqn 12.3–27, namely

$$\mathbf{p}_0 = -\mathbf{A}^{-T}\mathbf{N}_Q^T\mathbf{Q} \tag{12.4–10}$$

where $\mathbf{p}_0$ now has eight components, $\mathbf{A}$ is given by Eqn 12.4–2, and

$$\mathbf{N}_Q = \begin{bmatrix} 1 & x_1 & y_1 & x_1y_1 & 0 & 0 & 0 & 0 \\ 0 & 0 & 0 & 0 & 1 & x_1 & y_1 & x_1y_1 \\ 1 & x_2 & y_2 & x_2y_2 & 0 & 0 & 0 & 0 \\ 0 & 0 & 0 & 0 & 1 & x_2 & y_2 & x_2y_2 \\ \vdots & & & & & & & \\ 1 & x_n & y_n & x_ny_n & 0 & 0 & 0 & 0 \\ 0 & 0 & 0 & 0 & 1 & x_n & y_n & x_ny_n \end{bmatrix}$$

12.5 Transformation matrix

In Chapters 5 and 7, Sections 5.3 and 7.6, we saw that, for some members of framed structures, the local coordinate axes for the members were not in the same directions as the structure coordinate axes (or global coordinate axes, as some authors call them), and we derived transformation matrices to transform member stiffness matrices for local coordinates to member stiffness matrices for structure coordinates.

In finite element work, where the local coordinate axes are not in the same directions as the structure coordinate axes, it is also necessary to derive a transformation matrix $\mathbf{T}$ to express the element stiffness matrix in terms of structure coordinates, before we can assemble the overall stiffness matrix for the structure. Of the two types of elements —triangular and rectangular—which we have studied so far, transformation matrices are required only for rectangular elements. The reason is that, with triangular elements (Fig. 12.3–1) we can always select local coordinate axes which coincide with the structure coordinate axes; the displacement function in Eqn 12.3–1 will always satisfy the four conditions in Section 12.3 irrespective of the orientation of the local coordinate axes. However, this is not so with the rectangular element; the displacement function in Eqn 12.4–1 would not satisfy the condition of boundary compatibility (see condition 4, Section 12.3) unless the local coordinate axes are parallel to the sides of the element.

Hence the orientation of the element dictates the orientation of the local coordinate axes, and a transformation matrix is required when these axes are not in the same directions as the structure axes.

We shall derive the transformation matrix for the rectangular element in Fig. 12.5–1, in which x and y are the local coordinate axes and x' and y' are the structure coordinate axes.

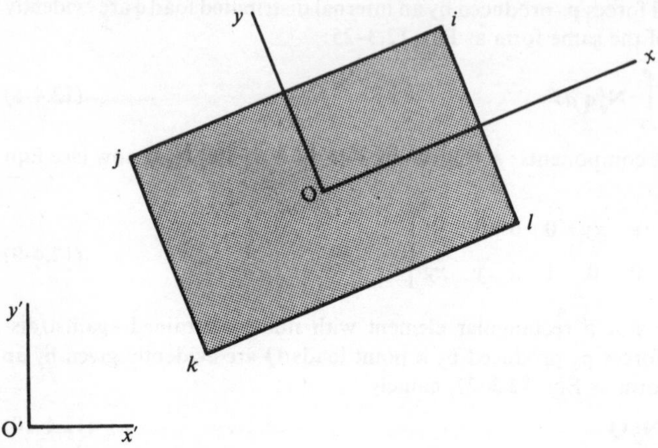

Fig. 12·5–1

Referring to local coordinates, the nodal forces $\mathbf{p}$ and the nodal displacements $\mathbf{\Delta}$ are related by

$$\begin{bmatrix} p_{xi} \\ p_{yi} \\ p_{xj} \\ p_{yj} \\ p_{yk} \\ p_{yk} \\ p_{xl} \\ p_{yl} \end{bmatrix} = \begin{bmatrix} K_{xi,\,xi} & K_{xi,\,yi} & K_{xi,\,xj} & \cdots & K_{xi,\,yl} \\ K_{yi,\,xi} & K_{yi,\,yi} & K_{yi,\,xj} & \cdots & K_{yi,\,yl} \\ & & & & \\ \vdots & & & & \\ & & & & \\ K_{yl,\,xi} & K_{yl,\,yi} & K_{yl,\,xj} & \cdots & K_{yl,\,yl} \end{bmatrix} \begin{bmatrix} \Delta_{xi} \\ \Delta_{yi} \\ \Delta_{xj} \\ \Delta_{yj} \\ \Delta_{xk} \\ \Delta_{yk} \\ \Delta_{xl} \\ \Delta_{yl} \end{bmatrix}$$

That is,

$$\mathbf{p} = \mathbf{K}\mathbf{\Delta} \tag{12.5–1}$$

Let p'_{xi} and p'_{yi} be the components of the nodal force at i in the directions of x' and y' respectively, i.e. p'_{xi} is the sum of the projections of p_{xi} and p_{yi} in the x'-direction, and p'_{yi} is the sum of the projections of p_{xi} and p_{yi} in the y'-direction. From Eqn 5.3–2, we have

$$\begin{bmatrix} p'_{xi} \\ p'_{yi} \end{bmatrix} = \begin{bmatrix} l_1 & m_1 \\ l_2 & m_2 \end{bmatrix}^T \begin{bmatrix} p_{xi} \\ p_{yi} \end{bmatrix} \tag{12.5–2}$$

where l_1 and m_1 are the direction cosines of x with respect to x' and y', and l_2, m_2 are those of y.

Equation 12.5–2, together with three similar equations at the other three nodes, can be grouped together in a single equation:

$$\mathbf{p}' = \mathbf{T}^T\mathbf{p} \tag{12.5–3}$$

where

$$\mathbf{T} = \begin{bmatrix} l_1 & m_1 & 0 & 0 & 0 & 0 & 0 & 0 \\ l_2 & m_2 & 0 & 0 & 0 & 0 & 0 & 0 \\ 0 & 0 & l_1 & m_1 & 0 & 0 & 0 & 0 \\ 0 & 0 & l_2 & m_2 & 0 & 0 & 0 & 0 \\ 0 & 0 & 0 & 0 & l_1 & m_1 & 0 & 0 \\ 0 & 0 & 0 & 0 & l_2 & m_2 & 0 & 0 \\ 0 & 0 & 0 & 0 & 0 & 0 & l_1 & m_1 \\ 0 & 0 & 0 & 0 & 0 & 0 & l_2 & m_2 \end{bmatrix}$$

The matrix $\mathbf{T}$ is called the **transformation matrix**; it transforms nodal force components in the x'- and y'-directions into statically equivalent components in the x- and y-directions.

By similar reasoning, we can relate the nodal displacements $\mathbf{\Delta}'$ in the x'- and y'-directions to nodal displacements $\mathbf{\Delta}$ in the x- and y-directions:

$$\mathbf{\Delta}' = \mathbf{T}^T\mathbf{\Delta} \tag{12.5–4}$$

where the transformation matrix $\mathbf{T}$ is identical to that in Eqn 12.5–3.

If we now pre-multiply Eqn 12.5–1 by $\mathbf{T}^T$, we have

$$\begin{aligned} \mathbf{T}^T\mathbf{p} &= \mathbf{T}^T\mathbf{K}\mathbf{\Delta} \\ &= \mathbf{T}^T\mathbf{K}\mathbf{T}\mathbf{T}^T\mathbf{\Delta} \end{aligned} \tag{12.5–5}$$

In Eqn 12.5–5 we are entitled to insert $\mathbf{TT}^T$ between $\mathbf{K}$ and $\mathbf{\Delta}$ because, as the reader should himself verify, the product $\mathbf{TT}^T$ is a unit matrix. (See Section 5.3.) Also, from Eqn 12.5–3, $\mathbf{T}^T\mathbf{p} = \mathbf{p}'$; from Eqn 12.5–4, $\mathbf{T}^T\mathbf{\Delta} = \mathbf{\Delta}'$. Hence, Eqn 12.5–5 may be written

$$\mathbf{p}' = \mathbf{T}^T\mathbf{K}\mathbf{T}\mathbf{\Delta}' \tag{12.5–6}$$

Since for structure axes, nodal forces $\mathbf{p}'$ and nodal displacements $\mathbf{\Delta}'$ are related by

$$\mathbf{p}' = \mathbf{K}'\mathbf{\Delta}' \tag{12.5–7}$$

in which $\mathbf{K}'$ is the element stiffness matrix for structure axes, it is readily seen by comparing Eqns 12.5–6 and 12.5–7 that

$$\mathbf{K}' = \mathbf{T}^T\mathbf{K}\mathbf{T} \tag{12.5–8}$$

In practical problems, $\mathbf{K}'$ is generated by standard computer procedure from Eqns 12.5–8 and 12.4–6. The direction cosines in the transformation matrix $\mathbf{T}$ are determined from the structure coordinates of the nodes of the element. Referring to Fig.

12.5–1, let the structure coordinates of the nodes i and j (i.e. their coordinates with respect to structure axes $x'y'$) be (x'_i, y'_i) and (x'_j, y'_j). Then

$$\begin{aligned} l_1 &= \frac{x'_i - x'_j}{\sqrt{[(x'_i - x'_j)^2 + (y'_i - y'_j)^2]}} \\ m_1 &= \frac{y'_i - y'_j}{\sqrt{[(x'_i - x'_j)^2 + (y'_i - y'_j)^2]}} \\ l_2 &= -m_1 \\ m_2 &= l_1 \end{aligned} \tag{12.5–9}$$

12.6 Assembling the structure stiffness matrix

In Chapter 7, Section 7.9, we explained how to assemble the structure stiffness matrix of a framed structure from the member stiffness matrices of the individual members. In finite element work, the structure stiffness matrix is assembled in essentially the same way from the element stiffness matrices.

Consider, for example, a structure such as the plate in Fig. 12.6–1. For illustration purposes, the structure is idealized into 11 elements simply connected at the nodes; in practice, of course, a much larger number of elements would be used, in order to obtain more meaningful results. Also, in Fig. 12.6–1, a combination of triangular and rectangular elements is used, to illustrate the procedure when such a combination occurs. All the nodes and elements are numbered; the nodes are numbered consecutively from 1 to 14 and the elements consecutively from 1 to 11. Here the nodes and the elements have been deliberately numbered in a haphazard manner in order to emphasize that in principle the order of numbering is immaterial, even though in practice we naturally prefer to number them in a systematic pattern whenever possible.†

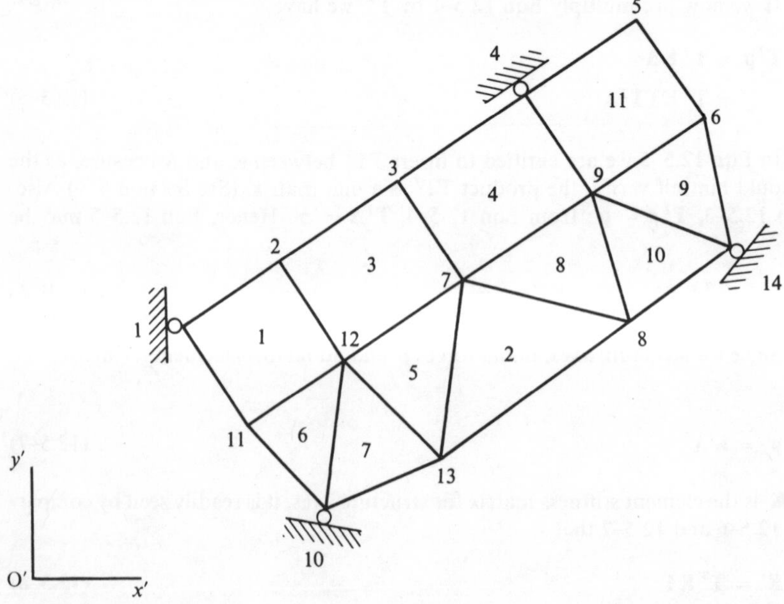

Fig. 12·6–1

† See also Section 13.4 on the band width of the stiffness matrix.

Since there are fourteen nodes, each having two nodal force components (in the x'- and y'-directions) the overall stiffness matrix $\mathbf{K}_s$ for the structure is of order 28 × 28. That is

$$\mathbf{K}_s = \begin{bmatrix} (K_s)_{x'1,x'1} & (K_s)_{x'1,y'1} & (K_s)_{x'1,x'2} & \cdots & (K_s)_{x'1,y'14} \\ (K_s)_{y'1,x'1} & (K_s)_{y'1,y'1} & \cdots & \cdots & (K_s)_{y'1,y'14} \\ (K_s)_{x'2,x'1} & (K_s)_{x'2,y'1} & \cdots & & \\ \vdots & & & & \\ (K_s)_{y'14,x'1} & (K_s)_{y'14,y'1} & \cdots & \cdots & (K_s)_{y'14,y'14} \end{bmatrix} \quad (12.6\text{–}1)$$

The 784 (i.e. 28 × 28) nodal stiffnesses of the structure stiffness matrix $\mathbf{K}_s$ are built up of contributions from the element stiffness matrices. For rectangular elements, the element stiffness matrices must be those for structure axes and not those for local coordinate axes, i.e. $\mathbf{K}'$ and not $\mathbf{K}$ should be used. For triangular elements, we can always select the structure coordinate axes as local coordinate axes, so that $\mathbf{K}'$ and $\mathbf{K}$ are identical.

To explain the way in which the individual nodal stiffnesses in $\mathbf{K}_s$ are built up, consider the nodal stiffnesses associated with displacements at node 7. A magnified view of the elements connected at node 7 is shown in Fig. 12.6–2, in which for clarity the elements are drawn separated from one another. Nodal stiffnesses such as $(K_s)_{x'2,x'7}$ and

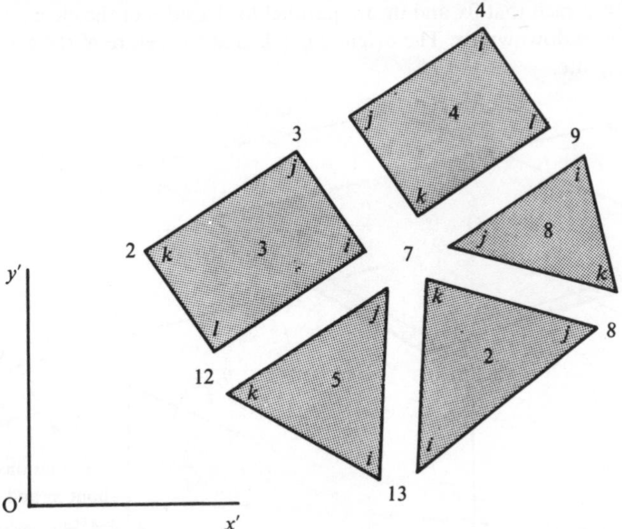

Fig. 12·6–2

$(K_s)_{x'2,y'7}$ are made up of contributions from the element stiffness matrix $\mathbf{K}'_3$ of element 3; $(K_s)_{y'3,x'7}$ and $(K_s)_{x'3,x'7}$ are made up of contributions from the element stiffness matrices $\mathbf{K}'_3$ and $\mathbf{K}'_4$ of elements 3 and 4; and $(K_s)_{x'7,x'7}$ or $(K_s)_{x'7,y'7}$ is each made up of contributions from element stiffness matrices of all the elements meeting at node 7, namely, elements 2, 3, 4, 5, and 8. Specifically,

$$\begin{aligned}
(K_s)_{x'2,x'7} &= (K'_{x'k,x'i})_3 \\
(K_s)_{x'2,y'7} &= (K'_{x'k,y'i})_3 \\
(K_s)_{y'3,x'7} &= (K'_{y'j,x'i})_3 + (K'_{y'j,x'k})_4 \\
(K_s)_{x'3,x'7} &= (K'_{x'j,x'i})_3 + (K'_{x'j,x'k})_4 \\
(K_s)_{x'7,x'7} &= (K'_{x'k,x'k})_2 + (K'_{x'i,x'i})_3 + (K'_{x'k,x'k})_4 + (K'_{x'j,x'j})_5 + (K'_{x'j,x'j})_8 \\
(K_s)_{x'7,y'7} &= (K'_{x'k,y'k})_2 + (K'_{x'i,y'i})_3 + (K'_{x'k,y'k})_4 + (K'_{x'j,y'j})_5 + (K'_{x'j,y'j})_8
\end{aligned} \quad (12.6\text{–}2)$$

and so on.

Referring to Fig. 12.6–1, if node 7 alone is displaced while all other nodes are held in position, then nodal forces occur only in nodes 2, 3, 4, 12, 7, 9, 13, and 8; hence, nodal stiffnesses such as $(K_s)_{x'11,x'7}$, $(K_s)_{y'6,x'7}$, etc., are zero. In other words, only those elements of the structure which have a node at node 7 will contribute towards the nodal stiffnesses associated with displacements of node 7.

The above explanation refers to node 7; but evidently the same method can be used to build up other nodal stiffnesses in the structure stiffness matrix.

Before we conclude this section, it may be helpful to discuss briefly the nodal reference letters *i*, *j*, *k*, etc., for the individual elements in Fig. 12.6–2. For a triangular element, the reference letters *i*, *j*, *k* can be designated to the three nodes in any way we like. For a rectangular element, the reference letter *i* can be arbitrarily designated to any node, but once *i* has been assigned to a node, it is desirable to proceed anticlockwise to assign *j*, *k*, and *l* to the remaining nodes, in order to preserve the convention for local co-ordinates as shown in Fig. 12.5–1; otherwise Eqn 12.5–9 for direction cosines will not be applicable.

12.7 Rectangular elements in bending

Consider a thin rectangular plate element *ijkl* of uniform thickness *t* (Fig. 12.7–1). Select local coordinate system 0*xyz* such that 0*x* and 0*y* are parallel to the sides of the element and the positive *z*-direction is downwards. The origin 0 is taken at the centre of the element, at mid-depth of the plate.

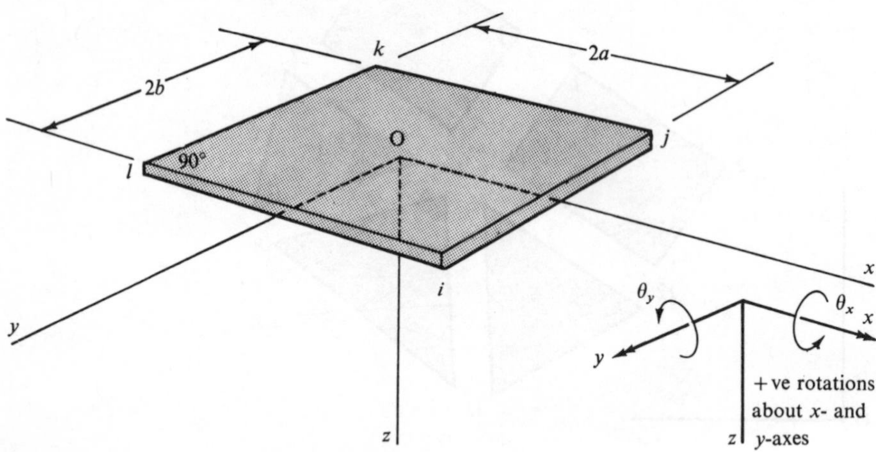

Fig. 12·7–1

It has been found[4] that a suitable displacement function is

$$w = \alpha_1 + \alpha_2 x + \alpha_3 y + \alpha_4 x^2 + \alpha_5 xy + \alpha_6 y^2 + \alpha_7 x^3 + \alpha_8 x^2 y$$
$$+ \alpha_9 xy^2 + \alpha_{10} y^3 + \alpha_{11} x^3 y + \alpha_{12} xy^3 \qquad (12.7\text{–}1)$$

where *w* is the deflection of the plate in the positive *z*-direction.

The displacements at each node are taken as the rotation about the *x*-axis, the rotation about the *y*-axis, and the deflection *w*. That is, at node *i*,

$$\mathbf{\Delta}_i = \begin{bmatrix} \theta_{xi} \\ \theta_{yi} \\ w_i \end{bmatrix} \qquad (12.7\text{–}2)$$

From Chapter 11 (Eqn 11.13–3) we have

$$\theta_x = +\frac{\partial w}{\partial y}; \qquad \theta_y = -\frac{\partial w}{\partial x} \tag{12.7–3}$$

The displacements at all four nodes of the plate element are denoted by

$$\boldsymbol{\Delta} = \begin{bmatrix} \boldsymbol{\Delta}_i \\ \\ \boldsymbol{\Delta}_j \\ \\ \boldsymbol{\Delta}_k \\ \\ \boldsymbol{\Delta}_l \end{bmatrix} = \begin{bmatrix} \theta_{xi} \\ \theta_{yi} \\ w_i \\ \theta_{xj} \\ \theta_{yj} \\ w_j \\ \theta_{xk} \\ \theta_{yk} \\ w_k \\ \theta_{xl} \\ \theta_{yl} \\ w_l \end{bmatrix} = \begin{bmatrix} \left(\dfrac{\partial w}{\partial y}\right)_i \\ -\left(\dfrac{\partial w}{\partial x}\right)_i \\ w_i \\ \\ \vdots \\ \\ \left(\dfrac{\partial w}{\partial y}\right)_l \\ -\left(\dfrac{\partial w}{\partial x}\right)_l \\ w_l \end{bmatrix} \tag{12.7–4}$$

Substituting in the displacement function for w in Eqn 12.7–1, we have

$$\boldsymbol{\Delta} = \mathbf{A}\boldsymbol{\alpha} \tag{12.7–5}$$

which is expressed in explicit form in Table 12.7–1.

Table 12.7–1

$$\boldsymbol{\Delta} = \begin{bmatrix} \theta_{xi} \\ \theta_{yi} \\ w_i \\ \theta_{xj} \\ \theta_{yj} \\ w_j \\ \theta_{xk} \\ \theta_{yk} \\ w_k \\ \theta_{xl} \\ \theta_{yl} \\ w_l \end{bmatrix} = \begin{bmatrix} \overset{1}{0} & \overset{2}{0} & \overset{3}{1} & \overset{4}{0} & \overset{5}{x_i} & \overset{6}{2y_i} & \overset{7}{0} & \overset{8}{x_i^2} & \overset{9}{2x_iy_i} & \overset{10}{3y_i^2} & \overset{11}{x_i^3} & \overset{12}{3x_iy_i^2} \\ 0 & -1 & 0 & -2x_i & -y_i & 0 & -3x_i^2 & -2x_iy_i & -y_i^2 & 0 & -3x_i^2y_i & -y_i^3 \\ 1 & x_i & y_i & x_i^2 & x_iy_i & y_i^2 & x_i^3 & x_i^2y_i & x_iy_i^2 & y_i^3 & x_i^3y_i & x_iy_i^3 \\ & & & & & & & & & & & \\ & & & & & \multicolumn{3}{c}{\text{As above, but use } x_j,\, y_j} & & & & \\ & & & & & & & & & & & \\ & & & & & & & & & & & \\ & & & & & \multicolumn{3}{c}{\text{As above, but use } x_k,\, y_k} & & & & \\ & & & & & & & & & & & \\ & & & & & & & & & & & \\ & & & & & \multicolumn{3}{c}{\text{As above, but use } x_l,\, y_l} & & & & \\ & & & & & & & & & & & \end{bmatrix} \begin{bmatrix} \alpha_1 \\ \alpha_2 \\ \alpha_3 \\ \alpha_4 \\ \alpha_5 \\ \alpha_6 \\ \alpha_7 \\ \alpha_8 \\ \alpha_9 \\ \alpha_{10} \\ \alpha_{11} \\ \alpha_{12} \end{bmatrix}$$

Hence

$$\boldsymbol{\alpha} = \mathbf{A}^{-1}\boldsymbol{\Delta} \tag{12.7–6}$$

The nodal forces are denoted by

$$\mathbf{p} = \begin{bmatrix} \mathbf{p}_i \\ \mathbf{p}_j \\ \mathbf{p}_k \\ \mathbf{p}_l \end{bmatrix} = \begin{bmatrix} T_{xi} \\ T_{yi} \\ p_{zi} \\ \vdots \\ T_{xl} \\ T_{yl} \\ p_{zl} \end{bmatrix} \tag{12.7–7}$$

where T_{xi} is the moment about the x-axis acting on the element at node i; T_{yi} is that about the y-axis; p_{zi} is the force in the z-direction at node i.

Note that since moments T_{xi}, T_{yi}, etc., act at nodal points, it is immaterial whether they are regarded as twisting moments or as bending moments. For example, T_{xi} may be interpreted as the bending moment about the x-axis acting on the face li (Fig. 12.7–1) at node i, or it may equally be interpreted as the twisting moment about the x-axis acting on face ij at node i.

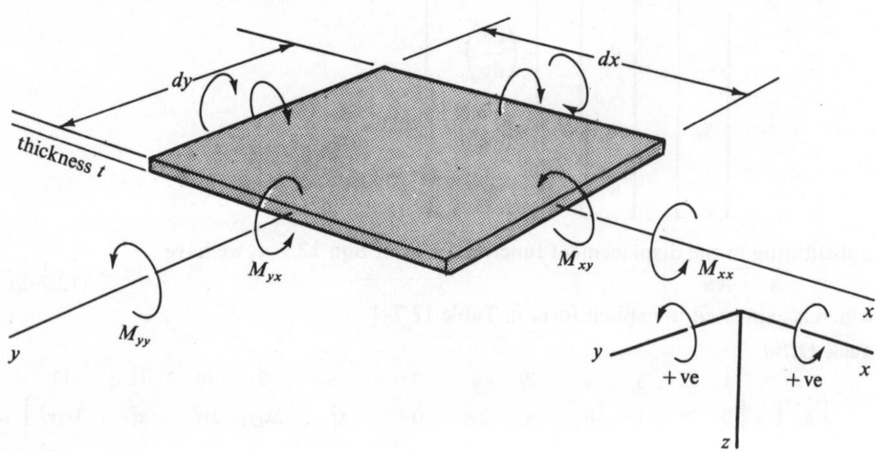

Fig. 12·7–2

Next consider the internal bending moments and twisting moments within the element $ijkl$. Figure 12.7–2 shows an element $dx\,dy$ within the element $ijkl$ in Fig. 12.7–1. In the figure,

- M_{yx} represents the bending moment per unit width acting on a face normal to the y-axis, tending to produce a positive rotation about the x-axis.
- M_{xy} represents the bending moment per unit width acting on a face normal to the x-axis, tending to produce a positive rotation about the y-axis.
- M_{xx} represents the twisting moment per unit width acting on a face normal to the x-axis, tending to produce a positive rotation about the x-axis.
- M_{yy} is similarly defined.

In Fig. 12.7–2 all moments labelled are shown acting in the positive directions, as defined by the conventional right-hand-screw rule for vectors. Eqns 11.13–8 in Chapter 11 show that

$$M_{yx} = -\int_{-t/2}^{t/2} \sigma_y z\,dz = +\frac{Et^3}{12(1-\nu^2)}\left\{\frac{\partial^2 w}{\partial y^2} + \nu\frac{\partial^2 w}{\partial x^2}\right\}$$

$$M_{xy} = +\int_{-t/2}^{t/2} \sigma_x z\,dz = -\frac{Et^3}{12(1-\nu^2)}\left\{\nu\frac{\partial^2 w}{\partial y^2} + \frac{\partial^2 w}{\partial x^2}\right\}$$

$$M_{xx} = -\int_{-t/2}^{t/2} \tau_{xy} z\,dz = +\frac{Et^3}{12(1-\nu^2)}(1-\nu)\frac{\partial^2 w}{\partial x \partial y}$$

$$M_{yy} = +\int_{-t/2}^{t/2} \tau_{yx} z\,dz = -\frac{Et^3}{12(1-\nu^2)}(1-\nu)\frac{\partial^2 w}{\partial x \partial y}$$

$$= -M_{xx} \tag{12.7–8}$$

That is

$$\begin{bmatrix} M_{yx} \\ M_{xy} \\ M_{xx} \end{bmatrix} = \frac{Et^3}{12(1-\nu^2)}\begin{bmatrix} 1 & -\nu & 0 \\ -\nu & 1 & 0 \\ 0 & 0 & \frac{1-\nu}{2} \end{bmatrix}\begin{bmatrix} \frac{\partial^2 w}{\partial y^2} \\ -\frac{\partial^2 w}{\partial x^2} \\ 2\frac{\partial^2 w}{\partial x\,\partial y} \end{bmatrix}$$

that is

$$\mathbf{M} = \mathbf{D}\boldsymbol{\kappa} \tag{12.7–9}$$

The matrix $\boldsymbol{\kappa}$ is called the **curvature matrix** for element $dx\,dy$, because

$$\frac{\partial^2 w}{\partial y^2} = \frac{\partial}{\partial y}(\theta_x \text{ of Eqn 12.7–3}) = \text{bending curvature about } x\text{-axis}$$

$$-\frac{\partial^2 w}{\partial x^2} = \frac{\partial}{\partial x}(\theta_y \text{ of Eqn 12.7–3}) = \text{bending curvature about } y\text{-axis}$$

$$2\frac{\partial^2 w}{\partial x\,\partial y} = \text{absolute magnitude of } \frac{\partial}{\partial x}(\theta_x) + \text{absolute magnitude of } \frac{\partial}{\partial y}(\theta_y)$$

i.e. sum of absolute magnitudes of the twisting curvatures about the x- and y-axes.

Since rotation is equal to curvature × arc length, we see that for element $dx\,dy$,

$$\text{bending rotation about } x\text{-axis} = \frac{\partial^2 w}{\partial y^2}dy$$

$$\text{bending rotation about } y\text{-axis} = -\frac{\partial^2 w}{\partial x^2}dx$$

$$\text{twisting rotation about } x\text{-axis} = \frac{\partial^2 w}{\partial x\,\partial y}dx$$

$$\text{twisting rotation about } y\text{-axis} = -\frac{\partial^2 w}{\partial y\,\partial x}dy$$

$$= -\frac{\partial^2 w}{\partial x\,\partial y}dy$$

Remembering that M_{yx}, etc., are moments per unit length, we see that the moments acting on $dx\,dy$ in Fig. 12.7–2 are:

$$M_{yx}\,dx, \qquad M_{xy}\,dy, \qquad M_{xx}\,dy, \qquad M_{yy}\,dx$$

The strain energy in the element $dx\,dy$ is the work done by these moments, i.e.

$$\frac{1}{2}\left\{(M_{yx}\,dx)\left(\frac{\partial^2 w}{\partial y^2}\,dy\right) + (M_{xy}\,dy)\left(-\frac{\partial^2 w}{\partial x^2}\,dx\right)\right.$$
$$\left. + (M_{xx}\,dy)\left(\frac{\partial^2 w}{\partial x\,\partial y}\,dx\right) + (M_{yy}\,dx)\left(-\frac{\partial^2 w}{\partial x\,\partial y}\,dy\right)\right\}$$

Noting from Eqn 12.7–8 that $M_{yy} = -M_{xx}$, we see that with $\mathbf{M}$ and $\boldsymbol{\kappa}$ as defined in Eqn 12.7–9, we have

$$\text{strain energy in element } dx\,dy = \tfrac{1}{2}\boldsymbol{\kappa}^T\mathbf{M}\,dx\,dy \tag{12.7–10}$$

Referring to Eqn 12.7–9, we can express the curvature matrix $\boldsymbol{\kappa}$ in terms of $\boldsymbol{\alpha}$, i.e.

$$\boldsymbol{\kappa} = \mathbf{B}\boldsymbol{\alpha} \tag{12.7–11}$$

where $\mathbf{B}$ is obtained by differentiating Eqn 12.7–1, i.e.

$$\mathbf{B}\boldsymbol{\alpha} = \begin{bmatrix} \dfrac{\partial^2 w}{\partial y^2} \\ -\dfrac{\partial^2 w}{\partial x^2} \\ 2\dfrac{\partial^2 w}{\partial x\,\partial y} \end{bmatrix} = \overset{\begin{matrix}1 & 2 & 3 & 4 & 5 & 6 & 7 & 8 & 9 & 10 & 11 & 12\end{matrix}}{\begin{bmatrix} 0 & 0 & 0 & 0 & 0 & 2 & 0 & 0 & 2x & 6y & 0 & 6xy \\ 0 & 0 & 0 & -2 & 0 & 0 & -6x & -2y & 0 & 0 & -6xy & 0 \\ 0 & 0 & 0 & 0 & 2 & 0 & 0 & 4x & 4y & 0 & 6x^2 & 6y^2 \end{bmatrix}} \begin{bmatrix} \alpha_1 \\ \alpha_2 \\ \alpha_3 \\ \alpha_4 \\ \vdots \\ \alpha_{12} \end{bmatrix} \tag{12.7–12}$$

From Eqns 12.7–11 and 12.7–6 we have

$$\boldsymbol{\kappa} = \mathbf{B}\mathbf{A}^{-1}\boldsymbol{\Delta} \tag{12.7–13}$$

and substituting this into Eqn 12.7–9, we have

$$\mathbf{M} = \mathbf{D}\mathbf{B}\mathbf{A}^{-1}\boldsymbol{\Delta} \tag{12.7–14}$$

We are now in a position to derive an expression for the element stiffness matrix $\mathbf{K}$ defined for the rectangular bending element by

$$\mathbf{p} = \mathbf{K}\boldsymbol{\Delta} \tag{12.7–15}$$

where $\mathbf{p}$ represents the nodal forces (Eqn 12.7–7) and $\boldsymbol{\Delta}$ represents the nodal displacements (Eqn 12.7–4).

Imagine that the element *ijkl* in Fig. 12.7–1 is subjected to nodal displacements $\boldsymbol{\Delta}$, which result in nodal forces $\mathbf{p}$ as given by Eqn 12.7–15, and internal moments $\mathbf{M}$ as given by Eqn 12.7–14. Next impose virtual nodal displacements $\overline{\boldsymbol{\Delta}}$, the 12 components of which can be arbitrary.

The virtual external work done by the real nodal forces is

$$\overline{\boldsymbol{\Delta}}^T\mathbf{p} = \overline{\boldsymbol{\Delta}}^T\mathbf{K}\boldsymbol{\Delta} \tag{12.7–16}$$

The virtual nodal displacements $\overline{\boldsymbol{\Delta}}$ are compatible with virtual internal curvatures given by Eqn 12.7–13, i.e.

$$\overline{\boldsymbol{\kappa}} = \mathbf{B}\mathbf{A}^{-1}\overline{\boldsymbol{\Delta}} \tag{12.7–17}$$

Hence from Eqn 12.7–10, the virtual strain energy in the element *ijkl* of Fig. 12.7–1 is

$$\int_{-a}^{+a}\int_{-b}^{+b} \bar{\boldsymbol{\kappa}}^T \mathbf{M}\,dx\,dy = \int_{-a}^{+a}\int_{-b}^{+b} \{\bar{\boldsymbol{\Delta}}^T \mathbf{A}^{-T}\mathbf{B}^T\}\{\mathbf{D}\mathbf{B}\mathbf{A}^{-1}\boldsymbol{\Delta}\}\,dx\,dy$$

(using Eqns 12.7–17 and 12.7–14)

$$= \bar{\boldsymbol{\Delta}}^T \mathbf{A}^{-T}\left\{\int_{-a}^{+a}\int_{-b}^{+b} \mathbf{B}^T\mathbf{D}\mathbf{B}\,dx\,dy\right\}\mathbf{A}^{-1}\boldsymbol{\Delta} \qquad (12.7\text{–}18)$$

Equating Eqn 12.7–16 with Eqn 12.7–18 and noting that the virtual displacements $\bar{\boldsymbol{\Delta}}$ are arbitrary (for example, the components of $\bar{\boldsymbol{\Delta}}$ can be all equal to unity), we see that

$$\mathbf{K} = \mathbf{A}^{-T}\left\{\int_{-a}^{+a}\int_{-b}^{+b} \mathbf{B}^T\mathbf{D}\mathbf{B}\,dx\,dy\right\}\mathbf{A}^{-1} \qquad (12.7\text{–}19)$$

where, of course, **A**, **B**, and **D** are respectively given by Eqns 12.7–5, 12.7–12, and 12.7–9.

The matrix

$$\int_{-a}^{+a}\int_{-b}^{+b} \mathbf{B}^T\mathbf{D}\mathbf{B}\,dx\,dy$$

is given in explicit form in Table 12.7–2.

Now that the explicit forms of both the matrix **A** and the matrix $\iint \mathbf{B}^T\mathbf{D}\mathbf{B}\,dx\,dy$ are known, standard computer procedures can be used to generate the element stiffness matrix **K** in Eqn 12.7–19.

Note that the matrix $\iint \mathbf{B}^T\mathbf{D}\mathbf{B}\,dx\,dy$ contains very few non-zero elements; this is because the origin of the local coordinates is chosen at the centroid of the element (Fig. 12.7–1). If, for example, the origin is placed at node k, say, then the matrix $\iint \mathbf{B}^T\mathbf{D}\mathbf{B}\,dx\,dy$ will contain more non-zero elements.

Compatibility. The displacement function in Eqn 12.7–1 satisfies the first three of the four conditions in Section 12.3, but does not fully satisfy condition 4, in that complete compatibility cannot be maintained along element boundaries.

Equation 12.7–1 shows that along a boundary, x or y is constant. For example, along boundary *ij* (Fig. 12.7–1) x is constant and the displacement function reduces to

$$w = C_1 + C_2 y + C_3 y^2 + C_4 y^3$$

where the four constants C_1, C_2, C_3, and C_4 are uniquely defined by the nodal displacements θ_{xi}, w_i, θ_{xj}, and w_j. Since these nodal displacements are common to adjacent elements, continuity of vertical deflection w exists along boundary *ij*, and for similar reasons along all other boundaries of the element.

However, it is not possible to prove that the slope $\theta_y = -\partial w/\partial x$ is continuous along *ij* or *kl*. Similarly, discontinuity of $\theta_x = \partial w/\partial y$ may occur along *jk* and *li*.

Despite the discontinuity of slopes along boundaries, it has been found[4] that the displacement function (Eqn 12.7–1) is satisfactory.

12.8 Various elements for two- and three-dimensional analyses

In Sections 12.3 and 12.4, the three-node triangle and four-node rectangle were discussed. After the reader has obtained experience in the use of these simple elements, he may wish to employ more sophisticated elements. This section lists the displacement functions suitable for various elements used in two- or three-dimensional analyses. In all these elements, the nodes transmit direct force only.

Table 12.7–2

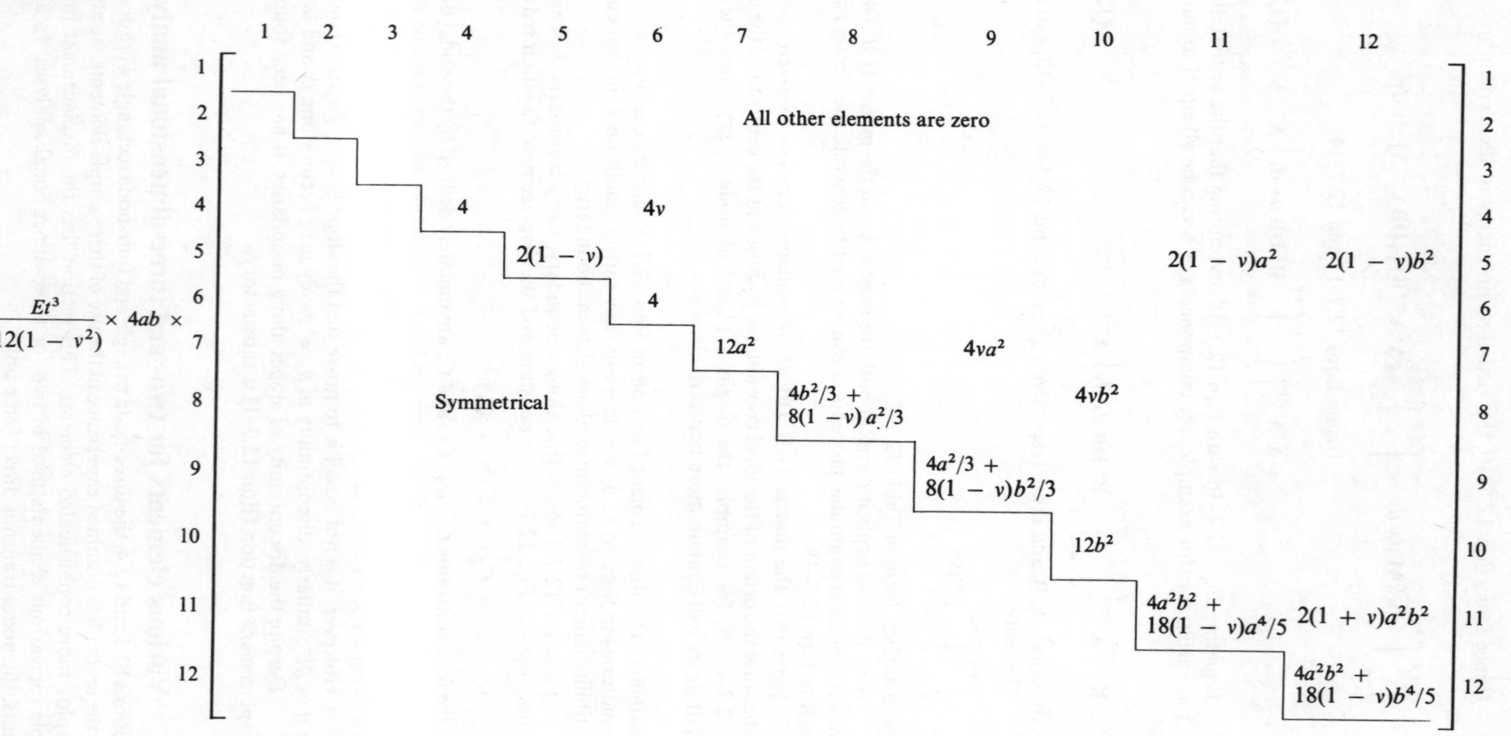

$\dfrac{Et^3}{12(1-\nu^2)} \times 4ab \times$

	1	2	3	4	5	6	7	8	9	10	11	12
1												
2												
3												
4				4		4ν						
5					$2(1-\nu)$						$2(1-\nu)a^2$	$2(1-\nu)b^2$
6						4						
7							$12a^2$		$4\nu a^2$			
8								$4b^2/3 + 8(1-\nu)a^2/3$		$4\nu b^2$		
9									$4a^2/3 + 8(1-\nu)b^2/3$			
10										$12b^2$		
11											$4a^2b^2 + 18(1-\nu)a^4/5$	$2(1+\nu)a^2b^2$
12												$4a^2b^2 + 18(1-\nu)b^4/5$

All other elements are zero

Symmetrical

six-node triangular element (Fig. 12.8–1)

$$\begin{bmatrix} u \\ v \end{bmatrix} = \begin{bmatrix} 1 & x & y & xy & x^2 & y^2 & 0 & 0 & 0 & 0 & 0 & 0 \\ 0 & 0 & 0 & 0 & 0 & 0 & 1 & x & y & xy & x^2 & y^2 \end{bmatrix} \begin{bmatrix} \alpha_1 \\ \alpha_2 \\ \vdots \\ \alpha_{12} \end{bmatrix} \quad (12.8\text{–}1)$$

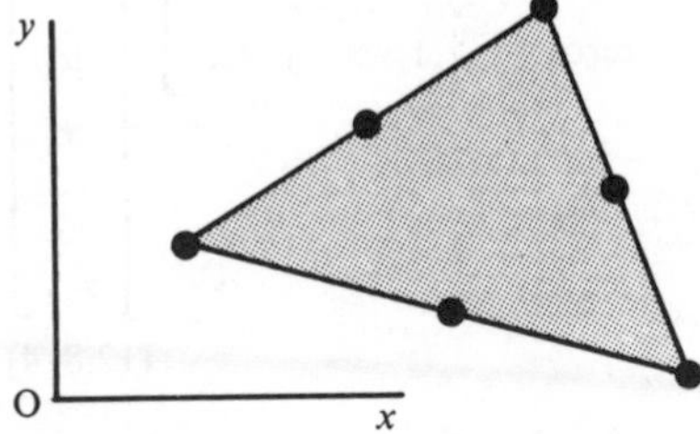

Fig. 12·8–1

eight-node rectangular element (Fig. 12.8–2)

$$\begin{bmatrix} u \\ v \end{bmatrix} = \begin{bmatrix} 1 & x & y & xy & x^2 & y^2 & x^2y & xy^2 & 0 & 0 & 0 & 0 & 0 & 0 & 0 & 0 \\ 0 & 0 & 0 & 0 & 0 & 0 & 0 & 0 & 1 & x & y & xy & x^2 & y^2 & x^2y & xy^2 \end{bmatrix} \begin{bmatrix} \alpha_1 \\ \alpha_2 \\ \vdots \\ \alpha_{16} \end{bmatrix} \quad (12.8\text{–}2)$$

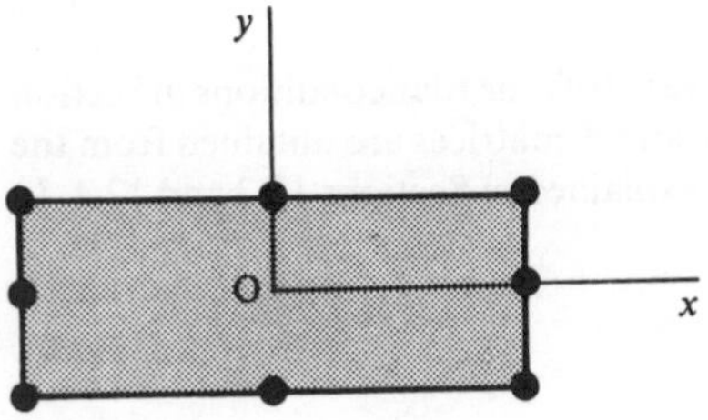

Fig. 12·8–2

four-node tetrahedron for three-dimensional analysis (Fig. 12.8–3)

$$\begin{bmatrix} u \\ v \\ w \end{bmatrix} = \begin{bmatrix} 1 & x & y & z & 0 & 0 & 0 & 0 & 0 & 0 & 0 & 0 \\ 0 & 0 & 0 & 0 & 1 & x & y & z & 0 & 0 & 0 & 0 \\ 0 & 0 & 0 & 0 & 0 & 0 & 0 & 0 & 1 & x & y & z \end{bmatrix} \begin{bmatrix} \alpha_1 \\ \alpha_2 \\ \vdots \\ \alpha_{12} \end{bmatrix} \quad (12.8\text{–}3)$$

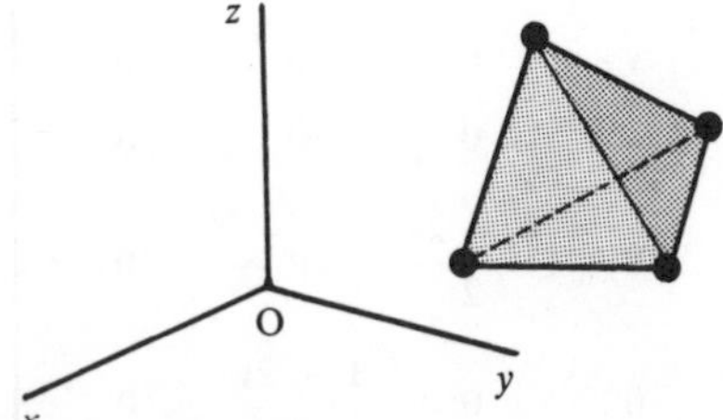

Fig. 12·8–3

As in the three-node triangle for two-dimensional analysis (Section 12.3), the **B** matrix for this element is independent of the coordinates x, y, z. That is, internal strains are uniformly distributed.

eight-node cuboid for three-dimensional analysis (Fig. 12.8–4)

$$\begin{bmatrix} u \\ v \\ w \end{bmatrix} = \left[\begin{array}{c:c:c} 1 \quad x \quad y \quad z \quad xy \quad yz \quad zx \quad xyz & \text{zeros} & \text{zeros} \\ \hdashline \text{zeros} & 1 \quad x \quad y\text{, etc.} & \text{zeros} \\ \hdashline \text{zeros} & \text{zeros} & 1 \quad x \quad y\text{, etc.} \end{array}\right] \begin{bmatrix} \alpha_1 \\ \alpha_2 \\ \alpha_3 \\ \alpha_4 \\ \vdots \\ \alpha_{24} \end{bmatrix}$$

3 × 24 matrix

(12.8–4)

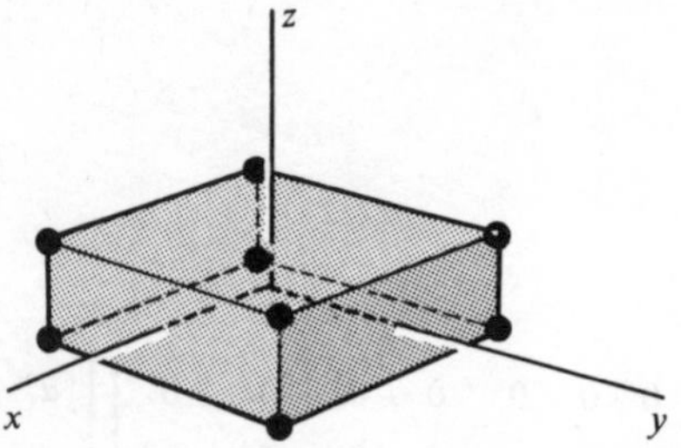

Fig. 12·8–4

All the displacement functions listed above satisfy the four conditions in Section 12.3. In using any of the above elements, the **A**, **B**, and **K** matrices are obtained from the given displacement function, using the procedure explained in Sections 12.3 and 12.4. In general, the relevant equations are of the form:

$$\boldsymbol{\varepsilon} = \mathbf{BA}^{-1}\boldsymbol{\Delta}$$

$$\boldsymbol{\sigma} = \mathbf{DBA}^{-1}\boldsymbol{\Delta}$$

$$\mathbf{K} = \mathbf{A}^{-T}\left\{\int_V \mathbf{B}^T\mathbf{DB}\,dV\right\}\mathbf{A}^{-1} \qquad (12.8\text{–}5)$$

For two-dimensional elements, **D** is as given in Eqn 12.3–12; for three-dimensional elements

$$\mathbf{D} = \frac{E}{(1+\nu)(1-2\nu)} \times \begin{bmatrix} 1-\nu & \nu & \nu & 0 & 0 & 0 \\ \nu & 1-\nu & \nu & 0 & 0 & 0 \\ \nu & \nu & 1-\nu & 0 & 0 & 0 \\ 0 & 0 & 0 & \dfrac{1-2\nu}{2} & 0 & 0 \\ 0 & 0 & 0 & 0 & \dfrac{1-2\nu}{2} & 0 \\ 0 & 0 & 0 & 0 & 0 & \dfrac{1-2\nu}{2} \end{bmatrix}$$

(12.8–6)

12.9 Computer flow charts

In previous sections of this chapter, the finite element method was presented in a form suitable for computer programming. Several flow charts are presented in this section. These flow charts should enable any reader who is familiar with the elements of programming to write his own computer programs for structural analysis.

Subsection 12.9(a) gives a flow chart for plane stress analysis, using triangular elements. Subsection 12.9(b) gives a similar one for rectangular elements. After the reader has mastered these flow charts he will not have much difficulty in writing his own flow chart for a combination of triangular and rectangular elements, or for rectangular elements in bending. Subsection 12.9(c) gives a fairly detailed flow chart for constructing the structure stiffness matrix. The steps in each flow chart are marked, and explanations regarding each step are given at the end of the flow chart.

(a) **Computer flow chart for plane stress analysis using triangular elements.** Consider an arbitrary two-dimensional structure, idealized as an assemblage of triangular elements (Fig. 12.9–1).

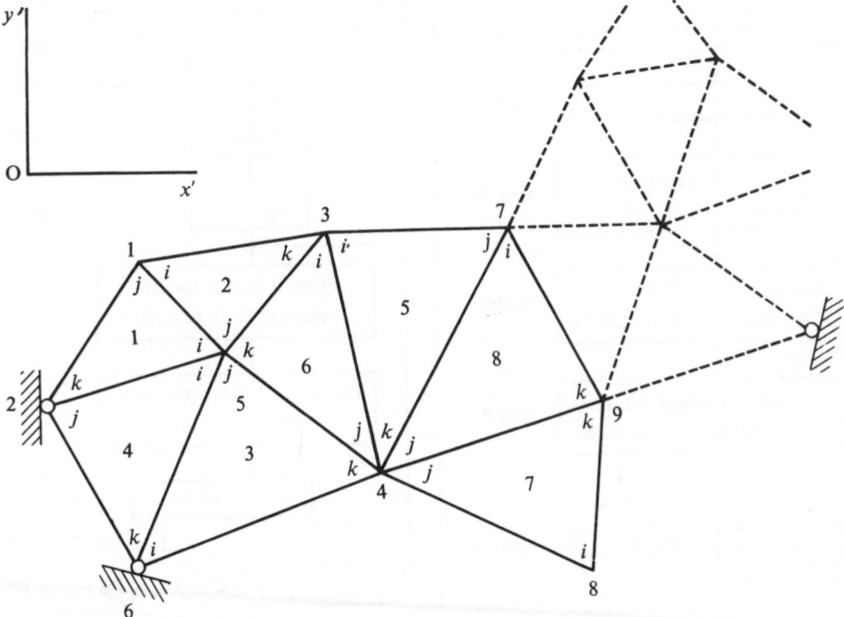

Fig. 12·9–1

Suppose there are a total of n_e elements in Fig. 12.9–1, with a total of n_n nodes. (In a practical problem, n_e and n_n would generally be quite large.) The nodes are numbered consecutively from 1 to n_n, and the elements are numbered consecutively from 1 to n_e, as explained in Section 12.6. It is also necessary to identify the nodal reference letters i, j, k of each individual element, by relating these letters to the node numbers; this is illustrated in Table 12.9–1.

Table 12.9–1

Element No.	1	2	3	4	5	...	n_e
Node *i*	5	1	6	5	3	...	
Node *j*	1	5	5	2	7	...	
Node *k*	2	3	4	6	4	...	

We are now ready to write the flow chart as follows:†

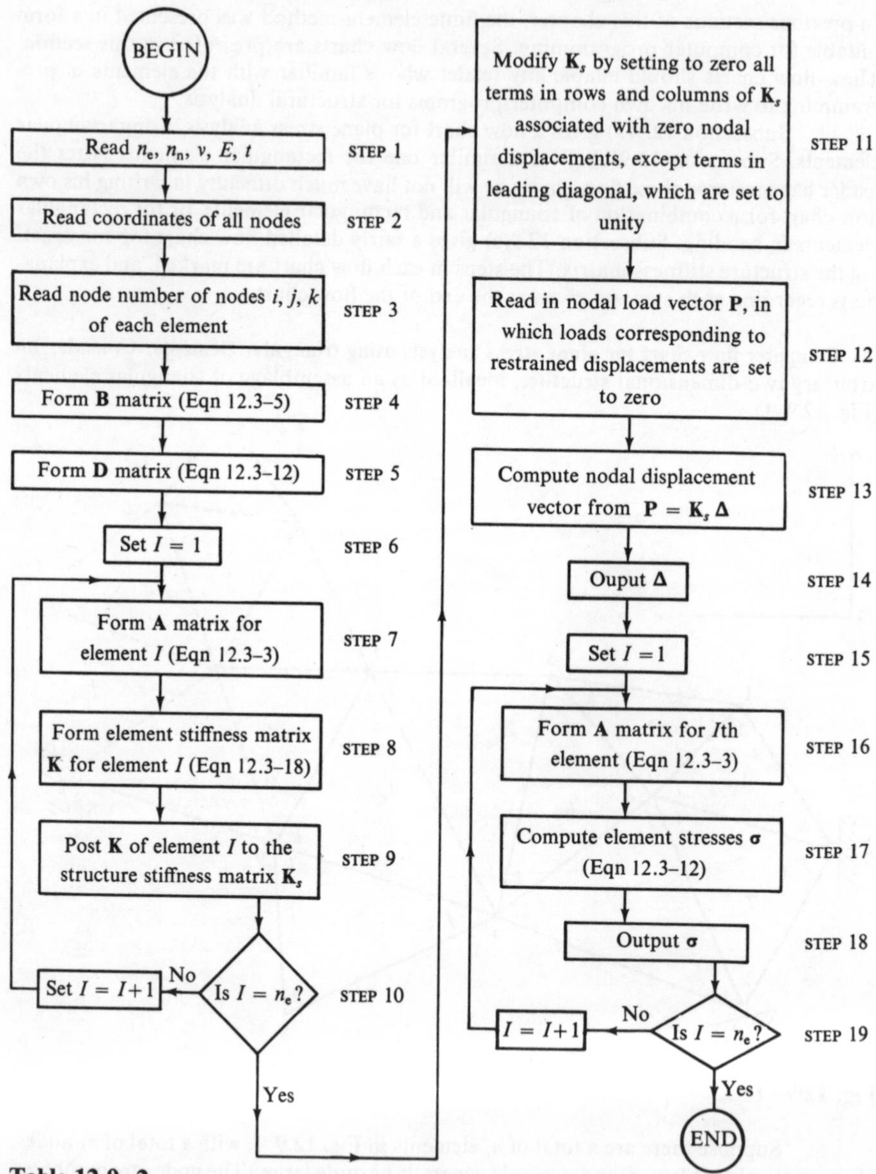

Table 12·9–2

In Step 1 of the above flow chart, we transfer into the computer as input data (a) the total number of elements in the structure, n_e, (b) the total number of nodes, n_n, (c) the Poisson's ratio of the material of the structure, v, (d) Young's modulus of the material, E, and (f) the thickness t of the structure.

In Step 2 we input the coordinates of every node. These are structure coordinates, but since for triangular elements we can always choose the structure coordinate

† Thanks are due to Dr P. J. Robins and Dr K. L. Taylor for valuable advice.

axes as local coordinate axes, there is no need to distinguish between the two types of coordinates.

In Step 3, we transfer the information in Table 12.9–1 to the computer, namely.

$$i(1) = 5, \quad j(1) = 1, \quad k(1) = 2$$

$$i(2) = 1, \quad j(2) = 5, \quad k(2) = 3 \text{ etc.}$$

where $i(1)$ denotes the node number for node i of the first element, $i(2)$ denotes the node number for node i of the second element, and so on.

In Steps 4 and 5 we set up the matrices **B** and **D** in the computer, in accordance with Eqn 12.3–5 and 12.3–12.

In Step 6, we set the index I to 1; that is we are beginning with element No. 1 (Fig. 12.9–1).

In Step 7 we set up the **A** matrix from Eqn 12.3–3, again using a standard computer procedure. Referring to Eqn 12.3–3, note that x_i, y_i, x_j, y_j, etc., for the Ith element are $x_{i(I)}, y_{i(I)}, x_{j(I)}, y_{j(I)}$, etc. The node numbers $i(I), j(I), k(I)$ were input to the computer in Step 3, for all values of I from 1 to n_e inclusive. Similarly, the coordinates of $i(I), j(I)$, and $k(I)$ were input in Step 2 for all values of I. Hence, for any element, the nodal coordinates x_i, y_i, x_j, y_j and x_k, y_k are available in the computer memory, for use in setting up **A** in accordance with Eqn 12.3–3.

In Step 8, we set up the 6×6 element stiffness matrix **K** from Eqn 12.3–18. Note that the matrices **A**, **B**, **D**, the thickness t, and the coordinates of the nodes i, j, k for each element, are already in the computer memory; the algebraic operations in Eqn 12.3–18 are executed by the computer using a standard procedure.

In Step 9, we assemble the structure stiffness matrix $\mathbf{K}_s$ from contributions of individual element stiffness matrices **K**. The mechanics of this step are explained in detail later in Subsection 12.9(c). At the conclusion of Step 9, all the 36 terms of **K** for the first element have been posted to appropriate positions in $\mathbf{K}_s$, and we set $I = 2$ (Step 10) and repeat Steps 7 to 9 for the second element, and so forth, until all n_e elements are dealt with; that is until $I = n_e$. Then the entire structure stiffness matrix $\mathbf{K}_s$ is assembled and we move on to Step 11.

Step 11 is the same as similar steps in the analysis of framed structures, as explained in Chapter 7, Sections 7.10 and 7.12; the structure stiffness matrix $\mathbf{K}_s$ as assembled by Steps 1 to 10 is that of an unrestrained structure. Obviously unless sufficient nodal restraints are prescribed to prevent rigid-body movement of the structure as a whole, it is impossible to analyse this system, because the nodal displacements of an unrestrained structure cannot be uniquely determined by the nodal forces. This physically obvious fact is represented mathematically by the stiffness matrix $\mathbf{K}_s$ being singular, i.e. $\mathbf{K}_s^{-1}$ does not exist. The modifications in Step 11 will convert $\mathbf{K}_s$ to a non-singular matrix.

Steps 12 to 14 are the same as similar steps for analysis of framed structures.

In Steps 14 to 18, the internal stresses are computed for each element in turn. In Step 16, $\boldsymbol{\sigma}$ is computed from Eqn 12.3–12, i.e.

$$\boldsymbol{\sigma} = \mathbf{DBA}^{-1}\boldsymbol{\Delta}$$

where $\boldsymbol{\Delta}$ here refers to the displacements of the nodes i, j, k of the element being dealt with for the time being, For the Ith element, we have

$$\boldsymbol{\sigma}_{I\text{th element}} = \mathbf{DB}\{\mathbf{A}^{-1}_{I\text{th element}}\}\begin{bmatrix} \Delta_x \text{ of node } i \text{ of } I\text{th element} \\ \Delta_y \text{ of node } i \text{ of } I\text{th element} \\ \Delta_x \text{ of node } j \text{ of } I\text{th element} \\ \Delta_y \text{ of node } j \text{ of } I\text{th element} \\ \Delta_x \text{ of node } k \text{ of } I\text{th element} \\ \Delta_y \text{ of node } k \text{ of } I\text{th element} \end{bmatrix}$$

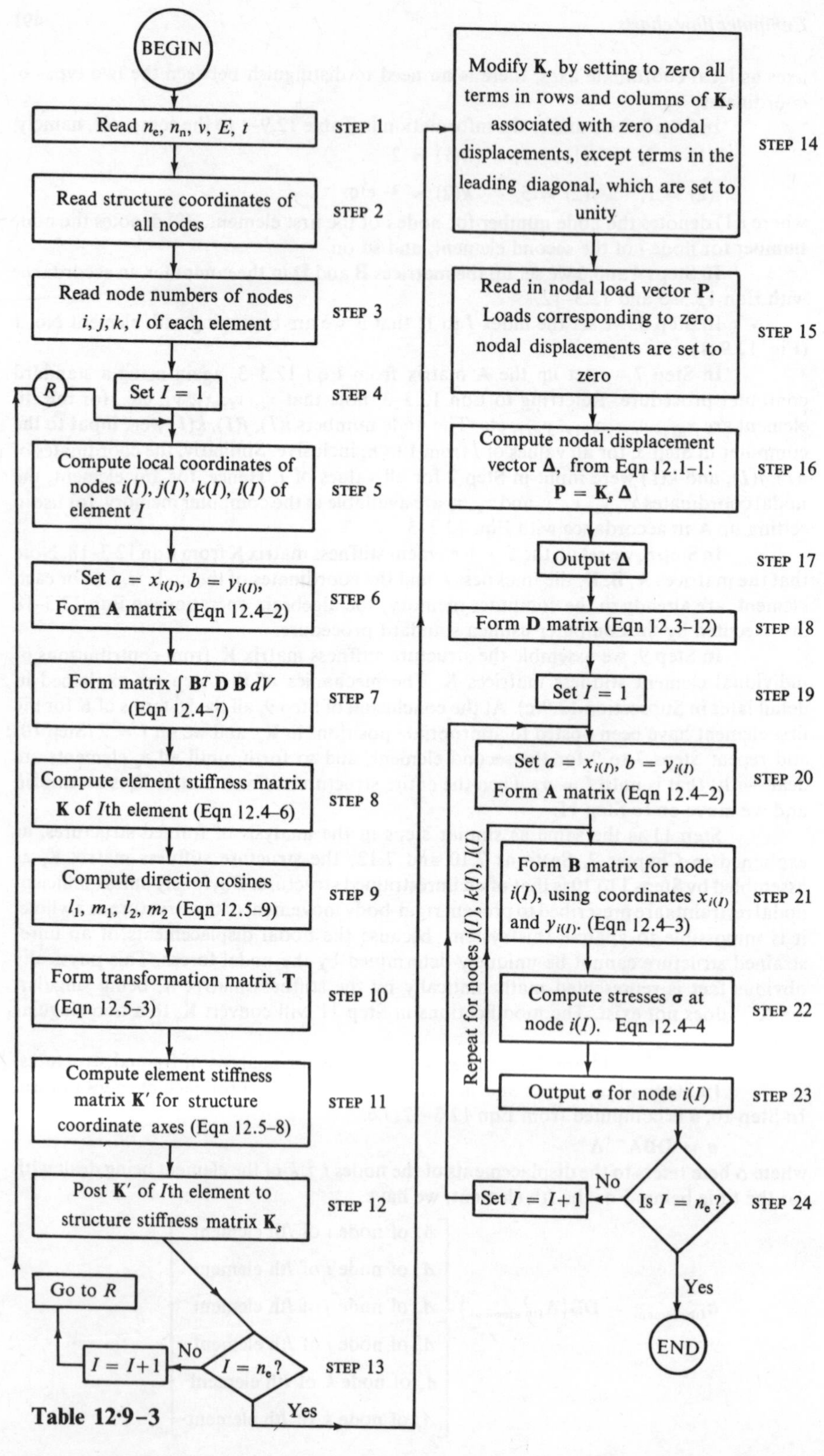
BEGIN
Read n_e, n_n, ν, E, t
STEP 1
Read structure coordinates of all nodes
STEP 2
Read node numbers of nodes i, j, k, l of each element
STEP 3
R
Set $I = 1$
STEP 4
Compute local coordinates of nodes $i(I)$, $j(I)$, $k(I)$, $l(I)$ of element I
STEP 5
Set $a = x_{i(I)}$, $b = y_{i(I)}$, Form A matrix (Eqn 12.4–2)
STEP 6
Form matrix $\int \mathbf{B}^T \mathbf{D} \mathbf{B} \, dV$ (Eqn 12.4–7)
STEP 7
Compute element stiffness matrix K of Ith element (Eqn 12.4–6)
STEP 8
Compute direction cosines l_1, m_1, l_2, m_2 (Eqn 12.5–9)
STEP 9
Form transformation matrix T (Eqn 12.5–3)
STEP 10
Compute element stiffness matrix K′ for structure coordinate axes (Eqn 12.5–8)
STEP 11
Post K′ of Ith element to structure stiffness matrix $\mathbf{K}_s$
STEP 12
Go to R
$I = I+1$
No
$I = n_e$?
STEP 13
Yes
Modify $\mathbf{K}_s$ by setting to zero all terms in rows and columns of $\mathbf{K}_s$ associated with zero nodal displacements, except terms in the leading diagonal, which are set to unity
STEP 14
Read in nodal load vector P. Loads corresponding to zero nodal displacements are set to zero
STEP 15
Compute nodal displacement vector Δ, from Eqn 12.1–1: $\mathbf{P} = \mathbf{K}_s \boldsymbol{\Delta}$
STEP 16
Output Δ
STEP 17
Form D matrix (Eqn 12.3–12)
STEP 18
Set $I = 1$
STEP 19
Set $a = x_{i(I)}$, $b = y_{i(I)}$ Form A matrix (Eqn 12.4–2)
STEP 20
Repeat for nodes $j(I)$, $k(I)$, $l(I)$
Form B matrix for node $i(I)$, using coordinates $x_{i(I)}$ and $y_{i(I)}$. (Eqn 12.4–3)
STEP 21
Compute stresses σ at node $i(I)$. Eqn 12.4–4
STEP 22
Output σ for node $i(I)$
STEP 23
Set $I = I+1$
No
Is $I = n_e$?
STEP 24
Yes
END

Table 12·9–3

The displacements of nodes $i(I)$, $j(I)$, and $k(I)$ had of course already been generated by the computer in Step 13, for all values of I from 1 to n_e.

(b) **Computer flow chart for plane stress analysis using rectangular elements.** With the background provided by Subsection 12.9(a) we shall present the flow chart in Table 12.9–3, followed by brief explanations of the various steps.

In the above flow chart for rectangular elements in plane stress, Steps 1 to 3 are essentially the same as corresponding steps for triangular elements in Subsection 12.9(a).

In Step 5 we compute the local coordinates of nodes of the Ith element (Fig. 12.5–1):

$$x'_{0(I)} = \tfrac{1}{4}\{x'_{i(I)} + x'_{j(I)} + x'_{k(I)} + x'_{l(I)}\}$$

$$y'_{0(I)} = \tfrac{1}{4}\{y'_{i(I)} + y'_{j(I)} + y'_{k(I)} + y'_{l(I)}\}$$

$$x_{i(I)} = \tfrac{1}{2}\sqrt{[(x'_{i(I)} - x'_{j(I)})^2 + (y'_{i(I)} - y'_{j(I)})^2]}$$

$$y_{i(I)} = \tfrac{1}{2}\sqrt{[(x'_{i(I)} - x'_{l(I)})^2 + (y'_{i(I)} - y'_{l(I)})^2]}$$

$$x_{j(I)} = x_{k(I)} = -x_{l(I)} = x_{i(I)}; \qquad y_{j(I)} = -y_{k(I)} = -y_{l(I)} = y_{i(I)}$$

where $x'_{0(I)}$ and $y'_{0(I)}$ are the structure coordinates of the centroid of the Ith element. They are also the structure coordinates of the origin of the local coordinate axes. Note that in Subsection 12.9(a) for triangular elements, it was not necessary to compute local coordinates. (Why?)

Steps 6 to 8 are self-explanatory. The algebraic operations are carried out with standard computer procedures.

The computation of direction cosines (Step 9) was not required in Subsection 12.9(a) for triangular elements, but is required here to construct the transformation matrix (Step 10) which transforms **K** into **K**′ (Step 11).

Step 12 is similar to Step 9 in Subsection 12.9(a)—for details, see Table 12.9–4.

Steps 13 to 17 are again similar to Steps 10 to 14 in that subsection.

Step 18 is similar to Step 5 in Subsection 12.9(a), but occurs comparatively late in the present flow chart, because it is only now that **D** is required as a complete matrix. Earlier, in Step 7, the elements of **D** were fused with other quantities in the 8 × 8 matrix $\int \mathbf{B}^T\mathbf{D}\mathbf{B}\, dV$.

Steps 19 to 24 are for generating the nodal stresses in the elements. Note that for rectangular elements, the **B** matrix is a function of the coordinates of the point at which stresses are sought; hence, for each element, Steps 21 to 23 have to be done four times—for the four nodes of the element. And Steps 20 to 23 have to be carried out n_e times—to cover all the n_e elements.

Note also that the stresses $\boldsymbol{\sigma}$ output in Step 23 are stresses σ_x, σ_y, τ_{xy}, i.e. they are in the directions of the local coordinate axes; they are not principal stresses nor are they stresses in the directions of the structure axes. Determination of principal stresses or maximum shear stresses was explained in Chapter 11, Section 11.5.

(c) **Computer flow chart for constructing the structure stiffness matrix.** In this subsection, a detailed flow chart is given for constructing the structure stiffness matrix of an assembly of triangular elements. This flow chart replaces Steps 6 to 10 of the flow chart in Subsection 12.9(a); mastery of this flow chart would also enable the reader to write a similar one to amplify Step 12 of the flow chart for rectangular elements in Subsection 12.9(b).

Before presenting the flow chart, we shall discuss a computer-oriented system of indexing nodal displacements. For explanation purposes, consider the arbitrary assembly of triangular elements in Fig. 12.9–1. Since at each node there are two displacements—one in the x'-direction and the other in the y'-direction, the total number of nodal displacements is twice the total number of nodes, i.e. the nodal displacement vector $\boldsymbol{\Delta}$ has $2n_n$ components. (Similarly, the nodal force vector **P** also has $2n_n$ components.) Of course, some of the nodes are restrained, and hence some of the $2n_n$ displacements are equal to zero.

We shall number the nodal displacements in the same order as nodes, and shall take the x' displacement before the y' displacement at each node. Thus, the x' displacement at node 1 will be called displacement 1, and the y' displacement at node 1 will be called displacement 2. Similarly, the x' displacement at node 2 will be called displacement 3 and the y' displacement at node 2 will be called displacement 4, and so on. At an arbitrary node J, therefore, the displacement in the x'- and y'-directions will be called displacements $2J - 1$ and $2J$ respectively. At the last node, i.e. node n_n, the two displacements would likewise be $2n_n - 1$ and $2n_n$ respectively.

Using the above indexing system, the structure stiffness matrix $\mathbf{K}_s$ can be written as (see also Eqn 12.6–1):

$$\mathbf{K}_s = \begin{bmatrix} (K_s)_{1,1} & (K_s)_{1,2} & (K_s)_{1,3} & \cdots & (K_s)_{1,2n_n} \\ (K_s)_{2,1} & (K_s)_{2,2} & (K_s)_{2,3} & \cdots & (K_s)_{2,2n_n} \\ (K_s)_{3,1} & (K_s)_{3,2} & (K_s)_{3,3} & \cdots & \\ (K_s)_{4,1} & (K_s)_{4,2} & \cdots & & \\ \vdots & & & & \\ (K_s)_{2n_n-1,1} & (K_s)_{2n_n-1,2} & \ddots & \cdots & (K_s)_{2n_n-1,2n_n} \\ (K_s)_{2n_n,1} & (K_s)_{2n_n,2} & \cdots & \cdots & (K_s)_{2n_n,2n_n} \end{bmatrix}$$

$$2n_n \times 2n_n \text{ matrix} \qquad (12.9\text{–}1)$$

Similarly, the element stiffness matrix for a triangular element can be written as (see also Eqn 12.1–4):

$$\mathbf{K} = \begin{bmatrix} K_{1,1} & K_{1,2} & \cdots & K_{1,6} \\ K_{2,1} & K_{2,2} & \cdots & K_{2,6} \\ \vdots & & & \\ K_{6,1} & K_{6,2} & \cdots & K_{6,6} \end{bmatrix}$$

$$6 \times 6 \text{ matrix} \qquad (12.9\text{–}2)$$

To construct the structure stiffness matrix $\mathbf{K}_s$ from the element stiffness matrices $\mathbf{K}$, consider a typical element I (Fig. 12.9–2); this element I could be any of the n_e elements in Fig. 12.9–1.

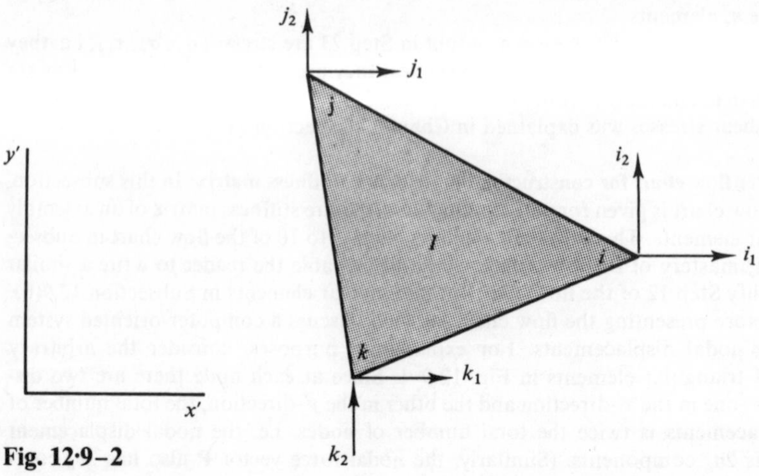

Fig. 12·9–2

The nodes i, j, k of this element have node numbers $i(I), j(I), k(I)$ which have of course been recorded in Table 12.9–1. The nodal displacements of this element are:

$$\begin{aligned}
i1 &= 2i(I) - 1, \text{ being } x' \text{ displacement at node } i \\
i2 &= 2i(I), \quad \text{ being } y' \text{ displacement at node } i \\
j1 &= 2j(I) - 1, \text{ being } x' \text{ displacement at node } j \\
j2 &= 2j(I), \quad \text{ being } y' \text{ displacement at node } j \\
k1 &= 2k(I) - 1, \text{ being } x' \text{ displacement at node } k \\
k2 &= 2k(I), \quad \text{ being } y' \text{ displacement at node } k
\end{aligned} \qquad (12.9\text{–}3)$$

The element stiffness matrix **K** (Eqn 12.9–2) contributes to the structure stiffness matrix $\mathbf{K}_s$ (Eqn 12.9–1) in the following manner:

$K_{1,1}$ contributes to $(K_s)_{i1,i1}$
$K_{1,2}$ contributes to $(K_s)_{i1,i2}$
$\vdots \qquad\qquad \vdots$
$K_{1,6}$ contributes to $(K_s)_{i1,k2}$
$K_{2,1}$ contributes to $(K_s)_{i2,i1}$
$K_{2,2}$ contributes to $(K_s)_{i2,i2}$ where displacements $i1$, $i2$, etc., are as defined by Eqn 12.9–3
$\vdots \qquad\qquad \vdots$
$K_{2,6}$ contributes to $(K_s)_{i2,k2}$
$\vdots \qquad\qquad \vdots$
$K_{6,1}$ contributes to $(K_s)_{k2,i1}$
$K_{6,2}$ contributes to $(K_s)_{k2,i2}$
$K_{6,6}$ contributes to $(K_s)_{k2,k2}$ (12.9–4)

We are in a position to write the computer flow chart (Table 12.9–4).

In Step 1 of this flow chart, we set the $2n_n \times 2n_n$ elements of the structure stiffness matrix to zero, using a standard computer procedure available in most computing centres. The object is to clear the computer storage for $\mathbf{K}_s$, to have it ready for storing data to be posted to it in subsequent steps of the flow chart.

In Step 1, we also set the index I to 1; that is, we are beginning with element No. 1 (Fig. 12.9–1).

In Step 3, we compute the displacement numbers for the displacements at the nodes of this Ith element. Referring to Table 12.9–1, it is seen that when $I = 1$, $i(I) = 5$, $j(I) = 1$, $k(I) = 2$. Therefore, in Step 3 with $I = 1$, we have

$$\begin{aligned}
i1 &= 2i(I) - 1 = 9, \qquad i2 = 10 \\
j1 &= 2j(I) - 1 = 1, \qquad j2 = 2 \\
k1 &= 2k(I) - 1 = 3, \qquad k2 = 4
\end{aligned}$$

In Step 4, we post those element nodal stiffnesses associated with displacement $i1$ (that is, displacement 9 in this case) to the relevant nodal stiffnesses in the structure stiffness matrix. Thus,

we post $K(1, 1)$ to $K_s(i1, i1)$, i.e. $K_{x'i,x'i}$ to $K_s(9, 9)$
we post $K(2, 1)$ to $K_s(i2, i1)$, i.e. $K_{y'i,x'i}$ to $K_s(10, 9)$
we post $K(3, 1)$ to $K_s(j1, i1)$, i.e. $K_{x'j,x'i}$ to $K_s(1, 9)$

and so on.

Note that we have written $K_s(i1, i1)$ instead of $(K_s)_{i1,i1}$, and $K_s(i2, i1)$ instead of $(K_s)_{i2,i1}$, etc.; this is because the stiffnesses $(K_s)_{i1,i1}$, $(K_s)_{i1,i2}$, $(K_s)_{i1,j1}, \ldots$, etc.

($2n_n \times 2n_n$ in all) are stored in the computer as elements of an **array**, a term which the reader will understand as soon as he has studied computer programming.

Similarly, Steps 5 to 9 are concerned with posting element nodal stiffnesses associated with nodal displacements $i2$ to $k2$.

At the conclusion of Step 9, all the 36-element nodal stiffnesses in the $\mathbf{K}$ of the first element have been posted to their respective nodal stiffnesses in $\mathbf{K}_s$. Then in Step 10, the computer checks whether I is equal to the total number of elements n_e. If not, I is set to $I + 1$ and Steps 2 to 9 are repeated, until all n_e elements are dealt with, and the construction of $\mathbf{K}_s$ is complete.

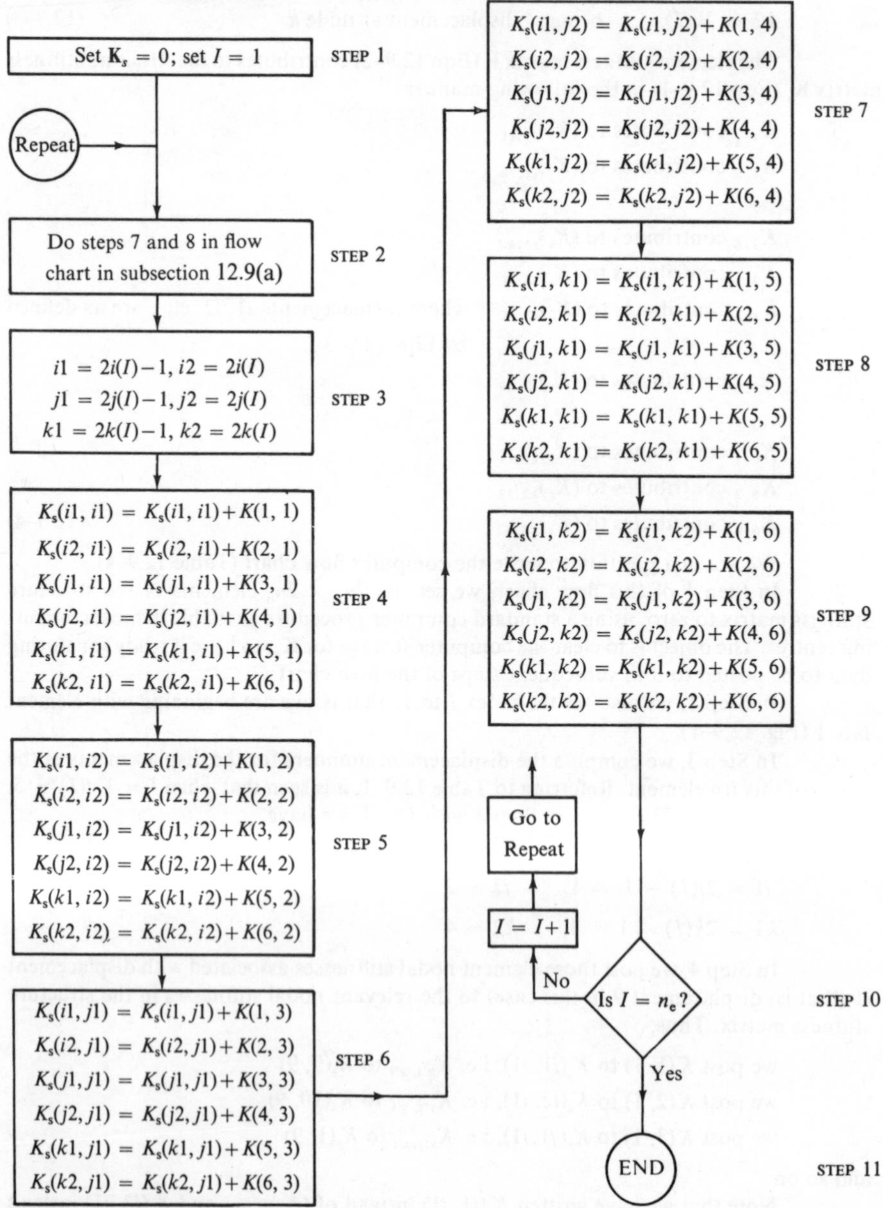

Table 12·9–4

Example 12.9–1. Prepare a flow chart to replace Step 3 to Step 9 of that in Subsection 12.9(c).

SOLUTION In Subsection 12.9(c), a flow chart is given for building up the structure stiffness matrix $\mathbf{K}_s$ from the stiffness matrices $\mathbf{K}$ of the various triangular elements. This flow chart has the advantage that it shows clearly how the computer can be instructed to carry out the operations involved in the building up of $\mathbf{K}_s$. It has the disadvantage that, if the number of degrees of freedom of the individual elements is large, then a large number of statements or computer instructions is required.

Table 13.1–2 in Chapter 13 gives a practical flow chart for skeletal structures. Using the same general approach, a flow chart can be drawn up for triangular elements, as shown below:

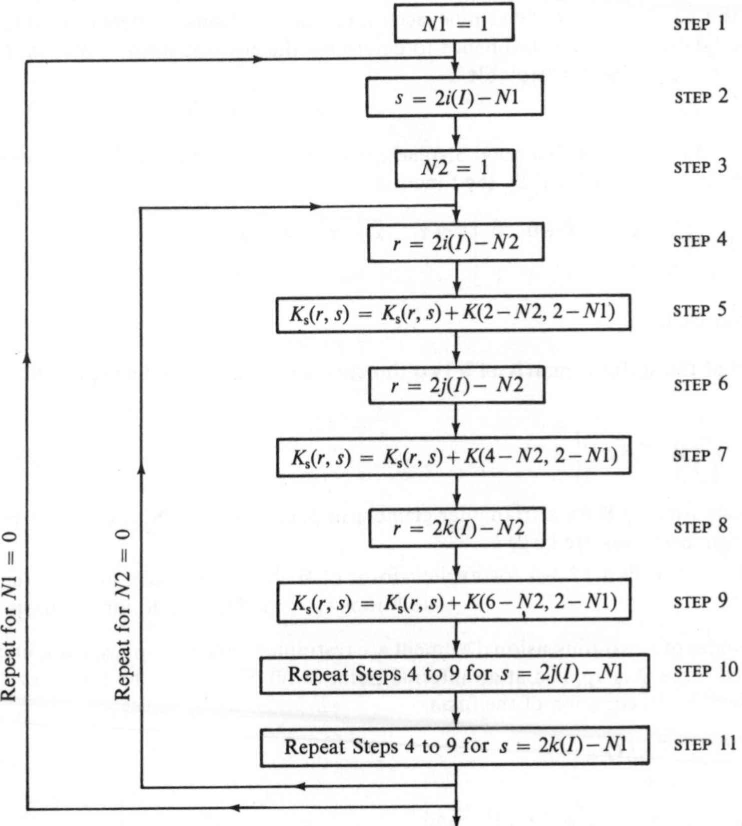
N1 = 1
STEP 1
s = 2i(I) − N1
STEP 2
N2 = 1
STEP 3
r = 2i(I) − N2
STEP 4
K_s(r, s) = K_s(r, s) + K(2 − N2, 2 − N1)
STEP 5
r = 2j(I) − N2
STEP 6
K_s(r, s) = K_s(r, s) + K(4 − N2, 2 − N1)
STEP 7
r = 2k(I) − N2
STEP 8
K_s(r, s) = K_s(r, s) + K(6 − N2, 2 − N1)
STEP 9
Repeat Steps 4 to 9 for s = 2j(I) − N1
STEP 10
Repeat Steps 4 to 9 for s = 2k(I) − N1
STEP 11
Repeat for N1 = 0
Repeat for N2 = 0

Table 12·9–5

Consider the element No. 1 in Fig. 12.9–1; that is, $I = 1$. The reader should verify that Steps 4 to 9 in the first cycle (with $N1 = 1$, $N2 = 1$) we post

$K(1, 1)$ of element No. 1 to $K_s(9, 9)$

$K(3, 1)$ of element No. 1 to $K_s(1, 9)$

$K(5, 1)$ of element No. 1 to $K_s(3, 9)$

in that order.

He should finally satisfy himself that Steps 2 to 11,† cycled as indicated for

(1) $N1 = 1, N2 = 1$,
(2) $N1 = 1, N2 = 0$,
(3) $N1 = 0, N2 = 1$, and
(4) $N1 = 0, N2 = 0$,

would do the work of Steps 3 to 9 of the flow chart in Subsection 12.9(c).

Problems

12.1. A three-dimensional continuum is to be analysed by the finite element method, using elements for which a suitable displacement function is not known. The analyst therefore plans to carry out precise laboratory tests on elements made of the same material as the continuum, to determine accurately the relationship between nodal forces and nodal displacements and hence to determine the element stiffness matrix. Is his approach to the problem acceptable?

Ans. No. See Example 12.3–1.

12.2. State the conditions which a good displacement function should satisfy, and show that these conditions are satisfied by the function

$$\begin{bmatrix} u \\ v \end{bmatrix} = \begin{bmatrix} 1 & x & y & 0 & 0 & 0 \\ 0 & 0 & 0 & 1 & x & y \end{bmatrix} [\alpha_1 \quad \alpha_2 \quad \alpha_3 \quad \alpha_4 \quad \alpha_5 \quad \alpha_6]^T$$

for a triangular element.

12.3. Show that the stiffness matrix of a two-dimensional element can be expressed in the form:

$$\mathbf{A}^{-T}\left\{\int \mathbf{B}^T \mathbf{D}\mathbf{B}\, dV\right\}\mathbf{A}^{-1}$$

State the explicit form of **B** for a triangular element in plane stress; is this explicit form applicable if displacements are large?

Ans. See Eqn 12.3–4 for explicit form of **B**. Not applicable, because the relations $\varepsilon_x = \partial u/\partial x$, etc., no longer apply.

12.4. The n nodes of a two-dimensional element are restrained against displacement, and a concentrated force **Q** is applied at an internal point (x, y). Show that the nodal force vector **p** is given by an equation of the form

$$\mathbf{p} = -\mathbf{A}^{-T}\mathbf{N}_Q^T\mathbf{Q}$$

where

$$\mathbf{p} = [p_{x1}\ p_{y1}\ \cdots\ p_{xn}\ p_{yn}]^T \quad \text{and}$$
$$\mathbf{Q} = [Q_x\ Q_y]^T$$

State the explicit form of $\mathbf{N}_Q$ for a rectangular element

Ans. $$\mathbf{N}_Q = \begin{bmatrix} 1 & x & y & xy & 0 & 0 & 0 & 0 \\ 0 & 0 & 0 & 0 & 1 & x & y & xy \end{bmatrix}$$

† Note that the complete statement in Step 10 should read 'Repeat Steps 4 to 9 for $s = 2j(\text{I}) - N1$, with the K elements in Steps 5, 7, and 9 replaced by $K(2 - N2, 4 - N1)$, $K(4 - N2, 4 - N1)$, and $K(6 - N2, 4 - N1)$ respectively.' Similarly, in executing Step 11, the above mentioned K elements would be replaced by $K(2 - N2, 6 - N1)$, $K(4 - N2, 6 - N1)$, and $K(6 - N2, 6 - N1)$ respectively.

12.5. Show that if the element coordinate axes and the structure coordinate axes are not parallel, then the element stiffness matrix **K** for element coordinates is related to the element stiffness matrix **K**′ for structure coordinates by an equation of the type

$$\mathbf{K}' = \mathbf{T}^T\mathbf{K}\mathbf{T}$$

Are such transformations necessary for (a) triangular elements, (b) rectangular elements?

Ans. (a) No. (Why not?) (b) Yes. (Why?)

12.6. The nodes of a rectangular element are held against displacements, and an in-plane force **Q** is applied at point (x, y), producing nodal forces $\mathbf{p}_0$.

(a) Explain whether the strain energy due to **Q** is a linear function of **Q**.

(b) If, while the element is sustaining the force **Q**, the nodes are given virtual displacements, explain why the virtual strain energy is zero.

Ans. (a) No strain energy. (b) See Example 12.3–2.

12.7. For a rectangular element in bending,

$$\mathbf{M} = \mathbf{D}\boldsymbol{\kappa}$$

where

$$\mathbf{M} = [M_{yx} \; M_{xy} \; M_{xx}]^T$$

and

$$\boldsymbol{\kappa} = \begin{bmatrix} \dfrac{\partial^2 w}{\partial y^2} & -\dfrac{\partial^2 w}{\partial x^2} & \dfrac{2\partial^2 w}{\partial x\, \partial y} \end{bmatrix}^T$$

(a) Explain, with the help of a diagram, the meaning of M_{xy} and M_{xx}; define their positive directions.

(b) State the explicit form of **D**.

(c) Express the matrix $\boldsymbol{\kappa}$ in terms of the coordinates of the point (at which the curvatures occur) and the nodal displacements.

Ans. See Section 12.7.

12.8. Write a computer flow chart for posting the element stiffness matrix **K**′ of a rectangular element in bending to the structure stiffness matrix $\mathbf{K}_s$.

Ans. Use Example 12.9–1 for guidance.

12.9. If the nodal displacements of an element are $\boldsymbol{\Delta}$, then the nodal forces **p** are given by

$$\mathbf{p} = \mathbf{K}\boldsymbol{\Delta}$$

where **K** is the element stiffness matrix.

(a) Explain why the set of forces **p** is always in equilibrium.

(b) Explain why such equilibrium is *not* a consequence of the use of the principle of virtual work in the derivation of **K**, but is a direct consequence of the fact that the displacement function allows for a rigid-body motion to occur without straining the element.

(c) Hence, or otherwise, show that if the displacement function in Eqns 12.3–1 does not contain the coefficients α_1 and α_4, then equilibrium of nodal forces cannot be guaranteed even though the element stiffness matrices **K** (Eqn 12.3–17) are determined by using the principle of virtual work.

References

1 Turner, M. J., Clough, R. W., Martin, H. C., and Topp, L. J. Stiffness and deflection analysis of complex structures. *J. aeronaut. Sci.*, **23**, No. 9, 805–823, 1956.

2 Argyris, J. H. and Kelsey, S. *Energy Theorems and Structural Analysis*. Butterworth, London, 1960.

3 Clough, R. W. The finite element in plane stress analysis. *Proceedings of the 2nd ASCE Conference on Electronic Computation*, 345–378. Pittsburgh, September, 1960.

4 Zienkiewicz, O. C. and Cheung, Y. K. The finite element method for analysis of elastic, isotropic and orthotropic slabs. *Proc. Instn. civ. Engrs.*, **28**, 471–488, 1964.

General references

Bathe, K.J. *Numerical methods in finite element analysis*. Prentice Hall, Englewood Cliffs, 1976.

Brebbia, C.A. and Conner, J.J. *Fundamentals of finite element techniques for structural engineers*. Butterworth, London, 1973.

Cheung, Y.K. and Yeo, M.F. *A practical introduction to finite element analysis*. Pitman, London, 1979.

Cook, R.D. *Finite element analysis*. John Wiley, New York, 2nd edition, 1981.

Whiteman J.R. *A bibliography for finite elements*. Academic Press, London, 1975.

Zienkiewicz, O.C. *The finite element method*. McGraw-Hill, Maidenhead, 3rd edition, 1977.

Chapter 13
Computer application

13.1 Flow chart for matrix stiffness method

The discussion of the analysis methods presented in the earlier chapters of this book has on purpose omitted reference to the details of how solutions to real problems are actually obtained, although it is accepted that the digital computer will be used in all except trivial cases. The purpose of this chapter is to fill in the details, and to study some of the computational problems which may arise. Flow charts are written so far as is possible in a manner independent of computer language or computer design, and no mention is therefore made of any special hardware or software facilities not generally available. Of course, when a program is actually written, it is inevitably in a particular language, and generally intended for use on a particular machine or range of machines, and some modifications would then be necessary. The flow charts of this and the following section assume that adequate general-purpose computer library subroutines are available for matrix handling, and special questions such as those of equation solution and banding in the stiffness matrix are left to Sections 13.3 to 13.6.

There is no essential difference between a stiffness analysis on a skeletal structure and a finite element analysis on a continuum, and the main steps in the flow chart of Table 13.1–1 could therefore be applied to either. There are some differences of detail, however, and the chart is written primarily to suit skeletal analysis. For the sake of generality all the fundamental steps have been listed, but it is likely that in many cases some would be omitted—for instance, bending properties are not required in a truss analysis, or torsion properties with a plane frame.

In the chart, the integer LC is used to count the load cases as they are handled, and I is used to count the members as the member stiffness matrices are assembled.

Notes on Table 13.1–1

(1) A sample set of data set out according to this scheme is to be seen in Example 13.1–1.

(a) The preliminary data would include a mention of the type of structure—space frame, grillage, etc.—the number of members, number of joints, number of load cases and common material properties such as Young's modulus and the shear modulus.

(b) The coordinates of each joint in structure axes are listed against the joint number. Joint numbers are used in preference to the letters of Chapter 7 for ease of handling on the computer.

(c) It frequently happens that many members in a structure have the same sectional properties, and it is helpful to detail each type used against a section code number.

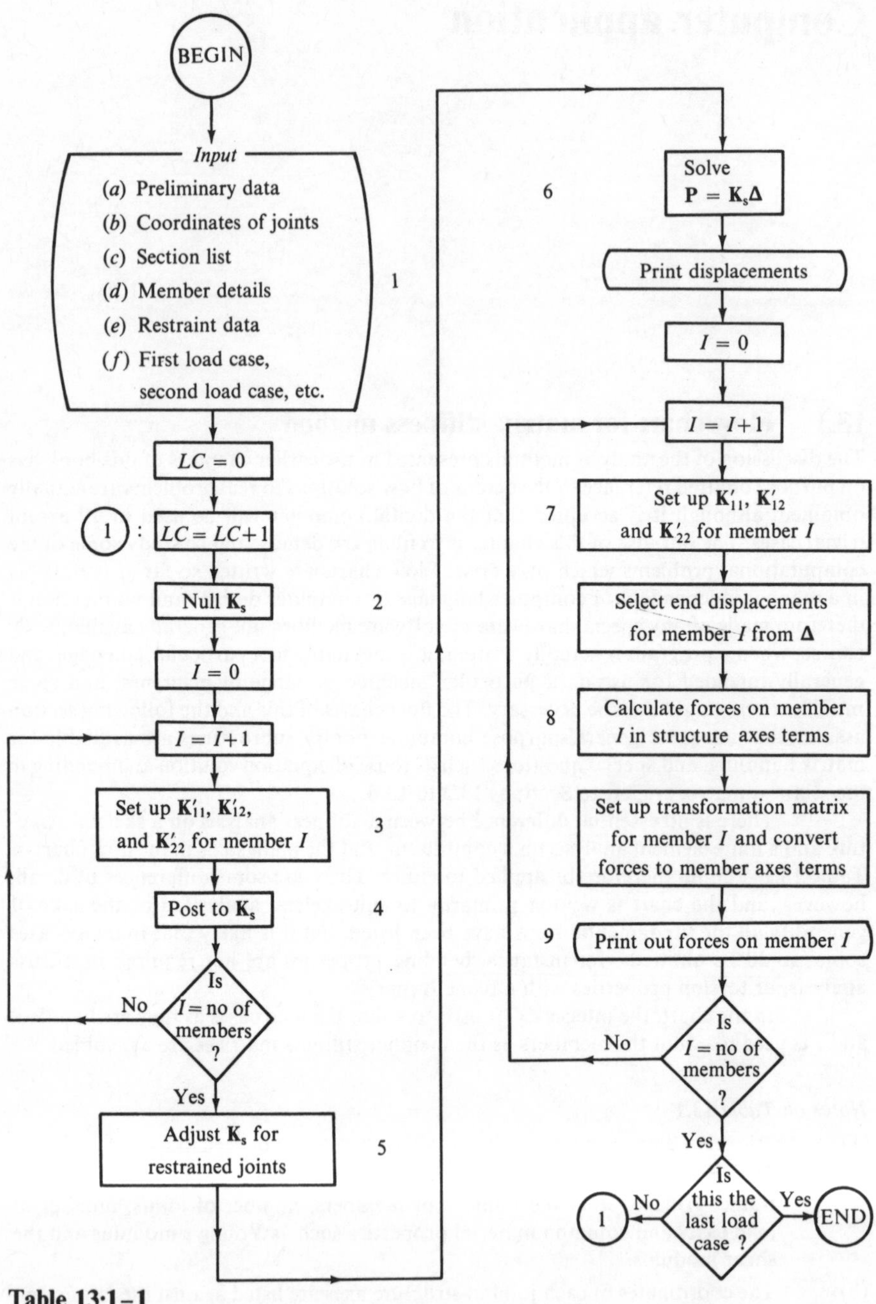

Table 13·1–1

(d) The member details listed here are a member code number, a section code number, the numbers of the joints the member connects, and some information on the end conditions—whether ends are pinned, and if so about which axis rotation is possible.

The projections of each member on the structure axes could be included in this list, thus making list (b) superfluous. Once the information under headings (b) and (d) has been entered, a useful check is to draw the lay-out of the structure on the computer's plotter where this facility is available, or on a visual display unit (VDU). This provides a very valuable check on the member connectivity, and it quickly shows up any error that might go unnoticed in a data table.

(e) This table lists the details of supports—whether they are encastré, pinned, roller, etc.

(f) The details of each load case are entered in turn. All cases may be read in together, or each one as it is required.

(2) The terms of the stiffness matrix $\mathbf{K}_s$ are all set to zero.

(3) Each member is taken in turn. The joint numbers at each end, together with the joint coordinates are used to calculate the member length and to set up the transformation matrix $\mathbf{T}$ (Eqn 5.3–16). The matrices $\mathbf{K}_{11}$, $\mathbf{K}_{12}$, and $\mathbf{K}_{22}$ (Eqn 7.18–1) are set up and converted to structure axis form by carrying out the multiplications $\mathbf{T}^T\mathbf{K}_{11}\mathbf{T}$, etc. (Eqn 7.6–5). The order and form of the matrices will depend on the structure type and end conditions detailed in (1a) and (1d) respectively (see Section 7.11).

(4) The matrices $\mathbf{K}'_{11}$, $\mathbf{K}'_{12}$, and $\mathbf{K}'_{22}$ calculated in Step 3 are posted into the relevant positions of the structure stiffness matrix $\mathbf{K}_s$ according to the rules discussed in Section 7.9. A subroutine POST for doing this will be discussed at the end of the section.

(5) The rows and columns of $\mathbf{K}_s$ corresponding to zero displacements at pinned and roller supports are set to zero. Encastré supports can also be treated in this way, with all the associated rows and columns being set at zero and unity being inserted in the leading diagonal positions. This process was discussed in Section 7.10 and illustrated in Section 7.12.

(6) As an alternative to solution of the equations $\mathbf{K}_s$ may be inverted, and if this is done the inverse can be used for all load cases. However, as inversion of an $n \times n$ matrix involves as many separate multiplications as solution of the set of n equations for n different loadings, solution is generally to be preferred.

(7) The matrices $\mathbf{K}'_{11}$, $\mathbf{K}'_{12}$, and $\mathbf{K}'_{22}$ are again set up for each member to allow calculation of member forces. These matrices could be stored from Step 3, but the amount of storage required would be quite considerable and the time saved would probably not be significant.

(8) Member forces are calculated first in structure axes (Eqn 7.6–5) and then converted to member axes (Eqn 5.3–4).

(9) It is sometimes convenient to store the member forces for each load case separately so that the effects of various load combinations can be studied.

The subroutine POST. The posting of the member stiffness matrices into the structure stiffness matrix must be carried out term by term and the flow chart for a subroutine POST for achieving this is shown in Table 13.1–2. The subroutine is designed to cater for any order of matrix $\mathbf{K}'_{11}$, etc., so is suitable for continuous beams ($\mathbf{K}'_{11}$, 1×1),† plane trusses ($\mathbf{K}'_{11}$, 2×2), plane frames, grillages, and space trusses ($\mathbf{K}'_{11}$, 3×3) and space frames ($\mathbf{K}'_{11}$, 6×6). It is assumed that all terms of $\mathbf{K}_s$ are zero before posting begins. The problem is illustrated in Fig. 13.1–1 for a space frame member joining joints numbered 7 and 12. In the figure the small squares represent positions for the individual terms of the matrix. The matrices $\mathbf{K}'_{11}$, etc., are of order 6×6. If it is assumed that the joint numbering is continuous and the first six joints therefore correspond to rows and columns 1 to 36 of $\mathbf{K}_s$, the first position occupied by each of the three matrices is $\mathbf{K}_s(37, 37)$ (meaning the location in the 37th row and 37th column of $\mathbf{K}_s$), $\mathbf{K}_s(37, 67)$ and $\mathbf{K}_s(67, 67)$. The final matrix $\mathbf{K}'_{21}$ is completed from $\mathbf{K}'_{12}$. The flow chart shows a jump to cater for the situation where end 1 of a member is encastré and numbered 0, as in this case only matrix $\mathbf{K}'_{22}$ is required (see for example Eqn 7.9–1).

† ($\mathbf{K}'_{11}$, 1×1) means that matrix $\mathbf{K}'_{11}$ is of order 1 by 1.

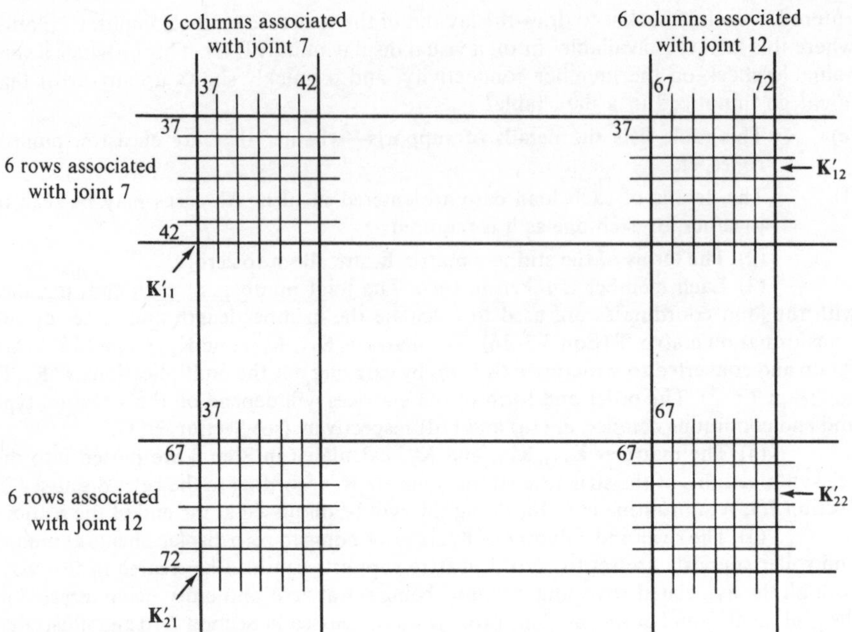

Fig. 13·1–1

Notes on Table 13.1–2

The member stiffness matrices are entered in $\mathbf{K}_s$ row by row.

(1) The terms of $\mathbf{K}'_{11}$ are added to values already in $\mathbf{K}_s$.
(2) The terms of $\mathbf{K}'_{12}$ are entered, but no summation is necessary.
(3) The positions in $\mathbf{K}_s$ required for $\mathbf{K}'_{21}$ are completed from $\mathbf{K}'_{12}$ as $\mathbf{K}'_{21} = \mathbf{K}'_{12}{}^T$.
(4) The terms of $\mathbf{K}'_{22}$ are added to values already in $\mathbf{K}_s$.

Example of input data—Example 13.1–1. The following example illustrates a possible lay-out of the data for the stiffness analysis of the plane frame of Fig. 13.1–2. For programming convenience most of the data is numerical, but some computer bureau programs make extensive use of alphabetic characters and statements (for code numbers and data table headings, for instance). They may also allow data items to be presented in a range of different units—metres may be suitable for joint coordinates while square centimetres may be most suitable for cross-sectional areas—or they may even be able to cater for a mixture of metric and non-metric data. There is no essential difficulty in the inclusion of such facilities but they may increase the size of the program very considerably. For simplicity, it is assumed here that all data is presented in a consistent set of units (kN and mm) and if this is done all results obtained are in the same set.

If a main-frame computer were to be used, the data for an analysis would normally be set up in a data file. The main program would then call this file to provide the necessary input. Likewise, output might be dumped to an output file, which could be printed out subsequently as desired. Frequently, input would be via a keyboard with a VDU, and the VDU could also be used to check output. Typical input is shown on the right hand side of Table 13.1–3. The notes on the left-hand side of the page are for guidance only.

Many analysis packages are designed to run on micro-computers. Here the program would have a data input stage where information and questions would appear on the computer VDU, and the user would type speciic answers—numeric and alphanumeric—as relevant. The headings might well be similar to those on the left-hand

Table 13.1–2. A flow chart for subroutine POST

N = Number of degrees of freedom per joint for the structure under consideration (1, 2, 3, or 6).
$\mathbf{K}_s$ = Stiffness matrix. All terms set to zero before the first member is handled.
$E1$ = Joint number at end 1 of the member.
$E2$ = Joint number at end 2 of the member ($E2 > E1$).
I and J = Two integer counters.

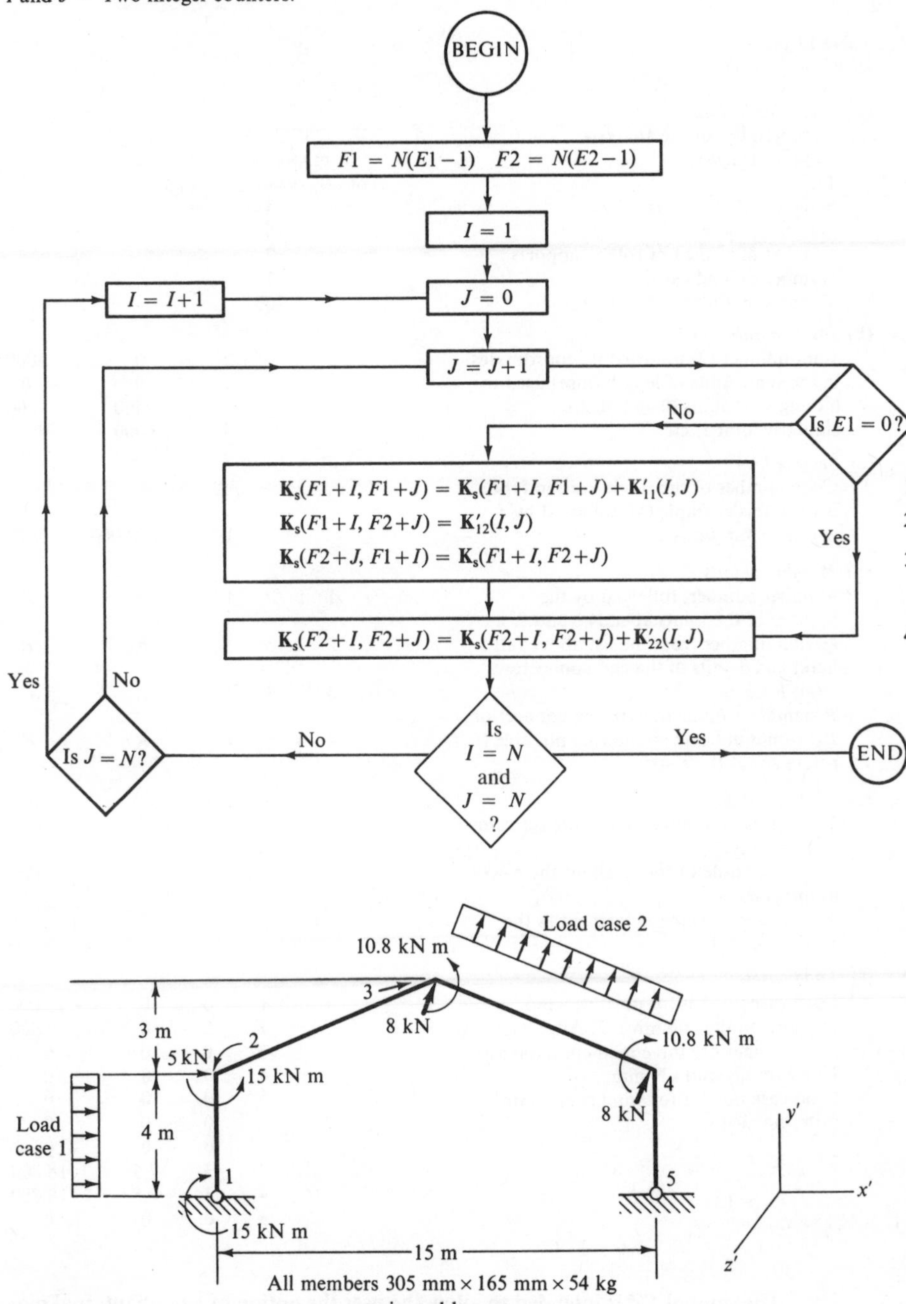

Fig. 13·1–2

side of Table 13.1–3. Sometimes graphics are used and the analyst may insert values into charts. A visual check on the layout of the structure may be possible on the screen, and sometimes output data can be viewed before selective printing.

The order of joint numbering can take any convenient form, but see Section 13.4 on banded matrices.

Table 13.1–3
(*See* Fig. 13.1–2)

(a) *Preliminary data-*

Code number of the analysis	17/23/A
Type of structure	PLANE FRAME
Title:	PORTAL FRAME ANALYSIS
Number of members	4
Number of joints	5
Number of pinned or roller supports	2
Number of load cases	2
Young's modulus (kN/mm^2)	200

(b) *Joint coordinates-*

Joint number (2) followed by coordinates	2	0	4000
in the same units of length (mm) used for	1	0	0
Young's modulus (0 and 4000),	3	7 500	7000
similarly joint 1, etc.	4	15 000	4000
	5	15 000	0

(c) *Section list-*

Code number of the single-section type used in this example (1), followed by *I* and *A* in mm units	1	116 860 000	6830

(d) *Member details-*

Member number, followed by the number of the joints at ends 1 and 2, a	1	1	2	1	*R*	*R*
section number from the list above (all 1 here) and details of the end connection	2	2	3	1	*R*	*R*
(all *R* here).	3	3	4	1	*R*	*R*
R signifies a moment-carrying connection, *PZ* would indicate an internal pin able to rotate about the z'-axis	4	4	5	1	*R*	*R*

(e) *Restraint data-*

The numbers of joints which are restrained in some way.	1	*PZ*
PZ—a pin able to rotate about the z'-axis as under *d*.	5	*PZ*
X—a roller allowing movement in the x'-direction, etc.		

(f) *Load cases-*

Load case no. 1 is for positive wind	1	1	0	0	−15 000
pressure on the column 1–2. After each		2	5	0	15 000
joint number are three load components.		3	0	0	0
Units are kN and kN mm.		4	0	0	0
Load case no. 2 is for wind suction on		5	0	0	0
rafter no. 3–4.	2	1	0	0	0
		2	0	0	0
		3	3	7.5	18 800
		4	3	7.5	−18 800
		5	0	0	0

The symbol *PZ* is intended to allow the user the option of having internal pins within a structure, and in this case the modified member stiffness matrices of Eqns 7.11–4 or 7.11–5 may be used. If a pin is internal, several members meet at the pinned joint, and so long as the modified $\mathbf{K}'$ matrices are not used for all such members the rows of zeros in $\mathbf{K}'$ are overlaid by non-zero terms in the assembly of $\mathbf{K}_s$. However, at a pinned support such as joints 1 or 5 in Fig. 13.1–2 there is one member only, and if the modified $\mathbf{K}'$ matrices were used in this case the result is a complete row and column of zeros in $\mathbf{K}_s$. This would lead the computer to conclude that $\mathbf{K}_s$ was singular and the equations consequently insoluble. Joints 1 and 5 are therefore regarded as rigid under heading (d), the presence of the pin being allowed for under (e).

13.2 Flow chart—elastic critical loads of plane frames

The flow chart of Table 13.2–1 is based on the method described in Section 9.12, and involves repeated stiffness method analysis of the structure at progressively increasing load factors. It was shown in Section 7.9 that the stiffness matrix $\mathbf{K}_s$ of a stable structure has the property of positive-definiteness, and a test for this property is performed at each load level. As soon as a load factor is used for which $\mathbf{K}_s$ ceases to be positive-definite and becomes singular it is known that the critical load has been reached. The test applied is to the determinant of the matrix which should be positive until singularity when it becomes zero, but in practice exact singularity cannot be obtained and the sign becomes negative corresponding to a state of unstable equilibrium.

The determinant is used here as a convenient quantity for which a standard computer library routine is likely to be available, but alternatives are discussed in Section 13.5. Some difficulties may occasionally be encountered in stability analyses, and these have already been discussed in Section 9.12.

Notes on Table 13.2–1

(1) The input data is very similar to that required by a stiffness analysis, the only difference from that of Table 13.1–1 being under item (a). The preliminary data must include an initial value for the load factor (LF), the increment to be applied to it (INC), and the required accuracy in the final result (ACC).

(2) N is used to count the number of loading cycles and DET 1, DET 2, and DET 3 are values of the determinant of the stiffness matrix. The axial forces in the members of the structure are not generally known before the analysis begins, and this chart assumes them all zero. Initial guesses might be read in as input data, but this is unlikely to provide much overall saving in computing time.

(3) A check on N might be inserted here to limit the number of load cycles in the event of any ill-conditioning preventing convergence.

(4) As the axial forces at each load level are initially only known approximately, a number of solutions (counted by I) are performed.

(5) The structure stiffness matrix is set up as described in Tables 13.1–1 and 13.1–2, but the member stiffness matrices used are those of Eqn 9.7–15 involving the stability functions. The functions are calculated as required using the current estimates of the axial forces.

(6) The repetition of the analysis performed at each load level should be terminated when the terms of the stiffness matrix converge to a steady state at successive cycles. The determinant (DET 2) is used here as a convenient quantity whose value depends on the stiffness matrix terms—if its values have reached a steady state, it is also likely that the terms of the matrix have done so. The alternative is to check the convergence of the individual displacements.

(7) The repeated analysis is terminated when the proportionate change in determinant is less than 0.1%. This limit is quite arbitrary, but seems to be reasonable in practice. The final value of determinant obtained is stored in DET 3 and its sign is used as the test of positive-definiteness.

(8) When the loading approaches the critical level, the stiffness matrix becomes increasingly ill-conditioned and successive values of the determinant are found to vary

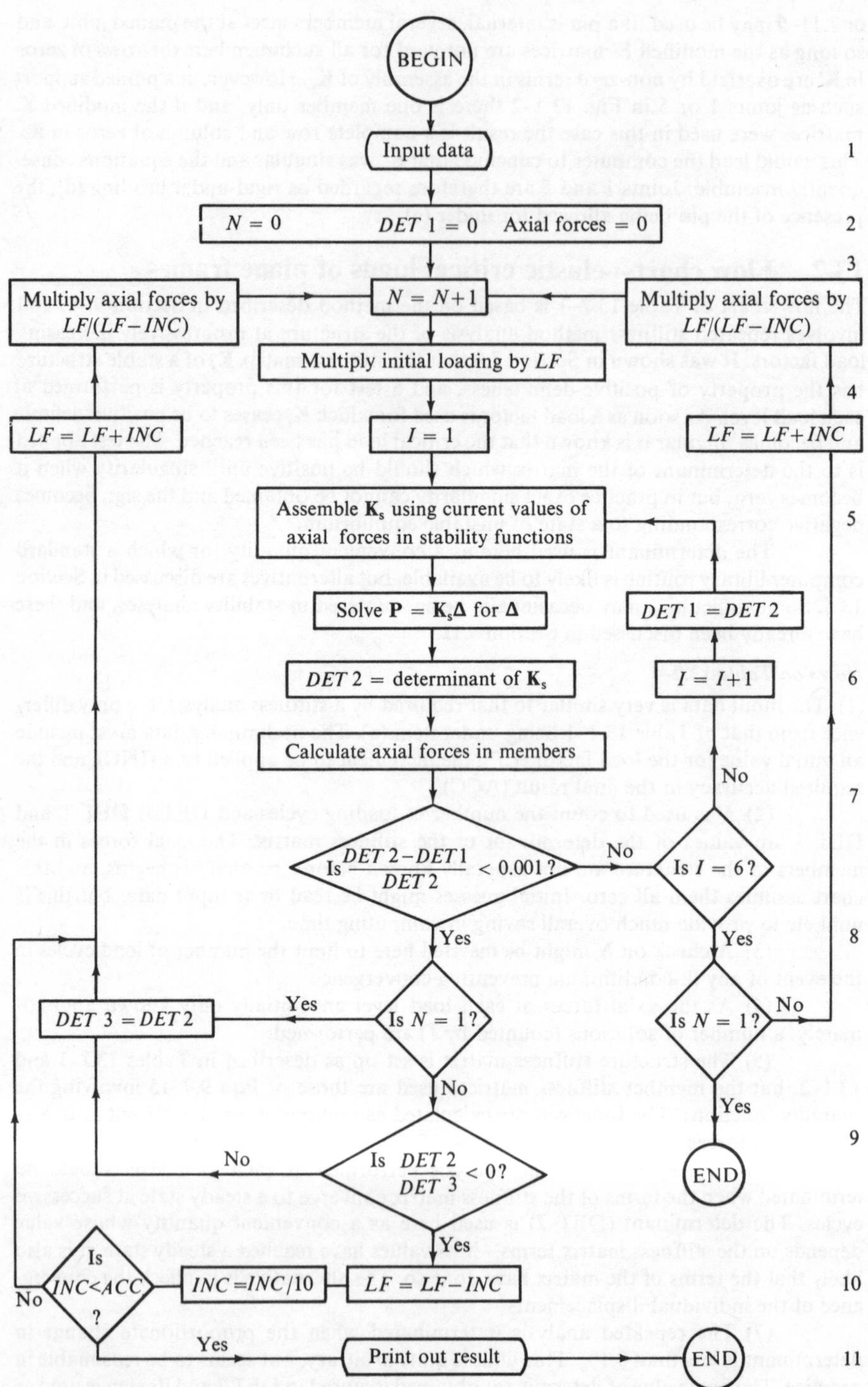
BEGIN
Input data
1
N = 0
DET 1 = 0
Axial forces = 0
2
3
Multiply axial forces by LF/(LF−INC)
N = N+1
Multiply axial forces by LF/(LF−INC)
Multiply initial loading by LF
4
LF = LF+INC
I = 1
LF = LF+INC
Assemble K_s using current values of axial forces in stability functions
5
Solve P = K_sΔ for Δ
DET 1 = DET 2
DET 2 = determinant of K_s
I = I+1
6
Calculate axial forces in members
No
7
Is (DET 2−DET 1)/DET 2 < 0.001?
No
Is I = 6?
Yes
Yes
8
Yes
DET 3 = DET 2
Is N = 1?
Is N = 1?
No
No
Yes
9
No
Is DET 2/DET 3 < 0?
END
Yes
Is INC<ACC?
No
INC = INC/10
LF = LF−INC
10
Yes
Print out result
END
11

Table 13·2–1

widely. In this case there is no virtue in continuing with repeated analyses, and the chart shows their termination at $I = 6$. If this termination is found necessary on the first load level attempted ($M = 1$) then obviously the initial loading is too near the critical level for any satisfactory analysis to be performed. In this case the analysis is stopped and must be repeated with a smaller set of loads. Otherwise the load factor can be increased, the axial forces increased by the same ratio as a good initial guess of their new values, and a further load level examined.

(9) When satisfactory convergence is achieved at any load level as described in Step 7 the value of the determinant obtained is compared with that from the previous load cycle. If no sign change is found then the load factor and axial forces are increased by one step and a new load level studied. (At the first load level this comparison is impossible, and is bypassed.)

(10) If a sign change is found in Step 9 (i.e. a change from positive to negative) then the critical value of load factor has been passed. The load factor is then decreased to its previous value, and increased again by smaller steps. If the reduced increment is below some prescribed accuracy (and great accuracy is not warranted) then the analysis is ended.

(11) Although output is shown only at this final stage it is generally desirable to trace the progress of the iterations by printing frequent intermediate results. Otherwise some of the snags discussed in Section 9.12 might pass unnoticed.

13.3 Solving large sets of linear equations

Although the worked examples on the stiffness method in Sections 7.12 and 7.14 required the solution of small sets of equations only, practical problems can lead to very large sets indeed. The solution of a space frame or a three-dimensional finite element grid involving six degrees of freedom at each of N nodes, requires the solution of $6N$ linear equations, and the storage of the complete stiffness matrix in the computer would require $36N^2$ or 14 400 locations for the modest number of 20 nodes. Many solution methods are available for linear equations, and where a small number only is involved the method chosen is immaterial. For large numbers, however, it is important to choose the most economical in computer time and storage, and that best suited to take advantage of any special properties the equations may possess.

The two classes of solution method available are termed direct and indirect methods. In the former a single set of operations is carried out on the equations and the results are obtained, while in the latter, solution is attempted by a series of successive approximations. A set of operations, usually much shorter than with a direct method, is repeated several times while the results either converge towards a steady level or show no definite trend. The only inaccuracy in the direct method is due to values being stored in the computer at each stage of the calculation to a limited number of decimal places, and fortunately structural problems are generally sufficiently well conditioned for this not to be a serious problem. The indirect methods, however, have the additional disadvantages that the rate of convergence may be slow, or the results may not converge at all, and for these reasons direct methods are to be preferred whenever possible.

A method which is particularly well suited to the solution of structural equations is Choleski's triangular decomposition method. Any square matrix **A** may be factorized or decomposed in the form:

$$\mathbf{A} = \mathbf{L}\mathbf{U} \tag{13.3–1}$$

where **L** and **U** are lower and upper triangular matrices and in the particular case of a symmetric matrix it can be arranged that:

$$\mathbf{U} = \mathbf{L}^T \quad \text{and} \quad \mathbf{A} = \mathbf{L}\mathbf{L}^T \tag{13.3–2}$$

The matrix **L** is unique for any given **A** and the leading diagonal terms are all positive if **A** is positive-definite (see Section 7.9).

solved for are $\mathbf{D}^{1/2}\mathbf{\Delta}$, and hence the required $\mathbf{\Delta}$ is obtained by pre-multiplying once more by $\mathbf{D}^{-1/2}$. Although $\mathbf{D}^{-1/2}$ is a square matrix, little additional storage is required as the terms may be stored in a single vector.

Table 13.3–1 Flow chart for Choleski decomposition

Matrix to be decomposed is $\mathbf{K}_s$ of order N

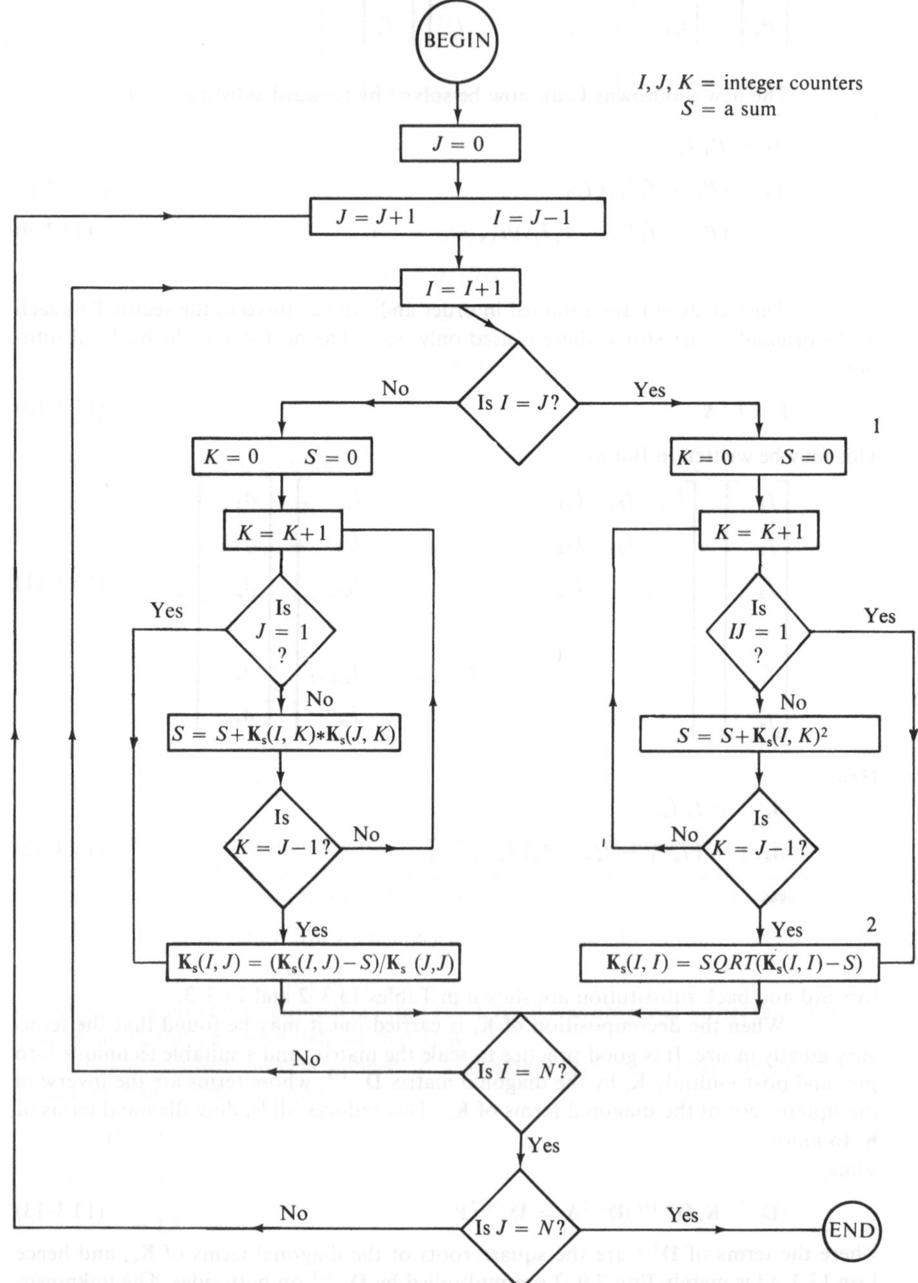

Notes on Table 13.3–1

(1) The right-hand loop calculates leading diagonal terms of **L** (Eqn 13.3–4) while the left-hand loop calculates the off-diagonal terms (Eqn 13.3–5).

(2) Once calculated the terms of **L** are stored in $\mathbf{K}_s$. Some of the terms of $\mathbf{K}_s$ remain in the upper triangular part, but they may be left there as these positions are not used in the forward or back substitution.

Table 13.3–2 Flow charts for forward substitution

The equation to be solved is $\mathbf{K}_s\mathbf{\Delta} = \mathbf{P}$ and the matrix $\mathbf{K}_s$ now holds the lower triangular matrix **L**.

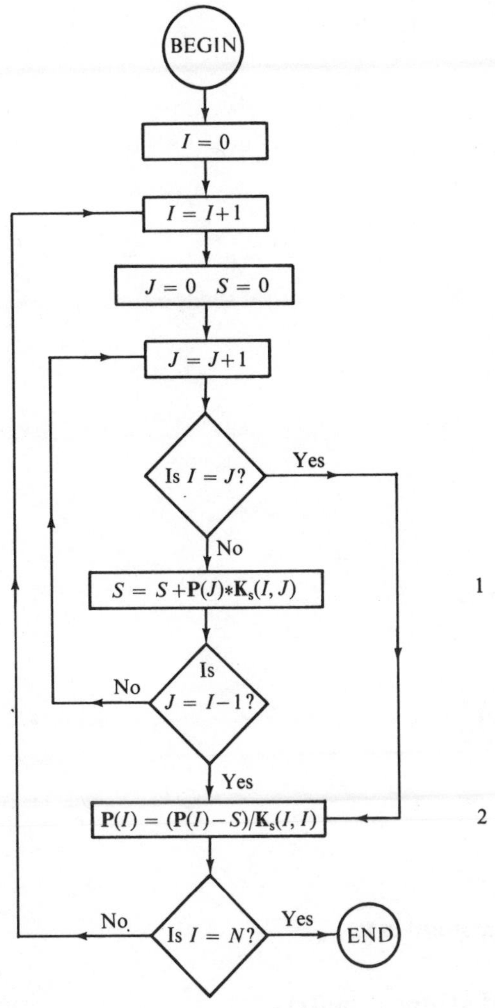

Notes on Table 13.3–2

(1) The pattern is seen in Eqn 13.3–9.

(2) Once calculated the *f* values (Eqn 13.3–9) are stored in **P**.

Table 13.3–3 Flow chart for back substitution

The lower triangular matrix L is stored in $\mathbf{K}_s$ and the intermediate values *f* are stored in P.

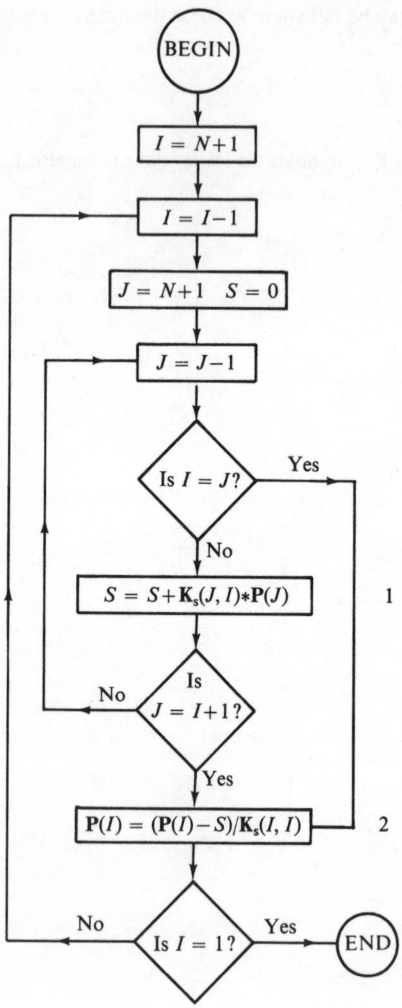

Notes on Table 13.3–3

(1) The pattern is seen in Eqn 13.3–12.

(2) Once calculated the unknowns **Δ** are stored in **P**.

13.4 Use of banded nature of stiffness matrix

The band width b of a symmetric matrix **A** is defined as the largest number of elements in any row from the diagonal to the extreme right-hand non-zero element, inclusive. Matrices for which b is much less than the order are said to be banded—the non-zero terms are grouped close to the diagonal—and with suitable node numbering most stiffness matrices and finite difference coefficient matrices show this property (see Section

7.9 and Table 11.14–1). If a symmetric, positive-definite banded matrix is decomposed by Choleski's method it can be concluded that:

1 The matrix **L** is banded, and is of the same band width as **A**.
2 At no stage of the decomposition is information required on more than b rows of $\mathbf{L}^T$. This can be seen by examination of the numbers of terms in the summation of Eqns 13.3–4 and 13.3–5.

For, if **A** = **LU** and **A** is banded, this may be written in detail as:

$$\overset{\leftarrow b=3\rightarrow}{\begin{bmatrix} a_{11} & a_{12} & a_{13} & & \\ a_{21} & a_{22} & a_{23} & a_{24} & \ddots \\ a_{31} & a_{32} & a_{33} & a_{34} & \ddots \\ & a_{42} & a_{43} & a_{44} & \ddots \\ & & \ddots & \ddots & \ddots \end{bmatrix}} = b \begin{bmatrix} l_{11} & & & \\ l_{21} & l_{22} & & \\ l_{31} & l_{32} & l_{33} & \\ l_{41} & l_{42} & l_{43} & \ddots \\ \vdots & & \ddots & \ddots \end{bmatrix} \overset{\leftarrow b \rightarrow}{\begin{bmatrix} l_{11} & l_{21} & l_{31} & l_{41} & \cdots \\ & l_{22} & l_{32} & l_{42} & \cdots \\ & & l_{33} & l_{43} & \cdots \\ & & & l_{44} & \cdots \\ & & & & \ddots \end{bmatrix}} \qquad (13.4\text{–}1)$$

The terms l_{11}, l_{21}, and l_{31} may be determined as in Section 13.3.

$$a_{41} = 0 = l_{41} \times l_{11}$$

then

$$l_{41} = 0 \quad \text{as} \quad l_{11} \neq 0$$

The repetition of this argument shows that the band width of **L** is the same as that of **A**, although the band is of course to one side of the diagonal. As with the unbanded case, computer storage locations of **A** may be overwritten by those of **L** as they are calculated.

On these conclusions could be based a solution process requiring only a very small amount of computer core store, or alternatively capable of dealing with very large sets of stiffness equations. A small part only of the stiffness matrix $\mathbf{K}_s$ would be set up initially—a portion of order b—and the first b rows of **L** could be calculated. These rows of **L** being no longer required could be output to a backing store, and further portions of $\mathbf{K}_s$ set up in the space made available. Such a process could solve any set of equations, however large, provided only that the computer fast store could accommodate the minimum of b^2 of storage with a small surplus for data storage, etc. The price that is paid is of course in transfer times, and it would be more economical to assemble as much of $\mathbf{K}_s$ as possible at one time.

Without recourse to the above type of solution considerable savings in computer storage can be made through use of the symmetric-banded nature of $\mathbf{K}_s$. If $\mathbf{K}_s$ is symmetric there is no need to store the complete matrix, and if it is banded there is no need to store the zero terms. The non-zero terms of the stiffness matrix of Fig. 13.4–1 could thus be stored in a rectangular array $\mathbf{RK}_s$ of order $n \times b$, requiring only nb storage locations instead of the n^2 required for $\mathbf{K}_s$.

Assembly of $\mathbf{RK}_s$ could be carried out in exactly the same manner as for a square matrix (Section 13.1), but the column numbering in all addresses would be amended. An item intended for location $i, j (i < j)$ of the square $\mathbf{K}_s$, would be posted instead to location $i, (j - i + 1)$. Similarly, in carrying out a Choleski solution this amended addressing scheme would be used, with the banded **L** matrix also stored in $\mathbf{RK}_s$. The application of this technique to the structure of Fig. 7.9–2, which has 18 degrees of freedom and a band width of 9 would lead to a computer storage saving of 50%, and the figure would be much more for many structures.

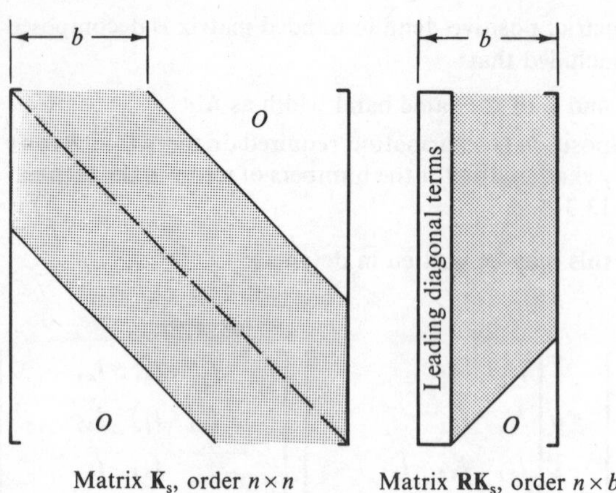

Fig. 13·4–1

Techniques for node numbering. It is clear from the foregoing discussion that there is great merit in numbering the nodal points in a structure, or finite difference or finite element grid, in a manner which leads to a minimum width of band. No infallible rules can be stated, but the general principle is to regard the structure as a strip and number transversely to the longitudinal centre line. For example, the structure of Fig. 7.9–2 is a high narrow tower, and the numbering is carried out along horizontal lines. The pin-jointed structure of Fig. 13.4–2 has no clearly defined length or breadth, however, and the numbering is carried out transversely to a diagonal line. The joints have been given numbers here, as this is what is used in the computer, despite the earlier convention of using capital letters. The order of the full stiffness matrix would be 24 × 24, with a band width of 7. Had the structure been numbered round the circumference, the matrix could have been full, with no banding at all. Three-dimensional structures are more difficult to deal with, and each must be considered on its merits, but the same general rule applies.

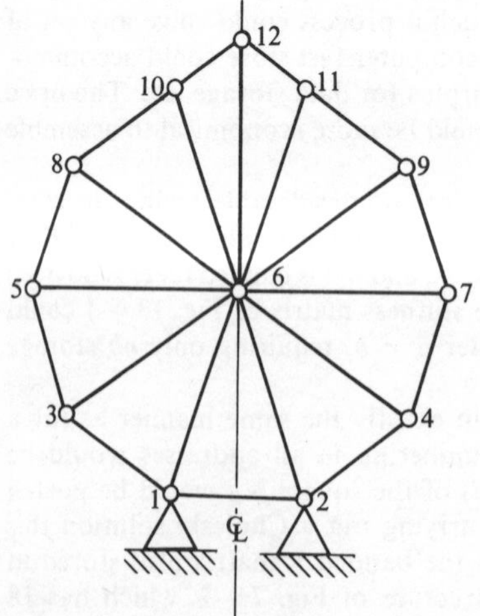

Fig. 13·4–2

13.5 Eigenvalues and eigenvectors

If **A** is a square matrix of order n, then there is at least one vector **x** and one corresponding scalar λ for which

$$\mathbf{Ax} = \lambda\mathbf{x} \tag{13.5–1}$$

x and λ are termed respectively an eigenvector and eigenvalue of matrix **A**. Introducing the unit matrix **I**, Eqn 13.5–1 could be rewritten

$$(\mathbf{A} - \lambda\mathbf{I})\mathbf{x} = 0 \tag{13.5–2}$$

This equation may be written in full as:

$$\begin{bmatrix} (a_{11} - \lambda) & a_{12} & a_{13} & \cdots & a_{1n} \\ a_{21} & (a_{22} - \lambda) & a_{23} & & \vdots \\ \vdots & & \ddots & & \vdots \\ a_{n1} & \cdots & \cdots & \cdots & (a_{nn} - \lambda) \end{bmatrix} \begin{bmatrix} x_1 \\ x_2 \\ \vdots \\ x_n \end{bmatrix} = 0 \tag{13.5–3}$$

where a_{11}, a_{12}, etc., are the individual terms of **A**, and x_1, x_2, etc., are the various unknowns. Equation 13.5–3 forms a homogeneous set of linear equations, whose determinant must be zero, or

$$|\mathbf{A} - \lambda\mathbf{I}| = 0 \tag{13.5–4}$$

This determinant could now be expanded, providing an nth order polynomial in λ, and although this would be a poor way of calculating λ, it does show that n values of λ exist. To each there will be a corresponding vector **x**. Equation 13.5–1 could therefore be rewritten as

$$\mathbf{AX} = \mathbf{XD} \tag{13.5–5}$$

where **X** is a square matrix of all the eigenvectors **x**, termed the modal matrix, and **D** is a diagonal matrix whose terms are the n eigenvalues. The eigenvectors of Eqn 13.5–3 cannot be calculated explicitly as the equations are homogeneous, but omission of one equation allows the ratios between the terms to be calculated (for an example see Section 9.13). The eigenvectors are usually stated in a normalized form with, for instance, the largest term or the square root of the sum of the squares of the terms, set to unity. It can be shown that if the matrix **A** is symmetric and positive-definite, all the eigenvalues are positive and real, and n eigenvectors exist.

The importance of eigenvalues and eigenvectors in vibration problems and principal stress analysis has already been discussed in Sections 10.5 and 11.5 respectively, but they are also of value in stability studies. As was shown in Section 9.11, a structure becomes unstable when its stiffness matrix becomes singular, and the test applied in that section was that its determinant would become zero. However, the modal matrix **X** is non-singular, and Eqn 13.5–5 can be rewritten

$$\mathbf{A} = \mathbf{XDX}^{-1} \tag{13.5–6}$$

and

$$|\mathbf{A}| = |\mathbf{X}||\mathbf{D}||\mathbf{X}^{-1}| = |\mathbf{X}||\mathbf{X}^{-1}||\mathbf{D}| = |\mathbf{D}|$$

as

$$|\mathbf{X}||\mathbf{X}^{-1}| = 1$$

Therefore, if the determinant of **A** is to be zero, that of **D** must be zero, and as the determinant of **D** is the product of its diagonal terms, then one of the eigenvalues must be zero also. It can be concluded that although the eigenvalues of a stiffness matrix are positive while the structure is stable, the values alter with increasing load until at instability one or possibly more eigenvalues become zero, and at this time these are the smallest of the eigenvalues. It is found that the smallest eigenvalue passes smoothly through zero as the neutral equilibrium condition is reached, and is followed by further values at each critical load. For stability purposes, therefore, it is only the smallest eigenvalue of the stiffness matrix which is generally required, whereas for vibration studies several of the lowest, and perhaps all of the eigenvalues, are of interest.

It is beyond the scope of this book to discuss in full the general question of the determination of eigenvalues and eigenvectors, but two methods will be described briefly without proof. The calculation of either the smallest or largest eigenvalue can be carried out by a very simple iterative technique, but this method cannot easily be extended to intermediate values. If values other than the extremes of the range are required, the full set must be calculated.

Solution of largest and smallest eigenvalues. An initial guess is made of the eigenvector ($\mathbf{z}_0$) corresponding to the largest eigenvalue of matrix **A**, and $\mathbf{z}_0$ is improved to $\mathbf{z}_1$ by the multiplication:

$$\mathbf{z}_1 = k_1^{-1}\mathbf{A}\mathbf{z}_0 \tag{13.5–7}$$

$\mathbf{z}_1$ is thus a new estimate of the normalized eigenvector (cf. Eqn 13.5–1), and k_1 is the largest term of $\mathbf{A}\mathbf{z}_0$. The general term after p iterations is

$$\mathbf{z}_p = k_p^{-1}\mathbf{A}\mathbf{z}_{p-1} \tag{13.5–8}$$

k_p converges to the eigenvalue of **A** of largest modulus $\lambda_{\max}$, and $\mathbf{z}_p$ is the corresponding eigenvector.

If the matrix $\mathbf{B} = \mathbf{A} - q\mathbf{I}$, where q is some scalar larger than $\lambda_{\max}$, then its eigenvalues may be determined from expansion of the determinant:

$$\begin{vmatrix} (a_{11} - q - \lambda_B) & a_{12} & \cdots \\ a_{21} & (a_{22} - q - \lambda_B) & \cdots \\ \vdots & \vdots & \ddots \\ & & (a - q - \lambda_B) \end{vmatrix} = 0 \tag{13.5–9}$$

(λ_B is an eigenvalue of B)

It can be seen from comparison with Eqn 13.5–3 that

$$\lambda_A = q + \lambda_B$$

(λ_A is an eigenvalue of **A**), or

$$\lambda_B = \lambda_A - q$$

The eigenvalues of **B** are thus equal to those of **A** less the quantity q, and therefore if $q > \lambda_{\max}$, the eigenvalue of largest modulus in **B** corresponds to the smallest eigenvalue of **A**, $\lambda_{\min}$. This is illustrated in Fig. 13.5–1 where it can be seen that if **A** is a positive-definite matrix all of whose eigenvalues are real and positive, the effect of q is to effectively shift the origin to the right-hand side of $\lambda_{\max}$.

The preceding iteration of Eqn 13.5–8 is now performed on matrix **B**, and the eigenvalue of largest modulus so found is corrected by q to give λ_{min} for **A**. The process is illustrated in the following Example 13.5–1.

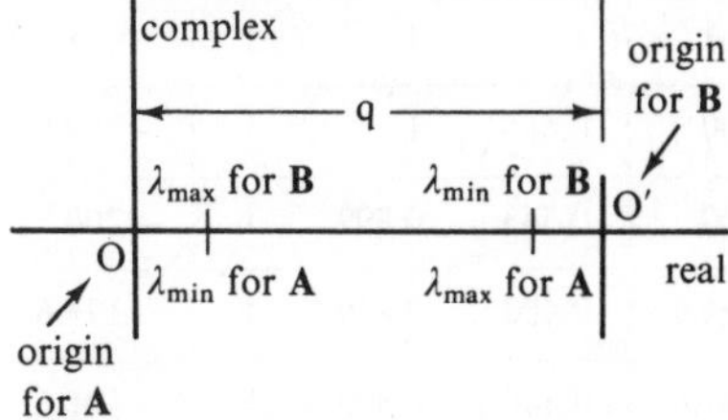

Fig. 13·5–1

Example 13.5–1. The largest and smallest eigenvalues of the matrix **A** used in Example 11.5–2 are required:

$$\mathbf{A} = \begin{bmatrix} 127 & -60 & 0 \\ -60 & 113 & -53 \\ 0 & -53 & 53 \end{bmatrix}$$

The starting eigenvector $\mathbf{z}_p$ is taken as $[1\ \ 1\ \ 1]^T$

p	$\mathbf{Az}_{(p-1)}$			$\mathbf{z}_p$			k_p
0				1	1	1	1
1	67	0	0	1	0	0	67
2	127	−60.0	0	1	−0.473	0	127
3	155.4	−113.4	25.0	1	−0.730	0.161	155
4	170.8	−151.0	47.2	1	−0.885	0.277	171
5	180.0	−175.0	61.6	1	−0.972	0.342	180
6	185.4	−187.9	69.7	−0.987	1	−0.371	−187.9
7	−185.2	191.9	72.6	−0.967	1	−0.379	191.9
8	−182.8	191.9	73.1	−0.952	1	−0.381	191.9
9	−181.0	190.3	73.3	−0.951	1	−0.385	190.3
10	−180.8	190.5	73.4	−0.950	1	−0.386	190.5
11	−180.8	190.5	73.4	−0.950	1	−0.386	190.5

The largest eigenvalue is therefore 191, with the corresponding normalized eigenvector being $[-0.95\ \ 1\ \ -0.39]^T$.

To calculate the smallest eigenvalue, the value of q is taken as 200, and the new matrix **B** is:

$$\mathbf{B} = \begin{bmatrix} 127-200 & -60 & 0 \\ -60 & 113-200 & -53 \\ 0 & -53 & 53-200 \end{bmatrix} = \begin{bmatrix} -73 & -60 & 0 \\ -60 & -87 & -53 \\ 0 & -53 & -147 \end{bmatrix}$$

The starting vector is again taken as $[1\ 1\ 1]^T$

p	$\mathbf{Bz}_{(p-1)}$			$\mathbf{z}_p$			k_p
0	1	1	1	1	1	1	1
1	−133	−200	−200	0.665	1	1	−200
2	−108.5	−179.9	−200	0.543	0.899	1	−200
3	−93.5	−163.7	−194.6	0.480	0.840	1	−194.6
4	−85.5	−154.9	−191.6	0.446	0.809	1	−191.6
5	−81.1	−150.2	−189.9	0.429	0.791	1	−189.9
6	−78.7	−147.9	−188.9	0.416	0.781	1	−188.9
7	−77.3	−145.9	−188.4	0.410	0.775	1	−188.4
8	−76.5	−145.1	−188.1	0.406	0.771	1	−188.1
9	−76.0	−144.5	−187.9	0.405	0.770	1	−187.9
10	−75.7	−144.3	−187.9	0.404	0.770	1	−187.9

The eigenvalue of **B** of largest absolute value is therefore −188, with the corresponding normalized eigenvector being $[0.40\ \ 0.77\ \ 1]^T$. The smallest eigenvalue of matrix **A** is therefore $-188 + 200 = 12$. The reader should check that the corresponding eigenvector is also $[0.40\ 0.77\ 1]^T$. Hence

Largest eigenvalue of $\mathbf{A} = 191$
Smallest eigenvalue of $\mathbf{A} = 12$

Although the convergence is quite rapid in this example, and slide-rule calculations are quite suitable, this is by no means always the case. The rate of convergence depends on the ratios between the various eigenvalues, and if either the largest or smallest is very close to another value, the process is unable to distinguish between them unless many iterations are performed with a large number of significant figures. This makes automatic computation essential.

Solution of all eigenvalues for symmetric, positive-definite matrices—L–R method. The positive-definite symmetric matrix $\mathbf{A}_1$ of order n is first decomposed to triangular form as in the Choleski method of equation solution (Section 13.3).

$$\mathbf{A}_1 = \mathbf{L}_1\mathbf{L}_1^T \tag{13.5–10}$$

The reverse multiplication of $\mathbf{L}_1^T\mathbf{L}_1$ is now carried out, and set equal to $\mathbf{A}_2$, or

$$\mathbf{L}_1^T\mathbf{L}_1 = \mathbf{A}_2$$

$\mathbf{A}_2$ is now decomposed, as before. Then

$$\mathbf{A}_2 = \mathbf{L}_2\mathbf{L}_2^T$$

and

$$\mathbf{L}_2^T\mathbf{L}_2 = \mathbf{A}_3$$

The general step being

$$\mathbf{A}_p = \mathbf{L}_p\mathbf{L}_p^T \quad \text{and} \quad \mathbf{L}_p^T\mathbf{L}_p = \mathbf{A}_{p+1}$$

The process is repeated until **A** becomes a diagonal matrix with the eigenvalues in descending order in the leading diagonal positions. The normalized eigenvectors may be obtained by solving Eqn 13.5–2 with each eigenvalue in turn.

When the method is programmed for the computer, the decomposition is carried out as in Section 13.3, and the multiplication $\mathbf{L}_p^T\mathbf{L}_p$ is carried out as follows:

$$\mathbf{L}_p^T\mathbf{L}_p = \mathbf{A}_{p+1}$$

or in full

$$\begin{bmatrix} l_{11} & l_{21} & \cdots & l_{n1} \\ & l_{22} & & \vdots \\ & & \ddots & \vdots \\ & & & l_{nn} \end{bmatrix} \begin{bmatrix} l_{11} & & & \\ l_{21} & l_{22} & & \\ \vdots & & \ddots & \\ l_{n1} & \cdots & & l_{nn} \end{bmatrix} = \begin{bmatrix} a_{11} & a_{12} & \cdots & a_{1n} \\ a_{21} & a_{22} & & \vdots \\ \vdots & & \ddots & \vdots \\ a_{n1} & \cdots & \cdots & a_{nn} \end{bmatrix}$$

$$a_{11} = l_{11}^2 + l_{21}^2 + l_{31}^2 + \cdots + l_{n1}^2$$

$$a_{22} = l_{22}^2 + l_{32}^2 + \cdots + l_{n2}^2$$

or

$$a_{ii} = \sum_{m=i}^{m=n} l_{m,i}^2$$

Also

$$a_{12} = l_{21}l_{22} + l_{31}l_{32} + l_{41}l_{42} + \cdots + l_{n1}l_{n2}$$

and

$$a_{13} = l_{31}l_{33} + l_{41}l_{43} + \cdots + l_{n1}l_{n3}$$

or

$$a_{ij} = \sum_{m=j}^{m=n} l_{m,i}l_{m,j} \qquad (i < j)$$

The condition $i < j$ ensures that the upper triangle is completed, and as the matrix $\mathbf{A}_p$ remains symmetric throughout the process, the lower half can be completed from symmetry. If this pattern is adopted, the new matrix $\mathbf{A}_{p+1}$ can be stored in the same locations in the computer as matrix $\mathbf{L}_p$.

It has been suggested at the beginning of this section that a calculation of the smallest eigenvalue can be used as a check on the singularity of the stiffness matrix during a stability analysis. However, the analysis requires that the equation

$$\mathbf{P} = \mathbf{K}_s\boldsymbol{\Delta}$$

should also be solved for displacements at each load level, and as equation solution destroys the original matrix, the eigenvalue check must be carried out first. The iterative process described alters $\mathbf{K}_s$ only in the leading diagonal terms (in forming **B**) and this by a fixed amount. $\mathbf{K}_s$ can therefore be corrected once the smallest eigenvalue has been calculated, and the equations then solved. If several eigenvalues are required, however, the stiffness matrix must be reassembled, as the *L–R* process destroys it.

Example 13.5–2. The three eigenvalues of the matrix **A** used in Example 13.5–1 will be determined using the *L–R* method. **A** is a stress tensor taken from Example 11.5–2, and although such matrices need not be positive-definite, it was seen in Example 13.5–1 that two eigenvalues were positive, and as the determinant is also positive, the third eigenvalue must also be positive. **A** is therefore positive-definite, and the *L–R* method can be used.

The analysis takes place from left to right, $\mathbf{A}$ being decomposed into $\mathbf{LL}^T$, and then a new $\mathbf{A}$ being set equal to $\mathbf{L}^T\mathbf{L}$.

$$\overset{\mathbf{A}}{\begin{bmatrix} 127 & -60 & 0 \\ -60 & 113 & -53 \\ 0 & -53 & 53 \end{bmatrix}} = \mathbf{A}_1 = \overset{\mathbf{L}_1}{\begin{bmatrix} 11.3 & & \\ -5.34 & 9.20 & \\ 0 & -5.76 & 4.45 \end{bmatrix}} \mathbf{L}_1^T$$

$$\overset{\mathbf{L}_1^T\mathbf{L}_1}{\begin{bmatrix} 156.4 & -49.1 & 0 \\ -49.1 & 117.8 & -25.6 \\ 0 & -25.6 & 19.8 \end{bmatrix}} = \mathbf{A}_2 = \overset{\mathbf{L}_2}{\begin{bmatrix} 12.5 & & \\ -3.93 & 10.13 & \\ 0 & -2.53 & 3.66 \end{bmatrix}} \mathbf{L}_2^T$$

$$\overset{\mathbf{L}_2^T\mathbf{L}_2}{\begin{bmatrix} 171.6 & -39.8 & 0 \\ -39.8 & 109.2 & -9.25 \\ 0 & -9.25 & 13.4 \end{bmatrix}} = \mathbf{A}_3 = \overset{\mathbf{L}_3}{\begin{bmatrix} 13.1 & & \\ -3.04 & 10.00 & \\ 0 & -0.92 & 3.54 \end{bmatrix}} \mathbf{L}_3^T$$

- -

$$\overset{\mathbf{L}_7^T\mathbf{L}_7}{\begin{bmatrix} 191.1 & & \\ -7.6 & 91.5 & \\ 0 & 0 & 12.4 \end{bmatrix}} = \mathbf{A}_8 = \overset{\mathbf{L}_8}{\begin{bmatrix} 13.80 & & \\ -0.55 & 9.56 & \\ 0 & 0 & 3.52 \end{bmatrix}} \mathbf{L}_8^T$$

$$\overset{\mathbf{L}_8^T\mathbf{L}_8}{\begin{bmatrix} 191.4 & -5.26 & 0 \\ -5.26 & 91.5 & 0 \\ 0 & 0 & 12.4 \end{bmatrix}} = \mathbf{A}_9$$

The development of $\mathbf{A}$ into a diagonal matrix can be seen from the steps shown, and although $\mathbf{A}_9$ still has non-zero off-diagonal terms, the eigenvalues are defined to three figures. It can be seen that the matrix $\mathbf{A}$ retains its symmetric and banded form at each stage.

Alternative means of checking singularity. An alternative to the use of eigenvalues in checking singularity in a stability analysis if a Choleski solution is used, is to observe the sign in the square-rooting operation required for the calculation of diagonal terms in the decomposed matrix. For a positive-definite matrix these signs should all be positive, but one will become negative as soon as the singular condition is passed. The method has the advantage that, apart from examining the terms as they are calculated, no additional operations are required, but has the disadvantage that it provides no information from which eigenvectors can be calculated.

13.6 Reduction in order of stiffness matrix—use of sub-structures

Figure 13.6–1 shows a series of mesh points on an elastic structure—they might be joints in a skeletal structure, or finite element node points in either two or three dimensions. If a stiffness analysis were performed on the structure, an equation of the type seen in Eqn 7.9–2 would be obtained:

$$\mathbf{K}_s \begin{bmatrix} \mathbf{\Delta}_A \\ \mathbf{\Delta}_B \\ \mathbf{\Delta}_C \\ \mathbf{\Delta}_D \\ \mathbf{\Delta}_E \\ \mathbf{\Delta}_F \\ \mathbf{\Delta}_G \end{bmatrix} = \begin{bmatrix} \mathbf{P}_A \\ \mathbf{P}_B \\ \mathbf{P}_C \\ \mathbf{P}_D \\ \mathbf{P}_E \\ \mathbf{P}_F \\ \mathbf{P}_G \end{bmatrix} \tag{13.6–1}$$

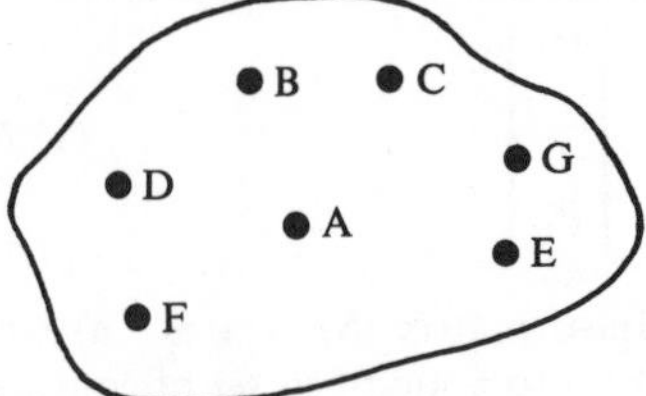

Fig. 13·6–1

Each term in the two vectors $\mathbf{\Delta}$ and $\mathbf{P}$ is a subvector of order appropriate to the type of structure, and the order of $\mathbf{K}_s$ would be 7 × 7 in equivalent submatrices.

Partitioning of the stiffness matrix would lead to:

$$\begin{bmatrix} \mathbf{S} & \mathbf{a} \\ \mathbf{a}^T & \mathbf{M} \end{bmatrix} \begin{bmatrix} \mathbf{\Delta}_A \\ \mathbf{\Delta}_B \\ \mathbf{\Delta}_C \\ \mathbf{\Delta}_D \\ \mathbf{\Delta}_E \\ \mathbf{\Delta}_F \\ \mathbf{\Delta}_G \end{bmatrix} = \begin{bmatrix} \mathbf{P}_A \\ \mathbf{P}_B \\ \mathbf{P}_C \\ \mathbf{P}_D \\ \mathbf{P}_E \\ \mathbf{P}_F \\ \mathbf{P}_G \end{bmatrix} \tag{13.6–2}$$

S and **M** are two symmetric matrices chosen such that on expansion of the left-hand side the rows of **S** multiply the unknowns $\mathbf{\Delta}_A$ to $\mathbf{\Delta}_E$, and those of **M** the unknowns $\mathbf{\Delta}_F$ and $\mathbf{\Delta}_G$. Although $\mathbf{K}_s$ is singular unless there are sufficient support restraints to prevent rigid-body movement, there is no reason why **S** or **M** should be. For if there were complete restraint at all the joints corresponding to either **S** or **M**, then the determinant of $\mathbf{K}_s$ would be the determinant of the other, and this must be positive as such restraint would be enough to prevent any rigid-body movement.†

Expansion of the expression on the left-hand side of Eqn 13.6–2 leads to

$$\mathbf{S} \begin{bmatrix} \mathbf{\Delta}_A \\ \vdots \\ \mathbf{\Delta}_E \end{bmatrix} + \mathbf{a} \begin{bmatrix} \mathbf{\Delta}_F \\ \mathbf{\Delta}_G \end{bmatrix} = \begin{bmatrix} \mathbf{P}_A \\ \vdots \\ \mathbf{P}_E \end{bmatrix} \tag{a}$$

† For example, if $\mathbf{\Delta}_A$ to $\mathbf{\Delta}_E$ inclusive are fully restrained, then $\mathbf{a} = \mathbf{a}^T = 0$ and $\mathbf{S} = \mathbf{I}$ (see Section 7.10); hence $|\mathbf{K}_s| = |\mathbf{M}|$.

and

$$\mathbf{a}^T\begin{bmatrix}\boldsymbol{\Delta}_A\\ \vdots\\ \boldsymbol{\Delta}_E\end{bmatrix} + \mathbf{M}\begin{bmatrix}\boldsymbol{\Delta}_F\\ \boldsymbol{\Delta}_G\end{bmatrix} = \begin{bmatrix}\mathbf{P}_F\\ \mathbf{P}_G\end{bmatrix} \tag{b}$$

Pre-multiplication of (a) by $\mathbf{a}^T\mathbf{S}^{-1}$ gives

$$\mathbf{a}^T\begin{bmatrix}\boldsymbol{\Delta}_A\\ \vdots\\ \boldsymbol{\Delta}_E\end{bmatrix} + \mathbf{a}^T\mathbf{S}^{-1}\mathbf{a}\begin{bmatrix}\boldsymbol{\Delta}_F\\ \boldsymbol{\Delta}_G\end{bmatrix} = \mathbf{a}^T\mathbf{S}^{-1}\begin{bmatrix}\mathbf{P}_A\\ \vdots\\ \mathbf{P}_E\end{bmatrix} \tag{c}$$

Subtraction of (c) from (b) leads to

$$(\mathbf{M} - \mathbf{a}^T\mathbf{S}^{-1}\mathbf{a})\begin{bmatrix}\boldsymbol{\Delta}_F\\ \boldsymbol{\Delta}_G\end{bmatrix} = \begin{bmatrix}\mathbf{P}_F\\ \mathbf{P}_G\end{bmatrix} - \mathbf{a}^T\mathbf{S}^{-1}\begin{bmatrix}\mathbf{P}_A\\ \vdots\\ \mathbf{P}_E\end{bmatrix} \tag{13.6–3}$$

Equation 13.6–3 thus has a reduced stiffness matrix $(\mathbf{M} - \mathbf{a}^T\mathbf{S}^{-1}\mathbf{a})$ which relates the displacements at two points only (F and G) to a modified set of equivalent loads. Partitioning of $(\mathbf{M} - \mathbf{a}^T\mathbf{S}^{-1}\mathbf{a})$ into those rows corresponding to $\boldsymbol{\Delta}_F$ and $\boldsymbol{\Delta}_G$ respectively would give

$$\begin{bmatrix}(\mathbf{K}'_{11})_{F-G} & (\mathbf{K}'_{12})_{F-G}\\ (\mathbf{K}'_{21})_{F-G} & (\mathbf{K}'_{22})_{F-G}\end{bmatrix}\begin{bmatrix}\boldsymbol{\Delta}_F\\ \boldsymbol{\Delta}_G\end{bmatrix} = \begin{bmatrix}\mathbf{P}_F\\ \mathbf{P}_G\end{bmatrix} - \mathbf{a}^T\mathbf{S}^{-1}\begin{bmatrix}\mathbf{P}_A\\ \vdots\\ \mathbf{P}_E\end{bmatrix} \tag{13.6–4}$$

This is a member stiffness equation exactly analogous to Eqns 7.6–5 and 7.6–6, and could therefore be used in the analysis of some much larger structure into which the body A, B ... G was connected at F and G only. For this reason the body A, B ... G is termed a substructure. The concept is particularly useful for analyses of large structures in which many identical parts each containing many nodes, such as lattice girders, for example, are connected to the remainder of the structure at a few nodes only. Once an analysis has been carried out on such a girder alone, which might in itself be a sizeable problem, the member can be treated in future cases as having no larger a stiffness matrix than a normal beam. It also allows the analyst to gain an overall picture of the main forces within a structure, without the picture being confused by too much detail—this can be filled in subsequently by local analyses if necessary.

The mathematical interpretation of Eqn 13.6–3 is that a proportion of the unknown quantities in the original problem have been eliminated as a group, and for this reason the process is given the name **block elimination**.

General references

Ashkenazi, V. Solution and error analysis of large geodetic networks. *Survey Review*, **19**, Nos. 146 and 147, 1967–8.

Fox, L. *An Introduction to Numerical Linear Algebra*. Oxford University Press, 1964.

Chapter 14
Plastic theory of structures

14.1 Introduction

In the foregoing chapters, the discussion has generally been restricted to structural behaviour in the elastic range, and the reader's attention was mainly directed to the stresses and deformations under working load. In this chapter, however, the behaviour of beams and frames in the plastic range will be examined; the aim is to develop a *plastic theory* for the determination of the ultimate-load capacities of structures.

Largely as a result of the work of Sir John Baker's team at Cambridge University[1], plastic theory has been established as a practical tool in structural design. Authoritative books[1–7] already exist which treat the subject in depth, and the aim of the present Authors is to present in a single chapter a clear account of those topics that are taught at undergraduate level at universities and polytechnics. Plastic theory can be applied to many types of structures, but, in keeping with the declared aim of the chapter, discussion will be restricted to beams and frames made of ductile materials, such as structural mild steel.

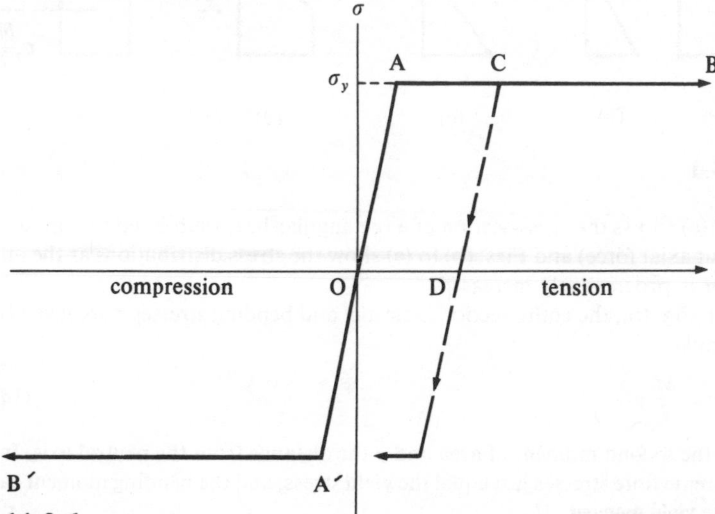

Fig. 14·2–1

14.2 The elastic-plastic stress-strain relation

Fig. 14.2–1 shows the stress-strain relation for an ideal **elastic-plastic** material (as in Fig. 9.3–5) which follows Hooke's law up to the yield stress σ_y, then yields plastically at constant stress. That is OA is a straight line having a slope equal to the modulus of elasticity E; AB is horizontal and extends without limit. If the stress is reduced at any point (C) in the plastic range the return path (CD) is a straight line parallel to OA. Both E and σ_y, and indeed the whole stress-strain relation, are the same for tension and for compression.

Admittedly, the idealized stress-strain relation in Fig. 14.2–1 is only a mathematical model, but it is a close approximation to the behaviour of structural mild steel as well as a reasonable first approximation to many continuously strain-hardening materials used in structural engineering. Typically, structural steel yields at a strain ε_y of the order of 0.1 %, and thereafter the strain increases at constant stress to at least ten times ε_y before strain-hardening occurs at a large strain of the order of 1.5%. The assumption of perfect plasticity after the yield stress is reached amounts to ignoring the effects of strain hardening and errs on the safe side.

In subsequent discussions (sections 14.3–14.5) we shall, unless otherwise stated, assume an elastic-plastic stress-strain relation.

14.3 Plastic bending without axial force

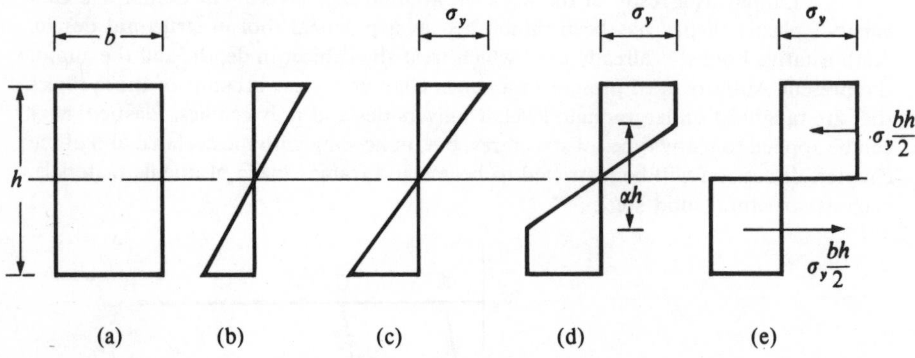

Fig. 14·3–1

Fig. 14.3–1(a) shows the cross-section of a rectangular beam subjected to pure bending (*i.e.* without axial force) and Figs. (b) to (e) show the stress distributions as the applied moment M is progressively increased.

In Fig. (b), the entire section is elastic, and bending stresses σ are given by the usual formula

$$\sigma = \frac{M}{I} y \tag{14.3–1}$$

where I is the second moment of area and y the distance from the neutral axis. In Fig. (c), the extreme fibre stresses just equal the yield stress, and the bending moment carried is called the **yield moment**, M_y:

$$\sigma_y = \frac{M_y}{I}\left(\frac{h}{2}\right) \quad \text{or} \quad M_y = \sigma_y \frac{I}{(h/2)} = \sigma_y \frac{bh^2}{6} \tag{14.3–2}$$

where $bh^2/6$ is, of course, the elastic section modulus Z.

As the moment is increased beyond M_y, the extreme fibre strain will continue to increase and the maximum strains will exceed the yield strain ε_y, but the maximum stresses remain constant at σ_y, as in Fig. (d). As the moment further increases, more and more fibres become plastic until finally (Fig. (e)) the whole cross-section is plastic; the bending moment at full plasticity is called the **plastic moment of resistance**, or simply the **plastic moment**, M_p. From Fig. (e)

$$M_p = \left(\sigma_y \frac{bh}{2}\right)\left(\frac{h}{2}\right) = \sigma_y \frac{bh^2}{4} \tag{14.3–3}$$

The stress distributions above are based on the proven assumption that plane sections remain plane in pure bending irrespective of whether stresses are elastic or plastic, provided the deformations are continuous.

In plastic analysis, the **plastic modulus**, Z_p, of a section is defined as the quantity which, when multiplied by the yield stress of the material, will give the value of the plastic moment M_p. It is seen from Eqn 14.3–3, that, for a rectangular section, $Z_p = bh^2/4$.

The ratio of the plastic moment M_p of a section to its yield moment M_y is called the **shape factor**. From Eqns 14.3–2 and 3, the shape factor for a rectangular section is

$$\frac{M_p}{M_y} = \frac{Z_p}{Z} = \frac{bh^2/4}{I/(h/2)} = 1.5 \tag{14.3–4}$$

since $I = bh^3/12$. It is thus seen that the shape factor is solely a function of the shape of the cross section; for example, it is 1.7 for a circular section and about 1.15 both for British Universal Beams and for American wide-flange beams.

Referring again to the stress distributions in Fig. 14.3–1, it should be noted that at any stage of the loading, the resultant force on the cross section must be zero, *i.e.*:

$$\int_A \sigma dA = 0 \tag{14.3–5}$$

Consequently, in elastic bending (Figs. 14.3–1(b) and (c)) the neutral axis must pass through the centroid of the cross-section; in plastic bending the neutral axis must divide the cross section into two equal areas (Fig. 14.3–1 (e)). For a symmetrical section, such as the rectangular section in Fig. 14.3–1, these two statements mean the same thing. However, for a non-symmetrical section, the position of the neutral axis changes as the moment increases from M_y to M_p as illustrated in Fig. 14.3–2.

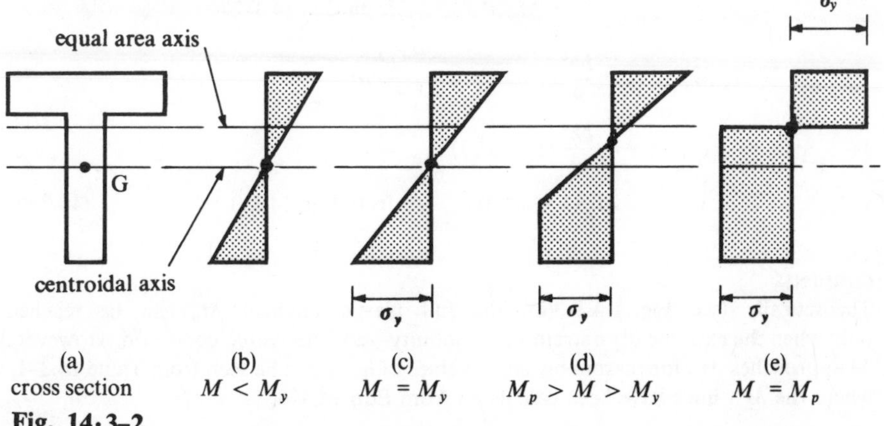

Fig. 14·3–2

In plastic analysis, the neutral axis is sometimes referred to as the **zero-stress axis.**

Example 14.3–1 Derive an expression for the bending moment in a rectangular section when the stress distribution is partly plastic and partly elastic (Fig. 14.3–3).

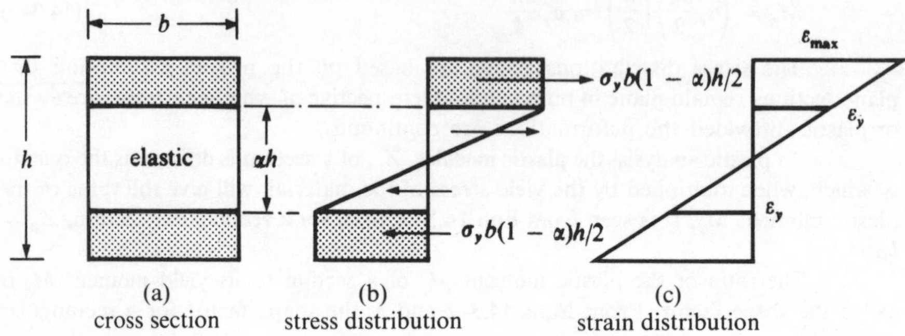

Fig. 14·3–3

SOLUTION Force in each plastic zone $= \sigma_y b\ (1-\alpha)h/2$

Moment lever arm for plastic zones $= \alpha h + (h - \alpha h)/2$

$$= (1+\alpha)h/2$$

$$\therefore M\ (\text{plastic zones}) = \left[\sigma_y b(1-\alpha)\frac{h}{2}\right]\left[\frac{(1+\alpha)h}{2}\right]$$

$$= \sigma_y \frac{bh^2}{4}(1-\alpha^2)$$

$$= M_p(1-\alpha^2) \quad \text{from Eqn 14.3–3}$$

$$M\ (\text{elastic core}) = \sigma_y\,[\text{elastic section modulus of elastic core}]$$

$$= \sigma_y \frac{bh^2}{6}\alpha^2$$

$$= M_y\,\alpha^2 \quad \text{from Eqn 14.3–2}$$

$$\therefore \text{Total } M = \sigma_y \frac{bh^2}{4}(1-\alpha^2) + \sigma_y \frac{bh^2}{6}\alpha^2$$

$$= \sigma_y \frac{bh^2}{4}(1-\alpha^2/3)$$

$$= M_p(1-\alpha^2/3) \quad \text{from Eqn 14.3–3} \qquad (14.3–6)$$

COMMENTS

Theoretically (see Fig. 14.3–3(c)) the full plastic moment M_p can be reached only when the extreme fibre strain ε_{max} is infinity—an impossible condition. However, M approaches M_p for reasonably small values of α, as can be seen from Table 14.3–1, where the M values have been calculated from Eqn 14.3–6.

Table 14.3–1 Relation between M and α

α	M
1	$0.667M_p$ †
1/10	$0.997M_p$
1/15	$0.999M_p$

† $M = 0.667\ M_p = M_y$ (Eqn 14.3–6)

Thus, if structural mild steel strain hardens at $\varepsilon_{max} = 15\ \varepsilon_y$(so that $\alpha = 1/15$), then the moment M would have reached 0.999 M_p by the time strain hardening occurs; even at $\varepsilon_{max} = 10\ \varepsilon_y$(so that $\alpha = 1/10$), M already reaches $0.997M_p$. Hence, for practical purposes, it is wholly acceptable to assume that the full plastic moment M_p can be developed. When the moment at a section of a beam reaches M_p, a **plastic hinge** is said to have formed at that section, because (see **Examples 14.3–2** and **3**) the local curvature approaches infinity and the rotation increases indefinitely under a constant value of the moment M_p.

Example 14.3–2 Sketch the moment-curvature curve for a rectangular section.

SOLUTION Refer again to Fig. 14.3–3 in **Example 14.3–1**. For M not exceeding the yield moment M_y, the curvature is:

$$\kappa = \frac{M}{EI} \tag{14.3–7}$$

which follows directly from the well-known relations $\sigma/y = M/I = E\ \kappa$.

For $M > M_y$, the flexural stiffness of the section is that of the elastic core only; Eqn 14.3–7 is therefore modified as

$$\kappa = \frac{M_e}{EI_e} \tag{14.3–8}$$

where M_e and I_e are respectively the resistance moment and the second moment of area of the elastic core. Referring to Fig. 14.3–3,

$I_e = bh^3\alpha^3/12$ and $M_e = \alpha^2\sigma_y bh^2/6$ as derived in Example 14.3–1.

Substituting into Eqn 14.3–8

$$\kappa = \frac{1}{\alpha}\left(\frac{2\sigma_y}{Eh}\right) = \frac{1}{\alpha}\kappa_y \tag{14.3–9}$$

where κ_y is the value of κ for α equal to unity and is hence the curvature for $M = M_y$. From Eqn 14.3–6

$$M = (1 - \alpha^2/3)\ M_p \tag{14.3–10}$$

Eliminating α from Eqns 14.3–9 and 10, we have

$$\frac{\kappa}{\kappa_y} = \frac{1}{\sqrt{3(1 - M/M_p)}} \tag{14.3–11}$$

Eqn 14.3–11 is plotted in Fig. 14. 3–4, which also shows an approximate curve for a typical British Universal Beam section or American wide-flange section.

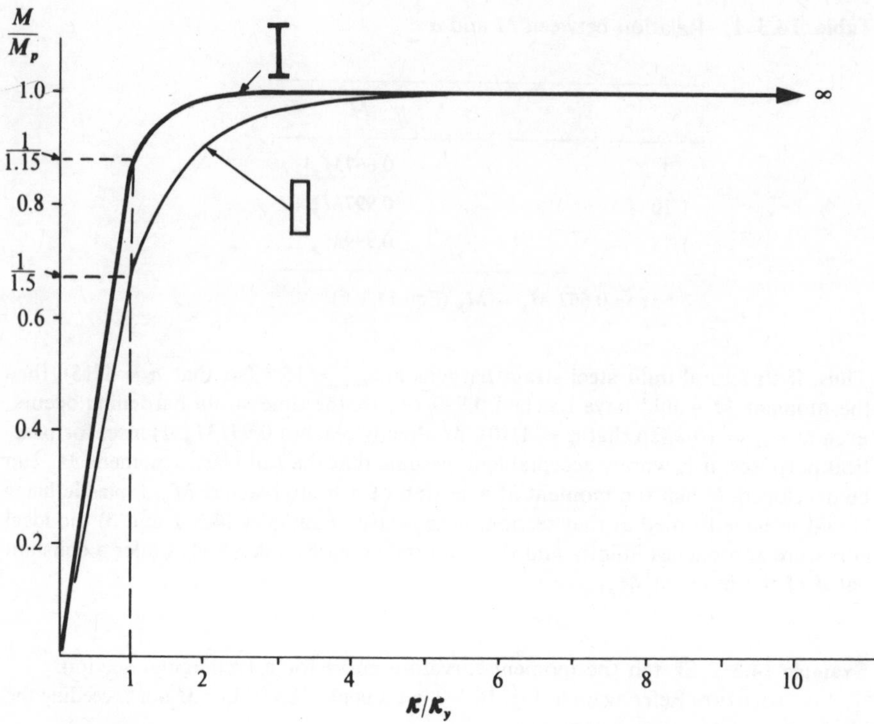

Fig. 14·3–4

In Fig. 14.3–4, the ordinate values 1/1.15 and 1/1.5 are respectively the reciprocals of the shape factors for the I section and the rectangular section; thus $M_p/1.15 = M_y$ for an I section, and $M_p/1.5 = M_y$ for a rectangular section. The moment-curvature relations are, as expected, linear up to $M = M_y$, after which curvatures increase rapidly with increase in moments and becomes very large as M approaches M_p.

Example 14.3–3 A simply supported beam of rectangular section $b \times h$ is loaded beyond the elastic range by a load Q at midspan.

(a) Determine the shape of the plastic zones.

(b) Determine the distribution of curvatures along the beam. What is the curvature at midspan?

(c) Explain how the deflections of the beam may be determined.

(d) If Q is increased to such magnitude that a plastic hinge just forms at the midspan section, repeat the solutions for (a), (b) and (c) above.

SOLUTION

(a) Referring to Fig. 14.3–5(a), for $M > M_y$ the shape of the plastic zones is completely defined by the coefficient α. From Eqn 14.3–6

$$M = M_p(1 - \alpha^2/3)$$

Therefore $\alpha = [3\,(1 - M/M_p)]^{1/2}$ (14.3–12)

where, for a typical section X-X, the value of M is given by the bending moment dia-

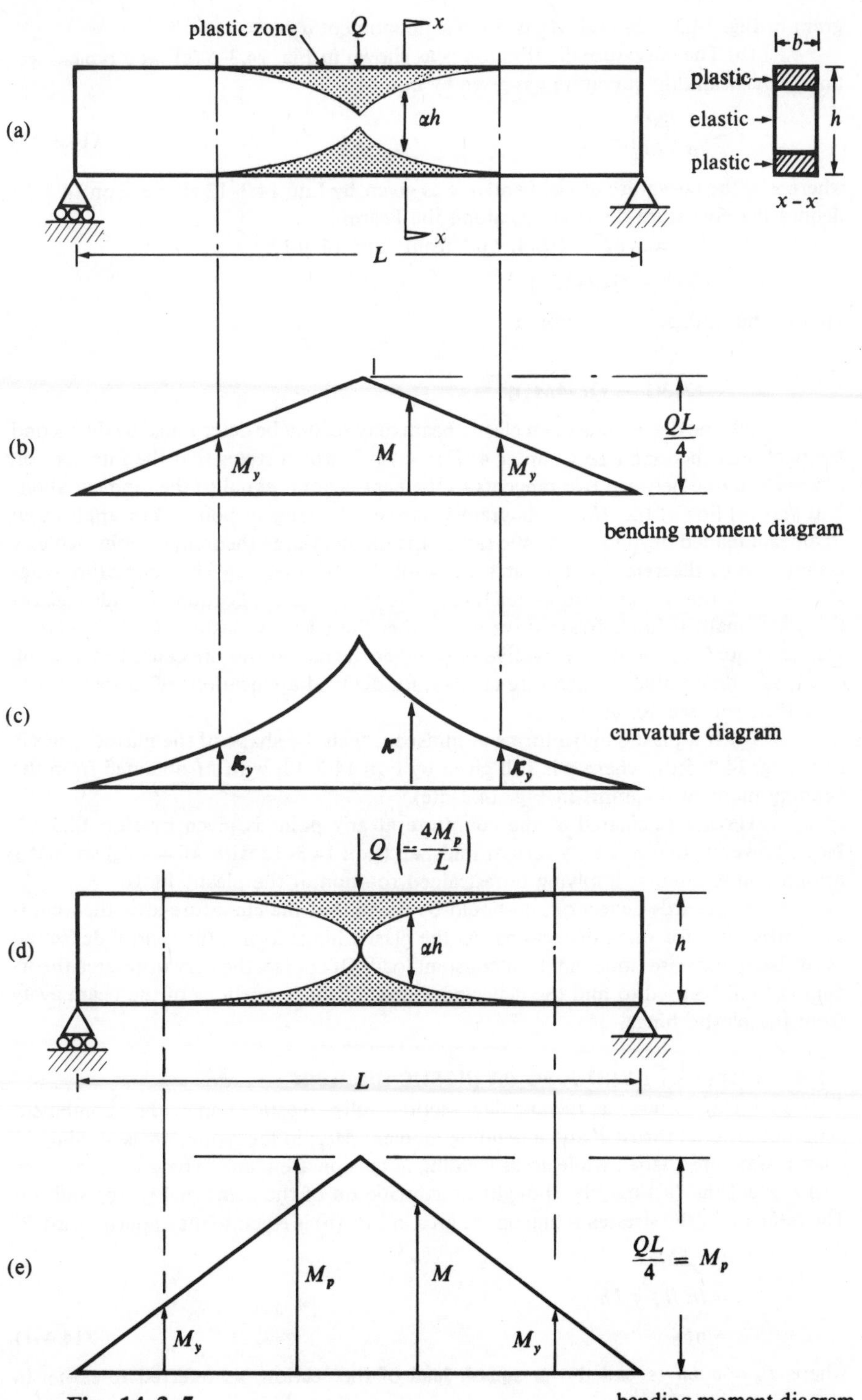
plastic zone
Q
x
αh
(a)
b
plastic
elastic
h
plastic
x - x
x
L
(b)
QL/4
M
M_y
M_y
bending moment diagram
(c)
κ
κ_y
κ_y
curvature diagram
Q (= 4M_p/L)
(d)
αh
h
L
(e)
QL/4 = M_p
M_p
M
M_y
M_y
bending moment diagram

Fig. 14·3–5

gram in Fig. 14.3–5(b) and M_p is the plastic moment $\sigma_y bh^2/4$.

(b) The curvature distribution is as shown in Fig. 14.3–5 (c); at a typical section of the beam, the curvature κ is given by Eqn 14.3–9:

$$\kappa = \frac{1}{\alpha}\left(\frac{2\sigma_y}{Eh}\right) = \frac{1}{\alpha}\kappa_y \tag{14.3–13}$$

where κ_y is the curvature at yield and α is as given by Eqn 14.3–12 above. Eqn 14.3–13 defines the curvature distribution along the beam.

At midspan $M = QL/4$, and from Eqn 14.3–12.

$$\alpha = [3\ (1 - QL/4M_p)]^{1/2}$$

Hence, the midspan curvature is

$$\kappa = \frac{2\sigma_y}{Eh[3(1 - QL/4M_p)]^{1/2}}$$

(c) The deflections of an elastic beam may readily be determined by the second moment-area theorem (see Chapter 4: Eqn 4.18–2) which states that the intercept on any vertical line between two tangents to the bent beam is equal to the moment about that vertical line of the "M/EI diagram" between the tangent points. For application to a beam loaded beyond the elastic range, the moment-area theorem is rephrased as a **curvature-area theorem**[8] by replacing the words "M/EI diagram" by "curvature diagram". Note that the curvature-area theorem is of general application, provided deflections are small: it holds irrespective of whether the beam is elastic, plastic or elasto-plastic. (Equally, it holds irrespective of whether the curvatures are caused by loading or by, say, differential temperature change; for detailed applications of the curvature-area theorem, see Reference 8.)

(d) If a plastic hinge forms at midspan, then the shape of the plastic zones is as in Fig. 14.3–5(d), where α is still given by Eqn 14.3–12, with M obtained from the bending moment diagrams in Fig. 14.3–5(e).

Having calculated α, the curvature at any point is given by Eqn 14.3–13. Note, however, that α is now zero at midspan (Eqn 14.3–12 with $M = M_p$) so that κ at midspan is infinity, implying unrestrained rotation at the plastic hinge.

As regards deflection, it should be noted that the curvature-area theorem is applicable only for small deflections. As the plastic hinge forms, the central deflection of the beam increases indefinitely at constant load. Of course, the curvature-area theorem may still be used to find the deflected shapes of those portions of the beam away from the plastic hinge.

14.4 Effect of axial load on plastic moment

Fig. 14.4–1(a) shows a rectangular section fully plastic under the combined action of an axial thrust P and a bending moment M'_p; in the figure, areas yielding in compression are shaded while areas yielding in tension are blank. The gross cross section $b \times h$ (Fig. (a)) may be thought of as made up of the areas in Fig. (b) and (c). The resultant of the stresses acting on the area in Fig. (b) is equal to the applied thrust P.

$$\begin{aligned} P &= \sigma_y bh_1 \\ &= (h_1/h)\ \sigma_y bh \\ &= nP_0 \end{aligned} \tag{14.4–1}$$

where $P_0 = \sigma_y bh$ is called the **squash load** of the section, as referred to earlier in Section 9.3. The ratio P/P_0 is called the **squash load ratio** and is usually denoted by n; for a rectangular section $n = h_1/h$, as shown in Eqn 14.4–1.

The resultant of the stresses acting on the areas in Fig. 14.4–1(c) is a couple equal to the applied moment M'_p. The reader should examine Fig. 14.4–1(a), (b) and (c), and satisfy himself that

$$M'_p = \sigma_y \frac{bh^2}{4} - \sigma_y \frac{bh_1^{\,2}}{4}$$

i.e. $M'_p = M_p - n^2 M_p$ (14.4–2)

where M'_p is the plastic moment of the section in the presence of the axial load P, and M_p is the plastic moment in pure bending. Dividing Eqn 14.4–2 throughout by σ_y

$$Z'_p = Z_p - n^2 Z_p \tag{14.4–3}$$

where Z'_p is the plastic second modulus in the presence of the axial load and Z_p that in pure bending. Eqn 14.4–2 shows that *the plastic moment of a rectangular section is always reduced by the presence of an axial load*—the reader should verify that (1) this is true irrespective of whether the axial load is tensile or compressive, and (2) this conclusion applies to any doubly symmetrical cross section.

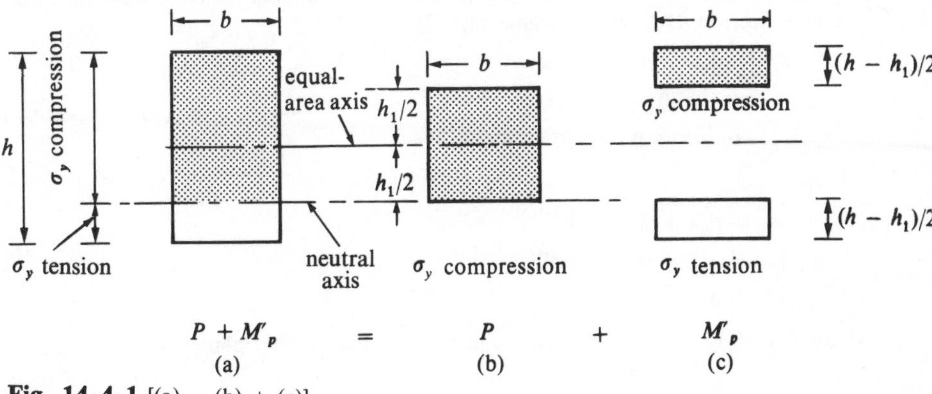

Fig. 14·4–1 [(a) = (b) + (c)]

For a mono-symmetrical section, such as a **T** section, we have to define carefully the term axial load. If an axial load is defined as one acting in the plane of the equal-area axis, it is clear from the analysis in Fig. 14.4–2 that an axial load will always reduce the plastic moment, by an amount equal to $\sigma_y \times$ [Z_p of area in Fig. 14.4–2(b)].

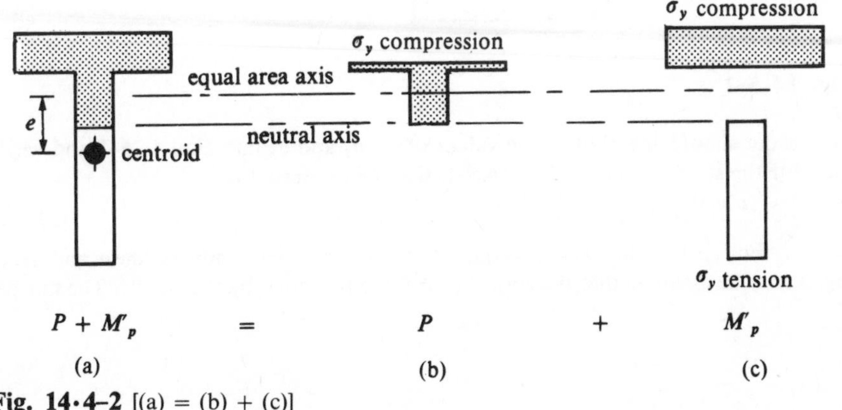

Fig. 14·4–2 [(a) = (b) + (c)]

If an axial load is defined as one acting through the centroid of the section, then it is effectively equal to a load acting in the plane of the equal-area axis plus an additional bending moment Pe, where e is the eccentricity of the centroid from the equal-area axis. If the sign of this additional moment is favourable then the plastic moment may appear to increase; this increase is, of course, purely illusory.

14.5 Effect of shear force on plastic moment

Under the action of combined shear and bending a typical element of a beam section is subjected simultaneously to shear and normal stresses. According to von Mises[9], the criterion for yield under combined stresses is:

$$(\sigma_1 - \sigma_2)^2 + (\sigma_2 - \sigma_3)^2 + (\sigma_3 - \sigma_1)^2 = 2\,\sigma_y{}^2 \tag{14.5–1}$$

where σ_1, σ_2 and σ_3 are the principal stresses acting on the element and σ_y is the yield stress in uni-axial tension; for a given material, σ_y may be determined from a tensile test.

Consider an element under the plane stress condition in Fig. 14.5–1. The principal stresses, σ_i $(i = 1,2,3)$ are given by the three roots of the determinantal equation (see Section **11.5** and **Example 11.5–2**):

$$\begin{vmatrix} (\sigma - \sigma_i) & \tau & 0 \\ \tau & (0 - \sigma_i) & 0 \\ 0 & 0 & (0 - \sigma_i) \end{vmatrix} = 0$$

that is,

$$\sigma_1 = \sigma/2 + \tfrac{1}{2}[\sigma^2 + 4\tau^2]^{1/2}$$

$$\sigma_2 = \sigma/2 - \tfrac{1}{2}[\sigma^2 + 4\tau^2]^{1/2}$$

$$\sigma_3 = 0$$

where σ_1 and σ_2 lie in the xy plane and σ_3 is normal to that plane.

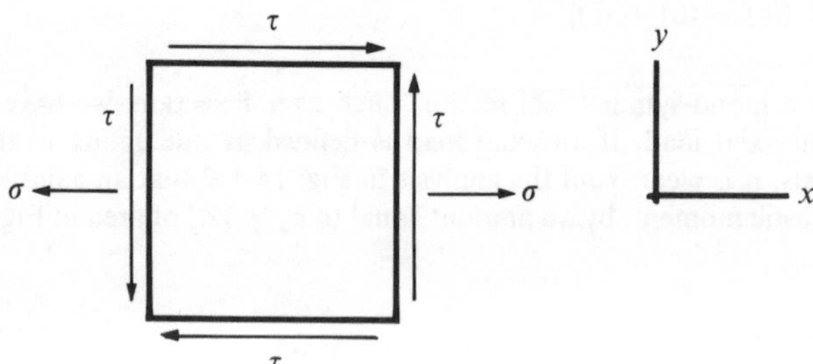

Fig. 14·5–1

The reader should substitute these values of σ_1, σ_2 and σ_3 into Eqn 14.5–1 and verify that, for the stress system in Fig. 14.5–1, the yield criterion is

$$\sigma^2 + 3\tau^2 = \sigma_y{}^2 \tag{14.5–2}$$

For an **I** section which becomes fully plastic under combined shear and bending, it may be assumed that the entire shear force is carried by the web[1, 2]. The normal stress in the flanges is then the full σ_y, while that in the web is σ, as given by Eqn 14.5–2.

For example, suppose the **I** section in Fig. 14.5–2(a) is fully plastic (that is, a plastic hinge has just formed) under the action of a shear force S and a moment M'_p; then the shear stress distribution is assumed to be that of Fig. 14.5–2(b) where

$$\tau = S/t_w h_w \tag{14.5–3}$$

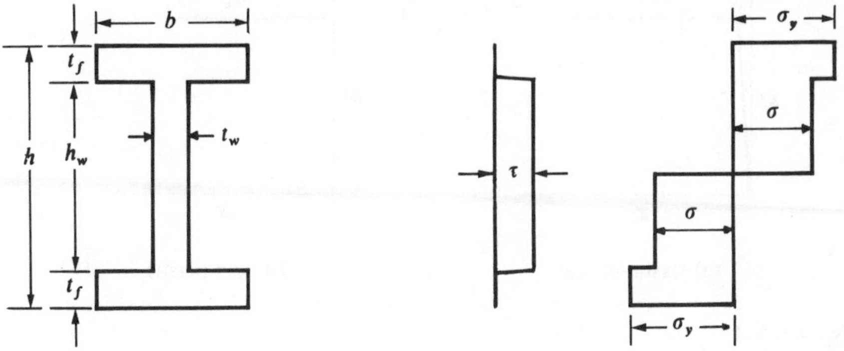

cross-section (a) shear stress (b) normal stress (c)

Fig. 14·5–2

The normal stress distribution is then as shown in Fig. 14.5–2(c), where σ is related to τ by Eqn 14.5–2. Therefore, the moment carried by the section is

$$M'_p = \sigma_y bt_f(h - t_f) + \sigma t_w h^2_w /4$$

$$= M_f + \frac{\sigma}{\sigma_y} M_w \tag{14.5–4}$$

where M_f is the full plastic moment of the flanges alone and $M_w(= \sigma_y t_w h^2_w/4)$ is that of the web alone; the normal stress σ in the web is, of course, computed from Eqn 14.5–2 using τ from Eqn 14.5–3.

So far only the **I** section has been considered; a discussion of plastic behaviour of the general cross section under combined shear and bending is outside the scope of this book, and the reader is referred to specialist texts[1, 6, 7].

14.6 Collapse loads and collapse mechanisms

The moment-curvature (M—κ) relations for the bending of beam sections were shown in Fig. 14.3–4. It can be shown that the load-carrying capacity of a frame or a beam depends only on the value of the plastic moment of resistance M_p and not on the complete M—κ relations. In plastic structural analysis, which is concerned with the determinat on of load-carrying capacities, the M—κ curve may be idealized as shown in Fig. 14.6–1(a); such an idealized **elastic-plastic** moment-curvature relation admits only of the elastic or the plastic state. (*This idealization amounts to taking the shape factor of the cross section as unity.*) In Fig. 14.6–1(a), the elastic curvatures are very small compared with the theoretically infinite curvature after plastic hinge formation. The

M—κ relation may therefore be further simplified to the **rigid-plastic** relation in Fig. 14.6–1(b).

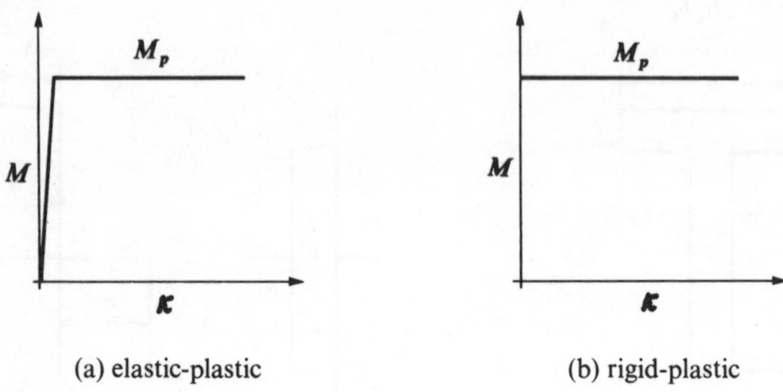

(a) elastic-plastic (b) rigid-plastic

Fig. 14·6–1

Indeed the fundamental theorems of plastic collapse (see Section **14.9**) refer, strictly speaking, only to rigid-plastic structures though in practice they may be applied to elastic-plastic structures provided the elastic deformations are not excessive.

In all subsequent discussions we shall, unless otherwise stated, assume a rigid-plastic moment-curvature relation. We shall further assume that the effects of axial load and shear force are negligible, and that local buckling, lateral buckling and shear failure are all prevented.

As a first example of the determination of the collapse load of a structure by plastic analysis, consider the two-span beam in Fig. 14.6–2(a), having a uniform cross section of plastic moment of resistance M_p. The beam supports a working load P, which is sufficiently low for stresses everywhere to be within the elastic range. The reader should verify that the bending moment diagram is as in Fig. 14.6–2(b). As the load P is progressively increased to say βP, the bending moment at section D reaches M_p and a plastic hinge forms there, as shown in Fig. 14.6–2(c). The beam is originally statically indeterminate and has one redundancy, but the formation of a plastic hinge removes that redundancy so that the beam in Fig. 14.6–2(c) is statically determinate. The bending moment diagram is now represented by abdc in Fig. 14.6–2(d)—the reader should verify this by statics. As the load is further increased, the bending moment at D remains constant at the full plastic value M_p, while that at B continues to grow until eventually, at a load of say λP, the moment at B also reaches M_p. The bending moment diagram is now ab'dc in Fig. 14.6–2(d), and there are two plastic hinges—at B and at D. Reference to the moment-curvature curve in Fig. 14.6–1(b) shows that the beam now undergoes unrestrained rotation at B and D; in other words, the structure has become a **mechanism**. (The conversion of structures into mechanisms by the formation of plastic hinges is further discussed in Comment (4) following **Example 14.6–4.**) The load λP at which the structure collapses as a mechanism is called the **collapse load** or the **ultimate load** and the factor λ which is the ratio of the collapse load to the working load, is called the collapse load factor, or simply the **load factor**. The collapse load, and hence the load factor, can easily be determined from Fig. 14.6–2(e). Consider a plastic hinge rotation ϕ at B; then from the geometry of Fig. 14.6–2(e), that at hinge D must be 2ϕ.

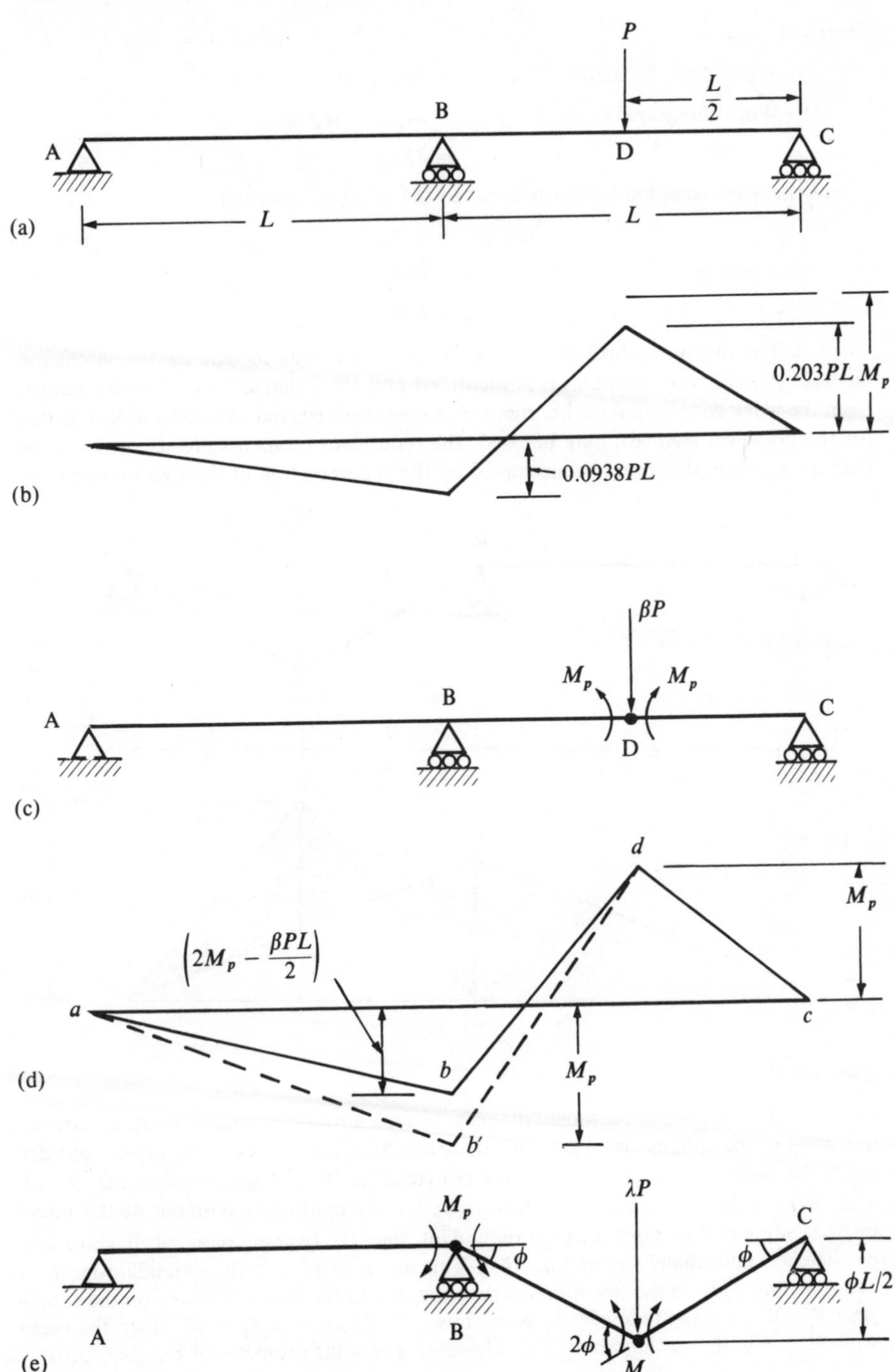
P
L/2
A
B
C
D
(a)
L
L
0.203PL
M_p
0.0938PL
(b)
βP
M_p
M_p
A
B
C
D
(c)
d
M_p
(2M_p − βPL/2)
a
c
b
M_p
(d)
b′
λP
M_p
C
φ
φ
φL/2
A
B
2φ
M_p
(e)

Fig. 14·6–2

Therefore

$$\text{Work done by load} = \lambda P(\phi L/2)$$

$$\text{Work dissipated in the hinges} = M_p\phi + M_p(2\phi) = 3M_p\phi$$

Hence the **work equation** (sometimes called the **collapse equation**) is

$$\lambda P\phi L/2 = 3M_p\phi \tag{14.6–1}$$

therefore

$$\lambda P = 6M_p/L$$

$$\lambda = 6M_p/PL$$

The above method of solution is called the **work method** or the **mechanism method**: the collapse mechanism is identified and the collapse load is then obtained from the work equation. Another method, sometimes referred to as the **statical method** (or the **graphical method**), may be used: the redundant moments are selected and the bending moment diagram is constructed by the superposition of the free-moment dia-

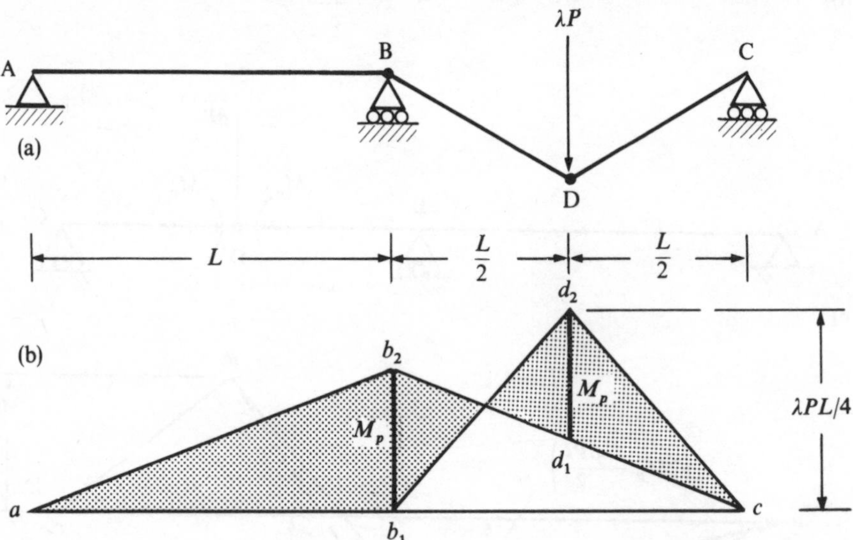

Fig. 14·6–3

gram on to the redundant-moment diagram in such a way that a mechanism is formed; the value of the collapse load is then calculated from statics. As an example, consider again the beam in Fig. 14.6–2(a)— for convenience the collapse mechanism of Fig. 14.6–2(e) is redrawn in Fig. 14.6–3(a). Suppose the continuity moment at the intermediate support B is selected as the redundant, then the bending moment diagram may be sketched immediately as in Fig. 14.6–3(b), in which ab_2c is the redundant-moment diagram, and b_1d_2c is the free moment diagram due to the load λP acting on the simple span BC. If the magnitude of b_1b_2 is so chosen that $b_1b_2 = d_1d_2 = M_p$, then the beam will collapse in the mechanism of Fig. 14.6–3(a). From the geometry of Fig. 14.6–3(b):

$$d_1d_2 = \lambda PL/4 - b_1b_2/2$$

That is $\quad M_p = \lambda PL/4 - M_p/2$

Therefore $\quad \lambda P = 6M_p/L \quad$ agreeing with Eqn 14.6–1.

COMMENTS (1) In the collapse mechanism in Fig. 14.6–2(e) (or Fig. 14.6–3(a)) the plastic hinges are shown to divide the beam into several rigid members, AB, BD and DC, which are assumed to remain straight so that all rotations take place in the plastic hinges. This is compatible with the rigid-plastic moment-curvature relation in Fig. 14.6–1(b). If a more realistic M—κ relation is used (see, for example, Fig. 14.3–4) the members AB, BD and DC would become curved under the action of the bending moments that exist there. However, the assumption that they remain straight in no way alters the fact that the work equation (Eqn 14.6–1) is 'exact'—see Section **14.9**.

(2) The work equation (Eqn 14.6–1) was derived from the concept of real work; the plastic hinge rotations (ϕ and 2ϕ in Fig. 14.6–2(e)) and the deflection of the loading point ($\phi L/2$) arose from the deformation of a real mechanism. However, since the load λP and the moments M_p at each of sections B and D form an equilibrium set while the displacement $\phi L/2$ and the rotations ϕ and 2ϕ form a compatible displacement set, we can at once form the **virtual work equation**:

Equilibrium set

$$\lambda P(\phi L/2) = M_p(\phi) + M_p(2\phi) \tag{14.6–2}$$

Compatible displacement set

The principle of virtual work was explained in Section **4.14**, and a further discussion of the principle as applied to mechanisms will be given in Section **14.9**. In the mean time it is sufficient to recognize that the so-called work equation is in fact a **virtual work equation** and that the **work method** may equally be called the **virtual work method.**

(3) Comments (1) and (2) above point to the fact that the collapse load is independent of the precise form of the moment-curvature relation; only the value of M_p enters into the work equation.

In analysing the two-span beam of Fig. 14.6–3 the load history has been traced on the assumption of zero initial stress: a plastic hinge formed at D, then another at B, and the structure collapsed as a mechanism. In fact, if the aim is only to determine the collapse load, then the initial stresses and the load history are immaterial; only the final collapse mechanism need be examined.

(4) The two-span beam in its collapse state satisfies three conditions:

(i) the **mechanism condition**

(ii) the **equilibrium condition**

(iii) the **yield condition**

That the mechanism condition is satisfied is obvious from Fig. 14.6–3(a)—the two plastic hinges turn the structure into a mechanism. That the equilibrium condition is satisfied can be verified from Fig. 14.6–3(b), where (and the reader should verify this) the bending moment distribution is in equilibrium with the load λP. By the yield condition is meant the condition that the bending moment (see Fig. 14.6–3(b)) nowhere exceeds the plastic moment of resistance M_p.

It can be proved that these three conditions are necessary and sufficient for the determination of the collapse load factor of a structure. Indeed, one of the fundamental theorems of plastic collapse is the **uniqueness theorem**, which states that if a bending moment distribution can be found which satisfies the three conditions of mechanism equilibrium and yield, then the load which corresponds to such a moment distribution, will be the true collapse load. The proof of the uniqueness theorem will be given in Section **14.9**.

In the rest of this chapter, we shall return to the uniqueness theorem from time to time.

Example 14.6–1. The propped uniform cantilever in Fig. 14.6–4(a) is of plastic moment M_p. Determine the valuè of P at collapse using (a) the work method (b) the statical method.

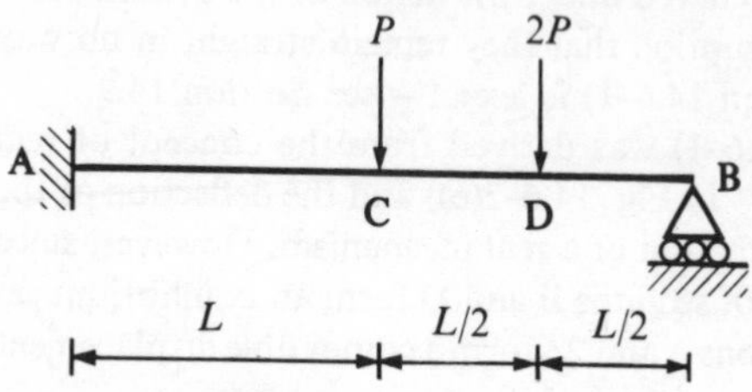

Fig. 14·6–4(a)

SOLUTION (see also comments at the end)

(a) **The work method** (or virtual work method)

There are three possible collapse mechanisms, as shown in Figs. 14.6–4(b), (c) and (d). We shall calculate the collapse load for each in turn.

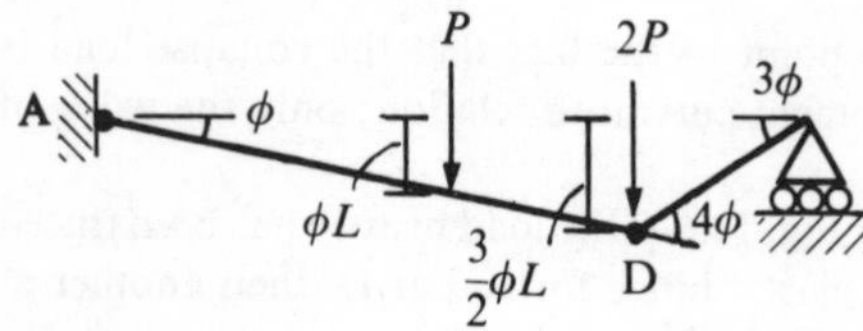

Fig. 14·6–4(b)

The work equation is

$$P(\phi L) + 2P(3\phi L/2) = M_p\phi + M_p(4\phi)$$

Therefore $$P = 5M_p/4L$$

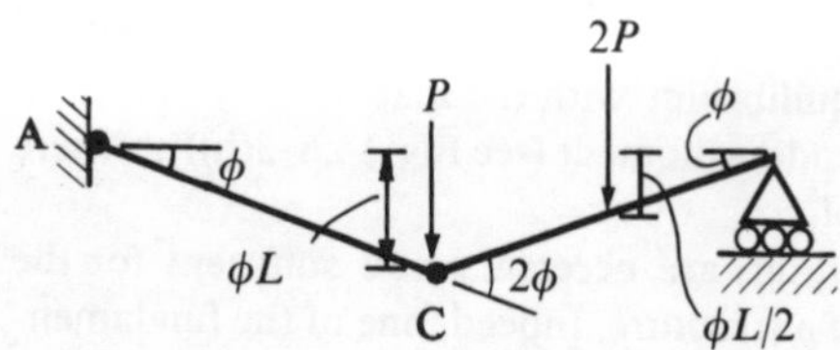

Fig. 14·6–4 (c)

$$P(\phi L) + 2P(\phi L/2) = M_p\phi + M_p(2\phi)$$

Therefore $$P = 3M_p/2L$$

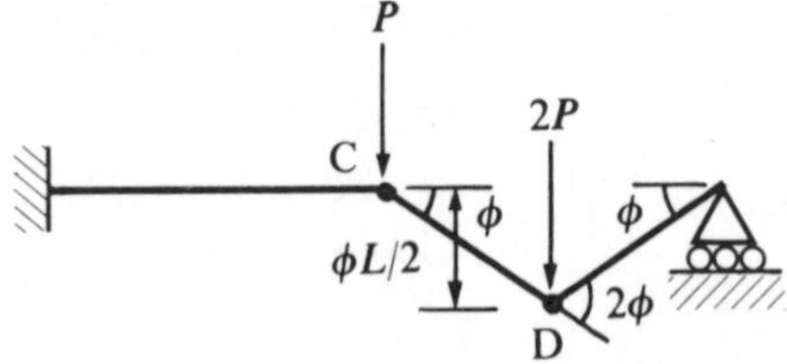

Fig. 14·6–4(d)

$$2P(\phi L/2) = M_p\phi + M_p(2\phi)$$

Therefore $\qquad P = 3M_p/L$

We have now examined all the three possible mechanisms and found that the one in Fig. 14.6–4(b) gives the lowest collapse load. This means that as the magnitude of P is gradually increased from zero, the collapse mechanism in Fig. 14.6–4(b) will be the first to form, when P reaches $5M_p/4L$. The other two mechanisms cannot form unless this one is prevented from forming, for example by strengthening the cross sections at points where hinges would have formed. We therefore conclude that $P = 5M_p/4L$ is the correct collapse value. (See Comment (2) below on the upper bound theorem.)

(b) **The statical method**

Figs. 14.6–4(e), (f) and (g) show three bending moment diagrams. Each diagram has been so drawn that the moment ordinate is exactly equal to the plastic moment M_p at two sections. In Figs. (e) and (f), $a_1\ a_2\ b$ is the moment diagram due to the redundant moment M_p at A, and $a_1c_2d_2b$ is the simple-span moment diagram due to the external loads P and $2P$. In Fig. (g) the moment M_p at section C has been selected as the redundant; that is, the triangle $a_1a_2c_2bc_1a_1$ is the redundant moment diagram (The reader should verify this. Hint: the moment M_p at C produces shear forces.), and $a_1a_3c_1d_2b$ is the moment diagram for the loads P and $2P$ acting on the beam with a hinge at C. The collapse values of P can be calculated from the geometry of the three bending moment diagrams.

The three values for P obtained by the statical method agree with those obtained by the work method. As before, we conclude that the lowest value, namely $P = 5M_p/4L$, is the correct one.

COMMENTS

(1) In the statical method above, we need not have calculated the collapse value of P for all the three bending moment diagrams. A closer examination will immediately reveal that both Figs. 14.6–4(f) and (g) violate the yield condition. In Fig. 14.6–4(f), the moment ordinate d_1d_2 exceeds M_p, which means that the collapse mechanism in Fig. 14.6–4(c) cannot occur unless plastic hinge formation is prevented at D by strengthening the cross section there. Similarly, in Fig. 14.6–4(g), a_2a_3 exceeds M_p; again, the mechanism in Fig. 14.6–4(d) cannot occur unless plastic hinge formation at A is deliberately prevented. The bending moment diagram in Fig. 14.6-4(e), on the other hand, satisfies the three conditions of mechanism (with plastic hinges at A and D), equilibrium (by the manner of its construction) and yield (since moment ordinates nowhere exceed M_p). Therefore we can at once conclude from the uniqueness theorem that the corresponding collapse load is the correct one; there is in fact no need to consider any other mechanisms.

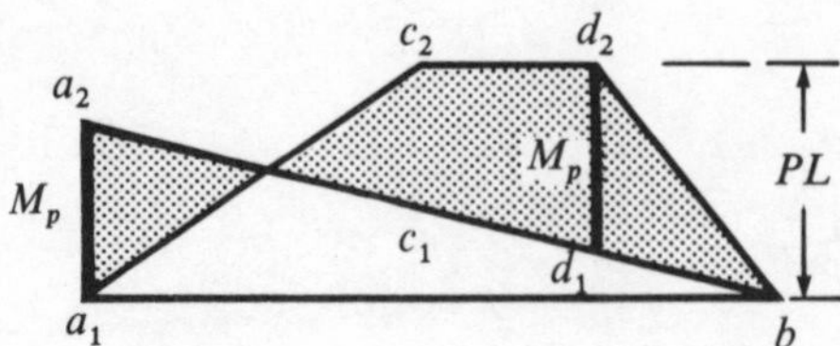

Fig. 14·6–4(e)

$$d_1d_2 \quad = PL - a_1a_2/4$$

Therefore $\quad M_p \quad = PL - M_p/4$

or $\quad P \quad = 5M_p/4L$

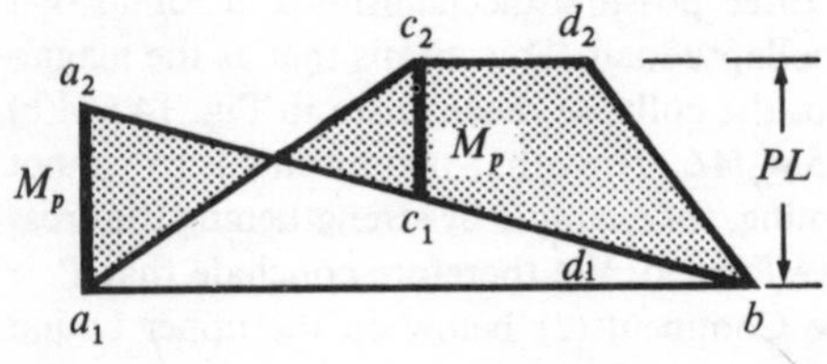

Fig. 14·6–4(f)

$$c_1c_2 \quad = PL - a_1a_2/2$$

$$M_p \quad = PL - M_p/2$$

Therefore $\quad P \quad = 3M_p/2L$

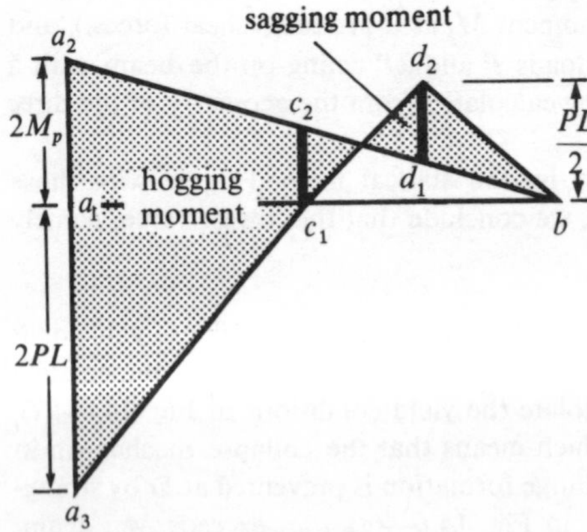

Fig. 14·6–4(g)

$$d_1d_2 \quad = PL/2 - c_1c_2/2$$

$$M_p \quad = PL/2 - M_p/2$$

Therefore $\quad P \quad = 3M_p/L$

(2) A fundamental theorem of plastic collapse is the **upper bound theorem** (sometimes called the **kinematic theorem**) which states that for a given structure subjected to a given loading, the magnitude of the loading which is found to correspond to any assumed collapse mechanism must be either greater than or equal to, but cannot be less than, the true collapse load. (The proof of the theorem will be given in Section **14.9**.)

Therefore, in an analysis we simply compute the collapse load for each possible mechanism and accept the lowest value as the correct one, as we did in the work method above. However, sometimes there is uncertainty about the number of possible mechanisms. A **statical check** is then necessary: the collapse bending moment diagram is drawn for the mechanism that gives the lowest collapse load; if the moment ordinate nowhere exceeds the plastic moment M_p then the uniqueness theorem guarantees that this mechanism will give the true collapse load. If M_p is exceeded somewhere, then the yield condition is not satisfied, and the search for a correct collapse mechanism must continue.

(3) The upper bound theorem is often referred to as the **unsafe theorem**, because, interpreted in a design sense, it states that the value of the plastic moment M_p obtained on the basis of an arbitrarily assumed collapse mechanism is smaller than, or at best equal to, that actually required. Consider, for example, the beam in Fig. 14.6–4(a). Suppose we have to determine the plastic moment of resistance required to carry the known loads P and $2P$. As we have seen in **Example 14.6–1**, the correct collapse mechanism (Fig. 14.6–4(b)) gives $M_p = 4PL/5$. The incorrect mechanisms in Figs. (c) and (d) give M_p as $2PL/3$ and $PL/3$ respectively.

(4) Another fundamental theorem of plastic collapse is the **lower bound theorem** (sometimes called the **static theorem**), which states that if a distribution of bending moments can be found such that the structure is in equilibrium under the external loading and such that nowhere is the plastic moment of resistance M_p exceeded, then the structure will not collapse under that loading—however 'unlikely' that distribution of moments may appear. The theorem is often referred to as the **safe theorem**. (The proof of this theorem will be given in Section **14.9**).

The use of the lower bound theorem is illustrated in **Example 14.6–2**.

(5) For future reference the three fundamental theorems of plastic collapse are displayed together below:

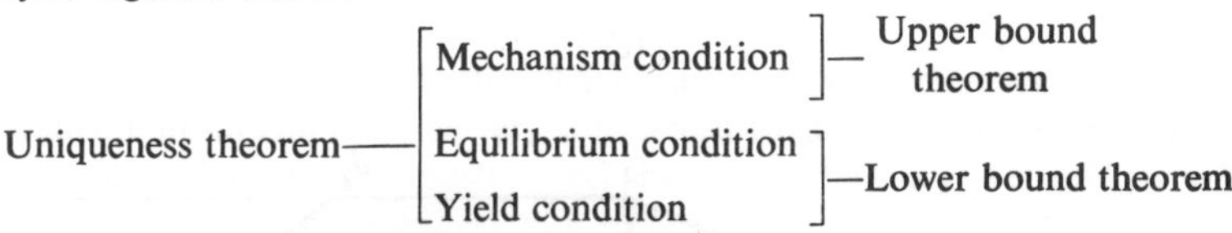

Fig. 14·6–5

Example 14.6–2. A propped uniform cantilever is to be designed to support the loads in Fig. 14.6–6(a). Explain how the lower bound theorem may be used to select a value of the plastic moment of resistance, M_p, which will guarantee that the beam will not collapse under the loading.

SOLUTION Fig. 14.6–6(b), (c) and (d) show three bending moment diagrams which are in equilibrium with the external loads P and $2P$. These are discussed in turn: **Fig. 14.6–6(b)**: The bending moment diagram is obtained by superposition of the simple-beam moment diagram $a_1c_2d_2b$ on to the diagram a_1a_2b due to the redundant moment M at the fixed end A. The value of M actually acting is, of course, not known,

Example 14.6–3. (Cambridge University Engineering Tripos: Preliminary Examination for Part I (1975).) Figs. 14.6–7(a) and (b) show respectively the dimensions of a propped uniform cantilever and its cross-section.

(a) Determine the plastic moment of resistance of the beam section if the yield stress of the material is 250 N/mm^2 in tension and in compression.

(b) If the load W may be applied at any position within the span, determine the minimum value of W that will cause collapse.

(c) If the load W is removed just after the formation of the collapse plastic hinges, sketch the shapes of the bending moment and shear force diagrams after the removal of the load.

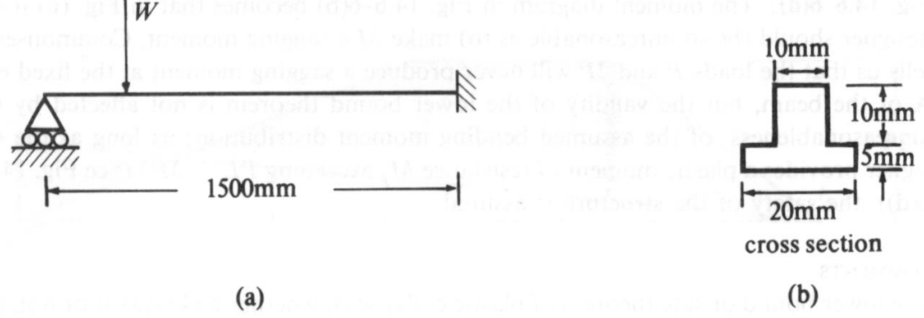

Fig. 14·6–7

SOLUTION

(a) The neutral axis divides the cross section into two parts of *equal area* as in Fig. 14.6–8(a).

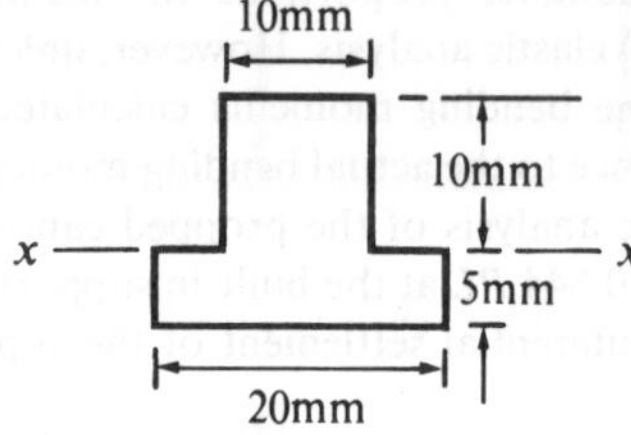

Fig. 14·6–8(a)

Therefore: Lever arm $= (10 + 5)/2 = 7.5$ mm

Plastic moment M_p $= 10 \times 10 \times 250 \times 7.5$

$= \underline{0.188 \text{ kN m}}$

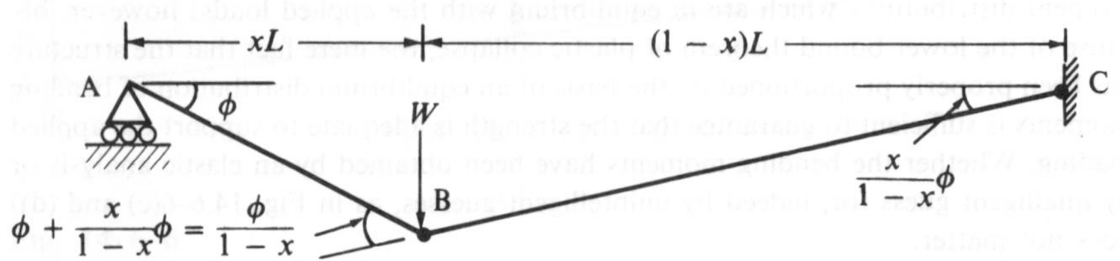

Fig. 14·6–8(b)

(b) Consider the mechanism in Fig. 14.6–8(b), where the position of the plastic hinge B is defined by the variable x.

The work equation is

$$W(xL\phi) = M_p\left(\frac{\phi}{1-x} + \frac{x\phi}{1-x}\right)$$

or

$$W = \frac{1+x}{x(1-x)}\,\frac{M_p}{L}$$

We do not know the value of x, but we do know that the required value of x is such as to make W a minimum.

$$\frac{dW}{dx} = 0 \text{ gives } x^2 + 2x - 1 = 0$$

therefore $x = 0.414$

and $W = \underline{730\text{N}}$

(c) In the absence of external applied load, the beam is acted on by support reactions only (see Fig. 14.6–8(c)).

Suppose the (unknown) support reactions are M and R. The only possible shape of the bending moment diagram is a triangle (Fig. 14.6–8(d)); the only possible shape of the shear force diagram is a rectangle (Fig. 14.6–8(e)).

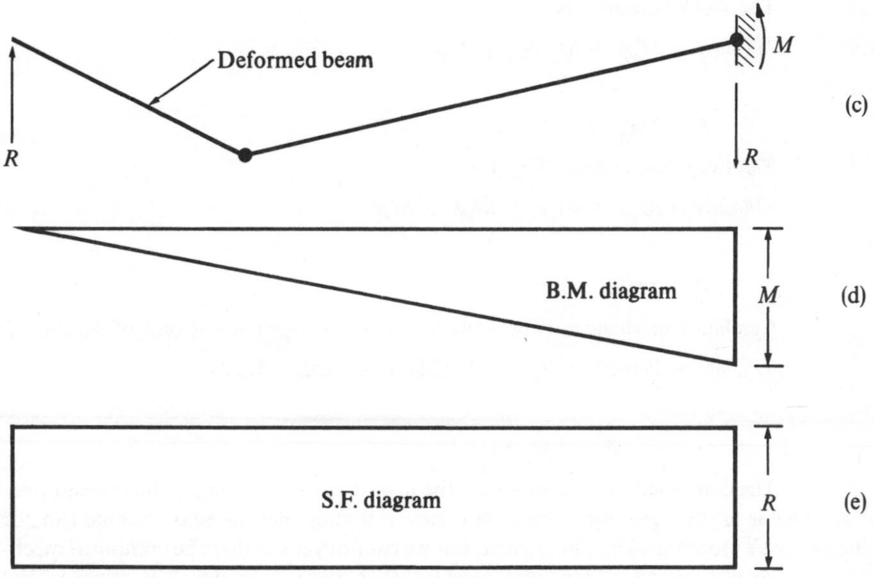

Fig. 14·6–8(c–e)

COMMENTS

(1) With reference to the mechanism in Fig. 14.6–8(b) it is interesting to note that (and the reader should verify this by elastic analysis) the maximum elastic moment at B is 0.174 WL and occurs when $x = 0.366\ L$, as shown in Fig. 14.6–9(a).

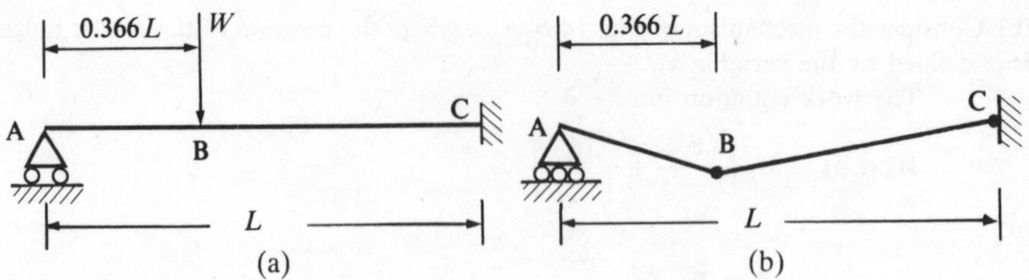

Fig. 14·6–9

However, the reader must not be tempted into thinking that the correct mechanism is that in Fig. 14.6–9(b). As we saw in part (b) of the solution above, the hinge B in the correct mechanism is at 0.414 L from A.

(2) With reference to Part (c) of the question, a common error is to think that the residual moment diagram has a 'kink', at the plastic hinge position.

Example 14.6–4. Fig. 14.6–10(a) shows a fixed-end portal frame of uniform plastic moment of resistance M_p. Determine the minimum value of P such that the loading shown would cause collapse.

SOLUTION We can identify three mechanisms of collapse, as shown in Fig. 14.6–10(b), (c) and (d):

Beam mechanism (Fig. (b))

The work equation is

$$2P(3a\phi) = M_p\phi + M_p(2\phi) + M_p\phi$$

or

$$P = 2M_p/3a$$

Sidesway mechanism (Fig. (c))

$$P(4a\phi) = M_p\phi + M_p\phi + M_p\phi + M_p\phi$$

or

$$P = M_p/a$$

Combined mechanism (Fig. (d)—see also Comment (2) at end of solution.)

$$2P(3a\phi) + P(4a\phi) = M_p\phi + M_p(2\phi) + M_p(2\phi) + M_p(\phi)$$

or

$$P = 3M_p/5a$$

The combined mechanism gives the lowest value of P, that is the lowest upper bound value of the three. What we can say now is that neither the beam mechanism nor the sidesway mechanism can be correct, but we cannot yet say that the combined mechanism is correct, because there may exist another mechanism which is more critical than that in Fig. (d).

As an exercise, the reader should carry out a statical check and verify that the bending moment diagram for the combined mechanism is as shown in Fig. (e)—*Hint*: see Comment (1) below. Fig. (e) shows that the portal frame in its collapse state in Fig. (d) satisfies the three conditions of mechanism, equilibrium, and yield. Hence, by the uniqueness theorem, the combined mechanism is the correct collapse mechanism and the corresponding value of P, namely, $P = 3M_p/5a$ is the correct answer.

COMMENTS

(1) If the reader has difficulties in drawing the bending moment diagram in Fig. 14.6–10(e), he should refer to the free-body diagram in Fig. (f):

$\sum$ Moments $= 0$ for member CD gives $V = 2M_p/3a$

$\sum$ Moments $= 0$ for CDE as a whole gives $H = M_p/2a$

P is, of course, $3M_p/5a$ as found earlier.

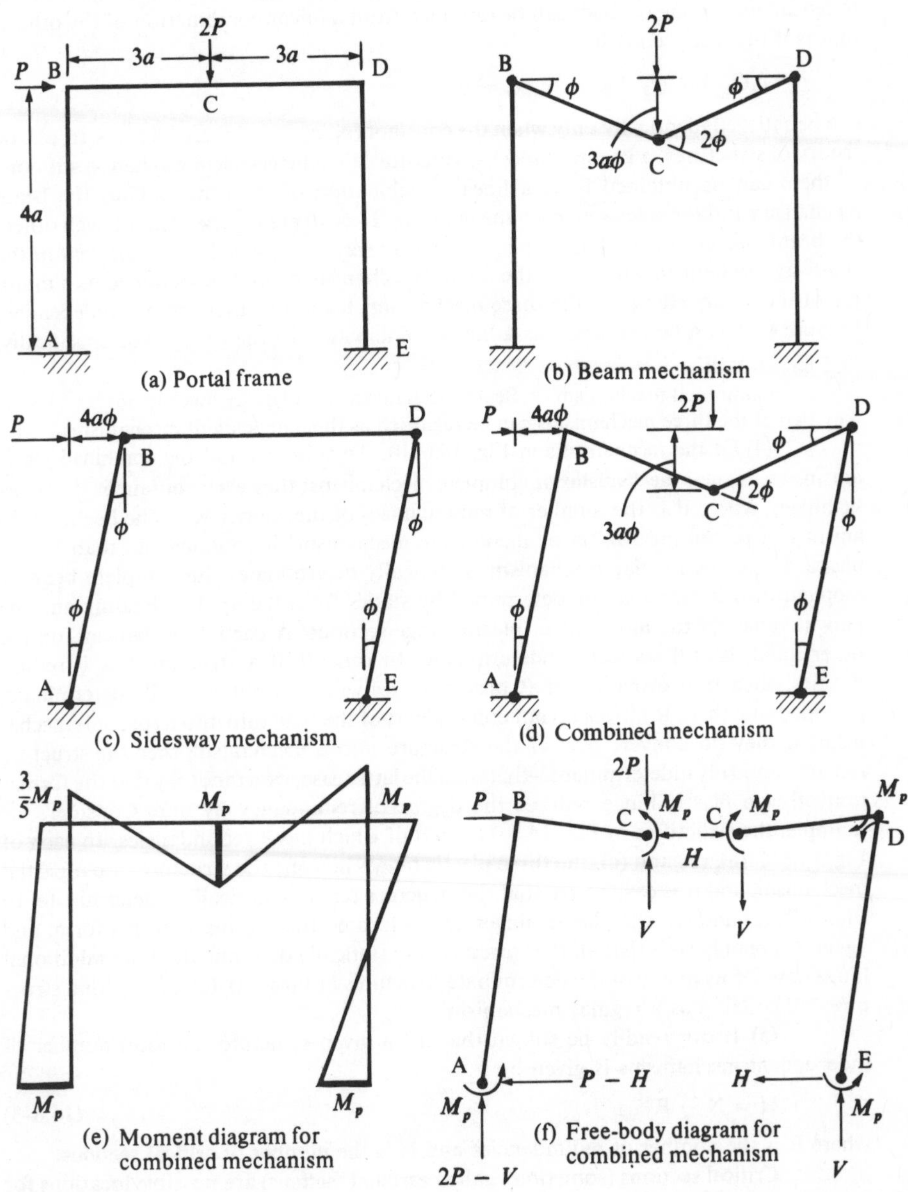

(a) Portal frame

(b) Beam mechanism

(c) Sidesway mechanism

(d) Combined mechanism

(e) Moment diagram for combined mechanism

(f) Free-body diagram for combined mechanism

Fig. 14·6–10

(2) In Fig. 14.6–10(d) the mechanism is referred to as a 'combined' mechanism, because it is obtained by combining the mechanisms in Figs. (b) and (c). Example 14.6–4 shows that it is possible to obtain a lower collapse load by combining two mechanisms to form a third mechanism. The combination of mechanisms has powerful applications in plastic analysis and will be dealt with in greater detail in Section 14.7.

(3) From Comment (2) above it can be seen that, although three mechanisms are shown in Fig. 14.6–10(b), (c) and (d), only two are independent, the third being obtainable by combining the first two.

In algebra, a set of quantities Q_1, Q_2, Q_3 . . . Q_n are said to be linearly independent if none of them can be obtained from a linear combination of the others; that is if the linear equation

$$c_1Q_1 + c_2Q_2 + \ldots c_nQ_n = 0$$

can be satisfied identically only when the constants c_1, c_2, . . . c_n are all zero. In plastic theory of structures, a group of mechanisms are called **independent mechanisms** if none of them can be obtained from a linear combination of the others. Thus the beam mechanism and the sidesway mechanism in Fig. 14.6–10 are independent of each other: the beam mechanism contains a plastic hinge at section C, which is not present in the sidesway mechanism, and hence the former mechanism cannot be obtained as a multiple of the latter. However, the three mechanisms taken together are not independent because each can be expressed as a linear combination of the other two. Specifically, in Fig. 14.6–10:

Combined mechanism = Beam mechanism + Sidesway mechanism

Any two of the three mechanisms can be regarded as the independent mechanisms.

(4) Of the mechanisms in Fig. 14.6–10, the sidesway and the combined mechanisms are **regular mechanisms** or complete mechanisms; they each contain R + 1 plastic hinges, where R is the number of redundancies of the framework. The beam mechanism is a **partial mechanism** or incomplete mechanism; it contains less than R + 1 plastic hinges. A regular mechanism is statically determinate; the complete bending moment distribution may be determined by statics from the applied loading and the known values of the moments at plastic hinge sections. A partial mechanism, on the other hand, is still statically indeterminate. In general, if a structure has R redundancies, then the formation of R plastic hinges will render it statically determinate, provided that these R hinges do not convert part of the structure into a (partial) mechanism; if they do convert part of the structure into a mechanism, then the structure remains statically indeterminate—that is, in the latter case, we cannot say that the formation of each plastic hinge reduces the degree of redundancy by one. Consider, for example, the structures in Fig. 14.6–11, each of which has 3 redundancies. In each of Fig. 14.6–11(a), (b) and (c), the three plastic hinges convert the structure into a partial mechanism, and it is easy to see that the structure remains statically indeterminate. In Figs. (d), (e) and (f), the plastic hinges are so located that no mechanisms form, and again it is readily seen that all the structures are statically determinate. If an additional hinge now forms in any of the determinate structures in Figs. (d), (e) and (f), that structure will collapse as a regular mechanism.

(5) It can readily be shown that for a given structure the total number of independent mechanisms is given by

$$M = \mathrm{N} - \mathrm{R} \qquad (14.6\text{–}3)$$

where R is the number of redundancies and N is the number of **critical sections.**

Critical sections (sometimes called **cardinal sections**) are possible locations for plastic hinges; thus for the portal frame in Fig. 14.6–10(a) the critical sections are A, B, C, D and E, which occur at the joints and at loading points. The bending moment diagram between two adjacent critical sections is a straight line; therefore a plastic hinge hinge cannot occur between critical sections (except in the particular case where the bending moment diagram between the sections is a rectangle of height M_p so that there is an infinite number of critical sections).

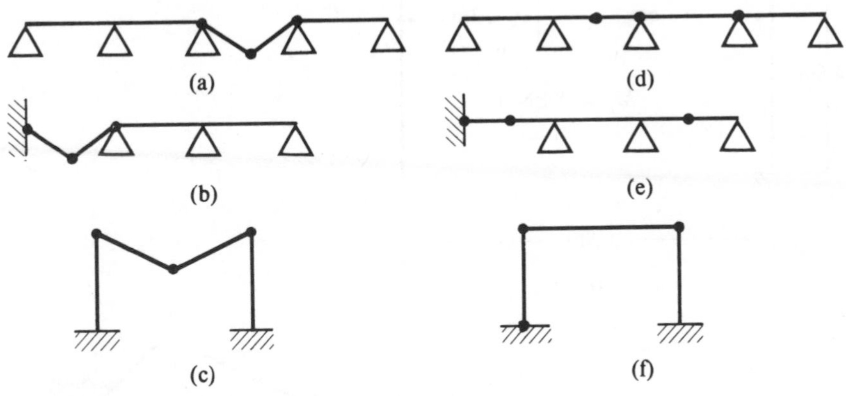

Fig. 14·6–11

To derive Eqn 14.6–3 consider a structure having R redundancies and N critical sections. The complete bending moment diagram is defined by the moment values at the N critical sections. Each virtual work equation (such as Eqn 14.9–1 or Eqn 14.9–2) written for such a mechanism represents a condition of equilibrium between the external loads and the bending moments at some of the N critical sections. Hence, M independent mechanisms, providing M independent equations, will enable the determination of M of the N bending moments required to define the complete bending moment diagram, so that N – M bending moment values still remain unknown. Since the number of redundancies is known to be R, we must have N – M = R; hence Eqn 14.6–3.

Example 14.6–5 (Contributed by Professor J Heyman.) The pitched-roof frame in Fig. 14.6–12(a) is of uniform members having plastic moment of resistance 90 kN m; loads V and H are applied as shown. Plot an **interaction diagram** from which may be read the positive values of V and H which will just cause collapse of the frame.

SOLUTION The frame has 5 critical sections and three redundancies; hence (see Eqn 14.6–3) there are two independent mechanisms.

We can immediately identify the pure sidesway mechanism in Fig. 14.6–12(b) and the part sidesway mechanism in Fig. 14.6–12(c); these are certainly independent (why?). Another possible mechanism is that in Fig. (d), which is obtained by adding together the mechanisms in Figs. (b) and (c)—see Comment (3) at the end of this Example.

For the collapse mechanism (b), the work equation is:

$$(4.5\theta)H = (4)(90)\theta$$

or $$H = 80 \qquad (14.6\text{–}4)$$

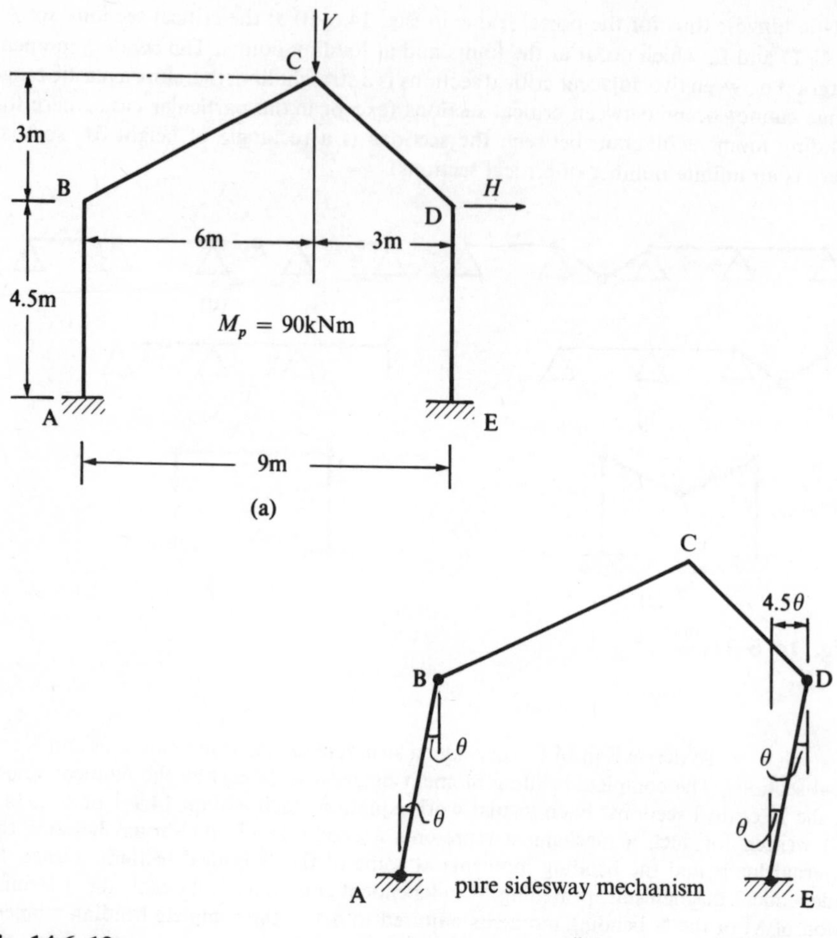

Fig. 14·6–12

For the collapse mechanism (c), the relative values of the hinge rotations and displacements are conveniently obtained by the method of **instantaneous centre** referred to earlier in Section **6.8**. Referring to Fig. 14.6–12(c) the locus of joint C′ is perpendicular to BC′ and that of joint D′ perpendicular to ED′. Therefore the instantaneous centre of rotation of the member C′D′ is located as the intersection I, of BC′ and ED′ produced. Consider a rotation θ of member BC′. Then

$$CC' = \theta(BC') = \phi(IC')$$

or
$$\phi = \left(\frac{BC'}{IC'}\right)\theta = \left(\frac{6}{3}\right)\theta = 2\theta$$

Similarly
$$DD' = \phi(ID') = \beta(ED')$$

or
$$\beta = \left(\frac{ID'}{ED'}\right)\phi = \left(\frac{4.5}{4.5}\right)\phi = \phi$$

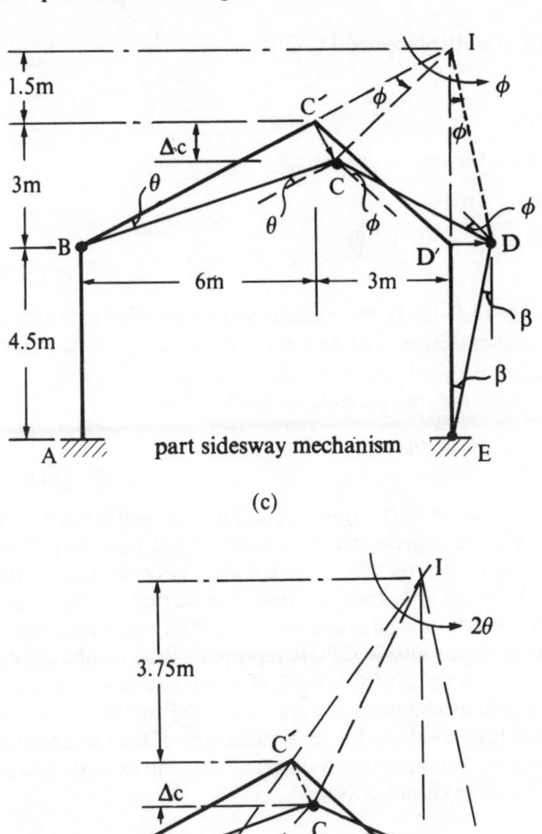

Fig. 14·6–12

Therefore the hinge rotations are

$$\theta_B = \theta$$
$$\theta_C = \theta + \phi = 3\theta$$
$$\theta_D = \phi + \beta = 4\theta$$
$$\theta_E = \beta = 2\theta$$

Also, the vertical component of the displacement C′C is

$$\Delta_C = (3)\,(\phi) = 6\theta$$

The horizontal displacement D′D is

$$\Delta_D = (4.5)\beta = (4.5)\,(2\theta) = 9\theta$$

The work equation is therefore

$$V(6\theta) + H(9\theta) = 90\,(\theta + 3\theta + 4\theta + 2\theta)$$

or

$$2V + 3H = 300 \qquad (14.6\text{–}5)$$

For the mechanism in Fig. 14.6–12(d), the instantaneous centre of rotation of the member C′D′ is located as the intersection, I of AC′ and ED′ produced. The reader should verify that, if $\theta_A = \theta$, then $\theta_C = 3\theta$, $\theta_D = 5\theta$, $\theta_E = 3\theta$, $\Delta_C = 6\theta$, and $D'D = 13.5\theta$. Therefore the work equation for this collapse mechanism is

$$V(6\theta) + H(13.5\theta) = 90(\theta + 3\theta + 5\theta + 3\theta)$$

or

$$6V + 13.5H = 1080 \qquad (14.6\text{–}6)$$

Using Eqns 14.6–4, 5 and 6, an interaction diagram may be plotted as in Fig. 14.6–13. Any combination (V,H) which is represented by a point on the boundary PQR would just cause collapse; if the point lies on PQ, then collapse is occurring by the mechanism of Fig. 14.6–12(c); if it lies on QR, then collapse is occurring by the mechanism of Fig. 14.6–12(d). A point inside the bounded region OPQR represents combinations of (V,H) which are safe; a point outside OPQR represents load combinations which cannot be carried by the frame. The point Q($V = 90$, $H = 40$) represents the load combination for which either of the two mechanisms in Figs. 14.6–12(c) and (d) can form. Similarly, the point R represents that load combination for which either of the mechanisms in Figs. 14.6–12(b) and (d) may occur. The interaction diagram also shows that, except when $H = 80$, the sideways mechanism (b) is not critical.

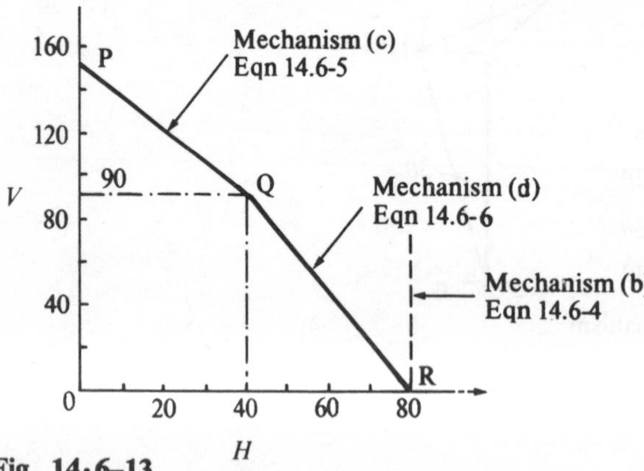

Fig. 14·6–13

COMMENTS

(1) Interaction diagrams of the type shown in Fig. 14.6–13 are always convex round the origin; they cannot contain re-entrant angles. Suppose, for example, two collapse mechanisms are represented by the lines PQX and YQR in Fig. 14.6–14(a). To prove that the boundary YQX, having a re-entrant angle at Q, is not acceptable, it is only necessary to show that for any load path it is not possible for the collapse load combination p′ on the line QX to be reached before the combination p on the line QR.

(2) Suppose there exists in addition to the mechanisms in Fig. 14.6–12 another collapse mechanism represented by the line ST in Fig. 14.6–14(b), then, following the argument in Comment (1) above, the point P (and indeed any point within the whole shaded area near P) would represent a load combination that cannot be carried by the frame. Similarly, if there is a collapse mechanism represented by the line JK, then Q would be outside the safe region. Thus, to establish that an interaction diagram is complete for a complex frame in which the collapse mechanisms are not self evident, and one or more of which might have been overlooked, one need only show that points such as Q are "possible" in the sense that the yield condition is not violated.

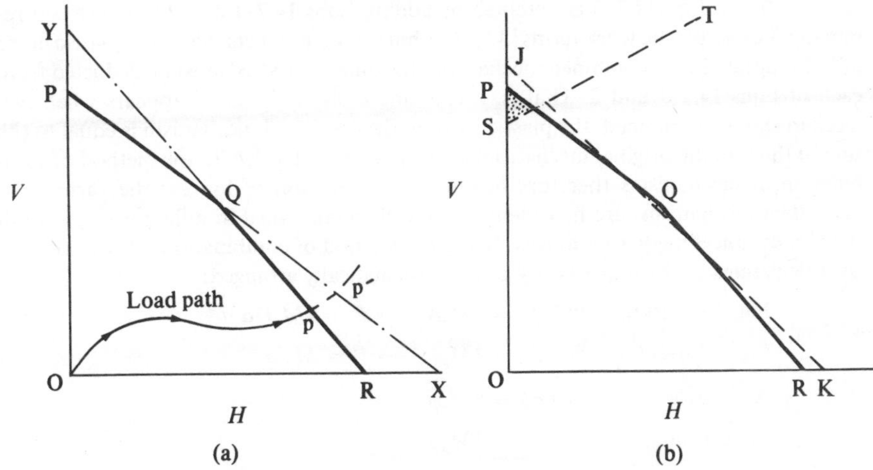

Fig. 14·6–14

(3) Finally, the reader should verify that the mechanism in Fig. 14.6–12(d) is not independent of those in Figs. (b) and (c), but is the result of adding together the latter two mechanisms. This being so, one would expect the work equation for the combined mechanism to be the sum of those for other two mechanisms. However, this is not quite true; the reader should add together Eqns 14.6–4 and 5 and verify that Eqn 14.6–6 results only if an amount $2M_p\theta_B$, which is equal to twice the plastic work term for the hinge B, is deducted from the total amount of plastic work absorbed in the frame. The reason for this deduction will be explained in the next section.

14.7 Combination of mechanisms

We have seen from Comment (5) at the end of Example 14.6–4 that if a structure has N critical sections and R redundancies, then there will be a total of N-R independent mechanisms of collapse; all other collapse mechanisms can be expressed as a combination of these N-R independent mechanisms. In that Example the two independent mechanisms in Fig. 14.6–10(b) and (c) were combined to form a third mechanism in Fig. (d); the collapse load obtained from the combined mechanism was lower than that obtained from either of the first two mechanisms. Example 14.6–4 in fact serves as a useful introduction to the method of the **combination of mechanisms**; it was then shown that the work equations for the three mechanisms were:

Beam mechanism (Fig. 14.6–10(b)):

$$2P(3a\phi) = [M_p\phi] + 3M_p\phi \qquad (14.7\text{–}1)$$

where the term $[M_p\phi]$ in square brackets represents the work absorbed by the plastic hinge rotation at B; it has been singled out here for a reason which will become clear later on.

Sidesway mechanism (Fig. 14.6–10(c)):

$$P(4a\phi) = [M_p\phi] + 3M_p\phi \qquad (14.7\text{–}2)$$

Combined mechanism (Fig. 14.6–10(d))

$$P(4a\phi) + 2P(3a\phi) = 6M_p\phi \qquad (14.7\text{–}3)$$

Thus, Eqn 14.7–3 is obtained by adding Eqns 14.7–1 and 2 together and removing the two plastic work terms $[M_p\phi]$ within square brackets. Since the plastic hinge at B disappears in the combined mechanism, the quantity $[M_p\phi]$ must be deducted from each of Eqns 14.7–1 and 2. That is, every time a plastic hinge disappears when two mechanisms are combined, the plastic work in the combined mechanism is equal to the sum of those in the original mechanisms less *twice* $M_p\phi$. The key to the method of combination of mechanisms therefore lies in the cancellation of hinges: the various independent mechanisms are first identified and then combined in different ways—with the aim of cancelling plastic hinges. Using the method of combination of mechanisms, the calculations for Example 14.6–4 can be systematically arranged:

Fig. 14.6–10(b): $2P(3a\phi) = 4M_p\phi \longrightarrow P = 2M_p/3a$

Fig. 14.6–10(c): $\underline{P(4a\phi) = 4M_p\phi} \longrightarrow P = M_p/a$

Adding : $10Pa\phi = 8M_p\phi$

Cancel hinge B: $\underline{\qquad - 2M_p\phi}$

Fig. 14.6–10(d): $10Pa\phi = 6M_p\phi \longrightarrow \underline{P = 3M_p/5a}$

When the work equations of two mechanisms are added together, the resulting equation will give a collapse load intermediate between the two original collapse loads; unless the combining of the mechanisms leads to the cancellation of a hinge, there is no possibility that the combined mechanism can give a lower collapse load than the lower of the two original loads.

In combining mechanisms it is often advantageous to make use of a special class of mechanisms known as **joint rotations**. Consider again the portal frame in Fig. 14.6–10; in Fig. 14.6–10(b) and (c), plastic hinges are shown to occur at the beam-column joints. Suppose we now differentiate between hinges in the beam and hinges in the column—such differentiation being particularly helpful when the beam and column sections have different plastic moments of resistance. If in Fig. 14.6–10(b) we choose to assume that all hinges occur in the beam, and in Fig. 14.6–10(c) we assume that all hinges occur in the columns, then the beam and sidesway mechanisms become as shown in Fig. 14.7–1(b) and (c), in which hinges do not form at the beam-column joints but at sections immediately adjacent to these joints. Note that the structure now has seven critical sections, as indicated by the short lines in Fig. 14.7–1(a); the number of redundancies is still 3 so that from Eqn 14.6–3

$$M = 7 - 3 = 4$$

That is, there are now four independent mechanisms. These are the beam mechanism (Fig. 14.7–1(b)), the sidesway mechanism (Fig. 14.7–1(c)), and the joint rotations at B and D (to be explained later).

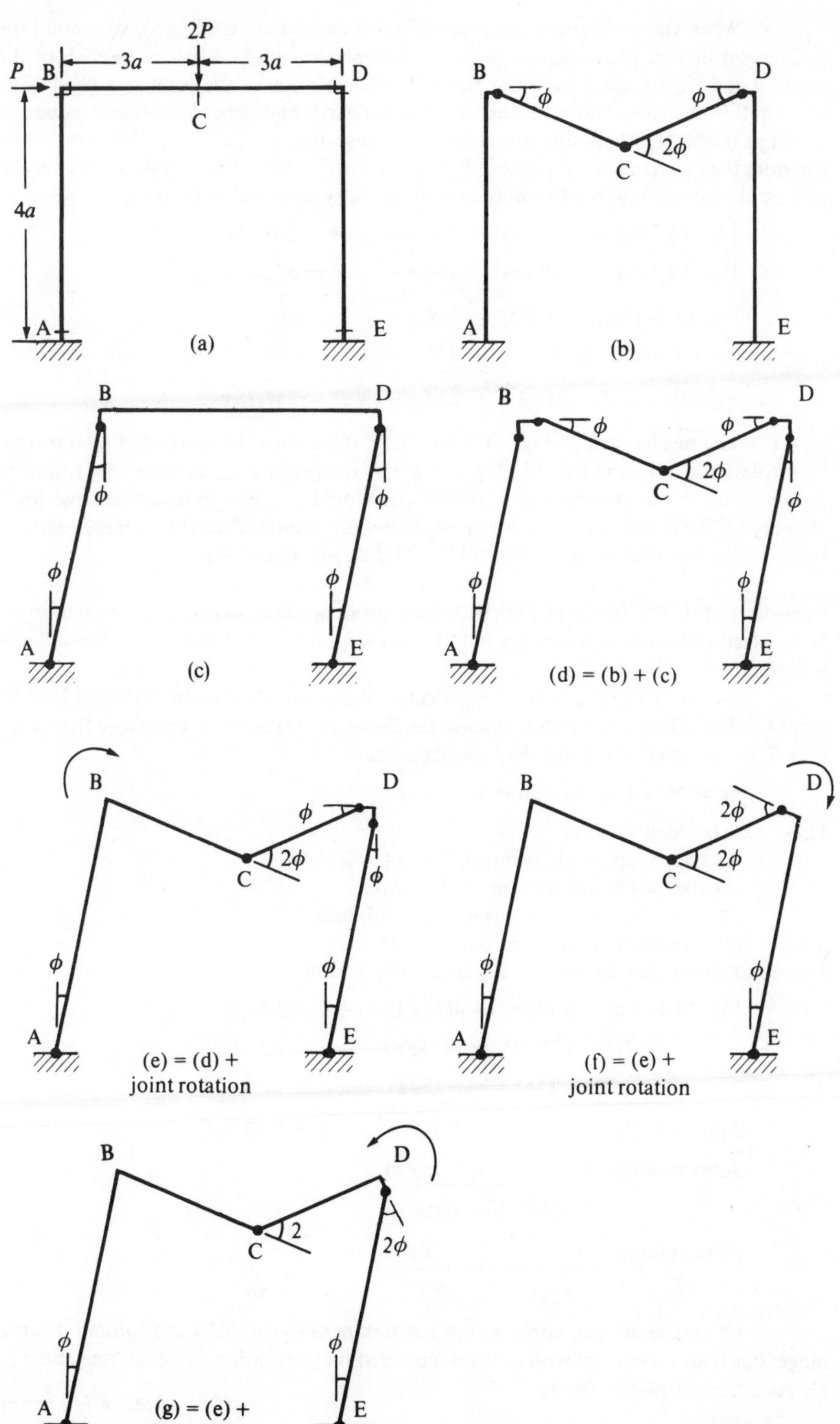
2P
P
B
3a
3a
D
C
4a
A
E
(a)
B
D
φ
φ
C
2φ
A
E
(b)
B
D
φ
φ
φ
φ
A
E
(c)
(d) = (b) + (c)
(e) = (d) +
joint rotation
(f) = (e) +
joint rotation
(g) = (e) +
joint rotation

Fig. 14·7–1

When the mechanisms in Fig. 14.7–1(b) and (c) are combined, we obtain the mechanism in Fig. (d), in which there are two plastic hinges adjacent to each of the joints B and D. A joint rotation at B, *i.e.* a local rotation of the joint as a rigid body will result in the cancellation of the two hinges near B and hence will give a reduction of $2M_p\phi$ in the total amount of plastic work absorbed in the frame. After this joint rotation, the mechanism is as shown in Fig. 14.7–1(e). Using the technique of joint rotations, the calculations for Example 14.6–4 may be re-arranged as follows:

Fig. 14.7–1(b): $2P(3a\phi) = 4M_p\phi \longrightarrow P = 2M_p/3a$

Fig. 14.7–1(c): $\underline{P(4a\phi) = 4M_p\phi} \longrightarrow P = M_p/a$

Fig. 14.7–1(d): $10Pa\phi = 8M_p\phi$

Joint rotation at B: $\underline{\qquad - 2M_p\phi}$

Fig. 14.7–1(e): $10Pa\phi = 6M_p\phi \longrightarrow \underline{P = 3M_p/5a}$

The mechanism in Fig. 14.7–1(e) may, if we wish, be converted by a further joint rotation to either of those in Figs. 14.7–1(f) and (g); of course, such joint rotations do not result in the cancellation of hinges and would not lead to lower collapse load. However, if the beam, say, has a lower M_p than the columns, then the collapse load will be lower for the mechanism in Fig. 14.7–1(f) than for that in Fig. (g).

Example 14.7–1 The frame in Fig. 14.7–2(a) carries working loads of values indicated. If the plastic moment of resistance is 15 kNm for each member, determine the collapse load factor.

SOLUTION The frame has 12 critical sections, as indicated by the short lines in Fig. 14.7–2(a). There are 6 redundancies (see Comment (1) at end). Therefore from Eqn 14.6–3, the number of independent mechanisms is

$M = \text{N} - \text{R} = 12 - 6 = 6$

These can be identified as

(1) the beam mechanism in Fig. 14.7–2(b)
(2) the beam mechanism in Fig. (c)
(3) the sidesway mechanism in Fig. (d) and
(4), (5) (6) the joint rotations at B, D, G.

The calculations can be set out systematically as follows:

Fig. 14.7–2(b): $6\lambda(4\phi) = 4(15)(\phi) \longrightarrow \lambda = 2.5$

(c): $12\lambda(2\phi) = 4(15)(\phi) \longrightarrow \lambda = 2.5$

(d): $\underline{6\lambda(4\phi) = 6(15)(\phi)} \longrightarrow \lambda = 3.75$

Fig. 14.7–2(e): $72\lambda\phi = 210\phi \longrightarrow \lambda = 2.92$

Joint rotation, B $\underline{\qquad - 2(15)(\phi)}$

$72\lambda\phi = 180\phi$

Joint rotation, D $\underline{\qquad - (15)(\phi)}$

$72\lambda\phi = 165\phi \longrightarrow \lambda = 2.29$

Of course, we can apply a joint rotation at G to eliminate the column (beam) hinge, but then we end up with a beam (column) hinge rotation of 2ϕ, so that there is no reduction in plastic work.

Fig. 14·7–2 ⟶

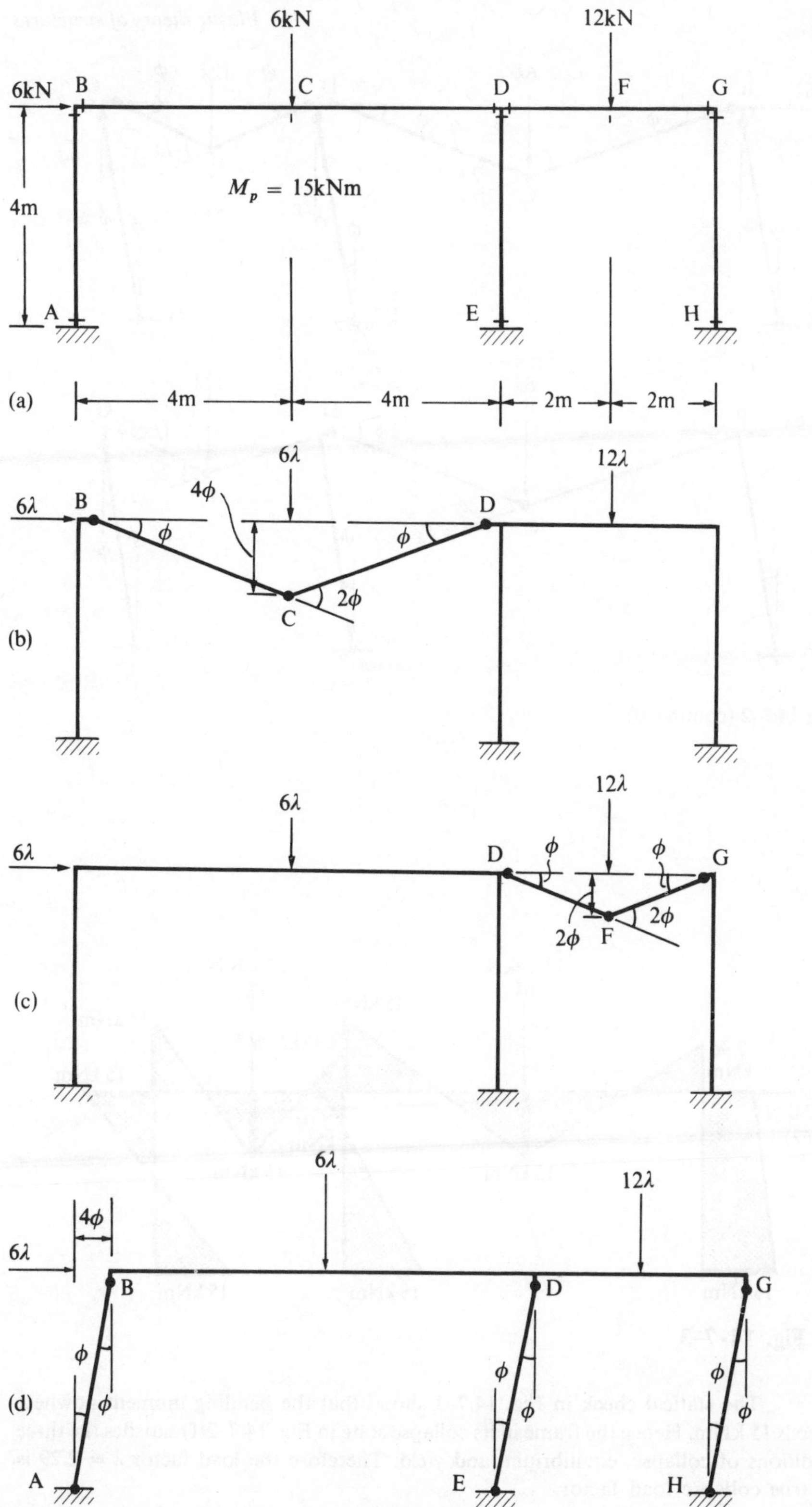
6kN
12kN
6kN
B
C
D
F
G
4m
$M_p = 15$kNm
A
E
H
(a)
4m
4m
2m
2m
6λ
12λ
6λ
B
4φ
D
φ
φ
C
2φ
(b)
12λ
6λ
6λ
D
φ
φ
G
2φ
2φ
F
(c)
6λ
12λ
4φ
6λ
B
D
G
φ
φ
φ
(d)
φ
φ
φ
A
E
H

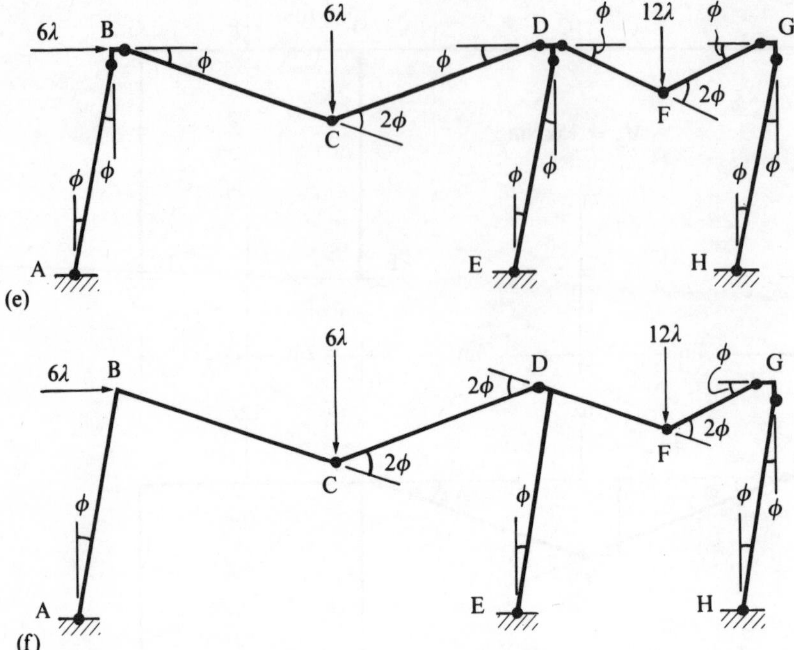

Fig 14·7–2 (continued)

Fig. 14·7–3

The statical check in Fig. 14.7–3 shows that the bending moment nowhere exceeds 15 kNm. Hence the frame in its collapse state in Fig. 14.7–2(f) satisfies the three conditions of collapse, equilibrium, and yield. Therefore the load factor $\lambda = 2.29$ is the true collapse load factor.

The bending moments in Fig. 14.7–3 are based on the freebody diagrams in Fig. 14.7–4, in which the bold arrows indicate known forces and known moments, such as the applied load and the moments at plastic hinge positions. The circled numbers indicate the order in which the calculations are made: thus,

Step 1: The shear force for the member FG is calculated as (15 kN m + 15 kN m)/2 m = 15 kN.

Step 2: The shear force at the end F of member DF is given by (27.48 kN − 15 kN) = 12.48 kN, where 27.48 kN is the known applied load and 15 kN is the shear force from Step 1. The bending moment at end D is therefore (12.48 kN) (2 m) − 15 kN m = 9.96 kN m

Step 3: For member CD, the shear force is (15 kNm + 15 kN m)/(4 m) = 7.5 kN.

Step 4: The shear force at end C of member CB is given by the difference of the applied force and the shear for CD; that is 13.74 kN − 7.5 kN = 6.24 kN. The bending moment at end B is therefore (6.24 kN) (4 m) − 15 kN m = 9.96 kNm.

Step 5: For member GH, the shear force is (15 kN m + 15 kN m) /(4 m) = 7.5 kN. The axial force is given by the shear force of member FG (see Step 1). The axial forces in member FG, DF and the force acting from the right of Joint D are all equal to the shear force in GH, namely 7.5 kN.

Steps 6 and 7: For Joint D, the moment on the left is the known plastic moment 15 kN m, that on the right is 9.96 kN m (from Step 2). Therefore, the moment acting at the bottom is (15−9.96)kN m = 5.04 kN m. Therefore, the bending moment at end D of member DE is also 5.04 kN m; this gives the shear force in DE as (5.04 kN m + 15 kN m)/(4 m) = 5.01 kN. Similarly, the shear force in the (very short) vertical member at Joint D is also 5.01 kN; hence the axial force in the short horizontal member on the left of Joint D is 5.01 kN m +

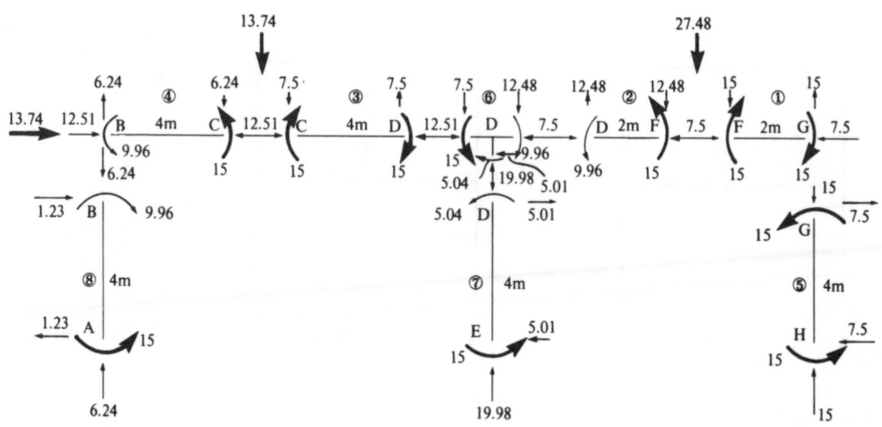

Fig. 14·7–4

7.5 kN m = 12.51 kN. The shear forces in the horizontal members at Joint D are 7.5 kN and 12.48 kN, from Steps 3 and 2 respectively; therefore the axial force in the vertical member at that joint is 7.5 kN + 12.48 kN = 19.98 kN, which also gives the axial force in DE.

The axial forces in CD and BC are therefore 12.51 kN.

Step 8: For member AB, the axial force is equal to the shear force in member BC (Step 4); the bending moment at end D is also obtained from Step 4 as 9.96 kNm. Hence the shear force in AB is (15 kNm − 9.96 kN m)/(4 m) = 1.23 kN. As a check, the axial force in BC plus the shear force in AB is 12.51 kN + 1.23kN = 13.74 kN, which equals the applied horizontal force at Joint B.

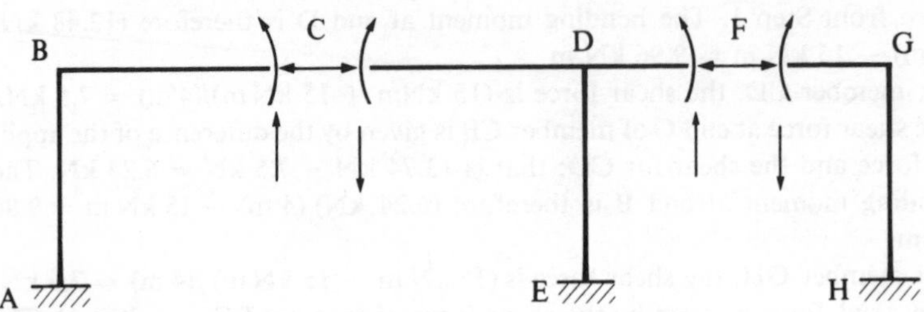

Fig. 14·7–5

COMMENTS

(1) In using Eqn 14.6–3 to determine the number of independent mechanisms it was stated that the frame had 6 redundancies. One effective method of finding the number of redundancies of frameworks is to make vertical cuts as illustrated in Fig. 14.7–5; each cut releases three redundancies (a moment, a shear and an axial force), and since two cuts are required to render the structure statically determinate, we conclude that the frame has six redundancies.

Similarly, the frame in Fig. 14.7–6 also has six redundancies.

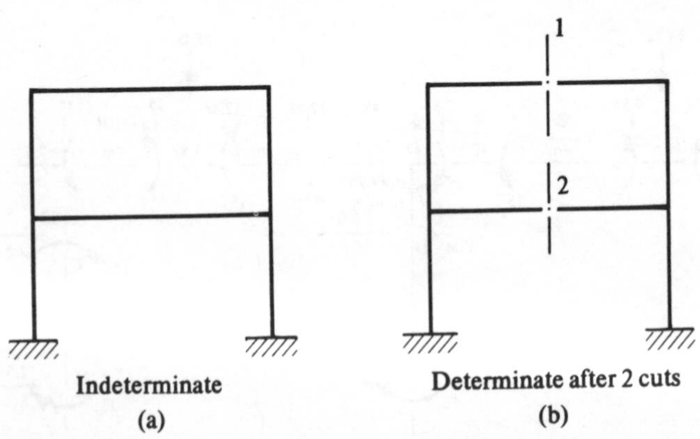

Fig. 14·7–6

(2) The frame in Fig. 14.7–2(a) collapses as a regular mechanism. Regular mechanisms being statically determinate, it was possible to draw the complete bending diagram using the equations of statics only. Sometimes a frame may collapse as a partial mechanism, which by definition is statically indeterminate; in such a case, the bending moment diagram at collapse cannot be completely defined—see Example 14.7–2.

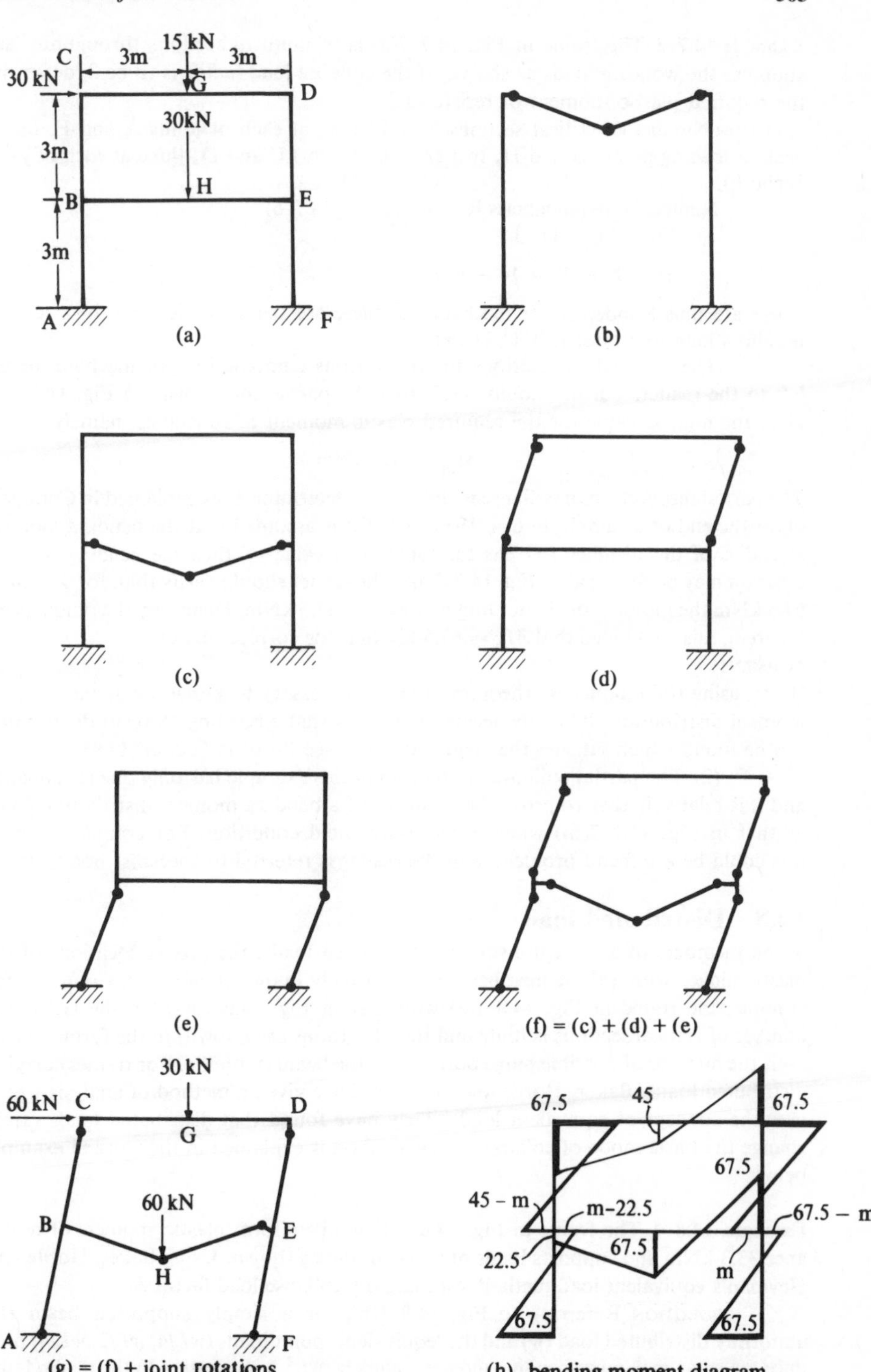
15 kN
C
3m
3m
30 kN
G
D
30kN
3m
H
B
E
3m
A
F
(a)
(b)
(c)
(d)
(e)
(f) = (c) + (d) + (e)
30 kN
60 kN C
D
G
B
60 kN
E
H
A
F
(g) = (f) + joint rotations at B and E
67.5
45
67.5
67.5
45 – m
m–22.5
67.5 – m
67.5
22.5
m
67.5
67.5
(h) bending moment diagram

Fig. 14·7–7

Example 14.7–2 The frame in Fig. 14.7–7(a) is of uniform sections throughout, and supports the working loads as shown. If the collapse load factor is to be 2, determine the required plastic moment of resistance.

SOLUTION Number of critical sections N = 14 (one at each of joints A and F; one at each of loading points G and H; two at each of joints C and D; three at each of joints B and E).

Number of redundancies R = 6 (see Fig. 14.7–6)

From Eqn 14.6–3

$$M = \mathrm{N} - \mathrm{R} = 14 - 6 = 8$$

There are thus 8 independent mechanisms; these are the 4 in Fig. 14.7–7(b), (c), (d), (e) plus 4 joint rotations, at B, C, D and E.

The detailed calculations for the various combinations of mechanisms are left to the reader, but he should verify that the partial mechanism in Fig. 14.7–7(g) gives the highest value for the required plastic moment of resistance, namely

$$M_p = 67.5 \text{ kNm}.$$

The partial mechanism of collapse is statically indeterminate, as explained in Comment (4) at the end of Example 14.6–4. However, if it is assumed that the bending moment at end E of the member ED has the (unknown) value m, then the bending moment diagram may be drawn as in Fig. 14.7–7(h). The reader should verify that, for $0 \leq m \leq$ 67.5 kNm the moment ordinates nowhere exceed 67.5 kNm. Hence, by the uniquencess theorem, it is concluded that $M_p = 67.5$ kN m is the correct answer.

COMMENTS

(1) In using the uniqueness theorem, it is not necessary to know the actual bending moment distribution; it is only necessary to show that a bending moment distribution can be found which satisfies the yield condition (see Proof in Section **14.9**).

(2) The partial collapse mechanism in this example has only one redundancy and it is relatively easy to prove the existence of a bending moment distribution (such as that in Fig. 14.7–7(h)) which satisfies the yield condition. For complex frames, this could be a difficult problem, and the reader is rererred to specialist books[1, 3, 6, 7].

14.8 Distributed load

When members of a structure support distributed loads, the precise locations of the plastic hinges within these members are not usually known in advance. Compare, for example, the frame in Fig. 14.8–1(a) with that in Fig. 14.6–10(a). In the latter, the number of critical sections is finite and their locations are known; in the former, however, the number of possible hinge positions in the beam is infinite. For frames carrying distributed loads, Baker, Horne and Heyman[1] have given a method of analysis which uses the concept of equivalent loads. They have found that distributed loads rarely change the basic mode of collapse. Their method is explained in the worked example below.

Example 14.8–1 The frame in Fig. 14.8–1(a) is of uniform plastic moment of resistance 33.3 kNm and supports loads of working values shown. Using Baker, Horne and Heyman's equivalent load method, estimate the collapse load factor λ.

SOLUTION Referring to Fig. 14.8–1(b), for a simply supported beam the uniformly distributed load (**w**) and the 'equivalent' point loads (**$wL/4$, $wL/2$, $wL/4$**) produce the same maximum bending moment, namely, $wL^2/8$. According to the method, the distributed load on the beam BD is replaced by the equivalent point loads:

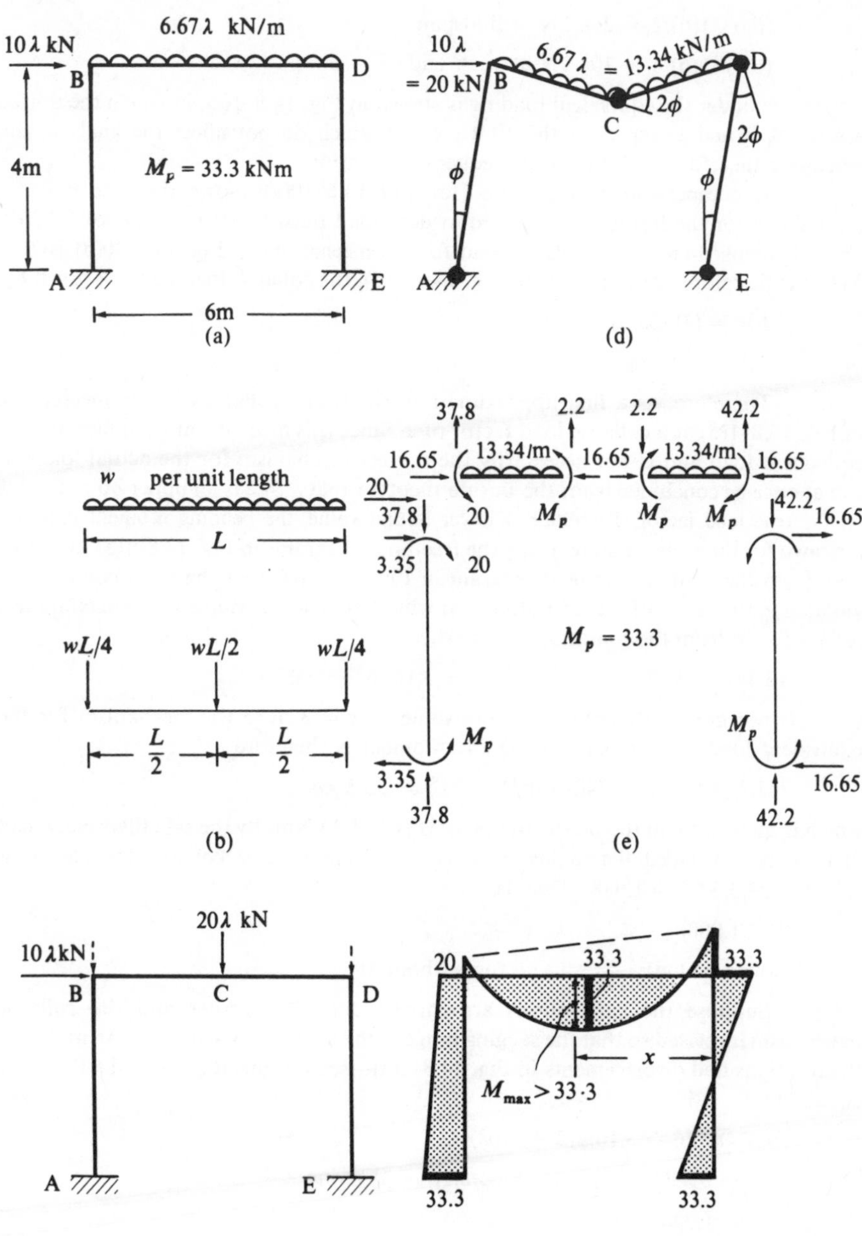

10λ kN
6.67λ kN/m
B
D
4m
M_p = 33.3 kNm
A
E
6m
(a)
10λ
= 20 kN
6.67λ = 13.34 kN/m
C
2φ
φ
(d)
w per unit length
L
wL/4
wL/2
wL/4
L/2
(b)
37.8
2.2
42.2
16.65
13.34/m
20
3.35
M_p
M_p = 33.3
(e)
20λ kN
10λkN
(c)
33.3
x
M_{max} > 33·3
(f)

Fig. 14·8–1

$$(6.67\lambda)(6)/2 = 20\lambda \text{ kN at midspan}$$

$$(6.67\lambda)(6)/4 = 10\lambda \text{ kN at each end.}$$

The frame under the equivalent loading is shown in Fig. 14.8–1(c), in which the dotted arrows at B and D represent the 10 kN loads which do not affect the analysis and which are therefore neglected in subsequent calculations.

A comparison of Fig. 14.8–1(c) and 14.6–10(a) shows that the collapse mechanism for the frame under the equivalent load must be that of Fig. 14.6–10(d). From Example 14.6–4, the collapse load for the mechanism in Fig. 14.6–10(d) is $P = 3M_p/5a$; therefore, for the frame in Fig. 14.8–1(d), the collapse load factor is given by

$$10\lambda = (3)\,(33.3)/5 = 20$$

or $$\lambda = 2$$

Therefore, as a first approximation, the frame collapses in the mechanism in Fig. 14.8–1(d) at a collapse load factor of 2. Since this mechanism is, in fact, for the equivalent load and is not necessarily the correct mechanism for the actual loads, it can at once be concluded from the unsafe theorem that $\lambda = 2$ is an upper bound value on the true load factor. To obtain a lower bound value, the bending moment diagram is drawn for the collapse state, using the free-body diagrams in Fig. 14.8–1(e). It can be seen from the bending moment diagram in Fig. 14.8–1(f) that the yield condition is violated in the beam BD; the location x at which the sagging moment is a maximum is readily found from the condition of zero shear:

$$13.34x = 42.2 \qquad \text{or} \qquad x = 3.16 \text{ m}$$

which is not greatly different from the value of $x = 3$ m in the mechanism for the equivalent load. The maximum sagging moment is therefore

$$(42.2)(3.16) - 13.34(3.16)^2/2 - 33.3 = 33.5 \text{ kNm}$$

which is quite close to the plastic moment M_p of 33.3 kNm. By the safe theorem, it can at once be concluded that a lower bound value on the true collapse load factor is $\lambda = 2 \times 33.3/33.5 = 1.988$. That is

$$\underset{\text{(lower bound)}}{1.988} \quad \leq \quad \lambda \quad \leq \quad \underset{\text{(upper bound)}}{2}$$

Suppose the calculations are carried one step further and the collapse mechanism is revised so that the sagging hinge in the beam occurs at $x = 3.16$ m. Using the rotations and displacements in Fig. 14.8–2(a)—see Comment (1) at end—the work equation is:

$$(6.67\lambda)(6)(6 - 3.16)\phi/2 + (10\lambda)(4\phi)$$
$$= 33.3[\phi + 6\phi/3.16 + 6\phi/3.16 + \phi] \text{ giving}$$
$$\lambda = 1.994$$

By the unsafe theorem this is an upper bound value, so that

$$\underset{\text{(lower bound)}}{1.988} \quad \leq \quad \lambda \quad \leq \quad \underset{\text{(upper bound)}}{1.994}$$

Of course, the process of successive approximation can be repeated: a bending moment diagram is drawn for the collapse state in Fig. 14.8–2(a), and a new lower bound value of λ determined; then from a revised value of x, a revised mechanism is investigated leading to a new upper bound value of λ, and so on. However, it is seen from this worked example that the process converges so rapidly that the answer $1.988 \leq \lambda \leq 2$ given by the equivalent-load mechanism is already very good; the second cycle of calculations, leading to $1.988 \leq \lambda \leq 1.994$ is hardly necessary.

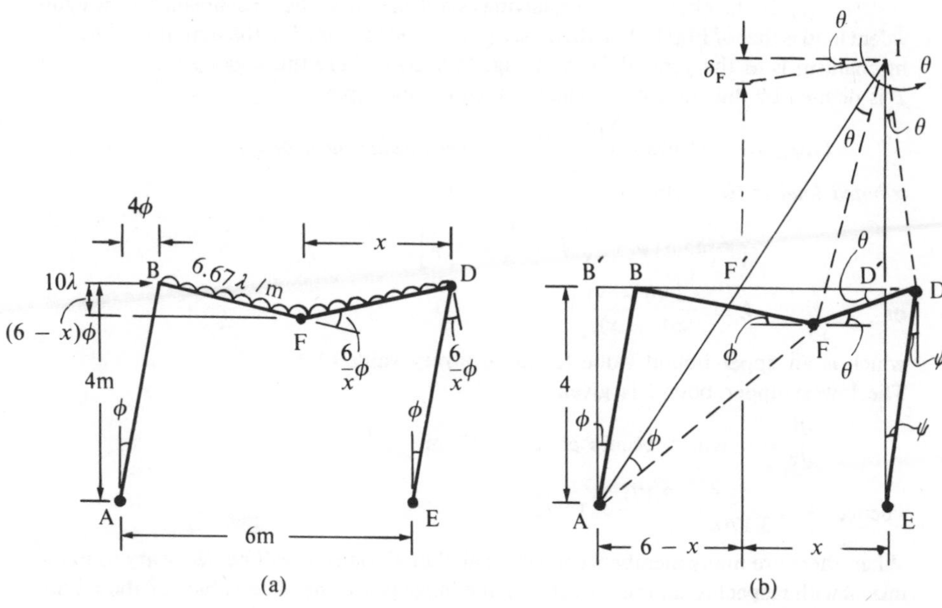

Fig. 14·8–2

COMMENT

(1) The displacements and rotations in Fig. 14.8–2(a) can be determined by the method of instantaneous centres explained earlier in Section **6.8**. Referring to Fig. 14.8–2(b)

$$\text{FF}' = (\text{IF}')\theta = (\text{AF}')\phi$$

$$\text{Hence } \theta = \frac{(\text{AF}')}{(\text{IF}')}\phi = \frac{(\text{B}'\text{F}')}{(\text{F}'\text{D}')}\phi = \frac{6-x}{x}\phi$$

$$\text{DD}' = (\text{ID}')\theta = (\text{ED}')\psi$$

$$\text{Hence } \psi = \frac{(\text{ID}')}{(\text{ED}')}\theta = \frac{(\text{ID}')}{(\text{AB}')}\theta = \frac{(\text{D}'\text{F}')}{(\text{B}'\text{F}')}\theta$$

$$= \frac{x}{6-x}\theta = \left(\frac{x}{6-x}\right)\left(\frac{6-x}{x}\right)\phi$$

$$= \phi \text{ as expected.}$$

Hinge rotation at F

$$= \theta + \phi = \frac{6-x}{x}\phi + \phi = \frac{6}{x}\phi$$

Hinge rotation at D

$$= \theta + \psi = \theta + \phi = \frac{6}{x}\phi$$

Horizontal displacement of joint B $= 4\phi$
Vertical component of displacement of joint F, δ_F

$$= x\theta = (x)\left(\frac{6-x}{x}\right)\phi = (6-x)\,\phi$$

(2) In the above Example, having established that the mechanism for the equivalent load is that of Fig. 14.8–1(d), it may be concluded[1] that for the actual loading the mechanism is of the general form in Fig. 14.8–2(a) where the sagging hinge position F is defined by the variable x. The work equation is then

$$(10\lambda)(4\phi) + (6.67\lambda)(6)\frac{(6-x)}{2}\phi$$ (*The displacement of the centroid of the distributed load is* $(6-x)\phi/2$.)

$$= 33.3\left[\phi + \frac{6}{x}\phi + \frac{6}{x}\phi + \phi\right]$$

or $$\lambda = \frac{3.33(6+x)}{x(8-x)}$$

which is an upper bound value for an arbitrary value of x.
The lowest upper bound is given by

$$\frac{d\lambda}{dx} = 0 \text{ which occurs at } x = 3.17 \text{ m}$$

whence $$\lambda = \frac{3.33(6+3.17)}{3.17(8-3.17)} = 1.995$$

When there are many members carrying distributed loads, it will be necessary to minimize λ with respect to all the x's defining the hinge positions. The values of the x's are then obtained from

$$\frac{\partial\lambda}{\partial x_1} = 0; \frac{\partial\lambda}{\partial x_2} = 0; \frac{\partial\lambda}{\partial x_3} = 0 \text{ etc.}$$

For a complex structure, the method is therefore rather tedious, but for simple structures it is very effective—see Example 14.8–2.

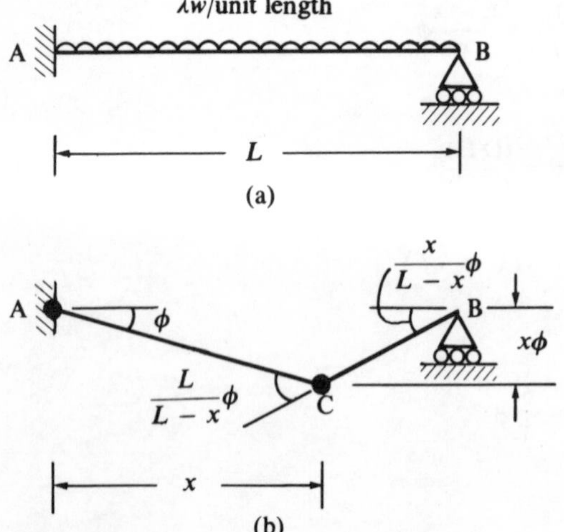

Fig. 14·8–3

Example 14.8–2. Fig. 14.8–3(a) shows a propped cantilever supporting a uniformly distributed load w per unit length. If the collapse load factor is to be λ, determine the required plastic moment of resistance M_p.

SOLUTION Let the unknown location of the sagging plastic hinge be defined by x, as in Fig. 14.8–3(b). The work equation is

$$\lambda\, wL\left(\frac{x\phi}{2}\right) = M_p\left[\phi + \frac{L\phi}{L-x}\right]$$

or

$$M_p = \lambda wL\,\frac{x(L-x)}{2L-x}$$

By the unsafe theorem, the value of M_p corresponding to an arbitrarily assigned value of x is less than, or at best equal to, the plastic moment required to withstand the load λw. Hence, M_p is maximised with respect to x:

$$\frac{dM_p}{dx} = 0 \text{ gives } x = \underline{0.586L}$$

$$\text{so that } M_p = \underline{0.0858\ wL^2}$$

COMMENTS

The above solution is based on the unsafe or upper bound theorem. An alternative solution, based on the safe or lower bound theorem is given in Example 14.8–3.

Example 14.8–3. Using the safe theorem, given an alternative solution to Example 14.8–2.

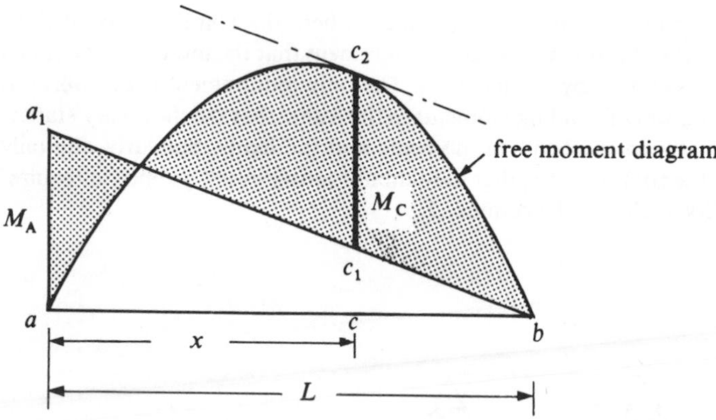

Fig. 14·8–4

SOLUTION Fig. 14.8–4 shows a bending moment diagram obtained by superposition: ac_2b is the simple-beam or free moment diagram for the uniformly distributed load λw per unit length, and aa_1b is the moment diagram for a hogging moment of arbitrarily assigned magnitude M_A at the built-in end of the beam. Thus, ac_2b is in equilibrium with a transverse load of λw per unit length and aa_1b is in equilibrium with zero transverse load; therefore the composite diagram, as shaded in Fig. 14.8–4, must also be in equilibrium with the load λw per unit length, whatever the value assigned to M_A.

Let M_C denote the magnitude of the largest sagging moment in the span. It can be concluded from the safe theorem that the beam will not collapse under the load, provided

$$M_p > M_A \quad \text{or} \quad M_C$$

whichever is (numerically) the larger. The value of M_p that satisfies this condition depends on the assigned magnitude of M_A. Referring to Fig. 14.8–4. the lowest value of M_p (= $0.0858\ \lambda wL^2$) is obtained if M_A is so chosen that $M_A = M_C$.

14.9 The fundamental theorems of plastic collapse

The proofs of the uniqueness theorem, the upper bound theorem and the lower bound theorem are given in this section. Extensive use will be made of the principle of virtual work (both in this section and in Section **14.10** on "Incremental collapse and shake-down"), and a brief discussion of the application of the principle to the mechanism type of virtual displacements is appropriate. Eqn 4.14–3 of Section **4.14** states that

$$\mathbf{P}^T\bar{\Delta} = \mathbf{F}^T\bar{\mathbf{e}}$$

and, as explained in the paragraph following that equation, if only bending moments need to be considered, then the quantity $\mathbf{F}^T\bar{\mathbf{e}}$ represents the summation of $Md\bar{\theta}$.

Thus if a rigid-jointed frame structure is considered with a mechanism of virtual displacements, in which all rotations are concentrated at the hinges while the rest of the structure remains rigid, the virtual work equation takes the form:

Equilibrium set

$$\Sigma P_j\Delta_j = \Sigma M_i\theta_i \qquad (14.9\text{–}1)$$

Compatible displacement set

in which M_i is the bending moment at the hinge i where the rotation θ_i occurs. As explained in Section **4.14**, it is not a necessary requirement that the moments M_i should be the actual moments caused by the loads P_j. The only requirement is equilibrium: M_i may be any distribution of bending moments in the frame that satisfies every statical test of equilibrium between the bending moments and the loads. Similarly, the only condition placed on the rotations θ_i is that they should satisfy every geometric requirement of compatibility with the deflections Δ_j.

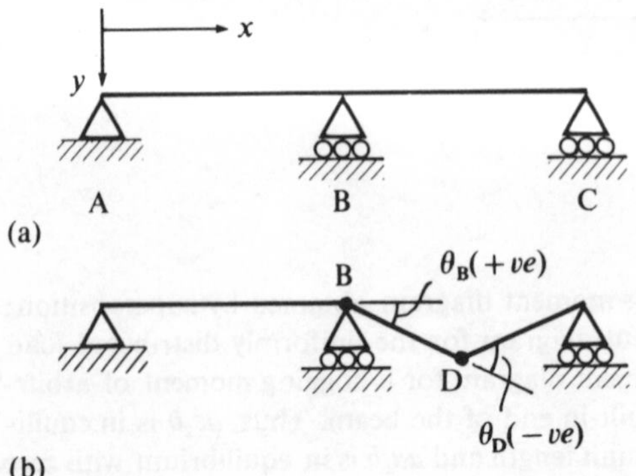

Fig. 14·9–1

Again, as explained in Section **4.14**, care must be taken to ensure that the **sign convention** for θ_i in Eqn 14.9–1 is consistent with that for M_i, so that a positive product $M_i\theta_i$ means that work is locally absorbed by the structure at the hinge section, just as a positive product $P_j\Delta_j$ means that work is done by the force P_j. Consider, for example, the continuous beam in Fig. 14.9–1(a). Suppose the sign convention adopted for bending moments happens to be 'hogging is positive'. Then for the virtual displacements defined by the mechanism in Fig. 14.9–1(b), the hogging rotation θ_B must be taken as positive, because for a positive (that is, hogging) moment at B, the rotation θ_B as shown would mean that energy is absorbed by the structure at that hinge; similarly, θ_D as shown must be taken as negative, so that for a positive moment at D energy is released by the structure as a result of that hinge rotation.

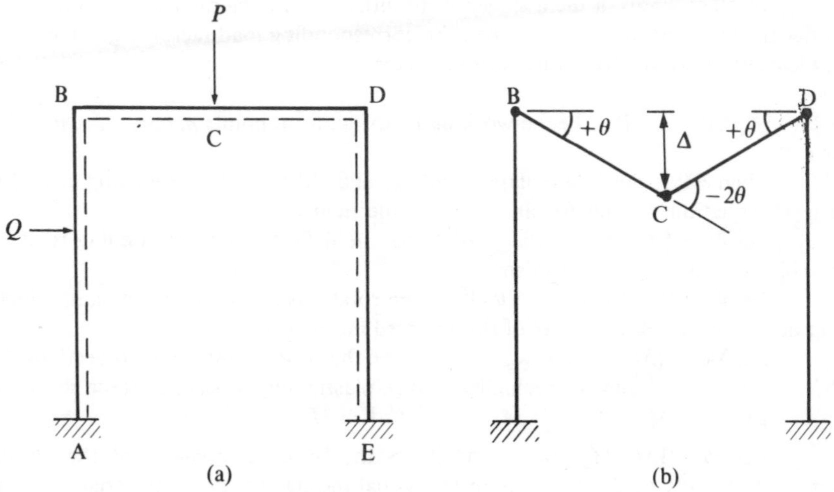

Fig. 14·9–2

Next consider the frame in Fig. 14.9–2(a). Suppose the sign convention chosen is such that bending moments that produce compression on the inside face (shown dotted) of the frame is positive. For the virtual-displacement mechanism in Fig. 14.9–2(b), the signs for the hinge rotations are then as shown. The virtual work equation is:

Equilibrium set

$$P(\Delta) + Q(0) = M_B(+\theta) + M_C(-2\theta) + M_D(+\theta)$$

Compatible displacement set (14.9–2)

where M_B,M_C and M_D denote the bending moments at the respective sections in *any* bending moment distribution that is in equilibrium with the loads (P,Q). *The equation is 'exact', irrespective of whether large shear and axial forces exist in the frame*—the virtual-displacement pattern has been so specified that such forces do not enter into the equation. *The equation holds irrespective of whether the frame is elastic or plastic.* As long as M_B, M_C and M_D belong to a distribution of bending moments that is in in equilibrium with the given loading (and the actual bending moments caused by the given loading is one such distribution), then Eqn 14.9–2 must be obeyed. In the particular case when Fig. 14.9–2(b) in fact represents a collapse mechanism, so that the moments M_B, M_C and M_D all have the known numerical value M_p, then in accordance with the sign convention in Fig. 14.9–2(a), $M_B = +M_P$, $M_C = -M_p$ and $M_D = +M_p$. Eqn 14.9–2 therefore takes the special form of a collapse work equation:

$$P\Delta = M_p|\theta| + M_p|2\theta| + M_p|\theta| \qquad (14.9\text{–}3)$$

that is, *in a work equation for a collapse mechanism, all the products $M_p\theta$ are positive.*

It is now possible to give the proofs of the plastic collapse theorems.

Upper bound theorem

The upper bound theorem, sometimes referred to as the unsafe theorem or the kinematic theorem, states that the load factor λ' obtained from the work equation written for any arbitrarily assumed mechanism is greater than, or at least equal to, the true collapse load factor λ.

In other words, if the assumed distribution of bending moments for the frame satisfies the mechanism condition, then the corresponding load factor is greater than or at least equal to, the true collapse load factor.

PROOF

Let $\mathbf{P} = [P_a P_b P_c \ldots P_m]^T$ be the working loads acting at points *a*, *b*, *c*, . . . *m* of the frame.

Then $\lambda\,\mathbf{P}$ is the true collapse loading, and $\lambda'\mathbf{P}$ the collapse loading given by the work equation written for an arbitrary mechanism.

Let $\mathbf{\Delta}' = [\Delta'_a\ \Delta'_b\ \Delta'_c \ldots\ \Delta'_m]^T$ be the deflections corresponding to $\mathbf{P}$ in the arbitrarily assumed mechanism.

Let $\theta' = [\theta'_A\ \theta'_B\ \theta'_C \ldots \theta'_N]^T$ be the rotations of the plastic hinges, which occur at sections A, B, C . . . N of the assumed mechanism.

Let $\mathbf{M}' = [M'_A\ M'_B\ M'_C \ldots M'_N]^T$ be the bending moments at sections A, B, C . . . N of the assumed mechanism; since plastic hinges occur at these sections,

$$|M'_A| = M_p,\quad |M'_B| = M_p \ldots |M'_N| = M_p$$

Let $\mathbf{M} = [M_A\ M_B\ M_C \ldots M_N]^T$ be the bending moments at the above mentioned sections A, B, C . . . N in the actual mechanism under the true collapse loading $\lambda\,\mathbf{P}$.

The work equation, written for the assumed mechanism, is

$$\lambda'\,\mathbf{P}^T\mathbf{\Delta}' = \mathbf{M}'^T\theta' \qquad (14.9\text{–}4)$$

Noting that, for the frame under the true collapse loading, $(\lambda\mathbf{P},\mathbf{M})$ is an equilibrium set, and that $(\mathbf{\Delta}', \theta')$ is a compatible set; the virtual work equation can be written:

Equilibrium set

$$\lambda\mathbf{P}^T\mathbf{\Delta}' = \mathbf{M}^T\theta' \qquad (14.9\text{–}5)$$

Compatible displacement set

Eqn 14.9–4 may be written as (if in doubt see Eqns 14.9–2 and 3):

$$\begin{aligned}\lambda'\mathbf{P}^T\mathbf{\Delta}' &= M'_A\theta'_A + M'_B\theta'_B + \ldots M'_N\theta'_N \\ &= M_p|\theta'_A| + M_p|\theta'_B| + \ldots M_p|\theta'_N| \end{aligned} \qquad (14.9\text{–}6)$$

In this equation, every term on the right-hand side is positive; hence $\lambda'\mathbf{P}^T\mathbf{\Delta}'$ is positive. Since load factors are taken as positive, the quantity $\mathbf{P}^T\mathbf{\Delta}'$ is positive.

Eqn 14.9–5 may be written as:

$$\lambda\mathbf{P}^T\mathbf{\Delta}' = M_A\,\theta'_A + M_B\,\theta'_B + \ldots M_N\,\theta'_N \qquad (14.9\text{–}7)$$

Subtracting Eqn 14.9–7 from Eqn 14.9–6,

$$(\lambda' - \lambda)\,\mathbf{P}^T\Delta' = (M_p|\theta'_A| - M_A\theta'_A) + (M_p|\theta'_B| - M_B\theta'_B) + \ldots + (M_p|\theta'_N| - M_N\theta'_N) \quad (14.9\text{–}8)$$

Since the moments $M_A, M_B, \ldots M_N$ are those in the actual mechanism, their magnitudes are less than M_p or at most equal to M_p. Specifically, if section A happens to be a plastic hinge section in the actual mechanism, then $|M_A| = M_p$ so that $M_A\theta'_A = M_p|\theta'_A|$; otherwise $M_A\theta'_A < M_p|\theta'_A|$ – if in doubt see Eqns 14.9–2 and 3. Therefore, each term on the right-hand side of Eqn 14.9–8 is either positive or zero; hence $(\lambda' - \lambda)\mathbf{P}^T\Delta'$ is either positive or zero. Since, as stated in the paragraph following Eqn 14.9–6, $\mathbf{P}^T\Delta'$ is always positive, we must have

$$\lambda' - \lambda \geq 0$$

i.e.

$$\lambda' \geq \lambda$$

Lower bound theorem

The lower bound theorem, sometimes referred to as the safe theorem or the static theorem, states that, if at any load factor λ'', a distribution of bending moments can be found which satisfies both the equilibrium condition and the yield condition, then λ'' is less than or at most equal to the true collapse load factor λ.

PROOF

The procedure is similar to that in the proof ot the upper bound theorem.

Let $\lambda\mathbf{P}$ be the true collapse loading and $\Delta(= [\varDelta_a\varDelta_b\varDelta_c \ldots \varDelta_m]^T)$ the deflections corresponding to $\lambda\mathbf{P}$.

Let $\theta(= [\theta_A\theta_B\theta_C \ldots \theta_N]^T)$ be the rotations at the plastic hinge sections under the true collapse loading, and $M(=[M_AM_BM_C \ldots M_N]^T)$ be the bending moments at the plastic hinge sections.

For an arbitrary loading $\lambda''\mathbf{P}$, let $\mathbf{M}''(= [M''_A M''_B M''_C \ldots M''_N]^T)$ be the bending moments at sections A,B,C . . . N, in a distribution of moments that satisfies both the equilibrium condition and the yield condition.

We can form the virtual work equation:

Equilibrium set

$$\lambda\mathbf{P}^T\mathbf{\Delta} = \mathbf{M}^T\boldsymbol{\theta}$$

Compatible displacement set

That is

$$\lambda\mathbf{P}^T\mathbf{\Delta} = M_A\theta_A + M_B\theta_B + \ldots M_N\theta_N$$
$$= M_p|\theta_A| + M_p|\theta_B| + \ldots M_p|\theta_N| \quad (14.9\text{–}9)$$

because A,B,C . . . N are plastic hinge sections in the actual mechanism under the true collapse loading $\lambda\mathbf{P}$.

We can also form the virtual-work equation:

Equilibrium set

$$\lambda''\mathbf{P}^T\mathbf{\Delta} = \mathbf{M}''^T\boldsymbol{\theta}$$

Compatible displacement set

That is

$$\lambda''\mathbf{P}^T\mathbf{\Delta} = M''_A\,\theta_A + M''_B\theta_B + \ldots M''_N\theta_N \qquad (14.9\text{–}10)$$

Subtracting Eqn 14.9–10 from Eqn 14.9–9

$$(\lambda - \lambda'')\,\mathbf{P}^T\mathbf{\Delta} = [M_p|\theta_A| - M''_A\theta_A] + [M_p|\theta_B| - M''_B\theta_B] + \ldots [M_p|\theta_N| - M''_N\theta_N] \qquad (14.9\text{–}11)$$

Since the moments $\mathbf{M}''$ satisfy the yield condition, $|M_A''| \leq M_p$, $|M''_B| \leq M_p \ldots |M_N| \leq M_p$, and each term on the right-hand side of Eqn 14.9–11 is either positive or zero, then

$$(\lambda - \lambda'')\,\mathbf{P}^T\mathbf{\Delta} \geq 0 \qquad (14.9\text{–}12)$$

However, in Eqn 14.9–9 every term on the right-hand side is positive and, since λ is positive, $\mathbf{P}^T\mathbf{\Delta}$ is positive. Therefore, from Eqn 14.9–12,

$$\lambda - \lambda'' \geq 0$$

or $$\lambda'' \leq \lambda$$

The uniqueness theorem

The uniqueness theorem states that, if at any load factor λ, a distribution of bending moment can be found which simultaneously satisfies (1) the mechanism condition, (2) the equilibrium condition and (3) the yield condition, then that load factor is the true collapse load factor.

PROOF

Since the bending moment distribution satisfies the mechanism condition, the upper bound theorem states that:

$$\lambda \geq \text{true collapse load factor}$$

Since the bending moment distribution satisfies the equilibrium condition and the yield condition, the lower bound theorem states that

$$\lambda \leq \text{true collapse load factor}$$

These two conditions can be satisfied simultaneously only if

$$\lambda = \text{true collapse load factor.}$$

COMMENTS

(1) Note that the uniqueness theorem does not guarantee that the collapse mechanism itself is unique. However, one thing is certain. If there are, say, two different collapse mechanisms, both of which satisfy the three conditions of mechanism, equilibrium, and yield, then the load factors for the two mechanisms must be the same—they must each be equal to the true collapse load factor.

(2) A corollary of the lower bound theorem is that the collapse load factor of a frame cannot be reduced by the addition of material to it or by the imposition of a restraint at a pin joint. To prove the corollary, let $\mathbf{M}$ be the distribution of bending moments in the original frame under the collapse load $\lambda\,\mathbf{P}$. By definition, $\mathbf{M}$ and $\lambda\,\mathbf{P}$ are in equilibrium, and $\mathbf{M}$ nowhere exceeds the local value of M_p. Next consider the modified frame under the same loading $\lambda\,\mathbf{P}$. The same distribution of bending moments $\mathbf{M}$ will of course still be in equilibrium with $\lambda\,\mathbf{P}$; also, since $\mathbf{M}$ nowhere exceeds M_P in the original frame it cannot possibly exceed that in the modified frame, because the addition of material to a section or the imposition of a restraint at a pin joint is each equivalent to increasing the local value of M_p. Thus, at a load factor λ we have found for the modified frame a bending moment distribution $\mathbf{M}$ which satisfies the equili-

brium and the yield conditions; by the lower bound theorem, λ is less than, or at most equal to, the true collapse load factor of the modified frame.

(3) Similarly, a corollary of the upper bound theorem is that the collapse load factor of a frame cannot be increased by the removal of material from the frame or by the removal of the rotational restraint at a joint.

(4) The three theorems of plastic collapse are, strictly speaking, only valid for rigid-plastic structures (see moment-curvature curve in Fig. 14.6–1(b)), which have zero displacement until the collapse load factor λ is reached. They are, however, applicable to elastic-plastic structures (see Fig. 14.6–1(a)) provided the effects of change of geometry and of instability are not important.

14.10 Incremental collapse and shakedown

Incremental collapse and shakedown is concerned with the progressive failure of structures under **variable repeated loading**, the number of cycles being low enough to prevent any question of fatigue arising. A theory will be developed for elastic-plastic structures having the moment-curvature relation OAB in Fig. 14.10–1; however, it will be shown that, with very little modification, the theory will be applicable to structures having the more realistic moment-curvature relation OA′B. As usual, it will be assumed that if the bending moment at a plastic hinge is reduced, the response at the plastic hinge section becomes elastic; that is, referring to Fig. 14.10–1, if at point B the moment is reduced, then the moment-curvature relation follows the straight line BC.

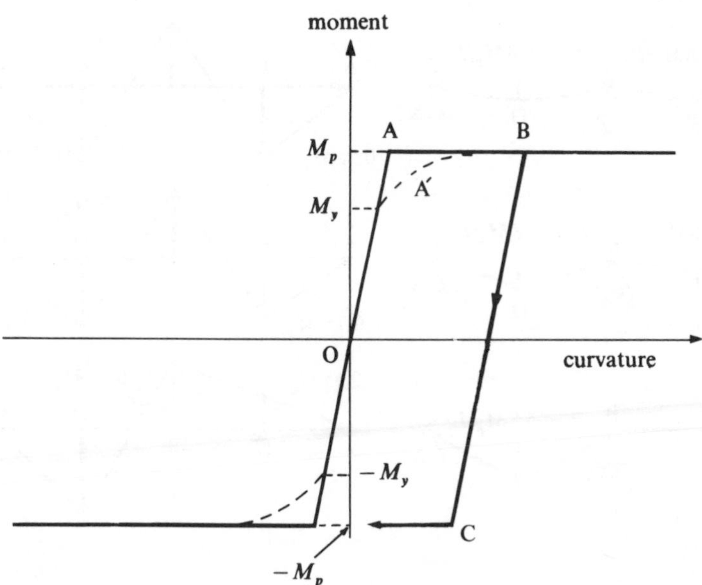

Fig. 14·10–1

As an example of a structure failing progressively under variable repeated loading, consider the behaviour of the two-span beam in Fig. 14.10–2 at the different loading stages:

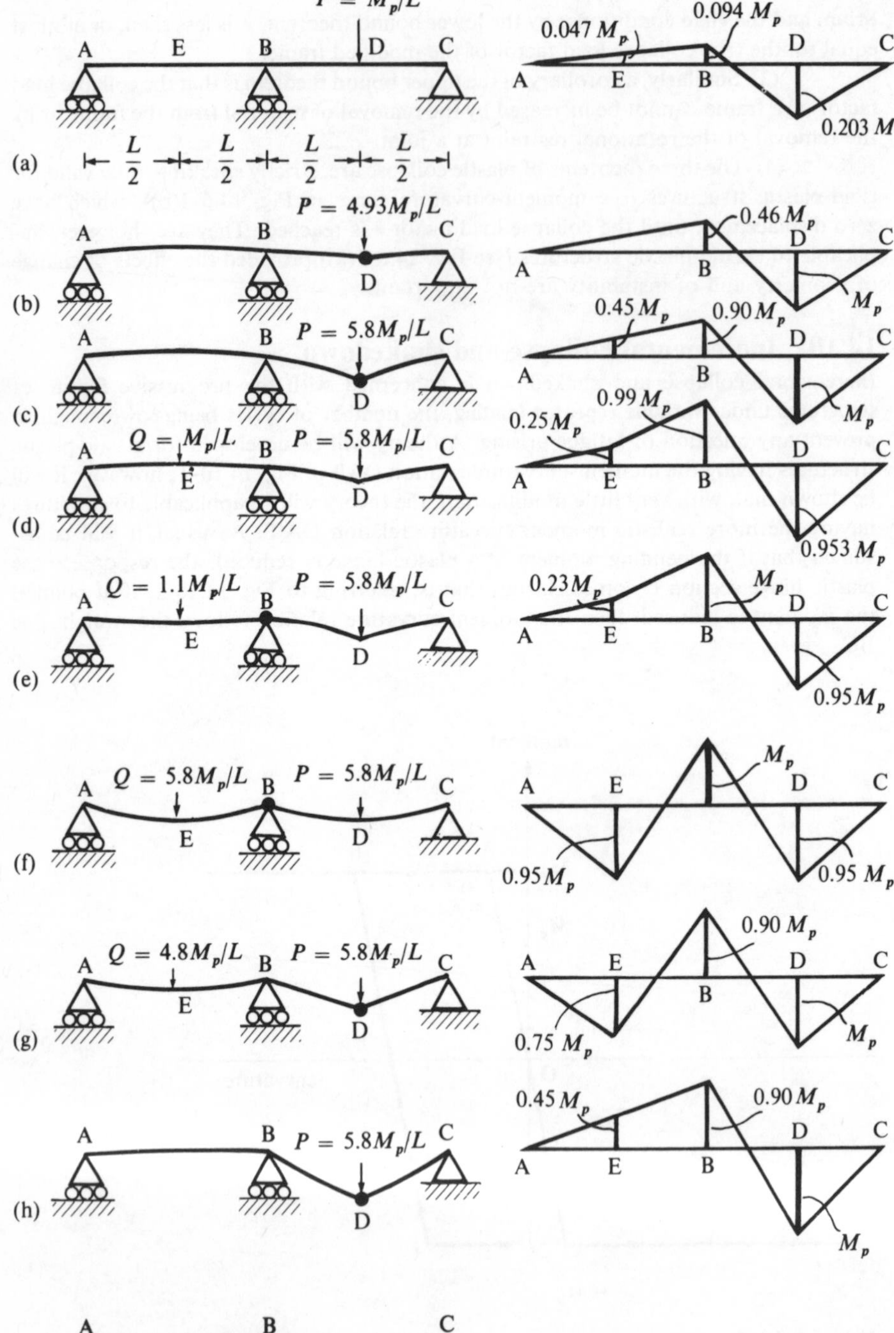

P = M_p/L
A E B D C
L/2 L/2 L/2 L/2
(a)
0.047 M_p
0.094 M_p
0.203 M_p
P = 4.93M_p/L
(b)
0.46 M_p
M_p
P = 5.8M_p/L
(c)
0.45 M_p
0.90 M_p
Q = M_p/L
(d)
0.25M_p
0.99 M_p
0.953 M_p
Q = 1.1M_p/L
(e)
0.23M_p
0.95M_p
Q = 5.8M_p/L
(f)
0.95M_p
0.95 M_p
Q = 4.8M_p/L
(g)
0.75 M_p
0.90 M_p
(h)
(i)

Fig. 14·10–2

STAGE 1 (Fig. 14.10–2(a)): A load P is gradually applied at D. For $P = M_p/L$ the bending moments are everywhere elastic.
STAGE 2 (Fig. 14.10–2(b)): When P reaches 4.93 M_p/L, the moment at D just reaches the plastic value M_p. That is, the bending moments are now 4.93 times those in Fig. 14.10–2(a).
STAGE 3 (Fig. 14.10–2(c)): When P is increased beyond 4.93 M_p/L, the plastic hinge at D rotates, leading to increased deflections. At $P = 5.8\ M_p/L$, the bending moments are as in Fig. (c). Note that as P is increased from 4.93 M_p/L to 5.8 M_p/L, the bending moment at D remains constant at the plastic value M_p. Therefore the bending moment diagram in Fig. (c) is equal to that in Fig. (b) plus that due to the load increment (5.8 M_p/L − 4.93 M_p/L) acting on the beam with a frictionless hinge at D. Specifically, the bending moment at the intermediate support section B is:

$$0.46\ M_p + (5.8\ M_p/L - 4.93\ M_p/L)(L/2) = 0.90\ M_p$$

STAGE 4 (Fig. 14.10–2(d)): A load Q is now gradually applied at E; the plastic hinge at D unloads and the response of the beam is purely elastic (see curve BC in Fig. 14.10–1). When Q reaches M_p/L, the bending moment diagram is as in Fig. (d), which is that in Fig. (c) plus the mirror image of Fig. (a).
STAGE 5 (Fig. 14.10–2(e)): When Q reaches 1.1 M_p/L, the hogging moment at B is just equal to M_p.
STAGE 6 (Fig. 14.10–2(f)): When Q is increased beyond 1.1 M_p/L, the plastic hinge at B rotates.
STAGE 7 (Fig. 14–10–2(g)): If the magnitude of Q is now reduced, the plastic hinge at B unloads and the beam responds elastically, until Q falls to 4.8 M_p/L, whereupon a sagging plastic hinge forms once again at D. As Q is further reduced, this hinge continues to rotate plastically, leading to increased deflections.
STAGE 8 (Fig. 14.10–2(h)): After the first cycle of applying and removing the load Q, the beam is left with permanent kinks at B and D. For each subsequent cycle, the process of "plastic rotation at D followed by plastic rotation at B" is repeated. Thus the result of the repeated loading is to cause **incremental collapse** as illustrated in Fig. 14.10–2(i). The collapse is incremental because the plastic hinges at B and D never exist simultaneously; otherwise the usual static collapse will occur.
Suppose the loads P and Q vary within a narrower range, say O to 5.05 M_p/L, as shown in Fig. 14.10–3:
STAGE 1 (Fig. 14.10–3(a)): As the load P is gradually increased, the sagging moment at D reaches M_p when P is 4.93 M_p/L; thereafter, plastic rotation occurs at the hinge D as P is increased to 5.05 M_p/L.
STAGE 2 (Fig. 14.10–3(b)): When Q is applied, the plastic hinge at D unloads and the response of the beam is elastic for the full range of Q from 0 to 5.05 M_p/L. In fact, for $Q = 5.05\ M_p/L$, the value of the hogging moment at B is just approaching M_p, so that no plastic rotation occurs.
STAGE 3 (Fig. 14.10–3(c)): When the magnitude of Q is reduced, the response of the beam is entirely elastic. The reader should verify that when Q is reduced to zero, the value of the sagging moment at D is just approaching M_p. That is for the complete cycle varying from zero to 5.05 M_p/L and back to zero, the beam's response is elastic.
STAGE 4 (Fig. 14.10–3(d)): if Q is now increased to 5.05 M_p/L the beam will again respond elastically to give the bending moments in Fig. 14.10–3(d). Indeed the loads P and Q may now be varied at random within the range 0 to 5.05 M_p/L and the behaviour of the beam will be purely elastic.

It will be recalled that when P was first applied (Fig. 14.10–3(a)), it produced

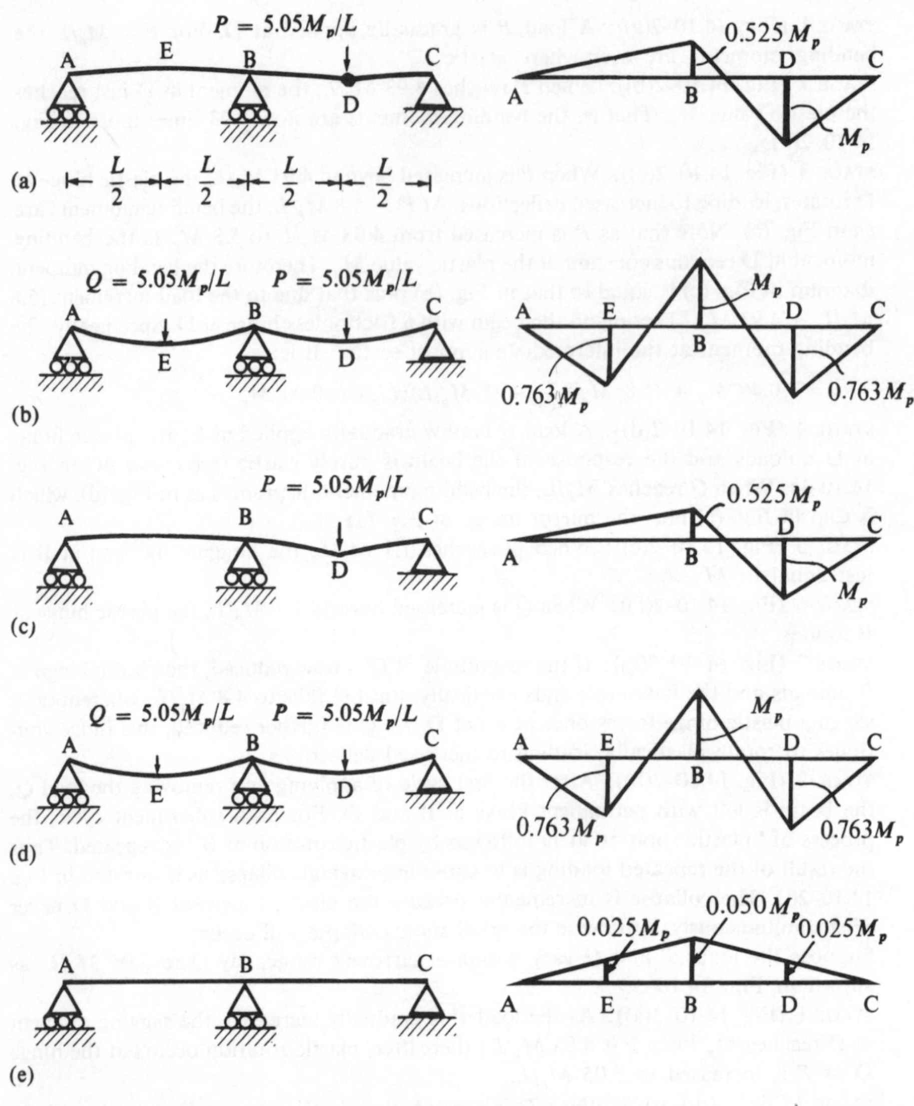

Fig. 14·10–3

plastic hinge rotation at D. However, it is now seen that once P has been taken up to 5.05 M_p/L, the subsequent response of the beam to random variations of the loads is elastic. Thus the beam has **shaken down** to elastic behaviour. A more detailed analysis (see Example 14.10–2) in fact shows that shakedown can occur provided the loads do not exceed 5.05 M_p/L That is if the ranges of loading are specified as

$$0 \le P \le \lambda M_p/L$$

$$0 \le Q \le \lambda M_p/L$$

then the **shakedown limit,** or **shakedown load factor,** is $\lambda = 5.05\ M_p/L$ for this particular example.

The reason why the beam in Fig. 14.10–3 shakes down while that in Fig. 14.10–2 does not can be understood by examining the bending moment diagram in Fig. 14.10–3(e), which is equal to that in Fig. 14.10–3(a) minus 5.05 times that in Fig. 14.10–2(a); that is, Fig. 14.10–3(e) shows the residual moments when the load P is taken up to $5.05\ M_p/L$ and then removed. It is these favourable residual moments that enable the beam to respond elastically to subsequent applications of the loads P and Q. For example, if the load $P = 5.05\ M_p/L$ is re-applied, the sagging moment at D will be

$$(5.05)(0.203\ M_p) - (0.025\ M_p) = M_p$$

where $0.203\ M_p$ is the elastic moment coefficient taken from Fig. 14.10–2(a) and $0.025\ M_p$ is the residual moment taken from Fig. 14.10–3(e). Therefore the moment at D just approaches M_p; if P is now removed, the beam responds elastically and the residual moments are again as in Fig. 14.10–3(e).

The residual moments when the maximum value of P and Q is $5.8\ M_p/L$ will now be examined. If the full loads P and Q are applied and then removed *monotonically*, so that the ratio P/Q remains constant during unloading, the residual moments are as shown in Fig. 14.10–4(c). If the loads are not removed monotonically, but Q is removed first and then P is removed, the residual moments become as shown in Fig. 14.10–4(f). A comparison of Figs. 14.10–3 and 4 shows that when the loading range for P and Q is 0 to $5.8\ M_p/L$, the residual moments change with the method of loading and unloading, but when the loading range is 0 to $5.05\ M_p/L$ they remain unchanged after the first cycle of applying and removing the load P. Indeed, as shall be seen later, whether a structure can shakedown or not depends on whether a stable set of favourable residual moments can be achieved.

Also, up to now the term residual moment has been used to mean the bending moment remaining in the structure after the removal of the load. However, Figs. 14.10–4(c) and (f) show that for a given structure carrying a given loading, the bending moment that will remain upon unloading can be different for different methods of unloading. Hence, to avoid ambiguity, the term **residual moment** will in subsequent discussions be defined by the equation:

$$m = M - \mathscr{M} \tag{14.10–1}$$

where m is the residual moment at the section under consideration, M is the **total moment** actually being produced at that section by the loading, and $\mathscr{M}$ is the **elastic moment** that would have been produced at that section by the same loading had the structure behaved elastically. Using Eqn 14.10–1, the residual moments for the beam loaded as in Fig. 14.10–4(a) are as shown in Fig. (c); similarly the residual moments for the same beam loaded as in Fig. 14.10–4(d) are as shown in Fig. (f). There is no ambiguity.

For a given set of loads W, say, the set of total moments M and the set of elastic moments $\mathscr{M}$ are each in equilibrium with the given loads. By associating these moments with an arbitrary mechanism, say (Δ, ϕ), the virtual work equations can be formed:

$$\sum W_j \Delta_i = \sum M_i \theta_i$$

$$\sum W_j \Delta_j = \sum \mathscr{M} \theta_i$$

It follows from Eqn 14.10–1 that

$$\sum m_i \theta_i = 0 \tag{14.10–2}$$

for all possible mechanisms; this is as one would expect, since the residual moments m are in equilibrium with zero external loads. (*Note*: Since residual moments are in equilibrium with zero external loads, any residual bending moment diagram must consist of straight lines between critical sections. Thus for a two-span beam, it must be a triangle, as in Fig. 14.10–4(f). See also **Example 14.6–3(c)**.)

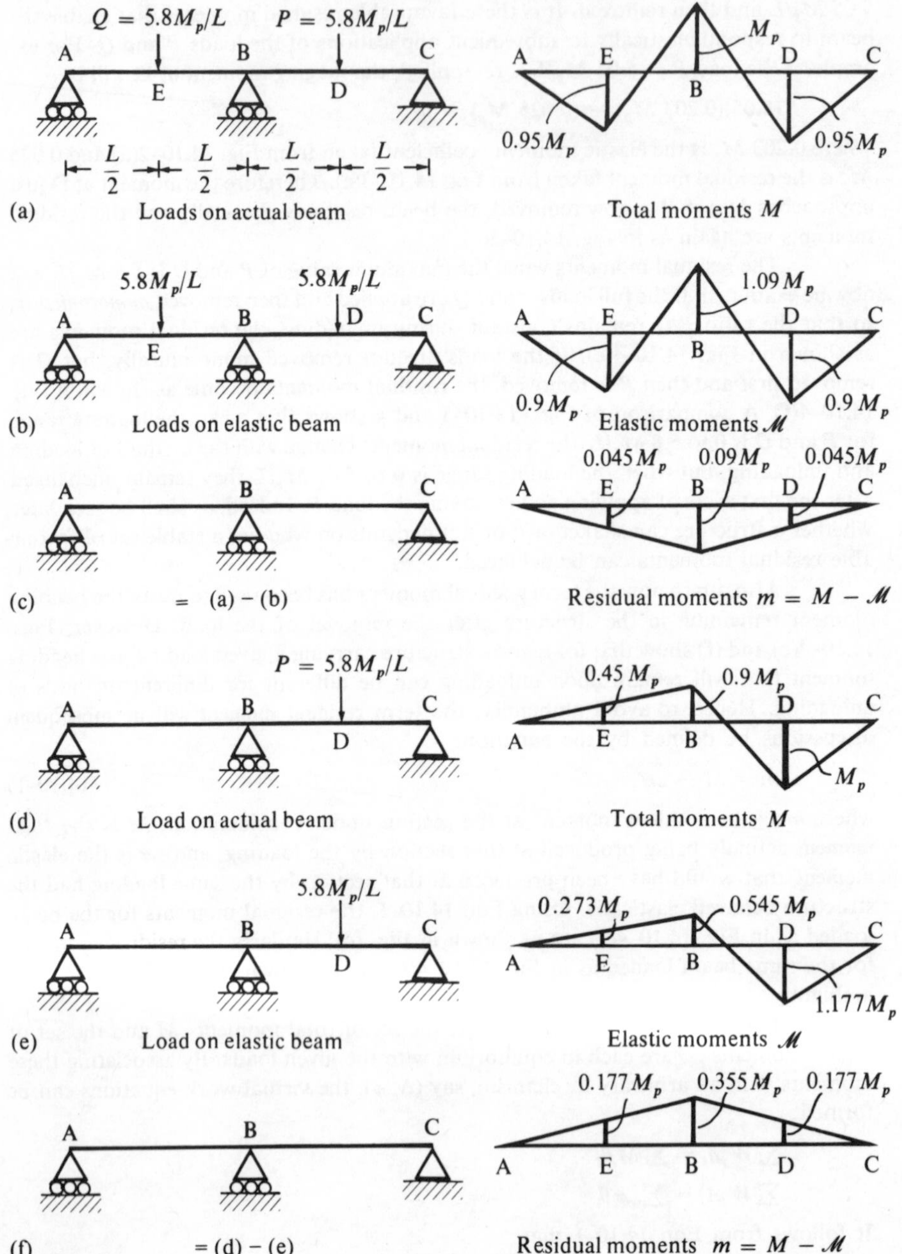

Fig. 14·10–4

Now consider a structure subjected to a set of loads each of which may vary at random within the prescribed maximum and minimum values. Let $\mathscr{M}_i^{\max}$ and $\mathscr{M}_i^{\min}$ denote the greatest and least values, respectively, of the elastic moment $\mathscr{M}_i$ that is calculated for the cross section i for all possible variations of the loads. The **shakedown theorem** then states that the structure will eventually shakedown if it is possible to find a distribution of residual moments $\bar{m}_i$ which will satisfy at every cross section i the conditions (see also Eqn 14.10–27):

$$\mathscr{M}_i^{\max} + \bar{m}_i \leq (M_p)_i \tag{14.10–3}$$

$$\mathscr{M}_i^{\min} + \bar{m}_i \geq -(M_p)_i \tag{14.10–4}$$

Note that it is not a necessary condition that the postulated residual moments $\bar{m}_i$ should be the same as the actual residual moments that will exist in the structure after it has shaken down; it is sufficient to be able to find any set of residual moments $\bar{m}_i$ that will satisfy Eqns 14.10–3 and 4.

To prove the theorem, let us assume that a particular distribution of residual moments $\bar{m}_i$ has been found that satisfies Eqns 14.10–3 and 4. Consider the quantity

$$U = \int \frac{(m_i - \bar{m}_i)^2}{2EI}\, ds \tag{14.10–5}$$

where m_i denotes the residual moment that actually exists at the cross section i at a given stage of the loading programme, ds is an elementary length of the member, EI is the flexural rigidity of the member at section i, and the integration is carried out over the whole structure; it is clear from Eqn 14.10–5 that U can never be negative.

Consider the effect on U of a small change in the applied loads. Let δM_i, $\delta \mathscr{M}_i$, and δm_i be the changes in the total moment, the elastic moment and the residual moment at section i. From Eqn 14.10–1

$$\delta M_i - \delta \mathscr{M}_i = \delta m_i \tag{14.10–6}$$

If the change δm_i in the residual moment is not zero (i.e. if δM_i differs from $\delta \mathscr{M}_i$) then plastic rotation must have occurred at one or more of those sections where the total moment happens to have reached $\pm M_p$. Let the change of rotation at a typical section k be denoted by $\delta\theta_k$. The change in U, which arises as a result of the changes δm_i, is given by Eqn 14.10–5:

$$\delta U = \int \frac{(m_i - \bar{m}_i)\, \delta m_i}{EI}\, ds \tag{14.10–7}$$

Now, the changes of rotation $\delta\theta_k$ in the plastic hinge sections are compatible with the changes of curvature $\delta M_i/EI$ in the rest of the structure. Also, the quantity $\delta \mathscr{M}_i/EI$ represents the changes of curvature which would have occurred if the structure had behaved elastically; hence the changes in curvature $\delta \mathscr{M}_i/EI$ are compatible with zero plastic-hinge rotations. Therefore the differences $\delta M_i/EI - \delta \mathscr{M}_i/EI$ must also be compatible with $\delta\theta_k$. Therefore, from Eqn 14.10–6, $\delta m_i/EI$ must be compatible with the actual changes in rotation $\delta\theta_k$ at the plastic hinge sections. Therefore, we can form the virtual work equation:

Geometrically compatible

$$\int (m_i - \bar{m}_i)\left(\frac{\delta m_i}{EI}\, ds\right) + \Sigma(m_k - \bar{m}_k)\, \delta\theta_k = 0$$

In equilibrium with zero external loads

(14.10–8)

Note that in the above equation each of the residual moment distributions m and $\bar{m}$ is in equilibrium with zero external loads (see Eqn 14.10–2); hence the difference $m - \bar{m}$ must also be in equilibrium with zero external loads.

It follows from Eqns 14.10–7 and 8 that

$$\delta U = -\sum(m_k - \bar{m}_k)\,\delta\theta_k \tag{14.10–9}$$

Suppose now that, at a particular section k,

$$m_k < \bar{m}_k \tag{14.10–10}$$

Then

$$\mathscr{M}_k^{\max} + m_k < \mathscr{M}_k^{\max} + \bar{m}_k$$

so that, from Eqn 14.10–3

$$\mathscr{M}_k^{\max} + m_k < (M_p)_k \tag{14.10–11}$$

where $\mathscr{M}_k^{\max} + m_k = M_k^{\max}$ represents the maximum possible moment that can actually occur at the section k under consideration; since $M_k^{\max}$ is less than $+(M_p)$, and since by hypothesis plastic rotation is occurring at the section k, then $\delta\theta_k$ must occur at the negative value of the plastic moment, $-(M_p)_k$, and that the sign of $\delta\theta_k$ must also be negative (see "sign convention" for rotations as explained in the paragraphs following Eqn 14.9–1). Therefore, in Eqn 14.10–9

$$(m_k - \bar{m}_k)\,\delta\theta_k > 0 \tag{14.10–12}$$

since both the quantities $(m_k - \bar{m}_k)$ and $\delta\theta_k$ are negative.

If, instead of supposing that $m_k < \bar{m}_k$ as in Eqn 14.10–10, it is assumed that

$$m_k > \bar{m}_k \tag{14.10–13}$$

it can be shown by a similar procedure that $\delta\theta_k$ is positive so that, again, Eqn 14.10–12 holds. The other possibility, that of

$$m_k = \bar{m}_k \tag{14.10–14}$$

leads to

$$(m_k - \bar{m}_k)\,\delta\theta_k = 0 \tag{14.10–15}$$

Reference to Eqns 14.10–3 and 4 in fact shows that for $m_k = \bar{m}_k$, the total moment actually acting at the section is just within the limits $\pm\,(M_p)$ so that no plastic hinge rotations occur; that is

$$\delta\theta_k = 0 \tag{14.10–16}$$

In other words, the structure behaves elastically if $m_k = \bar{m}_k$. We have thus shown that, whatever the relative values of m_k and $\bar{m}_k$, either Eqn 14.10–12 or Eqn 14.10–15 holds; it follows that

$$\delta U \leq 0 \tag{14.10–17}$$

that is, δU cannot be positive. If the structure behaves elastically so that no plastic rotation occurs, then $\delta\theta_k = 0$ (see Eqn 14.10–16); hence $\delta U = 0$ and U remains constant. Whenever plastic hinge rotation occurs, the change in U is negative. Since Eqn 14.10–5 shows that U cannot become negative, it follows that ultimately U must become zero or else settle down to some constant positive value. In either case, the structure approaches a state of shakedown.

Note that the validity of the proof of the shakedown theorem is unaffected by the presence of initial stresses in the structure. The presence of initial stresses merely means the presence of an intial set of residual moments, which may of course affect the order in which plastic hinges would form under the given loading, and may affect the number of load variations which have to take place before the structure shakesdown. However, an initial pattern of residual moments will be 'wiped out' by later yielding[3], and can have no effect on whether or not the structure can shakedown under the given loading.

With reference to Eqns 14.10–3 and 4, if a load factor λ is applied to the *range of loading*, then the greatest and smallest elastic moments will become $\lambda\mathcal{M}_i^{max}$ and $\lambda\mathcal{M}_i^{min}$ respectively. Suppose now it is possible to find a set of residual moments $\bar{m}_i$ which satisfies the following conditions at all cross sections i:

$$\lambda\mathcal{M}_i^{max} + \bar{m}_i \leq (M_p)_i \qquad (14.10\text{–}18)$$

$$\lambda\mathcal{M}_i^{min} + \bar{m}_i \geq -(M_p)_i \qquad (14.10\text{–}19)$$

The shakedown theorem then ensures that the structure will shakedown at the load factor λ, though of course it may also shakedown at some higher load factor. Therefore λ must be a **lower bound** on the correct shakedown load factor.

Example 14.10–1 Referring to the uniform two-span beam in Fig. 14.10–2, if the loads P and Q may each vary at random within the range 0 to $\lambda M_p/L$, determine *one* lower bound value of the shakedown load factor λ.

SOLUTION A convenient and useful lower bound on the shakedown load factor is immediately given by letting $\bar{m}_i = 0$. Eqns 14.10–18 and 19 then become

$$\lambda\mathcal{M}_i^{max} \leq (M_p)_i \qquad (14.10\text{–}20)$$

$$\lambda\mathcal{M}_i^{min} \geq -(M_p)_i \qquad (14.10\text{–}21)$$

Using the elastic bending moment diagram in Fig. 14.10–2(a), the following table is completed (sign convention: hogging moments are positive):

Section	Due to P		Due to Q		Combined Loading	
	$\mathcal{M}_i^{max}$	$\mathcal{M}_i^{min}$	$\mathcal{M}_i^{max}$	$\mathcal{M}_i^{min}$	$\mathcal{M}_i^{max}$	$\mathcal{M}_i^{min}$
B	$0.094\ M_p$	0	$0.094\ M_p$	0	$0.188\ M_p$	0
D	0	$-0.203\ M_p$	$0.047\ M_p$	0	$0.047\ M_p$	$-0.203\ M_p$
E	$0.047\ M_p$	0	0	$-0.203\ M_p$	$0.047\ M_p$	$-0.203\ M_p$

Table. 14·10–1

From the table, $\mathcal{M}_i^{max}$ is $0.188\ M_p$ and occurs at B; $\mathcal{M}_i^{min}$ is $-0.203\ M_p$ and occurs at D and E. Therefore Eqns 14.10–20 and 21 become

$$0.188\ M_p\,\lambda \leq M_p \qquad \lambda \leq 5.32$$

$$-0.203\ M_p\,\lambda \geq -M_p \quad \lambda \leq 4.93$$

Ans. $\lambda = 4.93$ is one lower bound to the correct shakedown load factor.

It was shown in Section **14.9** that in the case of static collapse, the load factor obtained from a collapse equation written for an arbitrary mechanism, is an upper bound on the correct value. The mechanism approach will now be applied to incremental collapse. Suppose a certain static collapse mechanism has rotations θ_i at cross-sections i, and that it is possible to find a set of residual moments $\bar{m}_i$ such that the following conditions are satisfied at all cross sections i:

$$\lambda \mathscr{M}_i^{\max} + \bar{m}_i = (M_p)_i \text{ if } \theta_i \text{ is } + \text{ve} \qquad (14.10\text{–}22)$$

$$\lambda \mathscr{M}_i^{\min} + \bar{m}_i = -(M_p)_i \text{ if } \theta_i \text{ is } - \text{ve} \qquad (14.10\text{–}23)$$

These ensure that incremental collapse will occur by the assumed mechanism at the load factor λ, unless it has already occurred by another mechanism at a lower load factor. Therefore λ given by Eqns 14.10–22 and 23 must be an **upper bound** on the shakedown load factor.

Using any consistent sign convention for moments and rotations (see paragraphs following Eqn 14.9–1), let θ_i^+ denote the hinge rotation at a section i where the sense of the rotation is positive, and $-\theta_i^-$ denote that where the sense of the rotation is negative. Therefore, when Eqn 14.10–22 and 23 are multiplied by the hinge rotations, either

$$\lambda \mathscr{M}_i^{\max} \theta_i^+ + \bar{m}_i \theta_i = (M_p)_i |\theta_i| \qquad (14.10\text{–}24)$$

or

$$-\lambda \mathscr{M}_i^{\min} \theta_i^- + \bar{m}_i \theta_i = (M_p)_i |\theta_i| \qquad (14.10\text{–}25)$$

where Eqn 14.10–24 is taken if the rotation is positive and Eqn 14.10–25 is taken if the rotation is negative; the product $(M_p)_i\theta_i$ is of course always positive (see explanation following Eqn 14.9–3). When Eqns 14.10–24 and 25 are summed for all the hinges of the assumed mechanism,

$$\lambda[\sum \mathscr{M}_i^{\max} \theta_i^+ - \sum \mathscr{M}_i^{\min} \theta_i^-] + \sum \bar{m}_i \theta_i = \sum (M_p)_i |\theta_i|$$

Since Eqn 14.10–2 states that $\sum \bar{m}_i \theta_i = 0$, it follows that

$$\lambda [\sum \mathscr{M}_i^{\max} \theta_i^+ - \sum \mathscr{M}_i^{\min} \theta_i^-] = \sum (M_p)_i |\theta_i| \qquad (14.10\text{–}26)$$

which gives an upper bound on the shakedown load factor. If the correct incremental collapse mechanism is used, then Eqn 14.10–26 will give the correct shakedown load factor.

Example 14.10–2 Using Eqn 14.10–26, obtain an upper bound on the shakedown load factor for the uniform two-span beam in Fig. 14.10–2, if the loads P and Q may each vary at random within the range 0 to $\lambda M_p/L$.

SOLUTION

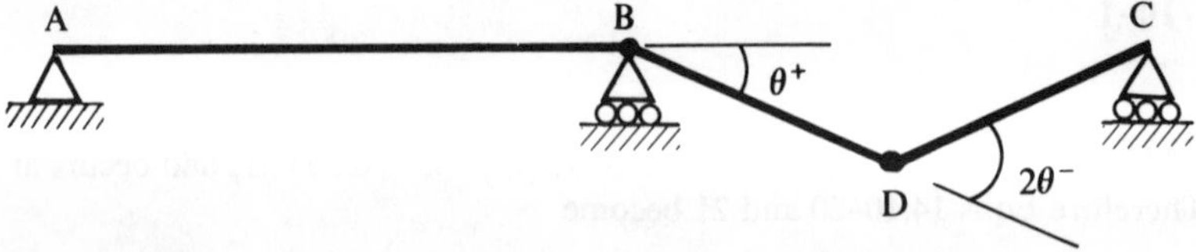

Fig. 14·10–5

Consider the mechanism in Fig. 14.10–5. The elastic moments in Table 14.10–1 are relevant. Substituting appropriate values into Eqns 14.10–26,

$$\lambda [(0.188 M_p)(\theta) - (-0.203 M_p)(2\theta)] = M_p(\theta + 2\theta)$$

whence $\lambda = 5.05$

Ans. $\lambda = 5.05$ is an upper bound on the shakedown load factor.

COMMENT For any arbitrary mechanism, Eqn 14.10–26 only gives an upper bound on the shakedown load factor. However, in this particular case it is seen that the mechanism in Fig. 14.10–5 is the correct incremental collapse mechanism; hence $\lambda = 5.05$ is the correct shakedown load factor.

Example 14.10–3 Using the method of **combination of mechanisms**, determine the shakedown load factor λ for the uniform frame in Fig. 14.10–6(a) if the load H can vary at random within the range 0 to 10λ kN and the load V can vary at random within the range 0 to 15λ kN. The plastic moment of resistance of each member is 50 kNm.

SOLUTION

Three collapse mechanisms can be identified. These are the beam mechanism in Fig. 14.10–6(b), the sidesway mechanism in Fig. (c) and the combined mechanism in Fig. (d).

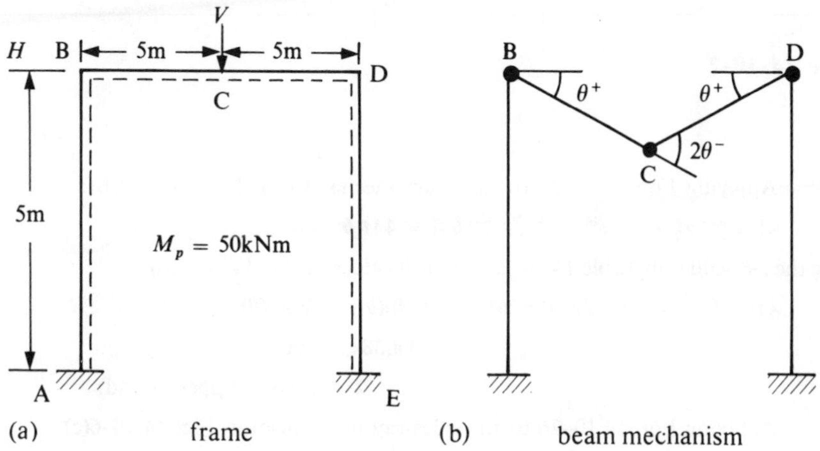

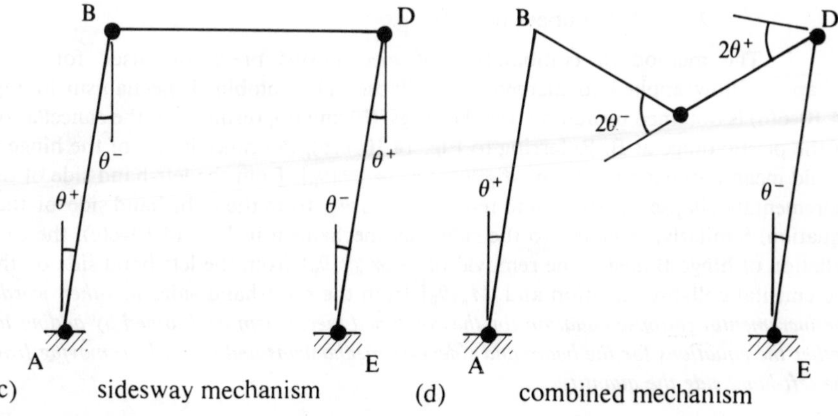

Fig. 14·10–6

The elastic moments due to the loads H and V are shown in Table 14.10–2, where the sign convention for bending moments is the same as that in Fig. 14.9–2; namely, a bending moment producing compression on that side of the member adjacent to the dotted line is designated a positive moment.

	Load V		Load H		Combined Loading		
Section	$\mathscr{M}_i^{max}$ (1)	$\mathscr{M}_i^{min}$ (2)	$\mathscr{M}_i^{max}$ (3)	$\mathscr{M}_i^{min}$ (4)	$\mathscr{M}_i^{max}$ (5)	$\mathscr{M}_i^{min}$ (6)	$\mathscr{M}_i^{max} - \mathscr{M}_i^{min}$ (7)
A	0	−7.50	15.63	0	15.63	−7.50	23.13
B	15.00	0	0	−9.38	15.00	−9.38	24.38
C	0	−22.50	0	0	0	−22.50	22.50
D	15.00	0	9.38	0	24.38	0	24.38
E	0	−7.50	0	−15.63	0	−23.13	23.13

Table 14·10–2

Applying Eqn 14.10–26 to the beam mechanism in Fig..14.10–6(b):

$$\lambda[\mathscr{M}_B^{max}\theta_B^+ - \mathscr{M}_C^{min}\theta_C^- + \mathscr{M}_D^{max}\theta_D^+] = 4M_p\theta$$

Using the $\mathscr{M}$ values in Table 14.10–2 and the θ values in Fig. 14.10–6(b).

$$\lambda[(15.00)(\theta) - (-22.50)(2\theta) + (24.38)(\theta)] = 4(50)(\theta)$$

$$84.38\lambda = 200$$

$$\lambda = 2.37 \text{(an upper bound)}$$

Applying Eqn 14.10–26 to the sidesway mechanism in Fig. 14.10–6(c).

$$\lambda[\mathscr{M}_A^{max}\theta_A^+ - \mathscr{M}_B^{min}\theta_B^- + \mathscr{M}_D^{max}\theta_D^+ - \mathscr{M}_E^{min}\theta_E^-] = 4M_p\theta$$

$$\lambda[(15.63)(\theta) - (-9.38)(\theta) + (24.38)(\theta) - (-23.13)(\theta)] = 4(50)(\theta)$$

$$72.52\lambda = 200$$

$$\lambda = 2.76 \text{(an upper bound)}$$

The method of combination of mechanisms previously used for static collapse is now applied to incremental collapse. The combined mechanism in Fig. 14.10–6(d) is obtained by adding together Figs. (b) and (c), resulting in the cancellation of the plastic hinge at B. Referring to Fig. 14.10–6(b), the cancellation of the hinge B would mean removing the term $\mathscr{M}_B^{max}\theta_B^+$, i.e. $\mathscr{M}_B^{max}|\theta_B|$, from the left-hand side of the incremental collapse equation, and removing $M_P|\theta_B|$ from the right-hand side of that equation. Similarly, referring to the sidesway mechanism in Fig. 14.10–6(c), the cancellation of hinge B means the removal of $-\mathscr{M}_B^{min}|\theta_B|$ from the left-hand side of the incremental collapse equation and $M_p|\theta_B|$ from the right-hand side. *In other words, the incremental collapse equation for the combined mechanism is obtained by adding together the equations for the beam and sidesway mechanisms and then (1) removing from the left-hand side the quantity*

$$(\mathscr{M}_B^{max} - \mathscr{M}_B^{min})|\theta_B|$$

which is the work term due to the range of elastic moments and (2) removing from the right-hand side the quantity

$$2M_p|\theta_B|$$

which is twice the plastic work at hinge B. Here $2M_P$ is 100 and, from Column 7 of Table 14.10–2, $\mathscr{M}_B^{max} - \mathscr{M}_B^{min} = 24.38$. We are now able to display the calculations as follows:

Beam mechanism:	$84.38\lambda = 200$	⟶	$\lambda = 2.37$
Sidesway mechanism:	$72.52\lambda = 200$	⟶	$\lambda = 2.76$
Adding:	$156.90\lambda = 400$		
Cancel hinge B:	$-24.38\lambda \quad -100$		
Combined mechanism:	$132.52\lambda = 300$	⟶	$\lambda = \underline{2.26}$

In this example, it is seen that there are only three collapse mechanisms. Each of the three λ's is an upper bound on the correct shakedown load factor, $\lambda = 2.26$ being the lowest upper bound is therefore the correct shakedown load factor.

COMMENTS

(a) For more complicated frames, it is sometimes difficult to be sure that all possible mechanisms have been investigated. To confirm that the answer obtained is correct, a statical check is necessary; methods are given in specialist texts[3,6,7].

(b) The shakedown load factor resulting from the analysis of any assumed mechanism of collapse, whether or not it is the correct mechanism, can never exceed the corresponding static collapse load factor for the same assumed mechanism. The shakedown load factor determined from the correct incremental collapse mechanism may be anything between, say, 60%, to 90% of the static load factor determined from the correct static collapse mechanism, though for realistic structures and loading conditions the former is unlikely to be less than 75% of the latter[6].

(c) Up to now it has been assumed that the structure is linear-elastic, so that the moment curvature relation is given by curve OAB in Fig. 14.10–1. It, referring to that figure, the moment curvature relation follows the more realistic curve OA′B with an elastic range $2M_y < 2M_p$, it is then necessary to add one further condition to Eqns 14.10–3 and 4:

$$\mathscr{M}_i^{max} - \mathscr{M}_i^{min} \leq 2M_y \tag{14.10–27}$$

which guards against the possible danger of **alternating plasticity**. Similarly, the additional condition

$$\lambda\,(\mathscr{M}_i^{max} - \mathscr{M}_i^{min}) \leq 2M_y \tag{14.10–28}$$

must be added to the Eqns 14.10–18 and 19 and to the Eqns 14.10–22 and 23. With these modifications, the shakedown theorem and the lower-bound and upper-bound arguments would be valid irrespective of whether the moment-curvature curve is OAB or OA′B in Fig. 14.10–1. However, it should be noted that for practical structural frames and realistic loadings, alternating plasticity is rarely of critical importance[3].

References

1 Baker, J.F., Horne, M.R. and Heyman, J. *The steel skeleton: 2. Plastic behaviour and design.* Cambridge University Press, 1956, 408 pp.

2 Baker, J.F. and Heyman, J. *Plastic design of frames: 1. Fundamentals.* Cambridge University Press, 1969, 228 pp.

3 Heyman, J. *Plastic design of frames: 2. Applications.* Cambridge University Press, 1971, 292 pp.

4 Calladine, C.R. *Engineering plasticity*. Chichester, Ellis Horwood, 1985, 318 pp.
5 Heyman, J. *Beams and frame structures*. Oxford, Pergamon Press, 1974, 136 pp.
6 Horne, M.R. *Plastic theory of structures*. London, Thomas Nelson and Sons, 1971, 173 pp.
7 Neal, B.G. *The plastic methods of structural analysis*. London, Chapman and Hall, 1977, 224 pp.
8 Kong, F.K. and Evans, R.H. *Reinforced and prestressed concrete*. Wokingham, Van Nostrand Reinhold, 3rd edition, 1987.
9 Timoshenko, S. and Goodier, J.N. *Theory of elasticity*. New York, McGraw Hill Book Co., 1970, p. 247.
10 Kong, F.K. and Charlton, T.M. The fundamental theorems of the plastic theory of structures. *Proceedings of the Michael R. Horne Conference on Instability and Plastic Collapse of Steel Structures* (Editor: L.J. Morris), Manchester, Sept 1983, London, Granada Publishing, 1983, pp. 9–15.

Problems

14.1 Explain whether the following statements are true or false[10]:

(a) In a statically indeterminate structure, the formation of the first plastic hinge will reduce the number of redundancies by one.

(b) If a statically indeterminate structure has R redundancies, then the formation of R plastic hinges will always render it statically determinate.

(c) The moment-area theorem (Eqn 4.18–2) is applicable to a structure loaded beyond the elastic range, provided the plastic hinges do not convert the structure or part of it into a mechanism. (The structure is made of an elastic-plastic material.)

(d) The uniqueness theorem states that for a given structure under a given loading, the collapse mechanism is unique.

(e) The lower-bound theorem states that if the bending moments actually produced by a given loading satisfy the conditions of equilibrium and yield, then the structure will not collapse under that loading.

(f) For a given structure if the bending moment distribution under a given loading is specified, then the bending moments that will remain after the removal of the loading are uniquely defined.

(g) For a rectangular section, the effect of an axial load is always to reduce the plastic moment of resistance, but for a T section, an axial load might sometimes increase the plastic moment of resistance.

(h) The assumption that plane sections remain plane under pure bending, while valid for linear elastic materials, is a very crude assumption for plastic bending.

(i) Strictly speaking, the upper and lower bound theorems of plastic collapse are applicable only if the structure is absolutely free of initial stresses and if no support settlements occur under load.

(j) The validity of the shakedown theorem is not affected by the presence of initial stresses or support settlements.

(k) If the following conditions (which are Eqns 14.10–3,4 and 14.10–27)

$$\mathscr{M}_i^{\max} + \bar{m}_i \leq (M_p)_i$$
$$\mathscr{M}_i^{\min} + \bar{m}_i \geq -(M_p)_i$$
$$\mathscr{M}_i^{\max} - \mathscr{M}_i^{\min} \leq 2M_y$$

are satisfied, then the shakedown theorem guarantees that the structure will shakedown after a finite number of load cycles.

(l) In using the upper bound theorem to determine the collapse load factor for a given structure under a given loading, a mechanism is assumed and the work equation $\lambda \sum W_j \Delta_j = \sum M_p |\theta_i|$ is written; we are entitled to write this equation because (Δ_j, θ_i) is a compatible displacement set and (λW_j, M_p) is an equilibrium set.

ANSWERS

(a) True

(b) As a general statement it must be rejected as false; see Comment (4) at the end of Example 14.6–4.

(c) False. Use curvature-area theorem; see Example 14.3–3, part (c).

(d) False. See Comment (1) following the proof of the Uniqueness theorem in Section **14.9.**

(e) The lower bound theorem does not require that the set of bending moments satisfying the equilibrium and yield conditions should be the actual moments produced by the loading.

(f) False. The bending moments that will remain after removal of the load depend on the method of removal of the load, unless the beam behaves elastically. See the paragraph preceding Eqn 14.10–1.

(g) Rectangular section: true; T section: depends on definition of axial load—see last two paragraphs of section **14.4.**

(h) False. That plane sections remain plane under pure bending follows from symmetry considerations and is just as valid for non-linear inelastic materials as for linear elastic materials. The only requirement is that deformations are continuous and the material isotropic.

(i) False

(j) True

(k) False. The shakedown theorem guarantees that the structure will eventually shakedown, but contains no statement regarding the number of load variations which has to take place before a condition of shakedown is actually reached.

(l) False. While (Δ_j, θ_i) is certainly a compatible displacement set, (λW_j, M_p) is not necessarily an equilibrium set; the fact that an equation has been written does not mean that the left-hand side and the right-hand side are in fact equal.

14.2

A multistorey frame consists of m bays and n storeys. Each horizontal member carries a vertical point load at midspan, and the frame is also acted on by a horizontal load at each floor level. Determine the number of independent mechanisms.

ANSWER

The total number of independent mechanisms is $2n(m + 1)$, of which $n(m + 1)$ are joint rotations. (*Hint*: No. of critical sections on top storey $= 4m + 1$; No. of critical sections on each intermediate storey $= 5m + 2$; No. of critical sections at ground level $= m + 1$ Therefore $N = (4m + 1) + (n - 1)(5m + 2) + (m + 1) = 5mn + 2n$; No. of redundancies (see Fig. 14.7–6 if necessary) $= 3mn$. Then use Eqn 14.6–3.)

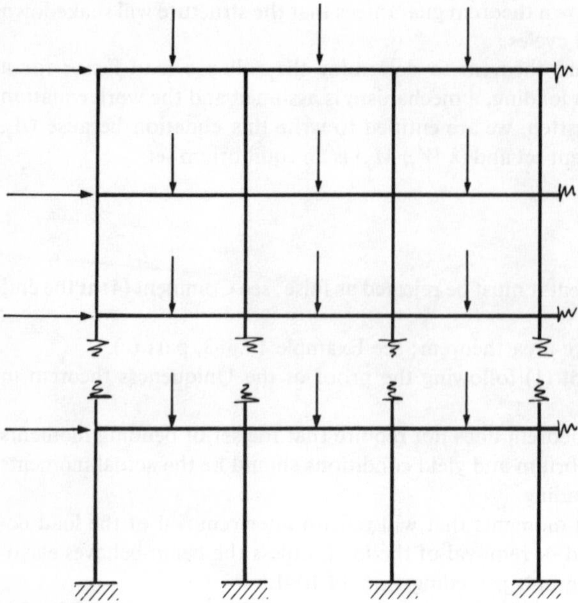

Problem 14·2

14.3

A propped uniform cantilever is of moment of resistance M_p, and it is known that the moment-curvature relation is elastic-plastic, as in Fig. 14.6–1(a) in Section 14.6. Point loads are applied at the third-span positions, and the value P of each load is gradually increased from zero. (The cantilever is initially unstressed.)

(a) Show that when the value of P reaches $3M_p/L$, a plastic hinge will form at B. For future reference, the deflections at C and D at this value of P are denoted by δ_C and δ_D.

(b) By writing work equations for assumed mechanisms, show that when the value of P reaches $4M_p/L$ a second plastic hinge will form at C so that the structure becomes a mechanism.

For future reference, the deflections at C and D, just at the instant when the second hinge begins to form at C, are denoted by Δ_C and Δ_D respectively.

(c) Explain why $(\Delta_C - \delta_C) = (\Delta_D - \delta_D)$.

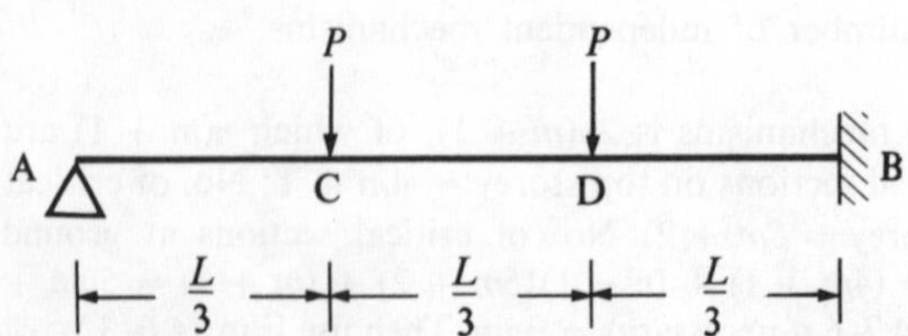

Problem 14·3

ANSWER

(c): The deflections $(\Delta_C - \delta_C)$ and $(\Delta_D - \delta_D)$ are equal because they are the deflections produced by the load increments M_p/L $(= 4M_p/L - 3M_p/L)$ acting on a *simply supported* beam AB. In other words, when P reaches $3M_p/L$ a plastic hinge forms at B; thereafter, increments of P are resisted by simple-beam action.

14.4

With reference to Problem 14.3(b), if, just after the formation of the second plastic hinge at C, the loads are removed, sketch the *shape* of the bending moment diagram after the removal of the loads.

ANSWER

Irrespective of the method of unloading, the bending moment diagram after load removal must be a triangle, as shown below. (If in doubt, see Example 14.6–3(c).)

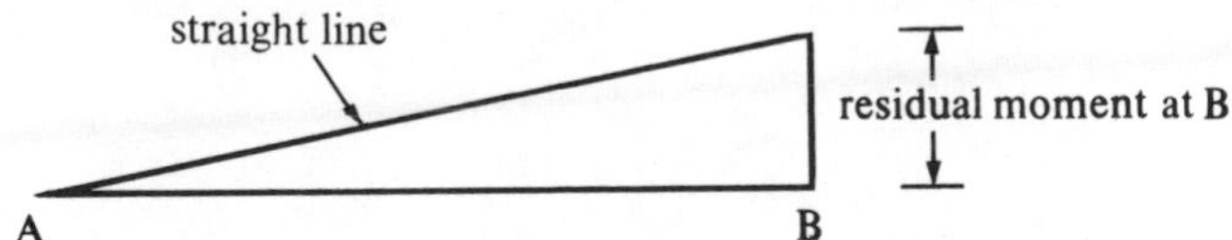

Problem 14·4

14.5

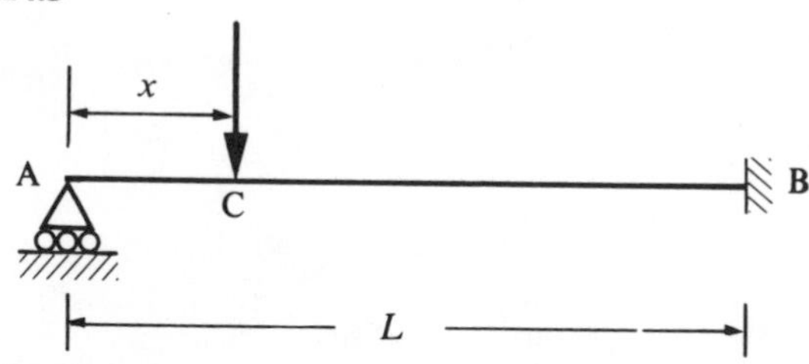

Problem 14·5

A propped cantilever is of uniform cross section having plastic moment of resistance M_p. It carries a concentrated load P which may be placed anywhere along the span length AB.

(a) Show that if collapse occurs at a load factor λ_c for a single traverse of the load P, then $\lambda_c = 5.83\ M_p/PL$.

(b) Show that if incremental collapse is to be avoided for multiple traverses at a load factor λ_s then $\lambda_s \leq 5.59\ M_p/PL$.

(*Hint for (a)*: See Example 14.6–3(b);
Hint for (b): For the load P at x from A, apply Eqn 14.10–26 to the corresponding static collapse mechanism. Then minimize λ_s with respect to x. If in difficulty see Reference 3, p.159.)

Appendix

Appendix 1

Table A1 Stability functions for compressive axial force ρ = 0.00 to 1.00

ρ	s		c		sc		$s(1+c)$		m	
0.00	**4.0000**		**0.5000**		**2.0000**		**6.0000**		**1.0000**	
		−132		25		33		−99		83
0.01	**3.9868**		**0.5025**		**2.0033**		**5.9901**		**1.0083**	
		−132		25		33		−99		85
0.02	**3.9736**		**0.5050**		**2.0066**		**5.9802**		**1.0168**	
		−132		25		33		−99		86
0.03	**3.9604**		**0.5075**		**2.0100**		**5.9703**		**1.0254**	
		−133		26		34		−99		88
0.04	**3.9471**		**0.5101**		**2.0133**		**5.9604**		**1.0343**	
		−133		26		34		−99		90
0.05	**3.9338**		**0.5127**		**2.0167**		**5.9505**		**1.0433**	
		−133		26		34		−99		92
0.06	**3.9204**		**0.5153**		**2.0201**		**5.9405**		**1.0525**	
		−134		26		34		−100		94
0.07	**3.9070**		**0.5179**		**2.0235**		**5.9306**		**1.0618**	
		−134		27		34		−100		96
0.08	**3.8936**		**0.5206**		**2.0270**		**5.9206**		**1.0714**	
		−135		27		35		−100		98
0.09	**3.8802**		**0.5233**		**2.0304**		**5.9106**		**1.0812**	
		−135		27		35		−100		100
0.10	**3.8667**		**0.5260**		**2.0339**		**5.9006**		**1.0913**	
		−135		28		35		−100		102
0.11	**3.8531**		**0.5288**		**2.0374**		**5.8906**		**1.1015**	
		−136		28		35		−100		105
0.12	**3.8396**		**0.5316**		**2.0410**		**5.8805**		**1.1120**	
		−136		28		36		−100		107
0.13	**3.8260**		**0.5344**		**2.0445**		**5.8705**		**1.1227**	
		−136		29		36		−101		110
0.14	**3.8123**		**0.5372**		**2.0481**		**5.8604**		**1.1336**	
		−137		29		36		−101		112
0.15	**3.7987**		**0.5401**		**2.0517**		**5.8504**		**1.1449**	
		−137		29		36		−101		115
0.16	**3.7849**		**0.5430**		**2.0553**		**5.8403**		**1.1563**	
		−138		29		36		−101		117

Table A1 (*contd*) ρ = **0.00 to 1.00**

ρ	s		c		sc		$s(1+c)$		m	
		−138		29		36		−101		117
0.17	**3.7712**		**0.5460**		**2.0590**		**5.8302**		**1.1681**	
		−138		30		37		−101		120
0.18	**3.7574**		**0.5490**		**2.0626**		**5.8200**		**1.1801**	
		−138		30		37		−101		123
0.19	**3.7436**		**0.5520**		**2.0663**		**5.8099**		**1.1924**	
		−139		30		37		−101		126
0.20	**3.7297**		**0.5550**		**2.0701**		**5.7998**		**1.2051**	
		−139		31		37		−102		129
0.21	**3.7158**		**0.5581**		**2.0738**		**5.7896**		**1.2180**	
		−139		31		38		−102		133
0.22	**3.7019**		**0.5612**		**2.0776**		**5.7794**		**1.2313**	
		−140		32		38		−102		136
0.23	**3.6879**		**0.5644**		**2.0813**		**5.7692**		**1.2449**	
		−140		32		38		−102		140
0.24	**3.6739**		**0.5676**		**2.0852**		**5.7590**		**1.2589**	
		−141		32		38		−102		143
0.25	**3.6598**		**0.5708**		**2.0890**		**5.7488**		**1.2732**	
		−141		33		39		−102		147
0.26	**3.6457**		**0.5741**		**2.0929**		**5.7385**		**1.2880**	
		−141		33		39		−103		151
0.27	**3.6315**		**0.5774**		**2.0967**		**5.7283**		**1.3031**	
		−142		33		39		−103		155
0.28	**3.6174**		**0.5807**		**2.1007**		**5.7180**		**1.3186**	
		−142		34		39		−103		160
0.29	**3.6031**		**0.5841**		**2.1046**		**5.7077**		**1.3346**	
		−143		34		40		−103		164
0.30	**3.5889**		**0.5875**		**2.1086**		**5.6974**		**1.3511**	
		−143		35		40		−103		169
0.31	**3.5746**		**0.5910**		**2.1126**		**5.6871**		**1.3680**	
		−143		35		40		−103		174
0.32	**3.5602**		**0.5945**		**2.1166**		**5.6768**		**1.3854**	
		−144		36		40		−103		179
0.33	**3.5458**		**0.5981**		**2.1206**		**5.6664**		**1.4033**	
		−144		36		41		−104		185
0.34	**3.5314**		**0.6017**		**2.1247**		**5.6561**		**1.4218**	
		−145		36		41		−104		190
0.35	**3.5169**		**0.6053**		**2.1288**		**5.6457**		**1.4408**	
		−145		37		41		−104		196
0.36	**3.5024**		**0.6090**		**2.1329**		**5.6353**		**1.4604**	
		−146		37		42		−104		202
0.37	**3.4878**		**0.6127**		**2.1371**		**5.6249**		**1.4806**	
		−146		38		42		−104		209
0.38	**3.4732**		**0.6165**		**2.1412**		**5.6145**		**1.5015**	
		−146		38		42		−104		216
0.39	**3.4586**		**0.6203**		**2.1454**		**5.6040**		**1.5231**	
		−147		39		42		−105		223
0.40	**3.4439**		**0.6242**		**2.1497**		**5.5936**		**1.5453**	
		−147		39		43		−105		230
0.41	**3.4292**		**0.6281**		**2.1539**		**5.5831**		**1.5684**	
		−148		40		43		−105		238
0.42	**3.4144**		**0.6321**		**2.1582**		**5.5726**		**1.5922**	
		−148		40		43		−105		246
0.43	**3.3995**		**0.6361**		**2.1626**		**5.5621**		**1.6168**	
		−149		41		44		−105		255
0.44	**3.3847**		**0.6402**		**2.1669**		**5.5516**		**1.6423**	
		−149		41		44		−105		264
0.45	**3.3698**		**0.6443**		**2.1713**		**5.5410**		**1.6688**	
		−150		42		44		−106		274

Table A1 (*contd*) ρ = **0.00 to 1.00**

ρ	s		c		sc		$s(1+c)$		m	
		−150		42		44		−106		274
0.46	**3.3548**		**0.6485**		**2.1757**		**5.5305**		**1.6962**	
		−150		42		44		−106		285
0.47	**3.3398**		**0.6528**		**2.1801**		**5.5199**		**1.7247**	
		−151		43		45		−106		295
0.48	**3.3247**		**0.6571**		**2.1846**		**5.5093**		**1.7542**	
		−151		44		45		−106		307
0.49	**3.3096**		**0.6614**		**2.1891**		**5.4987**		**1.7849**	
		−152		44		45		−106		319
0.50	**3.2945**		**0.6659**		**2.1936**		**5.4881**		**1.8168**	
		−152		45		46		−106		332
0.51	**3.2793**		**0.6703**		**2.1982**		**5.4775**		**1.8500**	
		−152		45		46		−107		346
0.52	**3.2640**		**0.6749**		**2.2028**		**5.4668**		**1.8846**	
		−153		46		46		−107		361
0.53	**3.2487**		**0.6795**		**2.2074**		**5.4562**		**1.9207**	
		−153		47		47		−107		376
0.54	**3.2334**		**0.6841**		**2.2121**		**5.4455**		**1.9583**	
		−154		47		47		−107		393
0.55	**3.2180**		**0.6889**		**2.2168**		**5.4348**		**1.9976**	
		−154		48		47		−107		411
0.56	**3.2025**		**0.6937**		**2.2215**		**5.4240**		**2.0387**	
		−155		49		48		−107		430
0.57	**3.1870**		**0.6985**		**2.2263**		**5.4133**		**2.0817**	
		−155		49		48		−108		450
0.58	**3.1715**		**0.7035**		**2.2311**		**5.4026**		**2.1267**	
		−156		50		48		−108		472
0.59	**3.1559**		**0.7085**		**2.2359**		**5.3918**		**2.1739**	
		−156		51		49		−108		496
0.60	**3.1403**		**0.7136**		**2.2407**		**5.3810**		**2.2234**	
		−157		52		49		−108		521
0.61	**3.1246**		**0.7187**		**2.2456**		**5.3702**		**2.2755**	
		−158		52		49		−108		548
0.62	**3.1088**		**0.7239**		**2.2506**		**5.3594**		**2.3304**	
		−158		53		50		−108		578
0.63	**3.0930**		**0.7292**		**2.2555**		**5.3485**		**2.3881**	
		−159		54		50		−109		610
0.64	**3.0771**		**0.7346**		**2.2605**		**5.3377**		**2.4491**	
		−159		55		50		−109		645
0.65	**3.0612**		**0.7401**		**2.2656**		**5.3268**		**2.5136**	
		−160		55		51		−109		683
0.66	**3.0453**		**0.7456**		**2.2706**		**5.3159**		**2.5819**	
		−160		56		51		−109		724
0.67	**3.0293**		**0.7513**		**2.2757**		**5.3050**		**2.6542**	
		−161		57		51		−109		769
0.68	**3.0132**		**0.7570**		**2.2809**		**5.2941**		**2.7311**	
		−161		58		52		−109		818
0.69	**2.9971**		**0.7628**		**2.2861**		**5.2831**		**2.8130**	
		−162		59		52		−110		873
0.70	**2.9809**		**0.7687**		**2.2913**		**5.2722**		**2.9003**	
		−162		60		53		−110		933
0.71	**2.9647**		**0.7746**		**2.2966**		**5.2612**		**2.9936**	
		−163		61		53		−110		999
0.72	**2.9484**		**0.7807**		**2.3018**		**5.2502**		**3.0935**	
		−163		62		53		−110		1074
0.73	**2.9320**		**0.7869**		**2.3072**		**5.2392**		**3.2009**	
		−164		63		54		−110		1156
0.74	**2.9156**		**0.7932**		**2.3126**		**5.2282**		**3.3165**	
		−165		64		54		−111		1248

Table A1 (*contd*) ρ = **0.00 to 1.00**

ρ	s	Δ	c	Δ	sc	Δ	$s(1+c)$	Δ	m	Δ
		−165		64		54		−111		125
0.75	**2.8991**		**0.7995**		**2.3180**		**5.2171**		**3.441**	
		−165		65		55		−111		136
0.76	**2.8826**		**0.8060**		**2.3234**		**5.2060**		**3.577**	
		−166		66		55		−111		147
0.77	**2.8660**		**0.8126**		**2.3289**		**5.1950**		**3.724**	
		−166		67		55		−111		160
0.78	**2.8494**		**0.8193**		**2.3345**		**5.1839**		**3.884**	
		−167		68		56		−111		176
0.79	**2.8327**		**0.8261**		**2.3400**		**5.1727**		**4.060**	
		−168		69		56		−111		193
0.80	**2.8159**		**0.8330**		**2.3456**		**5.1616**		**4.253**	
		−168		70		57		−112		213
0.81	**2.7991**		**0.8400**		**2.3513**		**5.1504**		**4.466**	
		−169		71		57		−112		237
0.82	**2.7822**		**0.8472**		**2.3570**		**5.1392**		**4.703**	
		−169		73		57		−112		265
0.83	**2.7653**		**0.8544**		**2.3627**		**5.1281**		**4.968**	
		−170		74		58		−112		298
0.84	**2.7483**		**0.8618**		**2.3685**		**5.1168**		**5.266**	
		−171		75		58		−112		338
0.85	**2.7312**		**0.8693**		**2.3744**		**5.1056**		**5.604**	
		−171		77		59		−113		386
0.86	**2.7141**		**0.8770**		**2.3802**		**5.0944**		**5.990**	
		−172		78		59		−113		446
0.87	**2.6969**		**0.8848**		**2.3862**		**5.0831**		**6.436**	
		−173		79		60		−113		520
0.88	**2.6797**		**0.8927**		**2.3921**		**5.0718**		**6.956**	
		−173		81		60		−113		614
0.89	**2.6623**		**0.9008**		**2.3981**		**5.0605**		**7.570**	
		−174		82		61		−113		737
0.90	**2.6450**		**0.9090**		**2.4042**		**5.0492**		**8.307**	
		−175		84		61		−113		901
0.91	**2.6275**		**0.9173**		**2.4103**		**5.0378**		**9.208**	
		−175		85		62		−114		
0.92	**2.6100**		**0.9258**		**2.4164**		**5.0264**		**10.333**	
		−176		87		62		−114		
0.93	**2.5924**		**0.9345**		**2.4226**		**5.0151**		**11.781**	
		−177		88		62		−114		
0.94	**2.5748**		**0.9433**		**2.4289**		**5.0036**		**13.711**	
		−177		90		63		−114		
0.95	**2.5570**		**0.9523**		**2.4352**		**4.9922**		**16.413**	
		−178		92		63		−114		
0.96	**2.5393**		**0.9615**		**2.4415**		**4.9808**		**20.466**	
		−179		93		64		−115		
0.97	**2.5214**		**0.9709**		**2.4479**		**4.9693**		**27.230**	
		−179		95		64		−115		
0.98	**2.5035**		**0.9804**		**2.4544**		**4.9578**		**40.728**	
		−180		97		65		−115		
0.99	**2.4855**		**0.9901**		**2.4609**		**4.9463**		**81.246**	
		−181		99		65		−115		
1.00	**2.4674**		**1.0000**		**2.4674**		**4.9348**		∞	

ρ	s		c		sc		s(1 + c)		m	
1.00	2.467		1.000		2.467		4.935		∞	
		−184		111		68		−117		
1.10	2.283		1.111		2.536		4.819		−7.902	
		−193		138		74		−118		
1.20	2.090		1.249		2.610		4.700		−3.847	
		−201		175		81		−120		
1.30	1.889		1.424		2.691		4.580		−2.495	
		−211		232		88		−123		
1.40	1.678		1.656		2.779		4.457		−1.818	
		−221		317		96		−125		
1.50	1.457		1.973		2.875		4.332		−1.411	
		−233		462		105		−128		
1.60	1.224		2.435		2.980		4.204		−1.138	
		−246		731		116		−130		
1.70	0.978		3.166		3.096		4.074		−0.944	
		−261				128		−133		
1.80	0.717		4.497		3.224		3.941		−0.798	
		−278				143		−135		
1.90	0.439		7.661		3.367		3.806		−0.683	
		−296				158		−138		
2.00	0.143		24.682		3.525		3.668		−0.5914	
		−319				177		−142		756
2.10	−0.176		−21.074		3.702		3.526		−0.5158	
		−343				199		−144		634
2.20	−0.519		−7.511		3.901		3.382		−0.4524	
		−374				226		−148		540
2.30	−0.893		−4.623		4.127		3.234		−0.3985	
		−408				256		−151		466
2.40	−1.301		−3.370		4.383		3.083		−0.3519	
		−449		697		295		−155		407
2.50	−1.750		−2.673		4.678		2.928		−0.3112	
		−499		442		340		−159		359
2.60	−2.249		−2.231		5.018		2.769		−0.2752	
		−560		303		397		−163		320
2.70	−2.809		−1.928		5.415		2.606		−0.2432	
		−636		220		469		−167		288
2.80	−3.445		−1.708		5.884		2.439		−0.2144	
		−731		165		560		−171		261
2.90	−4.176		−1.543		6.444		2.268		−0.1883	
		−856		127		680		−176		238
3.00	−5.032		−1.416		7.124		2.092		−0.1645	
				100		838		−131		218
3.10	−6.052		−1.316		7.962		1.911		−0.1427	
				79				−187		201
3.20	−7.297		−1.236		9.021		1.724		−0.1226	
				63				−192		187
3.30	−8.863		−1.173		10.395		1.532		−0.1039	
				51				−198		175
3.40	−10.908		−1.122		12.242		1.334		−0.0864	
				40				−204		164
3.50	−13.719		−1.082		14.849		1.130		−0.0700	
				31				−211		154
3.60	−17.867		−1.052		18.786		0.919		−0.0546	
				23				−218		146
3.70	−24.685		−1.028		25.386		0.701		−0·0399	
				16				−225		139
3.80	−38.173		−1.013		38.650		0.476		−0·0260	
				9				−234		133
3.90	−78.330		−1.003		78.574		0.242		−0·0127	
				3				−242		127
4.00			−1.000		∞		0.000		−0·0000	

Table A3 Stability functions for tensile axial force $\rho = 0.00$ to -2.00

ρ	s		c		sc		$s(1+c)$		m	
0.00	**4.0000**		**0.5000**		**2.0000**		**6.0000**		**1.0000**	
		1299		−235		−319		980		−749
−0.10	**4.1299**		**0.4765**		**1.9681**		**6.0980**		**0.9251**	
		1268		−213		−301		967		−626
−0.20	**4.2567**		**0.4553**		**1.9380**		**6.1947**		**0.8626**	
		1237		−193		−284		954		−531
−0.30	**4.3804**		**0.4360**		**1.9096**		**6.2900**		**0.8095**	
		1209		−177		−268		941		−456
−0.40	**4.5013**		**0.4183**		**1.8829**		**6.3841**		**0.7638**	
		1181		−162		−253		929		−397
−0.50	**4.6194**		**0.4021**		**1.8575**		**6.4770**		**0.7241**	
		1157		−149		−240		917		−348
−0.60	**4.7351**		**0.3872**		**1.8336**		**6.5687**		**0.6893**	
		1132		−137		−227		905		−309
−0.70	**4.8483**		**0.3735**		**1.8109**		**6.6592**		**0.6584**	
		1110		−127		−215		894		−275
−0.80	**4.9593**		**0.3608**		**1.7893**		**6.7486**		**0.6309**	
		1088		−118		−205		883		−247
−0.90	**5.0681**		**0.3490**		**1.7689**		**6.8369**		**0.6062**	
		1067		−110		−194		873		−223
−1.00	**5.1748**		**0.3381**		**1.7494**		**6.9242**		**0.5839**	
		1047		−102		−185		862		−203
−1.10	**5.2795**		**0.3279**		**1.7309**		**7.0105**		**0.5636**	
		1029		−95		−176		853		−185
−1.20	**5.3824**		**0.3183**		**1.7133**		**7.0957**		**0.5451**	
		1011		−89		−168		843		−170
−1.30	**5.4835**		**0.3094**		**1.6965**		**7.1800**		**0.5281**	
		994		−84		−160		833		−156
−1.40	**5.5828**		**0.3010**		**1.6805**		**7.2633**		**0.5125**	
		977		−79		−153		824		−144
−1.50	**5.6806**		**0.2931**		**1.6652**		**7.3458**		**0.4981**	
		961		−74		−146		815		−134
−1.60	**5.7767**		**0.2857**		**1.6506**		**7.4273**		**0.4847**	
		946		−70		−140		807		−124
−1.70	**5.8714**		**0.2787**		**1.6366**		**7.5080**		**0.4723**	
		932		−66		−134		798		−116
−1.80	**5.9645**		**0.2721**		**1.6232**		**7.5878**		**0.4607**	
		918		−62		−128		790		−108
−1.90	**6.0563**		**0.2659**		**1.6104**		**7.6668**		**0.4499**	
		905		−59		−123		782		−102
−2.00	**6.1468**		**0.2600**		**1.5982**		**7.7450**		**0.4397**	

Table A4 Stability functions for tensile axial force ρ = 0.00 to − 20.0

ρ	s		c		sc		$s(1 + c)$		m	
0.00	**4.0000**		**0.5000**		**2.0000**		**6.000**		**1.0000**	
								924		
−1.00	**5.175**		**0.3381**		**1.7494**		**6.924**		**0.5839**	
		972		−781				821		
−2.00	**6.147**		**0.2600**		**1.5982**		**7.745**		**0.4397**	
		841		−455		−991		742		−753
−3.00	**6.988**		**0.2145**		**1.4991**		**8.487**		**0.3644**	
		749		−297		−690		680		−473
−4.00	**7.737**		**0.1848**		**1.4300**		**9.167**		**0.3171**	
		680		−209		−505		629		−329
−5.00	**8.417**		**0.1639**		**1.3796**		**9.797**		**0.2842**	
		627		−156		−383		589		−245
−6.00	**9.044**		**0.1483**		**1.3413**		**10.385**		**0.2597**	
		583		−121		−300		554		−192
−7.00	**9.627**		**0.1362**		**1.3113**		**10.939**		**0.2405**	
		548		−97		−241		524		−155
−8.00	**10.175**		**0.1265**		**1.2872**		**11.463**		**0.2250**	
		519		−80		−198		498		−128
−9.00	**10.694**		**0.1185**		**1.2674**		**11.961**		**0.2122**	
		492		−67		−166		476		−109
−10.00	**11.186**		**0.1118**		**1.2508**		**12.437**		**0.2013**	
		471		−57		−141		457		−94
—11.00	**11.657**		**0.1061**		**1.2368**		**12.894**		**0.1919**	
		451		−50		−121		438		−82
−12.00	**12.108**		**0.1011**		**1.2246**		**13.333**		**0.1838**	
		434		−43		−106		424		−72
−13.00	**12.542**		**0.0968**		**1.2141**		**13.756**		**0.1766**	
		418		−38		−93		409		−64
−14.00	**12.960**		**0.0930**		**1.2048**		**14.165**		**0.1701**	
		404		−34		−82		396		−58
−15.00	**13.364**		**0.0895**		**1.1966**		**14.561**		**0.1644**	
		392		−31		−74		384		−52
−16.00	**13.756**		**0.0865**		**1.1892**		**14.945**		**0.1592**	
		380		−28		−67		373		−48
−17.00	**14.136**		**0.0837**		**1.1825**		**15.318**		**0.1544**	
		369		−25		−60		364		−44
−18.00	**14.505**		**0.0811**		**1.1765**		**15.682**		**0.1501**	
		360		−23		−55		354		−40
−19.00	**14.865**		**0.0788**		**1.1710**		**16.036**		**0.1461**	
		351		−21		−50		346		−37
−20.00	**15.216**		**0.0766**		**1.1660**		**16.382**		**0.1424**	

Appendix 2

End moments for beams carrying axial load together with a distributed load or a point load—see Section 10.8

ρ	*Distributed load* $f = m_1/(\frac{1}{12}wL^2)$	*Ratio $m_1/(WL)$ for a point load at position rL*								
		$r = 0.1$	0.2	0.3	0.4	0.5	0.6	0.7	0.8	0.9
2.0	1.6359	0.097	0.177	0.228	0.245	0.227	0.182	0.121	0.061	0.017
1.8	1.5223	0.094	0.169	0.214	0.227	0.209	0.166	0.110	0.056	0.015
1.6	1.4272	0.092	0.162	0.203	0.212	0.193	0.153	0.101	0.051	0.014
1.4	1.3463	0.090	0.156	0.192	0.199	0.180	0.142	0.094	0.047	0.013
1.2	1.2766	0.088	0.151	0.184	0.188	0.169	0.132	0.087	0.044	0.012
1.0	1.2159	0.087	0.146	0.176	0.179	0.159	0.124	0.082	0.041	0.011
0.8	1.1624	0.086	0.142	0.169	0.170	0.151	0.117	0.077	0.039	0.011
0.6	1.1150	0.084	0.138	0.162	0.162	0.143	0.111	0.073	0.037	0.010
0.4	1.0727	0.083	0.134	0.157	0.156	0.136	0.105	0.069	0.035	0.010
0.2	1.0345	0.082	0.131	0.152	0.150	0.130	0.100	0.066	0.033	0.009
0.0	1.0000	0.081	0.128	0.147	0.144	0.125	0.096	0.063	0.032	0.009
−0.2	0.9686	0.080	0.125	0.143	0.139	0.120	0.092	0.060	0.031	0.009
−0.5	0.9264	0.079	0.121	0.137	0.132	0.114	0.087	0.057	0.029	0.008
−1.0	0.8665	0.077	0.116	0.128	0.122	0.104	0.079	0.052	0.027	0.008
−2.0	0.7747	0.073	0.107	0.115	0.108	0.091	0.069	0.045	0.023	0.007
−5.0	0·6125	0.066	0.089	0.090	0.081	0.067	0.051	0.034	0.018	0.006
−10.0	0.4824	0.059	0.073	0.070	0.061	0.050	0.038	0.026	0.014	0.005
−20.0	0.3663	0.050	0.056	0.051	0.044	0.036	0.027	0.019	0.011	0.002

$m_1 = f(wL^2/12)$

$\rho = P/P_E$

(see Eqn 9.10–9)

Index